GENERAL RADIOTELEPHONE LICENSE HANDBOOK

Seventh Edition

Edward M. Noll is an accomplished author of numerous technical books, lessons, articles, and instruction manuals. He also devotes much of his time to circuit design and lecturing. He has taught radiocommunications at the university level and has served on the staff of several broadcast stations. His other SAMS books include: *Linear IC Principles, Experiments, and Projects*, *Wind/Solar Energy for Radiocommunications and Low-Power Electrical Systems*, *Oscilloscope Applications, and Experiments*, *Marine Radiotelegraph Operator Handbook*, *73 Dipole & Long-Wire Antennas*, *Solid-State Circuit Files* (Vols. 1 & 2), *Microprocessor Circuits* (Vol. 1), and *Broadcast Radio & TV Handbook*.

GENERAL RADIOTELEPHONE LICENSE HANDBOOK

Seventh Edition

by

Edward M. Noll

Howard W. Sams & Co., Inc.
4300 WEST 62ND ST. INDIANAPOLIS, INDIANA 46268 USA

SEVENTH EDITION
SECOND PRINTING—1984

International Standard Book Number: 0-672-21930-1
Library of Congress Catalog Card Number: 82-50660

Edited by: *Welborn Associates*
Illustrated by: *T. R. Emrick*

Printed in the United States of America.

Preface

The major objective of *General Radiotelephone License Handbook* is to prepare the prospective license holder for taking the required FCC examination. The thorough coverage in the license handbook can also provide a review and updating for a current license holder who wishes to become or is already a two-way radio technician. Remember that transmissions or tests during the installation, servicing, or maintenance of two-way radio units that may affect the proper operation of a two-way radio station shall be made by or under the immediate supervision and responsibility of a person holding a general radio operator license. Such a license is required for these activities in the private land-mobile radio services, aviation services, marine services, and citizens band services. A general radiotelephone license requires a passing grade for FCC Element III. The operator and/or technician of an international radio broadcast station must also have such a license.

Particular attention is given to solid-state fundamentals and circuits. There is a coverage of actual circuit types used in modern two-way radio equipment. Digital and microprocessor fundamentals are introduced, preparing you for the new way in radiocommunication systems.

An opportunity for the general radiotelephone license holder also exists in radio and tv broadcasting. Although a restricted radiotelephone operator permit (no license examination required) is the only license required to operate a station, the fact that you have a general license would look good on your resume. Furthermore a general radiotelephone license is required to install and maintain broadcast remote pick-up stations, broadcast aural STL and intercity relay stations, tv broadcast auxiliary stations, tv broadcast translators, broadcast low-power auxiliary stations, fm broadcast translator and booster stations, as well as broadcast experimental and development stations.

Two queries often posed by persons studying for the license examinations are: What am I qualified to do and what are my responsibilities as a license holder? This handbook attempts to answer these important questions as well as prepare you for the examination.

EDWARD M. NOLL

Contents

CHAPTER 1

CHAPTER 2

CHAPTER 3

CHAPTER 4

CHAPTER 5

CHAPTER 11

CHAPTER 12

CHAPTER 13

CHAPTER 14

CHAPTER 15

CHAPTER 16

CHAPTER 17

CHAPTER 18

CHAPTER 19

CHAPTER 1

Operator and Station Licensing

1-1. INTRODUCTION

Two-way radio systems, numbering in the millions, are installed, adjusted, repaired and operated by communications technicians having a *General Radiotelephone License.* Many of these radio units are operated by persons with a lower grade of license and, in some cases, by unlicensed persons under the supervision of the licensed operator of a base station. However, the installation, adjustment, and repair of two-way radio gear is handled by a general license holder. In fact, this grade of license is a prerequisite for making any adjustments that influence the operating frequency, power input/output, or modulation characteristics of the transmitter. In addition the operator of an international radio broadcast station must hold a general radiotelephone license. Such a license is also required for the operation of certain mobile equipment in the domestic radio and television broadcast field.

The purpose of this book is to assist the prospective license holder to obtain a general radiotelephone license. The coverage, although directed toward obtaining a license, is of such scope and detail it is also very useful to any presently licensed and active two-way radio technician as well as any prospective communications technician who wishes to enter this field. The first eight chapters of the book concentrate on fundamentals plus technical details on the various two-way radio services and equipment in a manner that helps you understand the responsibilities and knowledge required of the license holder. This information is particularly helpful in preparing for your license examination.

Typical two-way radio equipment is described. Most important, test equipment and test procedures are detailed. This latter information is particularly helpful in carrying out those adjustments which ensure compliance with FCC technical rules and regulations. Chapters 9 though 17 are in the form of self-evaluation questions and answers. They provide a review of the earlier chapters of the handbook as well as a review of some of the electronic basics that are not a part of the first eight chapters. Proper operating procedures along with *FCC Rules and Regulations* pertinent to two-way radio services are presented in Chapter 9.

The self-evaluation questions are only a guide to preparing for the general radiotelephone license examination. The actual FCC examinations are in the form of multiple-choice questions. These questions are not released and can be changed at will by the Federal Communications Commission. A sample multiple-choice test and answers are given in Chapters 18 and 19. The self-evaluation chapters sharpen you for the examination and give you a good idea of the basic subjects covered. *Study the entire book thoroughly* in preparing for the examination.

In summary, the book is an aid in securing the general license, discloses the responsibilities of such a license, and familiarizes you with the services you can render as a license holder. As you will learn, you can adjust and service, or supervise the adjustment and servicing, of a wide variety of radio stations.

1-1-1. Preparation for License Examination

The prospective general radiotelephone license holder must pass FCC Element III. There 100 multiple-choice questions. Passing grade is 75 percent.

After you feel prepared to take the examination you can contact the local district office of the Federal Communications Commission. These offices are listed at the end of this chapter. Application forms can usually be filled out prior to your trip to the district office. You will also receive an examination schedule that you might know when to appear for the examination.

As an aid in preparing for the examination the FCC releases a study syllabus. This syllabus is as follows:

TOPICAL OUTLINE FOR THE COMMERCIAL ELEMENT 3 OPERATOR EXAMINATION

A. RADIO ELECTRONICS FUNDAMENTALS

Concepts:

(1) Voltage
(2) Alternating Current, Direct Current
(3) Conductor, Insulator
(4) Open Circuit, Short Circuit
(5) Energy, Power
(6) Frequency, Wavelength
(7) Radio Frequency
(8) Audio Frequency
(9) Resistance
(10) Reactance
(11) Impedance
(12) Inductance
(13) Capacitance
(14) Impedance Matching
(15) Reactive Power
(16) Series and Parallel Resonance
(17) The "Skin Effect"
(18) Fields, Energy Storage, Electrostatic, Electromagnetic
(19) Exponential Charge/Discharge

(20) Induction; Magnetism
(21) Reluctance
(22) Hysteresis
(23) Permeability
(24) Mutual Inductance
(25) Magnetic Flux

Electrical Units:

(26) Metric Prefixes
(27) Volt, Millivolt, Microvolt, Kilovolt
(28) Ampere, Milliampere, Microampere
(29) Watt, Milliwatt, Kilowatt
(30) Hertz, Kilohertz, Megahertz, Gigahertz
(31) Ohm, Kilohm
(32) Farad, Microfarad, Picofarad
(33) Henry, Millihenry, Microhenry
(34) Decibel
(35) Joule
(36) Coulomb
(37) Volume Unit (VU)

Mathematical Relationships:

(38) Ohm's Law
(39) Kirchhoff's Laws
(40) Current and Voltage Dividers
(41) Electrical Power Calculations
(42) Series and Parallel Combinations; of Resistors, of Capacitors, of Inductors
(43) Turns Ratio; Voltage, Current, and Impedance Transformation
(44) Root-Mean-Square Value of a Sine Wave
(45) Resonant Frequency, Bandwidth, and "Q" of R-L-C Circuits
(46) Effective Radiated Power Calculations
(47) Time Constant for R-C and R-L Circuits
(48) Thevenin's Theorem
(49) Norton's Theorem
(50) Frequency and Period

B. RADIO WAVES, SIGNALS, AND EMISSIONS

The Communications Frequency Spectrum:

(1) Eight Major Bands of the Spectrum—VLF, LF, MF, HF, VHF, UHF, SHF, EHF
(2) Principal Allocations and Uses in Each Major Band

Radio Wave Propagation:

(3) Characteristics of Each Major Band
(4) Sky Wave ("Skip"); Ground Wave
(5) Ducting, Tropospheric Bending
(6) Absorption
(7) Line-Of-Sight
(8) Radio-Path Horizon
(9) Atmospheric Noise

Emissions and Modulation:

(10) Characteristics of Electromagnetic Radiation
(11) Wave Polarization
(12) Field Strength
(13) Amplitude Modulation
(14) Frequency Modulation
(15) Phase Modulation
(16) Single Sideband
(17) Double Sideband
(18) Carrier; Full, Reduced, Suppressed
(19) Sidebands
(20) Bandwidth
(21) Envelope
(22) Deviation
(23) Overmodulation
(24) Modulation
(25) FSK, AFSK

Signals:

(26) Frequency Translation, Multiplication, Mixing
(27) Fundamental, Harmonic
(28) Sine, Square, Triangle, and Sawtooth Waveforms
(29) Signal-to-Noise Ratio
(30) Analog, Digital, Binary, ASCII, BAUDOT
(31) Information, Bandwidth
(32) Data Rate, Baud Rate

C. ELECTRONIC PARTS AND COMPONENTS

Physical Appearance, Types, Characteristics, Applications, Failure Modes, and Schematic Symbols:

(1) Standardized Values (2) Tolerance
(3) Resistors; Carbon, Wirewound, Variable
(4) Capacitors; Disc, Electrolytic, Mica, Ceramic, Polystyrene, Tantalum, Air Variable, Trimmer
(5) Inductors; RF Chokes, Audio and Power Supply Chokes, Ferrite Beads, Stripline, Toroid
(6) RF, AF, and Power Transformers
(7) Fuses; Ratings, Slow-Blow, Chemical
(8) Batteries; Lead-Acid, Nicad, Alkaline, Cell Voltage, Capacity, Care, Recharging
(9) Vacuum Tubes; Diodes, Triodes, Tetrodes, Pentodes, Beam Power, Remote and Sharp Cut-Off Characteristic, Control Grid, Screen Grid Suppressor Grid, Plate, Cathode, Filament, Thermionic Emission, Plate Resistance, Amplification Factor, Transconductance
(10) Solid State Diodes; Junction, Rectifier, Zener, Varactor, Hot-Carrier, Pin
(11) Transistors; NPN, PNP, Junction, Unijunction, Power, Germanium, Silicon, Emitter, Base, Collector, Ratings and Parameters
(12) Field Effect Transistors; Enhancement, Depletion, MOS, CMOS, N-Channel, P-Channel, Gate, Source, Drain, Ratings and Parameters
(13) Silicon Controlled Rectifier, Triac
(14) Operational Amplifier Integrated Circuits
(15) Phase-Locked Loop Integrated Circuits
(16) TTL Digital Integrated Circuits
(17) CMOS Digital Integrated Circuits
(18) Light Emitting Diodes, Neon Lamp, Incandescent Pilot Lamp
(19) Panel Meters
(20) Quartz Crystals; Commercial Standard, Fundamental, Overtone
(21) Monolithic Crystal Filters, Ceramic Filters
(22) Helical Resonators
(23) Microphones; Dynamic, Condenser, Crystal
(24) Speakers, Output Transducers

D. PRACTICAL CIRCUITS AND TROUBLESHOOTING

Circuits—Characteristics and Functions:

(1) Power Supplies; Full-Wave, Half-Wave, Bridge, Center Tap, Filters, Mobile High Voltage and Inverters
(2) Filters; High-Pass, Low-Pass, Band-Pass, Band-Stop
(3) Voltage Regulators; Discrete, Integrated
(4) Block Diagrams of AM, SSB, and FM Transmitters and Receivers
(5) Amplifiers; Class A, AB, B, C
(6) Oscillators; Crystal, Other Common Types
(7) Modulators; AM, FM, Balanced, Phase
(8) Phase-Locked-Loop; Frequency Synthesizer
(9) Transmitter Final Amplifiers
(10) RF and IF Amplifier Stages; AVC and AGC

(11) Squelch; Carrier and Tone Actuated
(12) Detectors, Mixer Stages

Calculation of Voltages, Currents, and Power in Common Solid-State and Vacuum-Tube Circuits:

(13) Common Emitter Amplifier; Bias Network, Signal Gain, Input and Output Impedance
(14) Common Base Amplifier; Signal Gain, Input and Output Impedance
(15) Common Collector Amplifier; Signal Gain, Input and Output Impedance
(16) Common Source Amplifier; Signal Gain, Input and Output Impedance
(17) Cathode Follower Amplifier; Signal Gain, Input and Output Impedance

Troubleshooting Procedures:

(18) Changes in Circuit Performance Likely To Result from Failure of a Specific Component
(19) Circuit Components Likely To Be Faulty When Specific Changes in Circuit Performance Are Observed

E. COMMUNICATION EQUIPMENT AND INSTALLATIONS

FCC Type Acceptance and Specifications:

(1) Type Acceptance of Equipment
(2) Transmitter Carrier Frequency Tolerance
(3) Authorized Power and Emission; Emission-Bandwidth Designators, FCC Applications/Licensing

Transmitters, Receivers, Systems:

(4) Sub-Audible Tone Standards
(5) Microwave and Fixed Installations
(6) Installation Safety Precautions
(7) Prevention of Access by Unauthorized Persons
(8) Mobile Relay Stations (Repeaters); Timers
(9) RF and Remote Wireline Control Links
(10) Mobile Telephone (DPLMRS) Systems
(11) Selective Calling
(12) Trunked Systems
(13) Mobile Installations for VHF, UHF Including 800 MHz
(14) VHF Marine Equipment and Installations
(15) Alignment; Frequency, Power, and Modulation Checks
(16) Service-Maintenance Records

Antennas, Transmission Lines, and Auxiliary Equipment:

(17) Quarter-Wave Vertical Antenna
(18) Half-Wave Dipole Antenna
(19) Mobile Antennas
(20) Yagi Antenna
(21) Phased Vertical Array
(22) Vertical and Horizontal Polarization
(23) Antenna Gain, Beamwidth
(24) Radiation Patterns, Directivity, Major Lobes
(25) Coaxial Cable, Hardline
(26) Characteristic Impedance, Feedpoint Impedance
(27) Velocity Factor, Electrical Length
(28) Properties of Open, Shorted Sections of Transmission Line 1/8, 1/4, 3/8, and 1/2 Wavelength Long
(29) CB Phasing Harness
(30) Attenuation in Transmission Lines, Factors Affecting
(31) Duplexer, Cavity
(32) Diplexer
(33) Circulator

F. TEST EQUIPMENT AND MEASUREMENTS

Use of Test Equipment:

(1) Oscilloscope
(2) Multimeter
(3) Signal Generators
(4) Signal Tracer
(5) Reflectometer
(6) Field Strength Meter
(7) Non-Radiating Load; "Dummy Antenna"
(8) Watt Meter; Average, Peak Reading
(9) Frequency Counter
(10) Grid-Dip Meter
(11) Spectrum Analyzer; Spectrum Monitor
(12) Logic Probe
(13) Absorption Wavemeter
(14) CB Performance Tester
(15) Vacuum Tube Tester
(16) Transistor Tester
(17) Communications Monitor
(18) Performance Limitations of Test Equipment; Accuracy, Frequency Response, Stability

Measurement Procedures:

(19) Transmitter Power
(20) Deviation (FM)
(21) Percentage of Modulation (AM)
(22) Frequency, Drift
(23) Sensitivity; Sinad
(24) Spurious Signals
(25) Digital Logic Circuits
(26) Standing Wave Ratio
(27) Voltage
(28) Current
(29) Field Strength
(30) Resonant Frequency

G. ELECTROMAGNETIC COMPATIBILITY AND INTERFERENCE RESOLUTION

Recognition, Identification, Isolation, and Cure of Radio Frequency Interference

(1) Overload of Consumer Electronic Products by Strong RF Fields
(2) Interference to Consumer Electronic Products Caused by Radiated Harmonic Energy
(3) Disturbance in Consumer Electronic Products Caused by Audio Rectification
(4) Intermodulation Interference
(5) Receiver Desensitizing
(6) Cross Modulation Interference
(7) Capture Effect
(8) Vehicular Noise Suppression; Ignition Noise, Alternator Whine, Static

H. FCC RULES AND REGULATIONS

(1) FCC Station Licenses Required—Section 301, Communications Act of 1934
(2) Secrecy of Communications—Section 605, Communications Act of 1934
(3) The Rules and Regulations (Parts 0, 1, 2, 5, 13, 15, 17, 21, 81, 83, 87, 90, 94) 0.406
(4) Obedience to Lawful Orders 13.64
(5) Damage to Apparatus 13.65
(6) Unnecessary, Unidentified, or Superfluous Communications 13.66
(7) Obscenity, Indecency, Profanity 13.67
(8) False Signals 13.68
(9) Interference 13.69
(10) Exhibiting Signed Copy of Application 13.72
(11) Frequencies for Use in Distress and Search and Rescue Operations 83.352
(12) Frequencies for Calling 83.353
(13) Frequency for Aircraft in Distress 87.183(f)
(14) Station Authorization Required 90.113
(15) Modification of License 90.135

(16) Special Temporary Authority 90.145
(17) Station Construction Period 90.155
(18) Discontinuance of Station Operation 90.157
(19) Transmitter Measurements 90.215

I. RADIOTELEPHONE OPERATING PRACTICE

(1) Monitoring Before Transmitting; Defeat Sub-Audible Tone Squelch
(2) Selection of Transmitter
(3) Leaving Interval Between Transmissions for Others to Break
(4) Speaking into a Microphone (5) Phonetic Alphabet
(6) On-the-Air Testing Procedure (7) Length of Messages
(8) Radiotelephone Procedure Words; "Roger," "Over," "Out"
(9) Squelch Control/Volume Control
(10) International Maritime Distress Call Procedure
(11) VHF Marine Calling Procedure

A study of this syllabus discloses that a variety of basic, intermediate and advanced subjects are covered. To keep this handbook to a sensible size and cost it is assumed that the reader has a proper background in electrical and electronics basics. To be specific the handbook assumes you have a background that includes the topics of section A of the FCC Syllabus as well as topics 1 through 9 of section C.

General Radiotelephone License Handbook concentrates on solid-state technology (basic, intermediate, and advanced levels) as well as modern two-way radio equipment and test gear because these are the systems that you will encounter when employed in the field. However, in review form, many of the electrical and electronics basics are covered in the question and answer Chapters 10 through 17. Vacuum-tube data is given along with basic and more advanced vacuum-tube circuits. From these studies you can judge just how much additional review you need relative to the topics of section A and the early topics of section C of the syllabus.

A good study plan is to concentrate first on Chapters 1 through 9. Then begin your work on the question and answer chapters, Chapters 10 through 17. Try to answer each question before you read the followup discussion. This procedure will guide you in reviewing the first eight chapters as well as give you a clue to just how much you have to go back to a study of some electrical and electronics basics.

1-2 REFERENCES

The references that follow can be of help in advancing your career in radiocommunications or as an aid in better retaining electronic knowledge with practical experience. The latter three references in particular can be of assistance in obtaining practical experience at home or in a school laboratory situation. Reference 1 has been written for the communications technician who wishes to enter the radio and/or television broadcast field. It expands upon the radiocommunication and transmitter information given in this handbook. References 2 and 3 stress a better understanding of linear and digital integrated circuits by building up low-cost circuits on solderless circuit boards. Knowledge and practical skill in handling these devices are obtained. The fourth reference concentrates on microprocessor circuits and a chip-by-chip series of Democircuits permits you to build up a simple microprocessor system.

1. *Broadcast Radio and TV Handbook.*
2. Ed Noll's *Solid State Circuit Files,* Volume I, Bipolar Transistor, FET, and Linear IC Circuits.
3. Ed Noll's *Solid State Circuit Files,* Volume II, TTL and CMOS Circuits.
4. *Microprocessor Circuits,* Volume I, Fundamentals and Microcontroller.

Available from:

Howard W. Sams & Co., Inc.
4300 West 62nd St.
Indianapolis, Indiana 46268

1-3. LOWER-GRADE LICENSE

Many classes of two-way radio stations can be operated, but not serviced, by holders of a *restricted radiotelephone permit.* Such license holders may not adjust and service the radio station equipment in any manner that could in any way possibly result in improper transmitter operation. In general, for operating a radio station below 25 MHz, it is necessary that the person hold a commercial radio operator license or permit of any class. An unlicensed person, after being authorized to do so by the station licensee, may operate a mobile station during the course of normal rendering of service, while it is associated with and under the operational control of a base station of the same station licensee. For stations operating above 25 MHz, unlicensed persons may operate certain base and mobile stations after being so authorized by the station licensee.

1-4. FUNDAMENTAL SYSTEMS

In the two-way radio services there are three major station classifications—mobile, base, and fixed. A mobile station is one associated with a moving vehicle such as a truck, automobile, boat, or aircraft. A base station, often referred to as a land station, has a fixed position and is used for communicating with one or more mobile stations (and on occasion, with other base or fixed dispatch stations). The great majority of two-way radio systems come under the base-mobile classifications and usually include a single base station and one or more associated mobile stations.

A fixed station is one in a permanent location used to communicate with other fixed stations. This form of two-way radio is usually referred to as point-to-point communications. Normally, in such services, there are no facilities for communicating with mobile stations. Their principal objective is to convey information between two or more fixed locations. This differs from the base station, which also has a permanent position but is used for the purpose of communicating with mobile stations.

The most common arrangements of two-way radio stations are shown in Fig. 1-1. In the *simplex* arrangement of Fig. 1-1A, the

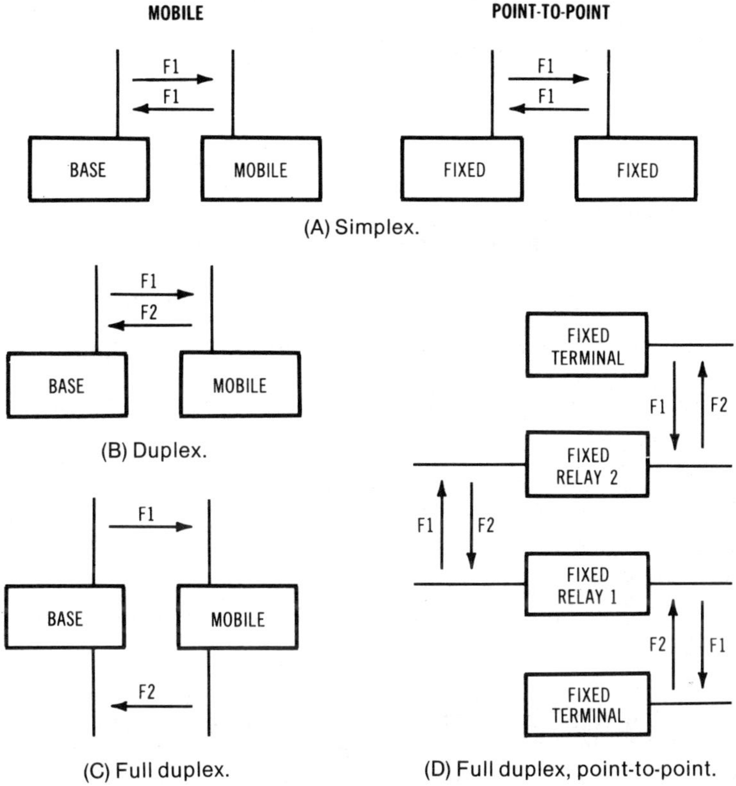

Fig. 1-1. Two-way radio systems.

base and mobile units operate on the same frequency (F1). A sequential "on-off" communication is established in which only one station can transmit at a given time, but can be heard by all other stations of the system. Each mobile station can hear both sides of a conversation between the base station and any mobile station. Likewise, communications can be established mobile-to-mobile when the stations are within range. Although not common, a simplex arrangement can also be used in a point-to-point radio system.

The *duplex* arrangement of Fig. 1-1B is a common two-way radio arrangement. The base station transmits on one frequency (F1), while the mobile stations use a different frequency (F2). In this arrangement the base station can communicate with all mobile stations, but the mobile stations cannot communicate directly with each other. Information to be conveyed between mobile units must be transmitted through the base station. Each mobile station receives only the base-station signal.

Although the duplex arrangement requires two transmitting frequencies (one for the base station and one for the mobile units), it is usually more satisfactory. It is possible to exert better control over the mobile units, and each mobile station is not confused by signals coming from other mobile stations (as with the simplex system).

The duplex arrangement also uses an "on and off" method of transmission. Quite often, frequency F1 is close to frequency F2. In fact, it is customary at each station to use the same antenna for transmitting and receiving. A relay or electronic switch, operating in conjunction with the transmit-receive switching of the station, changes the antenna between transmitter and receiver. It is not uncommon for the other segments of the station equipment to be used in both the "receive" and "transmit" positions.

A *full duplex* arrangement is shown in Fig. 1-1C. Operating frequencies F1 and F2 are well-separated; consequently, each station can transmit and receive at the same time. Each transmitter output is isolated and separated in frequency in order not to block the input of its companion receiver. With this arrangement, there can be simultaneous communications between the two stations as in line telephone conversations. If more than one mobile station is included, it will usually be on a different transmit frequency. The base station can use the same or a different frequency to establish communications with other mobile stations.

Full duplex is more common in point-to-point communication services than in mobile systems. In the typical arrangement of four stations shown in Fig. 1-1D, the transmit and receive frequencies of the various fixed stations are staggered, each station receiving on one frequency and transmitting on another. Highly directional antennas are employed, making it possible to establish duplex—and even full duplex—communications with minimum interference between stations. Notice that relay station 1 transmits on F1 and receives on F2, and that relay station 2 transmits on F2 and receives on F1. To permit the use of antenna systems with the highly directional characteristics needed to prevent interference, point-to-point stations are usually assigned frequencies which are in the microwave or upper half of the uhf spectrum.

1-5. STATION ASSIGNMENTS AND LICENSING

Each radio transmitter requires a station license. The transmitter operators do or do not require an operator's license, depending on the radio service they are associated with. For example, in most Maritime and Aviation Services it is necessary that the operator have a license. In fact, an operator's permit is required for practically all radio services that operate below 25 MHz. In the Citizens Band Radio Service a station license is required, but the operator is not licensed. Many of the radio services above 25 MHz, including the Private Land Mobile Radio Services, do not require an operator's license.

However, a distinction must be made between strictly operation and equipment maintenance. In most cases, the lower-grade operator licenses permit little else than the operation of those controls required in the normal handling of two-way radiocommunications. Adjustments or tests during or coincident with the construction, installation, servicing, or maintenance of a radio transmitter shall be made by or under the immediate supervision or responsibility of a person holding a commercial radio operator's license, either radiotelephone or radiotelegraph, as may be appropriate for the type of emission employed. Such a licensee is responsible for the proper functioning of the radio-transmitter equipment.

There are many point-to-point and mobile communications systems that can be operated, installed, adjusted, and maintained by general radiotelephone operators. In fact, practically all stations, except broadcast, radiotelegraph, and amateur stations, come under the authority of a general radiotelephone license holder.

Table 1-1. Major Two-Way Radiotelephony Bands

Medium frequency (mf)	1.6-25 MHz
High frequency (hf)	25-50 MHz
Very high frequency (vhf_1)	108-136 MHz
Very high frequency (vhf_2)	150-174 MHz
Ultrahigh frequency (uhf_1)	450-512 MHz
Ultrahigh frequency (uhf_2)	806-821 MHz
Ultrahigh frequency (uhf_3)	851-866 MHz

The major two-way radio assignments in the medium-frequency (mf) band are police radio, marine, and aviation, although there are some point-to-point relay assignments and a very limited number of land-mobile allocations. Because of the great emphasis on small boats, an active portion of the spectrum between 2 and 3 MHz has been allocated to small-ship stations operating in coastal and inland waters. There are numerous aeronautical station assignments, largely for the benefit of air-carrier services (passenger and cargo). Many two-way radio aeronautical assignments, particularly for private aircraft and airport facilities, are made in the vhf_1 band.

The upper portion of the hf spectrum is crowded with mobile assignments. It includes not only the very active Citizens Band, but a moderate percentage of the private land mobile radio services as well. It is a very active portion of the frequency spectrum insofar as land mobile-radio equipment is concerned. Each transmitter must in some way be linked to a general radiotelephone license holder.

Similar assignments are made in the two vhf bands. The 72-76

MHz band has been allocated largely to operational fixed-station use. Point-to-point relay systems have allocations in this spectrum. In time, most point-to-point allocations and some of the systems now operating on this band will use the superhigh frequency (shf) and the upper end of the uhf bands. A high degree of stability and freedom from interference can be obtained more readily in the microwave spectrum.

Aeronautical radio facilities dominate the vhf_1 spectrum. The assignments on the vhf_2 spectrum are similar to those of the hf band. Land mobile-radio assignments are predominant. However, aviation radio, railroad radio, and coastal radiomarine allocations are also available.

The allocations on the uhf band are similar to those of the vhf and hf bands. A number of fixed point-to-point allocations are becoming available, particularly at the high-frequency end of the uhf band. This band includes an impressive array of allocations for land vehicle, marine, and aviation facilities. Radio navigational aids (including radar) for aeronautical and marine services are served by uhf and higher microwave frequency assignments. Some mobile radio assignments are also made in the uhf television spectrum between 470–890 MHz.

1-6. ACTIVE TWO-WAY RADIO SERVICES

Four very active two-way radio services that require the services of a general licensed radiotelephone technician are private aviation, small-boat marine radiotelephone, private land mobile, and Citizens Band. The FCC is particularly concerned about transmitter adjustments and measurements. Of particular concern are power, frequency, and modulation characteristics. These are factors of concern when the transmitter is initially installed and when any change is made in the transmitter which may influence these parameters. Technical details are covered in Chapters 5, 6, and 7.

1-6-1. Private Land Mobile Radio Services

Frequencies are allocated for a variety of radio services under the main heading of public safety, special emergency, industrial radio and land transportation. Most allocations are in the frequency spectra of 25–50 MHz, 150–174 MHz, and 450–512 MHz. The spectra 806–821 MHz and 851–866 MHz are becoming increasingly active.

The radio services that come under public safety are local government, police radio, fire radio, highway maintenance, and forestry-conservation. The special emergency radio services provide allocations for medical services, rescue organizations, veterinarians, disaster relief organizations, school buses, beach patrols, isolated area radio service, communications standby, and emergency repair of public communications facilities. The categories that come under the industrial radio services are power, petroleum, forest products, motion picture, relay press, special industrial, business radio, manufacturers, and telephone maintenance. The land transportation radio services include motor carrier, railroad radio, taxicab radio, and automobile emergency.

In the above radio services there are approximately one million authorized stations. This provides some idea of the scope of the technical activities in the private land mobile radio services. Frequency assignments are at a premium and it is necessary to have local frequency-advisory boards that assist the FCC in making frequency assignments to minimize as much as possible interference among stations.

Technical standards are covered in detail in Chapters 4 and 5. However, the general scope of the material is introduced in the following paragraph under transmitter measurements as taken from Volume V of the FCC Rules and Regulations.

Sec. 90.215 Transmitter measurements.

(a) The licensee of each station shall employ a suitable procedure to determine that the carrier frequency of each transmitter authorized to operate with an output power in excess of two watts is maintained within the tolerance prescribed in Section 90.213. This determination shall be made, and the results entered in the station records in accordance with the following:

(1) When the transmitter is initially installed;

(2) When any change is made in the transmitter which may affect the carrier frequency or its stability.

(b) The licensee of each station shall employ a suitable procedure to determine that each transmitter authorized to operate with an output power in excess of two watts does not exceed the maximum figure specified on the current station authorization. On authorizations stating only the input power to the final radio-frequency stage, the maximum permissible output power is 75 percent for frequencies below 25 MHz and 60 percent of the input power for frequencies above 25 MHz. If a non-DC final radio-frequency stage is utilized, then the output power shall not exceed 75 percent of the input power. This determination shall be made, and the results thereof entered into the station records, in accordance with the following:

(1) When the transmitter is initially installed;

(2) When any change is made in the transmitter which may increase the transmitter power input.

(c) The licensee of each station shall employ a suitable procedure to determine that the modulation of each transmitter, which is authorized to operate with an output power in excess of two watts, does not exceed the limits specified in this part. This determination shall be made and the following results entered in the station records, in accordance with the following:

(1) When the transmitter is initially installed;

(2) When any change is made in the transmitter which may affect the modulation characteristics.

(d) The determinations required by paragraphs (a), (b), and (c) of this section may, at the opinion of the licensee, be made by a qualified engineering measurement service, in which case the required record entries shall show the name and address of the engineering measurement service as well as the name of the person making the measurements.

(e) In the case of mobile transmitters, the determinations required by paragraphs (a) and (c) of this section may be made at a test or service bench: Provided, That the measurements are made under load conditions equivalent to actual operating conditions; and provided further that after installation in the mobile unit the transmitter is given a routine check to determine that it is capable of being received satisfactorily by an appropriate receiver.

1-6-2. Marine Small-Boat Ship-to-Shore Radio

There are more than 350,000 stations authorized in the radiomarine services. A high percentage of these are compact ship-to-shore and ship-to-ship radiotelephones installed on small commercial and pleasure boats. In fact, any vessel, regardless of size, that is transporting more than six passengers for hire and is navigated in the open seas or any tidewater within the jurisdiction of the United States adjacent or contiguous to the open seas, must be equipped with an acceptable radio installation.

Small marine transceivers are operated in the 2–3 MHz and 156–158 MHz bands. These units are very compact and can be installed conveniently on small commercial and pleasure boats. Key allocations are given in Tables 1-2 and 1-3. In the case of the vhf assignments channel designators instead of specific frequencies are indicated.

The call and distress frequency for ship radiotelephone stations in the 2–3 MHz band is 2182 kHz. Channel 16 or 156.8 MHz is the calling and distress frequency for the vhf assignments.

Actual frequency assignments depend on the area in which you operate and it is the responsibility of dealers and technicians to provide guidance as to appropriate operating frequencies and facilities available in a specific geographic area of interest.

Typical vhf-fm marine transceivers are usually multichannel radios with as high as 55 transmit channels and 96 receive channels. There are also facilities for weather receive channels as well. Transmit power is the maximum permissible value of 25 watts, although many units have facilities for reducing power in order to minimize interference. Facilities for including a watch or channel 16 are also included.

Table 1-2. HF Intership Frequencies

Frequency (kHz)	Geographic area
2003	Great Lakes only.
2082.5	All areas.
2142	Pacific coast area south of latitude 42 degrees north, on a day only basis.
2203	Gulf of Mexico.
2638	All areas.
2670	All areas.
2738	All areas except Great Lakes and the Gulf of Mexico.
2830	Gulf of Mexico only.

Presently short-range marine communications now take place more frequently on the vhf rather than the mf bands. In fact, a vhf installation is compulsory even though mf equipment is installed on the boat.

The medium-frequency marine radiotelephone employs single-side band transmission rather than the old double-sideband, amplitude-modulation technique which required a greater operating bandwidth. Single-sideband operation requires only one-half of the emission bandwidth needed by standard amplitude modulation.

A typical 2–4 MHz radiotelephone might include eight or more crystal-controlled channels including the 2182 kHz emergency frequency. Again the selection of crystals for the channels requires guidance from the dealers and radiotelephone technicians of a given area. Assignments depend on the area facilities as well as the possible range of operation of the boat.

Table 1-3. VHF-FM Channel Designation

Channel Designators	TYPE OF COMMUNICATIONS Points of Communications	Commercial Vessels Channel Capability				Recreational Vessels Channel Capability				
		1/ 12	2/ 18	2/ 24	All	1/ 6	1/ 12	2/ 18	2/ 24	All
		Number of recommended channels in each group								
16 (Mandatory)	DISTRESS, SAFETY & CALLING Intership & ship to coast	1	2	2	1	1	1	2	2	1
06 (Mandatory)	INTERSHIP SAFETY Intership	1	1	1	1	1	1	1	1	1
65, 66, 12, 73 14, 74, 20	PORT OPERATIONS Intership & ship to coast	2	3	4	7		1	3	4	7
13	NAVIGATIONAL Intership & ship to coast	1	1	1	1		1	1	1	1
22	Liaison communications only with U.S. Coast Guard ship, coast, or aircraft stations	1	1	1	1	1	1	1	1	1
07, 09, 10, 11, 18, 19, 79, 80	COMMERCIAL Intership & ship to coast	3	4	6	8					
67, 08, 77, 88	COMMERCIAL—Intership	1	2	3	4					
09, 68, 69, 71, 78	NON-COMMERCIAL Intership & ship to coast					1	3	4	5	5
70, 72	NON-COMMERCIAL Intership						2	2	2	2
24, 84, 25, 85, 26, 86, 27, 87, 28, 88 3/	PUBLIC CORRESPONDENCE Ship to public coast	2	4	6	9	2	2	4	8	9
162.4MHz & 162.55 MHz	NOAA WEATHER SERVICE Ship receive only	1/	1/	1/	1/	1/	1/	1/	1/	1/

1/ It is recommended that in addition to the mandatory channels, each VHF ship station have both weather receive channels. A ship station equipped with a 6 or 12 channel transceiver that does not have one or two weather-receive-only channel(s) may wish to either employ one or two of the regular channel(s) for the reception of weather broadcasts or employ a separate weather receiver.

2/ It is recommended that ship stations which have two transceivers install channel 16 in both.

3/ Channel 88 is available for Public Correspondence in the Great Lakes and the Saint Lawrence Seaway only (§83.351(b)(72)).

The following applies to the adjustment of transmitting apparatus:

ADJUSTMENT OF TRANSMITTING APPARATUS

All adjustments of radio transmitting apparatus in any station subject to this part during or coincident with the installation, servicing, or maintenance of such apparatus which may affect the proper operation of such station, shall be performed by or under the immediate supervision and responsibility of a person holding a general operator license, who shall be responsible for the proper functioning of the station equipment; Provided, however, that:

(a) Only persons holding a radiotelegraph operator license shall perform such functions at radiotelegraph stations transmitting by any type of Morse code;

(b) Only persons holding a license, containing a ship radar endorsement shall perform such functions on radar equipment.

1-6-3. Aviation Radio Services

The aviation radio services set aside portions of the spectrum for radio communication and radio navigation facilities for aircraft operators, aeronautical enterprises, and organizations that require radio transmitting facilities for safety purposes and other necessities. The radio stations are allocated on the basis of a number of catagories-airborne- aeronautical advisory, aeronautical multicom, aeronautical enroute, aeronautical metropolitan, flight test, flying school, airport, aeronautical utility, aeronautical search and rescue mobile, aeronautical fixed, operational, radio navigational land, and civil air patrol.

Most of these stations require a licensed operator. Insofar as installation and maintenance are concerned a higher grade license is needed. If radiotelegraphy is employed, the operator must be the holder of an appropriate radiotelegraph license.

In general, communications in the aviation services shall be restricted to safe, expeditious, and economical operation of aircraft, and the protection of life and property in the air. There are many electronic aids to aircraft navigation; they can be considered in four general categories of communication, navigation, traffic control, and landing. Most of the modern-day aeronautical radio activity occurs in the frequency spectrum between 108 and 136 MHz. Radionavigation uses the spectrum between 108 and 118 MHz; air traffic control uses 118–136 MHz. Spotted throughout the spectrum are frequencies assigned to both aircraft and aeronautical ground stations. For example, in flying a private aircraft you will find the frequencies of the various ground stations given on navigational maps and/or charts. These aeronautical ground stations monitor certain aircraft frequencies and you can quickly establish contact enroute by setting your aviation radio to an appropriate frequency.

Many aviation radio units have dual-reception capability. Thus it is possible to receive a radio navigational signal and also maintain a two way radio contact with an aeronautical ground station.

Most modern flying is done via the vhf omnidirectional range stations. These are called *VOR* or *OMNI* stations. In the vhf spectrum there is largely static-free reception and the bending and false beams of the older lower-frequency radio range stations are not transmitted. A reliable directional pattern can be produced at the higher frequencies.

A typical installation on a small aircraft may be a transmitter and receiver that operate on a frequency indicated by a digital readout. The communicator can be used as a simple transceiver operating on the same frequency. However, duplex operation is also possible by proper settings with the transmit and receive frequencies differing. A strictly navigational receiver can also be included. Its associated meters indicate bearing and right-left positioning of the aircraft with respect to a VOR range station.

A low frequency receiver may also be installed to permit tuning to a low-frequency range station. Many aircraft include a *transponder* which sends out a coded signal. Thus at a ground station the aircraft can be seen and identified on a radar screen. Another addition could be a distance measuring (DME) unit. In operation it sends out an interrogating signal to a special range facility called a *VORTAC* station. The VORTAC station in turn sends back a signal to the aircraft unit that permits the pilot to determine the distance to the VORTAC station.

The frequency of 121.5 MHz is the universal simplex clear channel for use by aircraft in distress or condition of emergency. It is not assigned to aircraft unless other frequencies are assigned and available for normal communications needs. The channel is available as follows:

1. For emergency communications when circumstances beyond the control of a pilot prevent communication between the aircraft and ground stations on other regularly assigned channels.
2. For emergency direction-finding purposes.
3. For establishing air-to-ground contact by aircraft in distress, emergency, or when lost.
4. In connection with search and rescue operations, to provide a common channel for aircraft (either civil or military) not equipped to transmit on 123.1 MHz. This includes communications between aircraft, and between aircraft and ground stations. Stations having the capability should change to 123.1 MHz as soon as practicable.
5. To provide common frequency for survival communications and survival beacons (emission A2).
6. For air/ground communications between aircraft and ocean vessels for safety purposes when service on other vhf channels is not available.

The frequency 243 MHz is available to survival craft stations which are also equipped to transmit on 121.5 MHz.

Installation, maintenance and adjustments that influence the radiation from the transmitter must be made by an operator with a general radiotelephone license. If radiotelegraphy or other form of coded transmission is used, the operator must have an appropriate radiotelegraph license.

The various aeronautical ground stations must be operated by FCC-licensed operators with the exception of the FAA stations operated by government employees. Also certain land mobile stations do not require an operator; usually such vehicles are associated with an airport control station or other aeronautical ground station which is under the control of a licensed operator. The general operator requirements are:

GENERAL OPERATOR REQUIREMENTS.

(a) Except as provided for in FCC rules or as limited on the face of the operator license or permit, all stations in the Aviation Services shall be operated by persons holding any class of commercial radio operator license or permit issued by the Commission: *Provided*, That

(1) Only a person holding an appropriate license or permit shall operate aircraft stations

(i) Utilizing frequencies below 30 MHz not exclusively allocated to the aeronautical mobile service, or

(ii) Utilizing frequencies above 30 MHz not allocated exclusively to aeronautical mobile services and which are assigned for international use; and

(2) Only a person holding a general radiotelephone license shall operate an aircraft station at which the installation is not used solely for telephony, or at which the power is more than 250 watts carrier power or 1,000 watts peak envelope power.

(b) The licensed operator of a land or aeronautical public service station using telephony may permit other persons to transmit or to communicate under his direct supervision and responsibility over the facilities of the station in accordance with the terms of the station license.

TRANSMITTER ADJUSTMENTS AND TESTS BY OPERATOR.

All transmitter adjustments or tests during or coincident with the installation, servicing, or maintenance of a radio station, which may affect the proper operation of such station, shall be made by or under the immediate supervision and responsibility of a person holding a general radiotelephone or radiotelegraph first or second-class operator license who shall be responsible for the proper functioning of the station equipment: Provided however, That only a person holding a radiotelegraph first-or second-class operator license shall perform such functions at radiotelegraph stations transmitting by any type of the Morse Code for purposes other than automatically keyed station identification.

1-6-4. Citizens Band (CB) Radio Service

The Citizens Band (CB) Radio Service has had a phenomenal growth, as attested by the millions of transmitters now in operation. A segment of the frequency spectrum has been set aside for the Citizens Band Radio Service to provide a private, short-distance radiocommunications service for the business or personal activities of the licensees. Any person who is 18 years old or older may obtain a station license in this service if his application meets the requirements of the Citizens Band Radio Service. Partnerships, associations, trusts, or corporations meeting these requirements can be licensed including any state, territorial, local government entity, or any service organization or association, including civil defense. No operator license is required.

The Citizens Band Radio Service has many uses. It is particularly attractive because low-cost, efficient equipment is available and almost everyone is eligible for a license. The Citizens Band Radio Service is basically mobile in nature with a majority of installations made in trucks and autos. Travel conditions, highway instructions, and personal exchanges dominate the mobile conversations. There are local activities as well as continuous chatter among trucks and private cars on the interstate highways.

The family base-mobile installation is particularly popular. The base station (still considered a mobile installation in terms of FCC terminology) is installed at home, and a mobile unit is installed in the family car. When one member of the family is shopping or traveling to or from work, that member is able to maintain contact with home. In this manner delays, travel changes, and additional shopping information can be confirmed and arranged via two-way radio. Often the time of departure from one's business or place of employment is quite indefinite. At a reasonable distance from home, a mobile-to-home contact can be made to let the family know said person is on the way home.

Of course, the two-way radio is handy in case of a car breakdown, or other delays. In many localities it is not only possible to call home, but necessary assistance can be obtained by calling a station operated by a garage or service station. Channel 9 has been set aside for the exclusive use in requesting road help or for other emergencies. Citizens Band radio is often quite helpful on long trips because breakdowns on turnpikes often involve considerable delay if you have to await the routine scheduled trip of a repair truck along the highway. Citizens Band operation permits you to obtain a variety of services ahead of time such as lodging, repairs, location and traffic guidance, medical assistance, etc.

Citizens Band radio can be a special boon to a small businessman. Any type of pickup and/or delivery service can derive benefit in terms of reporting delays, breakdowns, changes in routing, and last-minute pickups or deliveries. A fuel-oil delivery service represents only one example of the many businesses that can use CB radio to advantage. Enroute trucks can be informed of call-ins from customers right after they have been completed and while the truck may be in the very vicinity of the customer's location. Parts jobbers in the service field use CB to advantage in making their deliveries to retail stores and service shops. Communications can even be maintained between the main store and outlying branches of various types of businesses.

Professional people can use Citizens Band radio for maintaining contact between car and office. Late call-ins or emergencies can be radioed to the doctor as he makes his rounds. Similarly, a veterinarian making his farm calls can keep in touch with his office.

Channel 9 may be used only for emergency communications or for traveler assistance.

CB channel designations and frequencies are as follows:

Channel	Frequency (MHz)	Channel	Frequency (MHz)
1	26.965	13	27.115
2	26.975	14	27.125
3	26.985	15	27.135
4	27.005	16	27.155
5	27.015	17	27.165
6	27.025	18	27.175
7	27.035	19	27.186
8	27.055	20	27.205
		21	27.215
9	27.065	22	27.225
		23	27.255
10	27.075	24	27.235
11	27.085	25	27.245
12	27.105	26	27.265

Channel	Frequency (MHz)	Channel	Frequency (MHz)
27	27.275	34	27.345
28	27.285	35	27.355
29	27.295	36	27.365
30	27.305	37	27.375
31	27.315	38	27.385
32	27.325	39	27.395
33	27.335	40	27.405

You must, at all times and on all channels, give priority to emergency communications.

A station license is required; no operator's license is needed. However, a general radio operator license is required for adjustments to radio transmitters during installation, testing, or servicing that may cause the transmitter to operate off frequency, or in a manner which may in other ways violate the rules.

FCC Rules and Regulations state:

COMPLIANCE WITH TECHNICAL REQUIREMENTS.

(a) Upon receipt of notification from the Commission of a deviation from the technical requirements of the rules in this part, the radiations of the transmitter involved shall be suspended immediately, except for necessary tests and adjustments, and shall not be resumed until such deviation has been corrected.

(b) When any station licensee receives a notice of violation indicating that the station has been operated contrary to any of the provisions contained in Subpart E of this part, or where it otherwise appears that operation of a station in this service may not be in accordance with applicable technical standards, the Commission may require the licensee to conduct such tests as may be necessary to determine whether the equipment is capable of meeting these standards and to make such adjustments as may be necessary to assure compliance therewith. A licensee who is notified that he is required to conduct such tests and/or make adjustments must, within the time limits specified in the notice, report to the Commission the results thereof.

(c) All tests and adjustments, which may be required in accordance with paragraph (b) of this section shall be made by, or under the immediate supervision of, a person holding a commercial operator license appropriate for the type of emission employed. In each case, the report which is submitted to the Commission shall be signed by the licensed commercial operator. Such report shall describe the results of the tests and adjustments, the test equipment and procedures used, and shall state the type, class, and serial number of the operator's license. A copy of this report shall be kept with the station records.

Several other important regulations are:

CB RULE 18. HOW HIGH MAY I PUT MY ANTENNA?

(a) If your antenna is installed at a fixed location, the antenna structure (whether receiving, transmitting, or both) must comply with either of the following:

(1) The highest point must not be more than 6.10 meters (20 feet) higher than the highest point of the building or tree on which it is mounted; or

(2) The highest point must not be more than 18.3 meters (60 feet) above ground.

(b) If your CB station is located near an airport, and if your antenna is more than 6.1 meters (20 feet) high, you may have to obey additional restrictions. The highest point of your antenna must not exceed one meter above the airport elevation for every hundred meters of distance from the nearest point of the nearest airport runway. Differences in ground elevation between your antenna and the airport runway may complicate this formula. If your CB station is near an airport, you may contact the FCC for a worksheet to help you figure the maximum allowable height for your antenna. Consult Part 17 of the FCC's Rules for more information.

CB RULE 20. HOW MUCH POWER MAY I USE?

(a) Your CB transmitter power output must not exceed the following values under any conditions:

AM (A3)—4 watts (carrier power)

SSB (A3J)—12 watts (peak envelope power)

(b) If you need more information about the power rule, see the technical rules in Subpart E of Part 95.

CB RULE 21. MAY I USE POWER AMPLIFIERS?

(a) You must not use or attach a linear or external radio-frequency power amplifier at any CB station in any way.

(b) There are no exceptions to this rule.

(c) The FCC will presume you have used a linear or other external rf power amplifier if:

(1) It is in your possession or on your premises; and

(2) There is other evidence that you have operated your CB station with more power than allowed by CB Rule 20.

(d) Paragraph (c) of this rule does not apply if you hold a license in another radio service which allows you to operate an external rf power amplifier.

CB RULE 41. HOW DO I HAVE MY CB TRANSMITTER SERVICED?

(a) You may adjust your own antenna to your CB transmitter and you may make "radio checks."

(b) Each internal repair and each internal adjustment to your CB transmitter must be made by, or under the direct supervision of, a person holding a general radiotelephone operator license.

(c) Except as provided in paragraph (d) of this section, each internal repair and each internal adjustment of a CB transmitter in which signals are transmitted must be made using a nonradiating ("dummy") antenna.

(d) Brief test signals using a radiating antenna may be sent to adjust a transmitter to an antenna or to detect or measure spurious radiation. These test signals may not be longer than one minute during any five minute period.

CB RULE 42. MAY I MAKE ANY CHANGES TO MY CB TRANSMITTER?

(a) You must not make or have anyone else make any internal modification to your CB transmitter.

(b) You must not operate a CB transmitter which has been modified by anyone in any way, including modification to operate on unauthorized frequencies or with illegal power.

1-7. STATION LICENSE

Radio transmitters in the two-way radio services may not be operated without a station license. Station license authorization is granted by the FCC. The necessary application form can be requested from the FCC in Washington, DC or from a district office (Table 1-4). In the private land-mobile radio services FCC form 400 is used.

A variety of technical and business information is required to compete the form. The business information requested verifies the eligibility of the applicant and includes a general description of the business and activities and how the radio system is to be employed in that activity. Business addresses must be given as well as the address of the transmitter site and the location of the transmitter control points. Antenna structure, height and other data must be indicated as well as a description of the possible area in which the

Table 1-4. Addresses of FCC Field Offices

Listed below are the addresses and telephone numbers of the FCC field offices. This list is alphabetical by state, and also includes the field offices in Puerto Rico and The District of Columbia (Washington, D.C.).

ALASKA, Anchorage 99510
Rm. G-63, U.S. Post Office &
Courthouse Bldg. P.O. Box 644
Phone (907) 265-5201

CALIFORNIA, Long Beach 90807
Rm. 501, 3711 Long Beach Blvd.
Phone (213) 426-7886

CALIFORNIA, San Diego 92101
7840 El Cajon Blvd.
Suite 405
Le Mesa, California 92041
Phone (714) 293-5460

CALIFORNIA, San Francisco 94111
323-A Customhouse
555 Battery St.
Phone (415) 556-7700

COLORADO, Denver 80202
Suite 2925, The Executive Tower
1405 Curtis St.
Phone (303) 837-4053

DISTRICT OF COLUMBIA 20554
(Washington, D.C.)
1919 M St., NW Rm. 411
Phone (202) 632-8834

FLORIDA, Miami 33130
919 Federal Bldg.
51 SW First Ave.
Phone (305) 350-5541

FLORIDA, Tampa 33602
Barnett Office Bldg. Rm. 809
1000 Ashley St.
Phone (813) 228-2605

GEORGIA, Atlanta 30309
Rm. 440, Massell Bldg.
1365 Peachtree St., N.E.
Phone (404) 881-7381

GEORGIA, Savannah 31402
Rm. 238, Federal Bldg. & Courthouse
P.O. Box 8004 (125 Bull St.)
Phone (912) 232-4321 Ext. 320

HAWAII, Honolulu 96850
7304 Prince Kuhio Fed. Bldg.
300 Ala Moana Blvd.
P.O. Box 50023
Phone (808) 546-5640

ILLINOIS, Chicago 60604
3935 Federal Bldg.
230 South Dearborn St.
Phone (312) 353-0195

LOUISIANA, New Orleans 70130
829 F. Edward Hebert Federal Bldg.
600 South St.
Phone (504) 589-2094

MARYLAND, Baltimore 21201
George M. Fallon Federal Bldg.
Rm. 819
31 Hopkins Plaza
Phone (301) 962-2727

MASSACHUSETTS, Boston 02109
1600 Customhouse
165 State St.
Phone (617) 223-6608

MICHIGAN, Detroit 48226
1054 Federal Bldg.
231 W. Lafayette St.
Phone (313) 226-6077

MINNESOTA, St. Paul 55101
691 Federal Bldg.
316 N. Robert St.
Phone (612) 725-7819

MISSOURI, Kansas City 64106
1703 Federal Bldg.
601 East 12th St.
Phone (816) 374-5526

NEW YORK, Buffalo 14202
1307 Federal Bldg.
111 W. Huron St. at Delaware Ave.
Phone (716) 856-5950

NEW YORK, New York 10014
201 Varick St.
Phone (212) 620-3435

OREGON, Portland 97204
1782 Federal Office Bldg.
1220 SW 3rd Ave.
Phone (503) 221-3097

PENNSYLVANIA, Philadelphia 19106
One Oxford Valley Office Bldg.
Room 404
2300 E. Lincoln Highway
Langhorne, PA 19047
Phone (215) 752-1324

PUERTO RICO, San Juan 00918 Rm. 747 Hato Rey Federal Bldg. & Courthouse Phone (809) 753-4567	**TEXAS, Houston 77002** 5636 Federal Bldg. 515 Rusk Ave. Phone (713) 226-4306
TEXAS, Beaumont 77701 323 Federal Bldg. 300 Willow St. Phone (713) 838-0271 Ext. 317	**VIRGINIA, Norfolk 23502** Military Circle 870 N. Military Hgw. Phone (804) 461-4000
TEXAS, Dallas 75242 Earle Cabell Federal Bldg. Rm. 13E7, 1100 Commerce St. Phone (214) 749-3243	**WASHINGTON, Seattle 98174** 3256 Federal Bldg. 915 Second Ave. Phone (206) 442-7610

station or units of the station normally will be operated. Other technical data including type of service, emission, and maximum input power requested must be shown.

The technician should see to the proper posting of licenses. As an aid to his customers the technician should keep an eye on the expiration dates of the various stations with which he is concerned. He can also keep an eye on license renewal dates making certain that the station owners renew within two months before the expiration dates.

CHAPTER 2

Transmission Characteristics

2-1. FREQUENCIES AND OPERATING CHARACTERISTICS

Two-way radio, mobile, and point-to-point services are distributed throughout the entire frequency spectrum. The FCC makes frequency assignments in accordance with the services to be rendered and the radio-program characteristics suitable for this service. The radio-frequency spectrum is apportioned into the following subdivisions:

vlf (very low frequency)....................................below 30 kHz
lf (low frequency)..30 to 300 kHz
mf (medium frequency)................................300 to 3000 kHz
hf (high frequency)................................3000 to 30,000 kHz
vhf (very high frequency)....................30,000 kHz to 300 MHz
uhf (ultrahigh frequency)..............................300 to 3000 MHz
shf (superhigh frequency)........................ 3000 to 30,000 MHz
ehf (extremely high frequency)..............30,000 to 300,000 MHz

The two lowest-frequency bands and a low-frequency segment of the medium-frequency band are used mainly in the maritime and aeronautical services. Up to 540 kHz, radiotelegraph transmissions are most common. Information is transmitted in the form of the International Morse code, and operators are required to have some form of commercial radiotelegraph license. Airway beacon and other stations that send out navigational signals use sections of this frequency spectrum, and the international distress frequency of 500 kHz (600 meters) is also located here.

In this frequency spectrum, the ground wave dominates (Fig. 2-1). As a result, a very reliable transmission medium exists, both for

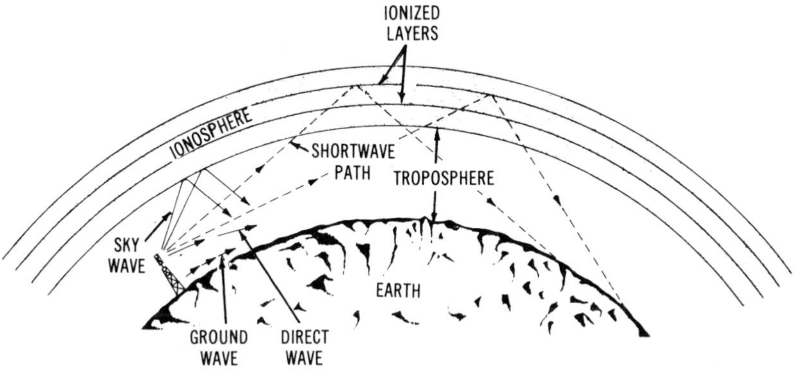

Fig. 2-1. Propagation of radio signals.

short- and for medium-distance communications.

If adequate transmitter power and receiver sensitivity are available (along with a large, efficient transmitting antenna), it is possible to establish very reliable long-distance communications. In recent years much experimental work has been done in this frequency spectrum toward developing reliable world-wide communications. Since very low frequencies will penetrate the ocean water more readily than higher-frequency waves, the former can be used for communicating with submarines.

In the medium-frequency spectrum between 300 and 3000 kHz, both the ground wave and the sky wave contribute to the net signal delivered to a receiving location. At the low end of the medium-frequency spectrum, the ground wave dominates the sky wave very much. Consequently, reliable ground-wave communications can be established with limited and occasional interference introduced from sky-wave components.

The radio broadcast band (am) starts at 540 kHz and extends to 1600 kHz. The ground wave serves as a reliable local and regional transmission medium for most of the broadcast band. During the night hours, especially in the winter, strong sky-wave components return to the earth. At varying distances from the transmitter, the ground waves and sky waves interact to cause fading, sometimes adding and sometimes subtracting. At the high-frequency end of the broadcast band, the ground-wave radiation is reliable over a shorter distance than it is at the low-frequency end. Particularly at night, the local ground-wave components are subject to interference from the sky-wave components originating at a substantial distance.

In this frequency spectrum, atmospheric noises are predominant. Summer transmission is plagued by lightning static, and the winter reception problem is interaction between the ground wave and sky wave.

In the frequency spectrum between 1.6 MHz and 5 MHz, there is a variety of two-way radio assignments. It is here that small-boat radio assignments are made. Small-boat installations must be installed, adjusted, and serviced only by general radiotelephone license holders.

The upper end of the medium-frequency spectrum provides reliable coverage up to 50 or 100 miles and under proper circumstances, even farther. For most small-boat and aviation two-way radio installations, the desired range is substantially less than 50 miles. Consequently, reliable performance is obtained with compact, low-power, single-sideband equipment.

Atmospheric noises are prevalent, and in mobile installations the problem of ignition interference must be considered. Sky-wave components bounced off the ionized layers sometimes travel a great distance. They then return to the earth at a high signal level and are capable of interfering with more localized communications. In some point-to-point communications systems, sky-wave transmission is advantageous in covering long distances (up to several hundred miles) on a reliable, scheduled basis.

Long-distance point-to-point communications and other services are assigned to the short-wave hf spectrum between 3 and 30 MHz. Various bands of short-wave broadcast stations occupy these frequencies. By using the sky-wave bounce from the ionized layers, these stations make reliable, world-wide communications possible.

The ionization of the various layers above the earth is a continuously shifting process. Careful observation and measurement over a period of years have permitted the development of consistent long-distance performance. These long-distance communications services—by proper choice of frequency, time, power, and directive antenna systems—are able to accurately pinpoint these strong signals to almost all corners of the earth.

Atmospheric static is strong at the low-frequency end of the spectrum, but gradually becomes weaker toward the 30-MHz limit. Conversely, ignition interference increases in intensity toward the high-frequency end of the short-wave spectrum.

Sunspot activities and aurora borealis, along with the consequent magnetic storms, have a decided influence on short-wave performance. Sunspot activities also have been cycled over the years and must be considered in short-wave radio transmission. The level and degree of ionization become a function of sunspot activity, varying from day to night, season to season, and year to year. On occasion, magnetic storms are so intense that all or large segments of the short-wave bands are rendered inoperative.

In addition to the point-to-point communications, many mobile services are crowded at the high-frequency end of the short-wave spectrum—all using equipment which must also be installed, adjusted, and serviced by or under the supervision of a general radiotelephone license holder. Many land mobile systems are allocated frequencies in the spectrum between 25 and 50 MHz. Also, there are maritime and aviation assignments in this spectrum.

At the high-frequency end of the short-wave spectrum, the direct wave (Fig. 2-1) predominates. The ground wave is attenuated to an insignificant level only a short distance from the transmitter. Sky-wave reflections are more sporadic. Thus, most communications are by way of the direct wave which travels in the immediate atmosphere between transmitter and receiver. The reliability of the direct wave makes this segment of the frequency spectrum ideal for mobile two-way radio systems. With proper facilities and sufficient power output, the reliable maximum range can be extended to 75 miles.

The popular Citizens Band assignments are made in the 27-MHz range. Millions of mobile stations are licensed to operate in this band. Power, modulation, and frequency adjustments must be made by a general radiotelephone license holder.

The vhf spectrum extends between 30,000 kHz and 300 MHz. In addition to many other less-publicized services, this segment contains our television and fm broadcast stations. Two frequency bands are used extensively for two-way mobile-radio systems, and a third band is assigned to point-to-point communications. All assignments, except broadcasting, come under the jurisdiction of a general radiotelephone license holder. A concentration of aeronautical two-way radio systems and other aircraft and marine navigational services are also assigned space in this frequency spectrum.

Direct-wave propagation (Fig. 2-1) is predominant in this spectrum. The ground wave drops to an insignificant value just a short distance from the transmitting antenna. The sky wave penetrates the atmosphere and the ionized layers before going off into space. In fact, in the vhf and uhf spectrums, guidance and telemetering signals are sent to missiles and satellites.

The lower half of the vhf spectrum is subject to some ionospheric reflection. Intense sunspot activities will result in a dense ionosphere and some reflection of vhf signals. Ignition and other sparking noises are strong at the low end of the vhf band, and gradually decrease toward the high end. Atmospheric noises seldom exist, or if they do, they are very low in intensity. Inherent receiver noises now become significant. In the vhf, uhf, and higher microwave spectra, receiver input-circuit noises are of primary concern. In most receivers, noise from the input stage predominates.

The immediate atmosphere (troposphere) exerts a great influence on the range and reliability of vhf transmissions beyond the horizon. Customarily, we refer to vhf and uhf transmission as being line of sight. However, there is some bending of the vhf wave by the atmosphere. This atmospheric bend, or refraction, causes the wave to travel beyond the strictly optical line of sight. How far the wave travels depends on how much it is bent (Fig. 2-2). Many meteorological factors influence the degree of bending—barometric pressure, temperature, humidity, and others. Discontinuities such as air-mass layers or temperature inversions in the upper atmosphere exert a great influence on the degree of bending. In fact, under extreme conditions the radio wave is confined in ductlike fashion

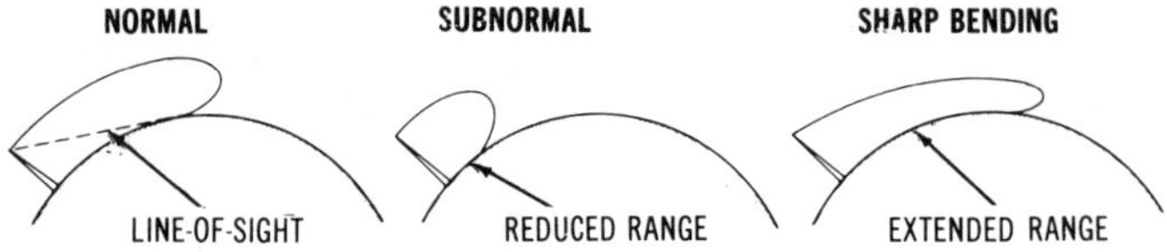

Fig. 2-2. Vhf propagation and tropospheric bending.

within these sharp discontinuities. Consequently, it can be propagated hundreds, and even thousands, of miles before returning to the earth.

Although these propagation phenomena are interesting and unusual, routine two-way radio communications must be more reliable. Consequently, communication systems are designed on the basis of line-of-sight range, plus a reasonable extension based on an average minimum amount of atmospheric bending. This distance is referred to as the radio-path horizon.

The uhf region between 300 and 3000 MHz represents an extension of the services provided in the vhf spectrum. Similar two-way radio assignments for land, marine, and aviation are made; and uhf television assignments occupy a good slice of this spectrum. Radio navigational aids, including radar, are also assigned here. Assignments for point-to-point microwave relay systems are made at the high-frequency end of the spectrum. In common terminology, that portion above 1000 MHz is called the microwave region.

The range of uhf transmission, in general, is less than that of vhf. However, the uhf wave, being more beam-like, can be reflected sharply by objects. For this reason, in metropolitan areas a uhf two-way radio system is often more satisfactory than its vhf counterpart—the ability of the wave to be reflected permits the signal to be bounced to a mobile unit surrounded by skyscrapers. Vhf, because of its greater range, seems to operate better in suburban and rural areas.

The uhf wave, particularly at the high-frequency end, can be packed into a concentrated pencil-like beam by the use of directional antennas. At this high frequency, the physical dimensions of a highly directional antenna are practical. As a result, this end is advantageous in point-to-point communications systems.

For transmission between two fixed points within line of sight, only two terminal stations are needed. If the communications system is to extend along a lengthy right-of-way (oil or natural-gas line, turnpike, railroad line, truck route), intermediate relay stations—either manned or automatic—can be used.

Microwave relay systems are in operation or are being planned for many industrial and commercial services. Here are more opportunities for the holder of a general radiotelephone license. Translators that carry television signals into remote areas use the uhf and microwave spectrums. Studio-transmitter links and remote-pickup equipment operate in the microwave spectrums.

The shf (superhigh frequency) spectrum extends between 3000 and 30,000 MHz. Radar and microwave relay services are assigned to this sector. However, much developmental work is being conducted here, to duplicate the services rendered in the uhf and vhf spectra, and some mobile operations are under test or development. Navigational devices in particular can take advantage of the very sharp physical dimensions. Licensing and technical requirements are somewhat more liberal on the many developmental frequencies.

In traveling through the atmosphere, radio (electromagnetic) energy is lost. This is called absorption. At higher frequencies, especially in the microwave spectra, this absorption becomes important. Absorption is significant in precipitation and, in heavy rain, there can be a considerable loss in transmission range, especially in the microwave ranges.

2-2. TYPES OF EMISSION AND MODULATION

The FCC classifies three major types of modulation—amplitude, frequency, and pulse. In an am system, as shown in Fig. 2-3, the information is conveyed by varying the amplitude of the resultant rf wave. An amplitude-modulated resultant wave is developed, formed by three signals—the original carrier; the upper sideband, the carrier plus the modulating frequency; and the low-frequency sideband, the difference between these two frequencies.

It is apparent that the bandwidth occupied by the signal in an amplitude-modulated system is set by the highest modulating frequency. If this frequency is 3000 hertz, the total bandwidth of the pure amplitude-modulated signal will be 6 kHz. If it is 10 kHz, the total bandwidth required will be 20 kHz.

In radio-broadcast systems, the highest audio-frequency component transmitted lies between 8000 and 12,000 hertz (although some of the high-fidelity broadcast stations transmit higher components). Thus, the bandwidth of broadcast stations lies somewhere between 15 and 25 kHz.

In am broadcasting, the frequency response is generally limited at the receiver. The if system in an average broadcast receiver is designed to pass audio components up to approximately 5000 Hz, in order to minimize interference between broadcast stations with operating frequencies that are close together. The bandwidth of the average am broadcast receiver is usually somewhat less than 10 kHz.

The low- and high-frequency audio components are of significance in the transmission of music and other program material. However, in terms of the intelligibility of voice communications, a low-frequency limit between 300 and 350 Hz and a high-frequency limit between 3000 and 4000 Hz are most satisfactory.

There are three advantages to confining the frequency response and bandwidth in voice communications. With a narrow bandwidth at the transmitting and receiving ends, the communication is less subject to heterodyne and sideband interference from other stations on the same or adjacent channels. Likewise, the narrowband signal does not radiate as many sidebands capable of interfering with other services on the same or adjacent channels. Moreover, a narrowband receiver minimizes the amount of static and man-made noise that enter, and its inherent noise is usually less than for the wide-band type. In summary, a narrowband system conserves the frequency spectrum, minimizes interference, and has a lower noise level.

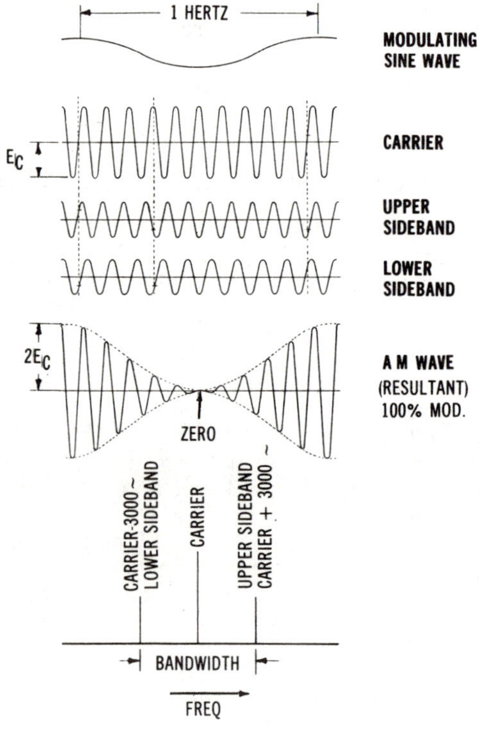

Fig. 2-3. The formation of an am wave.

In voice transmission, few significant frequency components are present above 3000 or 4000 Hz, and if present, they are usually low in amplitude and contribute very little to intelligibility. It is true that these overtones and high-frequency components do determine the quality and individualism of a person's voice. However, the intelligibility is not reduced when these high-frequency components are absent. Hence, most voice communications channels cut off at approximately 2500 to 3500 Hz. By so doing, there is a conservation of the frequency spectrum, along with a reduction in interference, and more economy in equipment.

Voice-frequency components below 300 Hz do determine the bass quality of a human voice, but again, they are not essential if intelligibility is the primary objective. The low-frequency components also contain the bulk of the voice power. Equipment designed for good low-frequency performance must be capable of handling the higher power levels contained in the low-frequency voice components. If these components are eliminated in the modulation process, the available power can be concentrated into the middle-frequency range. This is the range that has the most to do with good intelligibility. The removal of lows permits more effective use of the desired audio range, plus more economical equipment design because low-frequency performance and disturbances need not be considered.

The various types of transmission and emission are shown in Table 2-1. Notice that each type of transmission and modulation is given an identifying symbol. For example, regular am and double-sideband, full-carrier emission is given the symbol *A3*. These symbols are in common usage, particularly throughout all FCC publications. In making a specific am assignment for voice or other communications, a numerical prefix is often added to indicate the bandwidth assigned. For example, the symbol *6A3* indicates double-sideband amplitude modulation, the highest audio-frequency component being 3000 Hz. The total bandwidth would be 6 kHz.

2-2-1. Suppressed Carrier

Some voice-communication circuits, both base-mobile and point-to-point, use suppressed-carrier transmission, as shown in Fig. 2-4. In the am modulation process, the information to be conveyed between two points is contained in either sideband. The carrier itself is excess baggage—it really contains no useful information. Its presence does simplify receiver design, and the carrier can be put to work in the form of avc and afc. However, with appropriate receiver design it can be dispensed with, or at least transmitted at a much reduced level.

In conventional am systems with 100% modulation, two-thirds of the transmitted power is in the carrier and only one-third in the useful sidebands. For modulation percentages of less than 100% (usually the case except on peaks), the ratio of the power in the carrier to that in the sideband is substantially higher. As a result, a lot of transmitter power is wasted in the formation of the carrier. Furthermore, it is the interaction of carriers and sidebands that sets up the whistles and squeals which hamper communications under crowded conditions. Thus, if the carrier is reduced or suppressed, a source of interference is also removed.

Two other advantages of reduced-carrier transmission are the economies of transmitter design, and the fact that all useful available power can be concentrated into the information-carrying sidebands. Receiver design is more complex because a stable substitute for the carrier must be generated within the receiver.

Table 2-1. Various Types of Transmission and Emission

Type of modulation of main carrier	Type of transmission	Supplementary characteristics	Symbol
Amplitude modulation	With no modulation		A0
	Telegraphy without the use of a modulating audio frequency (by on-off keying).		A1
	Telegraphy by the on-off keying of an amplitude-modulating audio frequency or audio frequencies, or by the on-off keying of the modulated emission (special case: an unkeyed emission amplitude modulated).		A2
	Telephony	Double sideband	A3
		Single sideband, full carrier	A3H
		Single sideband, reduced carrier	A3A
		Single sideband, suppressed carrier	A3J
		Two independent sidebands	A3B
	Facsimile (with modulation of main carrier either directly or by a frequency-modulated subcarrier).		A4
	Facsimile	Single sideband, reduced carrier	A4A
	Television	Vestigial sideband	A5C
	Multichannel voice-frequency telegraphy	Single sideband, reduced carrier	A7A
	Cases not covered by the above, e.g., a combination of telephony and telegraphy.	Two independent sidebands	A9B
Frequency (or Phase) modulation	Telegraphy by frequency-shift keying without the use of a modulating audio frequency: one of two frequencies being emitted at any instant.		F1
	Telegraphy by the on-off keying of a frequency-modulating audio frequency or by the on-off keying of a frequency-modulated emission (special case: an unkeyed emission, frequency modulated).		F2
	Telephony		F3

	Facsimile by direct frequency modulation of the carrier.		F4
	Television		F5
	Four-frequency diplex telegraphy		F6
	Cases not covered by the above, in which the main carrier is frequency modulated.		F9
Pulse modulation	A pulsed carrier without any modulation intended to carry information (e.g., radar).		P0
	Telegraphy by the on-off keying of a pulsed carrier without the use of a modulating audio frequency.		P1D
	Telegraphy by the on-off keying of a modulating audio frequency or audio frequencies, or by the on-off keying of a modulated pulsed carrier (special case: an unkeyed modulated pulsed carrier).	Audio frequency or audio frequencies modulating the amplitude of the pulses.	P2D
		Audio frequency or audio frequencies modulating the width (or duration) of the pulses.	P2E
		Audio frequency or audio frequencies modulating the phase (or position) of the pulses.	P2F
	Telephony	Amplitude-modulated pulses	P3D
		Width- (or duration-) modulated pulses	P3E
		Phase- (or position-) modulated pulses	P3F
		Code-modulated pulses (after sampling and quantization).	P3G
	Cases not covered by the above in which the main carrier is pulse modulated.		P9

2-2-2. Single Sideband (SSB)

Still another form of am transmission with wide-acceptance in marine point-to-point and amateur communications is single-sideband, suppressed carrier (ssb). This type of transmission is symbolized as *A3J*. In single-sideband transmission, not only is the carrier suppressed or reduced substantially, but one sideband is also eliminated. Hence, as shown in Fig. 2-4, the required bandwidth of

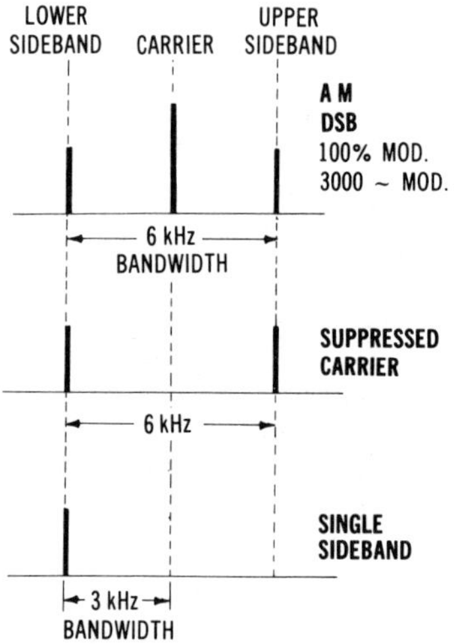

Fig. 2-4. Suppressed-carrier sideband transmission.

the signal has been cut in half. The total bandwidth of *3A3J* transmission thus is only 3 kHz.

In addition to the conservation of frequency in the spectrum, single-sideband transmission requires less power and further narrows the bandwidth to reduce interference. In two-way radio systems, single-sideband transmission is used widely in marine radiotelephone and for fixed point-to-point services. However, it can be expected in some, and anticipated in more and more, mobile installations. Radio amateurs have demonstrated its capabilities in mobile installations.

The power-saving feature of single-sideband transmission is obvious when we consider that, with conventional double-sideband 100% amplitude modulation, only one-sixth of the total power is contained in each sideband. Nonetheless, all the information to be conveyed is contained in one of the sidebands. With single-sideband transmission, all the available power of the transmitter can be concentrated into this one sideband. The narrow bandwidth and the absence of a carrier are very important interference-reducing characteristics.

Single-sideband transmission has become increasingly popular for transcontinental and intercontinental communications. It is somewhat less subject to ionospheric variables, and less troubled by selective fading, than other methods of transmission. With selective fading, some frequency segments of a transmitted signal fade more or less than others. This condition produces intermodulation distortion in the receiver and results in garbled speech. With only one sideband and a narrow bandwidth, such fading becomes less objectionable.

In many radio services single-sideband transmission is replacing conventional double-sideband amplitude modulation. This change-over has been proceeding on the radio amateur bands for some time. Sideband has almost completely replaced conventional am on most bands. Some of the fixed commercial services have also used sideband for a number of years. In fact, in some of the commercial services, particularly on the medium-frequency marine band, single sideband has become compulsory and am is being phased out.

In the 1.605- to 4.000-MHz spectrum, commercial A3 assignments are no longer available. This band has been changed over to single sideband as of Jan. 1, 1977.

There are three recognized sideband emissions (2.8A3H, 2.8A3A, and 2.8A3J). The 2.8 represents the highest permissible modulating audio frequencies. The 3 refers to the total bandwidth of 3 kHz. The H-form of emission refers to one sideband and full carrier. The A-form is one sideband and partial carrier, while the J authorization refers to one sideband and completely suppressed carrier. In certain portions of the medium-frequency spectrum even the H-type will be abolished after a certain cutoff date.

2-2-3. Frequency Modulation (FM)

In two-way mobile-radio systems, the use of frequency modulation has a number of advantages. The ignition systems of moving vehicles are a source of impulse noises, which appear as amplitude variations on the incoming rf signals. When am transmission is used, noise variations cannot be suppressed entirely, because of the hazard of also suppressing the desired amplitude variations of the incoming signals. In a frequency-modulation system, the desired information is in the form of a frequency deviation of the wave. Consequently, any amplitude variation introduced can be reduced almost completely without endangering the desired information carried by the incoming radio signal.

The very nearness of the noise source to the receiver, as in a moving vehicle, makes the frequency-modulation system attractive for mobile communications. In the uhf, vhf, and high hf bands, frequency modulation is the more common form for mobile communications. Although amplitude modulation is used for some services in these spectra, it is much more common in the short-wave and low-frequency half of the high-frequency (hf) band.

In an fm system, as shown in Fig. 2-5, the carrier is *frequency* modulated by the incoming signal. The frequency of the transmitted wave increases sinusoidally during the positive alternation of the modulating sine wave. The greatest upward frequency deviation (+5 kHz from the carrier) occurs at the positive crest of the modulating wave. As the modulating sine wave swings toward the zero axis, the frequency decreases sinusoidally to the carrier, or center, frequency. On the negative alternation, the frequency of the transmitted wave decreases with respect to the center frequency. The greatest downward swing away from the center frequency occurs at the crest of the negative alternation (−5 kHz from the center frequency).

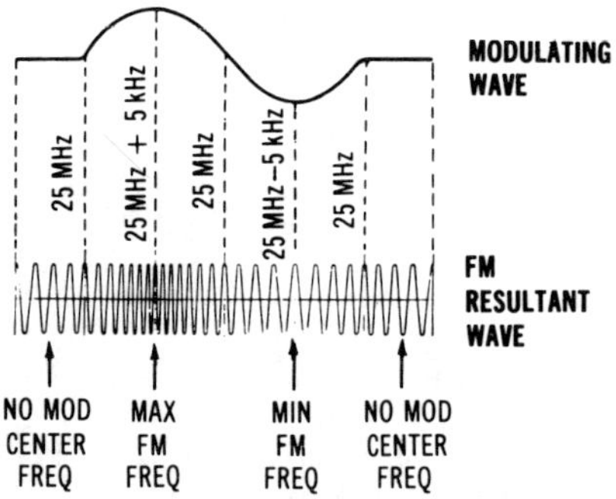

Fig. 2-5. Basic fm wave with a 25-MHz carrier.

The frequency-change limits of the fm wave determine the frequency deviation. In the example shown, the deviation is ±5 kHz. In assigning a channel for fm transmission, the FCC specifies 100% modulation as the maximum deviation permissible for the particular class of station. This maximum frequency deviation corresponds to the 100% modulation limit of an am wave which, as shown previously in Fig. 2-3, occurs when the wave rises to twice the original carrier amplitude at the positive crest of the modulating wave, and falls to zero at the negative crest.

This maximum fm modulation is not the same for each type of service. The maximum permissible deviation (100% modulation) is ±75 kHz for fm broadcasting stations. For the fm sound signal associated with television broadcasting, it is ±25 kHz. For most two-way fm radio systems, 100% modulation corresponds to ±5 kHz depending on the station classification. It is important that the general operator know the maximum deviation permitted for the type of equipment he must adjust or repair.

In an fm system, it might be assumed that the total transmission bandwidth is determined by the maximum deviation of the resultant wave. However, like the am wave, the fm wave is a composite of the carrier or center frequency and a number of sidebands. Unlike the am wave, however, which has only one pair of significant sidebands, the fm wave (Fig. 2-6) has one or more pairs of significant sidebands. They are a function of the modulation index which, at the highest modulating frequency, determines the number of pairs and hence the bandwidth of the fm transmission.

$$\text{Modulation index} = \frac{\text{Deviation}}{\text{Modulating Frequency}}$$

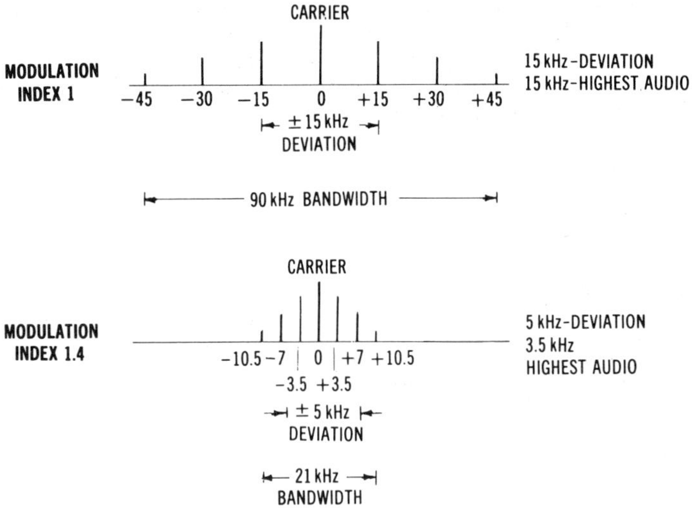

Fig. 2-6. Fm sidebands and distribution as a function of the modulation index.

The sideband pairs are separated from the center frequency by the frequency of the modulating wave and by its harmonic multiples (2,3,4,5, etc.). If an fm wave for a given maximum deviation has *three* significant sideband pairs, the total bandwidth is 2 × 3 × the highest modulating frequency. If the highest modulating frequency is 10 kHz, the total bandwidth required would be 60 kHz (2 × 3 × 10,000).

If the highest modulating frequency is reduced to 3 kHz, the total bandwidth required by the fm transmission will be only 18 kHz (2 × 3 × 3000). It is significant, just as in am practice, that reduction of the highest audio frequency reduces the bandwidth.

In fm broadcasting, particularly for high-fidelity transmission, the highest audio frequency is over 12 kHz. As we learned in communications, the best intelligibility is obtained by limiting the highest audio frequency to 3000–4000 Hz. Consequently, the fm wave used for communications can be crowded into a much narrower bandwidth than is required for fm broadcasting.

Another method of reducing the bandwidth of fm transmission is to reduce the amount of deviation. In an fm system the best noise rejection is obtained with the greatest deviation. For this reason, the narrowband system (nbfm) has a poorer noise rejection. However, the signal-to-noise ratio improves when the highest audio frequency is reduced. Inasmuch as the highest audio frequency for a voice communications service is much lower than required in broadcasting, it is possible to obtain a very fine signal-to-noise ratio despite the much more confined deviation of an nbfm system.

The symbols for fm emission are shown in Table 2-1. The two-way radio assignments are predominantly *F3*. A numerical prefix establishes the permissible bandwidth. For example, *18F3* signifies fm telephony with a total bandwidth of 18 kHz.

The general radiotelephone technician license applicant should also be familiar with some symbols of the other common forms of transmission. Symbol A1 is given to radiotelegraphy, in which the carrier is interrupted in accordance with a coded message (usually International Morse code). In this form of transmission, the carrier is turned on and off, as shown in Fig. 2-7.

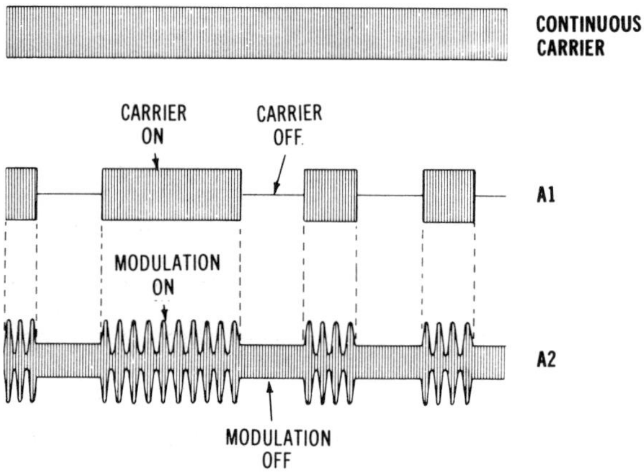

Fig. 2-7. A1 and A2 code transmission—Morse code letter L as an example.

Telegraph speed is often given in bauds. Based on the duration of the shortest code element the baud rate would be the number of these code elements transmitted per second. A Morse code speed of 25 words per minute has a baud rate of 20.

Another form of radiotelegraphy, common on the low- and very low-frequency ship bands, is A2. Here the carrier is transmitted continuously, but is modulated by an audio tone in accordance with the coded message. For example, an 800-Hz audio tone could be used. The tone is turned on and off to simulate the dots and dashes (on) and the spaces (off) of the International Morse Code.

Symbols F1 and F2 represent methods of sending frequency-modulated coded messages. Numbers 4 and 5 represent facsimile and television transmission, respectively. An A or F prefix is used, depending on whether am or fm is employed. In television broadcasting, A5 is used for picture transmission and F3 for sound.

Various types of pulse modulation can be noted in the last section of the chart. Pulse modulation techniques are used in advanced multiplex, telemetering, and computer data applications. Digital voice communications is a form of pulse transmission.

The baud also specifies rate in the transmission of digital data. In terms of serial data transmission, the baud rate is approximately the number of bits transmitted per second. This includes the actual data bits as well as so-called character-framing bits. For example, in a computer system of 63 characters with an 8-bit interval assigned to each character, the actual rate in bauds would be approximately 500 (63 × 8).

2-3. SIGNALS AND PROCESSING

In preparing a radio-frequency signal for transmission there are a number of modulation and processing steps. Most of these procedures will be described in detail in later chapters of the handbook. At this point it is important to become familiar with various signal terms and definitions used within the industry and by the FCC.

The fundamental radio-frequency wave is the *sine-wave.* This sine-wave at the transmit frequency is called the *fundamental* frequency. In the generation process, every attempt is made to form a pure single-frequency sine-wave carrier. In the generation and amplification process harmonic components can be formed. A *harmonic* is a sine-wave component with a frequency that is an integral multiple of the fundamental frequency. The second harmonic of a 600 kHz fundamental is 1200 kHz. The fifth harmonic of a 600 kHz fundamental wave is 3000 kHz.

In the processing of radio-frequency signals in transmitters and receivers, it is often necessary to multiply the signal frequency. In *frequency multiplication* the signal frequency is multiplied by an integral multiple. Such a stage is called a *frequency multiplier.* A times-two multiplier is called a *doubler* and multiplies the incoming frequency by a factor of two. A *tripler* stage would multiply the incoming signal by a factor of three. A 600 kHz signal applied to the input of a tripler would appear at the output with a frequency of 1800 kHz.

Another way of raising or lowering the frequency of a signal is known as *mixing.* In the mixing process two signals are heterodyned and the output is tuned to the sum or difference frequency. This technique is used in both transmitters and receivers. A 6 MHz signal mixed with a 5 MHz signal can be made to develop an output which is either 1 MHz (6 – 5) or 11 MHz (6 + 5).

The mixing process can be used to change over the frequency of a modulated radio-frequency signal in the same way. The entire baseband of modulating information can be changed up or down in frequency. By definition the *base band* refers to the spectra occupied by a carrier and all the associated modulating frequency components. These modulating components might be formed by a single voice frequency signal or even a number of subcarriers and their individual modulating components as well.

The mixing technique is also used in a process called *frequency translation.* In frequency translation an incoming carrier and its entire baseband is stepped up or down in frequency. An example of this technique is employed by television translators. In this application it could be used to provide television coverage to a small community in a mountain valley. A frequency translator on the mountain top would receive the signal from the television station, translate it directly to a different frequency spectra and re-radiate a signal down into the valley community.

The quality of a received signal is often evaluated in terms of *signal-to-noise ratio*. The signal-to-noise ratio is an amplitude ratio that compares the desired signal with the noise signal. Usually the ratio is given in decibels.

In modern radiocommunication systems many types of information are transmitted in addition to voice frequencies. *Information* refers to the total package that must be conveyed. *Data* transmission is the basic element-by-element means used to convey information. *Data rate* refers to the frequency at which the data elements are conveyed. In an amplification or modulation process the term *bandwidth* is the width of the frequency spectra needed to convey the desired information.

Several types of signals and codes that are used to convey information are analog, digital, binary, ASCII and Baudot. An *analog* signal is a continuous change of voltage or current representing the changes of a given situation. This might be an audio signal or the changes indicated by a physical variable such as temperature, velocity, rotation, etc. To prevent distortion such a signal must be processed by *linear amplifiers*. A *digital* signal uses discrete voltage levels to convey information and data. Most often there are two levels of voltage representing logic 1 and logic 0. The frequency, spacing, absence and presence of pulses and other pulse characteristics such as duration and spacing convey the desired information despite the fact that the digital pulse has only two specific voltage levels.

A *binary* signal is a two-level digital signal representing 0 and 1. In the binary system these two numbers can be used to represent any decimal number. Such a signal has a squared-wave pulse waveform.

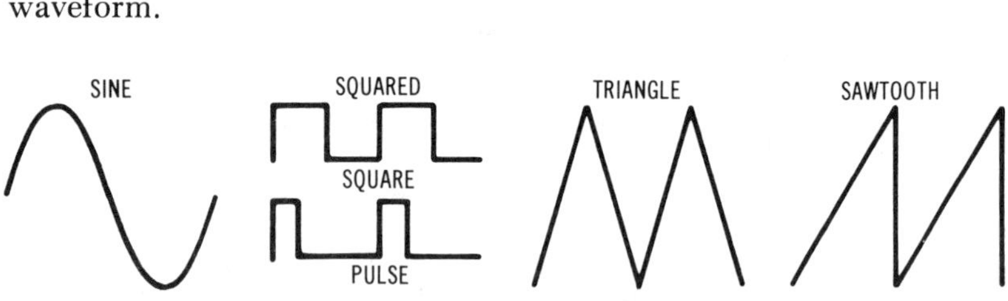

Fig. 2-8. Typical waveforms.

The *Baudot* code is a five- or six-bit (marks) code that is used to represent 32 or 64 numbers, letters and symbols. The code is used in radio teletype transmissions. *ASCII* is the abbreviation for American Standard Code for Information Interchange. It is an 8-bit code with one bit used for parity check and seven bits that convey 127 characters (numbers, letters and symbols). In addition to the *sine* wave, three other waveform types found in radiocommunications and radiocommunication test equipment are *square*, *triangle*, and *sawtooth*, as shown in Fig. 2-8. The true square wave and the pulse are basic to the formation of digital signals. They represent the two discrete voltage levels that can be used to represent logic 0 and logic 1. These two waveforms are also common in radiocommunication test instruments. Triangle and sawtooth waves are useful in digital signal processing and are also a part of test instrumentation, especially oscilloscopes.

CHAPTER 3

Solid-State Fundamentals

Semiconductor diodes, bipolar transistors, field-effect transistors, and integrated circuits are essential electron devices in modern two-way radio equipment. A knowledge of solid-state fundamentals and circuitry is mandatory if you are to install, tune, and service modern transceivers. A review and expansion of these subjects as related to two-way radio applications are the content of Chapter 3.

3-1. SEMICONDUCTOR THEORY

It is necessary to delve a bit into atomic theory to understand semiconductor operation. The difference between an insulator and a conductor is related to the number of free electrons. The greater the number of free electrons, the greater the conductivity of the material. These free electrons drift in a random manner throughout the conductor. However, with the application of an emf, free electrons are made to drift in a given direction producing a charge motion, or current.

Whether a material is a good conductor or a good insulator depends on its atomic makeup. Each element has its own specific grouping of electrons, both in number and in placement around the nucleus. These electrons may occupy one of several energy bands as shown in Fig. 3-1. Electrons of approximately the same energy level are grouped together in close-spaced orbits. Such a group is referred to as a shell. Shells are spaced a specific distance from the nucleus and contain a specific number of electrons. The number of shells and the number of electrons distinguish one element from another.

Electrons that orbit in shells near the nucleus have a low energy level and are tightly bound. Electrons in the outer shell have a high

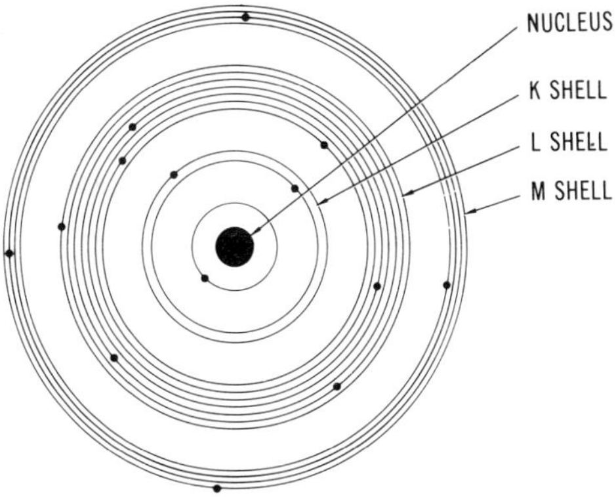

Fig. 3-1. Atomic makeup showing nucleus and energy bands.

energy content and are less tightly bound to the atom.

The important electrons of the outer shell are referred to as *valence* electrons. In the makeup of matter there are never more than eight electrons in the outer shell but there may be fewer than eight. When the element has exactly eight electrons it has a high order of stability and the valence electrons (electrons in the outer shell) are tightly bound to the atom.

Atoms having fewer than eight electrons are less stable and they are more ready to associate with the electrons of the outer shells of neighboring atoms. In general, atoms having shells of five, six or seven electrons are inclined to borrow additional electrons from other atoms. Atoms with one, two or three valence electrons are inclined to lose these electrons to other atoms. The interlocking of the valence electrons forms stable molecules of a substance, and under proper conditions the molecules form crystals.

Two important elements in the makeup of semiconductor diodes and transistors are germanium and silicon. As shown in Fig. 3-2 they have thirty-two and fourteen electrons, respectively. The first and second shells are full with two and eight electrons, respectively. This means that the third shell of the silicon atom contains four electrons.

When the first three shells of an atom are filled they contain two, eight, and eighteen electrons, respectively. In the case of germanium this means that the fourth shell also contains four electrons, making up a total of thirty-two. Therefore, both silicon and germanium have four valence electrons. As shown in Fig. 3-2, such an electron arrangement can, by the sharing of the four valence electrons with nearby atoms, form a stable crystalline structure. A stable crystalline construction acts as an insulator. Its conductivity is very poor because of the tightness with which the valence electrons are bound.

In the construction of semiconductor diodes and transistors the initial step is the developing of highly purified germanium or

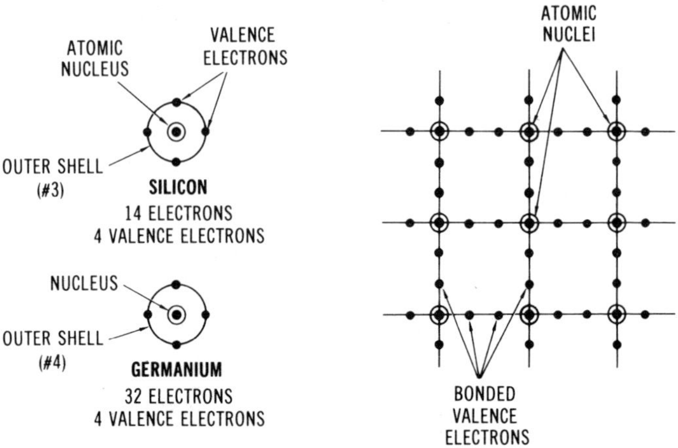

Fig. 3-2. Atomic structure of silicon and germanium.

silicon. These pure elements are referred to as *intrinsic* germanium or silicon crystals. Although the conductivity is poor such a crystalline construction is not really a perfect insulator because the bonding can be broken with high temperature or appropriate electrical potentials.

In the manufacture of semiconductor devices very minute yet accurate amounts of impurities are added to the intrinsic crystal. These impurities lower the resistivity of the crystal a specific amount. Therefore, a controlled amount of conduction is made possible. Such a crystal with the proper amount of impurity (doping) is called an *extrinsic* semiconductor. It has a conductivity that lies somewhere between the high conductivity of a conductor and the extremely low conductivity of an insulator.

3-1-1. P and N Crystals

Whether a semiconductor crystal is a p-type or n-type depends on the nature of the impurity. When an intrinsic crystal is doped with an element that has five valence electrons instead of four, the impurity atoms are inclined to bond with the intrinsic atoms. In so doing, as shown in Fig. 3-3, an extra electron exists for every combination of impurity and intrinsic atoms. This free electron is not bonded and will move about in random fashion. This type of crystal is referred to as an n-type and the motion of the electrons results in the movements of negatively charged particles. The semiconductor material formed does not have a negative charge because the total number of electrons in the substance equals the total number of protons. It is said, however, to have *electron carriers*.

A p-type semiconductor is made by doping the intrinsic crystal with an element that has three valence electrons. In this case the

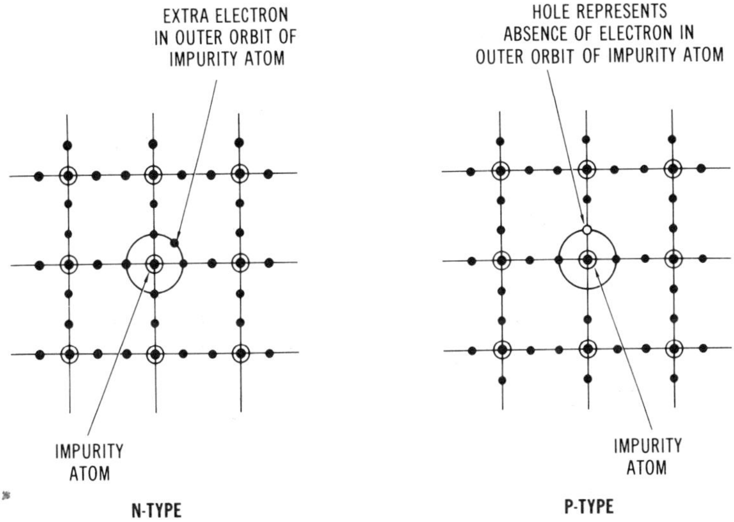

Fig. 3-3. Influence of impurity atoms in semiconductor crystal.

three electrons will be bonded with the intrinsic atoms. However, as shown in Fig. 3-3, at each impurity-intrinsic bond position there will be an electron vacancy. Such an empty charge position, at which there would be an electron if a complete bond had been established, is referred to as a *hole*. It represents the absence of an electron and therefore can be considered as a positive particle. Again this new semiconductor material has no charge because of the balance between electrons and protons.

In a p-type semiconductor material there is always a tendency for an electron from a neighboring atom to move into the empty spot or hole. In this case, there is a hole left in the bonded makeup of the neighboring atom. As a result there are free holes. In the p-type semiconductor the motion of charges throughout the material is a result of the random movement of holes (positive particles) instead of electrons (negative particles). Current that results from the random motion of holes is said to be supported by the motion of so-called *hole carriers*.

In summary, the type of doping of an intrinsic semiconductor material determines whether it becomes a p-type or n-type "extrinsic" semiconductor. The conductivity depends on the amount of doping.

Fig. 3-4C demonstrates charge flow in a p-type block. In a p-type block the current carriers are holes or positive particles. As shown, current direction in the external circuit is the same as for the n-type crystal. However, in the block itself it is the motion of holes (opposite from the direction of electron carriers in Fig. 3-4B) that provides the means of positive majority carriers. The temperature

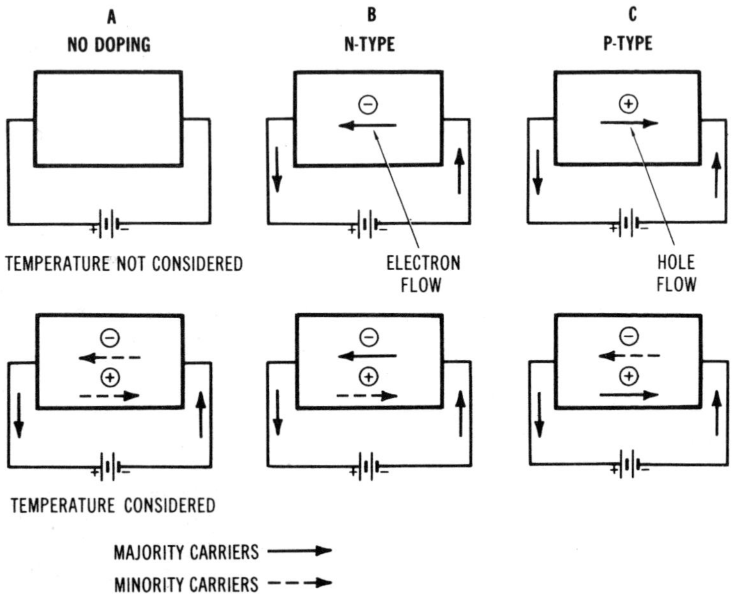

Fig. 3-4. Charge flow in semiconductor crystals.

breakup of bonds results in some free electrons as well. However, the number of free electrons is substantially less than the number of holes. Consequently, the temperature caused electron flow is the negative minority carrier conduction in p-type semiconductor material.

Minority current is a problem in the design of semiconductor and transistor circuits. In general there are fewer minority current carriers for silicon than germanium because there is tighter bonding in a silicon crystal. Thus, silicon devices are less temperature sensitive than germanium devices. The higher the doping, the greater is the number of electron or hole carriers. Often the current is spoken of as the movement of positive carriers or negative carriers instead of hole carriers or electron carriers, respectively. It is very important to keep in mind that in the case of n-type semiconductor material current is the result of the motion of electrons while in p-type material it is the result of the motion of holes.

The electron or hole carriers can be made to drift in a given direction to produce a current. When a battery is connected to an intrinsic-type crystal, there will be no charge flow if we neglect the influence of temperature. However, temperature does have a substantial influence on charge flow in a semiconductor crystal. Even at room temperature there is some breaking of valence bonds and, consequently, there is a random motion of holes and electrons. Hence, if we want to represent the small amount of current that exists as a result of temperature, it can be indicated as shown in the bottom block in Fig. 3-4A. The higher the temperature, the greater is the current that results from the breaking of the valence bonds.

Fig. 3-4B shows the current direction that results from the connection of a source of voltage to an n-type crystal. In the n-type there are electron carriers. Therefore, there will be an electron flow from the negative terminal of the battery through the semiconductor crystal to the positive terminal.

The influence of temperature on the current is shown in the bottom row of blocks of Fig. 3-4. This motion of plus charges is considerably less than the regular motion of electrons in the n-type crystal. Therefore, they are referred to as *minority carriers* while the main flow are *majority carriers*. In the p-type material, holes are the majority carriers, while electrons are the minority carriers.

3-2. SEMICONDUCTOR DIODES

Semiconductor diodes are manufactured in many forms for a wide variety of applications. The number of types seems almost limitless because of the ease with which characteristics can be controlled by shape, type of material, and degree of doping. Some diodes are used as detectors and others as rectifiers. Special forms are used as voltage regulators, switches, parametric amplifiers and mixers, oscillators, sensors, and voltage-variable capacitors.

3-2-1. The PN Junction

A semiconductor diode is formed by placing p- and n-type segments together to form a solid block of semiconductor material as shown in Fig. 3-5. The point at which the two segments touch is referred to as a pn junction. Such a junction permits current in one direction and blocks current in the opposite direction.

When a pn junction is first formed some of the holes of the p section drift into the n section while some of the free electrons of the n side drift into the p side. Thus, there is some neutralization of holes and electrons. As a result there is a space in the junction area referred to as the *depletion region* in which there are no current carriers (electrons or holes). This neutralization of carriers in the depletion region results in the development of a difference of potential, amounting to a few tenths of a volt between the p and n sections. The developed positive ionic potential on the n side repels any additional movement of holes from the p side and the negative potential on the p side repels any additional movement of electrons from the n side. This establishes a condition of equilibrium and produces what is referred to as the *barrier potential* of a pn junction.

When the diode junction is biased as in example A of Fig. 3-5, there is no charge flow and the junction is said to be *reverse-* or *back-biased*. When the circuit is connected, the electrons of the n side are attracted to the positive terminal of the battery. The holes of the p side, because they are positive charges, are attracted to the negative terminal of the battery. Due to the movement of current carriers away from the junction the width of the depletion area is increased and the barrier potential rises to the same value as the applied emf. The resistance of the diode in this case is very high and no significant external electric charge flows.

The semiconductor junction is said to have a high reverse resistance. Note the reverse bias is of the same polarity as that which symbolizes the barrier potential of the junction.

If the voltage applied to the pn junction is changed in polarity to that of Fig. 3-5B, the negative potential applied to the n side forces electrons toward the junction while the positive potential applied to the p side forces holes toward the junction. By forcing the majority carriers toward the junction the depletion area is reduced. If the applied emf is high enough to overcome the very low barrier potential there will be a continuous exchange of holes and electrons through the junction. The electron carriers of the n side move across the junction into the p side while the hole carriers in the p side move across the junction into the n side.

The electrons that leave the n side are continuously being

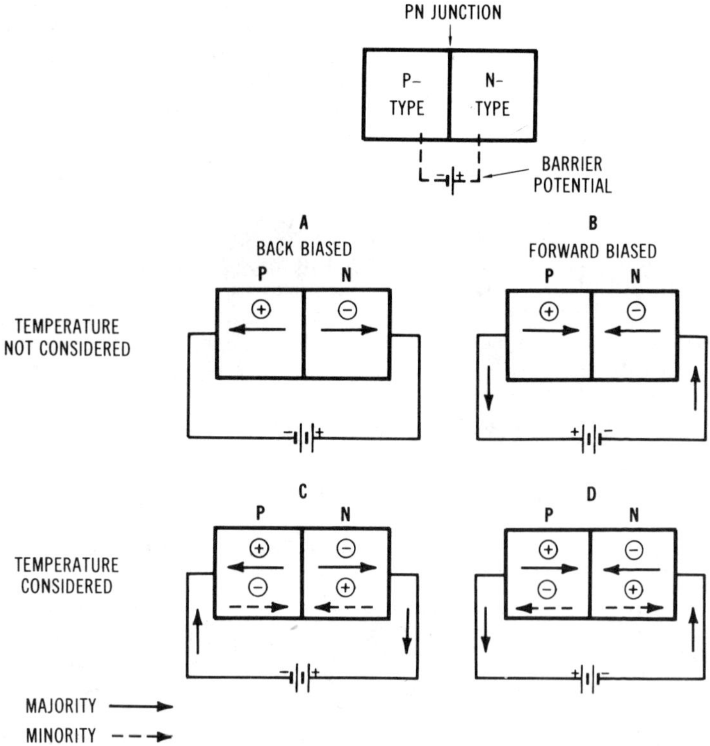

Fig. 3-5. The pn junction showing direction of majority and minority carrier movement.

replaced by the electrons from the source of voltage. Therefore, there is a continuous current through the junction and the external circuit. The diode resistance in this case is low and it is said that the forward resistance of the semiconductor junction is low. The junction is said to be *forward biased* or biased in the conducting direction. Note that this bias is such that the voltage applied to the p side is positive and that applied to the n side is negative. The forward bias is opposite in polarity to the symbol representing the barrier potential. Therefore, the barrier potential must first be overcome before the diode conducts in the forward direction.

The illustrations of Figs. 3-5C and 3-5D show temperature influence on the diode pn junction. When the diode is back biased, the majority carriers are forced away from the junction. However, the minority carriers, which are a result of the breakup of bonds because of temperature, are forced toward the junction. Thus, there is a movement of minority carriers across the junction and there is a small amount of reverse current. The quality of semiconductor diodes is checked by determining the ratio of the reverse to forward resistances.

At a low operating temperature there is little or no current for small values of reverse bias. If the reverse bias voltage becomes excessive, however, there is a wholesale breaking up of the electron bonds and the current rises sharply as shown. This is called the *avalanche* condition. This avalanche of current can become excessive and can destroy the pn junction. In some semiconductor designs the avalanche current is used to advantage.

At a high temperature a significant reverse current begins to flow even at low values of reverse bias, increasing as the back bias is increased. The high temperature conditions are shown by the dashed curve of Fig. 3-6. When the junction is severely overheated,

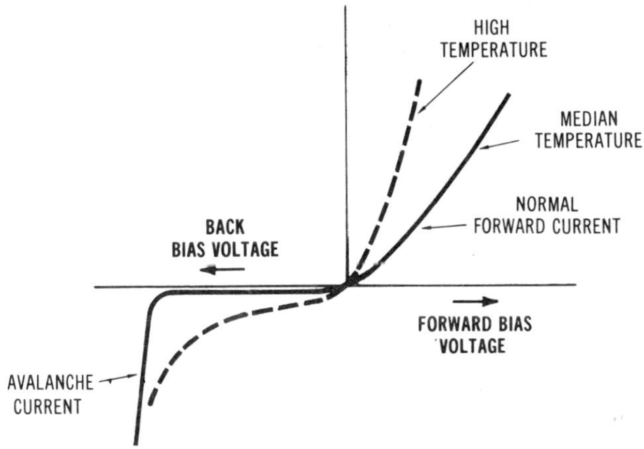

Fig. 3-6. Pn junction current-voltage characteristics.

the diode loses its unidirectional characteristic completely, and is no longer capable of operating as a pn junction. In general diodes are rated by the maximum forward average and peak currents and peak inverse voltage (piv).

When a diode is forward biased the majority carriers are forced toward the junction. In this case the minority carriers are attracted away from the junction. As a result the minority carriers make no contribution to forward current.

If the temperature effect is neglected the forward junction current increases with an increase in the forward bias. The current rise is not, however, exactly linear. With the high applied voltage more electron-pair bonds are broken, a higher operating temperature is produced and the current increase is accelerated. As shown in Fig. 3-6, the higher the operating temperature the faster the rate of forward current increase with an increase in the applied emf.

3-2-2. Special Semiconductor Types

In addition to the use of semiconductor diodes as rectifiers, mixers, detectors, etc., special types are used for modulators, voltage regulators, and other more specialized applications in two-way radio units. The zener voltage-regulator diode is one such type. Zener diodes are used widely to stabilize the supply voltages of transistor equipment.

The zener diode operates in the avalanche current region of the characteristics curve. This occurs when a high reverse voltage is applied to its pn junction. Both sides of the zener diode junction are heavily doped to obtain a relatively low breakdown voltage. Proper control of the doping sets this breakdown voltage at the desired level. Zener diodes are available with various voltage ratings.

A typical zener diode response is shown in Fig. 3-7. Note that at the avalanche point the current rises sharply. This tremendous current at one specific voltage is the condition that makes the zener diode an excellent voltage regulator. Once the diode is in the avalanche condition various amounts of current can be drawn through the diode, but the voltage across the diode will remain practically constant. The maximum current is limited, however, by the wattage rating of the diode.

Another semiconductor used in modern equipment for both frequency modulation and frequency-control applications is the voltage-variable capacitor, or *varactor*. The voltage-variable capacitor is constructed and doped in a manner that emphasizes the change in capacitance which results from variations in the depletion area of a pn junction. A typical response is shown in Fig. 3-8. By proper selection of an operating point an applied audio voltage can be made to cause the necessary change in capacitance to either frequency- or phase-modulate an rf signal.

In your study of the pn junction you learned that forward bias pushes the hole and electron carriers toward the pn junction and a high forward current results. With the application of reverse bias, however, the electron and hole carriers are attracted away from the

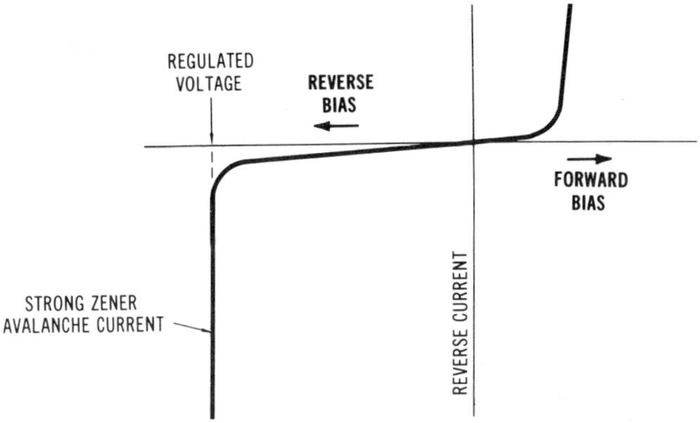

Fig. 3-7. Zener diode characteristics.

pn junction. The extent of the depletion area that results depends on the amount of back bias. In this mode of operation the depletion area and the pn end segments of the diode take on some of the characteristics of a capacitor. There are positive and negative particles present on the pn end segments just like the accumulation of charges on two capacitor plates. There is an appropriate electric stress between the two charges across the depletion area. Therefore, the depletion area acts as the dielectric of the capacitor.

When the back voltage is changed, the width of the depletion area changes in a like manner. As a result we are changing the effective spacing between the capacitor segments as well as the dielectric characteristics, producing a change in the diode capacitance. This change of capacitance is used to cause frequency modulation in modern communications transmitters. Similarly, the same characteristics may be used in automatic-frequency-control (afc) circuits of receivers.

A PIN diode operates as a variable resistance in vhf and higher-frequency circuits. Such a diode has highly doped p and n ends. These ends are separated by a relatively wide and lightly doped junction region.

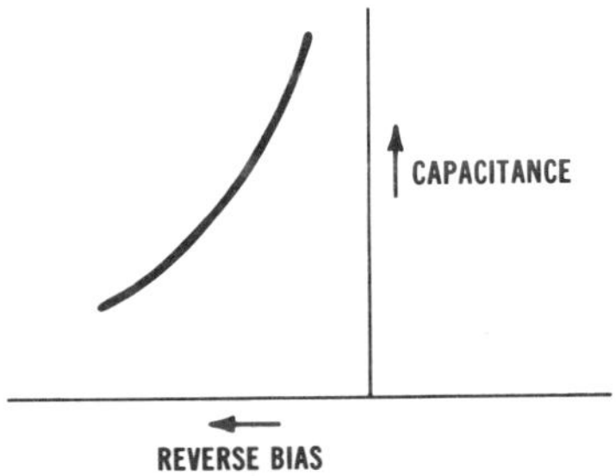

Fig. 3-8. Characteristics of a voltage-variable-capacitance diode.

In operation a dc bias current applied to the diode determines its resistance to the transfer of radio frequencies. When reverse-biased the junction resistance to radio frequencies is high; forward-biased, it is low. In this application the pin diode can be made to operate as a radio-frequency switch. In a suitable circuit, the bias can be made to vary in response to an agc bias. As a result the resistance to the transfer of a radio-frequency signal can be varied in accordance with the strength of a received signal.

A hot-carrier diode is a sensitive low-noise detector and mixer. It has a low junction capacitance and therefore performs effectively at vhf/uhf frequencies. As a mixer it has a high conversion efficiency and low noise content. The hot-carrier diode has a metal/semiconductor junction and there is a majority carrier flow with the application of even a very weak signal voltage.

3-3. BIPOLAR TRANSISTOR FUNDAMENTALS

A bipolar transistor is formed by combining two such pn junctions as shown in Fig. 3-9. The three segments are called

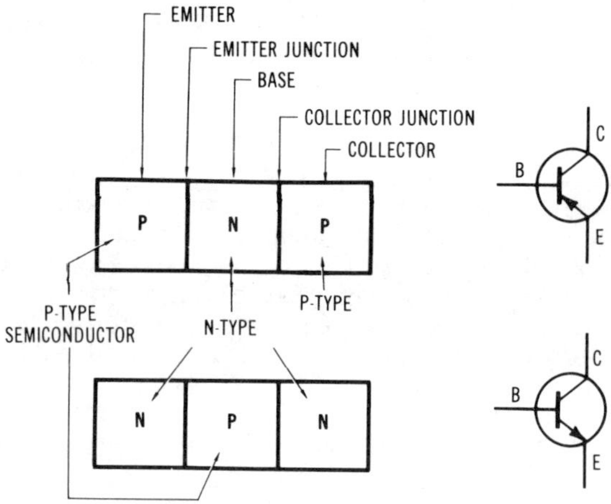

Fig. 3-9. Basic transistor construction and symbols.

emitter, *base*, and *collector*. Transistors come in two basic types, pnp and npn. In operation the junction between the base and the emitter (base-emitter junction or simply *emitter junction*) is forward biased while the junction between the base and the collector (collector-base junction or simply *collector junction*) is reverse biased. Proper biasing of the two types is shown in Figs. 3-10 and 3-11.

In operation, the emitter junction is forward biased. As a result there is a base-emitter current. The collector junction is back biased and, if we neglect the influence of the emitter junction, there would be no collector-base current. This would be the case if there were no emitter-junction bias or if the emitter junction were back biased. However, with the emitter junction forward biased, the carrier motion is such that there is a current through the collector junction. In fact, with the emitter junction forward biased and the

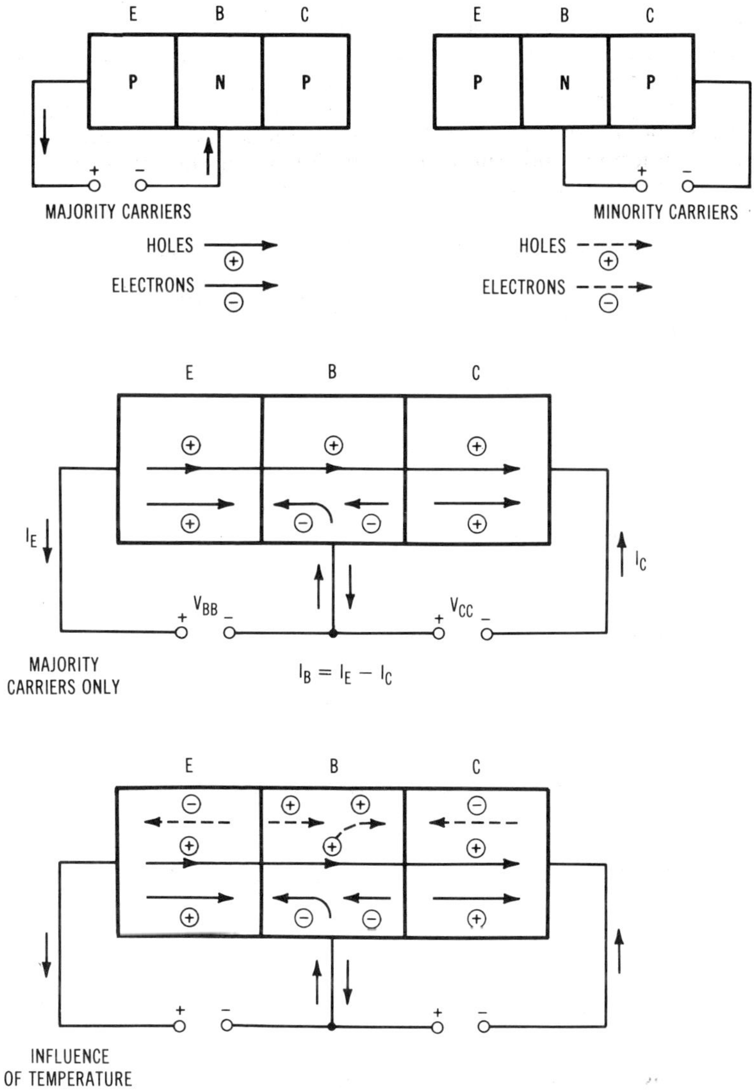

Fig. 3-10. Carrier movement in a pnp transistor.

collector junction back biased, the actual collector current is greater than the base current and the device functions as an amplifier.

The collector current is a result of the motion of holes through the base and across the collector junction for the pnp transistor; the motion of electrons over a similar path, in the case of an npn transistor.

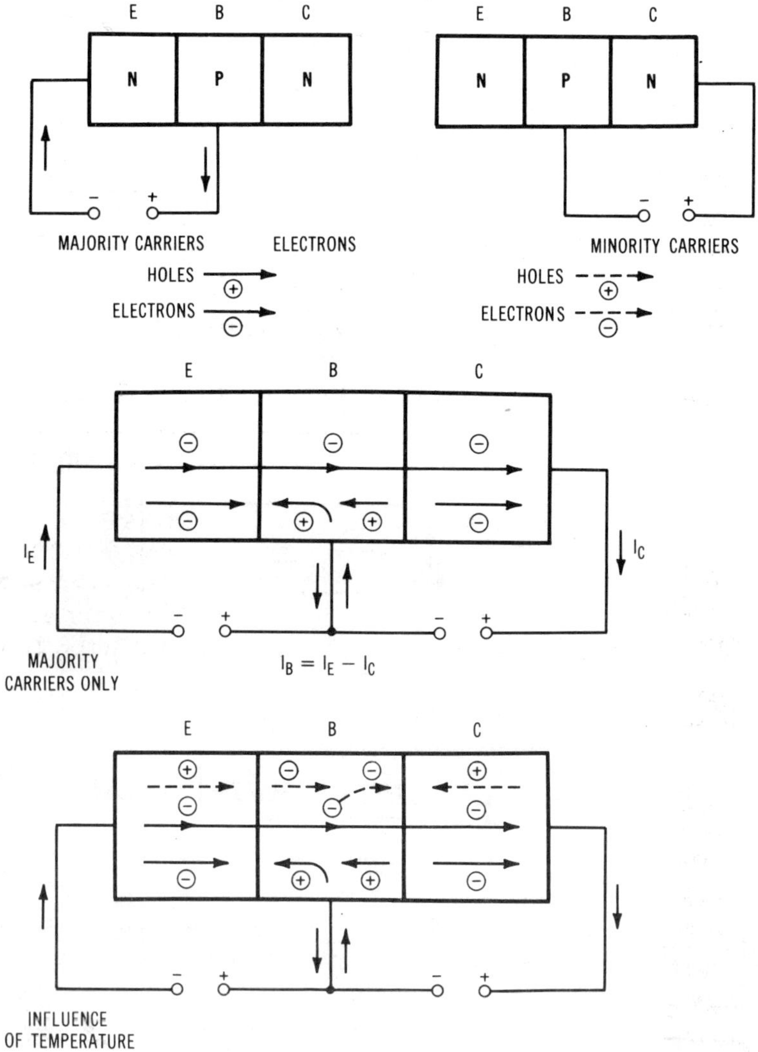

Fig. 3-11. Carrier movement in an npn transistor.

3-3-1. Majority Carrier Movement

Majority and minority carrier movements are illustrated in Figs. 3-10 and 3-11. In the case of the pnp transistor (Fig. 3-10), the two drawings at the top show the proper biasing of the transistor; the emitter junction is forward biased and the collector junction is back biased. With the two junctions biased correctly the majority carrier flows are illustrated in the center drawing.

When the emitter junction is forward biased, the electrons of the n-type base move to the emitter junction as do the holes of the p-type emitter. The holes moving across the junction are neutralized by base electrons.

For each electron that neutralizes a hole at the junction there is an electron supplied to the base from the bias source. Likewise at the terminal end of the emitter there must be an electron that leaves and enters the positive terminal of the voltage source. This continuous current through the junction and around the external circuit is referred to as the base-emitter current.

There are more holes injected across the emitter junction, however, than there are electrons to neutralize them. Since the base is very thin, these positive particles diffuse through the base to the collector junction. The holes have a positive charge and at the collector junction they are attracted to the negative potential of the collector. As a result they move through the collector junction into the collector element. The holes arriving in the collector element are neutralized by electrons from the collector supply voltage source. As electrons leave the negative terminal of the power source they must be replaced with electrons from the base element flowing into the positive terminal. This movement of charges is referred to as the collector current.

It is important to recognize that in terms of the base current flow, the emitter-base and collector-base currents are in opposite directions. Also it is important to understand that the hole flow to the collector junction is greater than the hole flow that is neutralized by electrons at the emitter junction. This is the result of a number of transistor characteristics. The emitter and the collector are heavily doped as compared to the base. Thus, there are few electrons in the base to neutralize the holes crossing the emitter junction, leaving many holes for diffusion to the collector junction. The base area is made very thin, therefore, the holes can diffuse quickly between the two junctions.

Note that the difference of potential between collector and emitter (two voltage sources in series) is higher than the base-to-emitter voltage (a single voltage source). Thus, the collector has a greater attraction for the injected holes than does the base.

The emitter current follows two branches. There is a flow (positive particles or holes) across the emitter junction and out of the base element. Another branch diffuses through the base and out of the collector element returning through the external circuit to the emitter. Thus, it can be stated that the emitter current:

$$I_E = I_B + I_C$$

The base current then is the difference between the emitter current and collector current (as shown by the opposite current arrows associated with the external base circuit).

$$I_B = I_E - I_C$$

$$I_C = I_E - I_B$$

Consequently, the collector current is greater than the base current. The ratio of the two currents (I_C and I_B) is known as the dc *beta* or dc current gain:

$$\beta = \frac{I_C}{I_B}$$

The dc beta is an important transistor characteristic. The collector current is less than the emitter current. The ratio of the collector current to the emitter current I_C/I_E is called the transistor *alpha*. It is always somewhat less than unity.

$$\alpha = \frac{I_C}{I_E}$$

As in the case of a single pn junction, the emitter-base current begins only after the barrier potential of the emitter junction is exeeded by the emitter-base forward voltage. In the case of a germanium transistor the barrier potential is approximately 0.2 volt; for a silicon transistor, approximately 0.7 volt. When the emitter junction forward-bias voltage is increased there is a rise in the emitter-base current. There is also a corresponding rise in the collector current. Fundamentally, a transistor is a current amplifier. The collector current follows the change in the base-emitter current.

In the pnp transistor, the more hole carriers that can be propelled across the emitter junction, the greater is the collector current. This motion of holes across the emitter junction is related directly to the base-emitter current. Thus, in considering transistor characteristics we often refer to *bias current* instead of *bias voltage*.

3-3-2. Diffusion Current

The movement of the carriers injected into the base (from the emitter junction) to the collector junction is referred to as a diffusion current. There is no difference of potential between the collector and the emitter sides of the base proper. The injected carriers (holes in the case of pnp transistor) appear with high concentration at the emitter junction region. Since there is a mutual repulsion between the charges, the random motion is such that the carriers will diffuse to a region of low concentration from an area of high concentration. In this manner the holes injected into the base diffuse in the direction of the collector junction. Once they arrive at the collector junction they are moved at high velocity in the manner of conventional current because of the difference of potential across this junction.

The thinner the base the shorter the time interval needed for the carriers to diffuse from the emitter to the collector sides of the base. Of course, the faster they are pulled through the collector junction the faster they must be replaced by charges from the emitter side. As a result the higher the collector voltage the more readily the charges are moved from emitter to collector junctions.

3-3-3. NPN Operation

The operation of an npn transistor is similar to that of a pnp except that the majority carriers are electrons rather than holes. To forward bias the emitter junction of an npn transistor it is necessary to connect the negative terminal of the voltage source to the emitter and the positive terminal to the base as shown in Fig. 3-11. To reverse bias the collector junction it is necessary to connect the positive terminal of the collector supply voltage to the collector and the negative terminal to the base.

When the emitter junction is forward biased there is a flow of electron carriers from the n-type emitter and a flow of holes from the p-type base toward the junction. As a result there are electron carriers injected into the base through the emitter junction. Some of these electron carriers are neutralized by the holes in the base. However, there is again a high ratio of injected negative carriers to base hole carriers. Therefore, electron carriers diffuse through the base to the collector junction. Here they are attracted by the positive potential applied to the collector and they flow through the collector into the output circuit.

Other relations are similar to those of the pnp transistor. The greater the number of electrons diffused to the collector junction in comparison to those neutralized in the base, the higher the current gain of the transistor. As a result the current relations for I_E, I_B, and I_C are identical to those of the pnp transistor. The exception is that the directions of current are opposite. Compare the external current arrows in Figs. 3-10 and 3-11, center diagrams.

3-3-4. Temperature Effect

The minority currents of a transistor are thermally induced the same as in the simple pn junction. The dashed lines in the bottom diagrams of Figs. 3-10 and 3-11 show the minority currents. In normal operation the emitter junction is forward biased. Therefore, the minority carriers that are present, electrons in the p-type emitter and holes in the n-type base, move away from the junction and the minority current is insignificant and not considered. Were the emitter junction reverse biased there would be no majority current but there would be some minority current because minority carriers would be repelled toward the junction.

In the normal operation of a transistor the collector junction is reverse biased. As a result the minority carriers, electrons in the p-type collector and holes in the n-type base are repelled toward the collector junction and there is a minority current. The minority current in the collector circuit when there is zero voltage present at the emitter junction is called the collector cutoff current (I_{CO}).

With normal operation of the transistor and a forward bias applied to the emitter junction there is a resultant collector current. However, there is also this minority collector current which is related to the effects of temperature.

In summary, the total collector current is a result of two components, the majority current (result of either the non-neutralized holes or electrons injected into the base from the emitter) and the minority current (thermal effect). Stated as an equation:

$$I_C = (I_E - I_B) + I_{CO}$$

One of the hazards of minority carrier flow is a distinctive effect called *thermal runaway*. Under certain circuit conditions the minority current can bias the emitter junction. Its direction is such that the emitter junction is forward biased even though no external forward bias is being applied. There will be some collector current through the emitter junction. As the junction is heated there will be more minority carriers formed and the minority current will increase and place a higher forward bias on the emitter junction. This results in additional collector current and further heating of the junction. The activity multiplies rapidly and the resultant high current and heating can destroy the transistor. Appropriate circuit design can reduce the possibility of thermal runaway.

3-4. TRANSISTOR CONFIGURATIONS AND OPERATION

The transistor can be connected in three basic transistor configurations. These are common base, common emitter, and common collector (emitter follower). In terms of the method of circuit connection they can be compared approximately with grounded grid, grounded cathode, and grounded plate (cathode follower) vacuum-tube stages. Thus, the collector, base, and emitter can be compared respectively with plate, control grid, and cathode of a vacuum tube (Fig. 3-12). Here the similarity ends and it is best to

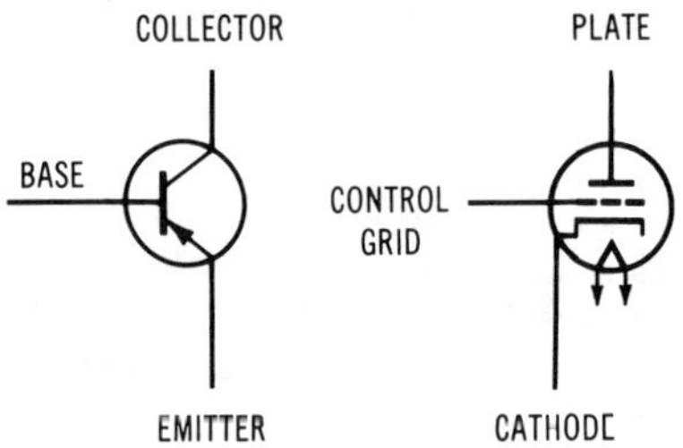

Fig. 3-12. Transistor and tube.

consider transistor and vacuum-tube functions separately with only occasional comparisons.

3-4-1. Common-Emitter Circuit

The most popular transistor amplifier configuration is the common-emitter circuit shown in Fig. 3-13. The circuit is shown for both pnp and npn transistors; they differ only with respect to the direction of electron flow and polarity of the bias voltages. Fig. 3-13A shows the use of two separate bias-voltage sources. Under other circumstances a single tapped-voltage source can be used as in Fig. 3-13B. In this arrangement the proper voltages can be obtained using supply voltage taps or resistive divider arrangements.

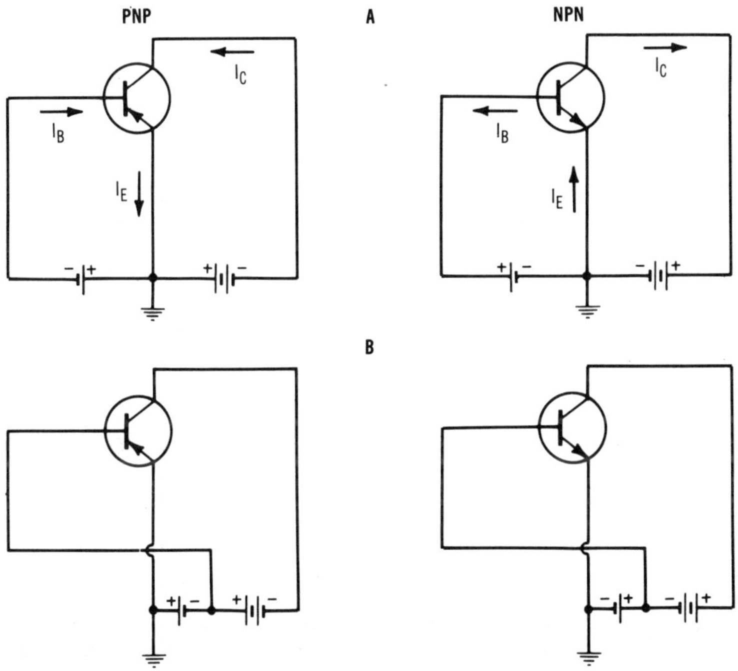

Fig. 3-13. The basic common-emitter circuit.

Notice the directions of electron flow and make a comparison with I_B, I_C, and I_E currents of Figs. 10 and 11. The following current relations are of significance:

$$I_C = I_E - I_B$$

In the common-emitter circuit the base current is the input current and it is customary to express collector current as:

$$I_C = \beta I_B$$

It is possible to establish a relation between alpha and beta. For example, expressions for I_C and I_E can be set down in terms of beta. As above, I_C is the product of beta and I_B. An expression for I_E in terms of beta would be as follows:

$$I_E = I_C + I_B = \beta I_B + I_B = I_B(\beta + 1)$$

The equation for alpha can then be set down and appropriate substitutions made to obtain alpha in terms of beta:

$$\alpha = \frac{I_C}{I_E} = \frac{\beta I_B}{(\beta + 1)I_B} = \frac{\beta}{\beta + 1}$$

It is possible to solve this equation for beta to obtain an expression in terms of alpha:

$$\beta = \frac{\alpha}{1 - \alpha}$$

In practice it is necessary to consider leakage current I_{CO}. Since this component of current flows in the emitter-base circuit through the emitter junction (refer to the current symbols of Fig. 3-14), it is

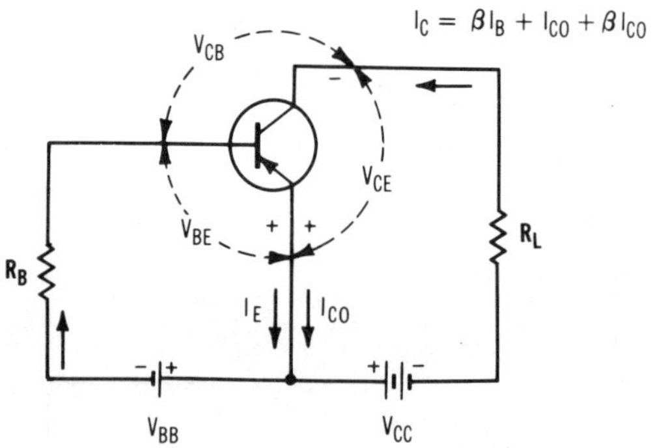

Fig. 3-14. Significant currents and voltages in common-emitter circuit.

amplified by a factor of beta. As a result the leakage current that flows in the collector circuit of the common-emitter configuration has a value of beta I_{CO}. Therefore, the total collector current is more accurately given with the following expression:

$$I_C = \beta I_B + I_{CO} + \beta I_{CO}$$

$$I_C = \beta I_B + I_{CO}(\beta + 1)$$

The factor I_{CO} which is the result of thermal energy discloses the very significant effect that temperature has on the collector current and operating conditions of a transistor stage. As long as I_{CO} is small in comparison to βI_B it will have only a limited influence on the operating conditions. As heating increases, it is possible that the I_{CO} component will double with each 10-degree rise in temperature.

Circuit configurations and bias arrangements are chosen so as to minimize the influence of temperature on the collector current and operating conditions of a transistor stage. When temperature has little influence on the stage it is said to have a high stability factor.

It is also important to give some consideration to the voltage drops around the transistor elements as shown in Fig. 3-14. In a transistor circuit the resistance of the emitter-base junction is very low. Thus, the amount of base-bias current, in the case of the common-emitter circuit, is dependent on the external series resistance (R_B). This is substantially larger than the resistance of the emitter junction. As a result the voltage drop from base to emitter (V_{BE}) is essentially a constant equal to the barrier voltage. In most bias calculations it is practical to substitute 0.7 volt for a silicon transistor and 0.2 volt for a germanium transistor.

The voltage across the collector junction is the difference in potential between the collector and base or (V_{CB}). It follows then that the difference of potential between the collector and the emitter (V_{CE}) becomes the sum of the previous voltages or:

$$V_{CE} = V_{CB} + V_{BE}$$

3-4-2. Common-Base Circuit

A fundamental common-base circuit is shown in Fig. 3-15. In this

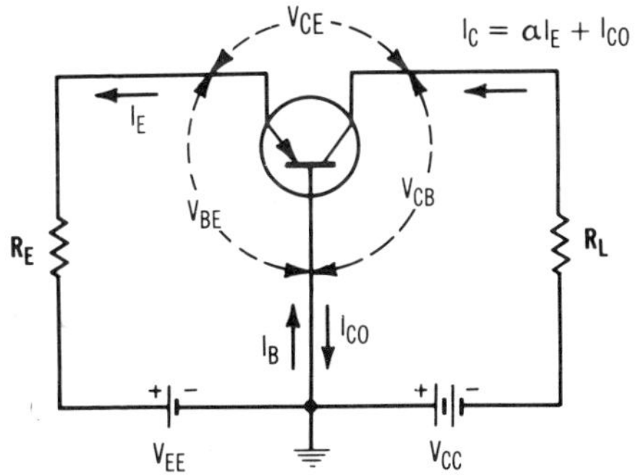

Fig. 3-15. Significant currents and voltages in common-base circuit.

illustration the emitter current is the input current and the collector current is again the output current. Since the emitter current is the sum of the collector and base currents, there can be no current gain in this type of circuit. In fact, the practical current gain approximates the value of alpha or:

$$I_C = \alpha I_E$$

A more exact expression for the collector current must consider the influence of temperature. Therefore:

$$I_C = \alpha I_E + I_{CO}$$

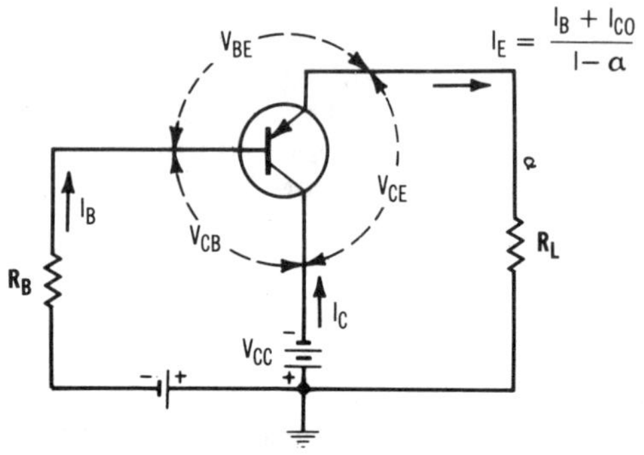

Fig. 3-16. Significant currents and voltages in common-collector circuit.

In the case of the common-base circuit the leakage current finds a direct path to the base. Thus, it does not bias the emitter junction as in the case of a common-emitter stage.

3-4-3. Common-Collector Circuit

The fundamental common-collector circuit is shown in Fig. 3-16. In this arrangement the input current is the base current while the output current is the emitter current. Again the emitter current is the sum of the base and collector currents and, therefore, current gain is possible. The equation for I_E in terms of I_B is derived by assigning a value of αI_E for I_C:

$$I_E = I_B + I_C$$

$$I_E = I_B + \alpha I_E$$

A more exact expression is obtained with the addition of the thermally generated leakage current that is also present in the emitter output circuit.

Solving for I_E:

$$I_E = \frac{I_B + I_{CO}}{1 - \alpha}$$

It must be pointed out that the previous basic equations consider only the direct currents in the three basic circuits. The common-emitter and common-collector circuits are capable of direct-current gain while the direct-current gain of the common-base amplifier is always less than unity.

3-5. BASIC AC OPERATION

Each transistor configuration (common-emitter, common-base, and common-collector) has different characteristics than the others.

Let us now consider each configuration in terms of ac amplification.

3-5-1. Common Emitter

A common-emitter circuit is shown in Fig. 3-17. An applied

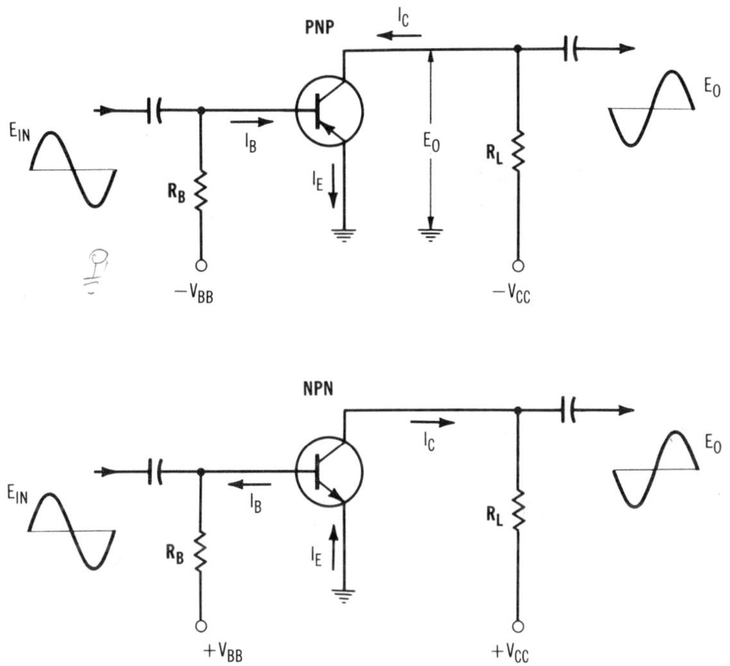

Fig. 3-17. Common-emitter circuits.

signal causes a voltage variation across the base resistor, R_B. This in turn varies the base current I_B about its operating point bias of 0.2 mA as shown in Fig. 3-18. As a result there will be a corresponding change in the collector current. The collector current change through the load (R_L) develops an output voltage component (E_O).

For a pnp stage the positive alternation of the input sine wave voltage decreases the forward bias and results in a decline in the base current. As a result, fewer positive carriers (holes) are injected into the base and the collector current declines. With the negative swing of the collector current there is less of a voltage drop across load R_L and the collector voltage (output voltage) becomes more negative. Hence, the output and input voltages are out of phase.

The negative input alternation provides additional forward bias for the emitter junction. As a result more positive carriers (holes) are injected into the base and there is a rise in the collector current.

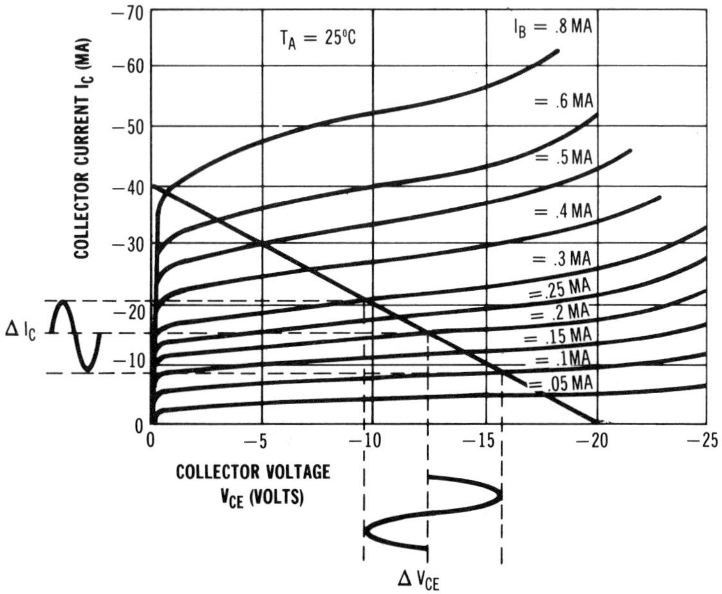

Fig. 3-18. Common-emitter circuit collector characteristics.

There follows a swing of the output voltage in the positive direction.

A similar relation exists between output voltage and input voltage for a common-emitter npn amplifier. For the npn circuit the positive alternation of the input sine wave increases the forward bias. Consequently more negative carriers (electrons) are injected into the base. As a result there is an increase in collector current. However, the current directions and bias polarities are opposite from those of the pnp type. Therefore, the increase in collector current results in a less positive voltage (negative swing) appearing across the output. Oppositely, for the negative alternation of the input wave there is a decrease in the forward biasing of the emitter junction. Fewer positive carriers (electrons) are injected into the base and there is a resultant decrease in collector current. Now the collector and output voltage swings positive to form the positive alternation of the output wave.

3-5-2. Graphical Analysis

The operation of a common-emitter stage can also be shown graphically as in Fig. 3-18. Again a 500-ohm load line is used. The collector supply voltage (V_{CC}) is 20 and the base supply voltage (V_{BB}) is such that the operating point base current is 0.2 milliampere. Thus, the operating-point collector current and collector voltage are 15 milliamperes and 12.5 volts, respectively. The operating point is so marked on the graph.

The transistor is a pnp type. When a positive input voltage is applied to the base the instantaneous bias is reduced causing a decrease in the base current, a negative input voltage results in a rise in the base current. Let us assume that the amplitude of the input voltage is such that the base current swings between 0.1 and 0.3 milliampere. Horizontal lines drawn from these two extremes to the collector current line show that the collector current will vary between 8.5 and 20.5 milliamperes. The collector current change swings the collector voltage between 9.5 and 16 volts. A value for the current gain is as follows:

$$\text{AC Gain } (h_{fe}) = \frac{\Delta I_C}{\Delta I_B}$$

$$= \frac{20.5 - 8.5}{0.3 - 0.1}$$

$$= 60$$

3-5-3. Common Base

In a common-base circuit, Fig. 3-19, the input and output

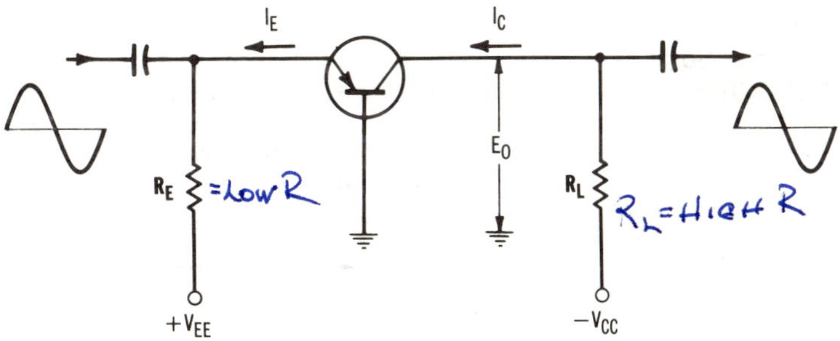

Fig. 3-19. Common-base stage.

voltages are in phase. A positive swing of the input increases the forward bias on the pnp emitter junction. Hence, a higher base current results and more positive carriers (holes) are injected into the base. Consequently, the collector current rises and the output voltage becomes less negative (swings positive).

During the negative alternation the forward bias decreases and fewer holes are injected in the base. The collector current falls and the output voltage swings negative.

It is to be noted that current gain is not possible with this arrangement because the input current is the emitter current (I_E). However, *voltage* and *power amplification* are possible because the output load resistance R_L is higher than the input resistance.

3-5-4. Common Collector

In the common-collector circuit, Fig. 3-20, the positive alternation of the input wave decreases the forward bias. Again

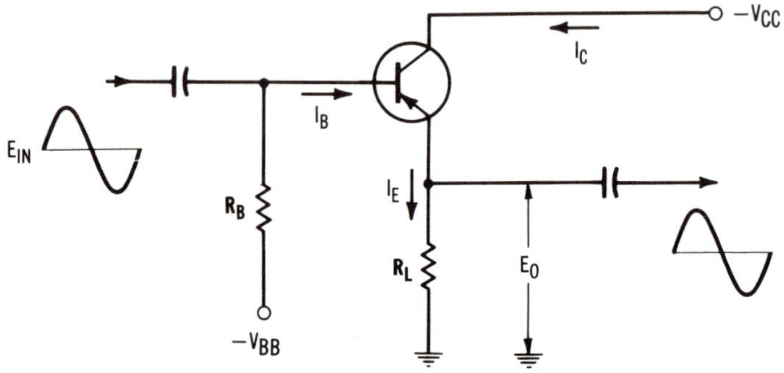

Fig. 3-20. Common-collector stage.

there is a decline in base current, injected holes, and collector current. With the decrease in the emitter current (a result of the decline in the collector current), there will be a positive swing of the emitter and output voltage. Oppositely, the negative alternation of the sine-wave input increases the forward bias and through transistor action the emitter current rises. Consequently, the emitter voltage swings more negative and there is a negative alternation of the output voltage.

The common-collector stage is capable of *current gain* because the base current is the input current while the emitter output current I_E is the sum of the collector and base currents.

This stage is often referred to as an emitter-follower because the emitter voltage tends to follow the voltage applied to the base.

3-6. SMALL-SIGNAL OPERATION

In the amplification of small signals the transistor can be considered to be an essentially linear device. In low-level or small-signal amplifiers the input signal variations are usually between the values of less than one microvolt up to some tens of millivolts. When a transistor is biased correctly and operates within this narrow input signal range, there is no significant distortion and the output and input signals are essentially identical in waveform. This condition applies to rf and if transistor amplifiers as well as low-level audio and video amplifiers and certain other small-signal amplifiers.

There are six factors of concern in evaluating transistor amplifiers. These are voltage gain (A_V), current gain (A_i), power gain (A_p), input resistance (R_{in}), output resistance (R_{out}), and phase. Some of these factors were discussed briefly in the previous topics.

The latter factor of phase was considered; you learned that input voltage and output voltage were in phase for both the common-base and common-collector configurations but out of phase for the common-emitter circuit. The three basic circuit configurations are repeated in Fig. 3-21 which gives special emphasis to the six factors just mentioned. These factors are also related to the actual load R_L placed on the stage as well as input-circuit constants. Each of the

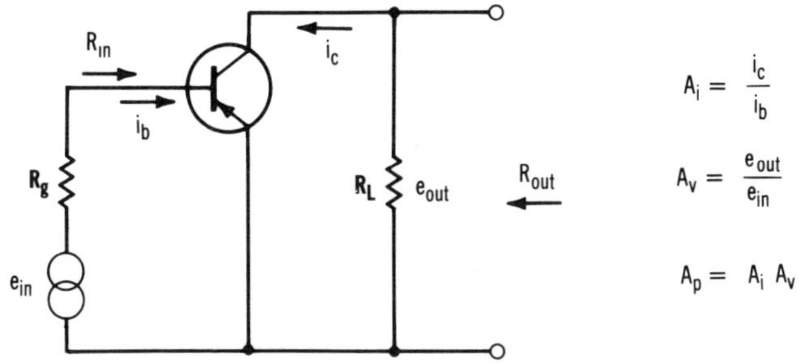

(A) Common-emitter.

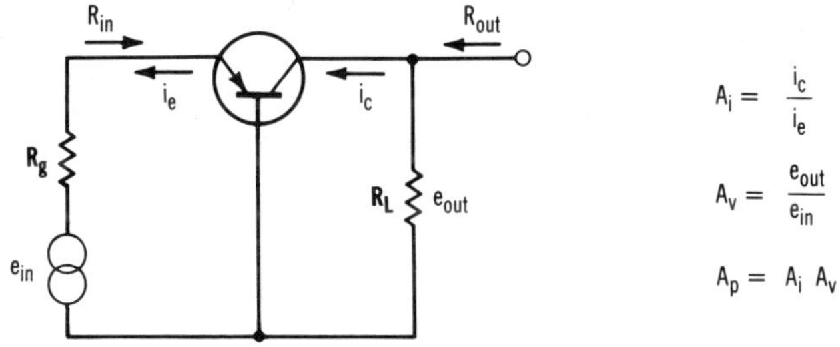

(B) Common-base.

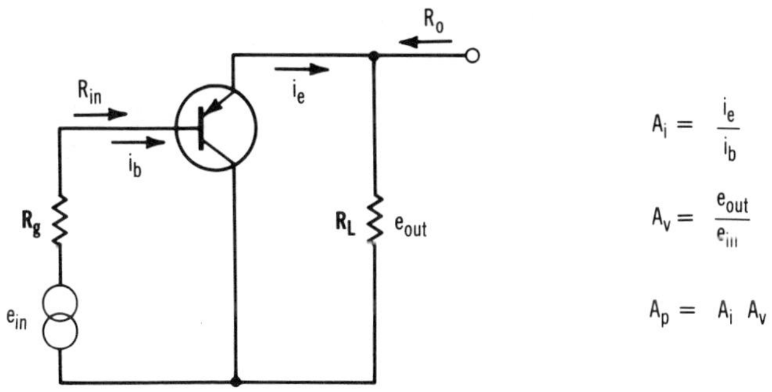

(C) Common-collector.

Fig. 3-21. Transistor configurations and ac equivalents.

first five factors listed will be considered in the following material. The relations apply only over the normal small-signal operating range of a given transistor type.

Common-base and common-emitter circuits are capable of approximately the same *voltage gain* A_v, with the common-base circuit having a slight edge, particularly at higher values of load resistance. The higher the value of the load resistance for either type of circuit the greater is the voltage gain. The voltage gain of the common-collector circuit is always less than unity.

The *current gain* (A_i) of common-collector and common-emitter configurations is approximately the same with the common-collector having a slight edge. Current gain decreases with an increase in load resistance for either type of circuit. The common-base stage has a current gain of less than unity.

The *power gain* (A_p) for all three transistor configurations is the product of the voltage and current gains or:

$$A_p = A_v A_i$$

The common-emitter stage has the highest power gain over the normal range of load resistance. For high values of load resistance, the common-base power exceeds that of a common-collector stage. However, at low values of load resistance, the common-collector stage has a higher power gain than the common-base circuit.

In general the power gains of both common-emitter and common-base circuits increase with a higher value of load resistance, while the power gain decreases with an increase in load resistance in common-collector stages.

The *input resistance* (R_{in}) to any transistor stage is low when compared to the input resistance of field-effect transistors and vacuum-tube circuits. The base-emitter junction is forward biased, thus, it has a low resistance. The *output resistance* (R_{out}) is substantially higher than the input resistance because it is across the reverse-biased collector junction.

The common-collector stage has the highest input resistance. The common-emitter circuit has an intermediate value of input resistance and the common-base circuit the lowest input resistance. The higher the value of the load resistance, the higher is the input resistance for a common-collector circuit. The input resistance of the common-base circuit gradually increases with an increase in load resistance while the input resistance of a common-emitter circuit gradually declines.

The use of feedback also influences resistive values. For the popular common-emitter circuit the use of negative feedback increases the input resistance substantially.

The common-base circuit has the highest output resistance and the common-collector stage the lowest. In a small-signal transistor circuit the output resistance is more dependent on the input circuit resistance than the output load resistance. In common-base and common-collector configurations the output resistance increases with an increase in the input or source resistance. This can be understood if you consider that the collector junction is reverse biased and the output conditions are a function of the carriers injected into the base which, in turn, are influenced by the source resistance.

The common-emitter configuration has the intermediate value of output resistance. This resistance decreases with an increase in the source resistance.

3-7. BIASING AND STABILIZATION

If a transistor is to operate in a linear amplifier circuit it must be properly biased and a suitable operating point must be found. The dc operating point depends on bias current, collector voltage and collector current. Obtaining the desired dc operating point requires the selection of the bias resistors and output load resistor to provide the proper input current and collector voltage.

The stability of a transistor amplifier is very important because if the operating point is allowed to shift with temperature changes, the circuit will introduce unwanted distortion. Besides the undesired effect of distortion, if a circuit is allowed to drift off the normal operating point with a rise in temperature, the physical safety of the transistor is endangered. The transistor could be destroyed by excessive collector current resulting from thermal runaway.

An input signal of a suitable magnitude causes the base current to vary equally on each side of the operating point, thus producing a collector current change and an output voltage variation across the load resistor, that is a reasonable copy of the voltage variation of the input signal. A major problem in the design of a transistor amplifier is keeping the dc operating point constant with changes in temperature. The operating point will shift with temperature because of the change in leakage current. This change in collector current can cause a shift in the dc operating point. If the collector cutoff current (I_{co}) value is small in comparison with the normal value of the operating collector current, it has little influence on the dc operating point.

The common-emitter circuit is especially vulnerable to the change in leakage current I_{co}. The leakage current in a common-emitter configuration is effectively multiplied by the factor of beta. However, the common-emitter circuit is used extensively because of other favorable operating characteristics.

The influence of temperature on common-emitter characteristics is shown in Fig. 3-22. The solid curves are for a particular standard operating temperature (usually 25°C). The dashed curves show the influence of an increase in transistor gain and/or leakage current with a rise in temperature.

For a given load resistance, represented by the load line of Fig. 3-22, it should be noted that the dc operating point has been shifted to a higher collector current (I_c) and a lower collector voltage (V_{CE}) when the temperature rises. In fact, for a given base-bias current the operating point has been shifted very near to the knee of the curve. When the operating point shifts to a point which is near the knee of the curve the positive alternation of an input current will become distorted. The increase in temperature, therefore, has

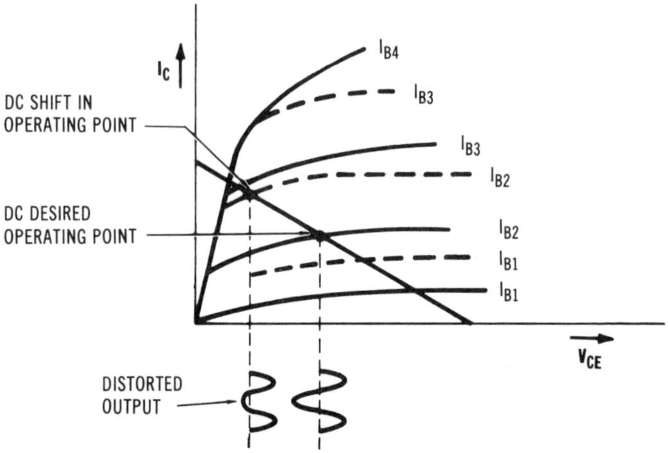

Fig. 3-22. Influence of temperature on common-emitter characteristics and dc operating point.

caused distortion in the form of a flattening of one alternation of the output wave.

In some cases the operating point can be shifted so that maximum safe transistor currents or power dissipations are exceeded. If the junction temperatures are increased in certain circuits there is also the danger of thermal runaway.

3-7-1. Bias Arrangements

Some typical bipolar transistor bias arrangements are given in Fig. 3-23. In the first arrangement of Fig. 3-23A a bias resistor (R_B) is connected between the supply voltage and the base. The ohmic value of the resistor is selected in accordance with the desired base current. Typical calculations are given later.

In the second arrangement of A an emitter resistor (R_E) is added. This resistor improves the bias stability. Now the actual base-bias current is determined by both resistors R_B and R_E.

In the circuit arrangement of B in Fig. 3-23, the base resistor R_B is returned to the collector voltage rather than the supply voltage. This factor must be considered in determining the value of the resistor to obtain a specific base-bias current. This circuit is also degenerative and improves the bias stability.

Example C of Fig. 3-23 is perhaps the most common type of biasing. In this case resistors R1 and R_B act as a bias current divider in conjunction with the base-to-emitter resistance of the transistor and resistor R_E (if present). Values for the three resistors are selected to establish good bias stability.

The arrangement of D in Fig. 3-23 is also common because multiple voltage supplies are common in modern transistor radio equipment. In this arrangement there is a separate emitter voltage

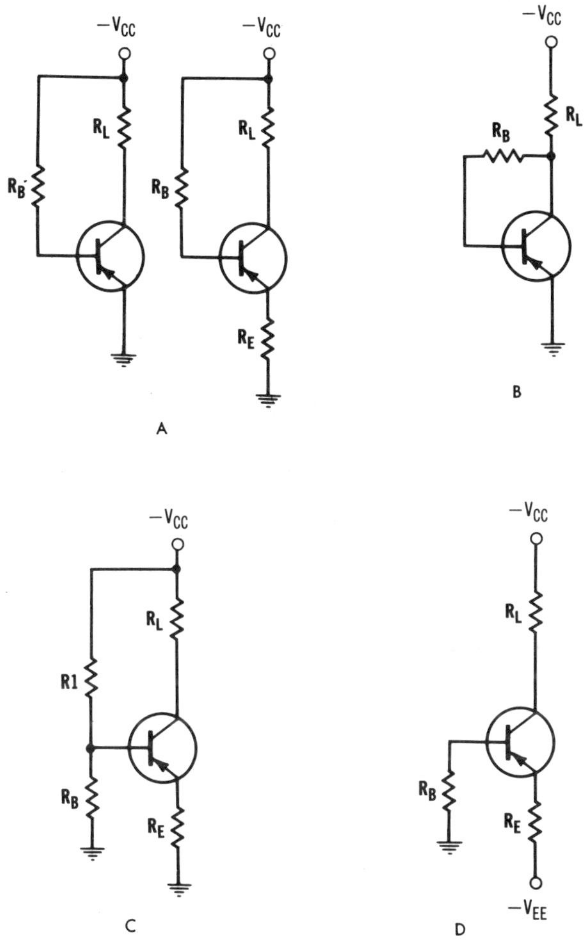

Fig. 3-23. Basic bias methods.

supply line that can be used to establish the appropriate base-bias current as determined by the ohmic values of resistors R_B and R_E.

3-7-2. Simple Bias Calculations

The simple circuit of Fig. 3-24 can be used to demonstrate base bias and other current relations in the common-emitter configuration. If we assume the supply voltage is 20 volts, load resistance 500 ohms, and collector current of 15 milliamperes, it is possible to determine the required value for resistor R_B for proper biasing of the stage.

If the beta of the transistor is 75, the base current can be determined by dividing the collector current by 75 or:

$$\beta = \frac{I_C}{I_B}$$

$$I_B = \frac{I_C}{\beta} = \frac{15}{75} = 0.2 \text{ milliampere}$$

The voltage drop across the base bias resistor (R_B) is the supply voltage minus the constant base-emitter barrier voltage of 0.7 volt. Therefore:

$$R_B = \frac{V_{CC} - V_{BE}}{I_B} = \frac{20 - 0.7}{0.2 \times 10^{-3}} = 96{,}500$$

The above operating point matches that shown previously in Fig. 3-18. It should be stressed that there is some spread in the value of the beta—it could be substantially more or less than 75. Under this condition some adjustment in the ohmic value of R_B is necessary if a desired collector current of 15 milliamperes is desired. A decrease in the ohmic value of R_B will increase collector current I_C. The emitter current (I_E) is, of course, the sum of the base current plus collector current, or 15.2 mA.

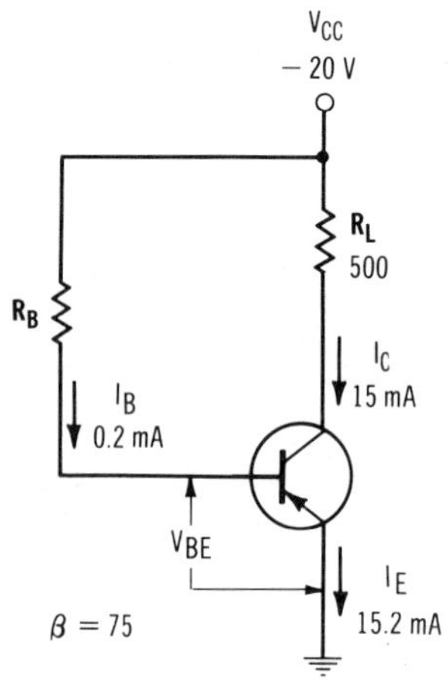

Fig. 3-24. Basic bias circuit.

3-7-3. Emitter Stabilization Resistor

The stability of a common-emitter circuit can be improved with the use of an emitter resistor (R_E). Such a circuit is shown in Fig. 3-25A. It might first appear that an emitter resistor alone could be used to obtain proper biasing of a transistor stage in the same manner as a cathode resistor in a vacuum-tube circuit. However, since the voltage drop across the emitter resistor is due to the emitter current the polarity of the voltage would develop a back bias for the base-emitter junction instead of a forward bias. In a fixed bias circuit using an emitter resistor it is still necessary to use some form of base bias.

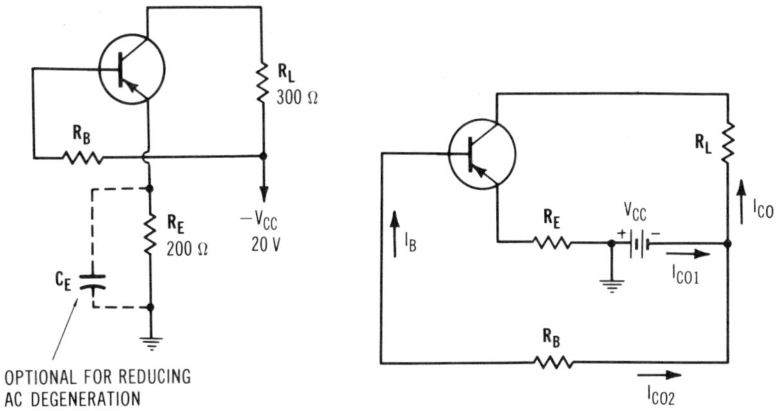

(A) Standard schematic notation.

(B) Rearranged for clarity.

Fig. 3-25. Stabilization of fixed bias with emitter resistor R_E.

The use of the emitter resistor, in conjunction with a base-bias circuit, improves the operating point stability because of the degenerative influence of the voltage developed across it. One of the contributing factors to the drift of the dc operating point is the effect of the leakage current as it flows through the emitter junction. If a higher percentage of the leakage current can be bypassed to the base rather than the emitter, the better the operating point stability will be. As shown in the equivalent schematic of Fig. 3-25B, the insertion of a resistor in the emitter circuit does this.

The leakage current in emitter resistor R_E is in such a direction that a compensating change will be made in the base-bias current. For example, an attempted rise in the leakage current (as a result of a temperature increase) develops a voltage across the emitter resistor that will decrease the base-bias current and bring about a correcting change in the collector current. The emitter resistor, therefore, exerts a degenerative influence on collector current. The emitter resistor because it is in series with the load resistor is a part of the collector-emitter output circuit.

If an ac signal is being applied to the stage there will be a corresponding ac voltage variation across the emitter resistor. This emitter voltage variation is in phase with the base voltage variation. Since these voltages are in series opposing across the base-emitter junction the effective input voltage is reduced. This ac degeneration can be reduced with the use of an emitter capacitor (C_E) that bypasses the ac voltage variation around emitter resistor R_E.

The following example demonstrates how the value of R_B can be determined:

The dc load on the collector-emitter circuit is to be 500 ohms. If a sample value of 200 ohms for R_E is used, it is necessary to reduce R_L to 300 ohms to maintain a 500-ohm dc load. Since V_{BE} is very low (0.7 volt), the sum of the voltage drops across R_E and R_B for practical applications equals the supply voltage. For a desired dc operating point of 0.015 ampere, the voltage drop across R_B must have the following value:

$$\begin{aligned} V_{RB} &= V_{CC} - V_{RE} \\ &= V_{CC} - I_E R_E \\ &= 20 - (0.015 \times 200) \\ V_{RB} &= 17 \text{ volts} \end{aligned}$$

Knowing the voltage drop across R_B, it is possible to determine the value of the base resistor required:

$$\begin{aligned} R_B &= \frac{V_{RB}}{I_B} \\ &= \frac{17}{0.0002} \\ R_B &= 85{,}000 \text{ ohms} \end{aligned}$$

3-7-4. Divider Bias

The most popular form of common-emitter biasing is the current-divider arrangement shown in Fig. 3-26. In this arrangement the divider current separates into two paths after R1; one path is through base resistor R_B, the second path through the series combination of emitter-junction and emitter resistor R_E. The current in the second path is the base current. The equivalent circuit shows the current paths.

If an emitter current of 0.015 ampere is assumed to be the same as used in the previous example, the voltage drop across R_E will be 3 volts. Since resistor R_B is in parallel with the series combination of the emitter-junction and emitter resistor R_E, the voltage drop across R_B must also be 3 volts if the base-emitter junction voltage is not considered. The required voltage drop across R1 will be 17 volts.

The divider bias arrangment permits the design of a highly stable common-emitter stage. The base resistor is kept low. For best stability use a current in resistor R_B that is greater than the dc base-bias current. However, there is a limit to which the value of a base resistor (R_B) can be reduced. The lower the base resistance is, the greater will be the load placed on any signal source.

Next determine the required values of resistors R1 and R_B for establishing the operating point. The desired base-bias current is 0.2 milliampere. Therefore, for good stability ten times this current should be shunted through base resistor R_B. This means that the

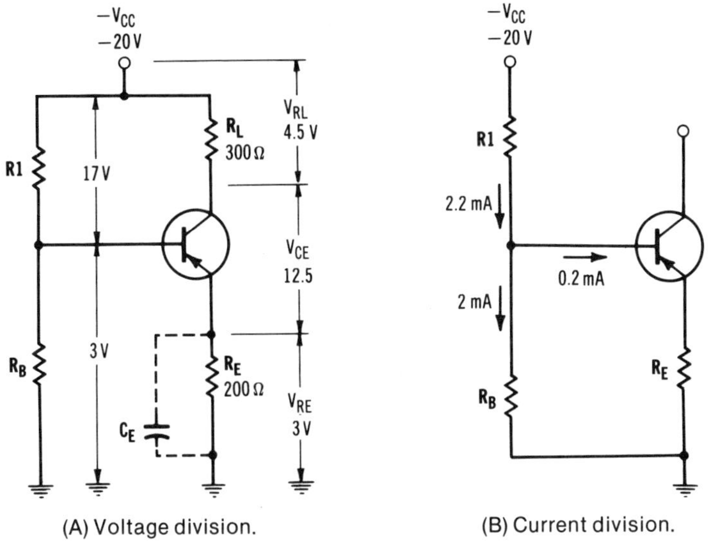

Fig. 3-26. Divider biasing arrangements.

current in R_B should be 2 milliamperes (Fig. 3-26B). Knowing this and the voltage drop across R_B, it is possible to determine the value of R_B as follows:

$$R_B = \frac{E_{RB}}{I_{RB}}$$

$$= \frac{3}{0.002}$$

$$R_B = 1500 \text{ ohms}$$

All the divider-circuit current is through resistor R1. Therefore, the total current in resistor R1 is 2.2 milliamperes (2 + 0.2). The voltage drop across this resistor is the difference between the supply voltage and the voltage across R_B or 17 volts (20 − 3). From this information it is possible to obtain the required value for resistor R1:

$$R1 = \frac{E_{R1}}{I_T} = \frac{17}{0.0022} = 7730 \text{ ohms}$$

3-8. AUDIO POWER AMPLIFIERS

Transistor stages can be biased class A, B, or C (Fig. 3-27). In class-A biasing the operating point on the transfer characteristic is

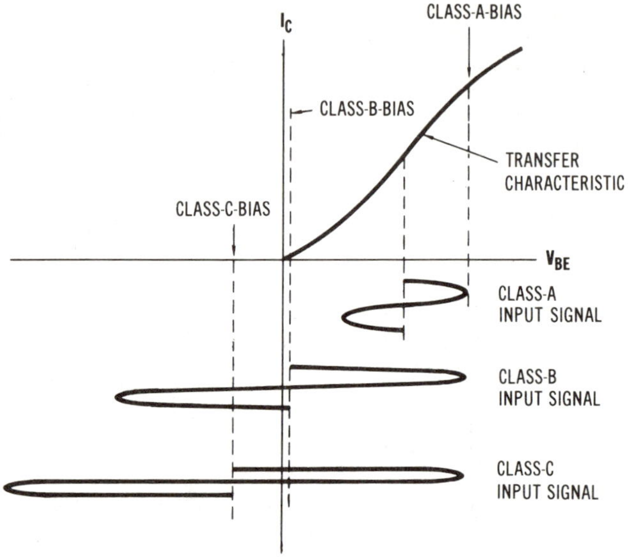

Fig. 3-27. Biasing modes for transistor.

such that the input signal swings over the linear portion of the transfer characteristic. Single-ended audio power amplifiers are biased in this manner.

In class-B biasing the stage is baised to cutoff. Of course that cutoff bias is equal to the barrier voltage. This is approximately 0.7 volt for a silicon transistor. Hence for true class-B operation it is necessary that a slight forward bias be employed. Thus for a class-B push-pull audio power amplifier there is a slight forward biasing equal to at least the barrier potential. Quite often an additional small amount of forward biasing is needed to minimize crossover distortion in the output.

When no bias is applied to a transistor stage it is biased class C because there is no collector current. Recall again that a small amount of forward bias is needed to overcome the barrier potential before there is any collector current flow. If desired additional back biasing can be used if the stage is to be biased even further beyond cutoff. This possibility will be considered in the description of class-C rf amplifier operation.

Two class-A audio power amplifier stages are shown in Fig. 3-28. Inasmuch as the transistor is basically a current amplifier some input power is needed. Therefore an input transformer is employed to match the output of the succeeding stage to the low-impedance input of the power amplifier. Class-A biasing is accomplished with resistors R1 and R_B. The circuit is temperature stabilized by emitter resistor R_E. This resistor keeps the operating point stabilized. An output transformer of proper turns ratio matches the low-

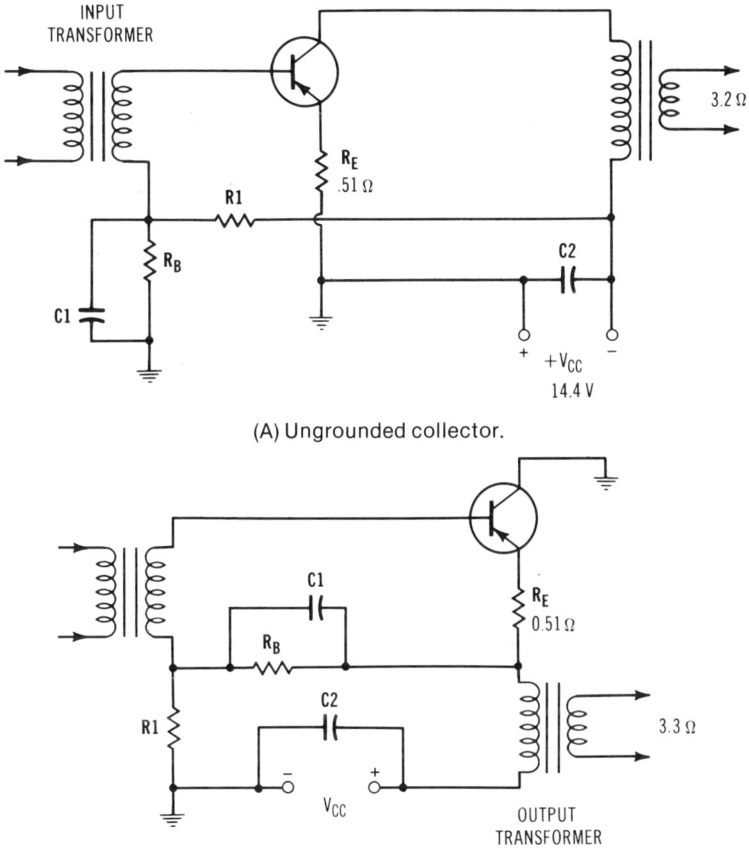

(A) Ungrounded collector.

(B) Grounded collector.

Fig. 3-28. Class-A power amplifier showing grounded- and ungrounded-collector arrangements.

impedance collector output to the low-impedance speaker system.

The circuits of Fig. 3-28 show both ungrounded and grounded collector arrangements. In a typical power transistor, the case of that device is usually connected electrically to the collector. The case in turn is often connected to a heat sink or chassis to provide transistor cooling in the form of radiation through the chassis or metallic heat sink that fits over the transistor case. Heat sinks are finned structures that function well as heat radiators. They are painted black to increase the efficiency of heat radiation.

If the transistor is used in a circuit that permits the collector to be operated at ground potential, the transistor case itself can be connected directly to the chassis or a grounded metallic heat sink.

This manner of connection applies to the circuit arrangement of Fig. 3-28B.

In other circuits the collector cannot be operated at ground potential. In such instances there must be an insulating washer to insulate the case of the transistor from the ground potential of the metallic heat sink. These washers have a low thermal resistance and will conduct the heat from the transistor case to the metallic surface.

The simplified circuit of Fig. 3-29 shows how an appropriate

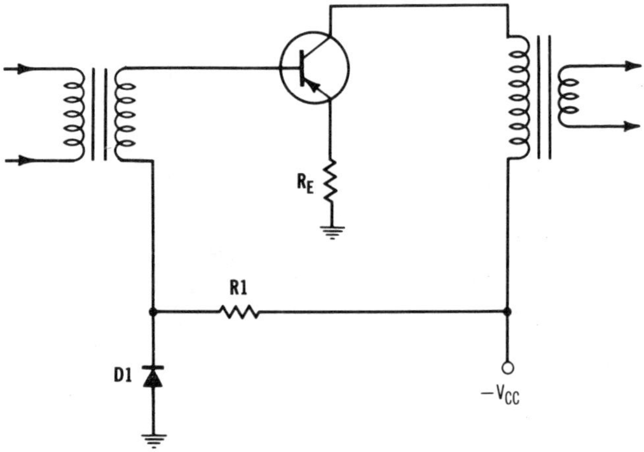

Fig. 3-29. Diode bias stabilization.

diode (D1) can be used to stabilize the bias of a transistor power stage. In this case the bias is established by resistor R1 and the resistance of the diode. The diode is an effective stabilizer because its constants are chosen to match the characteristics of the base-emitter junction of the transistor. In effect this base-emitter junction itself acts as a diode. When junction characteristics change with heating effects, there is a like change in the characteristics of the bias diode. Thus if the collector current tends to rise as a result of the heating of the base-emitter junction, the change in the characteristics of the external diode will follow and produce a compensating change in the base-bias current in resistor R1. Consequently the operating-point bias is stabilized and is less subject to temperature change.

A basic push-pull class-B amplifier using bipolar transistors is given in Fig. 3-30. A push-pull stage can be biased class B because one transistor will conduct on the positive alternation of the input wave and the other transistor on the negative alternation. As a result the complete sine wave is reconstructed in the output circuit.

In a theoretical case the class-B stage is biased to cutoff. However, the operating point is usually forward-biased a slight amount from this point to prevent crossover distortion in the

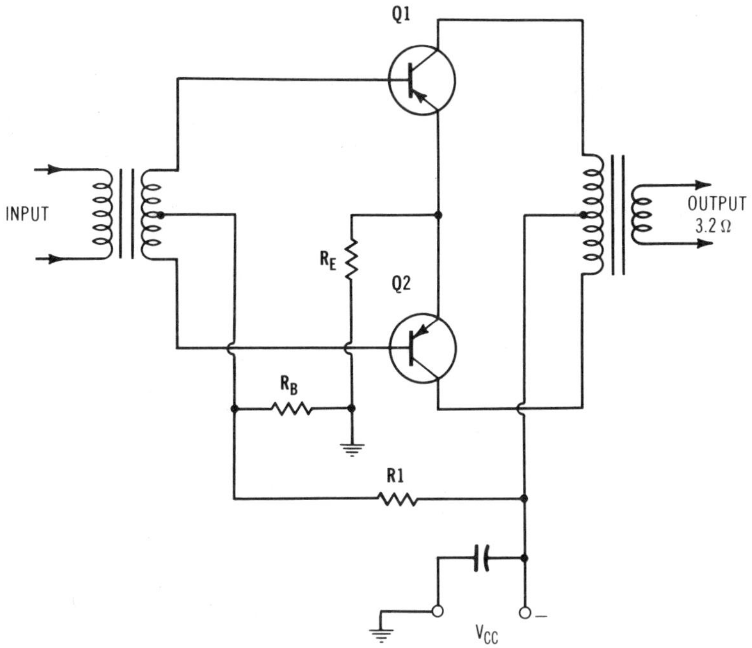

Fig. 3-30. Class-B push-pull amplifier.

output. The biasing arrangement of the circuit shown in Fig. 3-31 shows such an arrangement. The actual bias is determined by resistor R1 and the parallel combination of resistor R_B and the thermistor resistor.

The thermistor is used widely for bias-stabilizing transistor power amplifiers. It has a negative temperature coefficient; that is, its resistance decreases with an increase in temperature. A change in the thermistor resistance causes a lower forward-bias voltage to appear across the base and emitter junction with an increase in temperature. This reduction in forward bias compensates for the increase in conductance of the emitter junction with a rise in temperature.

The amount of temperature compensation is a function of the thermistor characteristic. Quite often the thermistor is placed in parallel with a bias resistor of specific value in order to provide an exact compensation.

3-9. TUNED TRANSISTOR AMPLIFIERS

Tuned amplifiers are used extensively in the receiver and transmitter sections of modern radiocommunications equipment. Neutralized and non-neutralized transistor tuned amplifiers are

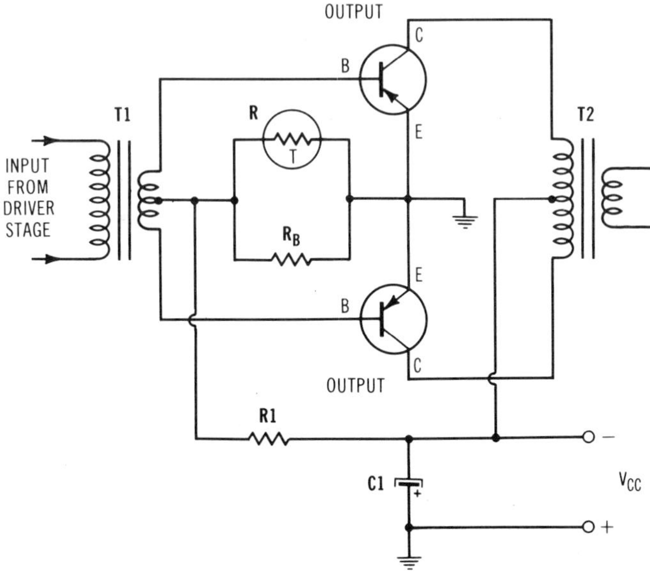

Fig. 3-31. Push-pull class-B stage using thermistor stabilization.

employed. Usually non-neutralized amplifiers are employed in the intermediate-frequency range while a neutralized amplifier is more likely to be found in the amplification of vhf or uhf signals.

When a tuned amplifier is designed for maximum gain conditions some neutralization is often advisable. Fewer stability problems and other favorable operating conditions are obtained when transistor stages are operated at lower gain. The lower-gain operation is obtained with the use of low-impedance loading at input and output or the use of intentional mismatch. The avoidance of neutralization is desirable from the standpoint of stability and freedom from feedback effects which can vary substantially with temperature.

In the case of the if amplifiers in a communication receiver the use of relatively low value tuned circuit impedance is not favorable from a standpoint of narrow-bandwidth requirements. However, it is helpful in terms of amplifier stability and agc operating conditions. As a result an optimum choice must be made. The amplifier design problem becomes less critical with the use of highly selective wave filters that determine the selectivity of the if system. Consequently, the if amplifier design can be slanted in favor of stability, and the added bandwidth will not harm the selectivity.

Several basic amplifier circuits as used in if application are shown in Fig. 3-32. The simplest form of if amplifier uses a single resonant circuit and a low-impedance untuned secondary winding matching the output of the first stage to the low-impedance input of the

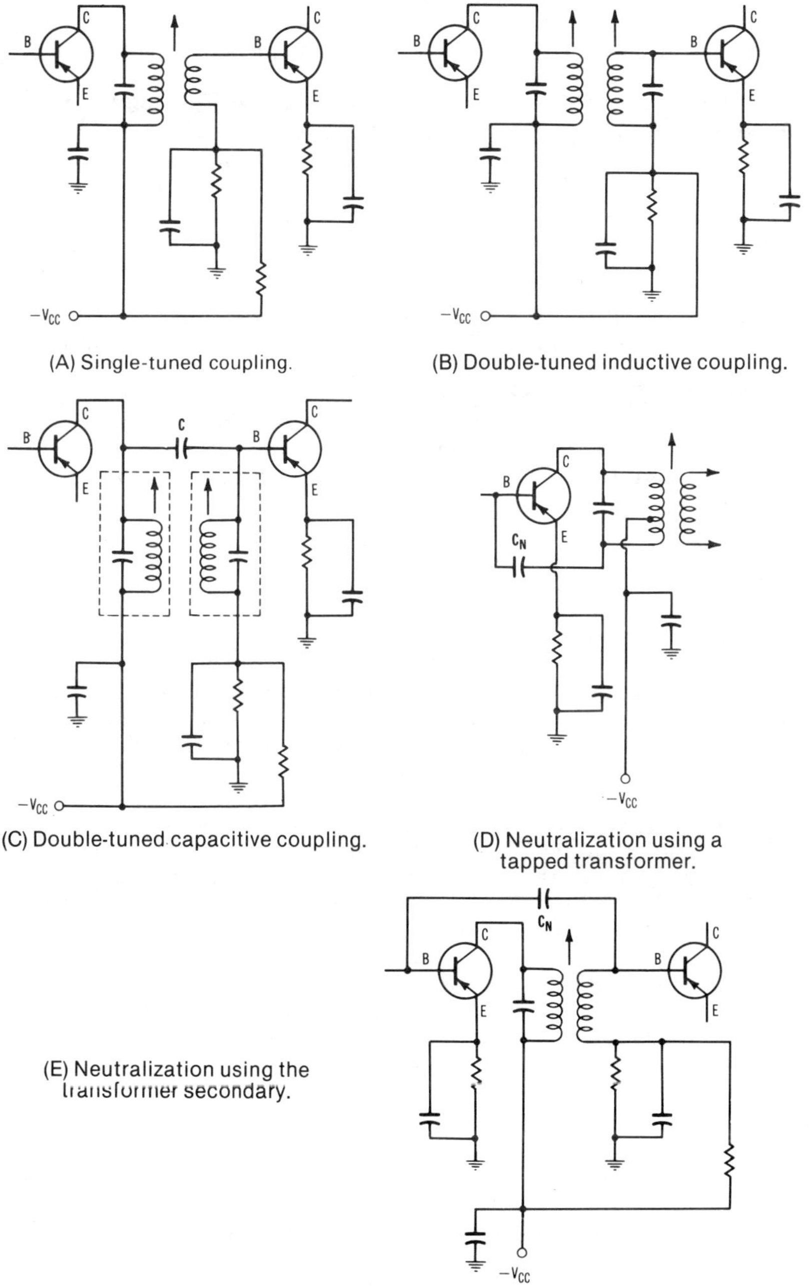

(A) Single-tuned coupling.

(B) Double-tuned inductive coupling.

(C) Double-tuned capacitive coupling.

(D) Neutralization using a tapped transformer.

(E) Neutralization using the transformer secondary.

Fig. 3-32. Tuned transistor amplifiers.

second (Fig. 3-32A). Examples in Figs. 3-32B and C show double-tuned interstage transformers. Fig. 3-32B is the more conventional form using inductive coupling between primary and secondary resonant circuits.

A more favorable arrangement, as shown in Fig. 3-32C, can be used to establish a good response curve as well as being more effective in blocking out strong off-channel signals. This circuit uses no inductive coupling. A small capacitor (C) provides the mutual coupling element between the two resonant circuits. The secondary inductor is sometimes tapped to permit appropriate match to the next stage, or a low-impedance secondary winding can be used.

The common-emitter stage is the most popular tuned amplifier transistor circuit configuration. It is to be noted that the biasing is similar to that employed in low-frequency audio circuits—a combination of base-divider biasing and emitter stabilization.

Figs. 3-32D and E show two common methods of neutralization used in vhf rf amplifiers and occasionally in high-frequency if amplifiers. In one arrangement the inductor is tapped near its low end and operates at rf-ground potential. Therefore an out-of-phase signal is made available between the bottom of the coil and the tap. This signal is fed to the base circuit through a neutralization capacitor (C_N). This out-of-phase component balances out the regenerated component that would cause amplifier oscillation. In Fig. 3-32E the required out-of-phase signal is made available between the secondary winding of the interstage transformer and fed back to the input base circuit through a neutralizing capacitor (C_N).

An example of a solid-state neutralized rf amplifier is shown in Fig. 3-33A. A double-tuned input resonant circuit contains low-impedance taps for matching the antenna to the base circuit of the transistor. The collector coil is also tapped to provide matching to the succeeding mixer. A low-impedance secondary winding could also be used. A tap on the primary provides the required out-of-phase feedback voltage for neutralization. In this particular rf stage the lower tap on the resonant circuit is at ac and dc ground potential. The pnp transistors in these circuits are biased properly with the application of a positive voltage to the base and emitter circuits.

No neutralization is needed in the two-stage amplifier of Fig. 3-33B. A single tuned resonant circuit is used for the input coupling. Double tuned transformers link both the first rf stage to the second and the output of the second stage to the mixer. Only the final resonant circuit is tapped to provide matching to the mixer. In configurations of this type sometimes intentional mismatching is employed to improve stability. The use of two stages instead of one develops the required overall gain and selectivity.

3-10. TRANSMITTER TRANSISTOR RF STAGES

Transistors are used in the rf section of a transmitter as oscillators, buffers, multipliers, and rf power amplifiers. Not all of

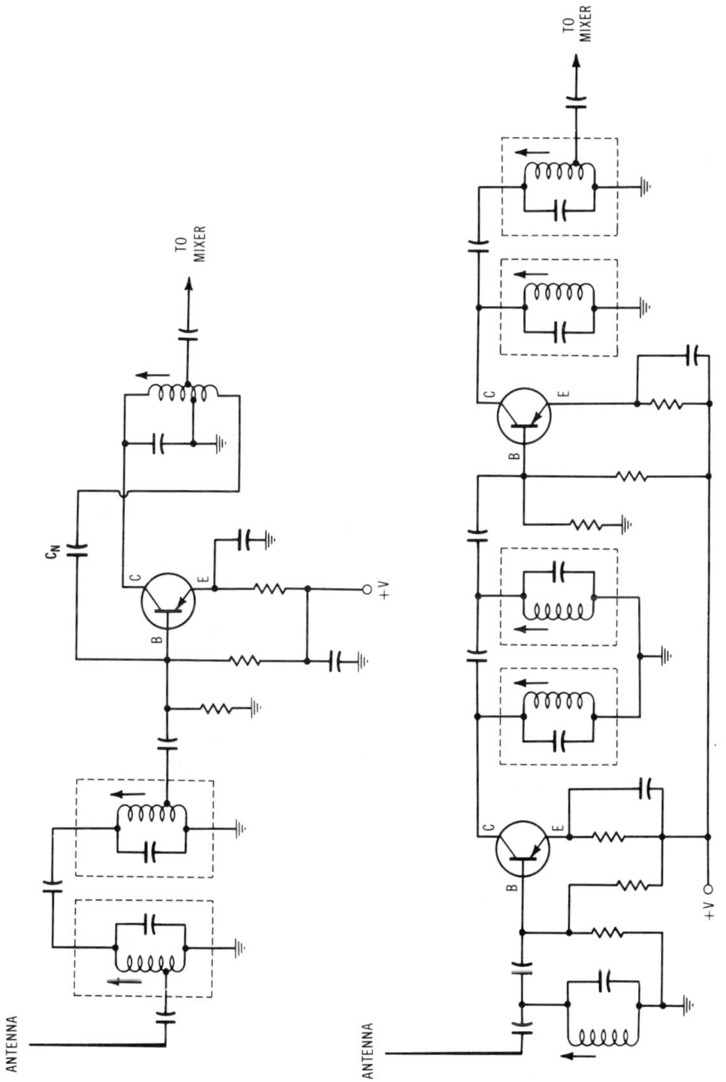

(A) Neutralized stage.

(B) Stage not neutralized.

Fig. 3-33. Basic rf amplifiers.

these stages operate class C in all transmitters. Quite often the low level stages are not operated class C because the advantage of high-efficiency class-C operation is nullified by the need for input driving power. Thus the overall efficiency is no better than that which can be obtained more conveniently with the use of class-A operation. However, in the driver-amplifier and power-amplifier stages of the transistor transmitter, class-C operation is preferable because of the more favorable efficiency obtained in the amplification of high-power signals.

Many low-level stages of the transmitter rf section are similar in circuit arrangement to the transistor tuned amplifiers used in the receiver, with the exception that less forward bias is used. Biasing is more nearly class B than class A. A typical stage is shown in Fig. 3-34. Double-tuned interstage coupling systems are used much less frequently than single-tuned types. Usually a tapped coil, split capacitor, or low impedance link is used to obtain impedance-matched coupling. Toroid transformers are especially popular.

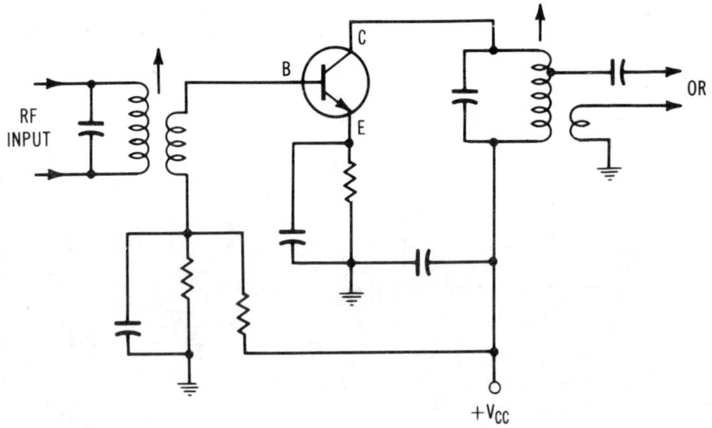

Fig. 3-34. Example of low-level rf amplifier for transmitter.

3-10-1 Basic Transistor Class-C Amplifier

A transistor operates readily as a class-C amplifier, because with no forward bias applied to its emitter-to-base junction, it is already biased class C. There is no collector current without some forward biasing. As you learned, germanium and silicon transistors require forward bias of approximately 0.2 volt and 0.7 volt, respectively, before there is collector current.

It is apparent from the above that transistors can be class-C biased by simply using no emitter-to-base junction bias as shown in Fig. 3-35. This amount of biasing is only slightly beyond the value of class-B cutoff bias and is used in low-level stages and is common for high-power vhf and uhf power amplifiers.

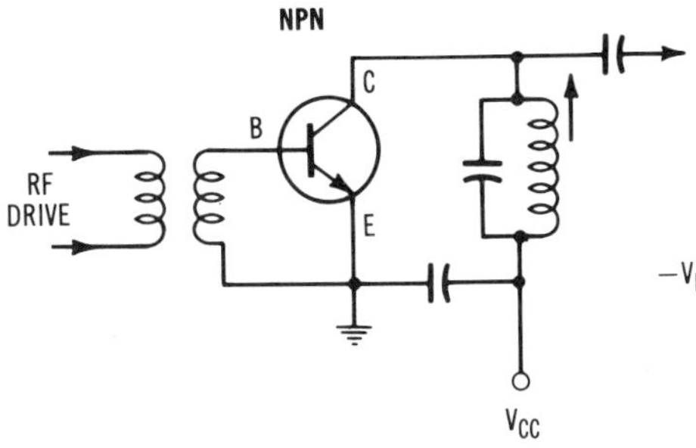

(A) Circuit.

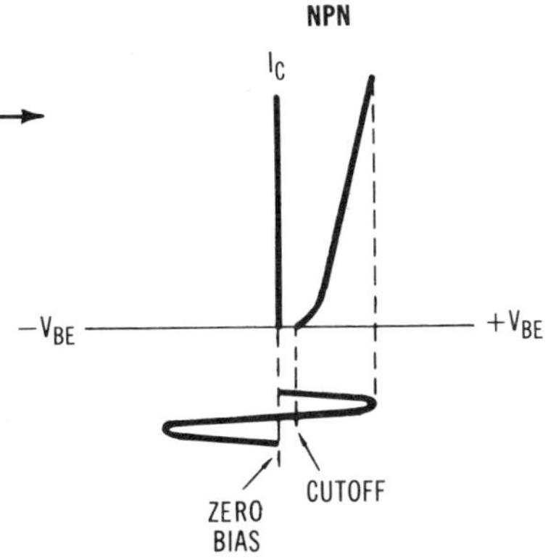

(B) Characteristics for npn transistor.

Fig. 3-35. Basic class-C amplifier.

The circuit arrangement of Fig. 3-36 shows how a simple resistor-capacitor combination in the base circuit permits biasing beyond cut-off. This reduces the angle of the collector current. In the npn stage, the more positive portion of the positive alternation of the input wave forward-biases the junction and produces base current. The direction of the base current is such that the charge placed on capacitor C_B is a back-bias. The emitter-to-base junction is reverse-biased as it should be for class-C operation.

The output current occurs in bursts that replenish the power in the output tank circuit. During the portion of the input wave that biases the emitter-to-base junction in the forward direction there is a short interval of collector conduction. At this time the resonant circuit capacitor is charged to its peak value. The oscillatory action in the resonant circuit reconstructs the complete rf cycle.

The *Q* of the resonant tank is significant. The unloaded *Q* should be as high as possible to obtain efficient operation of the tank circuit and to obtain maximum transfer of power from the tank to the load. The loaded *Q* of the tank circuit should be relatively low to permit effective transfer of power, but it must not be too low or the generation of harmonics will be excessive. This is quite a problem in transistor rf power amplifiers because of the inherent low impedance of the transistor. Thus suitable low-impedance resonant circuits and matching arrangements must be employed. Bifilar wound toroidal inductors are often essential.

The low impedance of the transistor has an advantage because neutralization is often not required. Quite often input impedance itself is very low and, along with the very low output impedance, reduces the tendency to self-oscillation. However, such a circuit must be protected from parasitic oscillations with appropriate radio-frequency chokes.

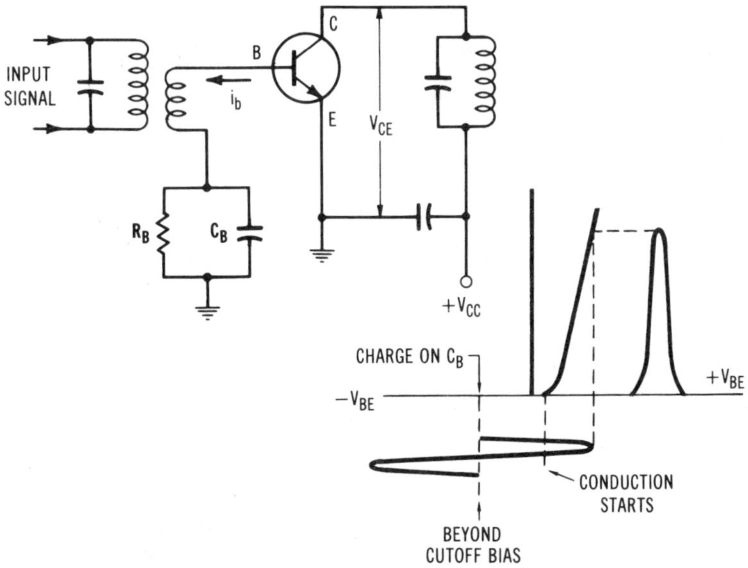

Fig. 3-36. Additional biasing with RC combination.

The pi-network (Fig. 3-37) is used widely in transistor power-amplifier circuits because of its convenient impedance matching capability and good harmonic rejection. It can be found in low- and moderate-power rf stages. For higher-powered stages it is often found in conjunction with other matching elements that ensure a proper match from the extremely low-impedance output of a high-powered transistor rf amplifier.

The pi-network can provide a proper match to a low-impedance antenna system. The output impedance of the transistor itself is often below this 50-ohm value and the function of the pi-network is to match the low-impedance output of the transistor stage to the higher impedance of the 50-ohm loading. Note the radio frequency chokes in both the input and output circuits. These prevent the supply voltage source and biasing networks from having an adverse influence on the loading and tuning of the input and output resonant circuits.

The tee-network tuning arrangement of Fig. 3-38 is popular in transistor power amplification. Its advantage is its ability to match the very low output impedance of a high-power rf amplifier stage to a significantly higher, but still relatively low, impedance load. The network consists of inductor L1 and capacitors C1 and C2. The network reflects a substantially lower impedance to the output of the transistor than the 50-ohm load into which the circuit must operate.

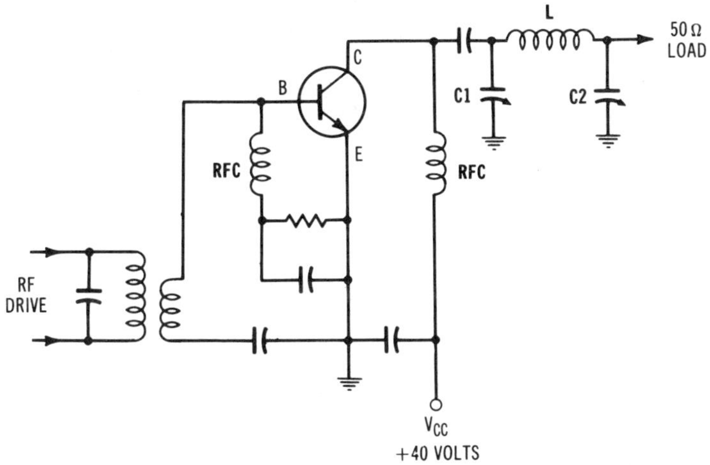

Fig. 3-37. Class-C stage with pi-network output circuit.

3-11. RF OSCILLATORS

Fundamentally an oscillator is an amplifier circuit in which the output is linked through a suitable network to its input. The radio-frequency oscillator contains a resonant circuit which determines the frequency of the oscillations generated. To cause oscillation a portion of the output signal that is fed back to the input must be in phase with an input change. Oscillations are sustained when the feedback or loop gain is greater than unity. The loop gain refers to the combined gain of the amplifier and the feedback network. In most oscillators the output variation is out of phase with the input and, to ensure in-phase feedback, the feedback network itself must produce a polarity shift. The manner of obtaining this feedback is the basic difference among the various types of oscillators.

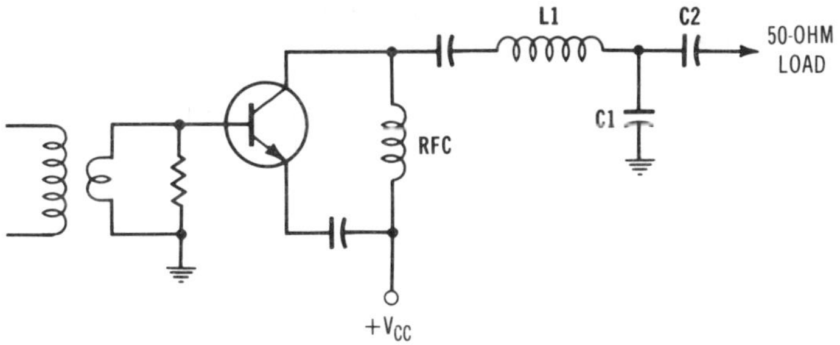

Fig. 3-38. Tee-network matching.

Two basic oscillator types are the *Colpitts* and *Hartley* circuits shown in Figs. 3-39A and B. Although they are used at both high and low radiocommunication frequencies, the Colpitts is somewhat more popular at the higher frequencies while the Hartley is more

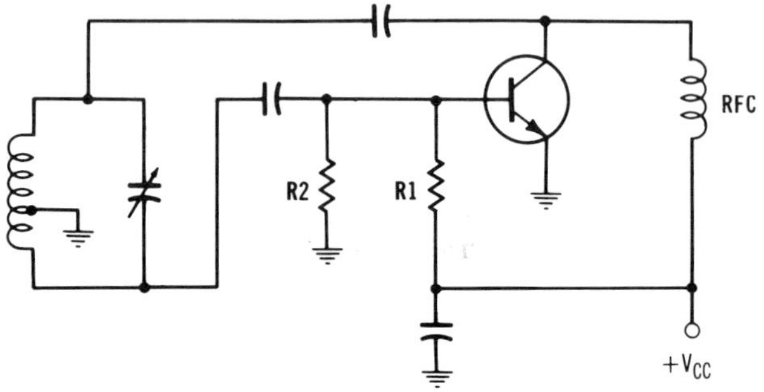

(A) Hartley circuit.

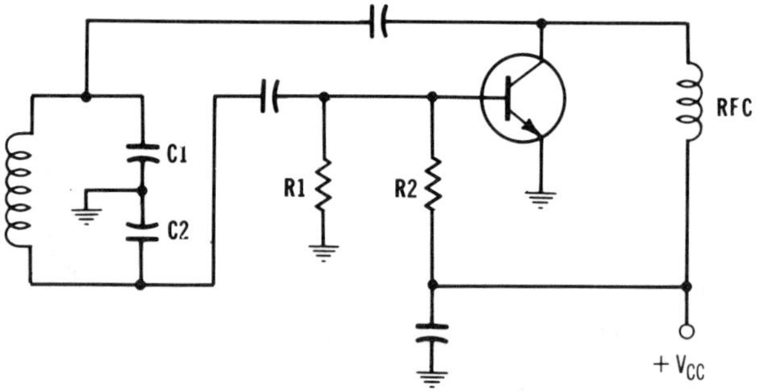

(B) Colpitts circuit.

Fig. 3-39. Basic oscillator circuits.

common at lower frequencies. In the Hartley oscillator the coil of the resonant circuit is tapped and usually placed at rf ground potential. This permits a small-amplitude out-of-phase rf voltage to be made available between the bottom of the coil and the tap. It is linked back to the input circuit. In practice the product of the network gain and the gain of the stage is made greater than unity to permit a buildup of the initially weak component and to compensate for circuit losses at the oscillating frequency. Thus, at the start of oscillation the feedback voltage is greater than the input voltage. However the gain of the stage falls off as the oscillation builds up. In so doing the greater-than-unity feedback causes the signal level to rise until the final maximum-magnitude level is attained. In effect the feedback to the input circuit is of just the proper level to continuously recycle the energy supplied to the resonant circuit and oscillations of constant amplitude are sustained.

The Colpitts oscillator (Fig. 3-39B) differs only in the manner in which the feedback voltage is obtained. In the Colpitts circuit the feedback voltage is picked off at the lower capacitor associated with the resonant circuit. It too provides a 180-degree shift and brings the output back in phase with the input voltage. The fraction of the input voltage that is fed back to the input is a function of the relative reactances of C1 and C2.

It should be noted that the oscillators are forward-biased a certain amount by resistors R1 and R2. Remember that the barrier potential must be overcome before the transistor draws any collector current. Thus with no bias the oscillations would not start. The bias resistors overcome the barrier potential and result in a start of collector current which then builds up to produce continuous oscillation.

A high-stability oscillator is shown in Fig. 3-40. The resonant frequency is determined by inductor L1, capacitor C and the combination of capacitors C2 and C3. The latter two capacitors are extremely high values and therefore a relatively low value of inductor L1 is needed to establish a given resonant frequency. The advantage of determining the oscillator frequency with a high C-to-L ratio is the great frequency stability that results. The large-value capacitors are so large in value that changes in transistor and external capacitive components have no significant influence on the operating frequency. The operating frequency is determined by the resonant circuit and not by stray external components.

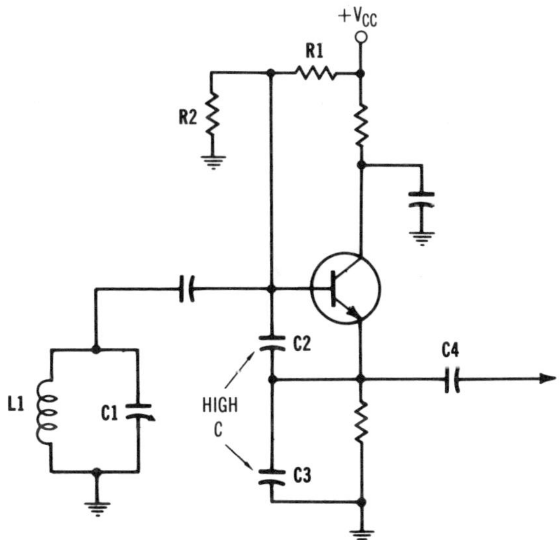

Fig. 3-40. High-stability oscillator.

Minimum loading of the oscillator circuit results when the output is removed from the low-impedance emitter circuit through an appropriate low-value coupling capacitor (C4). Resistors R1 and R2 provide the slight forward bias needed to start oscillations.

3-11-1. Crystal Oscillator

Certain crystalline materials produce a voltage when placed under mechanical stress. One such material is quartz, used in the manufacture of radiocommunication crystals. If the pressure on such a crystal is varied, an alternating voltage will be developed and conversely, when an alternating voltage is applied across the crystal, a physical vibration is set up in it. Quartz crystal vibrates freely, is very stable, and most important, it vibrates at a particular frequency determined by its size and structure.

A quartz crystal, when inserted into an oscillator stage tuned near its natural frequency, will vibrate strongly. In fact, a potential difference, much like the resonant voltage across a tank circuit, will be developed across the crystal. This voltage variation is supplied as feedback to the oscillator base (Fig. 3-41). A similar variation is

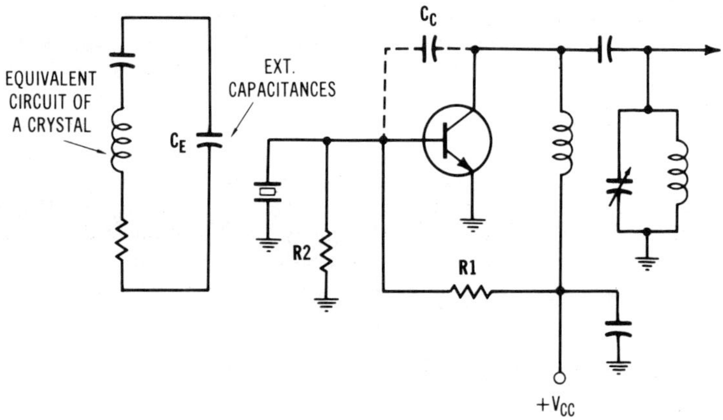

Fig. 3-41. Crystal oscillator circuit.

now set up in the collector tank circuit and is now fed back to the crystal to keep it in mechanical vibration. Inasmuch as the feedback is strongest at the mechanical-resonance frequency of the crystal, the frequency of oscillation is determined by the natural frequency of the crystal.

In fact, the characteristics of the crystal are quite similar to those of a resonant circuit as shown in Fig. 3-41. The crystal is the equivalent of a series resonant circuit having a high inductance and low capacitance. The crystal and the capacitance of its holder, along with the distributed external capacity shunted across the crystal, also form a parallel-resonant circuit. Since the series capacitance of the crystal is low, the external capacitance has only minimum influence on the oscillating frequency.

The feedback between output and input is provided by the collector-to-base capacitance (C_C) of the transistor. In some oscillator circuits, a small amount of external capacitance is connected between collector and base to establish the most favorable feedback condition.

The fundamental resonant frequency of a crystal is a function of its dimension and cut. The higher the frequency of operation the smaller is the crystal; a very-high-frequency crystal is very small, thin, and fragile. Fortunately crystals can be cut in such a manner that they are able to vibrate in a harmonic mode. Thus it is possible to obtain high-frequency output from a crystal that has lower frequency fundamental dimensions. Such a crystal is much less fragile than one cut for fundamental operation at a very high frequency.

The resonant frequencies of a crystal that correspond to the harmonic vibrations are referred to as *overtone* frequencies rather than harmonic frequencies. This differentiation is made because the overtone frequencies are close, but not exact, integrals of the fundamental frequency.

In general, overtone crystals are sold on the basis of their specific overtone frequency. The most popular overtone crystal is the third-overtone type; less common types are fifth- and seventh-overtone crystals. In a crystal oscillator designed for overtone operation the resonant frequency of the crystal circuit must be tuned to the appropriate overtone crystal frequency. The Q of the crystal at the overtone frequencies is high and good stability of the oscillator is assured. Usually an overtone crystal oscillator has a significantly lower output than that obtained from a fundamental oscillator circuit. Overtone crystals and oscillators are commonly used in communications equipment that is used in the vhf and uhf spectra.

3-11-2. Unijunction Transistor (UJT)

The unijunction transistor functions well as an unusual oscillator capable of generating very short duration low or high frequency pulses. The unijunction transistor consists of an n-type semiconductor bar with a moderately high resistance (Fig. 3-42A). Base 1 and base 2 connections are made at opposite ends of the bar. A small p-type junction is made along the bar and is called an emitter.

Understand that there is a voltage gradient along the bar that becomes progressively more positive from base 1 to base 2. When a voltage applied to the emitter exceeds the voltage gradient at the junction, the pn junction becomes forward-biased and there is a current flow. However when the conduction holes are injected from the emitter into the bar and the resistance drops between base 1 and emitter, a rapid increase in current is caused over this path.

When the emitter voltage is reduced to below the forward biasing

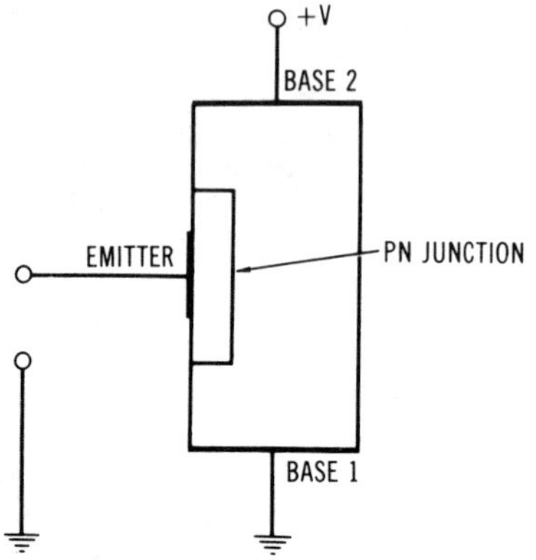

(A) Basic UJT.

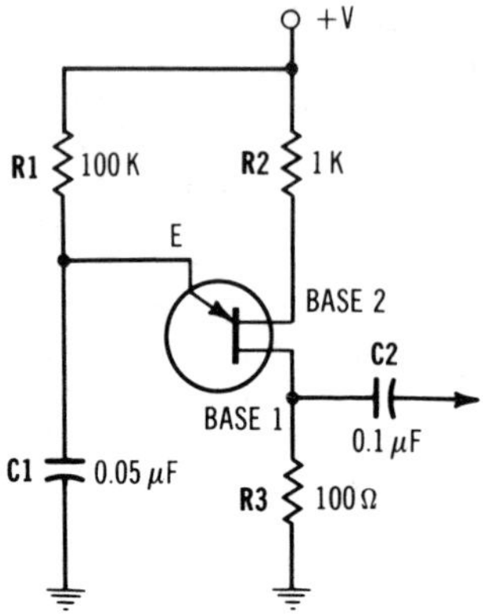

(B) Oscillator circuit.

Fig. 3-42. UJT plan and oscillator circuit.

value the pn junction ceases conduction and the current drops away to zero very quickly.

A typical ujt oscillator circuit is shown in Fig. 3-24B. When the circuit is turned on, capacitor C1 charges toward a positive value through resistor R1. This charging action finally supplies a potential to the emitter which is above the trigger value. Consequently, the unijunction goes into a high current state and a large current is present in the output resistor R3.

The resistance between the emitter and base 1 is now very low. Thus the capacitor is discharged quickly, causing the emitter voltage to drop below the unijunction conduction level. The junction is reverse-biased and the current in output resistor R3 drops very quickly to zero.

The frequency of the unijunction oscillator is determined largely by the time constant of resistor R1 and capacitor C1. This time constant is made very long if a low-frequency pulse train is to be obtained. Lowering the time constant will increase the output frequency. If the duration of the output pulse is to be increased for a given time constant, increase the value of C1 and decrease the value of R1 a like amount. If an adjustable frequency output is desired, resistor R1 can be a fixed resistor and potentiometer combination.

3-12. FIELD-EFFECT TRANSISTORS

A field-effect transistor (FET) is a three-element device consisting of source, gate, and drain (Fig. 3-43A). It is basically a single-junction (unipolar) device made up of a semiconductor bar (n-type material in the example) into which two facing strips of semiconductor material of the opposite type are diffused.

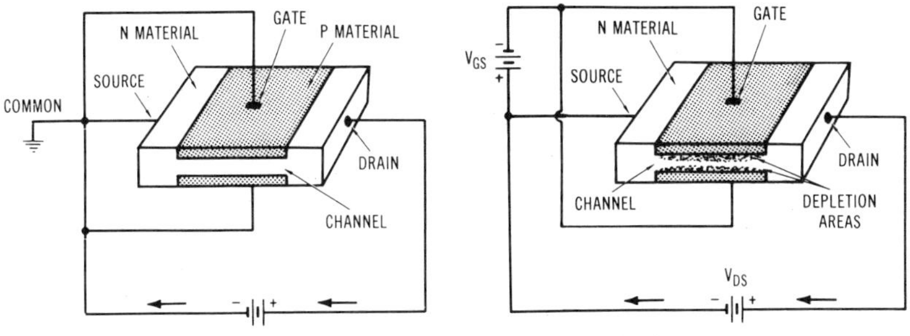

(A) Basic construction. (B) Biasing.

Fig. 3-43. FET construction and biasing.

In FET terminology the bar is called a channel while the element that controls the motion of charges along the channel is known as the gate. The common end of the channel is known as the source while the opposite end is called the drain. When the semiconductor bar is biased as in Fig 3-43A, there is a motion of electrons from source to drain through the bar or channel. In the fundamental operation of the FET, the gate is able to control this motion of charges along the channel. Let us consider how this is accomplished.

In a practical circuit the gate-to-channel junction is reverse biased with the source being used as the common connection, Fig. 3-43B. This reverse biasing of the junction causes a depletion region to extend into the channel. The presence of the depletion region in the channel causes an increase in the resistivity of the bar (decrease in conductivity). When the gate bias is increased there is a further penetration of the depletion region, a further rise in the channel resistance, and the motion of charges (drain current) is reduced. If the gate bias voltage is made high enough, the channel or drain current can be reduced to zero, thus producing cutoff.

The basic operation of an FET amplifier, along with a typical circuit and transfer characteristic, is illustrated in Fig. 3-44. An applied audio or other signal causes the gate voltage to vary about the gate bias set by the gate-bias voltage. As the gate voltage varies so does the extent that the depletion region varies back and forth in the channel causing a similar change in the channel resistance. Consequently the gate voltage variation causes a similar variation in the channel or drain current. In fact, as the gate voltage is made to vary with signal there results a substantial change in the drain current. The actual drain output voltage results from the variation of the drain current in the drain load resistor. The actual voltage gain of the stage is, of course, the ratio of V_O/V_{IN}.

A positive swing of the gate voltage produces an increase in the drain current. In turn there is a drop in the drain voltage (output

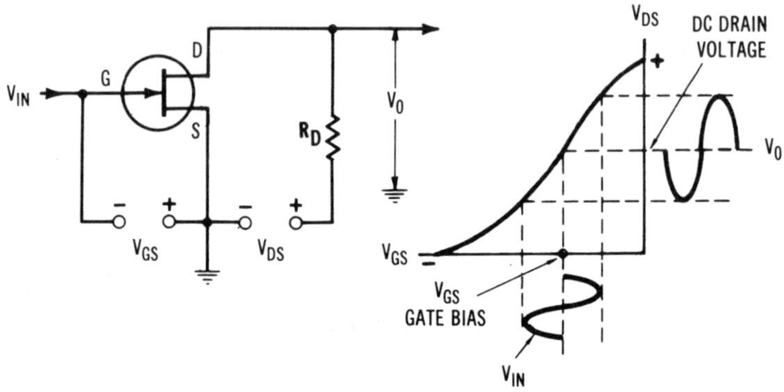

Fig. 3-44. Basic FET operation.

voltage swings negative). Conversely the negative alternation of the input voltage decreases the drain current and there is a positive swing of the drain and output voltages. Input and output voltages are said to be of opposite polarity, or out of phase.

In the example the channel was composed of n-type material and therefore the charge carriers (current) were electrons. The gate is composed of hole, or positive, carriers which neutralize some of the electron carriers to produce the depletion region that controls the resistance of the channel. This is known as an n-type junction field-effect transistor.

There are also p-type junction FETs. In this type the current in the channel is the result of the motion of hole charges. The gate has electron carriers and they in turn neutralize the hole carriers to form the depletion region that moves in and out of the channel in accordance with gate-voltage change. Operation is still the same with the channel current being a function of the gate-voltage change. The only difference is that the drain voltage is made negative instead of positive as in the case of the n-type junction FET.

3-12-1. Common-Source Amplifier

The field-effect transistor is basically a voltage amplifier rather than a current amplifier as is the bipolar transistor. There is no significant gate current, and, therefore, the input displays a high impedance to any signal source. The gate circuit responds to an input voltage change rather than an input current change. The output impedance too is moderately high and is higher than that displayed by the usual bipolar transistor.

A practical common-source FET amplifier circuit is given in Fig. 3-45. No external gate bias is necessary. Rather the presence of drain current in source resistor R2 develops the negative gate-to-source dc bias. Resistor R1 can have a high ohmic value because there is no significant gate current. This too aids in keeping the out-

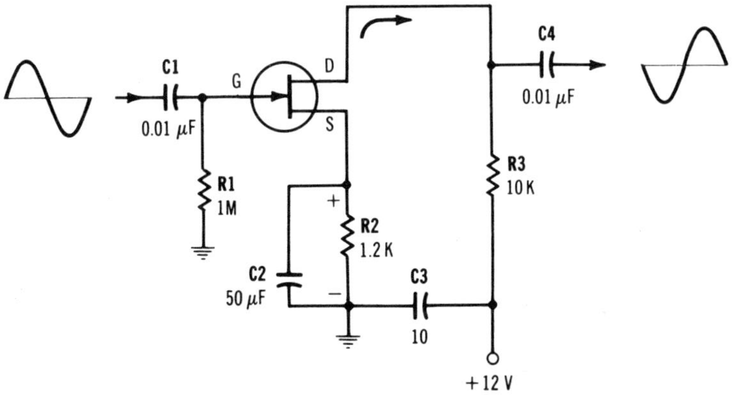

Fig. 3-45. Basic common-source amplifier.

put impedance high. The resistance of R2 is kept low in value in order to develop the voltage that will bias the transistor stage on the linear portion of its transfer characteristic. Typical values are given for a general purpose small-signal FET.

Resistor R3 is the drain load resistor across which the output signal is developed. Its value is selected to maintain the proper dc collector voltage and collector current.

Capacitor C2 prevents an ac voltage from developing across source resistor R2. This ac voltage is degenerative and would reduce the amplifier gain.

3-12-2. Common-Gate Amplifier

In the common-gate configuration (Fig. 3-46) a signal is applied between the source and common, and is removed between the drain and common. The Fig. 3-45 circuit was that of a common-source stage with the signal being applied to the gate and removed at the drain. The common-gate amplifier is capable of voltage gain but has a current gain that is less than unity. The common-source stage is capable of both voltage and current gain.

The common-gate stage has a very low input impedance and a very much higher output impedance. Therefore it can be used in any circuit where there must be a transformation between low input impedance and high output impedance at the same time there is a possible voltage gain. It is not a particularly common FET circuit although it does have some application in radio-frequency amplifiers.

Component values are approximately the same as those used in the common-source stage of Fig. 3-45. No gate resistor is used. Instead the gate is returned to the common. Signal is applied across the source resistor and this resistance along with the input resistance of the source circuit presents a low impedance to the source of signal. It is necessary to increase the value of the input coupling

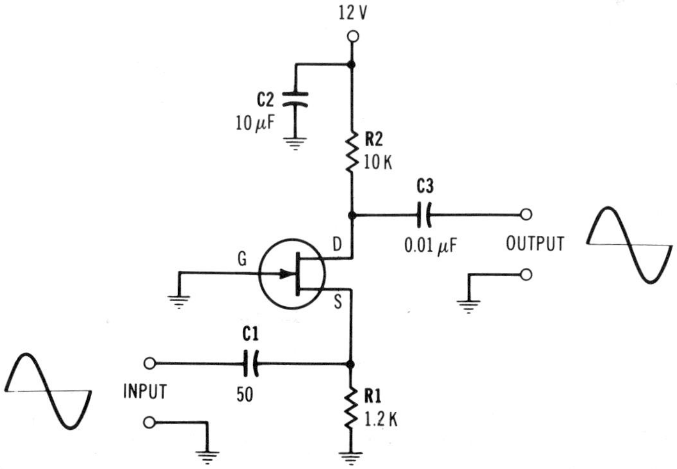

Fig. 3-46. Common-gate amplifier.

capacitor because of the much lower input impedance of the common-gate stage. This step maintains the frequency response to a low limit of at least 100 hertz. High-frequency response is well over 20,000 hertz.

When using the common-gate configuration input and output voltages are in phase. In the previous common-source circuit input and output voltages were 180 degrees related. The voltage gain is approximately the same as that obtained with the common-source stage. Maximum undistorted output can be obtained by proper regulation of the resistor R1 and R2 values.

3-12-3. Common-Drain (Source-Follower) Stage

In the *common-drain* configuration the input signal is applied to the base; the output is removed from the emitter (Fig. 3-47). The common-drain, or *source-follower*, stage is an effective impedance transformer. The input impedance is very high; output impedance is quite low. Thus the stage is capable of current gain but not voltage gain. In a properly designed stage the output voltage is just slightly less than the input voltage. Such a stage is used mainly in matching a high-impedance source of signal to a low-impedance load. It does so with a minimum loading effect on the source of signal.

The output voltage is in phase with the input voltage. Another advantage of the common-drain stage is its capability of handling a higher input signal level without distortion, while at the same time, it places a very minimum load on the source of the signal.

The practical source-follower stage of Fig. 3-47 uses a general-purpose FET. Resistor R1 in the gate circuit is returned to the junction of resistors R2 and R3. Therefore proper gate bias is applied to the stage and is that voltage developed across resistor R2. Resistors R2 and R3 in combination establish the proper drain voltage and serve as the output load resistance of the circuit.

The stage will operate over a considerable range of audio input signal levels and has a low-frequency response that extends down to at least 100 hertz. High-frequency response extends well above 20,000 hertz.

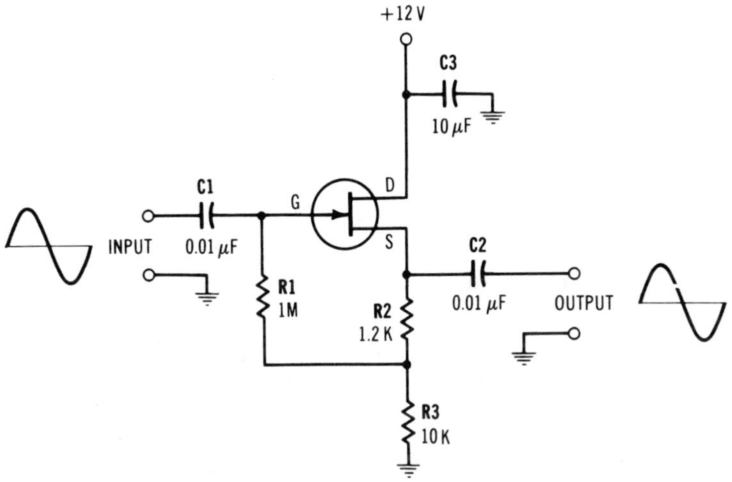

Fig. 3-47. Common-drain stage.

3-12-4. Stabilized Common-Source Amplifier

Two factors have a significant influence on FET bias requirements. One is the production spread, which may be such that the saturation current I_{DSS} may have a ratio of 3-to-1 or higher. Temperature drift is the second factor and involves both the warmup operating-point drift or a shift as the result of change in ambient temperature.

The changes in operating conditions as the result of production spread and temperature are pronounced when external gate bias is used. Self-biasing is much more self-adjusting because a shift in operating-point drain current is compensated for by the change in bias produced by the change of current in the source resistor. Additional operating-point stabilization can be obtained with a further increase in the ohmic value of the source resistor. Normally this would result in an improper bias, but a combination of gate-divider positive bias and a higher value of source bias can be used to set up the required operating point. Refer to Fig. 3-48.

Resistor R4 is increased in value to improve stabilization. However its ohmic value is too high for obtaining the correct operating bias. Resistors R2 and R3 are added to make appropriate

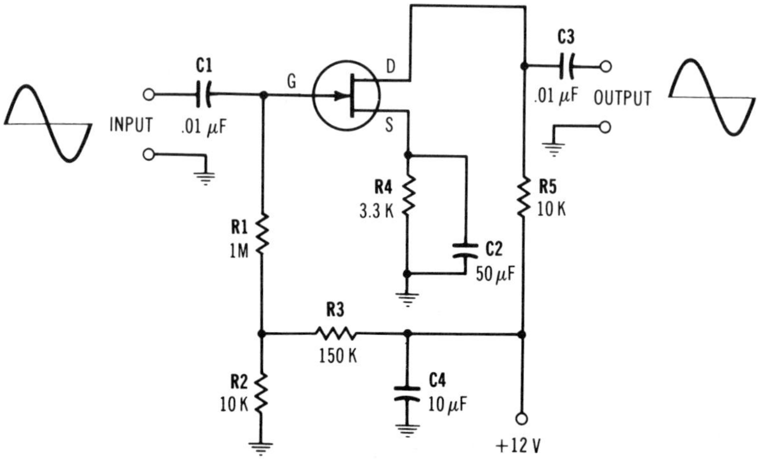

Fig. 3-48. Stabilized common-source amplifier.

compensation and the combination of the three resistors R2, R3, and R4 establish the correct and stabilized operating bias for the stage. Good stabilization is obtained by increasing the ohmic value of the source resistor three to five times. The ratio of the two resistors (R2 and R3) must then be set to establish correct bias.

3-12-5. Radio-Frequency Amplifier

The field-effect transistor is well suited to radio-frequency application. A low distortion and a low noise figure that extends up into the vhf and uhf spectra makes them ideal for use in radiocommunication receivers. Because of the high-impedance characteristic, simple circuit arrangements can be used.

A principal advantage of the FET in rf service is its low distortion. The field-effect transistor is a square-law device. When a single-frequency rf signal is applied to a square-law amplifier, its nonlinearity is such that three components develop in the output—fundamental frequency, second harmonic, and a dc component. The latter is not transferred to the next stage, and the second-harmonic component is so widely separated in frequency from the fundamental that tuned circuits prevent its transfer. Therefore, only the fundamental-frequency signal is conveyed to the next stage. The nature of its characteristics are such that additional higher-order harmonics are not developed. Also, the level of intermodulation distortion is reduced as compared to other devices, because there is less opportunity for the development of spurious beats with other signals.

Low-noise and low-distortion features also make the field-effect transistor an attractive mixer. In an FET mixer the sum or difference frequencies are developed with a minimum of cross-modulation difficulties. Associated circuits, as compared to bipolar transistors, are less complex and impedance matching and temperature instability are less of a problem.

An example of an un-neutralized simple stage is shown in Fig. 3-49A. The circuit is stabilized by shunting an appropriate resistive load across the FET output—either an actual resistor or a low-impedance load. This reduces the stage gain and increases the bandwidth.

Fig. 3-49B is that of a neutralized high-frequency FET amplifier. It has tuned input and output circuits along with an appropriate neutralizing inductor (L_N) that is connected between input and output resonant circuits.

A popular circuit for junction field-effect transistor amplifiers is the cascode arrangement of Fig. 3-49C. This circuit too requires no neutralization. The input stage is connected in a common-source configuration with unity gain; it is followed by a common-gate amplifier. The input stage has a high input impedance and is biased near or at zero. Therefore it provides a suitable load for a high-impedance small-signal source. Most important the input capacitance it presents to any such source is low.

The common-gate stage has a low input impedance. Consequently, a substantial load is placed on the drain circuit of the first stage, and the latter does not operate with gain. The gain is handled by the common-gate output stage which requires no neutralization.

The cascode combination has the gain of a single transistor stage. However, the circuit is unique in that good gain is obtainable, input and output impedance are high, and input and output voltages are in phase. Thus it is a very stable hf, vhf, and uhf amplifier.

3-13. MOSFET AMPLIFIERS

The structure of the *MOSFET* (metal-oxide semiconductor field-effect transistor) provides some unique characteristics and makes the MOSFET attractive for small-signal and low-noise radio-frequency amplifiers and mixers. The characteristics are favorable for use in other high-impedance applications and in chopper and switching circuits. In the structural arrangement (Fig. 3-50) the gate electrode acts as a control element just as it does in a junction field-effect transistor. However, that control is exerted whether the gate is swinging positive or negative. In the normal operation of a junction type the gate exerts an influence only when the junction is reverse biased.

In the MOSFET construction a thin gate layer is insulated from the semiconductor material that forms the channel by a thin layer of silicon dioxide. The electrical field that results when the gate is biased appropriately influences the active carriers in the source-drain channel. Since there is insulation instead of an actual junction between the gate and channel, the leakage current is very low and is little affected by temperature. A very high input impedance

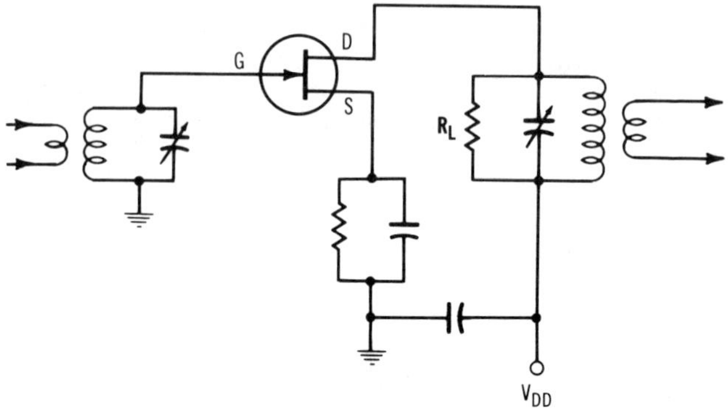

(A) Unneutralized.

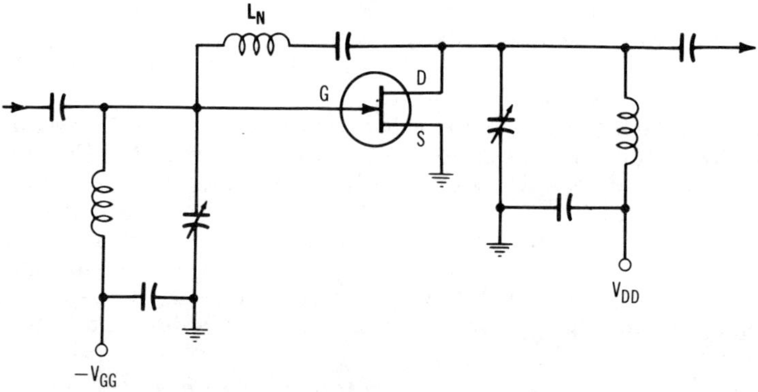

(B) Neutralized.

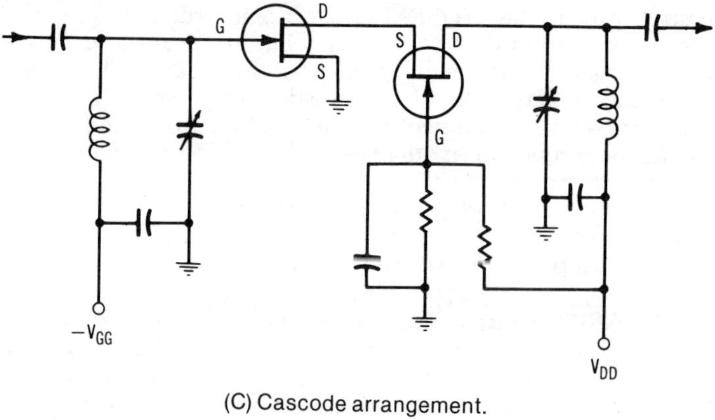

(C) Cascode arrangement.

Fig. 3-49. FET rf amplifiers.

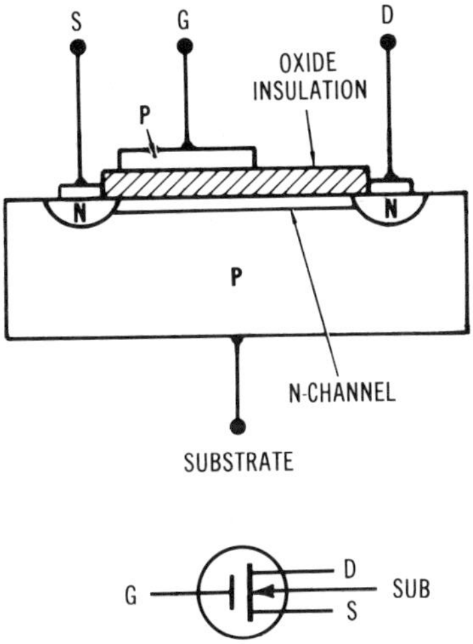

Fig. 3-50. Depletion MOSFET.

results. The gate and channel, separated by an oxide layer, are in effect the plates of a capacitor.

The arrangement of Fig. 3-50 shows the structure of a depletion-type MOSFET. In this example, the channel material is n-type and there are excess electrons. As the gate is made negative relative to the source, the number of electrons in the channel is depleted (reduced). This action is similar to that of a junction FET. However, during normal operation the gate of a junction FET may not be permitted to swing significantly past zero because the junction will then be forward-biased, thus decreasing the input impedance. The MOSFET has no such limitation. If the metal gate is made to swing past zero there is no increase in input current or decrease in input impedance, and the gate maintains its control of channel conductivity. This can be shown in the characteristic curves of Fig. 3-51. Note that the gate voltage can be made to swing either positive or negative.

A popular type of MOSFET construction is the enhancement-mode arrangement of Fig. 3-52. Note that in its basic structure there appears to be no actual channel between the source and the drain. However, the gate itself does span this region and when a positive bias is applied to the gate, a conduction path between the source and the drain is established. The conductivity of this path is a function of the charge on the gate. In this manner, the magnitude of the drain current is controlled.

The unusual characteristic of the enhancement-type MOSFET is that it is normally cut off (no drain current) until the application of the positive gate voltage. The application of a gate voltage causes the drain-source current to increase. The more positive the gate-source voltage is the higher is the drain current. This is shown in the characteristic curves of Fig. 3-53.

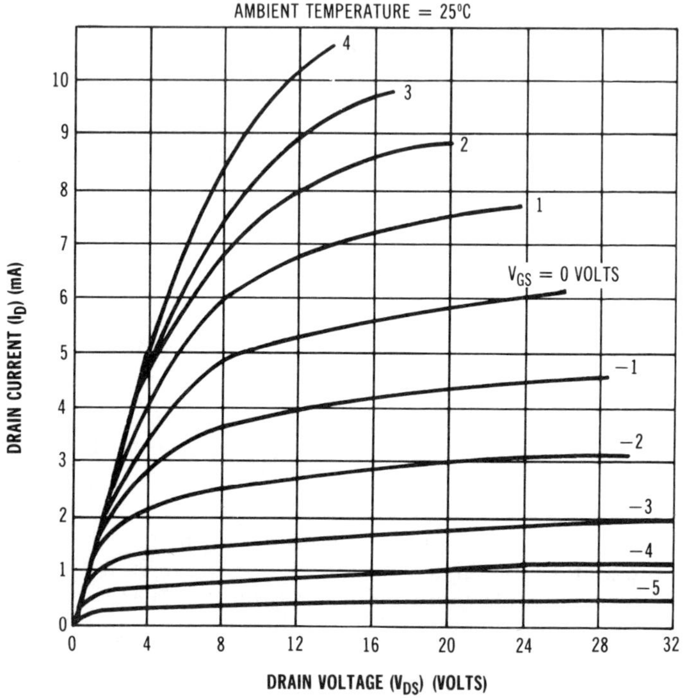

Fig. 3-51. Characteristic curves of a depletion MOSFET.

The high input impedance is both an advantage and a hazard of the MOSFET device. Since the input impedance is so high, a very tiny current or static charge is able to build up a high voltage on the gate which can possibly damage the device. Thus MOSFETs must be handled carefully and circuit arrangements must be such that excessive charges are kept off the gate. Another way of avoiding this hazard is by using gate-protective diodes built within the device (Fig. 3-54). These diodes are electrically connected between gate and source bypassing those voltage transients which could damage the device.

The two- or dual-gate arrangement provides a separate gate for signal and for applying an external control bias. The dual-gate configuration can be used conveniently in mixer and other balanced circuit arrangements. The dual-gate plan also makes possible a reduction in feedback capacitance obtained by ac grounding of the second gate. Thus the device can be used favorably in amplifier circuits without neutralization. Local-oscillator feedthrough to the antenna is reduced.

A typical vhf amplifier using a MOSFET transistor is given in Fig. 3-55A. Input signal is applied through an appropriate tuned circuit and matching arrangement to gate 1. Agc bias is applied to gate 2 and is able to regulate the gain of the rf stage in accordance with the magnitude of the incoming signal. No neutralization is used and, if desired, the output can be matched to a low-impedance load.

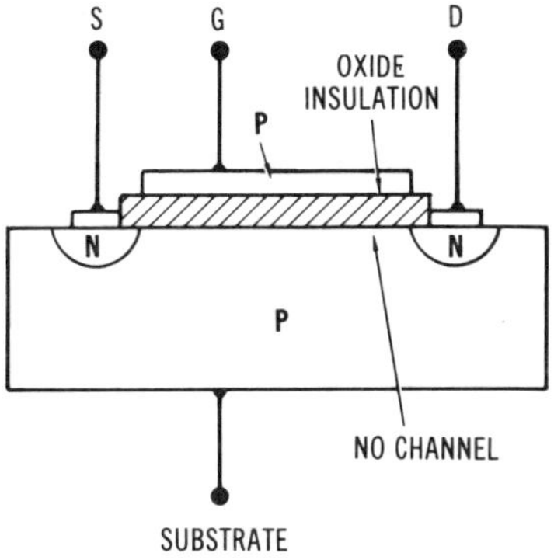

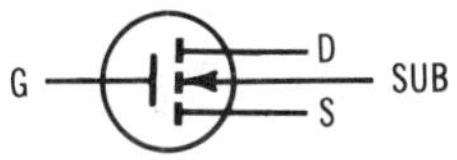

Fig. 3-52. Enhancement MOSFET.

A dual-gate mixer is shown in Fig. 3-55B. In this circuit, signal is applied to gate 1 while local-oscillator injection is made via gate 2. Circuit arrangement is simple and compact fine-performing mixers

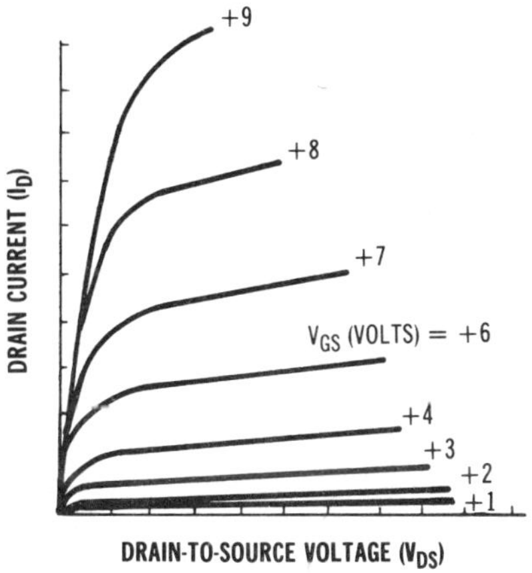

Fig. 3-53. Enhancement MOSFET curves.

can be constructed around dual-gate MOSFETs. Such circuits are found widely in the receivers of modern radiocommunications equipment.

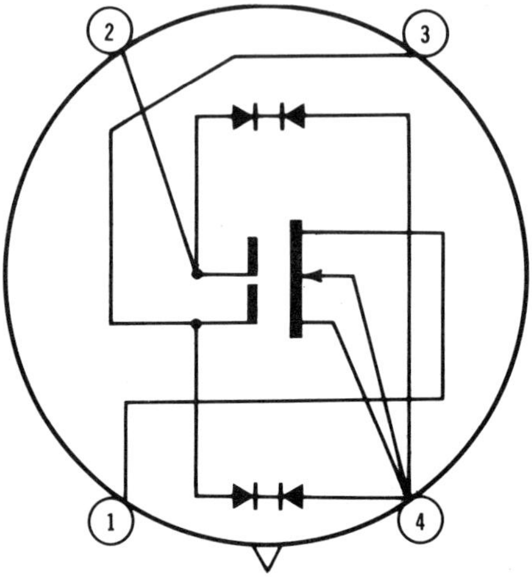

Fig. 3-54. Dual-gate MOSFET with protective diodes (40673).

3-14. FET OSCILLATORS AND CLASS-C AMPLIFIERS

Field-effect transistors have become more common in transmitters and two-way radio transceivers. Principal uses are likely to be in oscillator, low-level rf amplifier, multiplier, audio, modulator, and demodulator circuits. Of course, applications involving high-powered rf circuits can be anticipated.

Field-effect transistors have characteristics that are ideal for use in the crystal oscillators that appear in transmitters, transmitter test and calibration circuits, and receive sections of two-way—radio transceivers. A high-Q resonant circuit, important in the design of an easy-starting oscillator, is not loaded by a FET.

Either the Pierce or the Miller oscillator can be employed (Fig. 3-56). The Pierce oscillator has the advantage of not requiring any tuning; it can be operated over a wide frequency range simply by plugging in an appropriate crystal. The Miller circuit permits a higher output, and the drain (output) circuit is tunable.

As a solid-state oscillator, the field-effect transistor is superior to the bipolar type. A bipolar stage has a low impedance and can be troubled with feedback and starting problems if the crystal is not especially active. Also, the bipolar crystal oscillator must be forward-biased slightly or it will not start. In a field-effect circuit, biasing is a matter of control rather than of necessity.

If a low-impedance output is desired, the crystal oscillator can be followed by a source-follower or common-source amplifier. There is a minimum load on the crystal oscillator, and operation is little affected by any load that is connected to the output of the source follower.

A typical FET class-B or class-C amplifier is shown in Fig. 3-57A

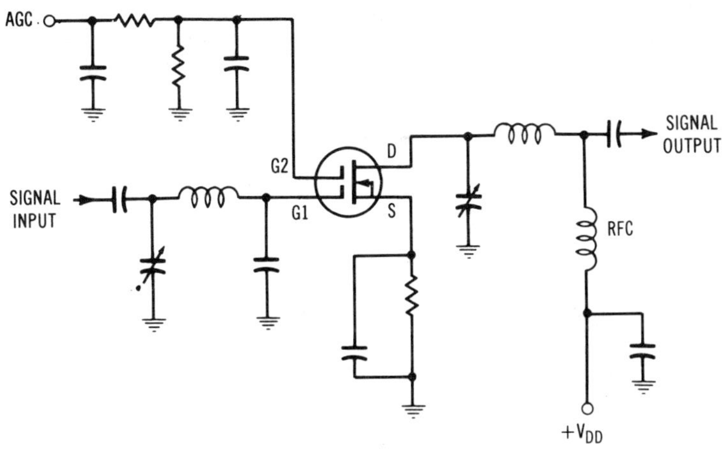

(A) Amplifier circuit vhf.

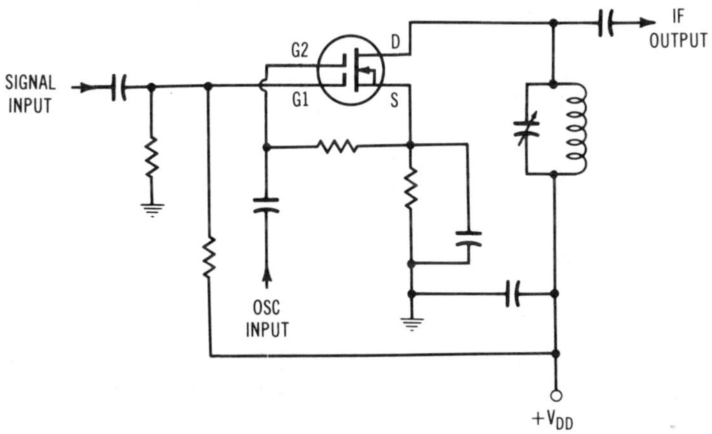

(B) Mixer circuit.

Fig. 3-55. MOSFET rf amplifier and mixer.

and B. External bias can be used. Biasing according to class of operation is shown in the transfer characteristic (Fig. 3-57C). For class-A operation the biasing is at the center of the linear portion of the transfer curve. To operate the amplifier class B it is necessary that the level of external bias match the cutoff bias of the transistor. A class-C stage is, of course, biased beyond cutoff.

The input impedance of an FET rf amplifier can be kept at its highest by making certain that the peak of the positive alternation of the input signal does not extend to the near-zero bias level that results in gate current. However, for more convenient and efficient operation and a somewhat lower input impedance, the stage may be operated in such a manner that the gate current is drawn at the peak of the positive alternation. In such an arrangement, one can

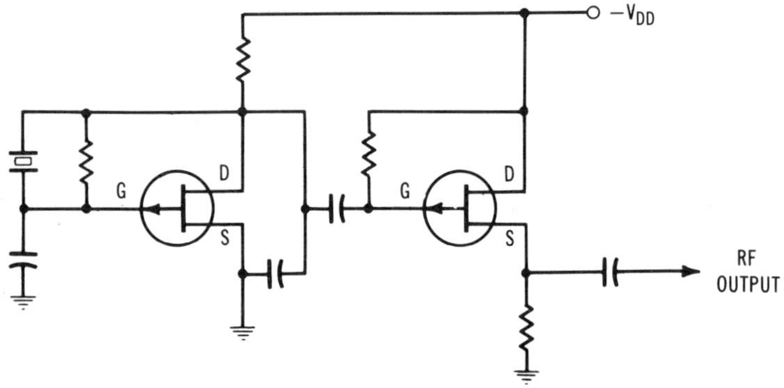

(A) Pierce crystal oscillator with source-follower buffer.

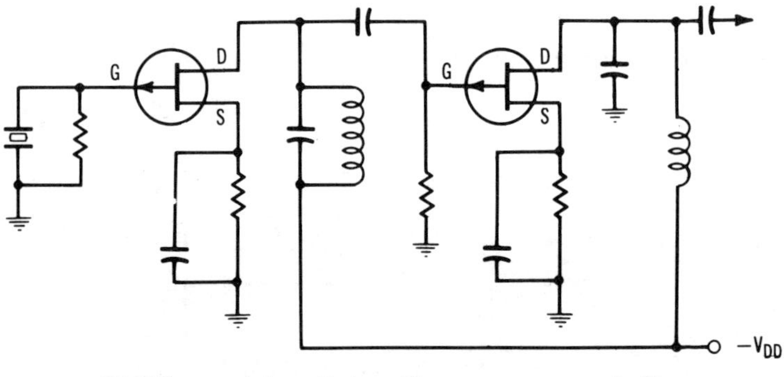

(B) Miller crystal oscillator with common-source buffer.

Fig. 3-56. FET crystal oscillators and buffers.

use gate current to establish the required class-B or class-C operation, using the circuit arrangement of Fig. 3-57B. The direction of the gate current is such that a negative bias is developed on the gate capacitor. With a proper time constant, this capacitor charge remains constant and serves as the dc gate bias for the stage. As mentioned, the input impedance is now lower, and somewhat more input power is required.

When the rf excitation is removed from its input, the drain current rises. It is advisable to use a protective source resistor which limits the drain current to a safe value in case rf excitation is lost.

The waveforms, shown in Fig. 3-58, demonstrate the operation of the FET class-C stage. The transistor draws drain current during

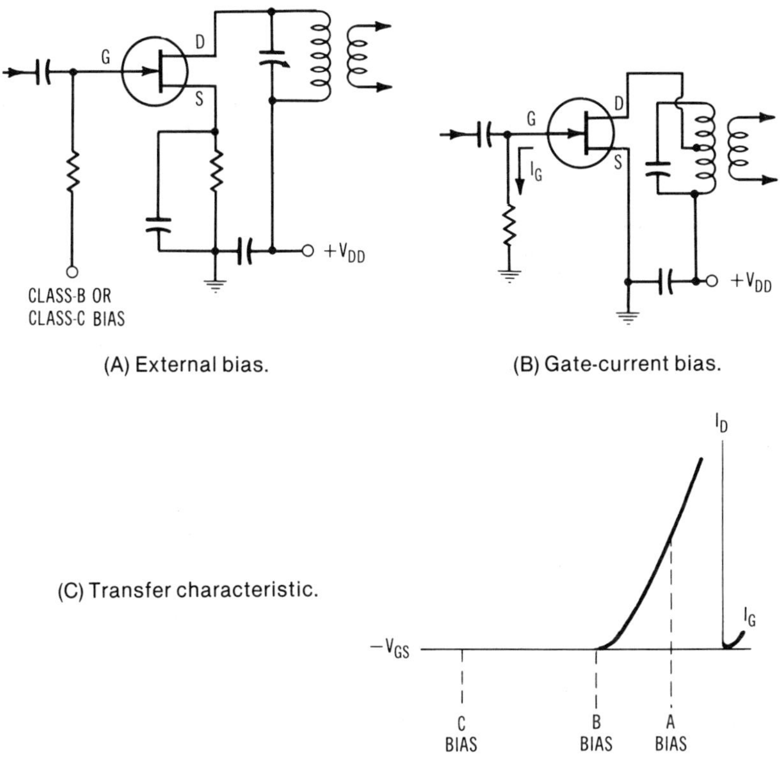

(A) External bias.

(B) Gate-current bias.

(C) Transfer characteristic.

Fig. 3-57. FET class-B and class-C amplifiers.

only part of the positive peak of the input wave. Thus, the drain current is drawn in bursts that supply electrical energy into the drain resonant circuit, charging the capacitor to a high negative value (sharp drop in drain voltage). However, a resonant circuit has a smoothing and energy-storing capability. When the bursts of drain current cease, the charge on the capacitor begins to fall off, and energy is transferred to the coil. After the charge reaches zero, the collapsing field of the coil releases its stored energy back into the tank circuit. The current reverses, and the capacitor is charged to an opposite polarity (positive), even though the drain current has stopped. The magnetic field around the coil is now of opposite polarity. When the capacitor loses its peak charge, the magnetic field again collapses, and a new cycle begins.

Soon after, a new burst of drain current is introduced because of the positive swing of the gate input. Again the capacitor in the drain tank circuit is charged to a maximum negative value, adding energy to the circuit. Although the rf amplifier operates class C and drain current is not present for the full input cycle, a good sine

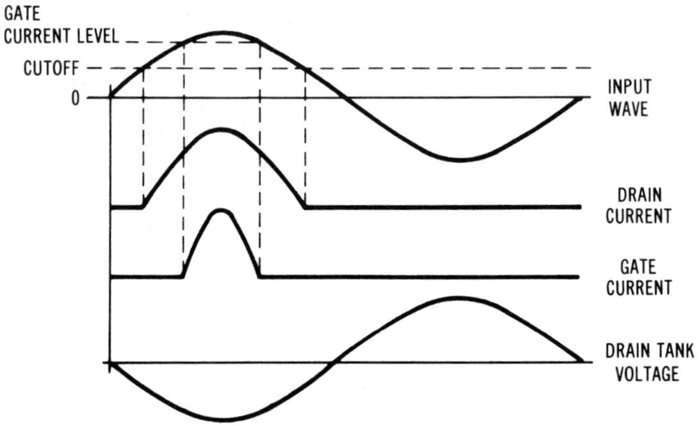

Fig. 3-58. Waveforms in a class-C neutralized amplifier.

wave is developed across the drain tank circuit when the proper choice of inductance, capacitance, and load has been made. The magnitude of the sine wave depends on the drain supply voltage. Over a substantial range, rf output voltage is a linear function of the drain supply voltage.

The positive peak of the input wave draws a small gate current, which develops a negative charge on the gate capacitor (C_G Fig. 3-57B). The current is such that the charge placed on the capacitor biases the gate beyond cutoff. Between intervals of gate current, the charge leaks off the capacitor through the gate resistor (R_G). If the time constant of $R_G \times C_G$ is high enough, a reasonably constant bias is maintained during the entire cycle of the input wave. This is equivalent to a dc voltage value that is significantly higher than the cutoff values. Of course, the capacitor is restored to full charge by the gate current at the crest of each positive input wave.

The length of time that drain current exists can be expressed as a number of degrees (angle) out of a full cycle (360 degrees). The angle of the drain current is a function of the desired efficiency, the output level, and the quality of the waveform. Values can occur between 90 and 150 degrees. The gate-current angle is significantly shorter because gate current is present only at the very peak of the rf input wave. This can be seen in the waveforms of Fig. 3-58.

3-14-1. VMOS Power FET

The conventional JFET and MOSFET transistors are fundamentally low-power devices. The regular FET is a horizontal-motion device (Fig. 3-59A) because the charge motion along the body or channel flows horizontally from source to drain. This limits the current density compelling the chip area to be large and the

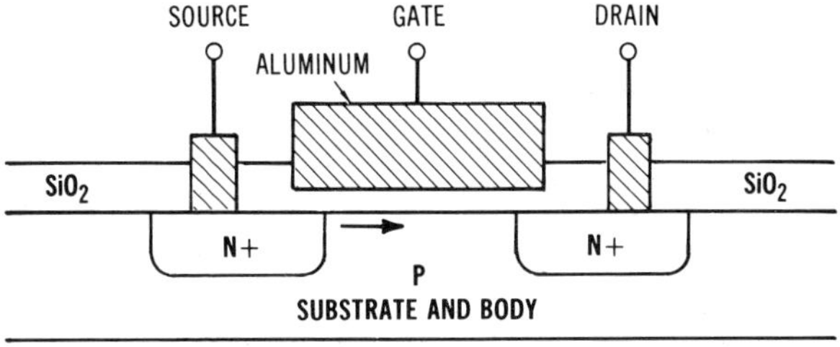

(A) Cross section of conventional MOSFET.

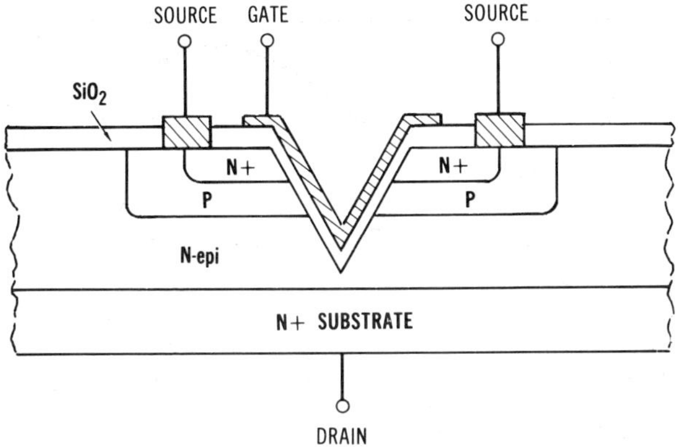

(B) Cross section of a VMOS channel.

Fig. 3-59. Structure of MOSFET and VMOS.

cost high if any significant power capability is to be obtained.

However this power limitation has been eliminated with the development of the VMOS (vertical metal-oxide semiconductor) field-effect transistor (Fig. 3-59B). The V-groove gate region and associated arrangement of elements permits a vertical motion of charges and a much greater current density is possible. It is an enhancement-mode transistor and requires that the gate and drain both be positive relative to the source if there is to be a motion of charges. Note the characteristics shown in Fig. 3-60. Before there is a drain current the gate voltage must be made positive. At zero volts and negative voltages the drain current is cut off.

In operation the positive gate sets up an n-type channel on both sides of the p-type body. Electrons flow from the source through the n-type channel and the epi (epitaxial) region to the substrate. The substrate serves as the drain connection for the VMOS device. The higher width-to-length of the VMOS channel plus other factors provide a much greater current density and a higher power-handling capability. The power capability is now comparable to that of bipolar devices without some of the disadvantages.

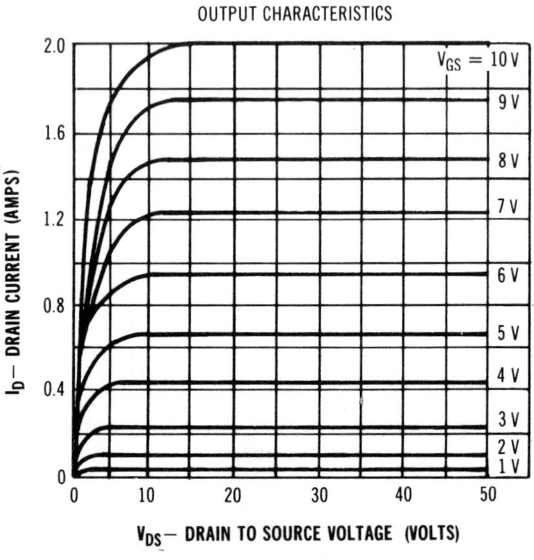

Fig. 3-60. VMOS characteristics.

The VMOS is a high-impedance voltage-drive device that provides good linearity and low distortion. A given device can function well as either a weak- or strong-signal device. As it heats, its characteristic is to draw less current instead of more. Consequently the thermal runaway hazard does not exist. It is not subject to secondary or mismatch breakdown. Furthermore there is no change in input capacity with changes in drive level and frequency.

A VMOS connected in a Pierce crystal-oscillator circuit is shown in Fig. 3-61. A good output is obtained using a 12-volt supply. Higher output can be obtained by increasing the drain voltage to 18–24 volts. The circuit differs from the conventional FET arrangement because the VMOS is an enhancement device and it is necessary to supply a starting bias by way of resistors R1 and R2. The value of resistor R3 can be regulated to control the output level and shape of the output wave. In general the lower the output, the more sinusoidal is the output waveform. The oscillator will operate on fundamental crystals from 1 to 25 MHz and possibly higher with adjustment of component values. A higher output can be obtained by using a resonant drain circuit instead of untuned.

A VMOS crystal oscillator and followup class-C amplifier for high-frequency operation is shown in Fig. 3-62. Output of the Pierce crystal oscillator is resistor-capacitor coupled to the gate of the output stage. No dc bias is supplied to this gate. Positive alternations of the signal from the crystal oscillator draw bursts of drain current through the output resonant circuit. A tapped

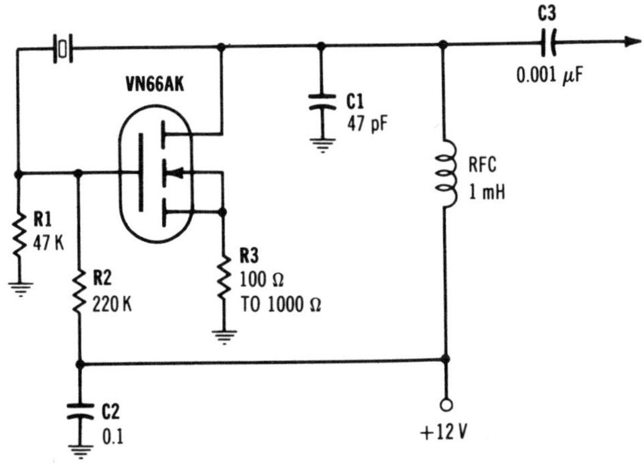

Fig. 3-61. VMOS crystal oscillator.

inductor provides an appropriate impedance match to the VMOS output. A low-impedance winding provides matching to an antenna system or followup low impedance load.

A typical VMOS vhf amplifier is given in Fig. 3-63. Tee-networks provide input and output matching from a low-impedance source and into a low-impedance load. The combination capacitive dividers provide impedance matching and, at the same time, in conjunction with the appropriate gate and drain inductors, resonate the input and output circuits to the desired vhf frequency. Supply voltage is applied to the drain through an appropriate 1-μH radio-

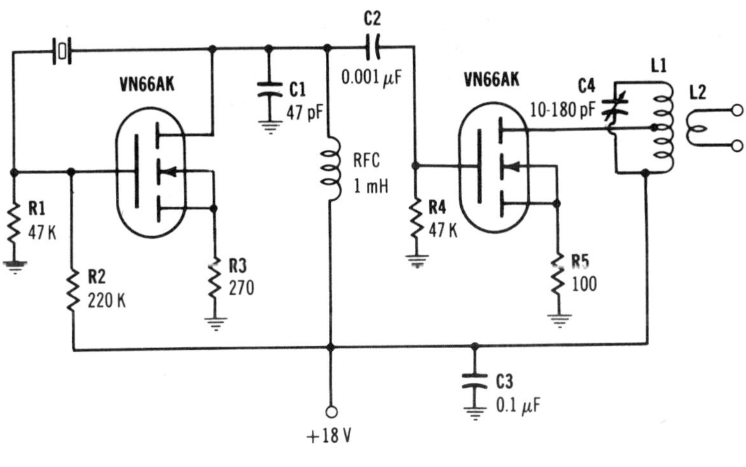

Fig. 3-62. VMOS oscillator-amplifier circuit for hf operation.

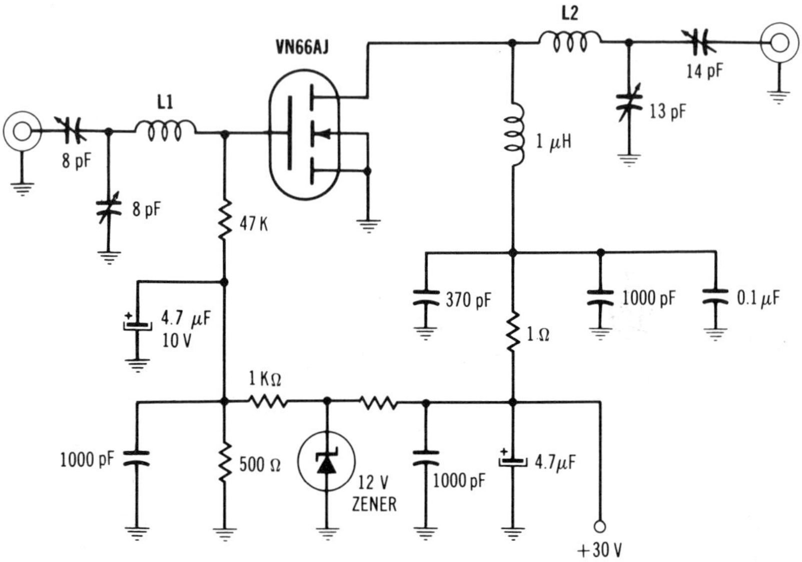

Fig. 3-63. VMOS vhf amplifier.

frequency choke. Optimum biasing for the gate is by way of a 12-volt zener diode that is linked to supply voltage point. An appropriate resistive divider then taps off the voltage required at the gate of the VMOS vhf amplifier.

3-15. INTEGRATED CIRCUITS

Integrated circuits are now being produced in the millions and are found extensively in two-way radio and test equipment. Inside each small package are components numbering from a few to tens of thousands of items. One device may function in a routine way in a simple system; another will perform a complex function in a complex electronic system. Each device may just be a simple amplifier or a complete system.

There are two basic types: linear and digital. Under the linear category are various types of amplifiers (radio-frequency, intermediate frequency, and audio-frequency), oscillators, modulators and demodulators, waveform generators, special arrays, etc. Linear types will be described in this chapter. There are also a great variety of digital integrated circuits. These will be covered in a later chapter on digital electronics.

There are two basic integrated circuit structures: monolithic and hybrid. The monolithic types are, by far, the more common with the entire integrated circuit completed as a single silicon chip. All circuit components are an inherent part of the structure. All components are inseparable in a continuous array of silicon atoms of proper polarization and differing degrees of impurity. Interconnections between components and connections to the output leads are handled by metalized patterns included in the manufacturing process. Thin wire leads connect the terminals to the pins of the integrated circuit casing, whether they are round, flat pack, or the most common, the dual in-line configuration (Fig. 3-64).

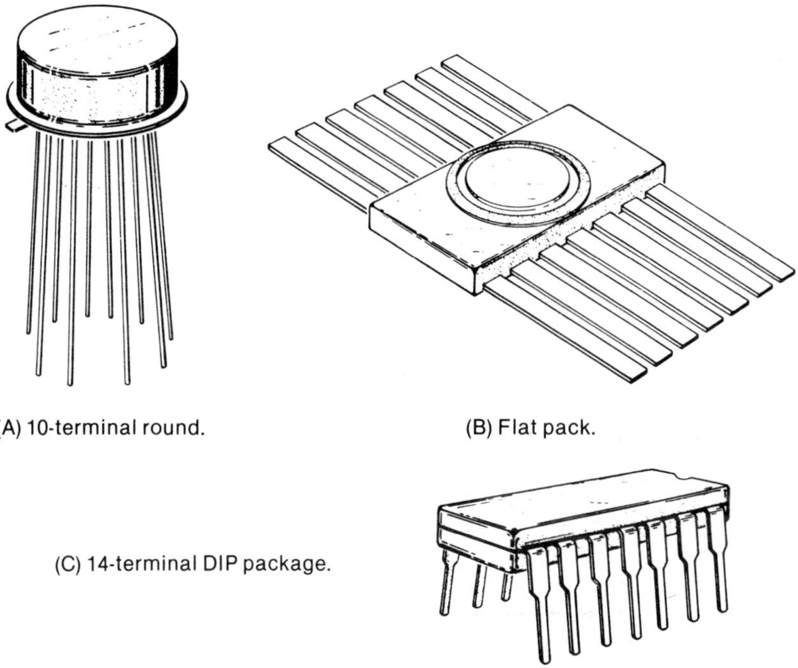

(A) 10-terminal round.

(B) Flat pack.

(C) 14-terminal DIP package.

Fig. 3-64. Integrated-circuit packages.

The structure of a monolithic IC is built around a single silicon specimen. More than one monolithic device can be incorporated in a single case. Electrical isolation among various devices and components is handled by polarization of the various layers and the deposition of special insulating strata. Isolating dielectric barriers separate individual monolithic components or groups of components both electrically and physically. In a hybrid structure, as the name implies, metallic depositions or wire bonds interconnect very tiny active and discrete components within a small case. A hybrid circuit can contain one or several individual monolithic structures plus discrete components in its design. Integrated circuit fabrication is simply an extension, elaboration, and miniaturization of the basic solid-state techniques. The transistor remains the key device; bipolar, FET, and MOSFET. Processed simultaneously along with these basic devices are diodes and resistors. Also there may be occasional capacitors, coils, and other components.

3-15-1. Differential Amplifiers

The differential amplifier is a key circuit of many ICs. The circuit (Fig. 3-65) is basically an emitter-coupled configuration. It has fine stability and a good rejection of undesired signal components. Being direct coupled, no interstage coupling capacitors are needed, providing a saving in space. Correct differential operation requires that the two collector resistances be identical in both ohmic value and general characteristics. In the fabrication of integrated circuits, these conditions are met quite readily and at low cost as compared to a similar amplifier designed to use discrete circuit components.

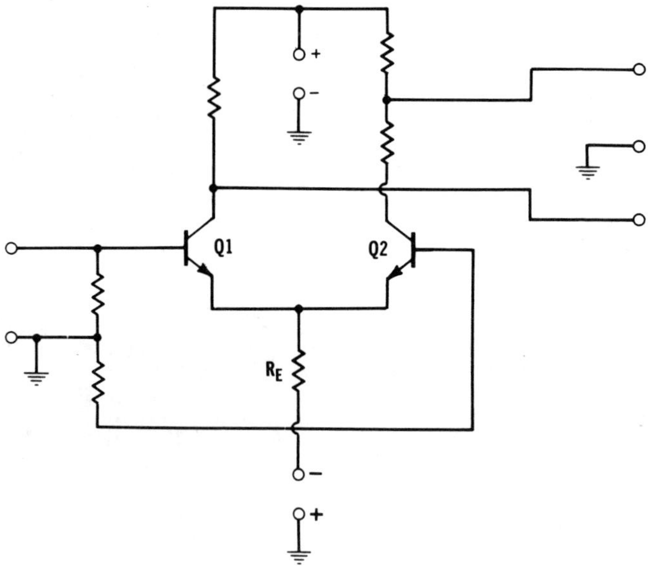

Fig. 3-65. Basic differential amplifier.

In basic operation, the differential amplifier responds to the signal difference that exists between the base inputs, developing equal-amplitude and out-of-phase collector signals. This type of input is called a *differential-mode* input signal.

When two equal-amplitude but same-polarity signals are applied to the base inputs, the ac signals across the common-emitter resistor are subtractive. In a situation of perfect balance, the differential amplifier thus performs in a bridge-like manner and there is no output from collector to collector. A much reduced output signal appears from each collector to common.

In-phase signals at the bases are referred to as common-mode input signals. This is usually in the form of any undesired signal, such as hum and interference components. Thus, another advantage of the differential amplifier is its ability to reject and reduce common-mode signals.

In the differential-mode operation, a signal applied to the base of transistor Q1 (Fig. 3-65) appears at the collector output of transistor Q1 and also across the common-emitter resistor. The latter signal component serves as the input signal for transistor Q2. Consequently, a signal of opposite polarity appears at the collector of transistor Q2. In effect, the differential amplifier acts as a phase splitter, developing two equal-amplitude but opposite-polarity signal components at the output.

The differential amplifier has a high order of dc stability, reducing the influence of supply voltage changes, temperature, etc. Supply voltage variations, in general, create a common-mode type of interference which the differential circuit inherently reduces. Thus, it is a stable dc amplifier and is very practical when designing the multistage circuits that are so common in ICs. Not only does the basic circuit have no need for interstage coupling capacitors, but the emitter bypass is also eliminated.

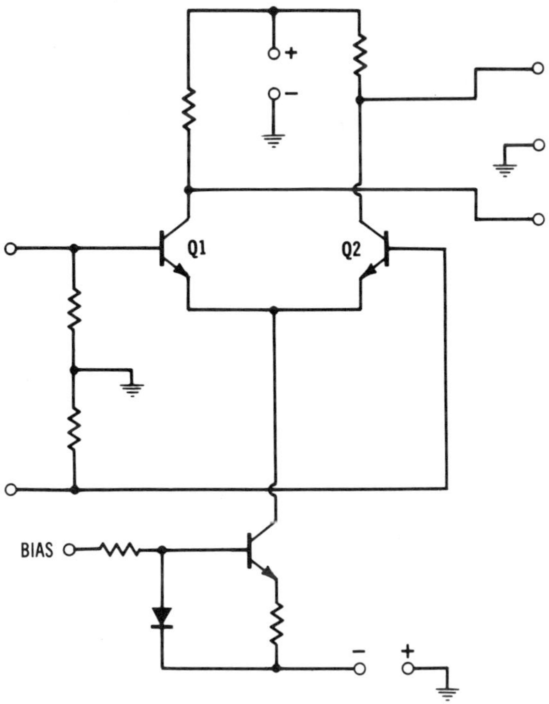

Fig. 3-66. Constant-current source using an active component.

In the rejection of common-mode signals, one depends on the degenerating effects of the common-emitter resistor. The higher the ohmic value of this resistance is, the greater is the rejection. Such increase is limited by supply voltage requirements and the difficulty involved in including high-value resistors in integrated circuits.

One answer to obtaining high stability and good rejection of common-mode signals is to include a constant-current emitter source composed of an additional active component rather than a high-value resistor. The fundamental arrangement is shown in Fig. 3-66. In this circuit, the combination of the transistor and its low-value emitter resistor acts as the high-resistance constant-current source. The presence of a common-mode signal on the differential transistors affects both voltages and junction resistances. A constant-current source holds the emitter and collector currents constant. In fact, the undesired voltage change appears totally across the constant-current source circuit, which is un-bypassed and, therefore, highly degenerative. Thus, the differential gain of the amplifier, in terms of the common mode, is greatly reduced.

Additionally, diodes in the base circuit of this constant-current source provide temperature compensation. When the characteristics of the base-emitter junction of the constant-current transistor and the diode junction are identical, there is exact compensation. An increase in the conductance of the base-emitter junction results from a rise in temperature. Since the compensating diode junction is physically near the transistor, there follows a similar change in its conductance. As a result, a compensation is made in the base bias, thus keeping the collector-emitter current constant. The basic circuit of Fig. 3-66 is very common in integrated circuits.

3-15-2. Darlington Configuration

The previous differential amplifiers have low input impedances. When a high impedance is required, Darlington circuitry is included in the integrated circuits. An additional transistor is included for each pair as shown in Fig. 3-67.

In the normal operation of a transistor, the emitter junction is forward-biased and conducts. Input resistance is low and approximates the product of beta times the emitter resistance:

$$R_{IN} = \beta R_E$$

To a degree, the input resistance can be increased by increasing the ohmic value of the emitter resistance. There is a resultant sacrifice in gain and a need for a higher supply potential. The Darlington-pair approach is to use the input resistance of a second transistor as the emitter resistance of the first (Fig. 3-67A). The input stage then operates with a highly degenerative emitter circuit and a resultant high input impedance. Both stages contribute an output with a net gain figure that is comparable to that obtained using a single transistor of the same type but operating with a much lower input resistance. Two such identical circuits are needed for the separate inputs of the differential configurations shown in Fig. 3-67B.

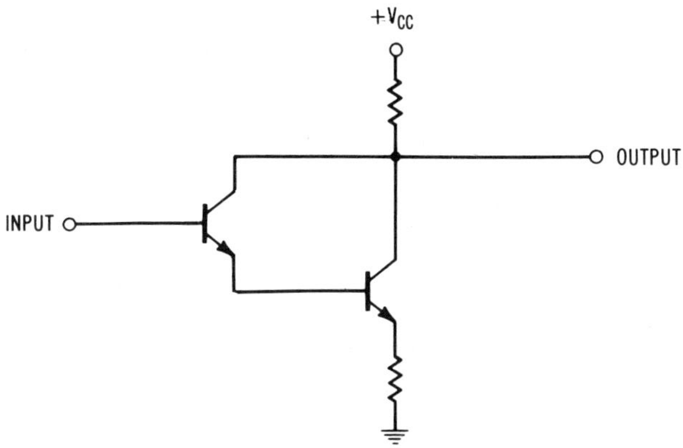

(A) Basic Darlington pair configuration.

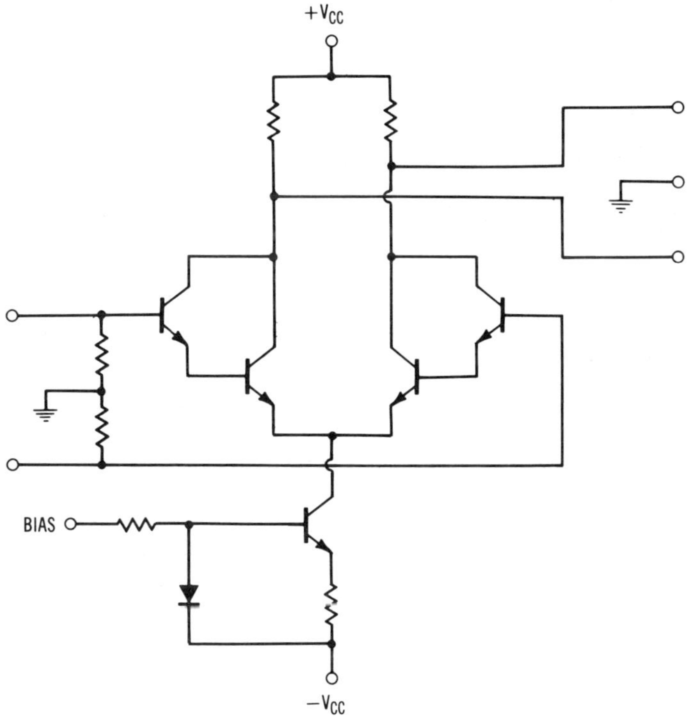

(B) Differential amplifier using Darlington pairs.

Fig. 3-67. Use of Darlington circuits in differential amplifiers.

The advantages of a differential amplifier match the needs of integrated circuits. The IC fabrication process permits the attainment of exceptional balance because all the components of the circuit are processed simultaneously. They have identical characteristics and can be placed near each other. They display similar temperature coefficients and maintain stable electrical characteristics over a wide temperature range. Good input and output isolation are inherent and no neutralization is required. This is basic to simple feedback systems.

Either no capacitors, or very few, are needed in differential circuitry. Large-value resistors can be avoided, and exact absolute resistor values are not critical, because differential amplifier performance depends mainly on resistance ratios.

The differential amplifier is a versatile configuration, lending itself well for use as an amplifier, oscillator, limiter, modulator, demodulator, mixer, or several other applications. Its ability to emphasize the desired signal and de-emphasize common-mode components was mentioned previously, along with its capability for highly linear amplification.

Differential amplifiers can be operated in various modes. The mode described above is referred to as the differential-input differential-output mode.

The differential amplifier has three additional operational modes (Fig. 3-68). In the circuit shown in Fig. 3-68A the signal is applied to the base of transistor Q1 and removed at the collector of transistor Q1. A positive input swing results in a negative swing between collector of transistor Q1 and common. This is known as the single-ended input, single-ended output, inverting mode.

Fig. 3-68B shows the input at the base of transistor Q1 and the output between the collector of transistor Q2 and common. The input swings positive and so does the collector voltage of transistor Q2. Output and input voltages are in phase. This operation is known as the single-ended input, single-ended output, noninverting mode.

In the final manner of operation (Fig. 3-68C), input voltage is applied from base to base and the output voltage can be removed from either of the collectors. This is known as differential input, single-ended output mode.

The various modes of operation require that gain be specified in one of three ways. First, differential-voltage gain is the ratio of collector-voltage change to the difference voltage between the two bases. Second, the double-ended differential-voltage gain is the ratio of collector-to-collector voltage change to the base voltage change. Finally, the single-ended differential-voltage gain refers to the ratio of collector-to-common change to the base-voltage change.

The internal schematic of a simple integrated circuit is given in Fig. 3-69. It consists of four transistors and six resistors that can be connected into any desired configuration including a differential amplifier. Separate base inputs are available and all emitters are joined together and connected to pin 4. An external high-value

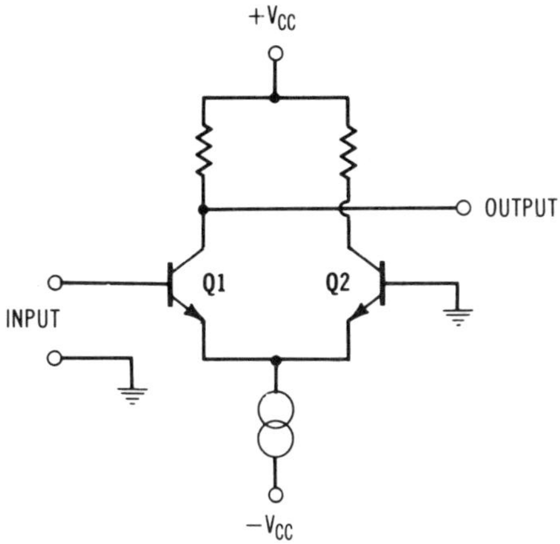

(A) Single-ended input, single-ended output, inverting mode.

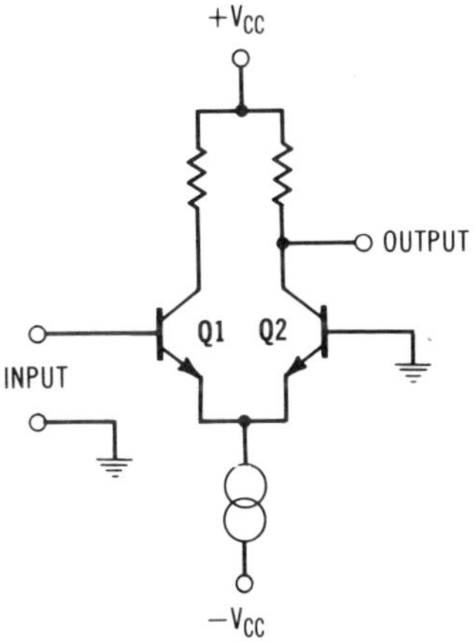

(B) Single-ended input, single-ended output, noninverting mode.

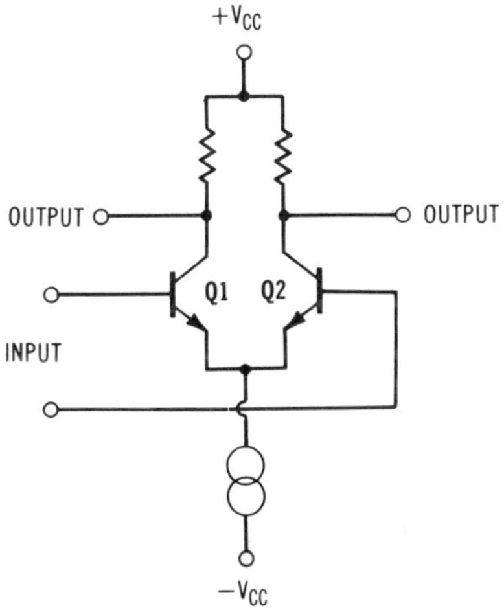

(C) Differential input, single-ended output mode.

Fig. 3-68. Operating modes of a differential amplifier.

resistor or a constant-current source can be connected to the emitters for stabilizing a differential amplifier.

The two symbols of Fig. 3-69B indicate the manner in which linear integrated circuits are often represented. The triangular symbolization is the more common and provides a simpler associated schematic diagram with the inputs and output isolated from each other in the resultant schematic layout.

A typical schematic diagram for a differential amplifier using this integrated circuit is given in Fig. 3-70. Base inputs 1 and 2 are paralleled and input signal from an audio generator can be applied. A differential-mode input is obtained by connecting bases 3 and 5 to ground through an appropriate capacitor. The two 1-megohm resistors provide biasing for the four bases. An emitter resistor is connected between pin 4 and ground.

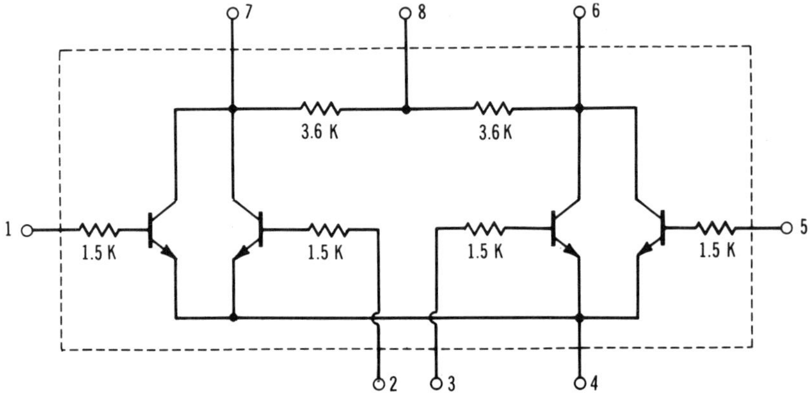

(A) Schematic diagram of IC.

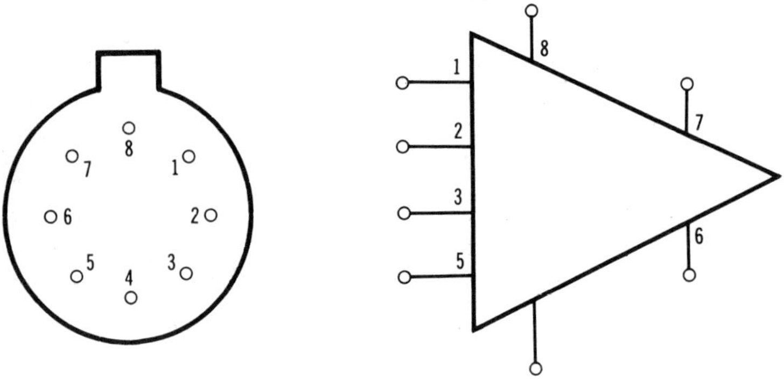

(B) Two symbols.

Fig. 3-69. Integrated circuit and symbols.

A balanced output is obtained between pins 6 and 7. An unbalanced output can be obtained between 6 and common or between 7 and common. Output can be selected according to whether the differential amplifier is to have an inverted or noninverted output signal.

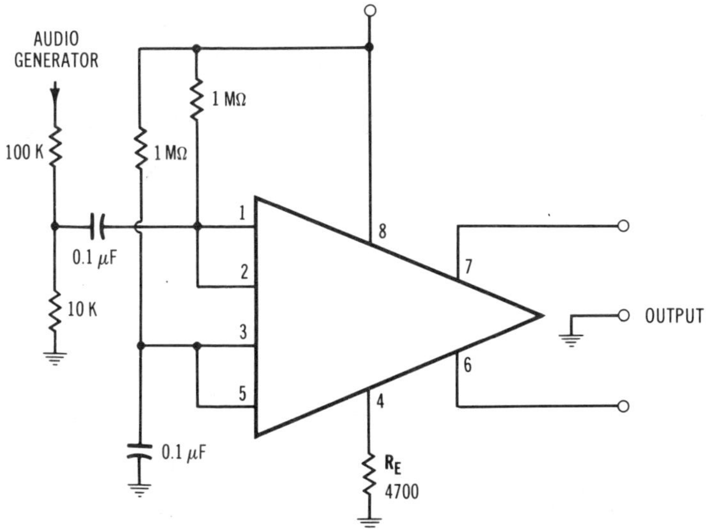

Fig. 3-70. Differential amplifier connections.

3-16. TYPES OF INTEGRATED CIRCUITS

An *operational amplifier* is a special form of linear integrated circuit with high dc and ac gain and high stability. Its many applications include digital as well as linear functions. Some of these are signal amplification, waveform generation and shaping, analog-digital conversion, impedance transformation, instrumentation, etc.

When the amplifier no-feedback gain (open-loop gain) is adequate and a feedback system is of correct design, the closed-loop gain and characteristics of the operational amplifier become a function of only the feedback components. Basically the relative ohmic values of two external resistors can be used to set the operating gain for a particular operational amplifier.

The operational amplifier or *op amp* is a multistage affair that includes the popular monolithic differential amplifier as well as other types of amplifiers, level shifting, and matching and voltage regulation stages. A typical operational amplifier is shown in Fig. 3-71. It consists of a pair of differential amplifiers and a cascaded single-ended output stage. The first differential amplifier uses transistors Q1 and Q2 along with constant-current transistor Q6. There is also a temperature-compensating diode (D1). The second amplifier consists of transistors Q3, Q4, Q7, and diode D2.

The feedback circuit of transistor Q5 reduces any common-mode error signal. It does this by evaluating the signal at the emitters of transistors Q3 and Q4. When the second differential amplifier is driven push pull, there should be a zero level at this point, indicating proper operation of both the first differential amplifier

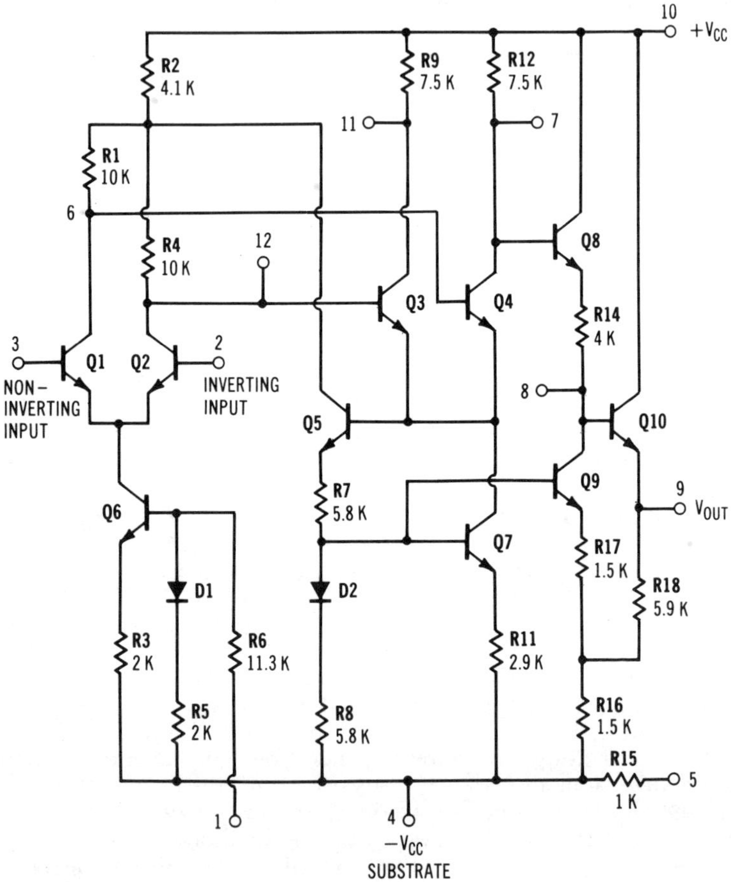

Fig. 3-71. Typical operational amplifier.

and the input system of the second. If an error voltage is present, a correcting voltage is developed across resistor R2 (in the differential collector circuit) by transistor Q5. This same transistor (Q5) also introduces an error bias into the constant-current transistor circuit (Q7) to bring a further reduction in common-mode signal.

The emitter circuit of this important stabilizing transistor (Q5) also supplies a dc error voltage to the base of transistors Q7 and Q9. Common-mode dc stabilization results. For example, a decline in supply voltage reduces the base bias at transistors Q7 and Q9. Since the collector of Q9 is linked to the base of transistor Q10 and to the emitter of transistor Q8, there is a similar change there as well as at the bases of transistors Q3 and Q4. The net result is to produce increases in collector voltages and compensate for the supply voltage decline. The feedback stabilization furnished by transistor Q5 offers excellent common-mode rejection, tolerance to supply voltage change, and high open-loop stability.

The output of the second differential amplifier supplies drive to the base of the emitter-follower transistor (Q8). In turn, transistor Q8 supplies a signal to the base of the single-ended emitter-follower output transistor (Q10). There is a limited amount of signal gain contributed by the output circuit as a result of the *bootstrap* (small amount of positive feedback) from the emitter of transistor Q10 to the emitter of transistor Q9. Therefore, transistor Q9 serves a dual purpose—as a constant-current source for the drive transistor and as a part of the bootstrap. The output system also provides a dc-level shift and the level at terminal 9 now corresponds to the no-signal input level.

The feedback plan of the operational amplifier provides high stability and characteristics that can be, if desired, related directly to the external feedback network. An operational amplifier has additional favorable attributes. The bandwidth of some extends uniformly from dc up to several hundred megahertz. Common-mode rejection is excellent and there is little dc offset or drift with temperature. If desired, the input impedance of some types can be made very high, and the power input and current requirements made insignificant.

Most operational amplifiers have single-ended outputs and push-pull inputs. There are some operational amplifiers with single-ended inputs, although these are not versatile and may have only a specialized function. An advantage of the differential input in this respect is demonstrated in Fig. 3-72. When signal is applied to one of the differential inputs (Fig. 3-72A), the output is not inverted (same polarity). If the signal is transferred to the second input, an

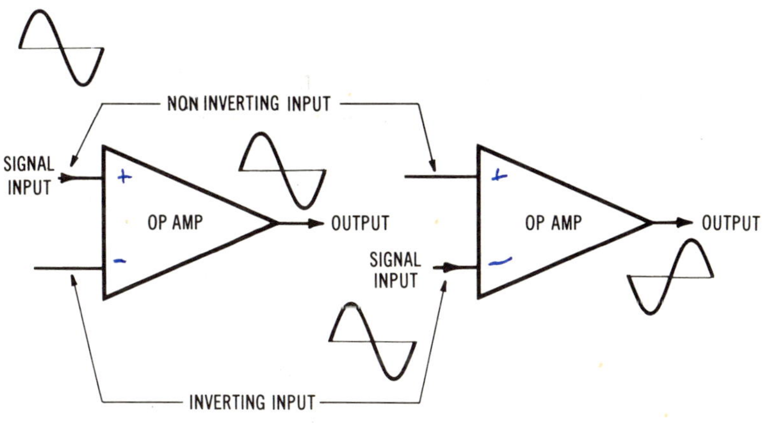

(A) Noninverting input.

(B) Inverting input.

Fig. 3-72. Operational amplifier inputs.

inverted (opposite polarity) signal appears at the output (Fig. 3-72B). In terms of an ac input signal, this means that an input can be selected to produce either an in-phase or an out-of-phase output signal. Ordinarily, the input terminals of operational amplifiers are labeled as either the inverting input (−) or the noninverting input (+).

Equivalent diagrams of inverting and noninverting operational amplifier configurations are given in Figs. 3-73A and B. These are closed-loop equivalents with a portion of the output transferred back to one side of the differential input by way of resistors Z_F and

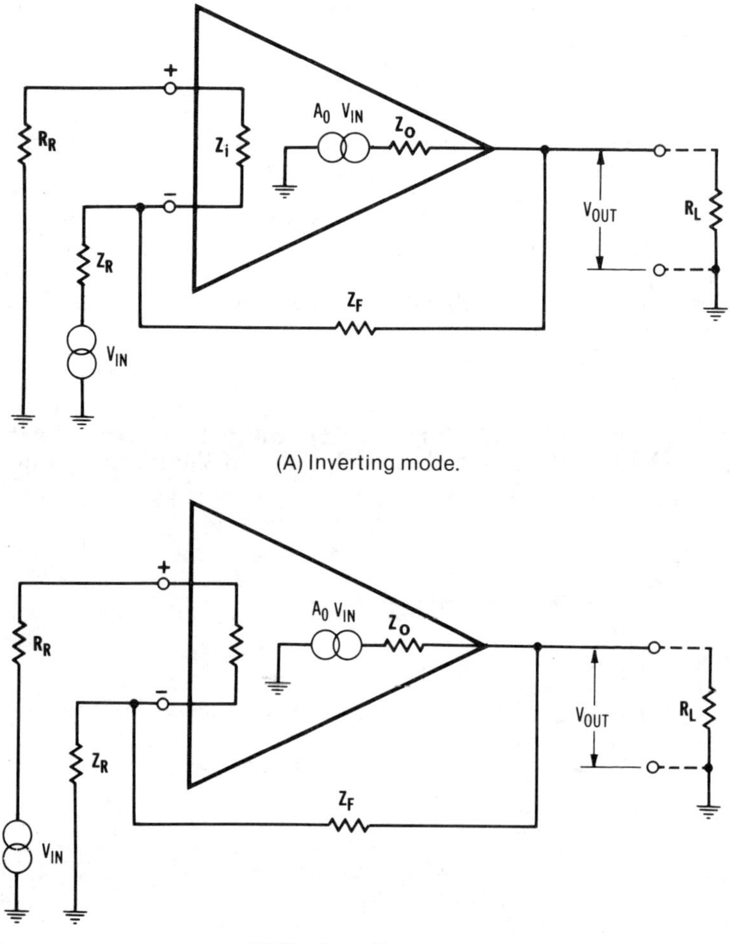

(A) Inverting mode.

(B) Noninverting mode.

Fig. 3-73. Equivalent diagrams of op amp configuration.

Z_R. Resistor Z_R, of necessity, must include the source impedance as well as bias components when used. Inasmuch as there is a differential input, the dc return path to common from both sides must be equal. Therefore, the ohmic value of resistor R_R must be made equal to the ohmic resistance Z_R and must consider the resistance of the signal source (V_{IN}).

The input impedance (Z_i) and output impedance Z_o are intrinsic to the operational amplifier. The output generator has a value of A_oV_{IN}. The term A_o is the open-loop differential voltage gain. (The closed-loop gain is lower than this value because of the influence of the feedback.)

The load resistance (R_L) is also shown although it is assumed in this discussion of the basic equivalents that its value is high in comparison to the output impedance of the operational amplifier. Therefore, it can for the moment be neglected.

The closed-loop gain expression is very simple.

$$A_V = \frac{V_{OUT}}{V_{IN}} = \frac{Z_F}{Z_R}$$

Note that the gain of the inverting configuration is entirely dependent on the values of the external feedback components. This is true only if certain requirements are met. The open-loop differential voltage gain must be very high. The intrinsic input impedance (Z_i) must be much greater than the value of Z_R and the paralleling feedback components. The intrinsic output impedance (Z_O) must be smaller than the feedback value (Z_F).

Similar operational requirements can be safely assumed for the noninverting configuration (Fig. 3-73B). Signal is applied to the noninverting side while feedback component must of necessity be applied to the inverted side of the differential input. The basic equation for the closed loop voltage gain is:

$$A_V = \frac{V_{OUT}}{V_{IN}} = \frac{Z_R + Z_F}{Z_R}$$

A knowledge of the configuration equivalents for inverting and noninverting operation permits an easier understanding of an operational equivalent circuit which includes the influence of the output load. This basic equivalent applies to both inverting and noninverting operation (Fig. 3-74). Within the equivalent, the terms Z_{IN} and Z_{OUT} will have different values for the inverting and the noninverting modes. The values for Z_{IN} and Z_{OUT} take into consideration the values of the feedback components. Even though the expression Z_R is included within the equivalent dashed block, one must make certain that its value includes the influence of the source impedance. For these reasons, the equivalent is appropriate for both configurations of an operational amplifier.

The voltage of the output generator is the product of the closed-loop voltage gain of the operational amplifier times the input

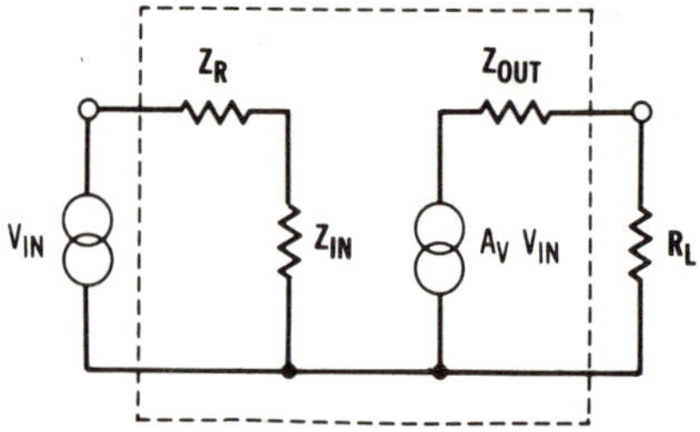

Fig. 3-74. Closed-loop equivalent, including output load.

voltage, or $A_V V_{IN}$. This voltage divides between the output impedance and the load. Hence, the output voltage itself becomes:

$$V_{OUT} = A_V V_{IN} \frac{R_L}{R_L + Z_{OUT}}$$

and the stage gain becomes:

$$\text{Voltage gain} = \frac{V_{OUT}}{V_{IN}} = \frac{A_V R_L}{R_L + Z_{OUT}}$$

If the ohmic value of the output impedance (Z_{OUT}) is low in comparison to the ohmic value of the load resistance, there is a further simplification of the equation for voltage gain. For the inverting mode:

$$\text{Voltage gain} = A_V \cong \frac{Z_F}{Z_R}$$

and for the noninverting mode:

$$\text{Voltage gain} = A_V \cong \frac{Z_R + Z_F}{Z_R}$$

In summary, the operational amplifier is a dc and ac amplifier of high stability. It can operate with a low gain or a high gain, depending on the values of three external resistors. In fact, the actual gain of the amplifier depends on external resistance ratios rather than internal components.

The two common input arrangements are shown in Fig. 3-75. Example A shows a noninverting amplifier with the amplified output voltage being of the same polarity as the input. Balanced operation occurs when the input resistances at the input terminals are identical. In the calculation of the input resistance it is always necessary to consider the influence of the feedback resistance. In the case of the noninverting amplifier, input 1 resistance is equal to R1. However, signal-input 2 resistance has an ohmic value which must match the parallel combination of resistors R1 and R2.

Typical gain and resistance values are listed in the charts. Note that gain increases when there is a higher ratio between the ohmic

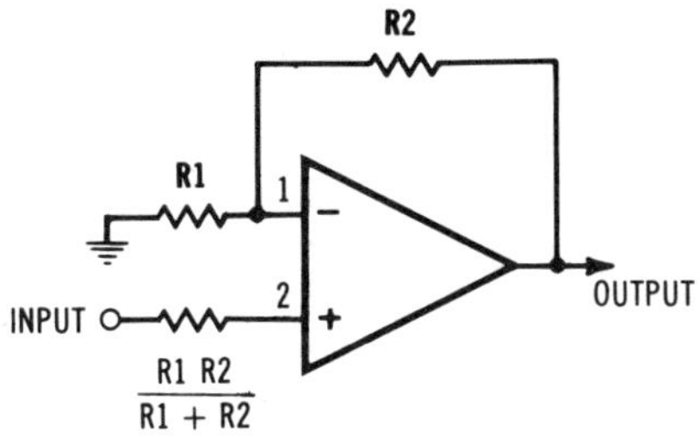

(A) Noninverting mode.

GAIN	R1	R2	BW
10	1 K	10 K	10 kHz
100	100	10 K	10 kHz

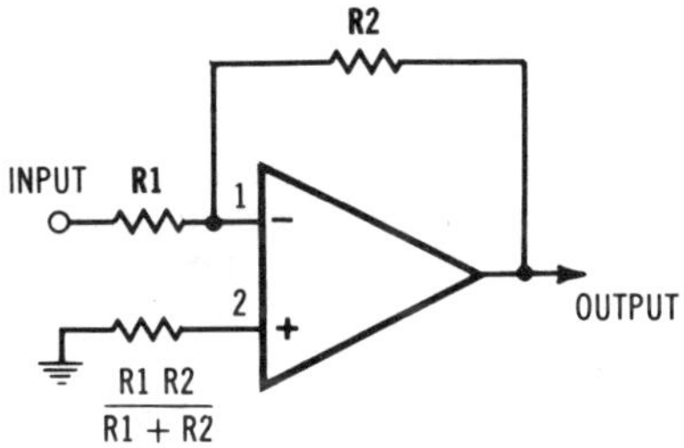

(B) Inverting mode.

GAIN	R1	R2	BW
10	1 K	10 K	100 kHz
100	1 K	100 K	10 kHz

Fig. 3-75. Operational amplifier operating modes.

values of resistors R2 and R1. This is to be anticipated because there is less feedback. Likewise, the bandwidth is reduced.

To obtain an inverted output as in Fig. 3-75 the signal is applied to the same input as the feedback component. Again the resistance values determine the stage gain. Stage gain approximates the ratio of feedback resistance R2 over the input resistance (R1). This can be verified from the chart by dividing the ohmic value of resistor R1 into that of resistor R2.

Circuits can be built around an 8-pin *minidip* 741 operational amplifier (Fig.3-76). The actual circuits are given in Fig. 3-77. A split voltage supply consisting of two 6-volt batteries is shown.

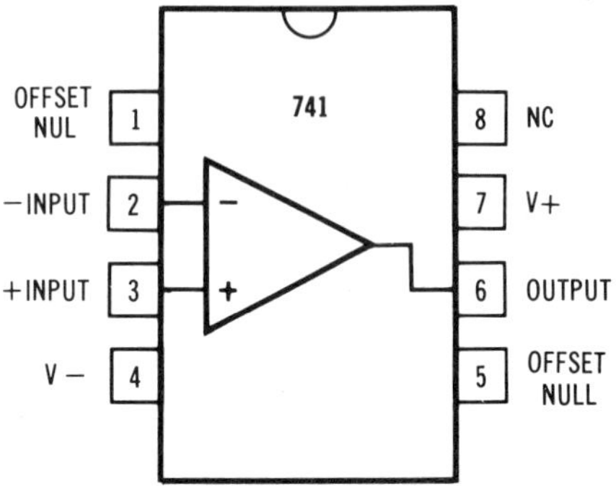

Fig. 3-76. Diagram of the 741 8-pin minidip IC.

Circuit A is a noninverting type with the signal being applied to the noninverting or + input of the operational amplifier. A sine wave input will result in a sine wave output of the same phase. The gain of the amplifier will be 10 as determined by resistors R1 and R2. If a positive rise of the dc voltage is applied to the input there will be a positive rise of the dc voltage at the output. A tenth of a volt rise at the input will show up as a 1 volt increase in the output.

Example B is that of an inverting operational amplifier. Now the signal is applied to the inverting or minus input of the operational amplifier. Now the signal is applied to the inverting or minus input of the operational amplifier. An amplified voltage that is 180 degrees related to the input voltage will appear at the output. Again the gain of the inverting amplifier is determined by resistors R1 and R2.

3-16-1. Audio Power Amplifier IC

Integrated circuit audio power amplifiers are used frequently in two-way radio equipment. An example of such a circuit is given in Fig. 3-78. The internal circuit is given in Fig. 3-78A while the pin-out diagram is shown in Fig. 3-78B. The input stage is a differential amplifier using Darlington pairs to raise the input resistance. In the circuit diagrams of C and D note that the input signal is applied to the noninverting input 3. Pin 2, or the inverting input, is connected to ground. Thus the input amplifier is fed in the differential mode.

The differential-input amplifier is followed by a Darlington-pair power output stage. The simple circuit arrangement of Fig. 3-78C provides a gain of 20. Additional gain can be obtained by connecting external components between pins 1 and 8. The use of the 10-μF capacitor between these two pins in Fig. 3-78B provides the maximum gain of 200. Note for this maximum gain operation that the 10-μF capacitor is connected across the internal 1.35K resistor which is part of the emitter system of the differential amplifier transistors.

A practical IC audio power amplifier is shown in Fig. 3-79. The feedback network between pins 1 and 8 consists of a resistor and a capacitor. In this manner the actual gain of the stage can be

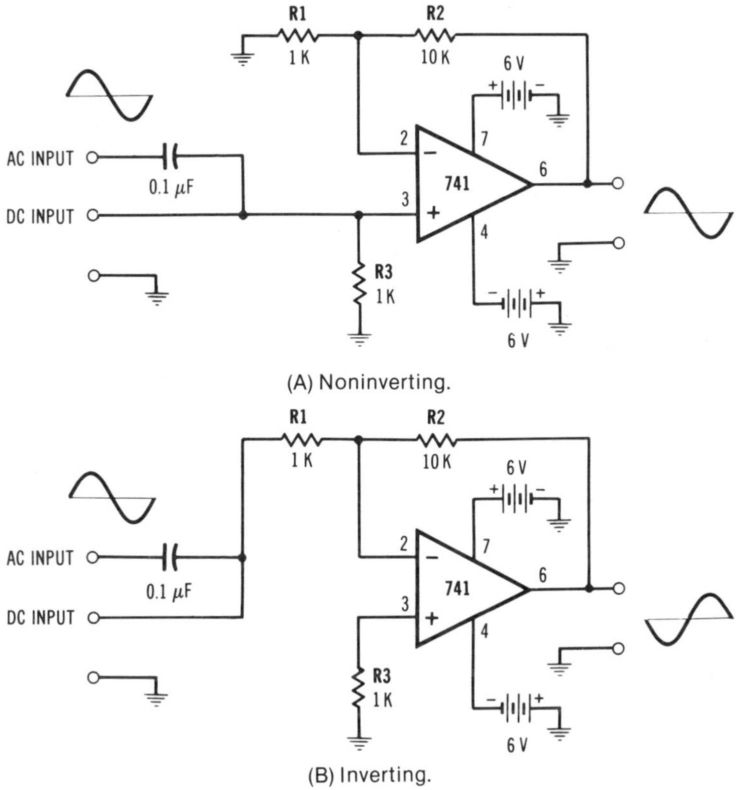

Fig. 3-77. Basic op amp circuits using the 741 IC.

decided and will be something less than the maximum value of 200 depending on the needs of the audio system. A decoupling capacitor (C4) is connected between pin 7 and ground and, in effect, is the supply voltage feed line. High-frequency rolloff is controlled by resistor R3 and capacitor C5.

3-16-2. FM IF Subsystem for Communications

Integrated circuits are available that contain a complete if system and fm detector along with an audio amplifier and other special circuits in one single chip. Such a unit can serve as a complete receiver subsystem for a communications package. The one shown in Fig. 3-80 includes a three-stage fm if amplifier/limiter channel, plus individual signal level detectors for each stage. This section is followed by a doubly balanced quadrature fm detector that can be used with either a single-tuned detector coil or double-tuned detector coil. The quadrature detector is followed by an audio

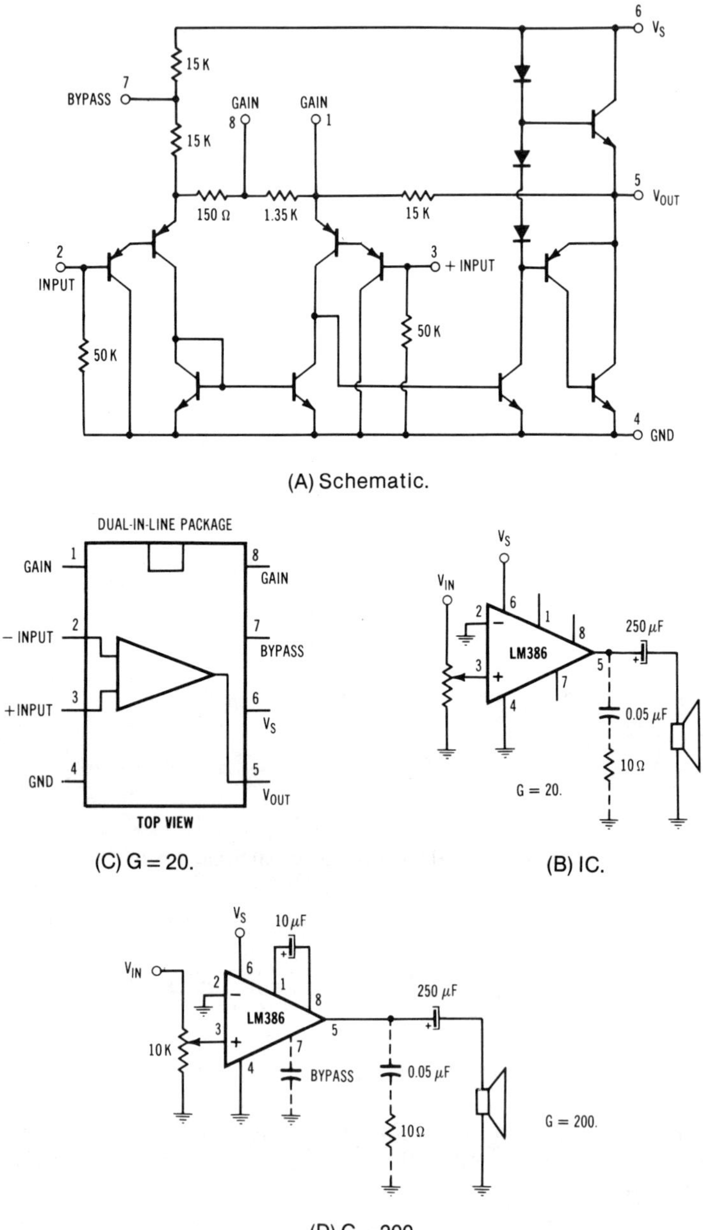

(A) Schematic.

(C) G = 20.

(B) IC.

(D) G = 200

Fig. 3-78. Audio power amplifier.

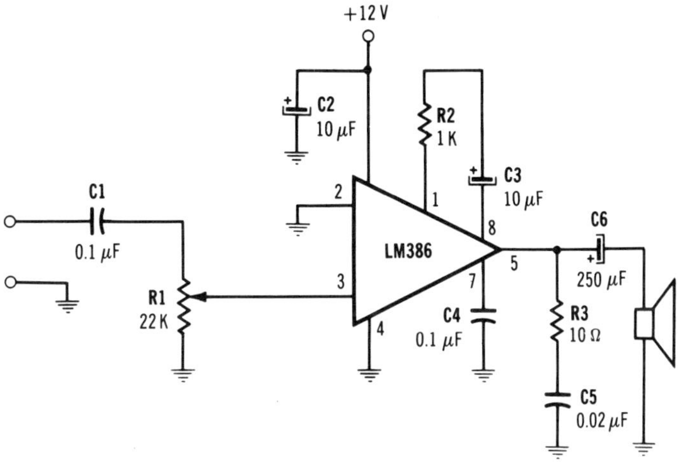

Fig. 3-79. Practical IC audio power amplifier.

amplifier with a typical 400-mV output level. The quadrature detector supplies drive to an automatic-frequency-control amplifier, the output of which can be used to hold the local oscillator on correct frequency.

The level-detector stages supply signal for a tuning-meter circuit as well as a delayed agc component for the rf amplifier of the tuner section of the communications receiver. A level detector is also associated with the quadrature detector. It supplies a drive component to the squelch system. The squelch output, in turn, is supplied to the muting stages of the audio amplifier. Such a squelch circuit silences the output of the receiver when there is no incoming signal above a level set by the muting sensitivity control.

Only the if input and the quadrature detector are tuned. A variety of frequencies for the if system can be chosen. Values indicated are for a 10.7-MHz intermediate frequency.

A complete circuit for the subsystem as it would be shown on a schematic diagram is given in Fig. 3-81. The various pin-outs can be compared to show how the output components are properly attached according to the function that takes place within the integrated circuit chip.

3-16-3. Phase-Locked Loop (PLL)

Many applications for phase-locked loops include frequency multiplication and division, signal conditioning and interference suppression, frequency synthesizers, data synchronizers, tracking filters, decoders, modulators and demodulators, etc.

The basic phase-locked loop (Fig. 3-82) has three main sections—phase comparator, low-pass filter, and voltage-controlled oscillator (vco)—all connected in a closed-feedback loop. Almost an

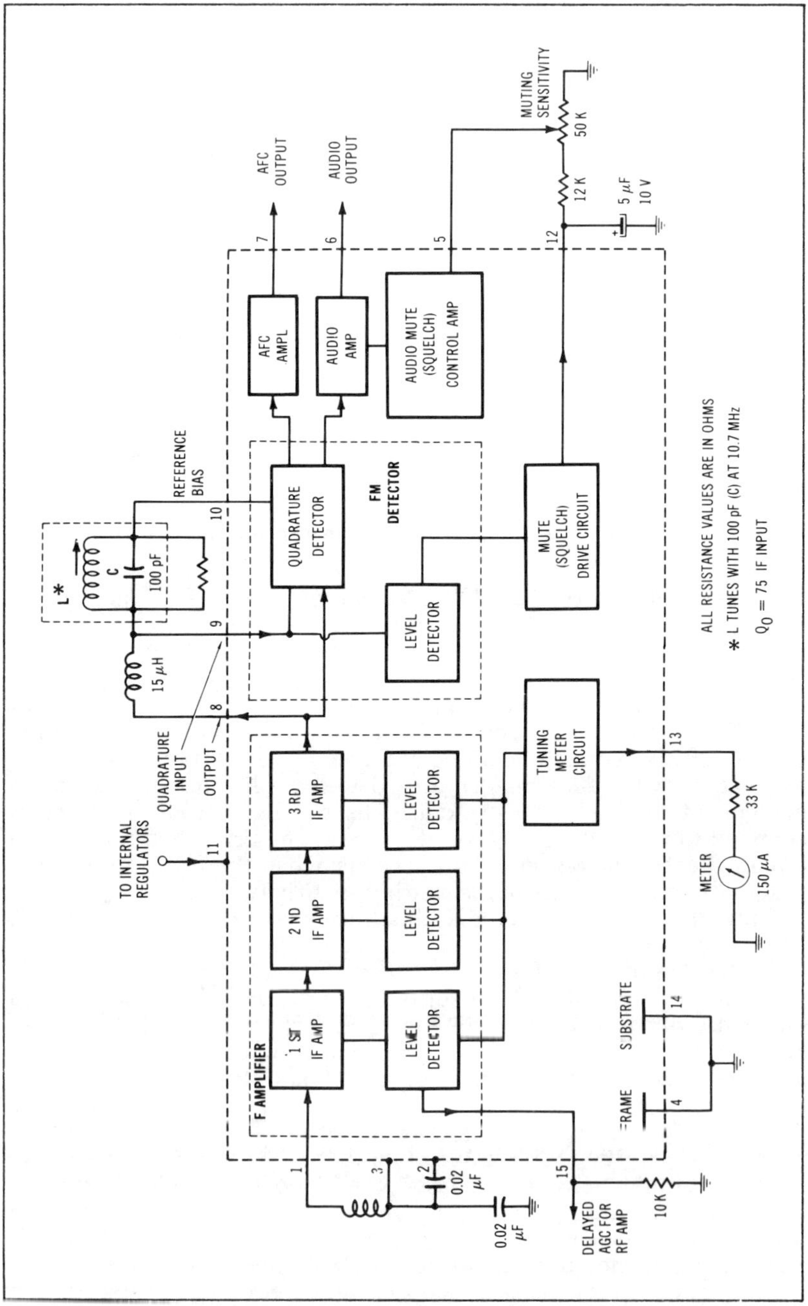

Fig. 3-80. A communications fm if subsystem.

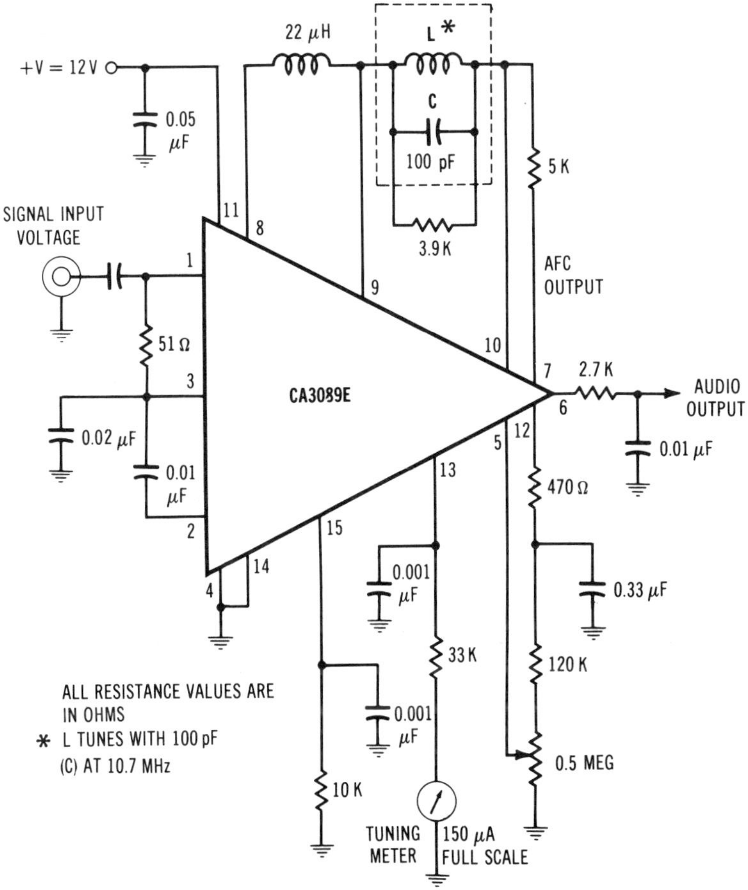

Fig. 3-81. Circuit diagram using the IC subsystem.

entire pll circuit can be included in an integrated circuit. Its major objective and big advantage in many applications is the ability to lock onto the frequency of any applied signal.

To understand the operation of the pll, let us assume first that the device is not locked onto the incoming signal, even though the two frequencies are reasonably close in frequency. The incoming signal and the locally generated signal from the vco (e_i and e_{osc}) are compared in the phase detector. The output of the phase detector will be a difference or beat frequency (e_X). Since it is low in frequency, output e_X is able to pass through the filter and pass to the voltage-controlled oscillator. The passband of the filter is such that the incoming and vco components do not pass. The feedback component (e_f) then shifts the frequency of the oscillator and locks

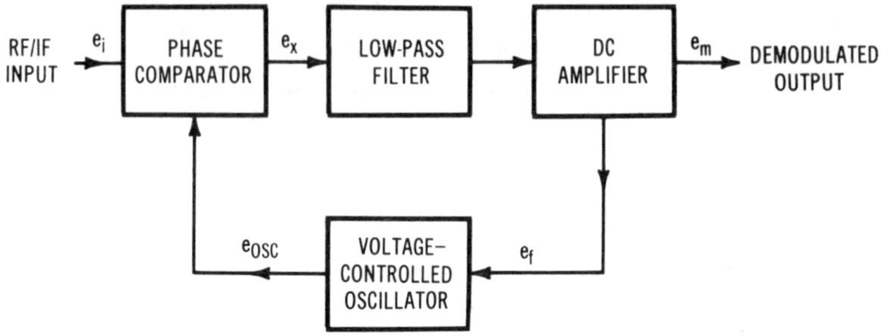

Fig. 3-82. Block diagram of a pll feedback loop.

it onto the incoming frequency.

Actually, it is a dc voltage that holds the oscillator on frequency because at the moment lock-in occurs, both the input signal and the vco signal are on the same frequency; thus, the output of the phase detector is zero. If either the incoming signal changes frequency or if the vco attempts to drift in frequency, there is an instantaneous shift in the phase between the two components in the phase detector. Immediately, there will be a dc output from the phase detector which will be either positive or negative, depending on the direction of the frequency separation. This dc output will cause a frequency-controlled voltage to be immediately applied to the voltage-controlled oscillator (vco) that will again equalize the two frequencies.

If it is the vco that attempts to drift in frequency, the dc voltage will prevent it from doing so. If it is the incoming signal that changed in frequency, the dc voltage will cause the vco to follow that change in frequency. From this it is apparent that the phase-locked loop can be made to hold its own frequency precisely or be made to follow the frequency of an applied signal. This action has many attractive attributes in various types of electronic systems.

The range of frequencies over which a pll can hold a lock is known as the lock-in range of the system. The lock-in range is always greater than the range of frequencies over which the lock can be acquired by the pll. The range of frequencies over which a lock can be acquired is known as the capture range of the pll system. Stated another way, when a pll is attempting to lock in on an incoming signal, it is able to do so if the incoming signal falls within its capture-frequency range. Once the lock is established, the pll is able to hold a lock over a greater range of frequencies than it needs for capture. The lock range represents a greater bandwidth than the capture range of the pll system. In practical operation, a lock will occur immediately when the incoming signal falls within the capture range of the pll system.

The loop filter is very important to the operation of the pll. It is true that the dc control voltage passes through the filter to the vco. However, the ac performance of the filter is also important. In fact, the bandwidth of the system performance is established by the loop filter. The bandwidth is selected in accordance with the desired lock-in or hold-in range and the time required to establish lock. Loop bandwidth must be kept narrow to minimize jitter that can be caused by external noise or interference components. However, the loop bandwidth must be adequate to pass desired components and establish an adequate capture range. In all circumstances, the loop bandwidth must be wide enough to accommodate the frequency components that make up the modulation. When used for the demodulation of an fm signal, one understands that the bandwidth should be greater than that required for the demodulation of a narrowband am signal.

A function block diagram of the *Signetics Corporation* NE561B pll integrated circuit is shown in Fig. 3-83. This device is an am/fm

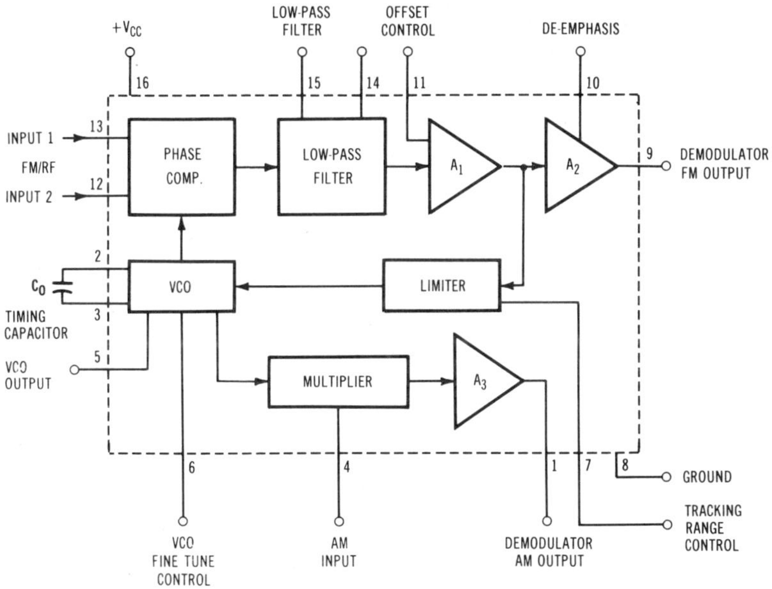

Fig. 3-83. Block diagram of the Signetics Corporation NE561B phase-locked loop.

demodulator system. The vco is an emitter-coupled multivibrator, the frequency of which can be controlled by an external capacitor (C_O). Its output is applied to the phase comparator along with the incoming fm radio-frequency signal. When the vco and the center frequency of the fm wave are matched, the dc component of the output of the phase comparator is a reference zero. If frequencies do not match, an error voltage results which is passed through the filter, a succeeding amplifier and limiter, and back to the vco.

The demodulated fm signal also passes through the filter amplifiers A_1 and A_2 to the fm output. A de-emphasis network can be attached to terminal 10 if required.

In the demodulation of an am signal, the signal is applied to the multiplier input. The pll system sets the vco on the exact carrier frequency. This vco component is also applied to the amplifier. Therefore, the multiplier acts as a product detector, and its output frequency is the difference between the am input signal and the locally generated carrier from the vco. The demodulated am signal is increased in level by amplifier A_3 and is recovered at terminal 1 of the device.

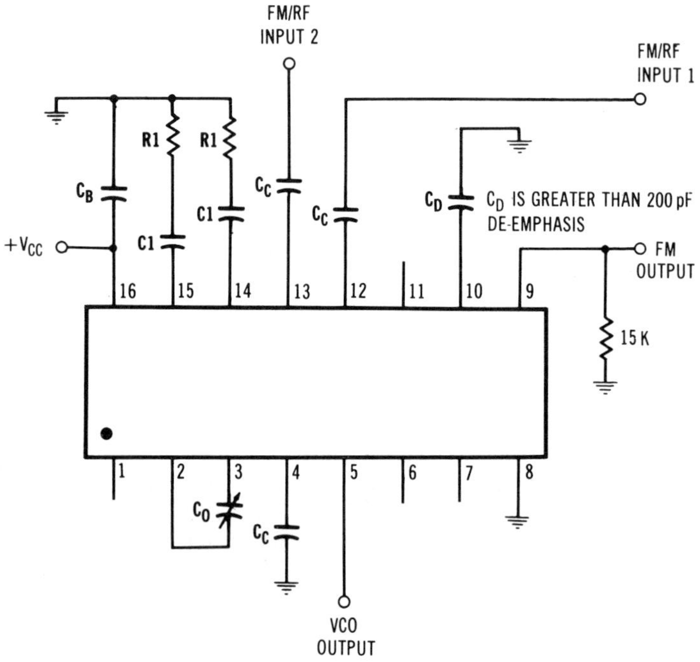

(A) Device and external circuitry.

FREQUENCY	C_O	R1	C1
10.7 MHz	30 pF	100 Ω	1000 pF
4.5 MHz	54 pF	50 Ω	2000 pF
455 kHz	650 pF	0	5000 pF
67 kHz	4600 pF	0	0.01 μF

(B) Component values for typical if frequencies.

Fig. 3-84. An fm pll demodulator.

A complete external circuit for using the device as an fm demodulator is shown in Fig. 3-84A. The fm signal is applied between terminals 12 and 13. Capacitor C_o determines the vco frequency. An approximate equation for determining its value is:

$$C = \frac{300}{f_o}$$

where,
C_o is the capacitance in picofarads,
F_o is the oscillator frequency in megahertz.

Values for typical intermediate frequencies are given in the chart of Fig. 3-84B. If the vco is to be used as a tunable local oscillator, this capacitor is made variable.

The values of the series resistor-capacitor combinations connected to terminals 14 and 15 determine the characteristics of the low-pass loop filter. The time constant of the filter is:

$$tc = C1(R1 + R_{in})$$

where,
R_{in} is the internal impedance and is approximately 6000 ohms.

A practical version of the equation would be:

$$C1 = \frac{1}{2\pi f_h R1}$$

where,
f_h is the desired bandwidth of the filter.

For example, frequency f_h might be 15 kHz for use in an fm receiver.

A capacitor can be used tc ground one side of the input to permit the fm signal to be applied single-ended and differential.

Capacitor C_D provides de-emphasis, and its value is based on the time constant:

$$tc = R_D C_D$$

Resistor R_D is 8000 ohms, the output impedance at the de-emphasis terminal. The time constant is whatever value needed for a particular fm demodulator service.

3-17. FREQUENCY SYNTHESIZERS

Frequency synthesizers are used extensively in setting the transmit and receive frequencies of modern two-way radio equipment. Other key frequencies for both transmit and receive mixing are also generated by the synthesizer system. Usually the synthesizer proper is an integrated circuit connected in a phase-locked loop configuration.

The basic functional plan of a frequency synthesizer is given in Fig. 3-85. In a synthesizer the output of the voltage-controlled oscillator is applied to a frequency divider instead of directly to a

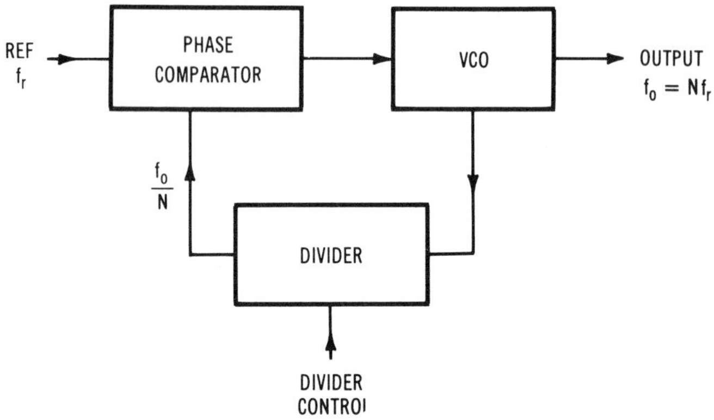

Fig. 3-85. Basic frequency synthesizer.

phase comparator. By changing the division ratio of the divider, it is possible to set the frequency of the voltage-controlled oscillator on some precise frequency.

In the phase comparator the output of the divider and usually a crystal-controlled reference frequency are compared. A correcting dc voltage at the output of the comparator is applied to the voltage-controlled oscillator and sets it on frequency. This locked frequency when applied to the divider will produce an output that will be identical to the reference frequency applied to the phase comparator.

In operation the output frequency is equal to the product of the divider ratio and the reference frequency. In the example of Fig. 3-85 let us assume that the divider ratio is 10 and the input reference frequency is 1 MHz. As a result the output frequency (f_o) is the product of (10×1) or 10 MHz. This output frequency (f_o) is also applied to the divider and the output of the divider is 1 MHz (10/10). Thus the voltage-controlled oscillator is locked in precisely on the operating frequency of 10 MHz. Any attempted change in the output frequency will set the phase-locked loop into operation and the dc correcting voltage at the output of the phase comparator will restore the voltage-controlled oscillator frequency to 10 MHz.

How can the frequency of the voltage-controlled oscillator be changed by the divider? Assume that the divider has a control system that establishes a count of 9 instead of 10. This would mean that output frequency f_o now becomes (9×1) or 9 MHz. In this case output frequency f_o is divided by 9 to produce, once again, a 1-MHz component at the output of the divider. This component is again compared with the 1-MHz reference frequency and the voltage-controlled oscillator is now held precisely on 9 MHz.

This is the basic technique used to establish an array of frequencies for a radio transceiver without the use of an array of crystals. For example, in a modern CB transceiver forty separate divider ratios that can be established with a digital divider control system can be used to precisely set the voltage-controlled oscillator on a specific frequency that will permit operation on a chosen one of the forty possible channels. Changing the channel selector will change the divider ratio in such a manner that operation will now be switched to the new channel specified on the digital channel readout.

Functional plan of a typical 40-channel CB transceiver using a frequency synthesizer is given in Fig. 3-86. Transistors are used throughout except for the integrated circuit that is a part of the phase-lock loop synthesizer and another integrated circuit for the audio output system used on both receive and transmit.

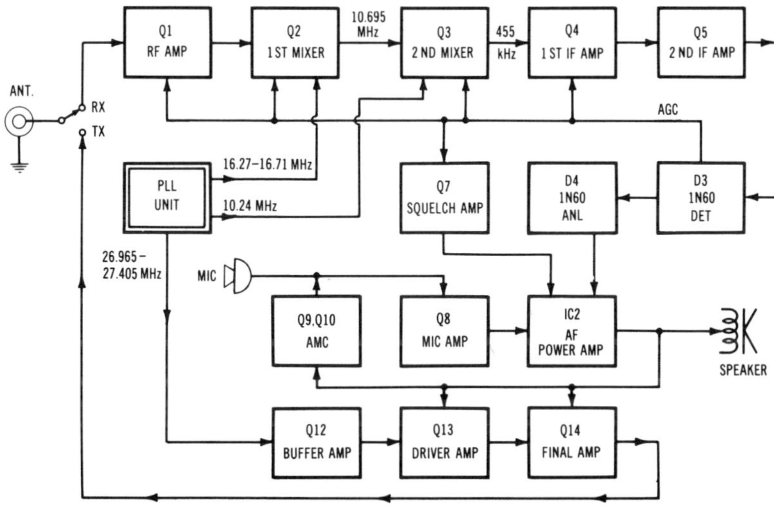

Fig. 3-86. Block diagram of a CB transceiver.

The pll unit generates three signal components. One of these is the transmit-frequency range and the other two are local oscillator injection components for the dual-conversion receiver.

Two crystal oscillators are used in the pll frequency synthesizer (Fig. 3-87). The vco, located at the center of the block diagram, generates a frequency range between 16.27 and 16.71 MHz depending on the selected channel. Its output is applied at that frequency to the first mixer of the receiver. Output also is applied to the transmitter mixer where it combines with the frequency generated by the crystal oscillator. The summation of these two frequencies produces a range of frequencies that falls within the assigned CB radio spectrum between 26.965 and 27.405 MHz.

Output is also taken from the vco and applied to yet another mixer through buffer Q204. In this mixer, the output is compared with a signal generated by a reference crystal oscillator. This

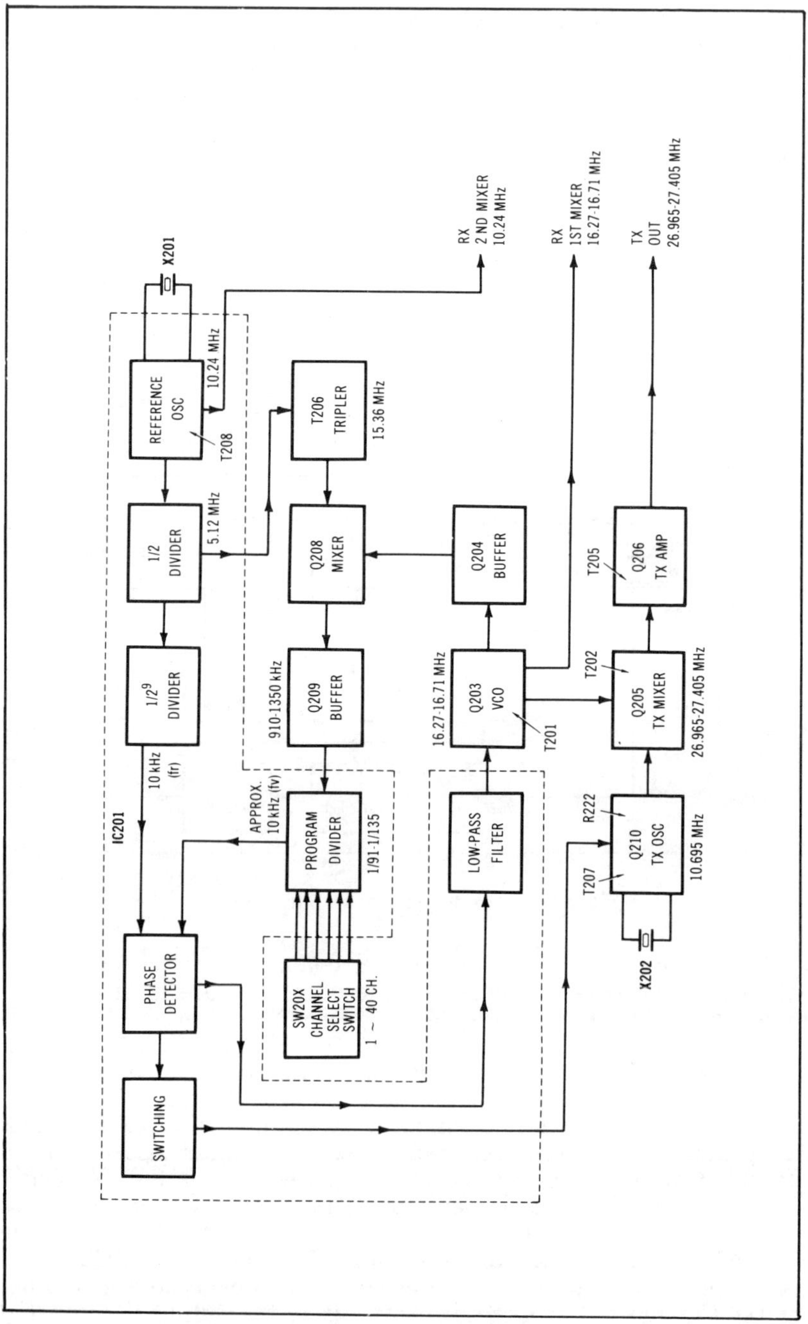

Fig. 3-87. Block diagram of a phase-locked loop synthesizer.

oscillator operates on 10.25 MHz and through a 2-to-1 divider develops a 5.12-MHz signal that is applied to tripler T206. Thus in the mixer the signal from the voltage-controlled oscillator and the reference crystal oscillator are mixed to produce a difference frequency that falls between 910 and 1350 kHz.

This latter signal is applied to the programmed divider. The actual division depends on the setting of the channel selector switch. The division is such that a 10-kHz signal is removed from the programmed divider. This is applied to the phase detector (top left in Fig. 3-87). The output of the 2-to-1 divider that follows the reference oscillator is further divided by a factor of 512 (2^9). Thus the 10-kHz reference frequency is also applied to the phase detector.

If there is any drift in phase between the two 10-kHz components there is a pulse excitation to the low-pass filter. The filter acts as a digital-to-analog converter and the generated information is applied to the vco as a correcting dc voltage. Thus the vco is held on frequency and that frequency depends on the setting of the channel selector switch. If the phase detector is inactive there is a special switching circuit that shuts off the 10.695-MHz transmit oscillator. Thus no spurious signal will be radiated by accident or by failure in the pll circuit.

Note that the output of the pll unit in Fig. 3-86 is supplied to the buffer-amplifier transistor (Q12) at the bottom left of the diagram. The output of this common-emitter amplifier is applied to a succeeding driver amplifier transistor (Q13). This stage is connected to the final amplifier. Thus it is the combined stages involving transistor Q13 and Q14 that develop the 4-watt output at a low-impedance level for application to antenna receptacle. This is the rf section of the transmitter.

The input microphone amplifier is located at the center (transistor Q8). Its output is applied through the squelch switching circuit to the integrated circuit audio amplifier modulator. By way of the secondary of a modulation transformer, the audio is applied to the dc supply-voltage to the collector circuits of both the driver amplifier (Q13) and the final amplifier (Q14). Both stages are collector-modulated to obtain full and linear modulation of the rf wave.

Note that the modulating wave at the output of the modulation transformer is also applied to the base of the automatic-modulation-control (Q9) at the lower left. Depending on the strength of the audio, a component is detected and filtered to develop a dc component applied to transistor Q10. It has a loading effect and limits the level of the microphone signal applied to the microphone amplifier. Thus overmodulation is avoided at the same time a high average level of modulation can be maintained.

3-18. WAVEFORM GENERATORS

Waveform generators are used in test equipment, timers and control circuits. Some of these use the popular 555 timer IC. The

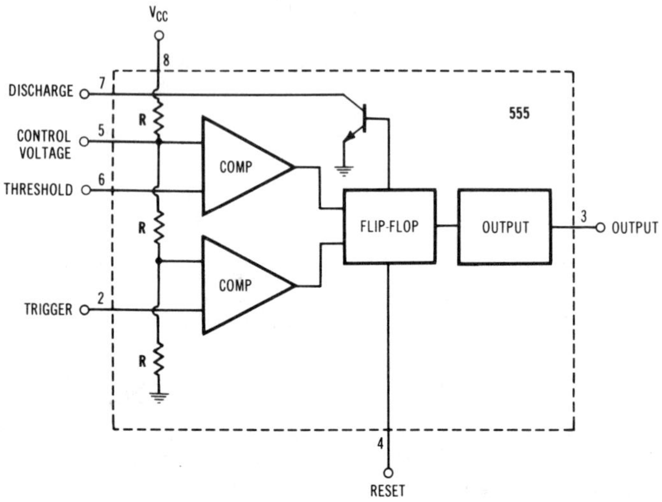

Fig. 3-88. Functional block diagram of the 555 IC timer.

timer consists of five basic circuits which are two comparators, a bistable flip-flop, a resistive voltage divider, a discharge transistor, and an output stage (Fig. 3-88). The three resistors of the divider are of equal value and the threshold comparator is referenced at two-thirds of the supply voltage while the trigger comparator is referenced at one-third of the supply voltage. Note that the output of the comparator is connected to the bistable flip-flop.

If the trigger voltage applied to the trigger comparator is dropped below one-third of the supply voltage, the comparator changes state and causes the flip-flop to drive the output to its high state.

The threshold input monitors the capacitive voltage of an external RC timing network (Fig. 3-89). When the capacitor voltage of this external circuit exceeds two-thirds of the supply voltage, the threshold comparator resets the flip-flop. In this case the flip-flop now drives the output to its low state. In this condition the discharge transistor is on and the external timing capacitor will be discharged. However, when the capacitor is discharged the comparator awaits another trigger pulse to cause the timing cycle to be repeated.

In a pulse-generator circuit the threshold and trigger inputs are tied together and connected to the external capacitor (C) as shown in the circuit of Fig. 3-89. Note how the external resistors are connected to the supply voltage and present a series path along with the capacitor from the supply voltage to common. The discharge circuit is connected to the junction of the two resistors while the threshold and trigger connections are made between the junction of the second resistor and the capacitor. A discussion of the timer con-

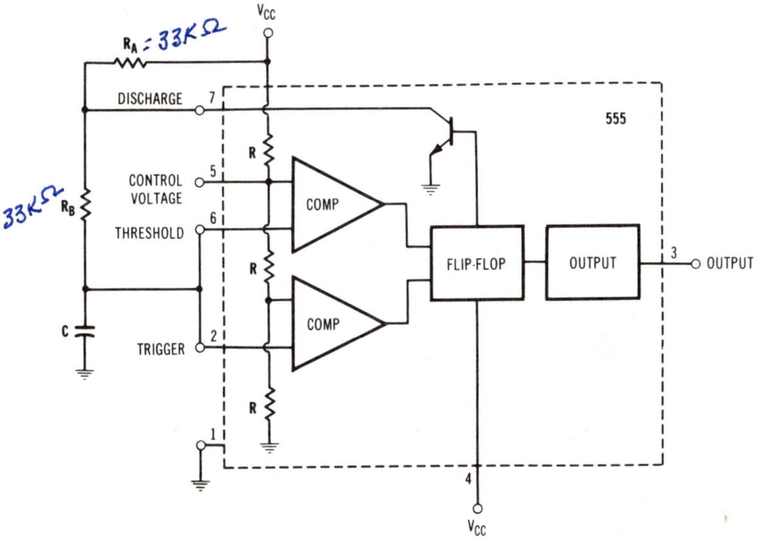

Fig. 3-89. External components for controlling timing of the 555 IC.

nected as an astable oscillator follows.

When the power is turned on the capacitor is in a discharge state. However, circuit activities cause the capacitor to charge toward the supply voltage level through resistors R_A and R_B. When the capacitor reaches the threshold level of 2/3 V_{CC}, the output drops low and the voltage on the capacitor discharges through resistor R_B and the discharge transistor of the timer. When the capacitor drops to the level of 1/3 V_{CC}, the trigger comparator triggers, causing the flip-flop to change over and a new charging cycle begins. Thus, the voltage on the capacitor varies between 1/3 V_{CC} and 2/3 V_{CC}. The output voltage at pin 3 is high (value of V_{CC}) during the charging interval and low (0 volts) during the discharge period.

A practical value for the discharge period would be:

$$t_{dis} = 0.693 R_B C$$

The charging time is longer and has a value of:

$$t_{ch} = 0.693(R_A + R_B)C$$

The total period of the charge and discharge activities becomes:

$$T = t_{ch} + t_{dis} = 0.693\ (R_A + 2R_B)C$$

The repetition rate or frequency is the reciprocal of the period, or

$$f = \frac{1}{T} = \frac{1.44}{(R_A + 2R_B)C}$$

The graph of Fig. 3-90 shows the relationship among capacitance, frequency of operation, and the quantity ($R_A + 2R_B$). Can you determine the approximate frequency of operation for the circuit of Fig. 3-91?

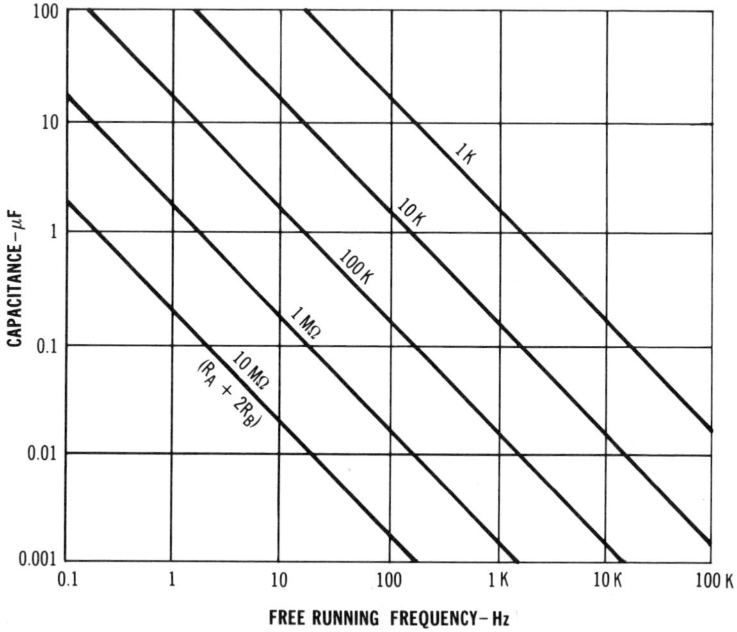

Fig. 3-90. Resistance, capacitance, and frequency relationship.

The quantity ($R_A + 2R_B$) is approximately 100 kΩ. The capacitor value is 0.01 microfarad. The 100-kΩ line crosses the 0.01 horizontal line between 1 and 2 thousand hertz. Use the frequency equation to obtain the exact value as follows:

$$f = \frac{1.44}{(.033 \times 10^6 + .066 \times 10^6)\ .01 \times 10^{-6}} = 1455 \text{ hertz}$$

In the circuit of Fig. 3-91 the pulse-repetition rate and period are determined by resistors R_A, R_B, and capacitor C1. The longer this time constant is, the lower is the frequency of the output pulse. A negative pulse appears at the output and has a duration determined by capacitor C1 and resistor R_B. The time interval between the trailing edge of the pulse and the leading edge of the next is set by the charge time constant of capacitor C1 plus resistors R_A and R_B.

If a shorter duration pulse of the same frequency is to be developed at the output it is necessary to decrease the value of resistor R_B and increase the value of resistor R_A. Thus, by

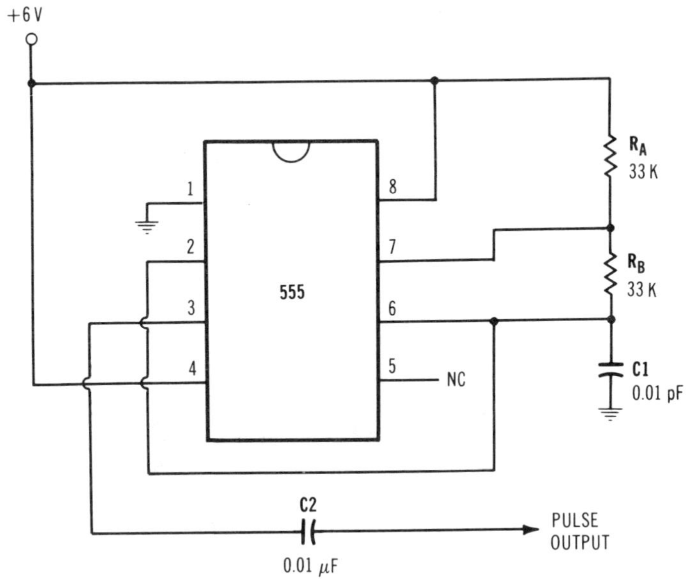

Fig. 3-91. Astable pulse generator.

controlling the relative values of the two resistors and the capacitor, a negative pulse of specific duration and repetition rate can be developed at the output.

Additional knowledge and practical experience with solid-state devices can be gained by constructing the simple demonstration circuits included in the solid-state circuit book of Reference 2, Chapter 1. Devices are low-cost and easy to obtain. Circuits are built on a solderless circuit board.

CHAPTER 4

Modulation Systems

In the two-way radio services two principal forms of modulation are used. These are *frequency modulation* and *amplitude modulation*. The former (fm) is usually used in the vhf and uhf spectra and is the dominant form of modulation in the private land mobile radio services. The latter form of modulation (am) is used more widely in the hf region in various radio services and also to some degree in the vhf and uhf region. Conventional double-sideband amplitude modulation is popular in the Citizens Band radio service. Single-sideband amplitude modulation (ssb) is found in the mf and hf marine radio services and other point-to-point services. These modulation systems are introduced in this chapter.

4-1. FM PRINCIPLES

Frequency modulation has several attractive advantages for radio communication use; it requires no significant audio power and has a very low susceptibility to noise and interference. Since a frequency-modulated wave is employed the amplitude variations contributed by noise can be limited sharply at the receiver. The ability to eliminate amplitude impulse noise is an important advantage when the equipment must be operated in close proximity to ignition systems and other noise sources, which radiate strong amplitude-varying rf signals. In a similar manner the beat variations (amplitude changes) between two interfering rf signals can be reduced. Thus by using a frequency-modulation system there will be less interference between stations operating on the same or adjacent channels. In fm systems the stronger station dominates the weaker station and pushes it into the background.

It is significant that fm transmission requires a wider emission bandwidth. The emission bandwidth is more that twice the highest modulating frequency. The greater the frequency deviation is (higher modulation index), the greater is the emission bandwidth for a given modulating frequency. However, bandwidth can be held down by establishing a limit for the highest audio frequency and by using a very low maximum-permissible frequency deviation. In the two-way radio services it is customary to limit the highest modulating audio frequency to 2500 hertz and the maximum deviation to ± 5 kHz.

In an fm system the frequency of the resultant rf wave is made to vary with the modulation. The extent of the frequency change, or deviation, varies with the amplitude of the modulating wave. The rate at which the frequency change varies with the frequency of the modulating wave.

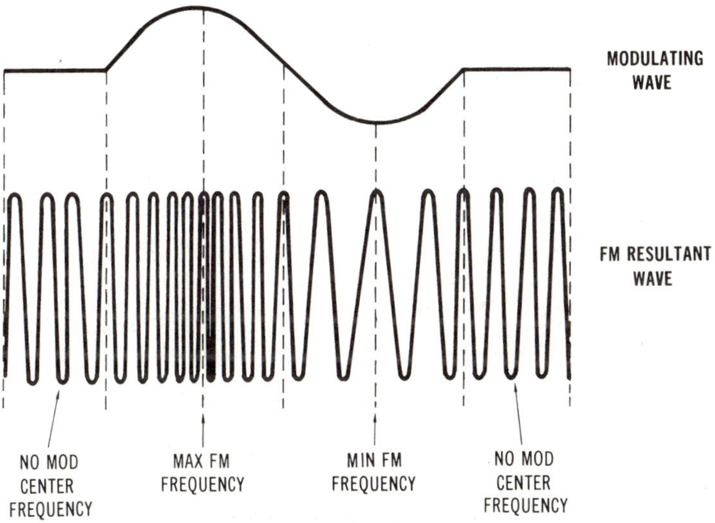

Fig. 4-1. The frequency-modulated wave.

As shown in Fig. 4-1 the actual amplitude of the resultant rf wave remains constant—only its frequency changes with the modulation. This latter characteristic means that any amplitude variations (noise) present in the resultant fm signal can be safely removed because the desired signal information has nothing whatsoever to do with amplitude change. This elimination of undesired amplitude variations is accomplished in the fm receiver.

4-1-1. Direct and Indirect Frequency Modulation

The basic theoretical principle of the frequency-modulation process can be demonstrated with a motor-driven capacitor. If such a capacitor had a linear frequency characteristic and was placed across the tank circuit of an oscillator, the frequency of the oscillator could be made to change with the capacitor rotor rotation. The oscillator frequency would, of course, be minimum when the capacitor plates are meshed fully and maximum when the capacitor plates are entirely out.

If the associated motor rotated the capacitor rotor at the rate of 100 revolutions per second, the frequency of the oscillator would also vary at the 100 cycle-per-minute rate. If the motor speed were increased or decreased there would be a similar change in the rate of frequency change. This would be the same as changing the frequency of the modulating wave in a practical fm system.

If the value of the capacitor is reduced, the range over which the frequency of the oscillator could change would be reduced. The lesser deviation would correspond to the application of a weaker modulating wave to the input of a practical fm system.

If it were practical for a rotating motor to follow the complex variations of speech, it could be used to generate an fm wave for a two-way radio system. However, a similar and practical deviation can be accomplished with the use of a transistor stage designed to place a variable capacitance or variable inductance across the tuned circuit of an oscillator. This type of circuit is called a *reactance modulator*.

An fm wave can also be formed by varying the phase of an rf wave in accord with the variations of a modulating signal. This technique is used extensively in fm two-way radio systems and the circuit is referred to as an *fm phase modulator*. The source of signal is usually a crystal oscillator, variable crystal oscillator (vxo), or variable frequency oscillator (vfo).

When an fm wave is formed by deviating the frequency of the oscillator with a reactance device it is called direct frequency modulation. The oscillator can also be any one of the above types.

In most fm units the fundamental oscillator operates at a much lower frequency than the assigned transmit channel frequency. A series of multiplier stages then follows the low-frequency oscillator to increase the frequency to the desired transmit channel frequency. As the frequency is multiplied, any deviation of the fundamental oscillator frequency will be increased by a like amount. To obtain a small linear frequency change, the actual oscillator deviation produced by the reactance circuit is quite limited. Assume the 12.95-MHz frequency of an oscillator is deviated by only 350 Hz and there is a multiplication factor of 12. The final transmit frequency will be 155.4 MHz (12 × 12.95) and the maximum deviation 4200 Hz (12 × 350). Somewhat greater deviation would result in the maximum permissible value of ±5 kHz. From this example it can be seen that frequency-multiplier stages increase the small frequency deviation of the fundamental oscillator to the permissible deviation desired.

Multiplier stages used in an fm transmitter can be conventional class-C amplifiers. The fm wave, which carries the modulating information, is in the form of a changing frequency. There is no danger of distortion because of the nonlinear amplitude characteristic of a class-C stage. The standard am and sideband modulation methods require linear amplification, after the modulation envelope has been formed, in order to retain the original modulation. In fact, the limiting activity of a class-C stage in terms of amplitude change is used to advantage in an fm system. In the fm process undesired amplitude variations may also occur simultaneously with a desired frequency change. These undesired amplitude variations are eliminated by the class-C multipliers, thus producing a constant amplitude fm signal output.

Very little audio power is necessary in the modulation of an fm wave as compared to a standard am system. The power in a resultant fm wave remains constant with modulation; it is simply distributed in differing amounts among carrier and sidebands, thus producing changing frequency but a constant amplitude output.

One of the problems with direct frequency modulation (Fig. 4-2A) is the center-frequency stability. It must be held constant to maintain the proper operating stability and prevent distortion. Both the oscillator and associated reactance device must be carefully designed for the most stable operation. A stable crystal oscillator or vfo is needed. An alternative is an afc system (Fig. 4-2B) that compares the center frequency of the transmitter to some crystal-controlled standard. A dc error voltage is developed at the output of the afc circuit and, through the reactance device, it alters the center frequency an amount necessary to correct any frequency drift.

A block diagram of an indirect frequency-modulation system is shown in Fig. 4-2C. In this case the rf carrier is phase modulated (pm). Phase (or angle) modulation, like direct frequency modulation, can be accomplished in a number of ways. Using a suitable correction arrangement, the phase-modulation process can be used to generate a true fm wave.

Most indirect systems use a crystal oscillator or vxo. Thus the center-frequency stability is good and an afc circuit is not required. A phase-modulation process is limited in the amount of deviation that can be obtained; however, this is an unimportant consideration in two-way radio systems because they are predominantly narrow-band limited-deviation systems.

In the phase-modulation process used in two-way commercial transmitters the audio variations continuously vary the phase of the rf carrier wave. They do so by placing a changing reactive component in the path of the rf carrier wave. This produces a changing phase shift. The higher the amplitude of the audio wave, the greater is the phase shift and the more the rf wave is delayed during one alternation. Conversely, on the opposite alternation of the audio wave the phase shift is of opposite sign and the rf wave is advanced instead of delayed. If the phase modulator speeds up the rf wave its period is reduced. This means it has a higher frequency (Fig. 4-3A). If the radio-frequency wave is delayed its period is increased and it has a lower frequency (Fig. 4-3B).

In practice the phase shift varies throughout the entire audio cycle reaching the maximum value only at the sine-wave peaks.

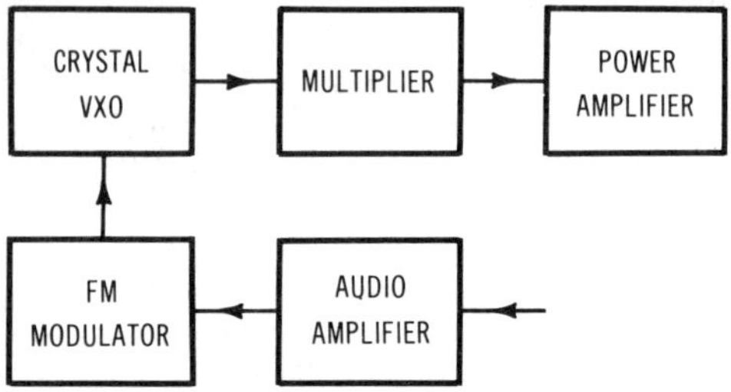

(A) Direct frequency modulation.

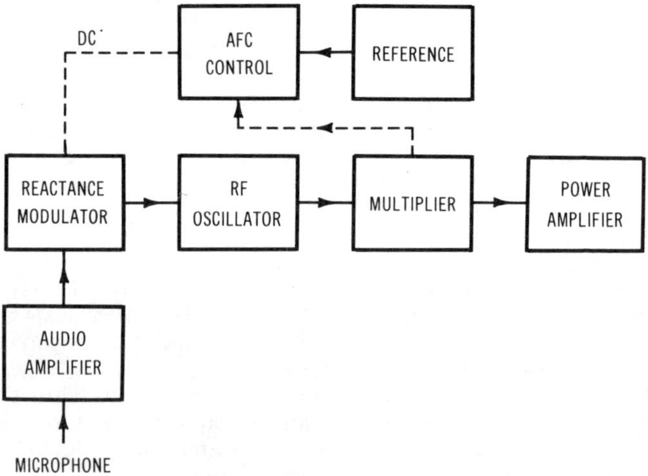

(B) Automatic-frequency-control system.

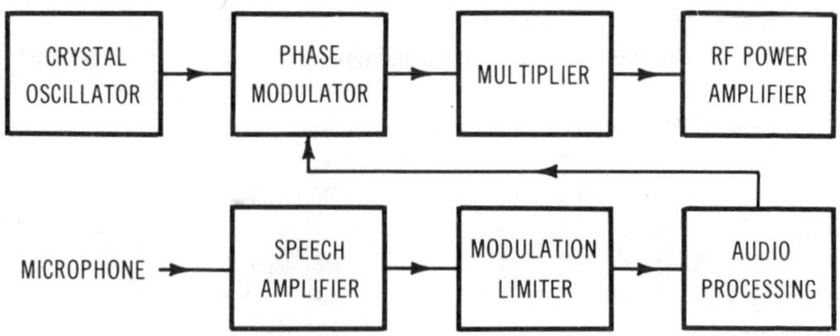

(C) Indirect fm system.

Fig. 4-2. Functional plans of fm systems.

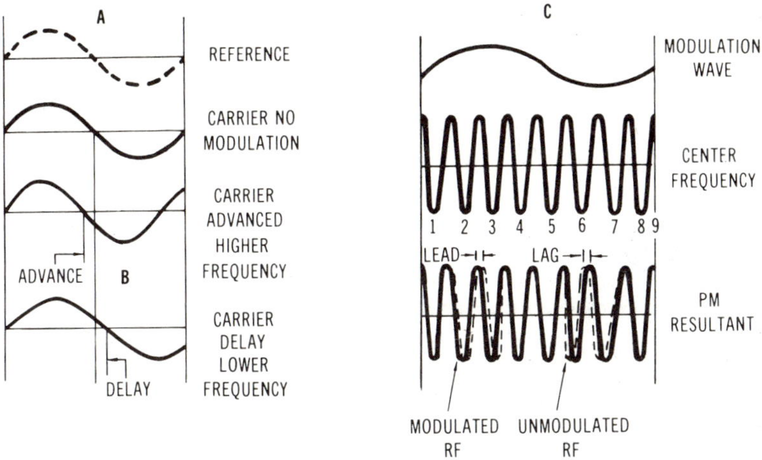

Fig. 4-3. Phase-modulation waveforms.

Positive and negative peaks produce the same phase shift but of opposite sign (lag and lead). The phase shift is zero when the audio wave passes through zero. Therefore the frequency of the rf wave deviates above and below the carrier value following the amplitude and polarity changes of the audio wave (Fig. 4-3C).

An increase in the modulating frequency increases the frequency deviation of the carrier. The amount of frequency modulation produced indirectly by phase modulation varies directly with the frequency of the modulating signal.

The amount of frequency deviation produced is relatively small. Usually a low carrier frequency is chosen to be phase modulated. After the fm wave has been generated by the indirect process it is applied to a series of multipliers. These multipliers increase both the frequency and frequency deviation. As in the direct fm system, the multipliers increase the deviation to the maximum permissible value allowed for a particular service.

There is one fundamental difference between frequency and phase modulation. In a direct fm system, the frequency deviation is constant for a given audio amplitude regardless of the audio frequency. The modulation index formula indicates that a change in audio frequency will cause a change in the modulation index even though the deviation remains fixed.

$$\text{Modulation index} = \frac{\text{frequency deviation}}{\text{modulating frequency}}$$

Thus in a direct fm system the modulation index decreases with modulating frequency, assuming a fixed deviation.

In a phase-modulation process, it is the modulation index that remains constant for a given amplitude regardless of audio frequency. Inasmuch as the modulation index is constant the deviation itself must vary with the audio frequency.

Frequency deviation = modulation index × modulating frequency

Actually the higher the audio frequency of a given amplitude is, the greater is the frequency deviation when using phase modulation.

It is apparent from the foregoing paragraphs that in generating a true fm wave using the phase-modulation process some form of compensating circuit must be used ahead of the modulator. Such a circuit is called a *predistorter* and is inserted between the audio source and the phase modulator.

The predistorter has a response that declines with an increase in frequency. Thus the output of the predistorter at a high audio frequency is less than the output at a lower frequency for the same input amplitude. Its response compensates for the fact that a phase modulator produces a greater deviation for a high-frequency modulating wave than for a lower-frequency one of the same amplitude. The compensation introduced by the predistorter is equal and opposite to that of the modulation process. Therefore, with predistortion it is possible to generate a true fm wave using the phase-modulation process.

In the usual fm two-way radio system there is an essentially narrow-band audio-frequency range, and any necessary predistortion can be handled very simply by proper control of overall amplifier frequency response. The associated modulation limiters and integrators inherently suppress the high-frequency components and prevent them from causing excessive deviation of the center frequency. Thus it is possible in some cases to handle predistortion without the actual use of a predistorter network.

4-2. COMPOSITION OF FM WAVE

The resultant fm wave, like an am envelope, is the algebraic summation of a number of individual rf waves. In the fm process, when modulating with a single-frequency tone, a number of sideband pairs are generated according to modulating frequency and deviation. This is unlike the am process which generates only a single pair of sidebands when modulation is by a single-frequency tone.

In a standard am system, the carrier remains constant in amplitude; however, the algebraic addition of the two sidebands and the carrier produces an amplitude-changing resultant (am envelope). In an fm modulation system the amplitude of the resultant wave remains constant while the amplitude of the carrier changes. The carrier amplitude change is such that when added to the various sideband pairs, a constant-amplitude but changing-frequency resultant is produced.

Actually the ratio of the powers in the sidebands and in the carrier changes with the modulating frequency and deviation. However, the resultant power output is always a constant. In fact, under specific conditions of modulating-wave amplitude and

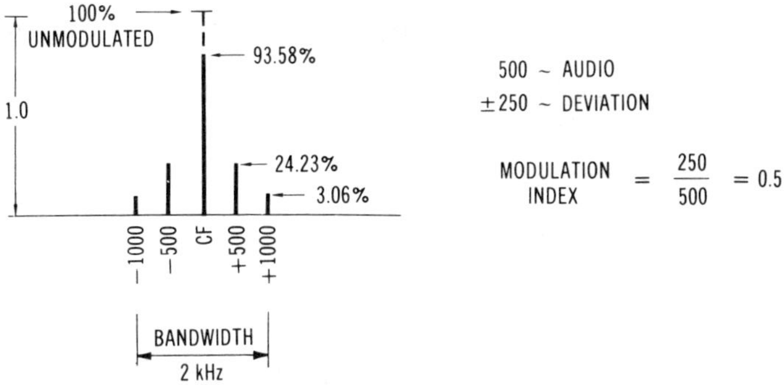

(A) ± 250-Hz deviation, 500-Hz audio modulation.

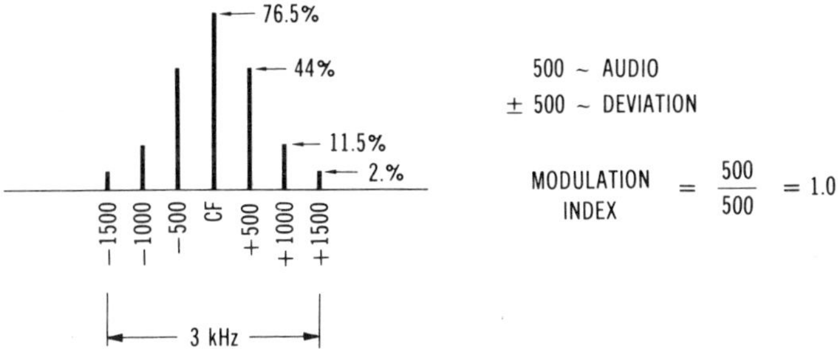

(B) ± 500-Hz deviation, 500 Hz audio modulation.

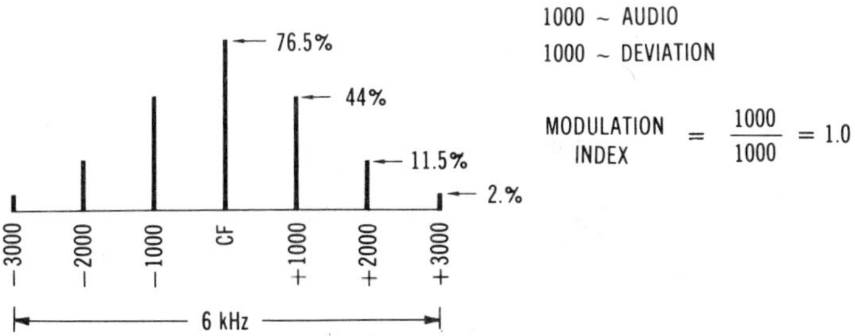

(C) ± 1000-Hz deviation, 1000-Hz audio modulation.

Fig. 4-4. Spectrum distribution of various fm waves.

frequency, it is possible that the carrier or center-frequency power can even fall to zero. In this case all of the power is in the sidebands. Under other conditions of frequency modulation, most of the power is in the carrier and only a small amount in the sidebands. In fact, the evaluation of shifts in power and relative changes of amplitude among the sidebands and carrier can be useful in checking the performance of an fm system.

The number of sideband pairs generated in the fm process

depends on the modulation index. The emission bandwidth, in turn, depends on the modulation index at maximum permissible deviation of the highest-frequency modulating wave. Some possible frequency distributions are illustrated in Fig. 4-4.

The wave equations and solutions that result when one rf wave is frequency-modulated by another lower-frequency wave are very complex. Such equations can be solved using *Bessel functions.* Despite the complex solutions, very practical mathematical results can be obtained using a table of Bessel functions. Table 4-1 provides a set of Bessel-function solutions that are applicable to the narrow-band frequency modulation used in two-way radio. The table gives the relative magnitudes of carrier (center frequency) and side frequencies for various modulation indexes. The unity reference represents the magnitude of the carrier when unmodulated. A sideband frequency is considered significant when its magnitude is more than 2 percent of the magnitude of the unmodulated carrier.

Table 4-1. Bessel Function Chart Applied to Carrier and Side Frequencies of an FM Wave

Modulation Index	Carrier (f_c)	$f_c \pm f_m$	$f_c \pm 2f_m$	$f_c \pm 3f_m$	$f_c \pm 4f_m$	$f_c \pm 5f_m$	$f_c \pm 6f_m$
0	1.00	0					
0.1	0.9975	0.0499					
0.2	0.99	0.0995					
0.3	0.9776	0.1483					
0.4	0.9604	0.1960					
0.5	0.9385	0.2423	0.0306				
0.6	0.912	0.2867	0.0437				
0.7	0.8812	0.329	0.0589				
0.8	0.8463	0.3688	0.0758				
0.9	0.8075	0.4059	0.0946				
1.0	0.7652	0.4401	0.1149	0.0196			
1.2	0.6711	0.4983	0.1679	0.0329			
1.4	0.5669	0.5419	0.2073	0.0505			
1.6	0.4554	0.5699	0.257	0.0725			
2.00	0.2239	0.5767	0.3528	0.1289	0.034		
3.00	0.2601	0.3391	0.4861	0.3091	0.1320	0.0430	
4.00	0.3971	0.066	0.3641	0.4302	0.2811	0.1321	
5.00	0.1776	0.3276	0.0466	0.3648	0.3912	0.2611	0.131

For example, when the fm modulation is such that the modulation index is 0.5, the magnitude of the center-frequency component is 93.85 percent of the unmodulated center-frequency level, the first pair of sidebands has a magnitude that is 24.23 percent of the unmodulated center frequency value and, finally, a second sideband pair has a magnitude of 3.06 percent of the unmodulated center-frequency value. If the modulating frequency were 500 Hz, there would be a pair of sidebands, one at each side of the center frequency, displaced by 500 cycles. Each sideband, upper and lower, would have a magnitude of 24.33 percent of the unmodulated carrier value. Furthermore, there would be a second pair of sidebands displaced by 1000 cycles (2 × 500) from the center frequency. These frequencies would have a magnitude of 3.06 percent. A simple spectrum distribution graph (Fig. 4-4A) can be used to illustrate relative magnitudes among the wave components and the required emission bandwidth for transmitting this fm signal.

If we increase the magnitude of the modulating wave and thus cause a greater deviation, the emission bandwidth will increase because an additional significant sideband pair will be transmitted (Fig. 4-4B).

The higher the audio frequency for a given index is, the greater is the separation between side-frequency pairs of significance, and the greater is the required emission bandwidth. Compare examples B and C in Fig. 4-4.

Collate these graphs with the information given in the Bessel Function table. The results prove that the emission bandwidth increases both with frequency and deviation. Therefore, in the narrow-band two-way radio services it is necessary to hold down the emission bandwidth by limiting both the extent of the deviation of the center frequency and the highest permissible audio frequency.

The maximum emission bandwidth assigned to a given radio service is a function of the maximum permissible deviation and the highest modulating frequency. The modulation index for this important operating condition is called the deviation ratio:

$$\text{Deviation Ratio} = \frac{\text{maximum permissible deviation}}{\text{highest modulating frequency}}$$

Calculation for the two-way radio deviation of ± 5 kHz based on a maximum permissible modulation frequency of 2500 Hz becomes:

$$\text{Deviation ratio} = \frac{5000}{2500} = 2$$

4-3. TYPICAL CIRCUITS

In the typical fm transmitter the modulation occurs at a low radio frequency. Narrow-band deviation is the rule and, consequently, many frequency modulators use a phase-modulator stage to obtain frequency modulation by the indirect means. Other transmitters employ direct frequency modulation of a crystal oscillator using a voltage-variable capacitor diode.

A basic bipolar phase modulator is shown in Fig. 4-5. Such a phase modulator is inserted in the radio-frequency path between the crystal oscillator source and the multiplier section of the transmitter.

The input voltage E_{in} follows two paths. One path is through capacitor C1 to the base of the phase modulator. This rf component is labeled E_B in the schematic and associated vector diagram. A

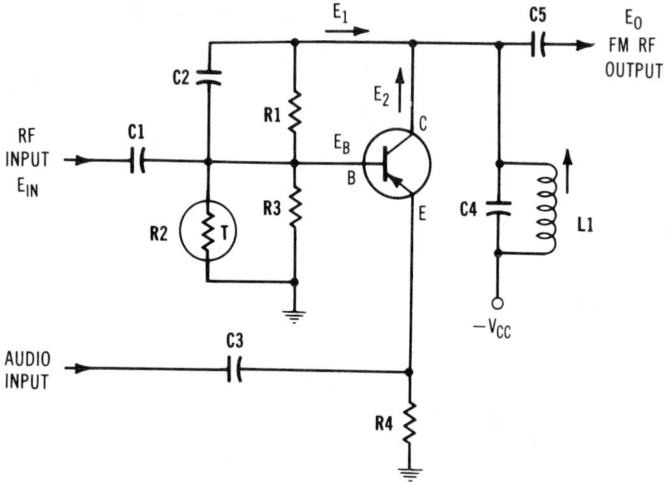

(A) Circuit.

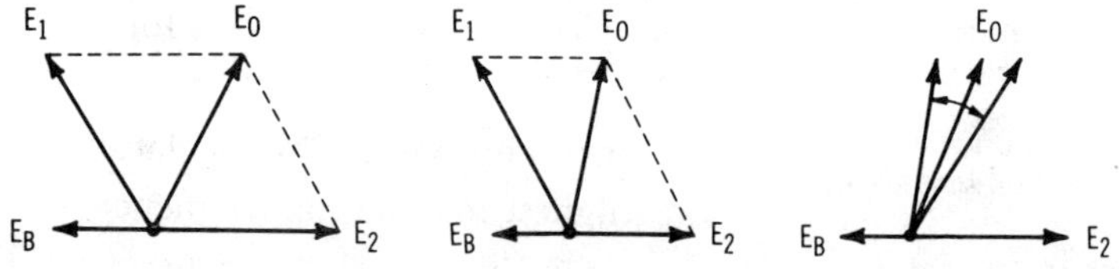

(B) Vector relationships.

Fig. 4-5. Transistor phase modulator.

second path is coupled to the output circuit by way of capacitors C1 and C2. This component is shown as voltage E_1. The rf component (E_B) supplied to the base appears inverted in the collector circuit of the phase modulator as voltage E_2. Since the transistor provides a polarity reversal, it would seem that the two components E_1 and E_2 present in the collector circuit would be of opposite polarity. However this is not the case because of the phase-shifting activity of the input capacitors which, along with other component values, have been selected to obtain an appropriate phase relation between the two rf components as shown in the vector diagram. They are neither in phase nor 180 degrees out of phase.

The vector diagram shows a possible relationship between the two components E_1 and E_2. In the output circuit the direct voltage E_1 remains constant in amplitude. Likewise the rf signal component (E_B) applied to the base is constant. However, the g_m of the stage varies with the applied modulating wave felt at the emitter.

Therefore the output voltage component (E_2) varies in amplitude with the modulating wave. As shown by the vectors, the amplitude of voltage E_2 varies with respect to the constant amplitude of voltage E_1.

The vector diagram of Fig. 4-5B shows how the phase modulation is caused. The net rf output voltage (E_o), or the resultant fm wave, is the vector sum of E_1 and E_2. The resultant voltage changes in phase as the amplitude of the E_2 component varies with the applied audio. The higher the amplitude of the applied audio is, the greater is the phase deviation and the greater is the resultant frequency modulation. The higher the frequency of the applied audio is, the more often the angle will swing between the two extremes.

The second vector diagram shows how the output voltage (E_o) has an angle that deviates further from voltage E_2 when the amplitude of E_2 declines.

The third vector diagram shows briefly what happens to the output voltage as the E_2 voltage varies in accordance with the amplitude of the audio input signal. This change in E_2 amplitude causes the angle of E_o to vary, following the changes of the audio wave. As mentioned, the audio component is applied to the emitter circuit, varying the base-emitter voltage in accordance with the audio modulation. The base-divider bias circuit includes a thermistor (R2) to stabilize the operating point against temperature change.

Transistors can also be used as variable reactance devices. The output capacitance of a transistor, the capacitance of the collector junction (C_o) specifically, can be made to vary quite readily. The collector junction is reverse-biased and the depletion area of the junction depends on the junction voltage. When a higher reverse voltage is applied to the junction the depletion area increases and reduces the junction capacitance. An audio signal applied to the base or emitter circuits can cause this capacitance to change. If the capacitance is shunted across a tuned circuit there will be a corresponding change in its resonant frequency (Fig. 4-6). The oscillator is connected in a common-base configuration with the small capacitor (C_s) providing the necessary feedback between collector and emitter circuit to sustain self-oscillation. The audio signal is supplied to the emitter and the resultant change in the collector-to-base voltage will cause the transistor-output capacitance (C_o) to vary and cause a change in the frequency of oscillation. This type of circuit is referred to as a self-reactance fm oscillator.

4-3-1. Voltage-Variable Capacitor

One of the effects of a changing voltage across a semiconductor diode is a change in capacitance. By suitable semiconductor design and doping, the change in capacitance can be emphasized and made to have a capacitance-versus-junction voltage characteristic that will permit it to operate as an efficient voltage-to-frequency converter.

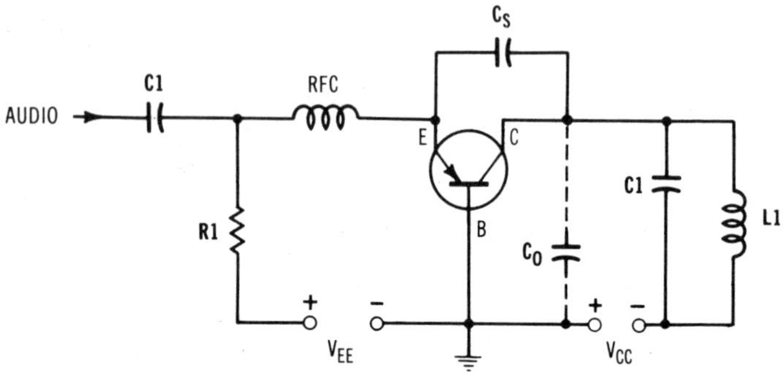

Fig. 4-6. Self-reactance modulator-oscillator system.

A voltage-variable capacitor placed across an oscillator tank circuit as shown in Fig. 4-7 can be used to control or vary the resonant frequency of the tuned circuit. As shown, the modulating voltage is supplied to the diode by way of an isolating rf filter. A voltage source sets the dc bias of the diode at a point on its voltage-capacitance transfer characteristic that ensures a linear deviation of the oscillator signal. As the bias point is made to vary with the positive and negative excursions of the modulating wave, there is a corresponding change in diode capacitance, and in the resonant frequency of the tuned circuit with which the diode is associated. A capacitive divider arrangement (capacitor C1 and voltage-variable capacitor D1) may be used to regulate the extent of the capacitance deviation for a given magnitude of audio input signal. Such an adjustment may be used to ensure a linear frequency deviation for a given maximum level of audio input signal.

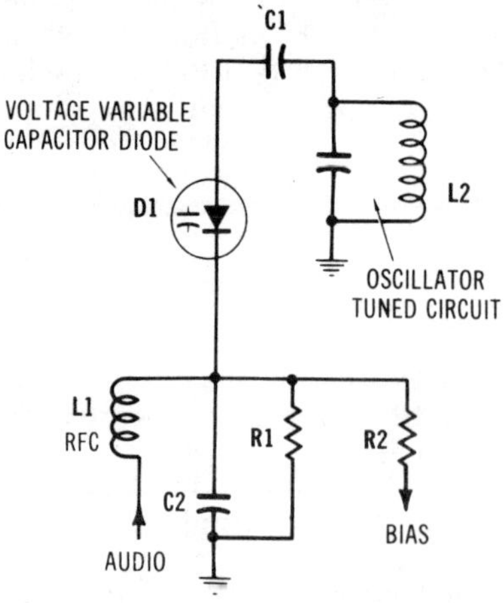

Fig. 4-7. Using voltage-variable capacitor diode to vary frequency of a resonant circuit.

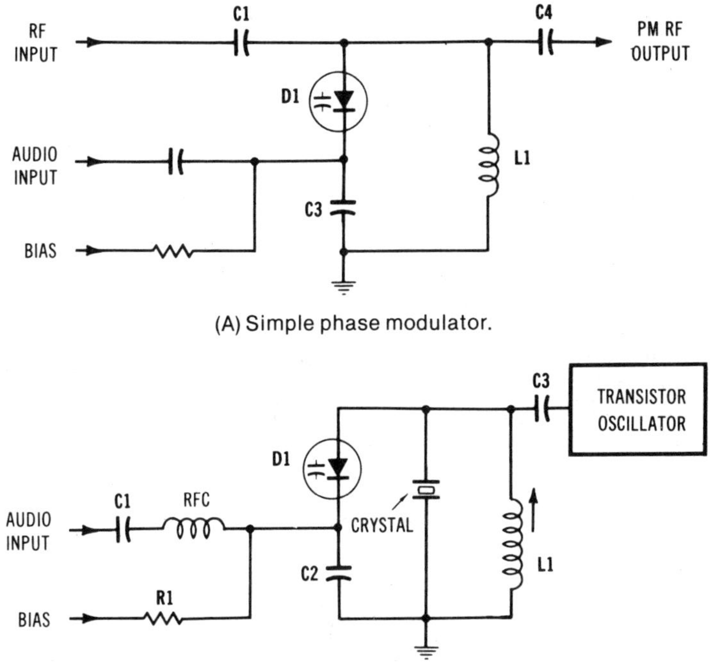

(A) Simple phase modulator.

(B) Direct-fm circuit.

Fig. 4-8. Variable-capacitance diode as a phase or frequency modulator.

The variable-capacitance diode can be used either to frequency- or phase-modulate an rf signal. In the simple phase modulator of Fig. 4-8A the variable-capacitance diode (D1) is a part of a resonant tank circuit that follows the oscillator. The coupling capacitor (C1) from the oscillator to the resonant circuit has a rather high reactance at the output frequency. Thus, the audio variation that is applied across the voltage-variable capacitor is able to change the resonant frequency of the tank circuit but not the oscillator frequency. However, the change in the resonant frequency causes a change in the phase angle of the rf signal delivered by the oscillator as it is developed across the tank circuit. This is to be anticipated because with a change in the frequency of the tuned circuit away from the frequency of the oscillator, the tuned circuit will present an impedance that now has a reactive component at the oscillator frequency.

The unmodulated resonant frequency of the tuned circuit depends on the LC constant and the capacitance contributed by the variable-capacitance diode (at the established dc bias). In the case of no applied modulation the oscillator frequency and the resonant frequency of the tuned circuit are one and the same. However, when an audio signal is applied to the variable-capacitance diode there will be a corresponding change in the diode capacitance. This will cause the resonant frequency of the tuned circuit to change. In effect the tuned circuit will now present a changing reactive component and the individual rf cycles will be shifted in phase. As in any form of phase modulator this change in phase will compress and expand the periods of individual rf cycles and frequency modulation will have been produced by an indirect means.

It is also possible to connect the variable-capacitance diode directly into the crystal circuit of an oscillator as shown in Fig. 4-8B. Often an overtone-crystal oscillator is used. The resonant circuit consists of inductor L2, capacitor C3, and the variable-capacitance diode (D1). An applied audio signal will cause a change in the capacitance and a limited deviation of the crystal oscillator frequency will result. By using an overtone oscillator, this change in frequency can be made substantial and narrow-band frequency modulation will result. The deviation, of course, can be increased by using succeeding multiplier stages.

A practical variable-capacitance frequency modulator used in conjunction with a Clapp-type crystal oscillator is shown in Fig. 4-9. The crystal operates in its series-resonant mode. Capacitors C_2 and C_3 are of high value and act as swamping capacitors. Their values are high enough to swamp out any capacitive changes that may be contributed by the transistor due to heating or other circuit changes.

The frequency of the oscillator is determined by the resonant characteristics of the crystal and the capacitance added in series by the variable-capacitance diode. In effect, the variable-capacitance diode is adding additional series capacitance to the series-resonant

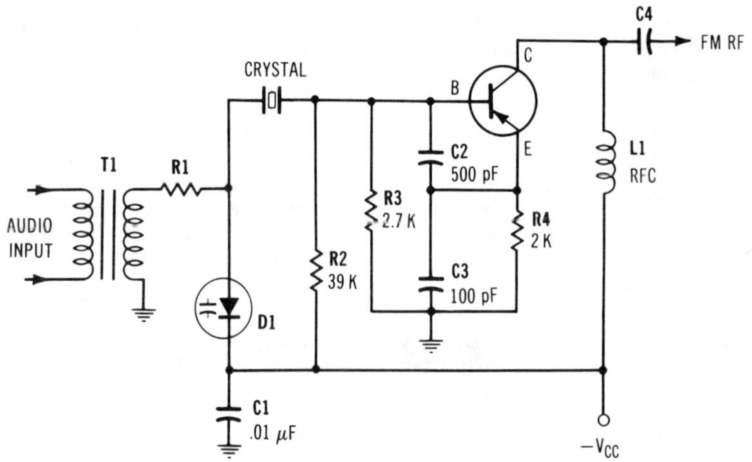

Fig. 4-9. Crystal oscillator with variable-capacitance-diode direct frequency modulation.

characteristic of the crystal when it is oscillating in its series mode. It is apparent, then, that any changes in the capacitance of D1 with the application of an audio wave will cause a corresponding deviation in the frequency of the crystal oscillator.

A two-diode phase modulator is shown in Fig. 4-10. The output of the transmitter crystal oscillator is supplied to the resonant input transformer T1; the fm component is derived at the output of resonant transformer T2. The capacitances of diodes D1 and D2 are varied by the voice-frequency components applied to the junction of resistors R1 and R2. A bias divider composed of resistors R3 and R4 set the diode bias for optimum linear deviation.

A small capacitor (C1) serves as coupling between the two transformers. This aids in keeping a precise center frequency at the same time the audio variations vary the reactances of the resonant circuits to produce a phase shift of the oscillator frequency.

The typical fm demodulator also uses a phase-shift principle in the manner in which it recovers the audio information from the received frequency-modulated signal. There are a variety of fm detectors. However, the conventional Foster-Seeley discriminator is basic to these types.

A basic discriminator circuit is shown in Fig. 4-11A. The operation of a discriminator is based on the phase relationship between two fm rf components that appear at its input. One component is coupled directly by way of capacitor C1 and appears across the discriminator inductor L1. Another rf component is coupled via the flux lines that link the primary and secondary resonant circuits. The secondary if voltage, when two coupled resonant circuits are employed, differs in phase 90° from the primary voltage, and, therefore, is also related by 90° to the if voltage that has been coupled directly across inductor L1. The 90° component is divided into two segments with reference to the secondary tap. One

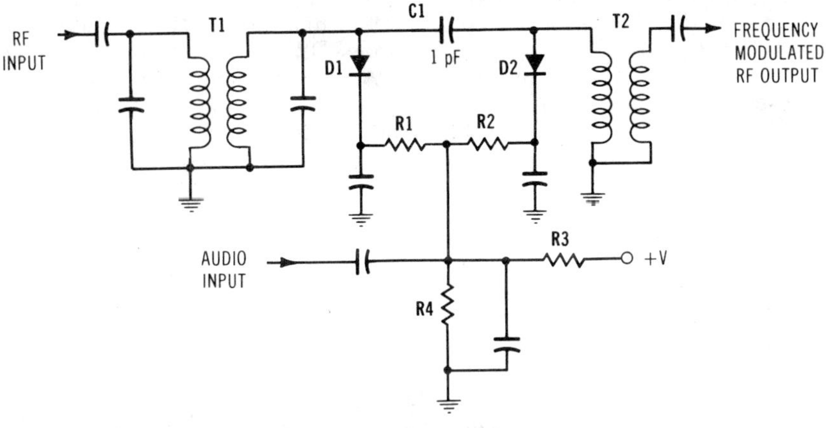

Fig. 4-10. Phase modulator using diodes.

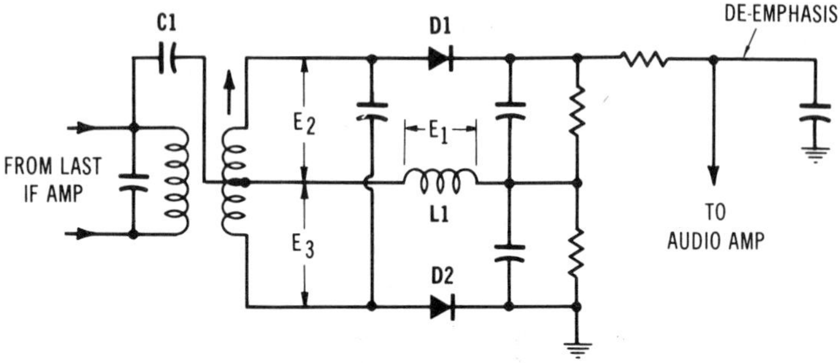

(A) Schematic.

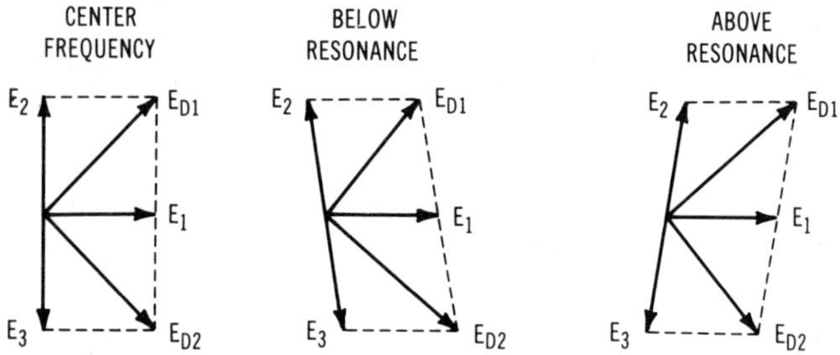

(B) Vectors.

Fig. 4-11. Basic discriminator circuit.

segment appears at the anode of diode D1, and the other appears at the anode of diode D2. In terms of the diode anodes, these two components are 180 degrees apart.

From the discussion, it is apparent that diode D1 is excited by two rf signals ($E_1 + E_2$), while diode D2 is excited by a signal component composed of ($E_1 + E_3$). At the frequency to which the discriminator transformer is tuned, these two signal components are of equal amplitude. Therefore they draw the same peak diode current, and, inasmuch as the diode currents occur in opposite directions in the output circuit, there is no output from the discriminator.

When the if signal deviates, the signal applied to the discriminator transformer does not remain on one frequency (center frequency). Rather it deviates with the modulation, swinging on each side of the center frequency. An off-resonance frequency applied to the discriminator transformer results in secondary components E_2 and E_3 which lead or lag the direct voltage (E_1) by more or less than 90 degrees. Therefore, when the incoming signal deviates on one side of the center frequency, a higher resultant voltage is applied to diode D1 than to diode D2. Conversely, when the signal swing is on the other side of center frequency, diode D2 is driven by a higher net voltage than diode D1. Refer to the vectors in Fig. 4-11B. When diode D1 draws the higher current, the net output of the discriminator is positive. Conversely, when the diode D2 draws the most current, the net output of the discriminator is negative.

How much positive or how much negative the output swings depends on just how far the incoming signal swings away from the center frequency. Thus, as the incoming fm signal deviates in frequency, the output voltage of the discriminator follows the frequency change, and a conversion is made from frequency change to amplitude change. The greater the deviation of the incoming signal is, the greater is the voltage range between maximum positive and maximum negative, as developed across the output.

The faster the rate of deviation of the incoming signal is, the faster will be the amplitude change at the output of the discriminator. In this manner, the modulating wave that is present as a frequency change of the incoming signal is converted back to the original amplitude variations of the modulating wave.

4-4. BLOCK DIAGRAM DESCRIPTION

A functional plan of an fm communications transceiver is given in Fig. 4-12. The transmission process begins with the generation of a crystal-controlled signal by the transmitter oscillator shown at the lower left. It is followed by the phase modulator that uses the phase-shift process to generate a frequency modulated wave indirectly.

The output of the microphone is applied to an audio amplifier and follow-up limiter. It is then applied to a roll-off filter that attenuates the high-frequency audio components sharply, limiting the audio bandwidth to a high of 2500 hertz. The filter is also necessary because of the tendency of a phase-modulation process to result in an fm deviation that is greater at higher frequencies. A preset deviation control is located between the limiter and the phase modulator. This control is often factory preset to prevent any deviation greater than ± 5 kHz.

Three multiplier stages follow to obtain a total frequency multiplication of 12. Hence the crystal-oscillator frequency is 1/12 of the transmit frequency.

Three rf power amplifiers follow building up the level of the signal to 25 watts output in the example of Fig. 4-12. A high attenuation low-pass filter follows. Its purpose is to reduce the harmonic output of the transmitter. Output is applied through the push-to-talk relay contacts to the antenna system.

On the receive mode the incoming signal is first applied to an rf amplifier and then to the first mixer. Proper bandwidth and selectivity is set by monolithic crystal filters that are located between the first mixer and the second mixer of the receiver. The bandpass of these crystal filters is only 16–20 kHz wide at the intermediate frequency of 10.7 MHz.

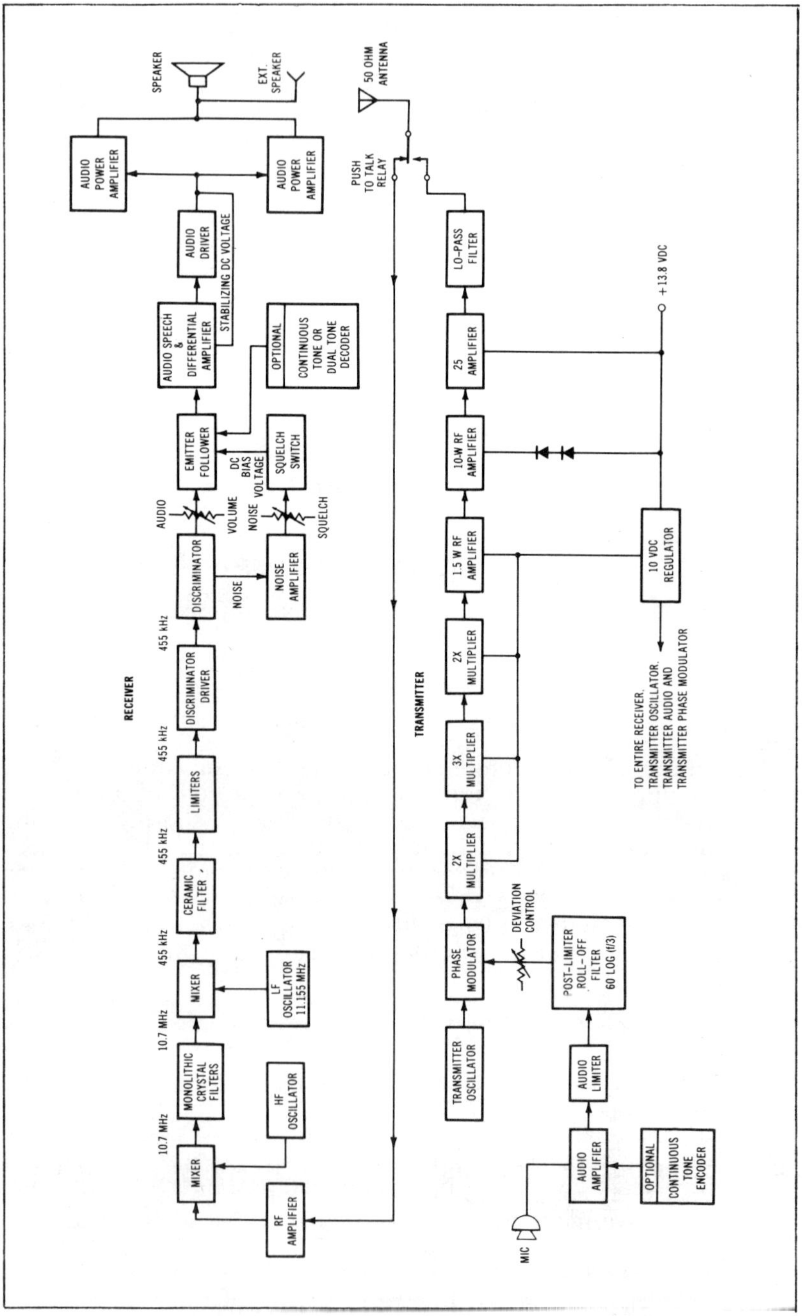

Fig. 4-12. Functional plan of fm communications transceiver.

The output of the second mixer is also followed by a critical ceramic filter. It has a bandpass of 16 kHz and excellent skirt selectivity. Therefore unwanted adjacent channel signals are rejected. A series of untuned intermediate-frequency and limiter stages follow to build up the level of the signal. Note that in the communications receiver the gain is handled by the untuned if amplifier. However, the selectivity is set by the dual-filter arrangement. Therefore operating constants of the amplifier and limiter stages do not influence the bandpass characteristics of the receiver. A discriminator driver and discriminator follow. The output of the discriminator is applied through the volume control to an emitter follower and a succeeding three-stage audio power amplifier ending in a class-B complementary-symmetry power output stage that drives the speaker.

A noise amplifier and squelch switch provide receiver squelch operation. Thus the squelch control can be used to cut down on the background noise level when there is no incoming signal. An incoming signal will break the squelch and the demodulated audio becomes audible.

Many modern fm communications transceivers also employ continuous-tone squelch systems (CTCSS). Such a system prevents the receiver from being activated except when the incoming signal contains a specific low-frequency tone that unsquelches the receiver. Unsquelching will occur only when the incoming tone is of the specific frequency which is used by the base station and other mobile stations of the communications system. Hence other systems operating on the same channel frequency do not unsquelch the receivers of your own particular system.

4-5. CONVENTIONAL AMPLITUDE MODULATION

In an amplitude-modulating system the radio-frequency wave is made to vary in amplitude in accordance with the audio variations. On the positive crest of the modulating wave the rf cycles at the output of the modulated amplifier, Fig. 4-13, are at maximum. When the modulating wave goes into its negative crest, the rf amplitude is a minimum. The rate at which the rf wave varies in amplitude depends upon the frequency of the modulating wave. The higher the frequency of the modulating wave is, the faster is the rate at which the rf wave varies in amplitude.

A basic am system is shown in Fig. 4-13. A constant-amplitude radio-frequency wave is applied to one of the inputs of the modulated amplifier. The audio signal through audio amplifier and modulator is also applied to the modulated amplifier. Actually the modulated wave causes the gain of the modulated amplifier to vary at an audio-frequency rate. Consequently the degree to which the radio-frequency wave is amplified depends on the modulating wave. Therefore the output of the modulated amplifier is a series of radio-frequency waves the amplitude of which varies and follows the changes in the modulating wave. Actually it is a mixing process

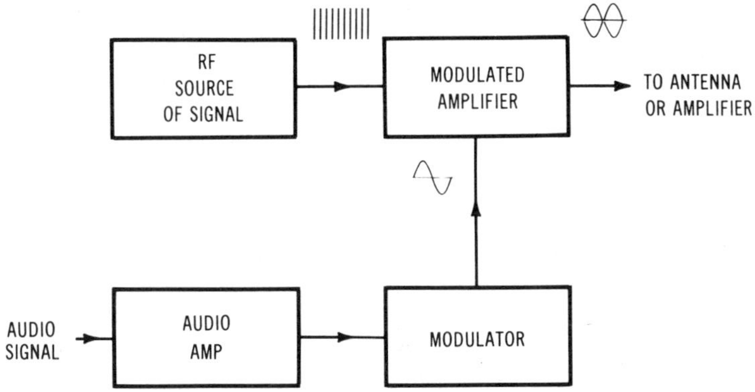

Fig. 4-13. Basic am system.

that takes place and the radio-frequency envelope at the output is the algebraic addition of a carrier of constant amplitude and two sidebands displaced on each side of the carrier frequency by an amount equal to the frequency of the modulating wave.

If a 30-megahertz wave is modulated by a 1000-hertz note there will be three waves that comprise the modulation envelope. These are the 30-megahertz carrier, the upper sideband at 30 megahertz plus 1000 hertz, and the lower sideband at 30 megahertz minus 1000 hertz. The extent to which the modulation envelope is modulated (depth of modulation) depends on the strength of the modulating wave.

Amplitude modulation is measured in percentage. When the amplitude of the rf voltage swings to twice the unmodulated value at the output of the modulated amplifier during a positive modulating crest and falls to zero on the negative crest, the carrier is said to be 100-percent modulated. This is shown in Figs. 4-14A

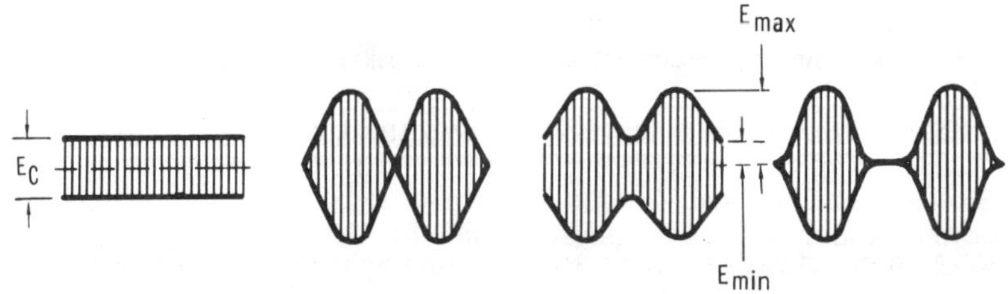

(A) Carrier—no modulation. (B) 100% modulation. (C) Less than 100%. (D) Over 100%.

Fig. 4-14. Modulation percentages.

and B. Full modulation of the carrier is important in obtaining a good transmission range and maximum demodulated audio at the receiver. For a lower-amplitude audio variation the modulation envelope is not filled out and appears as shown in example C. In actual practice 100-percent modulation is only obtained during the voice-frequency peaks. At other times the modulation percentage is less than this complete, or 100% value. In an amplitude-modulation system it is desirable to maintain as high an average modulation percentage as possible.

However, to obtain a linear and undistorted modulation characteristic the transmitter must not be overmodulated as shown in the waveform of example D in Fig. 4-14. Note that on the negative crest the wave flattens off to zero and remains so for a period of time. Thus the negative and positive crests of the envelope become asymmetrical and distortion is present. In fact, to prevent distortion the amplitude of the positive crest of the modulated rf sine wave must not be more than twice the amplitude of the unmodulated value. If it exceeds this value overmodulation is present. A formula often used to calculate modulation percentage is:

$$\text{Percent modulation} = \frac{E_{max} - E_{min}}{2E_c} \times 100$$

where,

E_{max} is the maximum amplitude of the modulation envelope,
E_{min} is the minimum amplitude of the modulation envelope,
E_c is the amplitude of the unmodulated carrier.

Since a modulating signal is seldom a pure sine wave but rather a combination of various frequencies of differing amplitudes it is often useful to determine positive and negative modulation percentages separately. Positive modulation refers to the increase in rf cycle amplitude above the unmodulated value; the negative modulation, to the drop in rf cycle amplitude below the unmodulated value. The positive and negative modulation percentage equations are:

$$\text{percent mod. (positive)} = \frac{E_c - E_{min}}{E_c} \times 100$$

$$\text{percent mod. (negative)} = \frac{E_{max} - E_c}{E_c} \times 100$$

Most modulating waves are not of the constant amplitude that maintains 100% modulation continuously. The average during voice modulation is considerably lower, and only on an occasional voice peak does the modulation reach 100%. Nevertheless an am system must be adjusted so that there will be no overmodulation on voice peaks. Since the voice peaks are only occasional, it is apparent that the average modulation must be lower than 100%. At the same time, for clarity of transmission and range, the average level of modulation should be kept as high as possible. Consequently two-way radio equipment usually includes special circuits that compress the voice peaks so they do not cause overmodulation and, at the same time, boost up the lower amplitude passages so that they are not too weak.

4-5-1. Carrier and Sideband Power

As mentioned previously, the modulation envelope is a composite waveform. During the modulation each rf cycle differs from the preceding and following ones. They are no longer pure sine waves. This is another way of stating that the resultant signal is no longer composed of a single wave (rf carrier) but, in the case of modulation by a single sine-wave tone, a carrier plus upper and lower sidebands. Actually the carrier, despite modulation, remains of constant magnitude and frequency. The only difference has been the addition of two sideband components.

The foregoing relationships indicate that the carrier power is unchanged with modulation. Any net addition in power is contributed by the two sidebands.

The instantaneous power of the modulation envelope varies continuously during the modulating cycle. At the peak of the envelope, assuming 100% modulation, the rf voltages are twice as high as during the unmodulated period. This doubling of the voltage represents a fourfold increase in power because the rise in power is equal to the square of the voltage. At the crest of the negative modulation, the resultant envelope falls to zero and the instantaneous power is zero.

When the power is averaged over the complete sine-wave cycle for 100% modulation, it is 50% higher than the unmodulated carrier power. One hundred percent modulation occurs whenever the amplitude of each sideband is one-half the amplitude of the

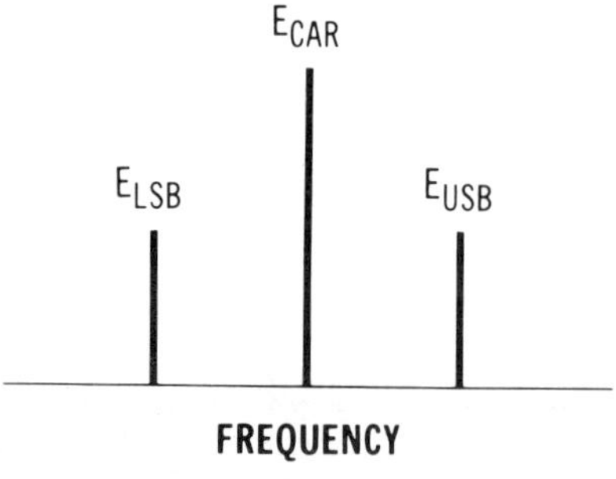

(A) 100% modulation.

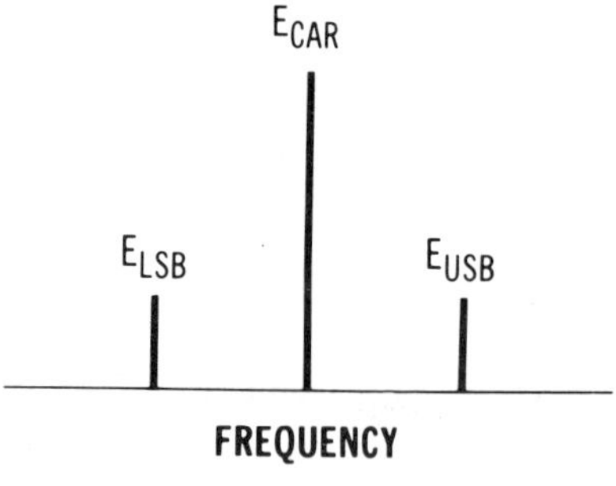

(B) 50% modulation.

Fig. 4-15. Carrier and sideband voltage levels.

unmodulated carrier as shown in Fig. 4-15. This means that each sideband contains one-fourth as much power as the carrier. Since there are two sidebands, they contain one-half the power of the carrier.

The modulating information to be transmitted is contained in the sidebands, each of which contain the same information. The carrier itself contains none of the information to be transmitted.

Let us now reconsider the power relationship in terms of the total power output. The total output (envelope power) is the sum of the power contained in the carrier and sidebands or:

$$P_o = P_{car} + P_{usb} + P_{lsb}$$

where,

P_o is the total output (envelope) power,
P_{car} is the power carrier,
P_{usb} is the upper sideband power,
P_{lsb} is the lower sideband power.

Assuming carrier power as unity:

$$P_o = 1 + \frac{1}{4} + \frac{1}{4} = \frac{3}{2}$$

Since the sidebands contain only one-half the power of the carrier, they represent only one-third of the total transmitted power. The carrier represents the remaining two-thirds.

$$\frac{P_{car}}{P_o} = \frac{1}{3/2} = \frac{2}{3}$$

$$P_{car} = \frac{2}{3} P_o$$

and

$$\frac{P_{sb}}{P_o} = \frac{1/2}{3/2} = \frac{1}{3}$$

$$P_{sb} = \frac{1}{3} P_o$$

It is apparent that amplitude modulation is a rather wasteful method of conveying information via a radio wave. All the information to be sent can be packed into one of the sidebands, representing one-sixth of the total transmitted power. The relationship is even more exaggerated for modulation of less than 100%—the prevailing condition in the usual voice communications. An example of 50% modulation is shown in Fig. 4-15B.

For 50% modulation the sideband amplitude is only one-quarter the carrier amplitude. Therefore each sideband contains but one-sixteenth the power in the carrier. Hence the power in both sidebands is only one-eighth the carrier power, a very small fraction of the total transmitted power. This relationship proves why a high average-modulation percentage is important in maintaining an effective transmission range and strong audio-frequency components at the receiver.

Efficiency factor is another consideration. Its value is:

$$\text{Efficiency factor} = \frac{\text{rf power output}}{\text{dc input power}}$$

$$= \frac{P_o}{I_C V_C}$$

The dc power input remains constant, whether the carrier is modulated or unmodulated. Likewise the carrier power output is a constant, and the rf carrier output is supplied by conversion from the power supply of the modulated amplifier.

However, the rf power output increases with modulation. What is the source of this additional power? You learned that the additional power with modulation is contained in the two sidebands. The sidebands are due to introduction of the modulating wave into the supply line via the modulator. Consequently, the rf sideband power is contributed by the audio-output system or modulator.

4-5-2. Modulator Power

With 100% sine-wave modulation, the rf power output of the transmitter is increased by 50%. To make this much additional power available, the audio power output of the modulator must be one-half the dc power input to the modulated amplifier. However, since only occasional voice peaks extend to 100% modulation, the average power output of the modulator can be less than 50% of the dc power input.

For 100% modulation, the audio peak voltage must be approximately equal to the supply voltage. Only under this condition will the voltage of the modulated amplifier swing to twice the supply voltage, a necessity for obtaining 100% modulation at the crest of the positive audio cycle. The negative sweep of the same modulating wave reduces the supply voltage to zero to produce the 100% modulated envelope trough.

To produce the desired peak audio voltage at the required power, the audio must be developed across a specific impedance. The impedance seen by the modulator output must equal the dc input resistance to the modulated amplifier. The dc input resistance is related to the dc component of supply current and voltage, or:

$$R_{in} = \frac{V_c}{I_c}$$

The modulation transformer or the output system of the modulator must be designed to match this impedance, just as an audio output transformer must have the proper turns ratio to match the impedance of a speaker voice coil.

4-6. TYPICAL AM CIRCUITS

A bipolar transistor can be amplitude-modulated by varying the supply voltage to one of its elements—base, emitter, or collector. However, in modern two-way radio equipment the dominant method is to use collector modulation (Fig. 4-16). The dc collector voltage is made to vary with the modulating signal by inserting the secondary of the modulation transformer in the supply voltage path. The collector voltage of the amplifier is increased or decreased by the audio modulating signal.

The unmodulated radio-frequency carrier is fed to the base of the modulated amplifier. The mixing of the two signals in the

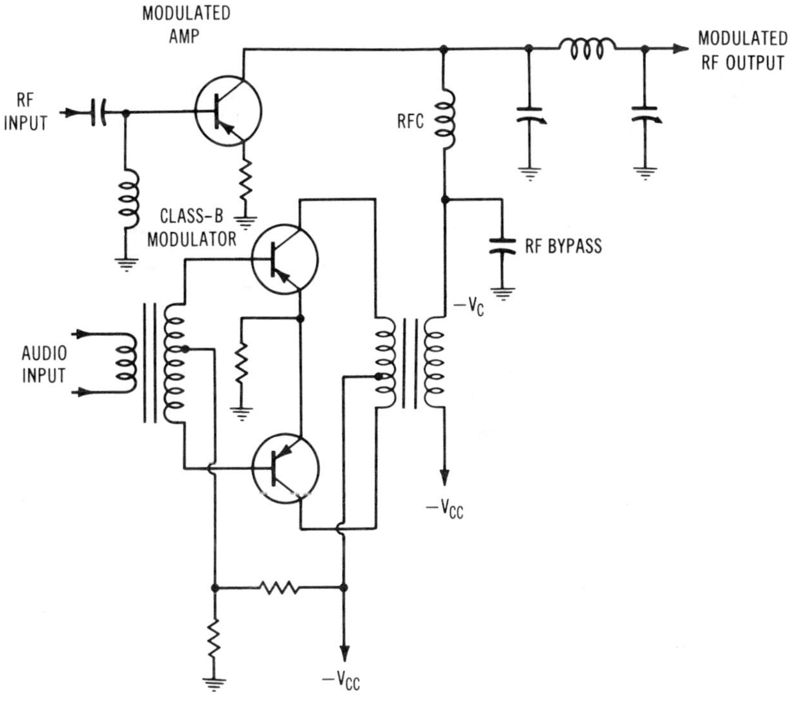

Fig. 4-16. Collector-modulation system.

modulation process results in the formation of a modulated rf signal across the collector tank circuit. In the modulated rf stage the tank circuit voltage varies linearly with the collector supply voltage; thus doubling the supply voltage doubles the tank rf voltage over a suitable operating range. If the modulated amplifier is operated from a 12-volt V_{CC} supply, the instantaneous collector supply voltage can be made to vary above and below 12 volts.

If the modulating wave is a pure sine wave this variation can be symmetrical. In bipolar transistor stages it is quite difficult to obtain absolute and symmetrical full 100% modulation. However, satisfactory and reasonably good amplitude modulation is obtained with a single stage.

The rate of variation of the supply voltage depends on the frequency of the modulating wave; the magnitude of the change depends on the amplitude of that wave. This change in the collector supply voltage becomes a replica of the modulating wave. If the variations on each side of the zero axis of the modulating wave are identical, the average dc collector voltage remains at 12 volts. Likewise in a perfect bipolar modulation system the dc collector current would also remain constant. This ideal operating characteristic is not always feasible in a simple bipolar modulation system.

The influence that a change in the collector voltage has on the operation of the modulated amplifier is shown in the waveform of Fig. 4-17. As the collector voltage changes in a class-C amplifier, so does the peak collector current drawn for each individual radio-frequency cycle. In fact, the peaks of the latter are a copy of the modulating wave as shown in the second waveform of Fig. 4-17.

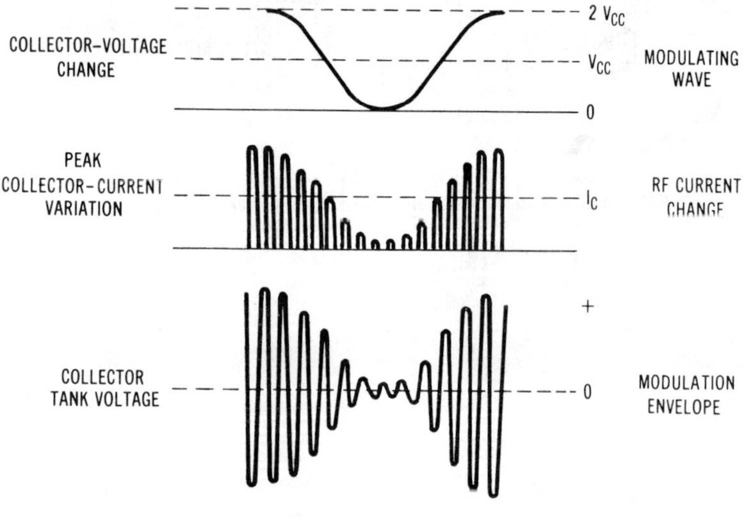

Fig. 4-17. Drain-modulation waveforms—100% modulation.

The peak collector current, in turn, determines the magnitude of the rf voltage developed across the collector tank circuit. The higher the maximum collector current is, the higher will be the amplitude of the corresponding rf voltage. It follows then that the rf collector voltage waveform is also a copy of the modulating wave. In this manner the standard am envelope is formed.

To develop the desired change in the envelope with full modulation capability, the collector voltage must be made to swing between twice the collector supply voltage and zero. The impedance seen by the modulator output equals the dc resistance of the modulated amplifier input. This dc resistance is quite low because of the low collector voltage and high collector current;

$$R_{in} = \frac{V_{cc}}{I_c}$$

It is this impedance that the output of the modulator (modulation transformer) must match. In so doing the proper load is reflected to the output of the modulator and it is possible to obtain a high level of modulation with good efficiency.

A more common system of amplitude modulation is that shown in Fig. 4-18. Linear and full 100% modulation is obtained. In this plan both the final and the driver are modulated. This arrangement is of particular help because the driver is then able to supply the necessary additional base drive to the final that can accommodate

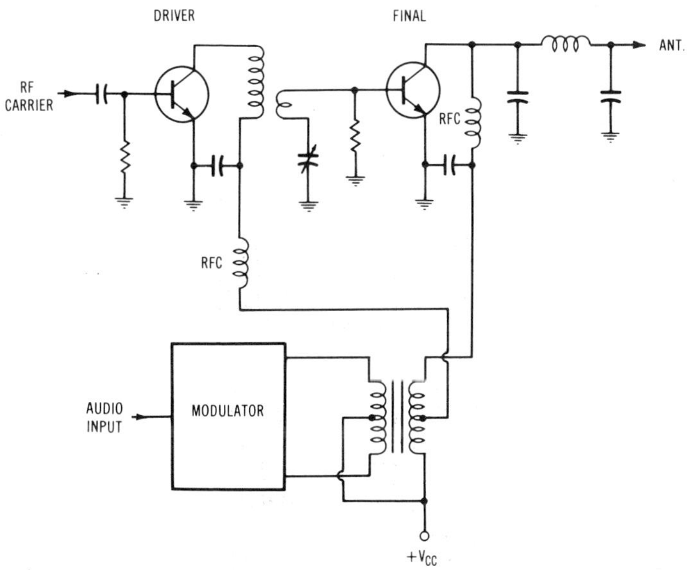

Fig. 4-18. Modulation of final and rf driver.

the 100% modulation peaks. The proper level of modulating wave to the driver to accomplish full modulation capability is set by the tapped position on the secondary of the modulation transformer. Note that the final rf amplifier obtains the full secondary voltage.

4-7. SINGLE-SIDEBAND MODULATION

The advantages of single-sideband modulation over standard amplitude modulation are narrower bandwidth, better signal-to-noise ratio, better signal-to-interference ratio, and more efficient use of available power. In standard amplitude modulation, a carrier and two sidebands are transmitted. As previously covered the carrier remains fixed in amplitude and carries none of the modulation, while the two sidebands carry identical modulation information. Thus it is possible to abandon the carrier and one sideband because all the desired data can be conveyed by the single remaining sideband.

Three specific processes are involved in the generation of a single-sideband signal. In the actual modulation process, the carrier itself must be suppressed, the undesired sideband removed, and the desired sideband converted in frequency and then amplified to a desired power level.

A simplified block diagram of a sideband transmitter is shown in Fig. 4-19. A crystal-controlled carrier generator is used to generate the basic carrier. This carrier is passed to a balanced modulator to which the audio signal is also applied. In a balanced circuit arrangement, the carrier is cancelled and two radio-frequency

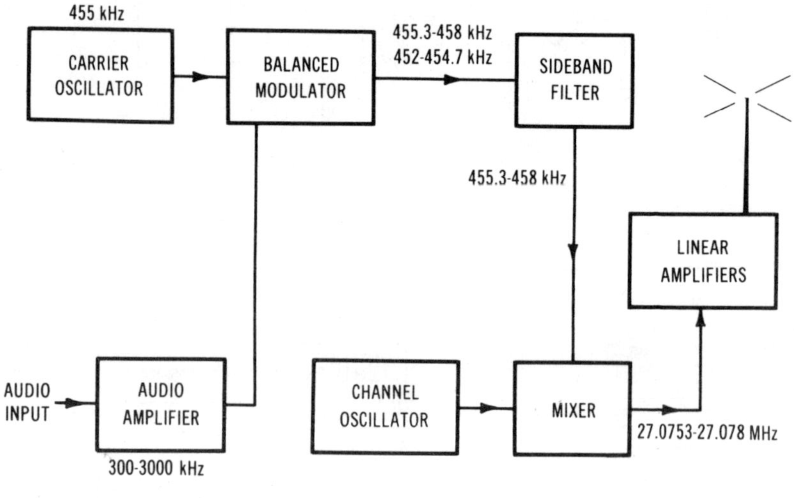

Fig. 4-19. Basic single-sideband transmitter.

sidebands are generated. If the source of carrier signal is a 455-kHz component, the sideband spectra for a 300- to 3000-hertz audio range would be as shown at the output of the balanced modulator. The two sideband spectra fall above and below the frequency of the absent carrier.

The second process in the generation of a single-sideband signal is removal of one sideband. It does not matter which sideband is removed because both carry like information. The choice depends on the basic system design. Sometimes a switching arrangement permits the generation of either one of the sidebands, upper or lower. In the filter method of sideband removal a carefully designed crystal or ceramic filter highly attenuates the sideband that is not to be transmitted (undesired sideband) but passes the desired sideband with a minimum of attenuation. Such filters have a highly selective response and must have a high order of stability.

The sideband signal must now be converted to the transmit frequency band. Inasmuch as it carries modulation, it must be converted in frequency and amplified in strength by linear circuits (harmonic multipliers and class-C amplifiers are not used). So-called linear mixers are used. Channel oscillator frequency component on mixing with the sideband signal produces an output frequency on the desired transmit band. In fact, the output circuit can be tuned to either a sum or difference frequency depending on whether the desired transmit band is above or below the sideband frequency. In the simplified example of Fig. 4-19, a high-frequency sideband is to be generated on CB channel 10 (27.075 MHz). The sideband spectra entering the mixer from the sideband filter has a frequency range between 455.3 and 458 kHz. When mixed with the channel oscillator frequency of 25.620 MHz, the output-frequency spectra of the mixer, for a 300- to 3000-hertz audio range would be 27.0753 to 27.078 MHz. Note that this is a range of frequencies between 300 and 3000 hertz above the channel-10 carrier frequency of 27.075 MHz.

Succeeding stages build up the weak upper-sideband signal present at the output of the mixer to the 12-watt pep limit set for sideband operation on the Citizens Band. This discussion simplifies the actual sideband transmission process. An actual single-sideband CB radio incorporates both transmit and receive functions in the sideband generation and demodulation process.

4-7-1. Waveform Relationships

A clear understanding of the makeup of various am waves can be obtained by considering Fig. 4-20. Such knowledge will permit a better understanding of the sideband-modulation envelope that is important when checking and adjusting sideband equipment. As you will recall, the standard am wave, when modulated by a single sine-wave tone, is a composite of three rf waves, a constant-amplitude carrier, and two side-frequency components which are above and below the carrier frequency by the frequency of the modulating wave. At 100% modulation, the two side frequencies are one-half the amplitude of the carrier.

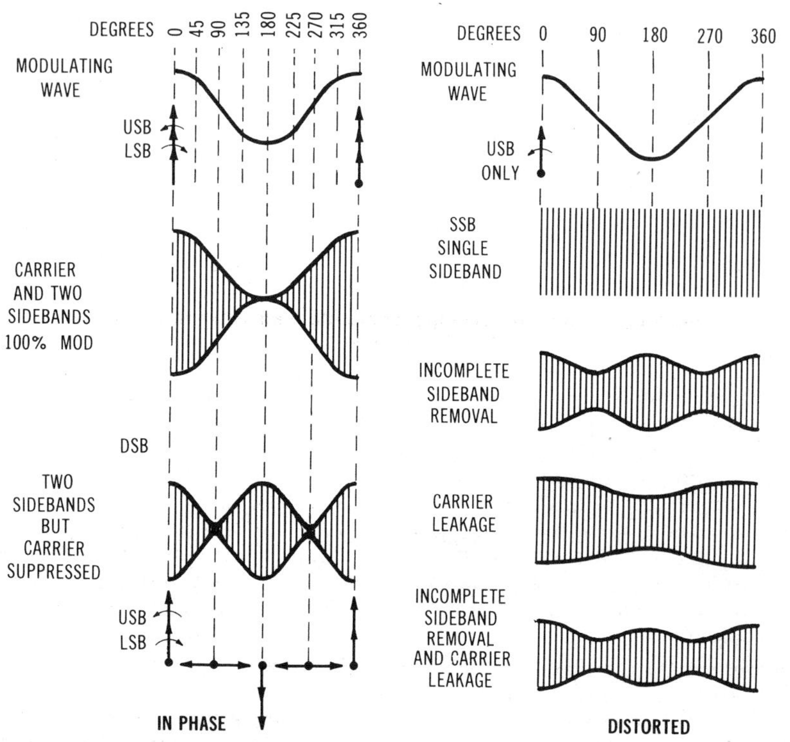

(A) In phase.

(B) Distorted.

Fig. 4-20. Typical sideband waveforms.

In the vector comparison shown in Fig. 4-20, all three components are in phase at the crest of the modulation envelope, and the peak amplitude builds up to twice the unmodulated signal value. Since the two side frequencies are not the same and also differ from the carrier frequency, the phase relationship among all three components changes throughout the modulating cycle. Note in Fig. 4-21 that the carrier is established as a zero-reference phase. On the positive crest of the modulating wave, the carrier and both sidebands momentarily are all in phase. Hence the resultant modulation-envelope signal rises to twice the unmodulated-carrier value. When the modulating sine wave is passing through zero, the two side frequencies are related 90 degrees to the carrier and 180 degrees to each other. Therefore, the side frequencies cancel and the net amplitude of the amplitude-modulated wave is the same value as its carrier.

On the negative peak of the modulating wave, the two side frequencies are again in phase with each other but 180° out of

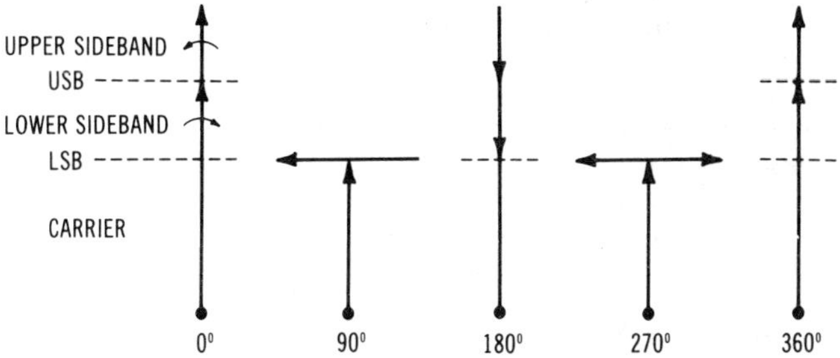

Fig. 4-21. Vector relationships for standard am wave through one modulating-wave cycle.

phase with the carrier. Now there is a complete cancellation and the net envelope output drops to the zero point of its cycle.

In summary, it is the changing phase relationship among the three components that causes the amplitude variations in their resultant amplitude-modulated envelope. Of course, the intermediate levels of the modulation envelope also depend on the instantaneous phase relationship among the three waves. Actually the side frequencies when following the amplitude changes of the modulating wave can be considered to rotate vectorially around the zero-reference phase set by the carrier. They rotate in opposite directions, one being above and one being below the carrier frequency.

Consider next the type of resultant modulation envelope obtained when the carrier is suppressed and the two side frequencies remain (Fig. 4-20A). As shown in Fig. 4-22A, the side frequency

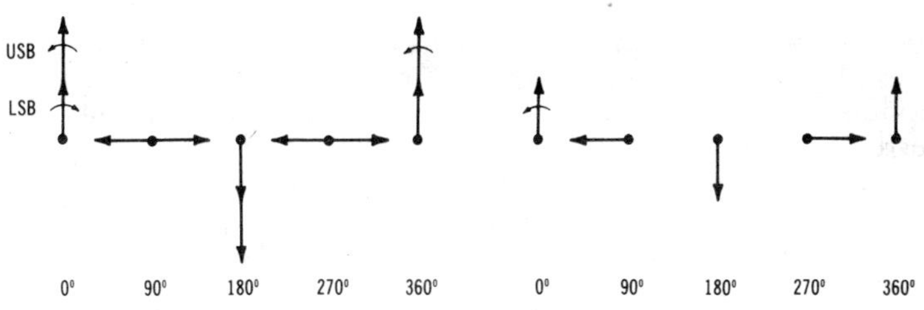

(A) Double sideband. (B) Single sideband.

Fig. 4-22. Vector relationships for dsb and ssb components.

components shift in phase relative to each other. Even though the carrier magnitude is zero, the side frequencies can be considered to revolve in phase with respect to the zero carrier-reference phase. As the modulating wave goes through its cycle, the side frequencies at times becomes out of phase and at other times exactly in phase during the modulating wave period. Actually the modulation envelope goes through two cycles during the one cycle of the modulating wave. The peak amplitude of the envelope rises to twice the individual sideband magnitude. It falls to zero when the two vectors are exactly out of phase.

The magnitude and vector relationships for a single-sideband signal are shown in Figs. 4-20B and 4-22B. Now there is only one rotating vector. In fact, only one rf component is generated as the modulating wave goes through its cycle. Hence a constant-amplitude resultant is produced.

The sideband modulation envelopes are important in assessing the performance of a sideband system. The nature of any distortion on these envelopes points up the various characteristic defects and maladjustments. In fact, the oscilloscope is an important tool for adjusting sideband systems. When the sideband system is functioning properly and is modulated by a single-tone frequency, the output rf wave is of constant amplitude as shown in Fig. 4-20B. If the undesired sideband is not completely removed there will be a double-frequency ripple on the sideband output waveform as shown. Likewise if there is carrier leakage into the output there will also be a frequency variation on the envelope that corresponds to the frequency of the modulating wave. Of course, the combination of the two defects produces the resultant wave shown at the bottom of Fig. 4-20B.

4-7-2. Power Relationships

In addition to the reduced bandwidth compared to that of a double-sideband system, a single-sideband system also conserves power. A mathematical example will prove this saving. A standard am transmitter with a rated carrier of 4 watts radiates 1 watt per sideband with single tone 100% modulation. The carrier power is 4 watts and the total sideband power is 2 watts. Peak power is, of course, 4 times the carrier value under this circumstance and reaches 16 watts, because the rf output voltage is twice the unmodulated carrier voltage during 100% modulation peaks. Nevertheless, the sideband power output is only 1 watt.

In order to generate a 12-watt sideband signal using the single-sideband modulation method, the peak envelope power is 12 watts because all of the energy is contained in the transmitted sideband. Furthermore only half the frequency spectra is required as compared to a double-sideband signal. Consequently two sideband signals can be made to fit in the same frequency spectrum as a single, double-sideband signal. Thus there is not only conservation of power but also conservation of frequency spectra.

4-7-3. CB Sideband Methods

Citizens-Band single sideband transceivers include both sideband and standard am capabilities. Most of the circuits are used interchangeably on am and sideband. Others must be switched in and out as a changeover is made between single sideband and standard amplitude modulation.

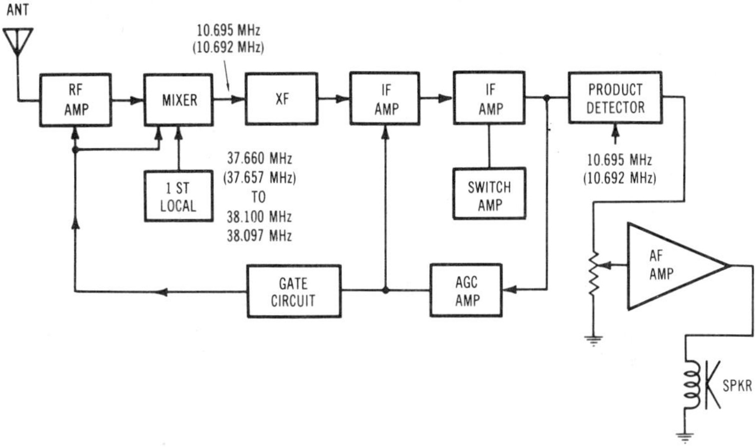

(A) An ssb receiver.

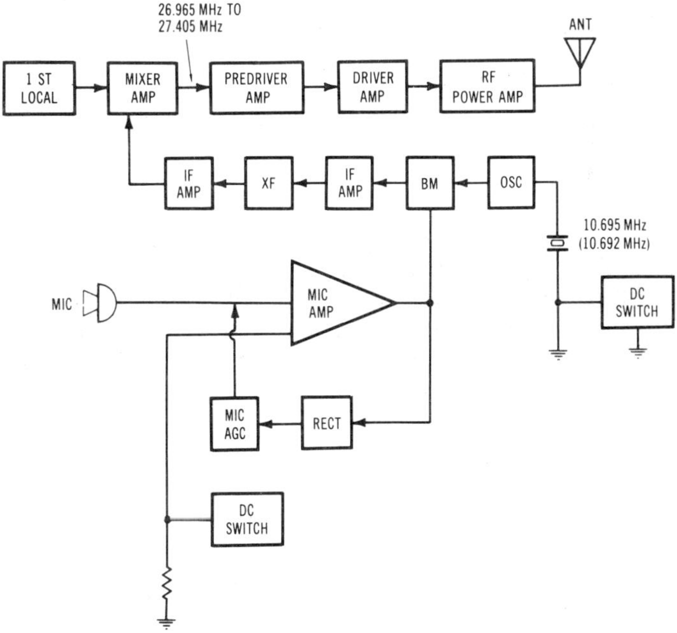

(B) An ssb transmitter.

Fig. 4-23. Block diagrams of ssb receiver and transmitter.

A functional block diagram of a CB receiver and transmitter is shown in Fig. 4-23. The receive signal is picked up by the antenna and is applied to the mixer through an rf amplifier stage. A local-oscillator signal is derived from a phase-locked loop circuit. The frequency range of the local oscillator injection signal varies between the limits indicated depending on whether upper- or lower-sideband reception is desired. The lower-sideband frequencies are indicated in parentheses. The output of the mixer is applied to the sideband filter which has a 3-kHz bandwidth. The carrier frequency is one of the two frequencies indicated depending on whether upper- or lower-sideband emission is being received.

A multistage if amplifier builds up the signal and applies it to the product detector. The demodulating frequency for the product detector is also derived from a phase-locked loop circuit. Volume control and an integrated-circuit audio amplifier follow. One of the if stages as well as the rf amplifier and mixer are under the control of an agc circuit system.

In the transmission process, the audio signal is derived from the microphone amplifier (integrated circuit) and is applied to the balanced modulator. The level-control circuit is associated with the audio section to maintain a high average modulation level without exceeding the peak level that will produce audio distortion or flat-topping in the modulation process. Appropriate switching is also necessary depending on whether the audio system is used for sideband or amplitude modulation.

The carrier signal is derived from an appropriate crystal oscillator which is a part of the pll circuit. The frequency of the crystal oscillator is changed by a switch depending on the choice of upper-sideband or lower-sideband emission.

The output of the if amplifier is applied to the sideband filter and on to a followup if amplifier that builds up the level of the sideband signal. The function of the balanced modulator is to produce a double-sideband signal without carrier. The followup filter then removes the undesired sideband. Therefore a single-sideband signal alone is applied to the mixer amplifier. The injection frequency for the mixer amplifier is derived from the pll circuit.

The output of the mixer is tuned to the difference frequency which falls in the Citizens-Band frequency range. A series of linear rf amplifier stages build up the sideband signal to the 12-watt pep level before application to the antenna system.

A functional plan of the pll section of the transceiver is shown in Fig. 4-24. The basic theory of the phase-locked loop was discussed previously. The programmable divider and associated circuits are shown at the bottom; the voltage-controlled oscillator and associated circuits, at the top center.

Let us begin with the phase detector (P/D) located at the immediate right of the programmable divider. In this circuit, the output of the programmable divider is compared with a reference signal derived from oscillator No. 2. Oscillator No. 2, at the lower

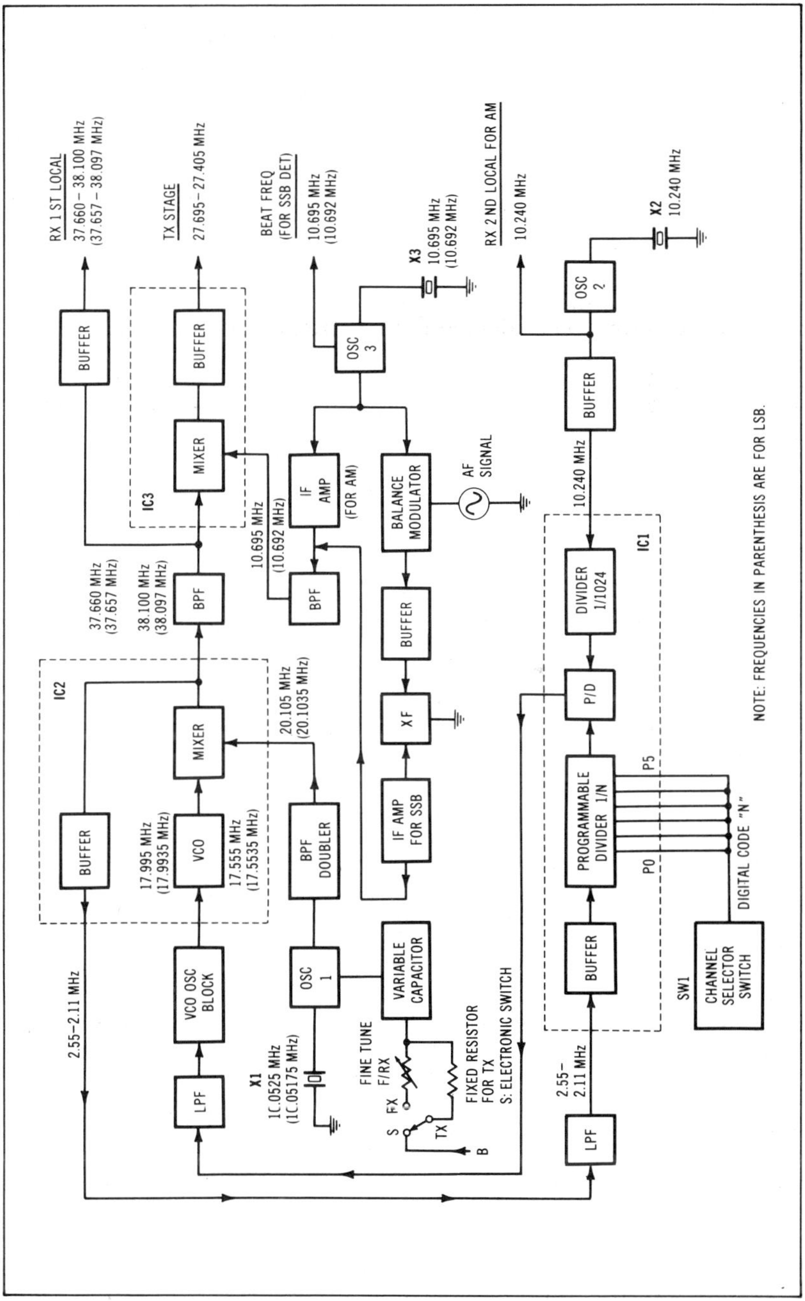

Fig. 4-24. Block diagram of GE Model 3-5825A pll section.

right, operates on 10.240 MHz. The signal is applied through a buffer to a divider chain that has a ratio of 1/1024. Therefore the frequency applied to the phase detector is 10 kHz. When the CB transceiver is on correct frequency for a given channel, the output of the programmable divider is also 10 kHz.

If there is a frequency differential, the output of the phase detector is applied through a low-pass filter to the frequency control circuit of the voltage-controlled oscillator (vco). The purpose of the phase detector is to keep the vco on correct frequency.

The comparison signal is derived from a mixer at the output of the vco and is applied through a buffer and low-pass filter to the input of the programmable divider. The frequency range of this feedback signal, depending on the channel received, falls between 2.11 and 2.55 MHz. This is the signal frequency applied to the programmable divider. However, the channel selector switch, along with its digital code and LED display, sets the proper divider ratio for a given channel. As a result, the output of the programmable divider, when everything is functioning properly, is 10 kHz. When switching channels, this divider will also change. Meanwhile an appropriate change is made in the setting of the vco because of the correcting voltage supplied by the phase detector.

Let us next consider the four signals that are supplied by the pll circuit. First a transmit frequency must be developed for both am and sideband operation. Then a local oscillator signal must be generated to permit both sideband and am mixer activity. A third signal must be developed for use in the sideband detector. The fourth, and final, signal supplied by the pll circuit is the second local-oscillator frequency for am demodulation. This transceiver uses single conversion for sideband reception and double conversion for amplitude modulation reception.

Oscillator No. 1 at the middle left in the diagram generates a 10-MHz component that is applied to a doubler and bandpass filter, and then to a mixer that is a part of the vco integrated circuit. Since the vco operates in the 17.5- to 18-MHz range, the sum frequency at the output of the mixer is in the 37- to 38-MHz range as shown in Fig. 4-24. It is applied to another mixer that is a part of integrated circuit IC3.

The third crystal oscillator, shown at the right center, generates a frequency of either 10.692 or 10.695 MHz depending on the selected sideband. For sideband modulation it is applied to the balanced modulator and associated sideband circuits as described previously for Fig. 4-23. This is noted again at the very center of Fig. 4-24. The output of the if amplifier passes through a bandpass filter to the mixer that is a part of integrated circuit IC3. The output of this mixer is tuned to the difference frequency and produces a frequency range between 27.965 and 27.045 MHz, which is the Citizens-Band radio spectra.

The output frequency of the vco section is that frequency required by the first local-oscillator injection signal. This is shown at the very top right in Fig. 4-23. The output oscillator (No. 3) also generates the frequency component that is applied to the product detector for single-sideband demodulation. These three signals shown at the top right are needed for operation for both the transmitter and the receiver for single-sideband operation.

In am transmission, the same frequency range is required for the transmit mode. Likewise the same frequency range is required for the first local-oscillator injection. However, to obtain a frequency component for the double-conversion process, still another frequency must be generated. This is supplied by oscillator No. 2 at the bottom right and has a frequency of 10.24 MHz. Note at the right center of Fig. 4-23 that for am transmission the output of oscillator No. 3 passes through an if amplifier and goes directly to the bandpass filter and on to the mixer associated with integrated circuit IC3. This generates the basic transmitter frequency range for am transmission.

4-8. SIDEBAND CIRCUITS

The first step in the generation of a single-sideband signal is the formation of sidebands along with the removal of the carrier. These two objectives can be accomplished simultaneously. Diodes, transistors, and integrated circuits are used in modern balanced-modulator circuits. Considerable help in understanding the operation of a balanced modulator can be gained by first reviewing the operation of a diode as a modulator. The diode is used frequently as an am detector or demodulator. In this application, the modulated signal is applied to the diode input. By rectifying the radio frequency cycles the lower-frequency modulating wave can be recovered in its output circuit.

A diode can be made to operate in the opposite fashion; that is, a weak audio signal and a stronger radio-frequency signal can be supplied to its input. The same nonlinear operation or rectifying action causes sum- and difference-frequency components to be developed in the output circuit. These are the familiar upper- and lower-sideband frequencies. The waveforms shown in Fig. 4-25 demonstrate the mixing or modulation process. When the radio-frequency signal alone is applied to the diode modulator, there is diode current during the positive alternations of the input cycle. Hence, positive bursts of current develop through the load. Inasmuch as the load is a resonant circuit a sine wave is reconstructed in the output. This action is similar to the influence of resonant circuits on the output of class-B or class-C rf amplifiers.

When an audio wave is applied to the input along with the radio-frequency wave there is an algebraic summation. The peak amplitudes of the diode current bursts now depend on the net diode anode voltage at the crest of each radio-frequency cycle. This peak voltage varies up and down with the modulating wave. As a result, the peak diode current varies correspondingly. This peak current diode change is comparable to the peak current variation of a modulated class-C amplifier. Again the resonant circuit, because of its energy storing ability, reconstructs the negative alternation of the

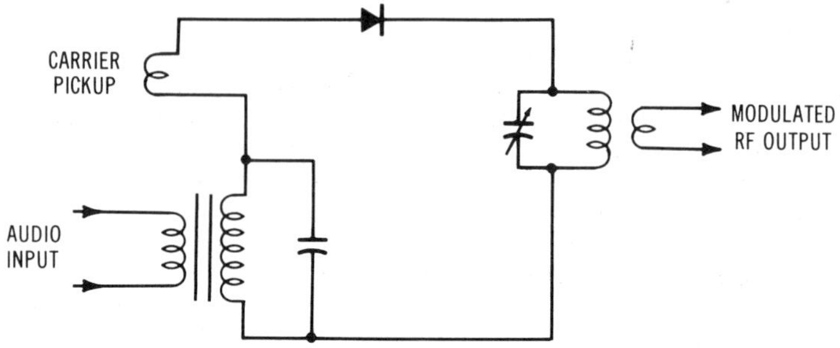

(A) Modulator circuit.

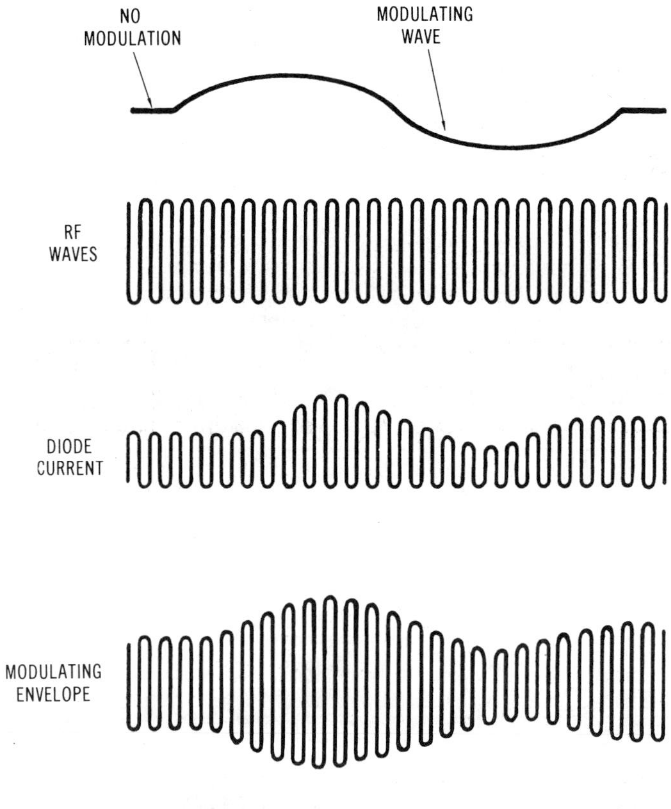

(B) Diode-modulator waveforms.

Fig. 4-25. Diode-modulator circuit operation.

output variation, forming the familiar amplitude-modulation envelope.

The input wave is the simple combining of two separate signals—high-frequency radio wave and low-frequency audio wave. The output wave results from nonlinear mixing or heterodyning and is composed of three radio-frequency components—the carrier plus two side frequencies. The audio wave is filtered out by the low impedance of the output circuit.

Diode modulation has poor efficiency. However, at a low signal level, combinations of diodes in a number of circuit arrangements can serve admirably as stable balanced modulators. Diodes have a low order of modulation linearity. Nevertheless by using a modulating wave of much lower amplitude than the radio-frequency wave (low percentage modulation), a linear output can be obtained (modulation envelope variations correspond faithfully to the modulating wave). By arranging combinations of diode modulators it is possible to generate the sideband frequencies at the same time the radio-frequency carrier itself is suppressed.

Two basic and common diode balanced modulators are the bridge and ring circuits shown in Figs. 4-26 and 4-27, respectively. Since radio-frequency and audio waves are supplied to the various diodes, modulation takes place. Thus carrier and sideband currents are produced. However, the carrier current components cancel with relation to the output system.

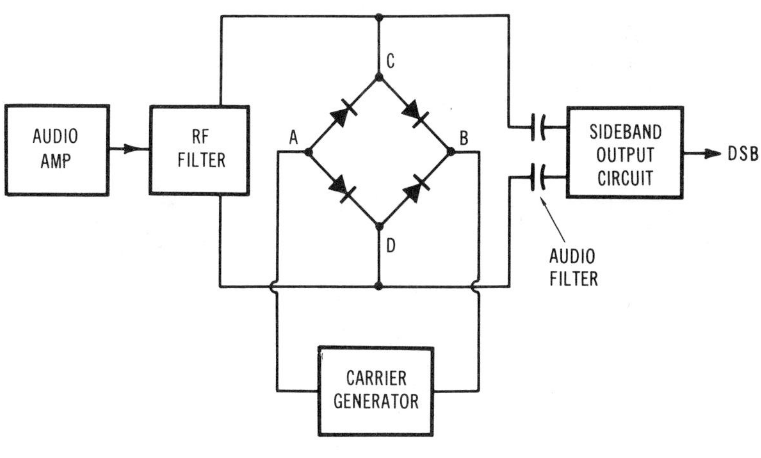

Fig. 4-26. Bridge-balanced-modulator operation.

Let us consider the operation of the bridge circuit in Fig. 4-26 when only the radio-frequency carrier is applied. When the radio-frequency cycle swings negative, point A on the bridge is then negative with respect to point B. Consequently four diodes are back-biased and there is no current. If the four diodes have the

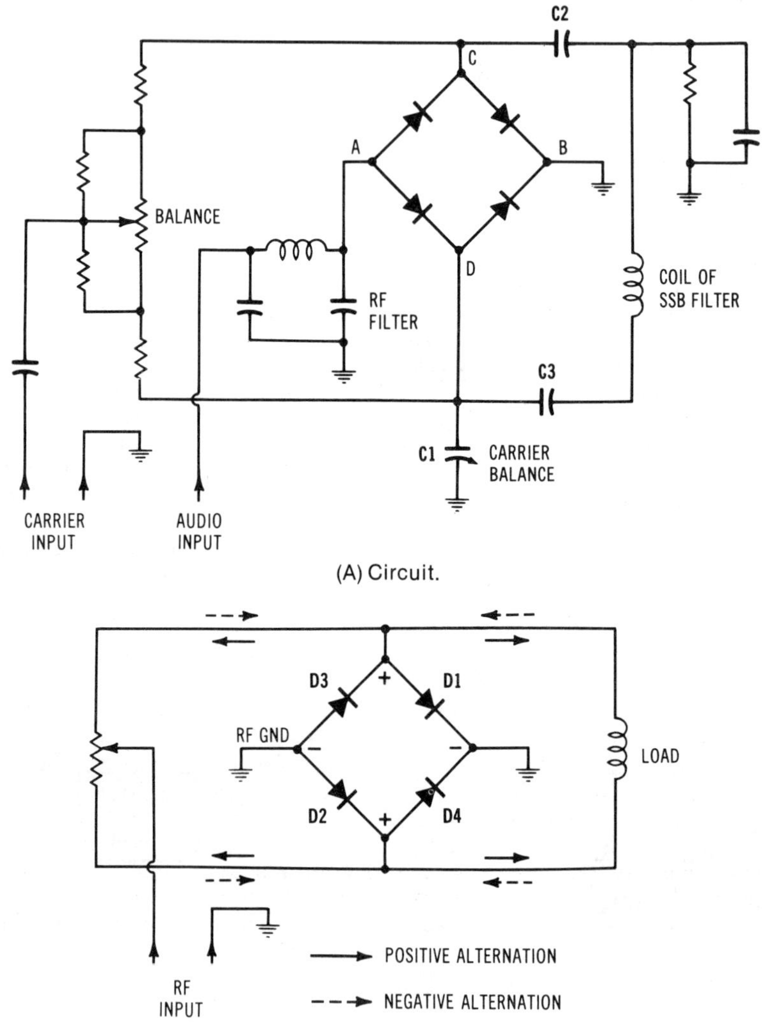

(A) Circuit.

(B) Simplified circuit.

Fig. 4-27. Ring-balanced modulator operation.

same reverse resistance, the bridge will be balanced and there will be no difference of potential between points C and D. Consequently there will be no current in the output circuit.

When the radio-frequency wave swings positive, all four diodes become forward biased. Hence, their resistance is low and there will be current. However, if they all have the same resistance, the bridge will remain balanced. Therefore no difference of potential develops between points C and D, and there will be no carrier current in the output circuit. Thus, it is apparent that neither the positive nor negative alternations of the radio-frequency wave cause an output voltage, and so the carrier is suppressed. However, the radio-frequency wave does appear across each of the four diodes and rectifier action takes place with the diode currents being switched on and off by the alternations of the radio-frequency wave.

Diode activities differ when a modulating wave is applied simultaneously with the carrier. Inasmuch as the audio wave is made substantially lower in amplitude than the radio-frequency wave, the diode conducts only during the positive alternation of the radio-frequency carrier. The amount of current present, as in the simple diode modulator, depends on the instantaneous sum of the combined radio-frequency and audio-frequency waves. As a result, the peak radio-frequency current will vary with the modulating wave as in any form of amplitude modulation.

The influence that the audio voltage has on the bridge is important. This voltage is applied between points C and D. As a result, the bridge is unbalanced—the degree of unbalance depending on the magnitude of the audio signal. The imbalance allows side-frequency currents in the output circuit. It is apparent then that the bridge, although balanced at the carrier frequency, will be unbalanced at off-carrier frequencies. As a result, the sideband frequencies develop in the output circuit. Audio-frequency components are blocked from the output circuit by the small series capacitors and the low shunt impedance of the output tank circuit at audio frequencies.

The ring-balanced modulator shown in Fig. 4-27 delivers up to twice the output of the other diode circuits. This is typical of the type modulator commonly used in Citizens Band transceivers.

Again the carrier is applied in balanced fashion to appear at the same phase between points C and common and D and common. The modulating wave is applied between points A and B of the diode ring. As shown in the simplified drawing of Fig. 4-27B, radio-frequency carrier current would be present in the output circuit except that the two paths are opposing. With proper balance, the current cancels and produces no carrier output.

The presence of the audio wave unbalances the diode operation. As mentioned previously, the carrier wave does appear across each diode, and the application of audio causes a heterodyning action. The side frequencies are produced and developed across the output because of the imbalance that the audio wave introduces into the ring circuit. In the ring-balanced modulator, side frequencies are developed during both alternations of the rf wave. This is unlike the bridge circuit described previously, where current occurred only with the positive alternations and less output was available. In diode balanced-modulator circuits, carrier suppression of 40 dB and higher can be attained by careful design and adjustment.

The radio-frequency carrier is introduced through a resistor network which includes a carrier-balance control. Any slight performance differentials in the ring circuit can be compensated for with this control. Thus it is possible to adjust the circuit for a minimum carrier output. Stray and differential capacities may also disturb the ring balance. Consequently, an additional carrier-balance control is included in the form of adjustable capacitor C1.

The audio wave is introduced via a low-pass filter that blocks the radio-frequency components from entering the audio system. In the output circuit, capacitors C2 and C3 have very high reactance at audio frequencies. Consequently, the audio components are blocked from the followup sideband filter.

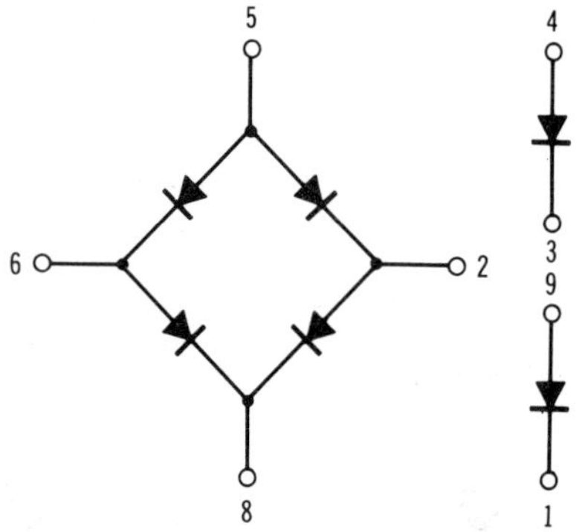

Fig. 4-28. Pin-out diagram of RCA CA-3019 diode array.

Integrated circuits are popular as balanced modulators because of their high level of carrier rejection. Bridge- or ring-type balanced modulators can be constructed from the popular RCA CA3019 diode array. A pin-out diagram of this integrated circuit is shown in Fig. 4-28. By proper selection of diodes, either a bridge- or ring-type modulator can be constructed.

A single-sideband generator with good carrier suppression is the balanced-differential amplifier shown in Fig. 4-29. As a function of the modulating signal, sideband outputs of 0.5- to 1-volt rms can be obtained. Such a balanced modulator circuit will operate with a carrier input of about 100 millivolts and an audio drive of about 0.5-volt rms. Internal circuit configuration and external circuit plan for a double-sideband suppressed-carrier generator are shown in Fig. 4-29A and B.

Two differential amplifier pairs are included in the integrated circuit and incorporate individual transistors in their common-emitter circuits to supply constant-current bias. A second transistor is included in each leg to inject the modulating signal. The carrier is applied in a differential-mode fashion to pairs of differential transistors. Outputs of these differential pairs are out of phase; under true balance the net carrier voltage is near zero. Out-of-phase audio is applied across the bases of the two modulation-insert transistors located in the emitter legs of the differential pairs. Upper- and

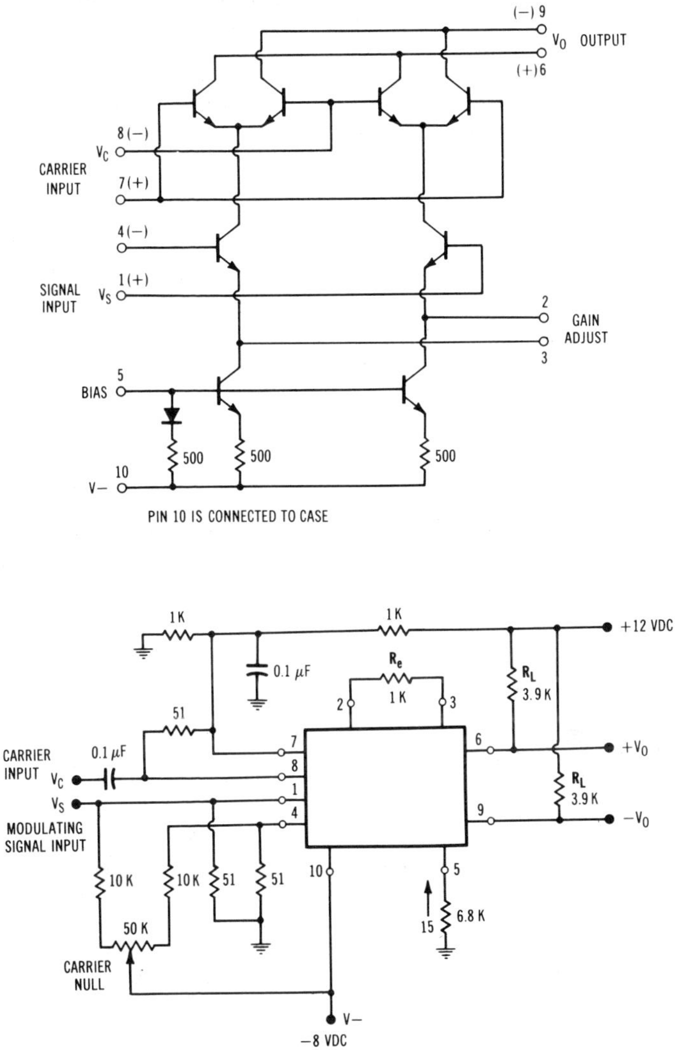

Fig. 4-29. Integrated circuit (Motorola MC-1596G) used as a balanced modulator.

lower-side frequencies develop across the output while the modulating wave itself is cancelled.

Note in Fig. 4-29B showing the external circuit that the carrier is applied between pins 8 and 7 and the modulating signal between pins 1 and 4. Bias for these two latter transistors is obtained from the negative 8-volt source connected to the arm of the carrier null potentiometer. This bias, because of its link to the emitters of the

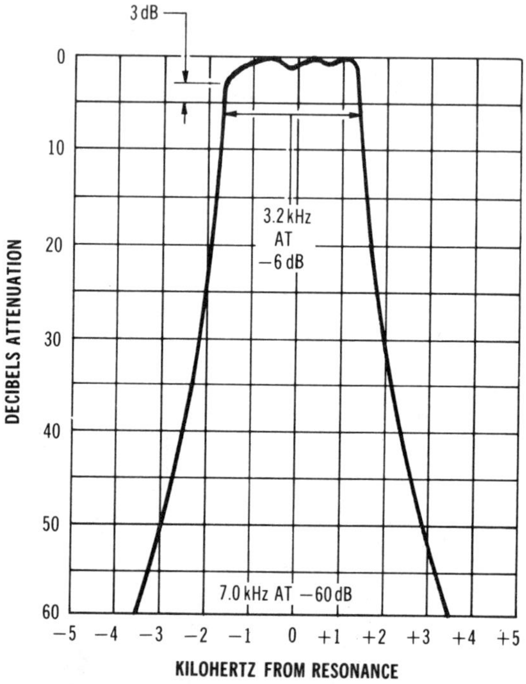

Fig. 4-30. Typical sideband-filter response.

differential pairs, sets their bias level and permits an appropriate adjustment for balancing the carrier to zero.

A balanced output is available between pins 6 and 9; and single-ended output can be derived from either pin and common. The carrier rejection is excellent using the proper levels of carrier and modulating wave.

In the filter-type sideband generator, a filter follows the balanced modulator. The output of the balanced modulator, you may recall, consists of two sidebands with the carrier suppressed. The sideband filter removes the undesired sideband. Sideband filters can be made of either crystal or ceramic material. The ceramic filter, however, is commonly used.

A typical filter-response curve is shown in Fig. 4-30. Note that it has very steep skirts and a bandwidth of only approximately 2.5 to 3.2 kHz. A 3-kHz response is typical. This narrow-frequency response means that the filter has good frequency rejection on one side of the carrier frequency but passes with minimum attenuation the side frequencies located on the other side of the carrier frequency. Hence, if an upper-sideband signal is to be transmitted, the sideband filter will pass the frequency spectrum of the upper sideband but will reject the lower-sideband frequency components.

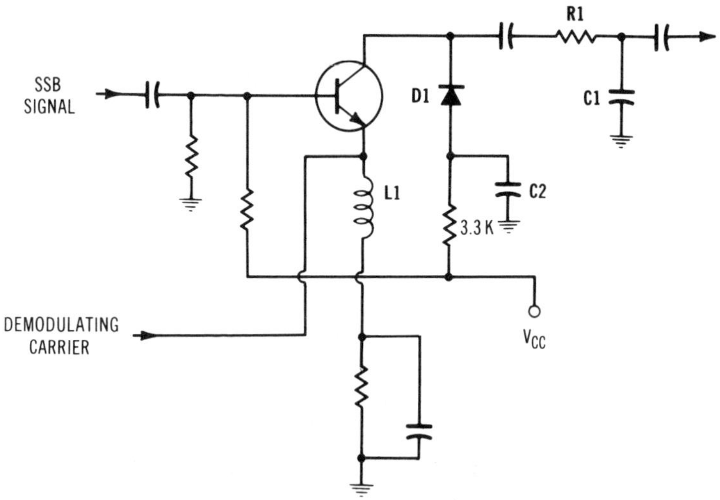

Fig. 4-31. Sideband product detector.

The sideband-product detector of a transceiver is shown in Fig. 4-31. In a product detector, two radio-frequency signals are mixed and the output circuit develops a difference frequency. The two rf components are filtered out. The single-sideband signal is applied to the base circuit of the transistor. A demodulating carrier for the product detector is applied to the emitter circuit, developing across inductor L. When the transceiver is set for sideband operation, diode D1 conducts and the 3.3K collector load resistor is active. The demodulated audio develops across the circuit and is coupled through a resistor-capacitor network to the audio system. Resistor R1 and capacitors C1 and C2 filter out the radio-frequency components. These three components also have values that limit the higher audio spectrum so that frequency signals above the desired voice-frequency range are attenuated.

CHAPTER 5

Digital and Microprocessor Electronics

Digital electronics has become an important part of modern radio-communications, augmenting the analog and linear circuits that comprise the complete two-way radio system. Digital emphasis is on control, switching, and display functions. Additional applications involve much of the test equipment used for two-way radio testing and service, such as frequency counters, voltmeters, and communications monitors. More elaborate radiocommunication systems and some test facilities incorporate microprocessors.

5-1. DIGITAL BASICS

A first step in understanding digital circuits is to learn just a bit about the binary numbering system. In a binary system, you can count with only two numerals, zero and one. In both the binary and decimal languages, *nothing* is zero or 0. Likewise the quantity *one* is the same for both languages and is written as one or 1. From this point on, things differ. In decimal language, the quantity two is written as 2; in binary language, it is written as 10. Very often the binary code used is in the form of a 4-bit series, and the quantities zero, one, and two are set down as follows:

zero = 0000
one = 0001
two = 0010

The 4-bit expression for zero is telling us that there is no eight, no four, no two, and no one. For the quantity one, it is saying that there is no eight, no four, no two, and one one. Likewise, for the quantity two, the expression says that there is no eight, no four, one two, and no one.

In 4-bit binary language, the quantity nine is written as 1001, telling that there is one eight, no four, no two, and one one. One 8 plus one 1 equals the decimal quantity 9. Table 5-1 makes a comparison between the decimal and binary presentations. It also shows the 4-bit presentation method alone with the quantity or weight represented by each digit of the four-bit sequence. Higher-value decimal quantities can be represented in binary form by adding more bits. For example, the quantity 16 in binary form using a 5-bit sequence would be expressed as 10000; decimal 17 would be 10001; and so on.

Table 5-1. Decimal-Binary Equivalents

Decimal	Binary	Four-Bit	Weight 8, 4, 2, 1
0	0	0000	(0) 8, (0) 4, (0) 2, (0) 1
1	1	0001	(0) 8, (0) 4, (0) 2, (1) 1
2	10	0010	(0) 8, (0) 4, (1) 2, (0) 1
3	11	0011	(0) 8, (0) 4, (1) 2, (1) 1
4	100	0100	(0) 8, (1) 4, (0) 2, (0) 1
5	101	0101	(0) 8, (1) 4, (0) 2, (1) 1
6	110	0110	(0) 8, (1) 4, (1) 2, (0) 1
7	111	0111	(0) 8, (1) 4, (1) 2, (1) 1
8	1000	1000	(1) 8, (0) 4, (0) 2, (0) 1
9	1001	1001	(1) 8, (0) 4, (0) 2, (1) 1
10	1010	1010	(1) 8, (0) 4, (1) 2, (0) 1
11	1011	1011	(1) 8, (0) 4, (1) 2, (1) 1
12	1100	1100	(1) 8, (1) 4, (0) 2, (0) 1
13	1101	1101	(1) 8, (1) 4, (0) 2, (1) 1
14	1110	1110	(1) 8, (1) 4, (1) 2, (0) 1
15	1111	1111	(1) 8, (1) 4, (1) 2, (1) 1

Various codes based on the binary 1 and 0 concept have evolved to meet the requirements of digital equipment operation and objectives. A simple code that has been extensively used is known as the binary-coded-decimal (BCD) method. Four binary bits are employed. Each digit position has a definite value (weight) in the order 8, 4, 2, 1.

In the true binary case, it is simply 8,4,2,1 as shown in Table 5-1. Conversion to decimal values involves the simple addition of the weights of the digits as shown previously. For special needs, there are various other types of codes, some involving more than four digits or characters. In the case of BCD, however, a simple 4-bit number is used to express only decimal quantities zero through nine. Although a 4-bit number can designate higher numbers (up to 15 as shown in the chart), the BCD code restricts each 4-bit character to decimal numbers from zero to nine.

When a higher number is to be indicated in binary form using the BCD code, additional 4-bit characters are conveyed. For example, the number 35 in the BCD code becomes 0011 0101.

3 5
0011 0101

Note that the first 4-bit character is the binary representation of three while the second is the representation of the decimal quantity five. Thus, the number 6751 would be written:

0110 0111 0101 0001

The intriguing part of the binary presentation is that numbers can be represented by combinations of binary 0 and binary 1. Furthermore, it is very easy to understand how to use a simple switch to see a binary 1 or a binary 0. For example, a closed switch could represent a binary 1; the same switch when open, a binary 0. A sequence of switches could then be used to demonstrate a binary counting system (Fig. 5-1). The closed position of the switch in the circuit is customarily called the on position; the open position is the off position. In digital vernacular, the closed position could designate true or binary logical 1, while the off position of the switch would be designated false or logical 0. In representing the decimal quantity twelve, switches 1 and 2 would be closed, while switches 3 and 4 would be open, setting up the sequence of 1100.

5-2. LOGIC CIRCUITS

In an actual digital system, the switches would be electronic and not mechanical. They operate at very high speeds causing many operations to occur simultaneously and at a high repetition rate.

Some considerable confusion exists because of the several ways of designating a logical 1 and a logical 0. In general, the terms yes and no, true and false, as well as 1 and 0, are identical. In actual circuits, logical 1's and 0's are represented by voltage levels. In a negative logic circuit the most positive voltage level (high) is 0 and the most negative level (low) is 1. Conversely, in a positive logic circuit, the most positive level (high) is defined as a logical 1, while the most negative voltage level (low) is a logical 0.

It follows, then, that in a positive logic circuit, the following terms are one and the same:

1 and 0
yes and no
true and false
high and low

Conversely, for negative logic circuits the identities are:

1 and 0
no and yes
false and true
low and high

The most common method used is positive logic.

In digital circuits using positive logic, inputs and outputs are either high (logic 1) or low (logic 0). There is no in between. Circuits act like fast switches changing over between high and low at the output in response to high and low conditions at their inputs.

The building-block of modern digital circuits is the so-called *logic*

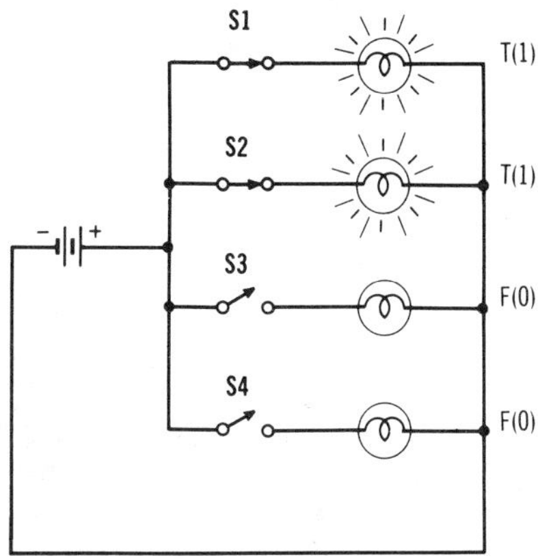

Fig. 5-1. Binary number representation using switches and lamps to show the decimal number 12.

gate. Although gate circuits can be built using discrete components they are usually in the form of multiple gates that are a part of a digital integrated circuit. The two most common families of digital integrated circuits are TTL (transistor-transistor-logic) and CMOS (complementary metal oxide semiconductors). The TTL device responds to voltages between 0 and +5 volts. Maximum supply voltage to a TTL device is +5 volts. The logic 0 voltage falls between 0 and 0.8 volt; the logic 1 between 2.4 and 5 volts. Thus a dc voltmeter or popular logic probe can be used to ascertain if input and output levels are at logic 0 or logic 1.

A CMOS digital integrated circuit will also operate at the same level as the TTL. However, supply voltages may extend between 4 and 15 volts. In most applications logic 0 would fall between 0 and 0.8 volt. The logic 1 value would be between 3 and 15 volts depending upon the supply voltage.

In addition to gates there are other digital ICs that perform a variety of functions. Some of these will also be discussed in this chapter.

5-3. LOGIC PROBE

A key test instrument in locating faults in digital circuits is the logic probe (Fig. 5-2). The logic probe contains light-emitting diodes (LEDs) which turn on according to the logic of the point in the digital circuit to which the logic probe is connected. If it is touched to a logic 1 point the high LED will come on; a logic 0 point, the low LED will light. The instrument can be used to check out either TTL or CMOS circuits.

A pulse LED also indicates the presence of square waves or pulses in the digital system. Each time the input changes to the logic probe the pulse LED is activated for 0.1 second. When observing low-frequency, low duty-cycle signals, the pulse LED provides an immediate indication of the pulse activity in the circuit. Also by observing the high and low LEDs the polarity of the pulse train can

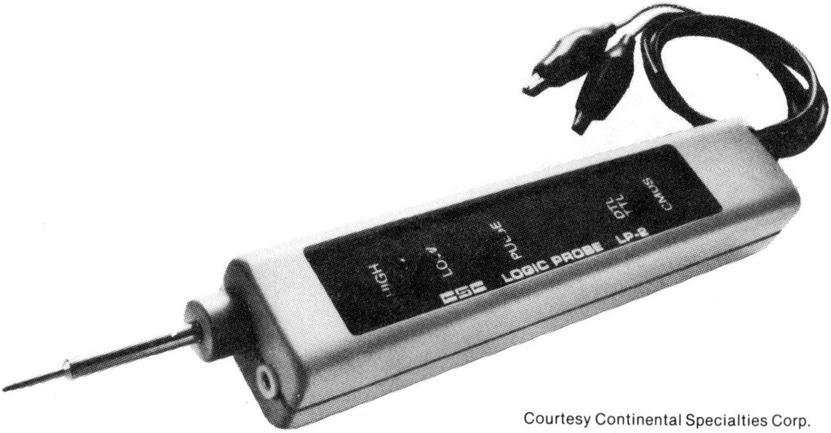

Courtesy Continental Specialties Corp.

Fig. 5-2. Logic probe.

be ascertained. If the high LED is on, the signal is normally high and pulsing low. If the low LED is on the signal is normally low and pulsing high. High-frequency signals cause the pulse LED to flash at a 10-hertz rate. The incorporation of a pulse stretcher in the logic probe makes this possible.

The logic probe can be moved from circuit to circuit in a digital system and by checking the logic at key test points proper or faulty operation can be ascertained. Measurements can be made under steady-state dc condition as well as with pulses present in the digital system.

5-4. TTL AND CMOS

The two most common digital integrated circuits are classified as TTL or CMOS types. A TTL digital IC incorporates bipolar transistors while the CMOS type is basically a MOSFET type. The term CMOS refers to *Complementary Metallic-Oxide Semiconductor*. The complementary designation refers to the fact that both p-channel and n-channel MOSFETs are used in the construction.

The TTL device is basically low impedance and supply voltage is confined to a +5 volts. The CMOS operates over a supply voltage typical of any MOSFET device. It is a high-impedance device and requires less power than a TTL. However, it must be handled more carefully and circuit design can be a bit more critical.

A typical TTL inverter is shown in example A of Fig. 5-3. As many as six of these inverters may be found in a single TTL inverter integrated circuit. Usually the circuit is referred to as an inverter gate. A logic 1 applied to its input results in a logic 0 output. A logic 0 input results in a logic 1 output. Typically logic 0 corresponds to 0 volts and logic 1 to +5 volts.

The first bipolar transistor in the circuit has a low-impedance

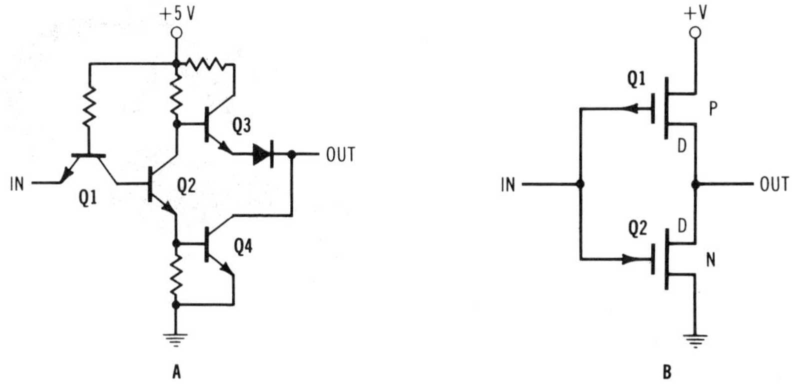

Fig. 5-3. TTL and CMOS inverter gates.

emitter input and a collector that is direct-coupled to the base of a phase inverter. The input transistor acts as a diode gate transferring the logic of the input to the base of the inverter. Inverted input logic is applied to the base of the top output-transistor pair; in-phase logic, to the base of the lower totem-pole output transistors.

In the totem-pole circuit the top transistor actually acts as the collector resistor of the bottom transistor. This connection speeds up circuit response to fast changes in input logic. The lower transistor is, of course, connected as a common-emitter type and its output logic is inverted as compared to its input logic. Therefore the overall device functions as a logic inverter.

A basic CMOS inverter circuit is given in Example B. The top FET is a p-channel type; the bottom, an n-channel. When the input logic is near ground or logic 0, the bottom n-channel type is nonconducting. However, the top p-channel does conduct and the output at the joined drains is near the supply voltage, or logic 1.

When the input is set to logic 1, the top FET becomes nonconducting. However, the bottom FET does conduct and the output at the joined drains is near ground, or logic 0. As in the case of a TTL as many as six of these logic gates can be mounted in a single integrated-circuit inverter.

There are six basic logic circuits that come under the general heading of logic gates. These are inverter, AND gate, NAND gate, OR gate, NOR gate, and Exclusive /OR gate. These gates are found built into digital integrated circuits with usually four or more gates of the same type included in the same chip.

5-4-1. Inverter

A digital inverter has an output logic which is the inverse of the input logic. Output Y is the complement of input A (Fig. 5-4A). Figs. 5-4A and 5-4B show the inverter symbol, truth table, and diagram.

A	Y
0	1
1	0

(A) Inverter symbol and truth table

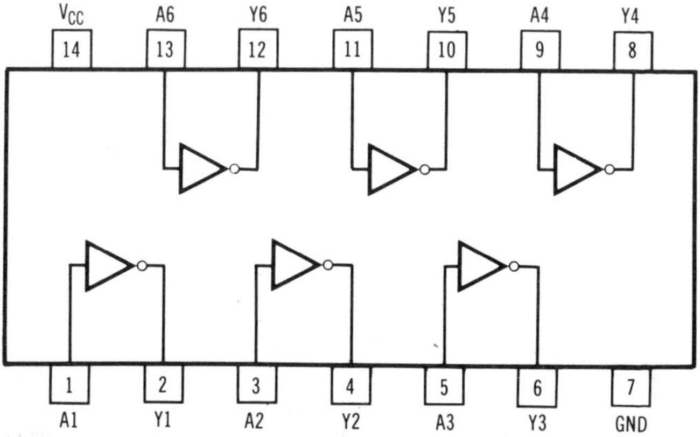

(B) Six inverters are included in one IC.

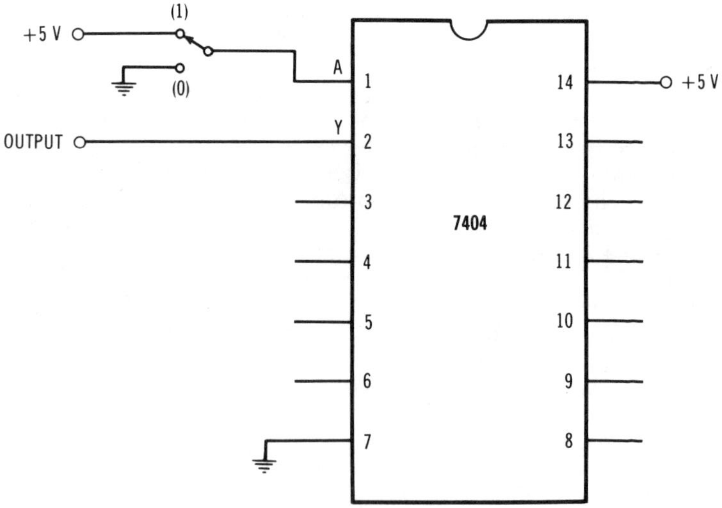

(C) Circuit.

Fig. 5-4. Symbol, pinout, truth table, and circuit of 7404 hex inverter.

It can be seen that if the input logic is 0, the output logic is 1. An input logic of 1 drives the output to logic 0. A clock pulse or other sequence of pulses applied to the input A will appear inverted at the output Y. The truth table is a manner of expressing the logic conditions of a digital circuit. It shows the logic output for every possible combination of input logic.

Data is given for the TTL 7404 hex inverter. There are six independent inverters included in this single IC chip. Note that there are twelve separate pins for the various inputs and outputs. Supply voltage is connected to pin 14, while pin 7 is grounded.

A simple demonstration circuit using just one of the inverter gates in the 7404 is given in Fig. 5-4C. The switch permits a connection of logic 1 or logic 0 to input A. If the switch were connected to the supply voltage it would be the same as connecting a logic 1 to the input. Under this condition the output at Y would be logic 0. A logic probe like the one shown in Fig 5-2 can be used to make the check. Also a dc voltmeter could be connected at the output pin 2 indicating whether the output logic was 0 (approximately 0 volts) or logic 1 (approximately +5 volts). Note the supply voltage connected to pin 14 and the ground to pin 7.

If a series of positive pulses were applied to input A, they would appear at the output Y as negative pulses, showing the inversion characteristic of an inverter stage.

5-4-2. Two-Input AND Gate

Schematic diagram, symbol, and truth table of a TTL 7408 AND gate are given in Fig. 5-5. Note the symbol for the AND gate and

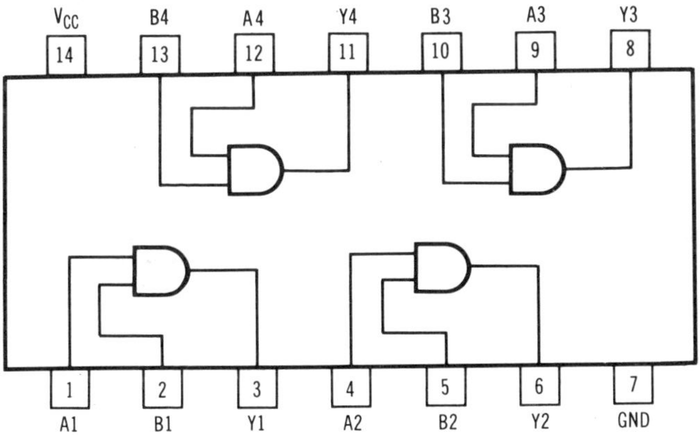

(A) Four AND logic gates in one IC.

(B) Logic gate symbol and truth table.

A	B	Y
0	0	0
0	1	0
1	0	0
1	1	1

Fig. 5-5. Diagram, symbol, and truth table of the 7408 AND gate

how it differs from the triangle symbol of the inverter. In an AND circuit, input gates A *and* B must be at logic 1 before there can be a logic 1 output. If either or both of the inputs are logic 0 (low), the output Y is logic 0 (low). These facts are again given in the truth table which shows the output logic state for every possible combination of input logic states. Every possible combination of A and B is given, showing the resultant output state.

The TTL 7408 contains four individual AND logic gates as shown in the schematic arrangement. Supply voltage positive is connected to pin 14; ground, to pin 7.

Truth table verification can be obtained with the simple circuit of Fig. 5-6 using one of the AND gates in the 7408. A logic probe or

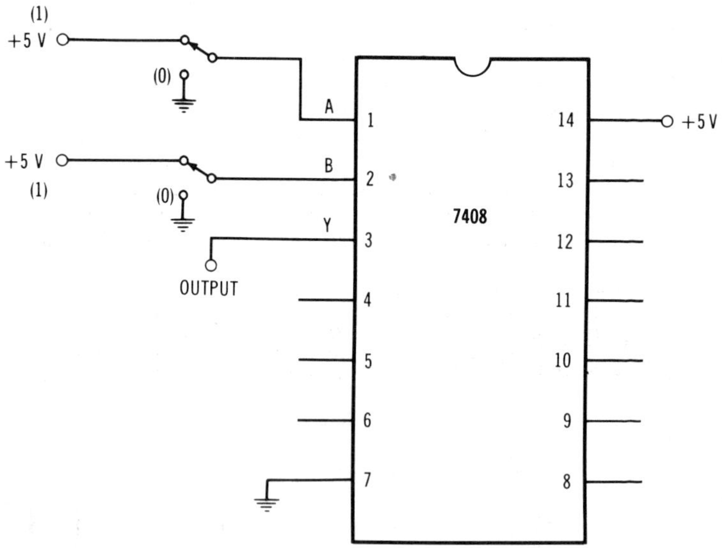

Fig. 5-6. Practical AND circuit.

dc voltmeter can be used to check the output logic at Y. When both switches are connected to logic 1 as shown, the output Y will also be logic 1. Any other combination of the two switches will result in a logic 0 at the output.

An AND gate circuit as you might find on a schematic diagram is given in Fig. 5-7. If a series of pulses, such as might be derived from a clock pulse generator, were applied to input A, their presence at the output would depend upon the setting of the logic at input B. Were the logic at input B set to logic 1, as shown, the pulses will appear at the output Y with the same polarity (no inversion). However, if the switch at B is set to logic 0 there will be no output at Y. Recall that in AND gate operation both inputs must be at logic 1 before there can be a logic 1 output.

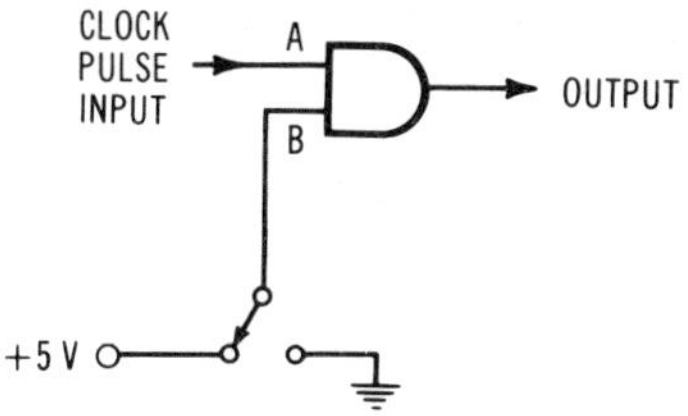

Fig. 5-7. AND gate schematic.

It should be mentioned that it is customary practice in presenting diagrams of logic circuits that the supply voltages are *not* indicated. It is assumed in this case that the proper potentials are applied to pins 14 and 7 as shown previously in Fig. 5-6.

It should be stressed that the input to gate B is usually a pulse waveform too and is not the simple switch arrangement of Fig. 5-7. The waveforms of Fig. 5-8 show such a possibility. Assume that clock pulses of a particular repetition rate are applied to input A as shown. A lower-frequency and longer-duration series of pulses is applied to gate B which drives gate B between logic 1 and logic 0. Note that when gate B pulse is at logic 1, the clock pulses applied to gate A appear at the output. Conversely when gate B is set to logic 0 by the applied gating pulse, there is no output pulse present at output Y. This shows how a gate pulse can be used to gate on

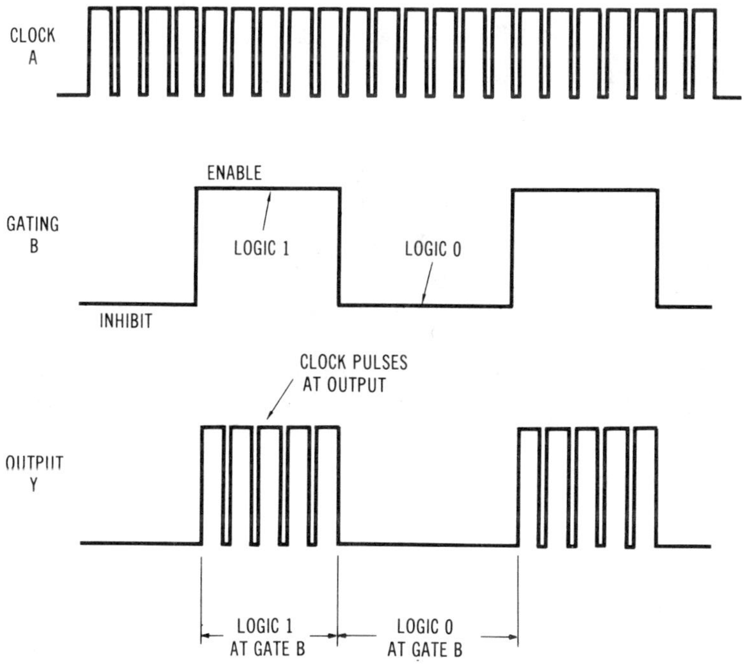

Fig. 5-8. Gated clock pulses AND gate

(enable) and off (inhibit) the AND gate, allowing or not-allowing the pulses at gate A to appear at output Y.

5-4-3. NAND Logic Gate

A schematic, symbol and truth table for a NAND gate are given in Fig. 5-9. Again the integrated circuit houses four individual NAND gates.

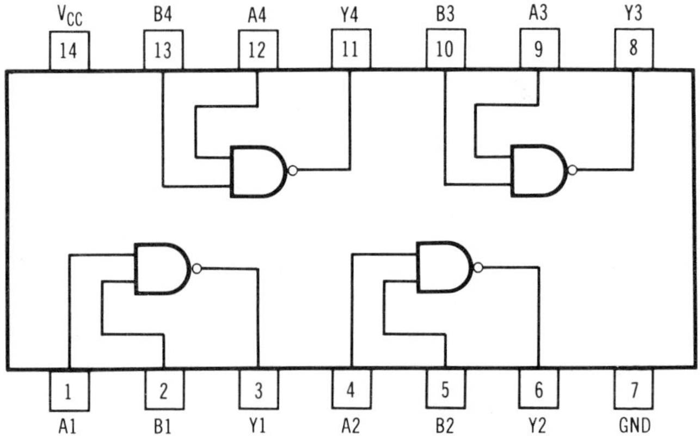

(A) Four NAND gates included in one IC.

A	B	Y
0	0	1
0	1	1
1	0	1
1	0	0

(B) Symbol and truth table.

Fig. 5-9. Diagram, symbol, and truth table of the 7400 quad NAND gate.

In a NAND gate configuration when both or either inputs are at logic 0, the output is at logic 1. If both inputs are high at logic 1, the output is low at logic 0. In fact, the NAND gate is referred to as a negated AND gate.

In comparing Fig. 5-9 with Fig. 5-5, note that the output logics in the Y column of the truth tables are complements. Note too that the two symbols are very similar except for the very small circle at the Y output of the NAND gate. A truth table can again be verified using the simple circuit of Fig. 5-10 and an appropriate logic probe or dc voltmeter.

When input data or a clock pulse is applied to input A there will be no output if the gate B is at logic 0. If gate B is set to logic 1, however, there will be an output signal. However in the case of the

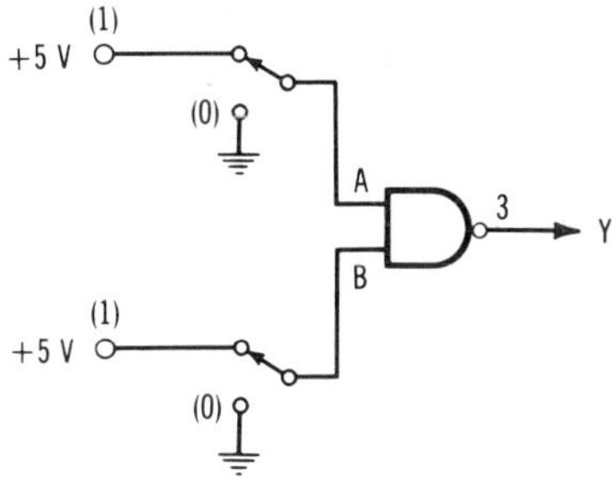

Fig. 5-10. Practical NAND circuit.

NAND gate the output signal will be the complement of the input data. Thus the NAND gate inverts the input signal.

When a combination of clock and gating pulses is applied to the input as shown in Fig. 5-11 there is a different output than that obtained when using an AND gate. Note again when the gating pulse at input B is a logic 0 there is no output. However, under the logic 1 condition, there is an output. This output is a series of inverted clock pulses as shown in Fig. 5-11.

In the previous examples only two inputs were shown. However, logic gates can have more than two inputs and quite often do. An

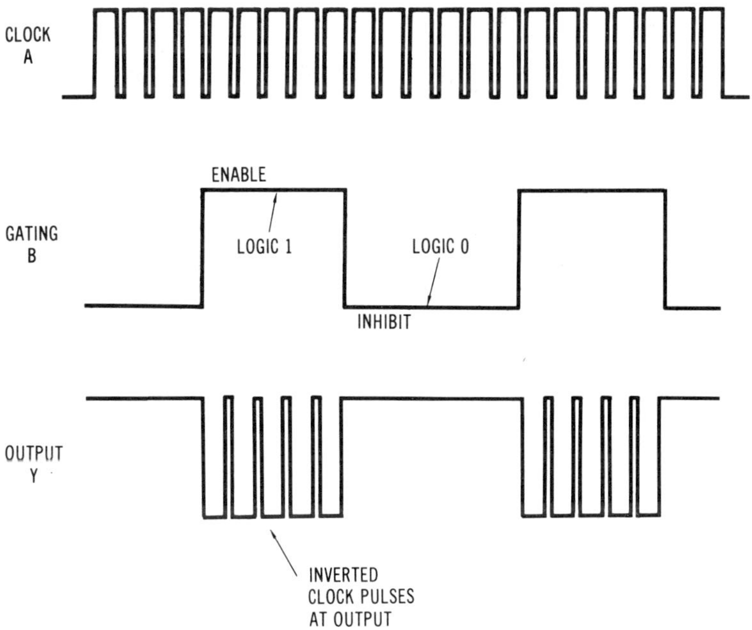

Fig. 5-11. Gated clock pulses NAND gate.

example of the symbol and truth table for a three-input NAND gate is given in Fig. 5-12. Note again that all the possible conditions for A, B and C have been set down. The resultant output is indicated in the Y column. Again for the NAND type of logic circuit the output is logic 0 when all inputs are logic 1. When all, two, or one input is at logic 0, the output is at logic 1.

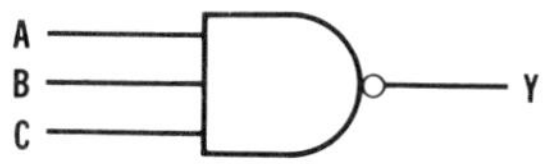

A	B	C	Y
0	0	0	1
0	0	1	1
0	1	0	1
0	1	1	1
1	0	0	1
1	0	1	1
1	1	0	1
1	1	1	0

Fig. 5-12. Three-input NAND gate.

5-4-4. Two-Input OR Gate

Diagram, symbol, and truth table for an OR gate are given in Fig. 5-13. In an OR gate circuit a logical 1 output results when a logic 1 is applied to both or either input gates. The output is logic 0 when both inputs are logic 0. Compare the truth table of the OR gate with that of the AND gate. In the case of the AND gate the output is logic 1 when input A *and* input B are logic 1. In case of the OR gate the output is logic 1 when input A *or* input B is logic 1.

If a clock pulse or other input data is applied to input A and input B is at logic 0, the input data is transferred to the output without inversion. Conversely, a logic 1 state at input B blocks the data applied to input A, and it does not appear at the output.

A gating signal applied to input B can be used to switch the output data on and off at the rate of the gating pulse. However, the transfer is made to the output whenever gate B is set at logic 0. This is the converse of the activity of an AND gate as detailed previously.

5-4-5. Two-Input NOR Gate

Diagram, symbol, and truth table for a NOR gate are given in Fig. 5-14. Note that the NOR gate symbol is the same as that for an OR gate with the exception of the small circle at the output. In NOR gate circuit operation when either or both inputs are at logic 1 the output is logic 0. The output is logic 1 only when both inputs are at logic 0. Compare the NOR and OR gate truth tables, observing that

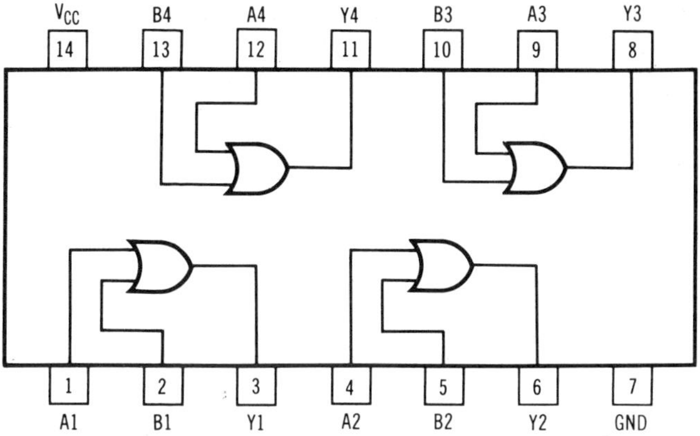

(A) Four OR gates included in one IC.

(B) Symbol and truth table of OR gate.

A	B	Y
0	0	0
0	1	1
1	0	1
1	1	1

Fig. 5-13. Diagram, symbol, and truth table of the 7432 OR gate.

the Y columns are complementary. The NOR is in fact referred to as a Negated OR gate, hence the name of NOR gate.

If the input data or clock pulse is applied to A gate and a logic 1 to B gate there will be no output. If a logic 0 is applied to the latter gate, however, the input signal appears in inverted form at the output. The output is, in fact, the complement of the input.

If required, a gating pulse can be applied to the B input. In this case the output is switched on and off and clock pulses appear in the output whenever the input gating pulse swings to logic 0. Output pulses are, of course, opposite in polarity from the input pulses applied to gate A.

5-4-6. Two-Input Exclusive/OR Gate

The Exclusive/OR gate is unique as shown in Fig. 5-15. When gates A and B are both set to logic 0 the output is logic 0. Furthermore when gates A and B are both set to logic 1, the output is also logic 0. If logic A is zero and logic B is 1, or logic A is one and logic B is zero, the output is logic 1. Symbol and truth table are given.

The Exclusive/OR gate is unusual in its response to input data or clock pulse. Assume that input data is being applied to gate input A. If gate input B is set to logic 0 the input data is transmitted directly to the output and is of the same polarity. However, if gate

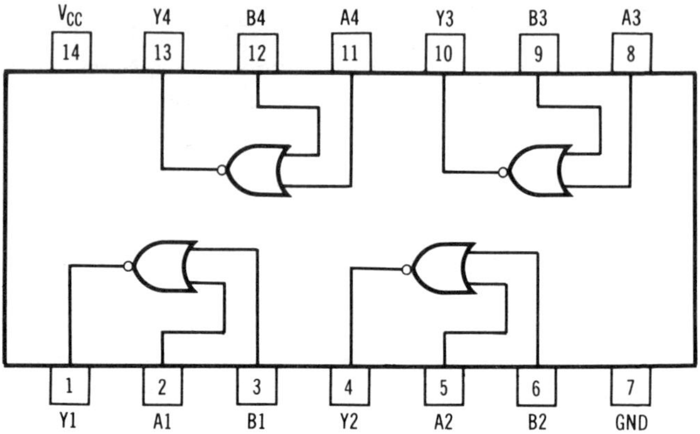

(A) Diagram.

A	B	Y
0	0	1
0	1	0
1	0	0
1	1	0

(B) Symbol and truth table of NOR gate.

Fig. 5-14. The 7402 NOR gate.

B is set to logic 1 the input data will also appear in the output but will be in inverted form.

It would be advisable to review all gate circuits. Pay particular attention to their symbols and truth tables. You may wish to set them down side by side and make appropriate comparisons. Memory can be aided by considering the relationships between AND and OR gates, AND and NAND gates, and finally OR and NOR gates.

5-5. LEDs AND 7-SEGMENT DIGITAL DISPLAYS

The light-emitting diode (LED) is a semiconductor diode that emits light when forward-biased (Fig. 5-16A). A series resistor is used to limit the diode current to a safe value. The most common LED color is red; yellow and green versions are available.

LEDs are used widely in digital and analog electronic systems. They are reliable and have a long life. In TTL systems the supply voltage is usually 5 volts. An appropriate resistor must be placed between the 5-volt source and any LED that is operated in the circuit. Resistor values usually fall in the 100- to 470-ohm range. In most cases the LED current should not exceed 50 mA and an actual current of 20 to 30 mA is typical. Common series resistor values are 180, 220, 270, and 330 ohms. Of course, the lower value resistors provide the brighter, glow. In CMOS circuits using higher-value

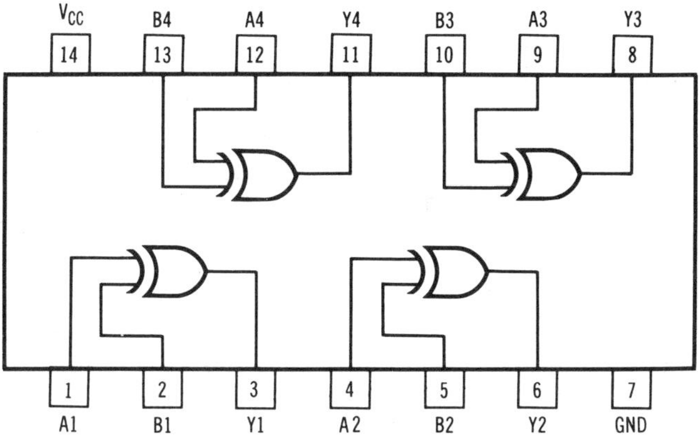

(A) Four EXC/OR gates in one IC.

(B) Symbol and truth table for EXC/OR gate.

A	B	Y
0	0	0
0	1	1
1	0	1
1	1	0

Fig. 5-15. Diagram, symbol, and truth table of the 7486 EXC/OR gate.

supply voltages. Ohm's law can be used to calculate the higher value of series resistance required.

Illustration B shows how an LED circuit in conjunction with a bipolar transistor can be used as a simple logic probe or lamp monitor. The LED circuit has a low impedance and could load down a particular higher resistance digital or analog circuit. The purpose of the bipolar transistor is to provide current amplification at the same time it presents a very much reduced load to the circuit to which the probe is connected. A very low input base current will result in the significant current needed to light the LED connected in the collector circuit.

This circuit can be used as a very simple probe for checking or monitoring TTL circuits. If the probe is connected to a logic 1 point (+5 volts), the LED will light. If connected to a logic 0 point (0 volts) it remains unlit.

The LED is the basis for digital number displays. A 7-segment display consists of seven LED slits, two or more of which are illuminated to display the numerals 0 through 9 (Fig. 5-17). For example, if segments f, g, b, and c are lit the number 4 will be displayed. The illumination of segments a, b, g, c, and d reproduces as the numeral 3. This typical 7-segment display plugs into a 14-pin DIP mount.

Anode voltage can be supplied to either pin 3 or pin 14. A

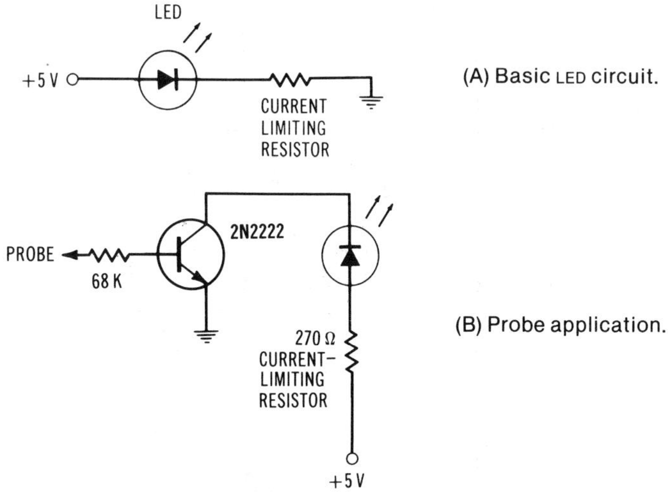

Fig. 5-16 LED probe.

specific LED slit is illuminated by connecting its associated pin to ground through the appropriate current-limiting resistor. Note in Fig. 5-17 that the individual slits are connected to pins 1, 2, 7, 8, 10, 11, and 13. Decimal point illumination is made via pin 6.

The simple circuit arrangement of Fig. 5-18 shows how the number 5 is illuminated. In this case slits a, f, g, c and d must be

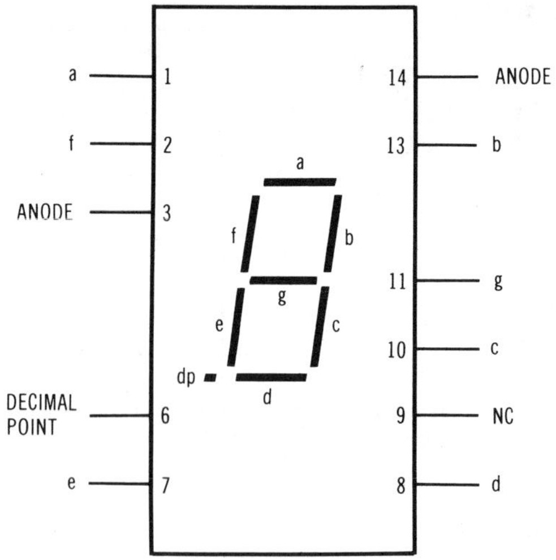

Fig. 5-17. Seven-segment display.

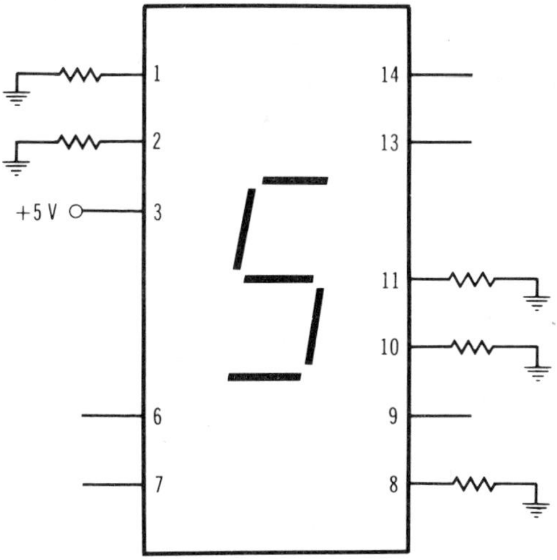

Fig. 5-18. Wiring for displaying the numeral 5.

connected to ground through current limiting resistors. This involves physical connections to pins 1, 2, 10, 11 and 8.

To operate the 7-segment display electronically a 7-segment decoder is required. The TTL 7447 integrated circuit, Fig. 5-19, is an example of such a decoder. It is able to convert a positive logic BCD input signal to a proper sequence that lights a 7-segment display.

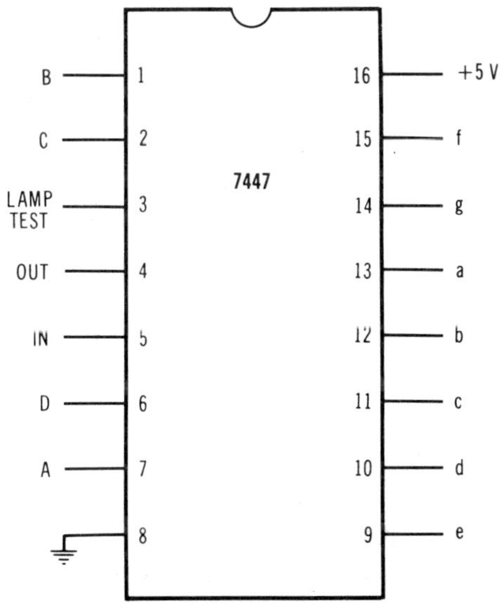

Fig. 5-19. 7447 BCD-to-7 segment decoder.

Recall that the BCD code is a 4-bit digital signal that represents the numerals 0 through 9 as a proper sequence of logic 0 and logic 1 bits. Refer back to Table 5-1. It is this coded signal, when supplied to pins 1, 2, 6 and 7, that provides the proper 7-segment illumination. The individual seven segments that connect to the 7-segment display involve pins 9 through 15. A schematic showing the decoder and 7-segment display properly interconnected is given in Fig. 5-20. Note the links between decoder output and display input by way of current-limiting resistors. If the digital logic signal applied to the 7447 is 0100 the b, c, f, and g output of the decoder will be placed at logic 0 and current will flow in the appropriate segments to illuminate the numeral 4.

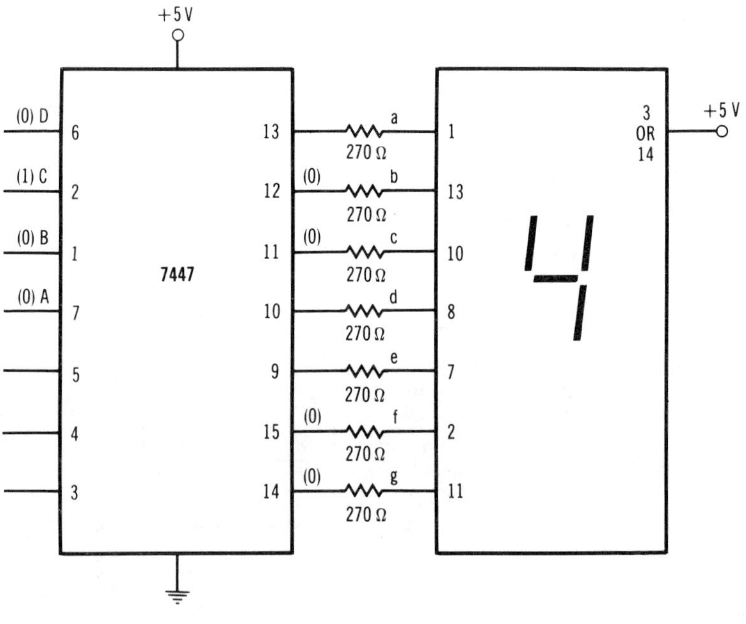

Fig. 5-20. Decoder and display connected together.

Truth tables are given in Fig. 5-21. The first column is the actual decimal numerals. The second set of columns shows the BCD code for the various numerals. It is this code that is supplied to the input of the 7447 decoder. The final set of columns shows the individual segments of the display. When an individual segment is connected to logic 0 by the decoder it will glow.

Pin 3 of the decoder (Fig. 5-19) can be used to test all of the segments of the display. When it is connected to logic 0 (grounded) all the slits will glow if they are operating properly. Pins 4 and 5 are used for blanking purposes and to provide a link to other 7447 devices that are connected when a multiple readout display is

NUMBER	D ("8")	C ("4")	B ("2")	A ("1")	a	b	c	d	e	f	g
0	0	0	0	0	0	0	0	0	0	0	1
1	0	0	0	1	1	0	0	1	1	1	1
2	0	0	1	0	0	0	1	0	0	1	0
3	0	0	1	1	0	0	0	0	1	1	0
4	0	1	0	0	1	0	0	1	1	0	0
5	0	1	0	1	0	1	0	0	1	0	0
6	0	1	1	0	0	1	0	0	0	0	0
7	0	1	1	1	0	0	0	1	1	1	1
8	1	0	0	0	0	0	0	0	0	0	0
9	1	0	0	1	0	0	0	0	1	0	0

Fig. 5-21. BCD-to-decimal decoding.

required. For example, if a 4-section numeral display is desired each individual display has its own decoder.

In a digital display system the decoder is preceded by a counter. The purpose of the counter is to convert a series of input pulses (clock) to the BCD output code. The TTL 7490 is an example of such a decade counter (Fig. 5-22). The decade counter can be considered to have ten stable states made available at pins 8, 9, 11, and 12. These ten states match the 4 bits of the BCD code through the sequence from 0 through 9. After a complete sequence has been made there is a return to the 4-bit 0000 that represents the numeral 0.

The purpose then of the decade counter is to count in sequence between 0 and 9 as the pulses arrive at pin 14 of the 7490. For each ten pulses coming in, in sequence, the BCD output will follow the

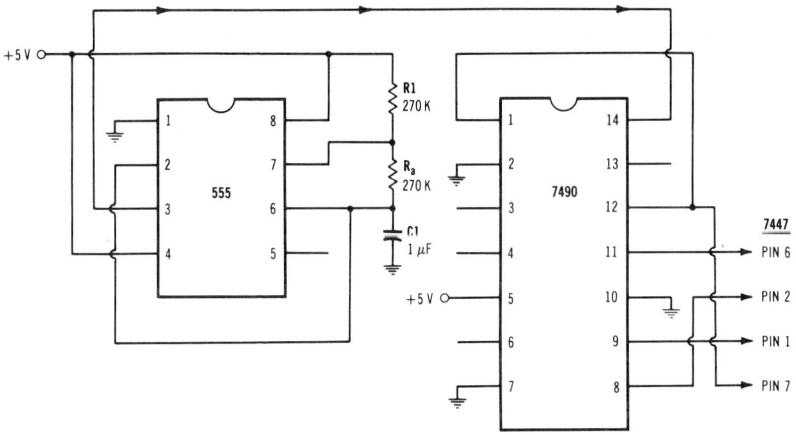

Fig. 5-22 BCD decade counter and clock generator.

sequence between 0000 and 1001 of the BCD code. After this sequence of ten pulses, the next arriving pulse will result in a 0000 output and the count sequence will resume again.

The Fig. 5-22 circuit includes a 555 clock pulse generator which is generating the sequence of pulses applied to the input pin 14 of the decade counter. If you combine Figs. 5-20 and 5-22 into an actual circuit you will come up with a demonstration counter that will continue to sequence between 0 and 9 and back to 0 and so on, as long as the clock generator is in operation.

In an actual digital display system other circuits and pulse sources are required. It may be that some analog quantity has to be measured. If so, circuits must be incorporated that will illuminate the display only for a given numeral as a function of the information being interpreted by the digital display system.

The circuit of Fig. 5-23 shows how a NAND gate can be used to control the number of clock pulses or other data that is applied to a counter. If the duration of the gate pulse is known one can

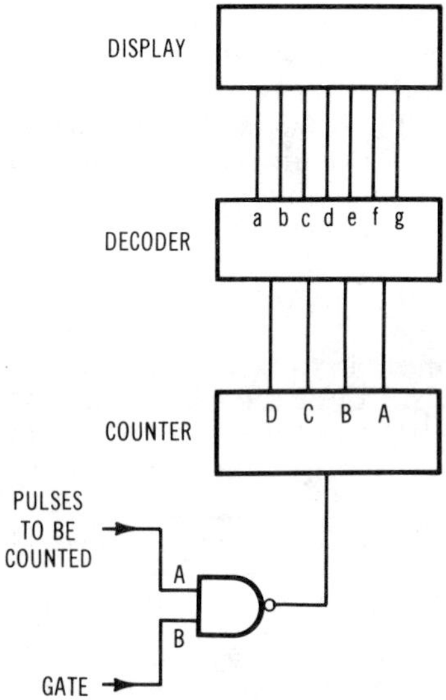

Fig. 5-23. Use of NAND gate to enable the count sequence.

determine how many pulses are applied to the counter during this interval of time. For example, if the duration of the positive gate pulse applied to input B is one second and the counter counts up to 7 during this interval of time it means that the pulses to be counted are arriving at a rate of 7 per second. This is the first step in understanding how a frequency counter operates.

Recall that for the operation of a NAND gate the application of logic 1 information to gates A and B results in an output of logic 0. Refer back to the waveforms of Fig. 5-11. When the gate pulse at B is positive (logic 1) the gate is said to be *enabled* and the pulses applied to gate A will appear in the output. They will be of negative polarity which is correct for applying the information to the counter.

The sequence of pulses applied to the counter results in the generation of the BCD code signals that correspond to the number of pulses that are applied from the gate. This BCD signal is decoded and converts to the 7-segment signal that operates the LED display. When the gate B signal is at logic 0 the gate is said to be *inhibited* and no pulses applied to gate A are able to reach the counter.

This very simple counter is only capable of counting between 0 and 9 and is just a single-digit display. If a higher count is desired, individual counters can be added in series. If the series of pulses applied to the first counter exceed ten in number the excess pulses leave the first counter and are supplied to the second counter. In fact, one pulse is sent to counter 2 for each ten pulses that are applied to counter 1. The combination of the two counters in series will permit a total count of 99.

A third counter can be added to the chain and will become active if there is an excess of 99 incoming pulses. Thus with a three digit display the total possible count jumps to 999. A fourth counter permits a maximum display 9999 and so on.

An example of such a four-digit display is given in Fig. 5-24. Note that the output of the count gate at the lower right is applied to counter No. 1. Furthermore there are interconnections that join counters 1 through 4 in series. In such an arrangement 4 digits will be displayed with the first counter displaying the lowest significant digit and count No. 4 the maximum significant digit.

If we assume that the count gate pulse applied to the count gate is 1 second in duration, and the signal input has a frequency of 2564 hertz, this is the number of pulses that will be applied to the counter chain in the one second period. Therefore the display at counter 4 will read 2, that at counter 3 will read 5, that at counter 2 will read 6, and counter 1 will display numeral 4.

If the frequency of the signal input changes a different number of pulses will be applied during the 1-second interval of the count gating pulse. Therefore the display will read a new frequency, either higher or lower, depending on the direction and extent of the frequency change at the signal input.

Consider the operation of the major blocks of the 4-digit display shown in Fig. 5-24. It is important that the count gating pulse be of an exact frequency and duration. Therefore a crystal-oscillator system is used to generate this calibrated pulse. Usually the crystal oscillates at a high frequency and a series of dividers brings down the frequency to the desired clock frequency. Timing and control circuits generate the various pulses required by the frequency counter. Three basic pulses are usually generated with specific durations and time positions. These are shown in simplified form in Fig. 5-25.

The count gate pulse was mentioned previously and is applied to the count gate along with a signal processed from the input signal.

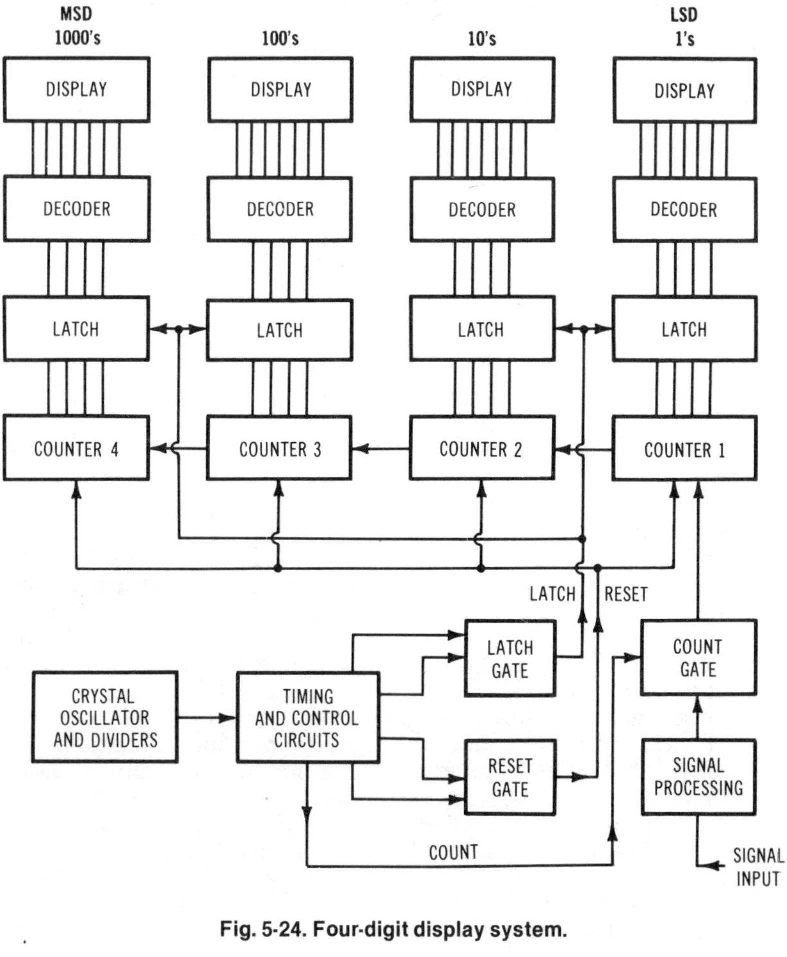

Fig. 5-24. Four-digit display system.

The input signal, more than likely, will be a sine wave. This sine wave is converted to a square wave by the signal processing block. This square wave has the same frequency as the input sine wave and is applied to the count gate.

The count gate permits a certain number of these pulses to pass through to its output as a function of the duration of the calibrated gating pulse. In our previous discussion we mentioned that the count gating pulse had a duration of 1 second. It could be some other known duration depending on the frequency range to be measured by the counter. A typical figure might be a duration of 100 milliseconds. At any rate it is an exact and known quantity and permits a certain number of the processed input pulses to pass through to counter 1. These pulses pass along the counter chain.

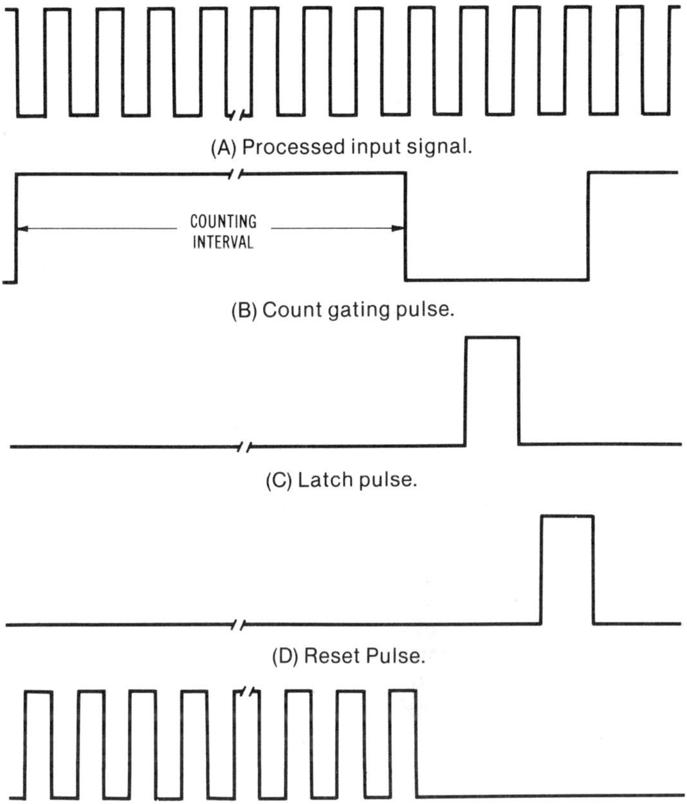

(A) Processed input signal.

(B) Count gating pulse.

(C) Latch pulse.

(D) Reset Pulse.

(E) Gated pulses to counter.

Fig. 5-25. Key digital display system waveforms.

They are said to *ripple* through the counter and the series connection of counters is referred to as a *ripple counter*.

As the pulses ripple through the counters, the entire operation would be displayed as the numbers run up on each display, finally reaching the desired final count for each individual section of the four. This would result in an annoying display and it would be more satisfactory if the only numerals displayed would be the last ones representing the actual frequency of the incoming signal. Blanking out of the run-up of the displays is handled by a series of latches and a latching pulse as shown in Fig. 5-24. The pulse counting in each of the individual counters keeps building up to the desired value as they are applied to the individual latch circuits. However, the final number is not passed to the decoder until a latch pulse is applied to each of the individual latches. Despite the fact that the ripple counters are very active during the count interval, only the final count number is seen, because this data is not passed to the decoder until the latch pulse arrives. Therefore the actual number that is displayed is the final count and you do not see the actual counting process visually on the displays.

After the numbers are displayed for a specific interval of time, a reset pulse arrives and restores each of the individual counters to its zero, and a new cycle of activity begins. This activity repeats itself at a fast rate and a continuous display appears and, at the same time, a change in the signal input frequency is recorded immediately. Thus the counter very quickly follows each change in the input frequency.

The waveforms of Fig. 5-25 tell us much about the operation of the display system. Waveform A is the processed input signal, the frequency of which is to be measured. Waveform B shows the count gating pulse. Counting occurs when this pulse is at logic 1. After a precise interval of time the count gating pulse goes to logic 0 and the counting ceases. At this time the latching pulse (waveform C) arrives and passes the last count number of each chain on to the decoder and to the 7-segment display. After the count has been displayed for an interval of time, the reset pulse arrives and quickly returns everything to zero, ready to begin a new count.

Waveform E shows the pulses that are gated to the first counter. Note that this activity occurs during the duration of the count gating pulse. No pulses are gated to counter 1 during the times that the latching and reset pulses are active. All of these activities occur at a fast rate but the eye sees a continuous display that does not flicker or show a count run-up.

5-6. DIGITAL IC CIRCUITS

There are many types of digital ICs with numerous applications and functions. Several important CMOS types and associated circuits are described.

5-6-1. D-Type Flipflop

A flipflop is a bistable two-element device. It has two stable states. An appropriate input signal causes a changeover from one state to the other.

In a so-called D-type flipflop (Fig. 5-26A) the logic applied to the D input is transferred to the Q output whenever the clock makes a positive transition (edge) from logic 0 to logic 1. The D abbreviates a delay activity because the logic at the D input is held until the clock signal makes the positive transition. Note that the flipflop has two outputs, Q and not-Q. They are complementary outputs. When Q is logic 1, the not-Q output is logic 0 and vice versa.

Assume input D is set to logic 0. The output at Q and not-Q does not change until the clock input changes from logic 0 to logic 1. When this happens the Q output becomes logic 0 and the not-Q output logic 1. If the D input is now set to logic 1, the circuit will

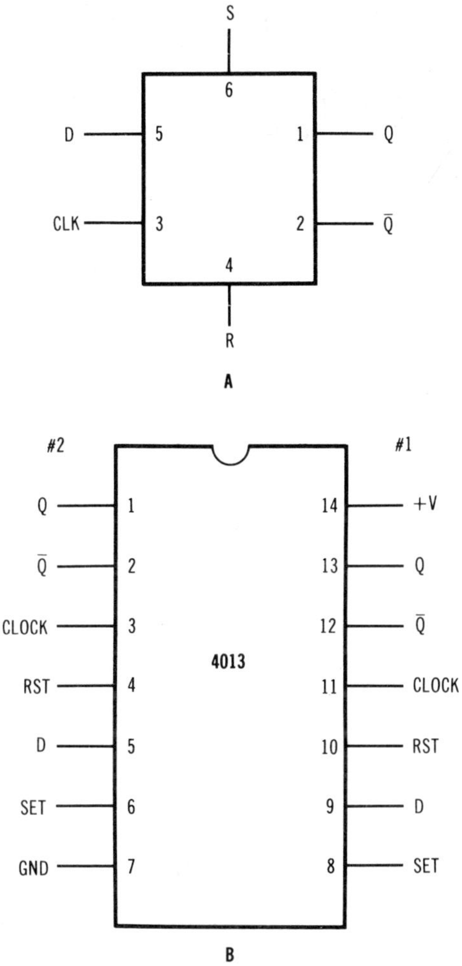

Fig. 5-26. Dual D-type flipflop.

flipflop upon arrival of the next positive transition of the clock from 0 to 1. At this moment the Q output changes to logic 1; the not-Q output to logic 0.

The set and reset potentials also influence the output. A positive set results in a logic 1 at Q and a logic 0 at not-Q, regardless of previous output logics. Conversely a positive reset results in a logic 1 at not-Q output and a logic 0 at Q output. These latter set and reset functions are dominant and overrule the output logic that has been established by the D logic.

An example of a D-type, flipflop is the 4013 digital IC. In fact, it contains two separate flipflops in the same package, Fig. 5-26B. A test circuit for checking out the operation of the flipflop is given in Fig. 5-27A. Four switches permit you to set the logic of the D, clock, set, and reset inputs. A transistor and LED permit you to observe the logic at the Q output.

As shown, a logic 1 is applied to the D input. The set and reset inputs are at logic 0 because of the link to ground through the 10K resistors. If you momentarily connect reset switch S4 to logic 1, the Q output will be logic 0 (LED off) and the not-Q output will be logic 1. Now momentarily connect the clock input to logic 1. Q output is 1 (LED on); not-Q output is 0. Connect switch S2 to ground,

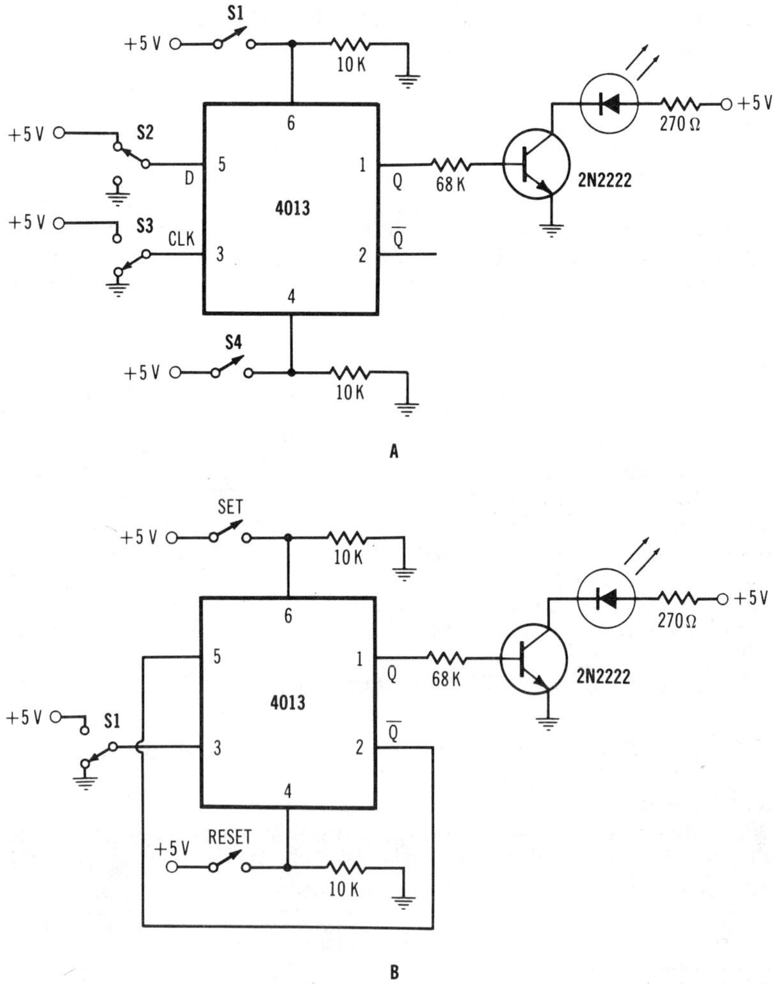

Fig. 5-27. Type-D flipflop circuits.

applying a logic 0 to the D input. The output will not change. However, if switch S3 is momentarily connected to logic 1, the output will flip and the LED will turn off. Connect switch S2 back to logic 1. Nothing will change until switch S3 is momentarily connected to logic 1. Then the output will flip again. Whenever the LED is off it can be overruled by momentarily connecting switch S1 to logic 1. This is the set function.

The basic circuit can be made to flipflop on a continuous basis by making a feedback connection between the not-Q output and the D input as shown in Fig. 5-27B. Under this condition there will be a change in the output logic every time the clock voltage swings positive. Now the logic at the not-Q output changes the D input each time the output logics change. In other words, the not-Q output takes the place of switch S2 in circuit A. Each time switch S1 is momentarily connected to logic 1, the output logic will change.

Assume that the Q output is logic 0. The first time switch S1 is momentarily connected to logic 1, the Q output will become logic 1. The next time switch S1 is momentarily connected to logic 1, the Q output changes to logic 0. The next time switch S1 is connected momentarily to logic 1, the Q output will be back at logic 1. Note that it required two positive voltage changes of the clock input to produce a complete cycle at the Q output (1 to 0 and back to 1). There is one output cycle for each two input cycles. If a square wave were applied to the clock input, the square wave at output Q would be one-half the frequency of the input square wave. The same applies for the output of the not-Q output. In this application the D-type flipflop is dividing the input frequency by a factor of 2. It is said to be acting as a *binary counter*.

It is possible to connect the output of one flipflop or binary counter to the clock input of the next. In this case the total division would be 4 (2 × 2). The combination is said to be acting as a divide-by-four counter. Four such binary counters connected in a series string would produce a total division of 16 (2 × 2 × 2 × 2). Furthermore by the proper choice of feedback links among the individual binary counters it is also possible to establish other counts such as 5 or 10 or some other whole count. There are many applications for flipflops and counters in digital systems.

5-6-2. Hexadecimal/BCD Counter

The CMOS 4029 is a versatile digital IC counter. The pinout diagram is given in Fig. 5-28. The package contains a series of D-type binary counters. Depending on the logic applied to pin 9, the total count can be made 16 (hexadecimal) or 10 (BCD). The logic applied to pin 10 determines whether there is an up-count or a down-count. In the case of BCD setting, then, the count can proceed 0 *up* to 9 or 9 *down* to 0.

The connections for division-by-ten counting (decade BCD) and division-by-16 (hexadecimal) counting are shown in Fig. 5-29 along with associated tables that show the DCBA binary output logic as a

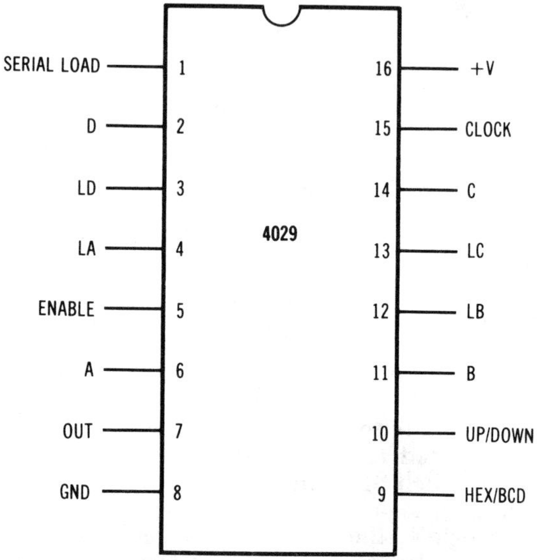

Fig. 5-28. Versatile counter CMOS chip.

function of the arriving clock pulse. Circuits are connected for up-counting. If down-counting is desired pin 10 must be connected to ground (logic 0).

There are many possible counting methods. Our accepted method is decade counting which involves numbering to the base 10. Binary counting is, of course, to the base 2 as introduced earlier in the chapter. Another popular counting method in computers and microprocessor systems is called hexadecimal meaning counting to the base 16. For that reason the counter just described has applications in hexadecimal systems because of its division-by-sixteen capability.

In the hexadecimal system both numbers and letters are used. In counting from 0 to 16 in hexadecimal it would be 0-1-2-3-4-5-6-7-8-9-A-B-C-D-E-F. In hexadecimal all decade numbers between 0 and 15 can be represented as a single symbol. Equivalents of binary, decade and hexadecimal counting are shown in Table 5-2. Additional details on counting methods can be found in Reference IV.

5-6-3. CMOS Decoder/Driver

The Motorole MC14495B is a versatile driver for a 7-segment display. In fact, it is a combination latch, decoder and driver in a single package. Furthermore the driver output can be supplied directly to the 7-segment display. No current-limiting resistors are required. Another attractive advantage is that it can be used for both BCD and hexadecimal displays. The decoding is such that the

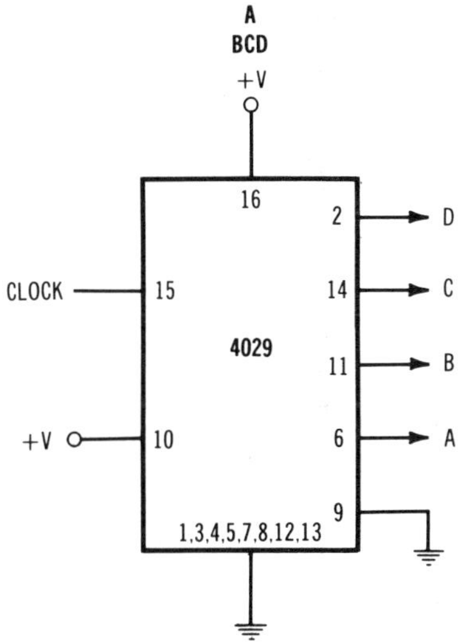

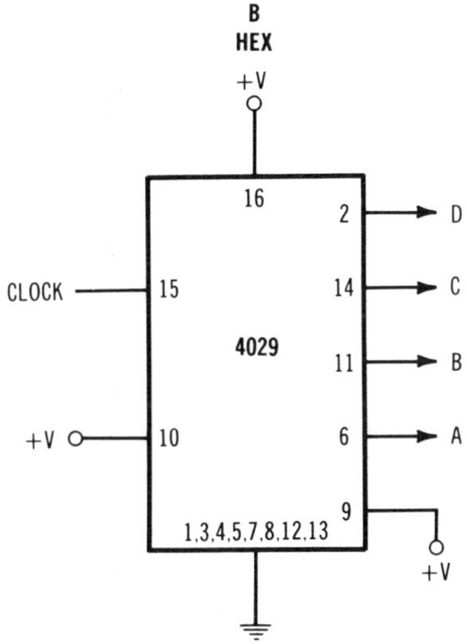

IN CLK	OUT DCBA
0	0000
1	0001
2	0010
3	0011
4	0100
5	0101
6	0110
7	0111
8	1000
9	1001
10	0000
11	0001
12	0010
13	0011
14	0100
15	0101
16	0110
17	0111
18	1000
19	1001
20	0000

(A) BDC.

IN CLK	OUT DCBA
0	0000
1	0001
2	0010
3	0011
4	0100
5	0101
6	0110
7	0111
8	1000
9	1001
10	1010
11	1011
12	1100
13	1101
14	1110
15	1111
16	0000
17	0001
18	0010
19	0011
20	0100

(B) Hex.

Fig. 5-29. Connections for BCD and Hex counting.

Table 5-2. Count Systems Equivalents

Decimal	BCD	Binary	Hex
0	0000	00000000	0
1	0001	00000001	1
2	0010	00000010	2
3	0011	00000011	3
4	0100	00000100	4
5	0101	00000101	5
6	0110	00000110	6
7	0111	00000111	7
8	1000	00001000	8
9	1010	00001001	9
10	0001 0000	00001010	A
11	0001 0001	00001011	B
12	0001 0010	00001100	C
13	0001 0011	00001101	D
14	0001 0100	00001110	E
15	0001 0101	00001111	F
16	0001 0110	00010000	10
17	0001 0111	00010001	11
18	0001 1000	00010010	12
19	0001 1001	00010011	13
20	0010 0000	00010100	14
21	0010 0001	00010101	15
22	0010 0010	00010110	16
23	0010 0011	00010111	17
24	0010 0100	00011000	18
25	0010 0101	00011001	19

same 7 segments of the display can also be used to reproduce the hexadecimal numbers A through F. The actual display capability is shown in Fig. 5-30. The A, C, E, and F are upper case. However, hex upper case B and hex upper case D are displayed as lower case b and d.

Input and output are shown in the pinout diagram. A latch pulse can be applied to pin 7. Its leading edge will latch whatever binary number is being applied to the DCBA input. This number will be displayed so long as the latch remains high. When the latch is made low, the display will continue to count. An LED connected directly to pin 4 will turn on whenever the six numbers A, B, C, D, E, and F are displayed.

An interesting demonstration circuit is given in Fig. 5-31. The clock can be derived from a 555 timer using components that will provide a very-low-frequency output. The output should be a conventional 7-segment common-cathode display. Note that the DCBA output lines of the counter are connected to the DCBA inputs of the 4495. The three switches can be used to demonstrate the versatility of the circuit. Switch S1 permits you to switch between up-count and down-count. As connected the circuit will up-count. Switch S2 selects BCD or hex counting. As connected in the diagram it will count BCD. Switching over to logic 0 (+V) will result in hex counting and the display of the numbers A through F. Switch S3 provides a latching demonstration. As connected there is no latching. When switch S3 is connected high (+V) the binary DCBA signal applied to the 4495 input at that moment will be locked and displayed continuously. Changing switch S3 back to low will restore normal counting.

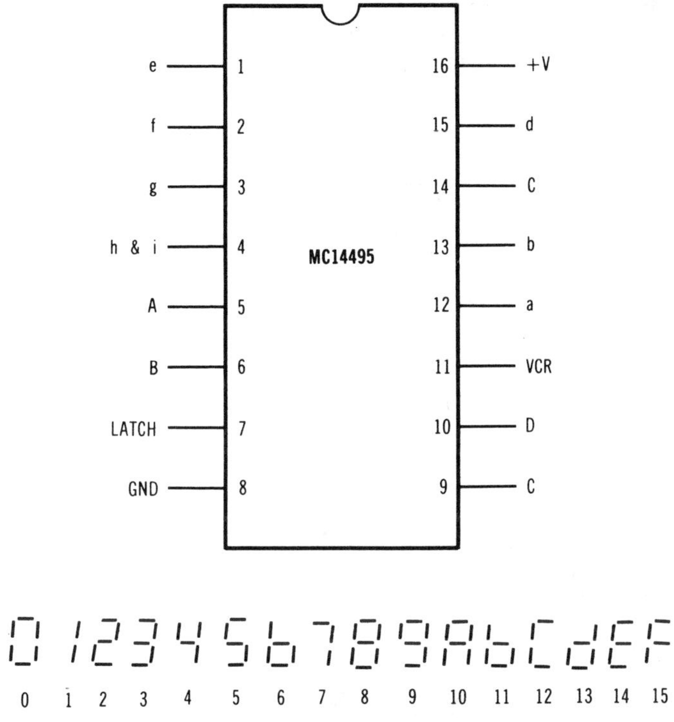

Fig. 5-30. Decoder/driver pinout diagram.

5-6-4. CMOS Data Selector

A data selector has several inputs that supply data to a single output on a scheduled basis which is determined by the binary addressing supplied to its control circuit. The 4512 chip is an example of such a selector. There are eight inputs (Fig. 5-32) that can be selected one by one according to the 3-bit binary signal applied to pins 13, 12, and 11. The highest number that can be represented by a 3-bit binary is 7, establishing proper selection for the maximum number of 8 inputs. Thus the input from 0 through 7 is selected in sequence when the 3-bit binary signal steps from 000 to 111. Of course, inputs can be selected at random according to the changing values of the CBA applied.

Output is removed at pin 14. If the logic at input 3 were 1 when the CBA code is 011, a logic 1 appears at output pin 14. Were the logic 0; the output would be 0. If the CBA address is changed to 110, the output at pin 14 would match the logic present at input 6.

The 4512 data selector is said to have a tristate output. Pin 15 provides tristate enabling. When pin 15 is low, the selector operates in normal fashion. However, if pin 15 is switched high, the output pin 14 is switched to a high-impedance state. Thus the three possible outputs are 0, 1, and high-impedance disconnect. An advantage of a tristate output is that it can be connected to a bus line along with similar circuits having tristate outputs. As a result there is minimum loading and interaction among chips connected to the bus line. Only the chips which are to be activated at the moment are placed on line with the tristate enabling facility.

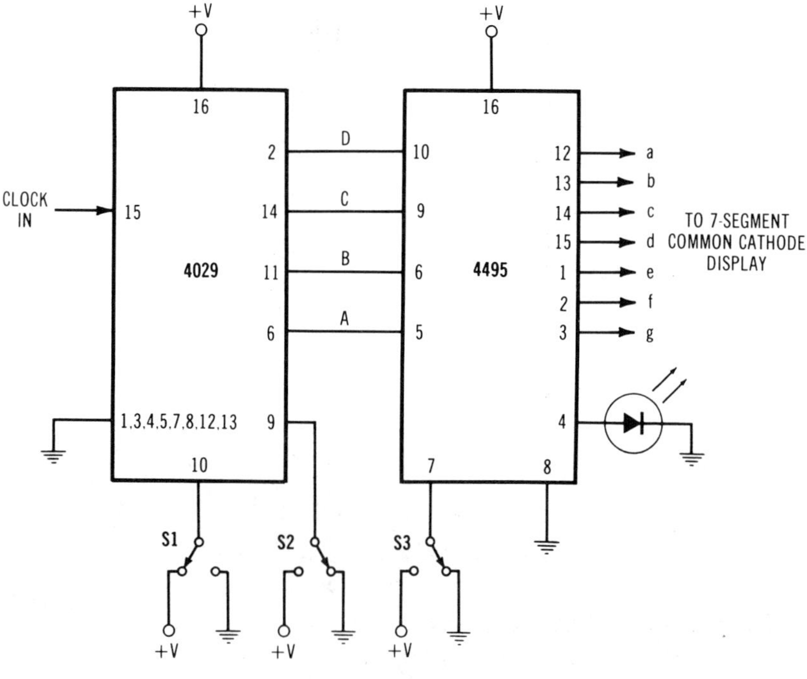

Fig. 5-31. Counter/decoder combination.

Circuit connects are shown in Fig. 5-33. The eight inputs are shown at the top left. With all the switches closed, a logic 0 is being applied to each input. Whenever a switch is opened for a particular channel, that input sees a logic 1 because of the connection of the input through its resistor to the supply voltage.

Output is removed at pin 14. A transistor and LED will cause the LED to turn on whenever the output is logic 1. Switches S1 through S3 can be used to provide a specific binary address. Tristate enabling can be checked with switch S4.

To transfer the logic at input 5 to the output you must use the CBA address of 101. By opening and closing the switch associated with input 5, the LED at the output will turn *on* for a logic 1 input and *off* for a logic 0 input. When the input switch of input 5 is open the LED is on.

If you now connect switch S4 high the LED will turn off, indicating that a disconnect has been made at the pin 14 output. For this setting of switch S4 there is really no output (neither 1 nor 0) and the transistor and LED have been completely disconnected from the 4512.

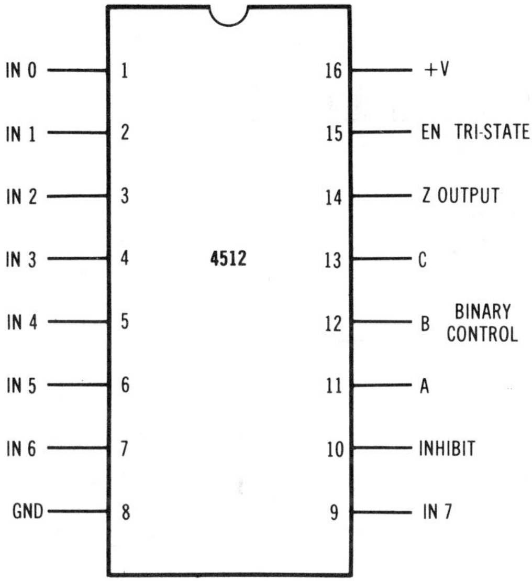

Fig. 5-32. Eight-channel data selector.

5-6-5. CMOS Addressable Output Latch

The performance of the output latch is complementary to the output of a data selector. In this circuit the input data is channeled to a specific output. The output selected is again determined by a binary addressing system. Furthermore the logic supplied to a specific output is latched and remains there until it is cleared by a reset pulse.

An example of an 8-output latch is the CMOS 4099 (Fig. 5-34). There are eight outputs and a single data input. The output is selected by the 3-bit binary control. Data is transferred to the output only when the WRITE/DISABLE pin 4 is low. Reset pulse is applied to pin 2.

Assume that there is a logic 1 on the data input when the binary address is 101. If at the moment pin 4 is also low a logic 1 will appear on the number 5 output, pin 14. Once latched this output data will not change regardless of the logic applied to pin 4. Thus one or more outputs can be activated, and, in effect, serial input data is presented as parallel output data. Outputs can be deactivated by setting the chip disable or reset (pin 2) high.

A demonstrating circuit is shown in Fig. 5-35. A group of switches on the left permit setting of the data logic, CBA

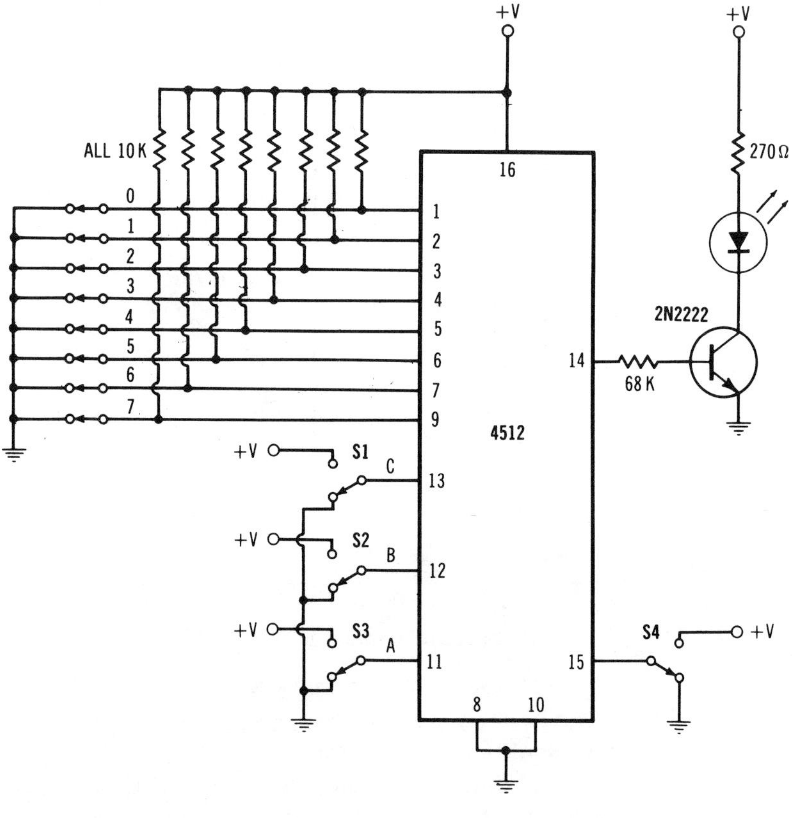

Fig. 5-33. Data selector circuit.

addressing, WRITE/ENABLE and reset. The output level of the 4099 is low and is built up with the use of an amplifying inverter. In this case the output of the inverter supplies current to individual LEDs.

When the circuit is connected exactly as shown the address is 000, the WRITE activity is enabled and, therefore, the 0 output LED turns on. This output remains high even though the switch S5 is set to logic 1. If another CBA address is now used such as 111, the number 7 LED will turn on as soon as switch S5 is returned to low. Now both the 0 output and the #7 output LED's are on. Output can be deactivated by setting switch S6 to high. Of course, in an actual circuit these activities would take place very quickly because of the digital signals being applied to the data, address inputs, as well as the WRITE and reset inputs.

Additional knowledge and practical experience with TTL and CMOS solid-state devices can be gained by constructing the simple demonstration circuits included in the *Solid-State Circuit Book* of Reference III, Chapter 1. Devices are low cost and easy to obtain. Circuits are built on a solderless circuit board.

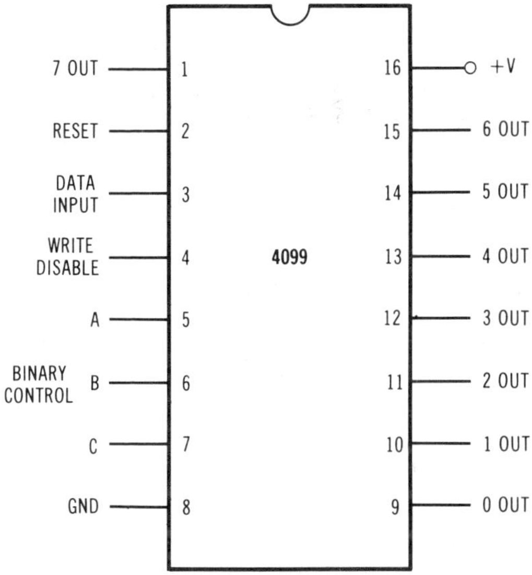

Fig. 5-34. Addressable output latch.

5-7. MICROPROCESSOR SYSTEMS

In elaborate and not so elaborate radiocommunication systems the microprocessor has become a popular control and organizing device. It is capable of monitoring complex facilities and making appropriate changes when necessary. Microprocessors are also found in test equipment. This section gives you a brief introduction to the major sections of microprocessor systems.

5-7-1. Microprocessing Simplified

The microprocessor unit (MPU) is the control center of a microprocessor control system or a microcomputer. Its awesome capabilities are housed within a small integrated circuit chip. Size varies from a standard 16-pin chip up to a 64-pin configuration. To carry out its activities the MPU must be surrounded by other devices, within or without, that can supply its input needs and respond to its output controls. A very simple diagram of an MCU plan is given in Fig. 5-36.

Each MPU type has its own so-called instruction set. Depending on size and power, a processor may have anywhere from sixteen, to scores, to hundreds of individual instructions that it is capable of carrying out. At a given instant a proper instruction is selected by a program stored in memory.

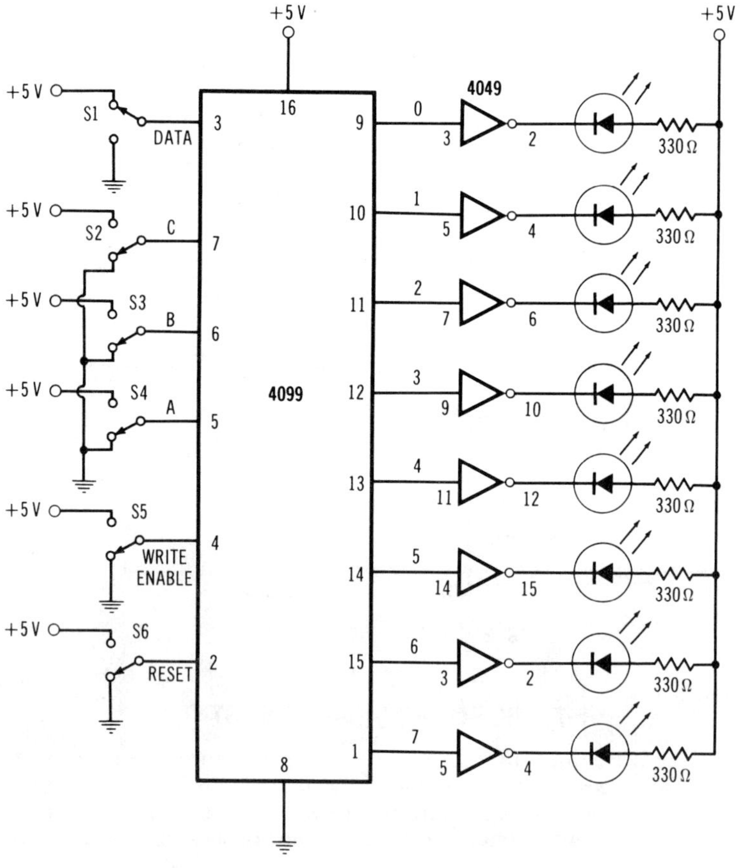

Fig. 5-35. Addressable output circuit.

First a programmer with a knowledge of the capabilities of the MPU instructions sets down such a program which the system will fulfill in an orderly sequence to meet the needs of a microcomputer, radiocommunication, or commercial control system for which it has been designed. The program is then stored in the memory. The system can be made to run the program whenever necessary or it can be made to go through the program endlessly depending on application.

The timing or sequencing of the microprocessor activities is under control of a clock. The clock is a pulse generator that permits very fast switching of the activities of the MPU and associated chips. Clock rates may be as high as 2 million pulses per second. This gives you some idea of the speed with which a system can go through a program and the speed with which information can be evaluated and result in output activities. No human could begin to react so quickly to the vast quantity of information that can be supplied, evaluated, and responded to. It may take a long time to prepare a program and check the validity of such a program. However, once set in operation, the speed with which information can be gathered and acted upon is phenomenal.

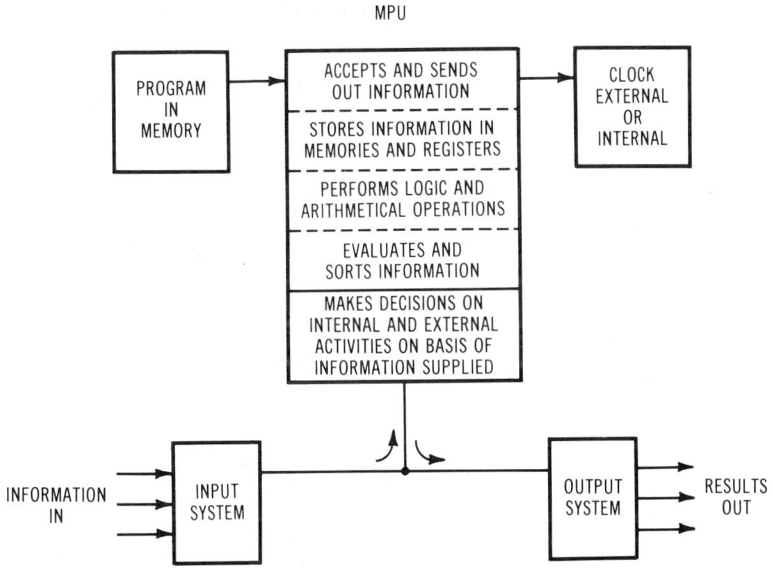

Fig. 5-36. MPU simplified plan.

In an MPU system there must be an input system to gather the desired information and an output system to carry out the dictates of the MPU evaluation.

In general, an MPU accepts and sends out information. Incoming information is placed in registers for short storage and immediate use. There are also memories that provide somewhat longer storage facility to allow enough time for manipulation. Each microprocessor includes an arithmetical logic unit (ALU) that solves logic functions and performs mathematical steps just like a calculator. Information must be evaluated, sorted, and distributed inside of, and outside of, the microprocessor proper. On the basis of the information supplied decisions must be made and proper internal and external activities must be initiated. No sorcerer ever matched the magic semiconductor brew in these tiny black boxes.

A Work Example

A simplified plan of an MPU in a work situation is shown in Fig. 5-37. In this application, the microprocessor system is to monitor,

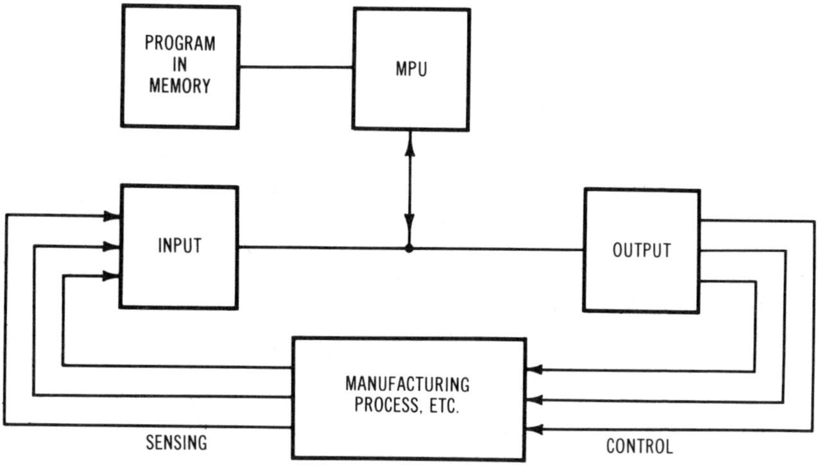

Fig. 5-37. The MPU at work.

measure and control a manufacturing procedure. There are input sensing lines and output control lines that extend to the manufacturing machinery. The sensing devices are called sensors or transducers. They are capable of measuring the device being manufactured in various stages of assembly, monitor the operation conditions of the machinery, count the number of items being produced, as well as many other parameters as a function of the manufactured item and manufacturing procedures. There may be three sensors or scores of them, or hundreds of them, depending on the complexity and exactness of the task. There can be as many controls over the manufacturing procedures as are required.

Displays can be associated with the process indicating where defects arise and what corrections are being made. Faults can be anticipated and corrections made automatically before a single bad part is produced. In the case of a serious fault a procedure or the entire process can be stopped automatically by the microprocessor system.

In a critical, or precision procedure, a tremendous advantage of a microprocessor control system is the speed with which it responds to a situation. An operating maladjustment can be identified and corrected long before a human operator would recognize that such an adjustment was needed, not even considering the time that the operator would require to react and make the necessary correction.

Sensors of many types are available. Their responsibility is to change over a reading or measurement to an electrical representation which is applied to the input of the microprocessor system. Likewise, the output of the system is an electrical one that manipulates switches and electronic controls to bring about the electrical and mechanical changes that the microprocessor has deemed necessary.

In addition to being the decision-maker of all computers, a microprocessor has many other applications. Their application in manufacturing, mentioned previously, is just one. They can operate a radiocommunication terminal, a broadcast transmitter or a color television camera. Sublime to silly, they can control the study of Saturn's rings or be the brain of an elaborate pinball machine. They can be part of an electronic toy or an automatic pilot. Present and future applications are so many that the microprocessor is sure to stay.

5-7-2. Bits and Bytes

Microprocessors and microprocessor systems are supplied with input signals in *bits*, evaluate and act upon information in bit form, and deliver output signals in the form of bits. A bit is short for binary digit. A binary digit or bit can only be one of two quantities, a 1 or a 0. Information presented in bit form is called data. A simple example of the use of bits is in order.

Each microprocessor system has a memory which locks up the program procedure. The memory information is stored in many locations like the rows and columns of the individual compartments of an elaborate shelf. However, only two types of information are stored: these are 1's and 0's. Each storage space has a specific address. Even the address is expressed in binary form.

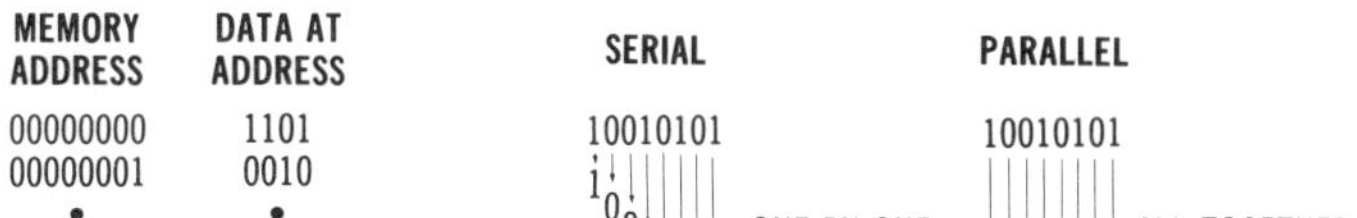

Fig. 5-38. Bits and bytes.

For example the address of the first row in a microprocessor program may be 00000000. The second program address would be 00000001. Each one of the 0's and 1's in an address is called a bit. The 8 bits of a complete address are called a *byte*. Refer to Fig. 5-38.

Not only is the memory address given in bits but the data stored at each address is given in bits. Let's assume that the first address stores four bits of information. The data stored might be 1101. This quantity when delivered to the microprocessor might cause it to perform one of the activities of its instruction set. If the data stored were 1011, some other instruction would be performed by the microprocessor.

In the second storage place or memory address you might find the data 0010. This might cause the microprocessor to respond in a specific way to the instructions just given previously. It might pick up some incoming digital data or deliver some digital data to the output. As in the case of the memory address each 0 and 1 stored by the memory is called a bit. The 4 bits of the data together are called a byte. If you wish to be more specific you state address bit and address byte, or data bit and data byte.

Binary information is conveyed from one circuit to another in one of two ways. These are serial and parallel. In a serial transmission system the byte is conveyed one bit at a time as shown in Fig. 5-38. For example the quantity 10010101 would be conveyed in the sequence 1-0-0-1-0-1-0-1 on one line.

If the information is to be conveyed in parallel form, the entire byte would be conveyed at one time. It is apparent that with the latter method, information can be conveyed at a much faster rate and a microprocessor will respond more quickly to operating conditions. However, the path of transmission would require eight individual conducting paths. In the serial conveyance method only a single line would be needed. Sometimes the latter is the more economical and simpler procedure when speed of action and evaluations must not be made as quickly.

5-7-3. More Essentials of a Microprocessor System

The principal activities of a microprocessor system were introduced in Figs. 5-36 and 5-37. However, additional circuits and signals are required if these activities are to be implemented. Additional essential circuits and interconnections are given in Fig. 5-39.

You learned previously that a memory must be addressed. There must be a means of moving the activity step by step from one memory byte position to another. This sequencing is accomplished with a counter as shown at the top right. Usually the counter is a part of the MPU proper although it is shown externally in Fig. 5-39. The counter is driven by the MPU clock which acts as the timing center for all MPU activities. If the memory has an 8-digit address there must be eight lines connected between the counter and the memory. Such a multiconductor interconnection is called a bus. Bus lines are drawn as shown in Fig. 5-39.

Address information goes out to other parts of the system as well. In particular the input/output (I/O) system must be addressed in such a manner that a particular input is chosen at a particular time and a particular output is selected at a particular time. This impor-

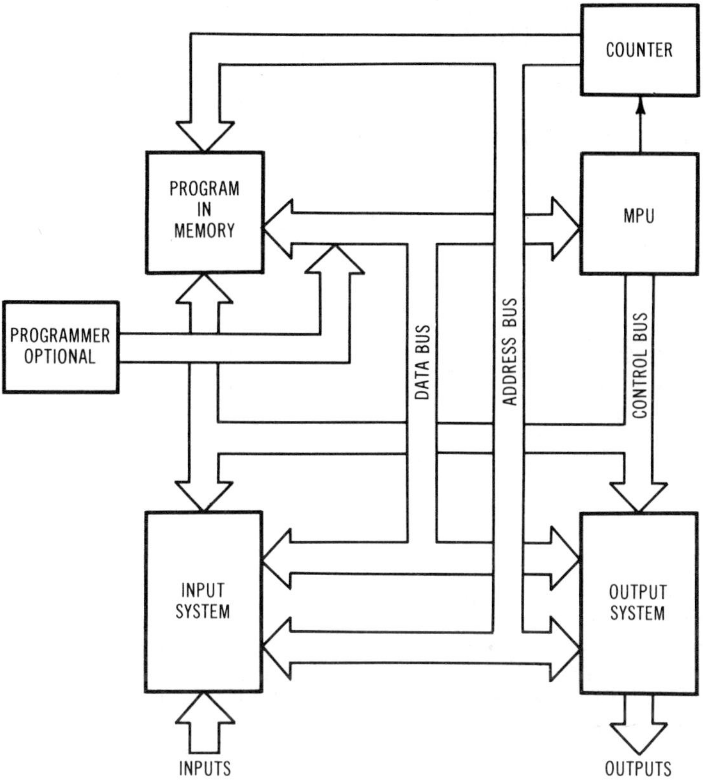

Fig. 5-39. Functional plan of microprocessor system.

tant address information is said to be distributed among the sections of a microprocessor system by an *address bus*. Note the interconnections among counter, memory, input system and output system are shown. Address information flows in a single direction out of the MPU to the various circuits where it is needed, inside of and outside of the MPU.

Data lines or *data bus* must also interconnect the MPU with the memory and I/O systems. The data bus line is a bidirectional facility because data must be exchanged among the devices of the microprocessor system. As the program is sequenced by the step-by-step change of memory address the information in each memory byte is released to the data bus. In Fig. 5-36 this was shown as a single line connecting the memory and the MPU. In all but the very simplest of processors a data bus interconnects the memory and MPU. Also the data bus connects to the I/O systems as well as other devices depending on the complexity of the system. Thus there are several types of information exchanged among units of the data bus line. Instructions go out from memory to the MPU in accordance with the MPU instruction that the program selects. Another memory byte will provide the information as to where particular data can be collected and how it is to be executed by the MPU. Thus data is delivered from the input system to the MPU. Furthermore data must leave the MPU and go to the output system. The data bus is indeed a two-way path.

In a design microprocessor system, or in a system that permits the program to be changed by the user there must also be a data bus connection between the programmer and the memory. Such a facility permits programs to be programmed, changed, and checked out for performance validity. Memories are used to store the digital 1's and 0's of a program. Actually there are three basic memory types--permanent, temporary, and a semipermanent version that can be erased with suitable facilities and replaced with a new program. A temporary memory chip is called a *RAM* for *Random Access Memory*. The data stored in a RAM is lost whenever its supply voltage is removed. A permanent memory chip is called a *ROM* for *Read Only Memory*. Information stored cannot be destroyed by loss of supply voltage. ROMs are prepared by the chip manufacturer according to the programs submitted by the users. An example of a semipermanent type is the EPROM or *Erasable Programmable Read Only Memory*. An ultraviolet light can be used to erase a program from this type and then the chip can be reprogrammed.

Operations in a microprocessor system must be synchronized and special information must be sent out to the various segments as a function of the particular instruction activated in the MPU. This is a function of the information sent along the conductors of the control bus. One such type of information is the clock pulse which goes out to synchronize operations in other circuits with the sequencing of the program memory. A pulse must go out to select input or output operation, depending on whether data is to be delivered to the MPU from the input system or supplied to the output system. There are other special pulses sent out in accordance with the activated instruction set of the MPU.

The typical microprocessor is rated on the basis of the number of lines on the data bus. It is said to be a 1-bit, 4-bit, or even a 64-bit microprocessor. The three most common values are 4 bit, 8 bit, and 16 bit. The higher the number of bits, the higher is the so-called power of the microprocessor. What this really means is that data can be transferred and acted on faster with the parallel movement of a high number of bits. Also a system is able to respond to a greater volume of data because of the higher number of bits.

5-7-4. Simplified Program Sequence

In a microprocessor system there are two activity phases. These are known as *fetch* and *execute*. In the *fetch phase* the instructions or operational code (OP CODE) is brought from memory and enters the MPU. In a typical microprocessor system the fetch would involve the time interval of one clock. The *execute phase* might require one or more clock periods depending on the nature of the information required to execute the instructions. In the execute activities the data must be gathered that is to be acted on. This data is known as the OPERAND. In the usual programming activity the address of the OPERAND must first be obtained from the memory. After the address is obtained then the OPERAND must be gathered from that address. This particular type of execute then would require two clock periods. The combination fetch and execute phases in this case would equal the time interval of three clock periods. Remember that the OP CODE is the instructions to be followed and the OPERAND the data upon which the instructions are to be performed. All of the above activities are timed and synchronized by the clock pulses.

Usually there are two clock pulses of the same frequency but differing polarity. One might be referred to as clock and the other as inverted clock. The two clock components are of the same frequency but may differ in polarity and duration.

Two clock components are shown in Fig. 5-40. The clock pulses are not square waves. Thus in comparing the A and B clock intervals the negative portion of the B clock begins ahead of the positive portion of the A clock. The A clock has its trailing negative transition occur ahead of the positive transition of the B clock. These relative timings are set to properly augment the fetch and execute activities.

The activities for the three clock pulse intervals of the previous example are shown. The first positive transition of the A clock

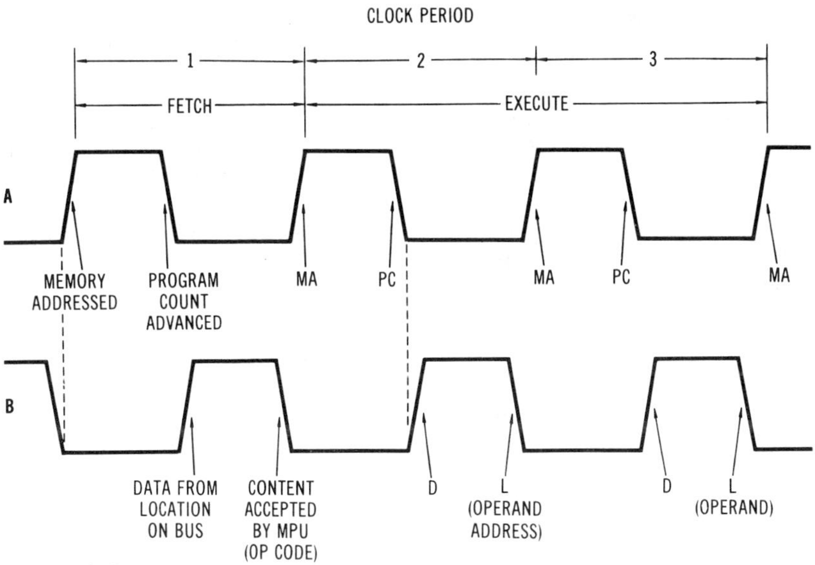

Fig. 5-40. Program activities.

addresses the first position in the memory. The trailing transition of the A clock simply advances the count in preparation for the next memory address. The first positive transition of the B clock places the data present in the first memory address on the data bus. The negative transition of clock B causes the MPU to accept this data. The data in this case is the OP CODE. This OP CODE data is now latched into the MPU and causes it to evaluate what must be done and then to execute the instruction. This concludes the activity of the first clock period and the appropriate OP CODE has been fetched.

Next the second memory position is addressed and according to OP CODE instructions it is necessary to determine the address of the OPERAND. The positive transition of the second B pulse causes this address to be placed on the data bus. The negative transition of the same pulse latches the OPERAND address to the MPU.

The third line of memory is addressed by the positive transition of the third clock A pulse. As in previous cases the program count is advanced by its trailing edge. The positive transition of the B clock third pulse places the OPERAND data on the data bus line. Its trailing edge latches the OPERAND to the MPU and the appropriate operation is performed in response to the OP CODE instruction supplied to the MPU during the fetch period. Note that the execute time required two clock periods. The speed with which the instruction was fetched and executed is revealing. Assuming a clock frequency of 500 kHz or 0.5 MHz, the clock period would be 2 microseconds (1/.5). Thus the entire procedure required only 6 microseconds.

The previous instruction OP CODE may have told the MPU to store this OPERAND in the ALU register (sometimes called an accumulator). Other instructions are now supplied from the program asking the MPU to add this stored data to the data supplied to one of the inputs of the microprocessor system. After the addition the new result is stored in the accumulator.

Now the MPU might be provided with instructions to compare this sum with the data on some other input line. The instructions might tell the MPU to place a logic 1 on output line 2 if the two quantities are of the same value. If they are not the same, then the MPU should put a logic 1 on output line 4 and so on.

In the above manner a short program of perhaps 12 instructions or a longer one with 200 instructions can be sequenced at a high rate of speed. The program can be repeated many times per second, demonstrating how quickly a microprocessor system can respond to input changes, transcribing them to output decisions.

5-7-5. Interfacing

Previous paragraphs covered the fundamental makeup of the microprocessor system. Such a system has input and output facilities that must be interfaced with a variety of input sensors. Likewise its output must be interfaced with the systems and devices under its control.

On the input side (Fig. 5-41) there might be a mechanical or digital switching panel that is used to supply the microprocessor system with specific operational sequences. The microprocessor installation could be involved in the measurement of electrical and/or physical parameters to which the system must respond. So-named sensors or transducers are able to evaluate a variety of parameters such as voltage, temperature, current, frequency, velocity, pressure, etc. Their primary responsibility is to form an electrical output current that is a measure of the parameters to be evaluated. Since a microprocessor is a digital device, these analog signals must be converted to digital data. This changeover is handled by an analog-to-digital converter or ADC. These interfacing circuits, then, process a great variety of possible information to digital representations that can be acted on by the microprocessor system.

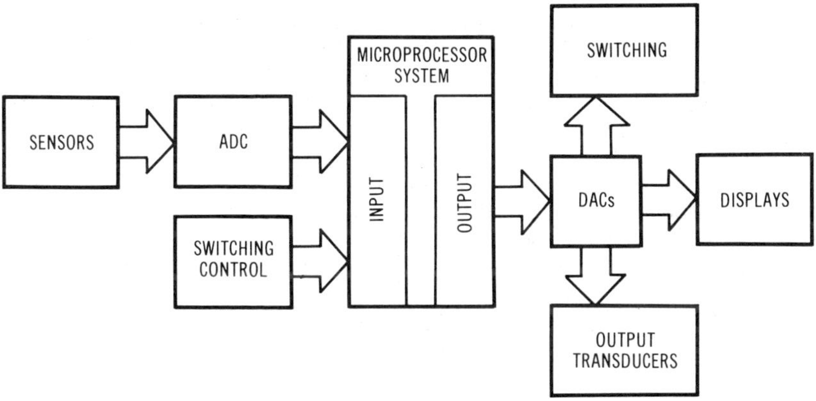

Fig. 5-41. Interfacing methods.

Digital information is also developed at the output of the microprocessor system. As a result it must often be supplied to a digital-to-analog converter, or DAC. The analog data is then used to operate a visual display, switch electrical/electronic devices on and off, change a frequency, operate an electromechanical valve, activate emergency equipment, etc. Interfacing is indeed an important segment of microprocessor control systems.

Additional knowledge and practical experience with solid-state devices can be gained by studying and building the demonstration circuits included in the microprocessor fundamentals book of Reference IV, Chapter 1.

CHAPTER 6

Test Equipment and Methods

Test instruments—both simple and complex—are needed to adjust, monitor, and troubleshoot the transmitters and receivers of two-way radio systems. For some transmitter measurements, the accuracy must be well within the two-way radio performance tolerances established by the FCC. Moreover, not only must the test equipment be of good quality, but it must be used very carefully as well.

The FCC is concerned primarily with the operating frequency and its stability, the power output, modulation level of the communications transmitter, and freedom from spurious radiation. The test equipment must be capable of making these measurements accurately. Otherwise, it would be difficult to determine whether the transmitter is complying with FCC technical standards. These tests must be performed by a licensed general radiotelephone technician.

The owner of a two-way radio system is also interested in additional performance criteria. He is concerned with reliable, intelligible, and interference-free communications. To him, transmitter and receiver performance are of equal importance. The base and mobile transmitters must convey a good signal within the area served by the installation. Transmitter power output and modulation levels must not be permitted to decline and should be held near, but not beyond, the maximum limits set by the FCC. The receiver sensitivity and stability must be maintained at a high level. It is preferable to keep the equipment in such condition that is can be used for long periods without retuning and service.

The operators of such equipment are not expected to have technical knowledge beyond turning the equipment on and off and switching from transmit to receive and back. It is imperative then that the technician in charge of the system know how to use the test equipment in a manner that will keep the system in top operating condition.

6-1. REQUIRED TEST EQUIPMENT

Two or three pieces of test equipment are needed to check out a transmitter to make sure it complies with FCC technical regulations. One such instrument is an accurate frequency counter or meter, used to determine if the transmitter is set exactly on the assigned frequency. Of course, such an instrument also permits the operator to tune the transmitter to its assigned frequency.

The same meter is generally used for checking frequency stability of the transmitter. Once a transmitter is set on its assigned frequency, its frequency drift may be no more than that permitted by FCC regulations. Generally, this requirement is stated as a percentage of the assigned frequency.

Modulation meters are required for checking the percentage of modulation, to make certain it is no greater than that permitted by FCC regulations. In an fm system, 100% modulation represents a maximum frequency deviation of ±5 kHz for most installations, and for other services a maximum deviation of ±2 kHz or ±15 kHz. Again, the modulation meter not only is useful in making certain that 100% modulation is not exceeded, but also permits the technician to tune the transmitter for adequate deviation.

Coverage and reliability are improved if the transmitter is made to deviate to its maximum limit on modulation peaks. Most transmitters include special circuits which prevent instantaneous modulation peaks in excess of 100%.

For am communications systems, the amplitude modulation must not exceed 100%. A test setup can be arranged to measure the actual amplitude-modulation percentage. Here again, the modulation level should be adjusted as high as possible (average above 70%) for the greatest coverage and reliability. At the same time, modulation peaks should not exceed 100%. In am transmitters, too, special circuits are usually incorporated to prevent modulation peaks from overmodulating the transmitter.

An accurate power-output meter is helpful in deriving the best performance from each transmitter. The assigned power output should not be exceeded. However, reliability and coverage can be improved if the transmitter operates at its assigned power output. A power-output meter is helpful in the attainment of maximum power output.

Most communications transmitters are FCC-approved and do not require an accurate power-output measurement. If the transmitter is operated according to its accompanying instructions, it can be assumed to be operating at the rated power output. Such transmitters, when operated in accordance with the tune-up instructions, are not capable of delivering more than the rated output. When the dc input power is normal, one can be reasonably certain the rated power output of such a transmitter is being obtained. The maximum permissible power is presented in terms of power input to the final stage.

Whether this type of transmitter or the one that requires power-output measurements is employed, a power-output indicator is useful—if for no other reason than making certain the transmitter is matched to the antenna system. When the two are matched, the antenna radiates almost the total power that the transmitter is capable of delivering.

Some radio-frequency indicators can be inserted into the transmission path, between transmitter and antenna, to measure power output and the standing-wave ratio on the transmission line. This type of instrument is very useful in measuring radiation from the line, and in making certain that maximum energy is transferred between the transmitter output and the antenna.

Simpler test equipment such as radio-frequency output indicators, absorption wavemeters, dip oscillators, and others are useful in transmitter tune-up and receiver adjustments, and in the localization of trouble in the two-way radio system.

An accurate signal generator is particularly important in receiver alignment. Many of these receivers have an extraordinary sensitivity, down to 1 microvolt or less. This sensitivity is possible only if the circuits have been critically aligned and the receiver carefully tuned. Keep in mind that many of these receivers are fixed-tuned—the user has no way of tuning in the signal if it is absent or mushy.

Some standard pieces of electronic test equipment such as a vom, dvm, capacitor checker, oscilloscope, and other component testers are helpful, too.

6-2. THE ABSORPTION WAVEMETER

The basic radio-frequency indicator and frequency meter is the absorption wavemeter shown in Figs. 6-1 and 6-2. It is hardly more than a calibrated resonant circuit which, when coupled near the source of radio-frequency energy, can withdraw some of the energy. This meter will absorb maximum energy when tuned to the same resonant frequency as the source.

Some form of radio-frequency indicator such as a lamp, neon bulb, or diode and dc-meter combination can be used to indicate the relative strength of the energy absorbed by the resonant circuit of the wavemeter. When the wavemeter is tuned to the frequency of the radio-frequency source, the indicator will read maximum; but if tuned to either side of the resonant-energy frequency, the meter reading or lamp brilliancy will decrease.

If the resonant frequency of the wavemeter tank circuit is known from the dial setting, the frequency of the radio-frequency energy can be determined. Usually the capacitor dial of the wavemeter is calibrated in frequency. Therefore, the setting of the pointer on the calibrated scale, when the capacitor is tuned for maximum rf indication, shows the frequency of the rf signal being measured. In

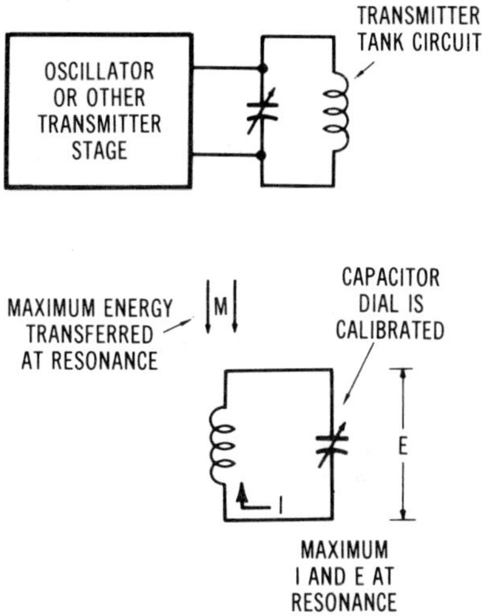

Fig. 6-1. Principle of the absorption wavemeter.

other absorption wavemeters, the dial is calibrated from 0 to 100, and the exact resonant frequency of the absorption tank circuit is determined from a chart.

Wavemeters are usually equipped with replaceable coils so they can be made to operate on different frequency bands, with a separate dial or calibration chart for each band.

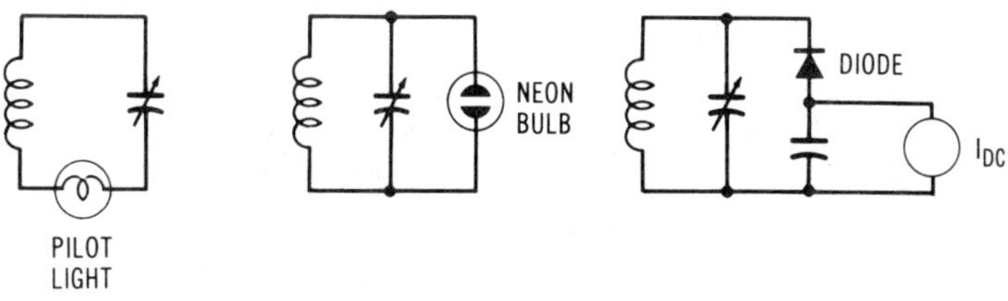

Fig. 6-2. Indicators for absorption wavemeters.

6-2-1. Using the Wavemeter

Energy from the tank circuit of the unit under test is coupled into the resonant circuit of the wavemeter. Use loose coupling to minimize loading the circuit under test, and couple the absorption wavemeter only close enough to the radio-frequency source to permit a usable rf indication (lamp glow or dc meter readings).

To provide minimum loading and additional convenience, some absorption wavemeters include a low-impedance pickup coil. The main meter coil remains part of the wavemeter proper and is not coupled close to the radio-frequency energy being measured. The small pickup loop is coupled to the source of energy and transfers the small amount needed into the absorption-wavemeter circuit (Fig. 6-3).

The absorption wavemeter is important in making approximate frequency measurements while a transmitter is tuned, and, in particular, for checking multiplier harmonic outputs. It also serves as a good indicator of the strength of the radio-frequency energy. When set at a fixed position from the energy source, the influence of tuning on the radio-frequency output can be noticed immediately.

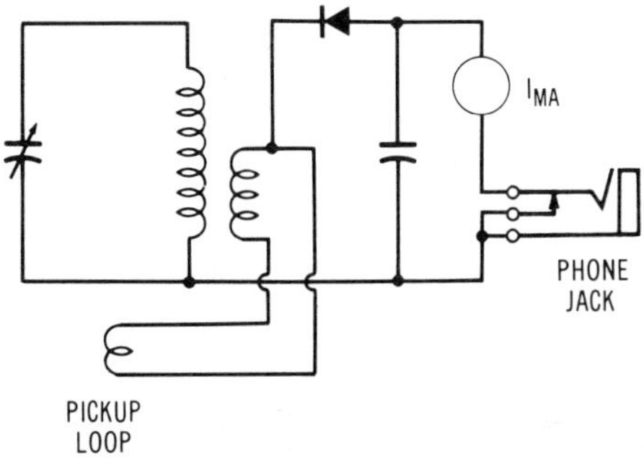

Fig. 6-3. Absorption wavemeter with pickup loop and phone jack.

Most absorption wavemeters also include a phone jack into which a pair of headsets can be plugged and any am modulation on the rf signal can be heard. Hence, when the absorption wavemeter is used to check out rf amplifiers that convey a modulated rf signal, the quality of that modulation can be tested approximately, using the detector circuit that is part of many absorption wavemeters. For proper demodulation of an fm radio-frequency signal, some form of fm detector would have to be associated with the wavemeter.

6-2-2. Wavemeter as an RF Indicator

The wavemeter principle can also be used as a sensitive radio-frequency output indicator. In fact, it is no problem to attach a small antenna to a wavemeter, as shown in Fig. 6-4, to make it even more sensitive to the radio-frequency output of a transmitter.

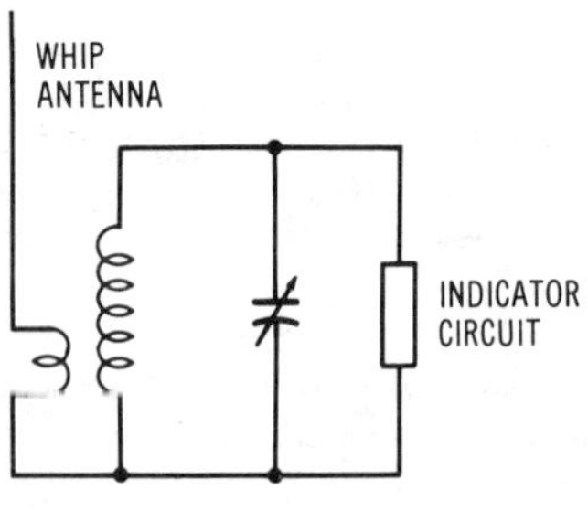

Fig. 6-4. Use of the wavemeter.

If the wavemeter is tuned to the operating frequency of the transmitter and placed somewhere in its immediate field, a strong indication can be obtained. It is in effect a *field-strength indicator*. With the meter positioned a fixed distance from a transmitter, various tuning adjustments can be made to maximize the rf output. In fact, the wavemeter, if placed in the field of the transmitter antenna, permits a rather good indication of the tuning of the antenna system and the effectiveness of energy transfer from transmitter to antenna.

6-2-3. Tuning Meter

A nonresonant rf indicator is shown in Fig. 6-5. This is a very popular instrument for field checking the power output and tuning of communications transmitters. It consists of a crystal diode and a sensitive dc meter. The indicator is a simple rectifier and filter

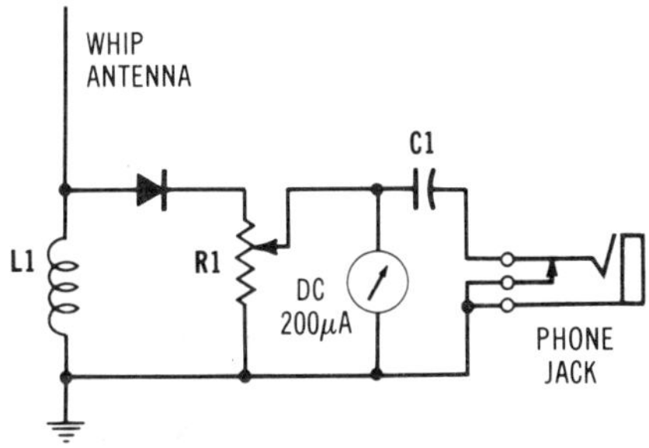

Fig. 6-5. A typical rf-field indicator.

circuit. Capacitor C1, inductor L1, and potentiometer R1 function as a filter to smooth out the unidirectional detector pulses. Thus, the steady current through the dc meter is a function of the signal strength at the input. To improve the sensitivity of the device, particularly when hf and vhf signals are being checked, a short antenna can be attached. Sensitivity can be adjusted with potentiometer R1. The amount of resistance inserted into the circuit influences the amplitude of the current through the dc meter.

A tuning meter of this type, when placed near the transmitter antenna, will indicate the relative power output of the transmitter, and the effect of any transmitter adjustment on the power output can then be noted.

It must be stressed that this type of rf meter, being untuned, is sensitive to an extremely wide frequency range. For this reason, a meter of this type cannot be used to check the transmitter frequency.

6-3. DIP OSCILLATORS

The dip oscillator is a useful test instrument for checking transmitter circuits. Most dip oscillators can also be used as absorption wavemeters.

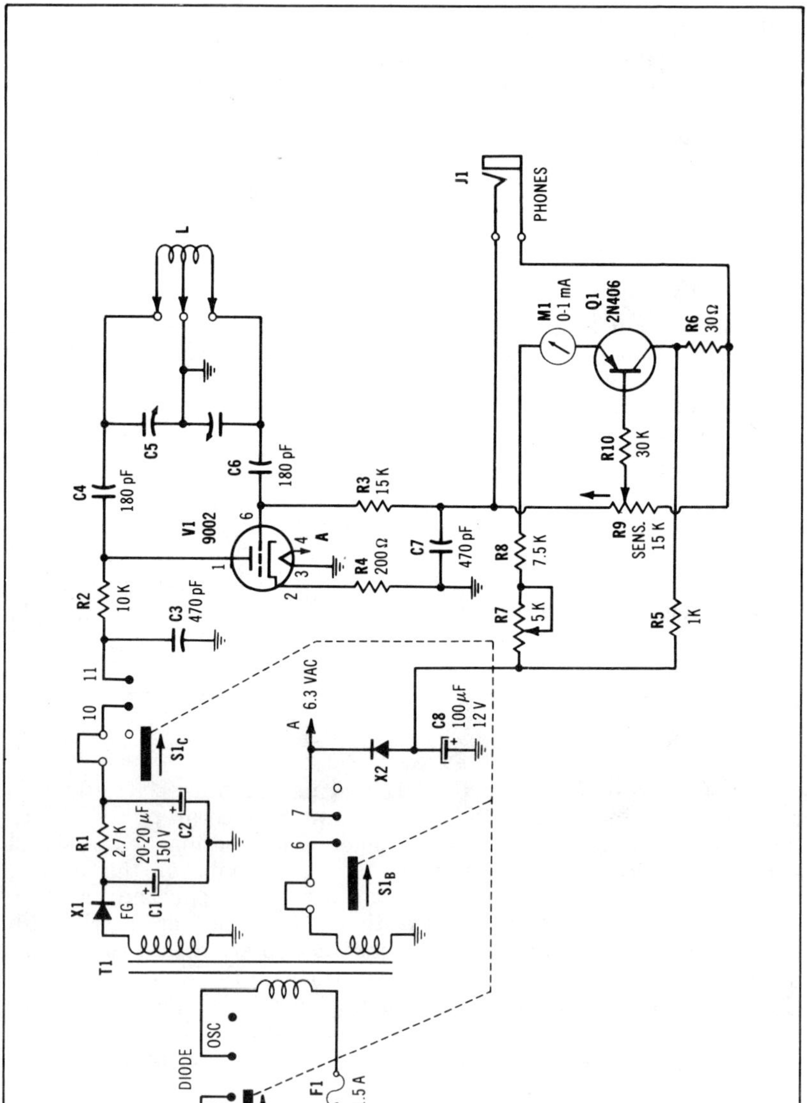

Fig. 6-6. Millen grid-dip oscillator.

The Millen grid-dipper uses a high-frequency triode connected in a Colpitts circuit, as seen in Fig. 6-6. Seven plug-in coils permit operation over a frequency range between 1.7 and 300 MHz. With use of the logging scale and an included calibration curve, the unit also functions between 165 kHz and 1.7 MHz. Plug-in coils provide overlapping frequency ranges between the two frequency extremes.

The meter sensitivity is increased with the use of a solid-state dc amplifier, transistor Q1. In the operation of a grid-dip oscillator, the loading effect of the external resonant circuit being measured causes a drop in the oscillator grid current. A tiny change in dc current is amplified by the transistor, resulting in a useful meter reading.

The dial of the variable capacitor of the grid-dip oscillator is calibrated in frequency. When used with the appropriate calibrated coil, the oscillator frequency can be read with reasonable accuracy.

Millen also has an all solid-state dipper available using a FET oscillator. It has a frequency range that extends between 1.6 and 300 MHz. Operation is from a 9-volt battery.

As a dip oscillator, the test unit can be used to determine the resonant frequency of a de-energized radio-frequency tuned circuit. When a resonant circuit is brought close to the oscillating tank circuit, the tuned circuit under measurement absorbs some of the energy from the oscillator tank circuit. (Refer to Fig. 6-7A.) Inasmuch as energy is removed from the tank circuit of the

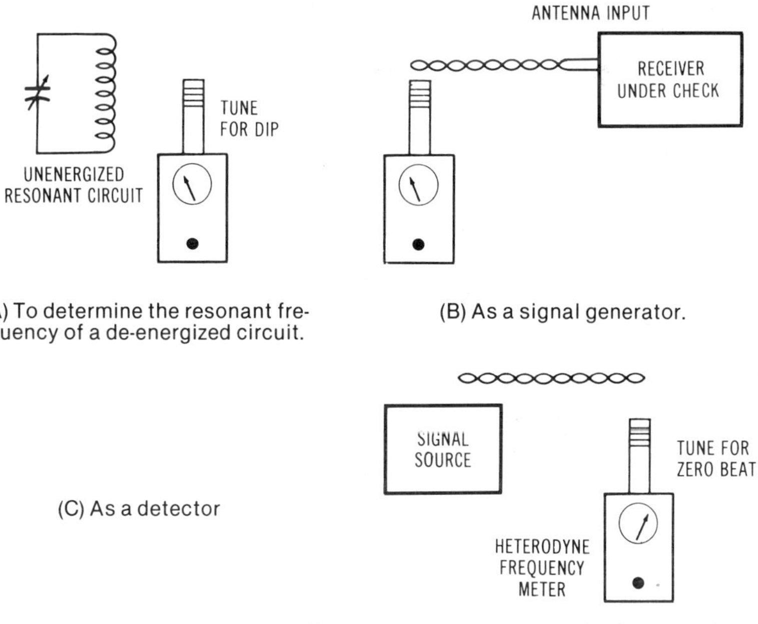

(A) To determine the resonant frequency of a de-energized circuit.

(B) As a signal generator.

(C) As a detector

Fig. 6-7. Using the dip meter.

oscillator, the oscillator feedback into the grid circuit decreases. As a result, the grid-current meter reading drops.

When the dip oscillator is set on the exact frequency of the resonant circuit under measurement, the meter will dip to its minimum value. This setting indicates that the dip oscillator is generating a signal of the same frequency as the resonant tuned circuit under check.

It is apparent that the dip oscillator is very helpful in determining the resonant frequency of various types of tuned circuits. Be it a small, weak-signal resonant circuit of a receiver or the larger, higher-powered resonant circuit of the transmitter final amplifier, the resonant frequency can be determined without any signal present. In transmitter circuits the dip oscillator can also be used to track down spurious resonant conditions that cause parasitic oscillations.

A dip oscillator can also be utilized as a signal generator, as shown in Fig. 6-7B. Therefore, it can be used in signal-tracing a communications receiver and, if the oscillator can be calibrated very accurately, for alignment work to a limited extent.

The dip oscillator can also be employed as an oscillating detector, similar in action to a heterodyne-frequency meter. In this application, as shown in Fig. 6-7C, the dip oscillator is brought near the source of rf energy. A pair of headphones, inserted into the phone jack, will pick up a beat note when the frequency of the grid-dip oscillator is brought near the signal-source frequency. If this note is zero-beat, the dip oscillator is set to the same frequency as the source of the signal. In this application the dip oscillator has been used for determining the frequency of an unknown signal source.

6-3-1. Dip Oscillator Calibration

The oscillating-detector principle of the dip meter can in itself be used to calibrate the meter accurately, as shown in Fig. 6-8. In this check, it is coupled near a crystal-controlled frequency standard. Whenever the dip oscillator is tuned to the crystal frequency or to one of its harmonics, a beat note will be heard. At the beat-note position, dip calibration should be checked. Some dip oscillators provide a calibration control so the oscillator can be reset if it has drifted.

Still another method of checking the calibration of a dip oscillator is to tune a communications receiver to a WWV frequency (2.5, 5, 10, 15, 20, and 25 MHz), as in Fig. 6-8B. The dip oscillator is then tuned to the same frequency. When a zero beat is heard *at the receiver output*, the dip oscillator has been set to the same frequency as the incoming WWV signal. The dial calibration can be checked at this point.

6-3-2. Dip Oscillator as an Absorption Wavemeter

When the oscillator is turned off, the dip oscillator becomes an absorption wavemeter. The tuned circuit and device act as a

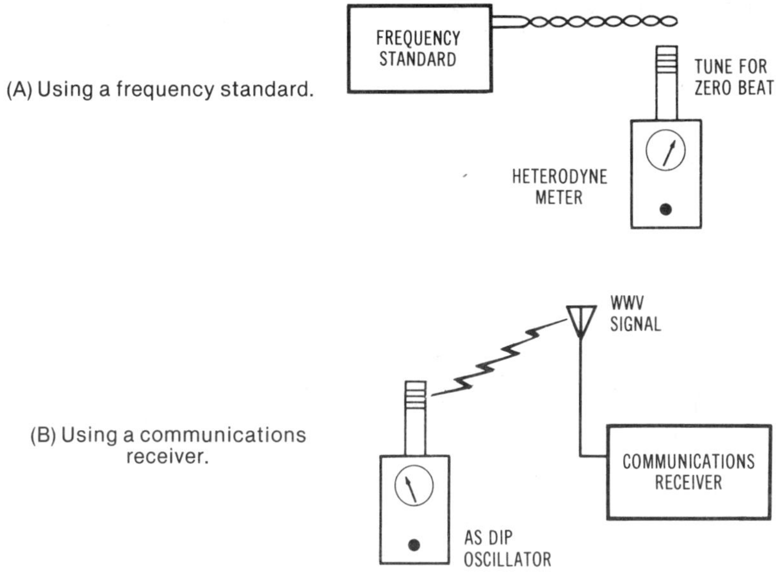

Fig. 6-8. Calibrating the dip meter.

detector and the meter becomes part of the diode-load circuit. When coupled near and tuned to the frequency of an rf energy source, the meter will display a current increase. The operation is similar to that discussed for absorption wavemeters.

The dip oscillator has many applications in testing and tuning transmitter circuits. The obvious applications are for tuning tank circuits and checking the output. Other applications include use as an rf indicator in neutralizing various stages of the transmitter, or in tracking down troublesome parasitic oscillations. It is also helpful for peaking various types of resonant traps used in transmitters to prevent parasitic oscillations, or for preventing the transfer of strong harmonic components from stage to stage and from the transmitter to the antenna system. Finally, the dip oscillator is handy for checking antenna performance. It can be used in bringing the antenna system to resonance and in minimizing standing waves on the transmission line.

6-4. POWER OUTPUT

To minimize interference and to be fair to all users of two-way radio equipment, transmitters must adhere to certain performance standards in the form of specific FCC technical rules and regulations. These standards cover frequency of operation, frequency stability, modulation level, and power output. Suitable test instruments are available for testing these transmitter characteristics. They must be capable of making measurements within the tolerances established by the FCC.

The maximum power output of the transmitter must comply with the maximum ratings established by the FCC for a particular service. As mentioned previously, the power output is set indirectly by the maximum power input restrictions of the FCC. Power-output measurements are not required for approved units. It is assumed that if the accompanying instructions are followed with regard to the power input, the output will not be excessive.

6-4-1. Power Meters

An rf power-output meter is helpful to the communications technician who does much work in this field. It permits him to determine if a given transmitter does function efficiently, and helps him tune up the transmitter in a manner which will give the owner as strong a signal as possible without exceeding the FCC maximum input rating.

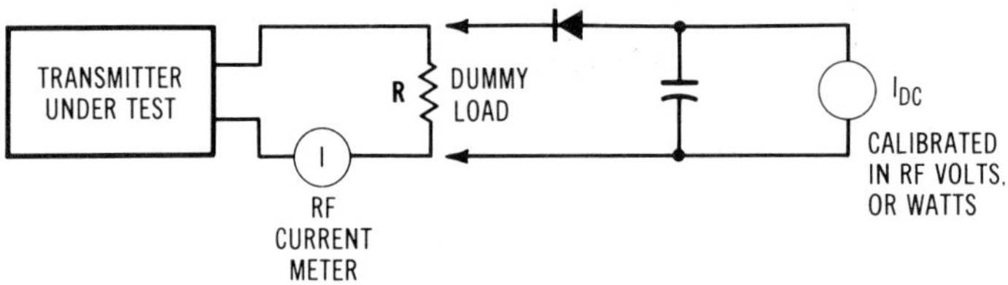

Fig. 6-9 Rf power-output measurement.

Power output can be measured by attaching a dummy load to the transmitter output. The dummy load should have the same impedance as the antenna system into which the transmitter normally works. An rf ammeter (Fig. 6-9) can be inserted into the dummy-load circuit. Power output will be:

$$P = I^2R$$

A calibrated rf voltmeter can also be used to measure the rf voltage across the dummy load. In this case the rf power output is:

$$P = E^2/R$$

An rf voltmeter is generally a crystal diode rectifier and a sensitive dc meter calibrated to measure rf voltage.

6-4-2. Commercial RF Power Meters

A problem in making rf power measurements is that the circuit operating conditions change when the test instruments are inserted across the transmitter output or into the transmission-line path between the transmitter and antenna system. Commercial instruments, referred to as insertion-type rf wattmeters, provide the answer to this problem.

Courtesy Bird Electronics Corp.

Fig. 6-10. Rf wattmeter.

Insertion-type wattmeters are the ones made by Bird Electronics Corporation and are shown in Figs. 6-10 and 6-11. Units are designed specifically for line insertion between a 50-ohm transmitter output and a 50-ohm antenna system. As shown in Fig. 6-11, it is inserted between the transmitter output and the coaxial transmission line going to the antenna system. Various plug-in termination elements are made available. By insertion of the proper elements or by switching seven wattage ranges can be measured—0–2.5, 0–10, 0–25, 0–50, 0–100, 0–250, and 0–500.

The instrument responds to the direction of wave travel and is capable of measuring the power in the forward wave traveling between transmitter and antenna, as well as the power lost in the reflected wave traveling back toward the transmitter because of a mismatch in the antenna system. It is apparent that the rf wattmeter is helpful not only in making power measurements, but also in tuning the transmission line and antenna system for a minimum standing-wave ratio to ensure the most efficient transfer of power from transmitter to antenna.

The wattmeter and coupling system are shown schematically in Fig. 6-11. The rf energy, passing between transmitter and antenna, is sampled by a pickup element and applied to a crystal diode and rectifier. A rectified and filtered dc current is then linked by a meter cable to a current meter with a sensitivity of 30

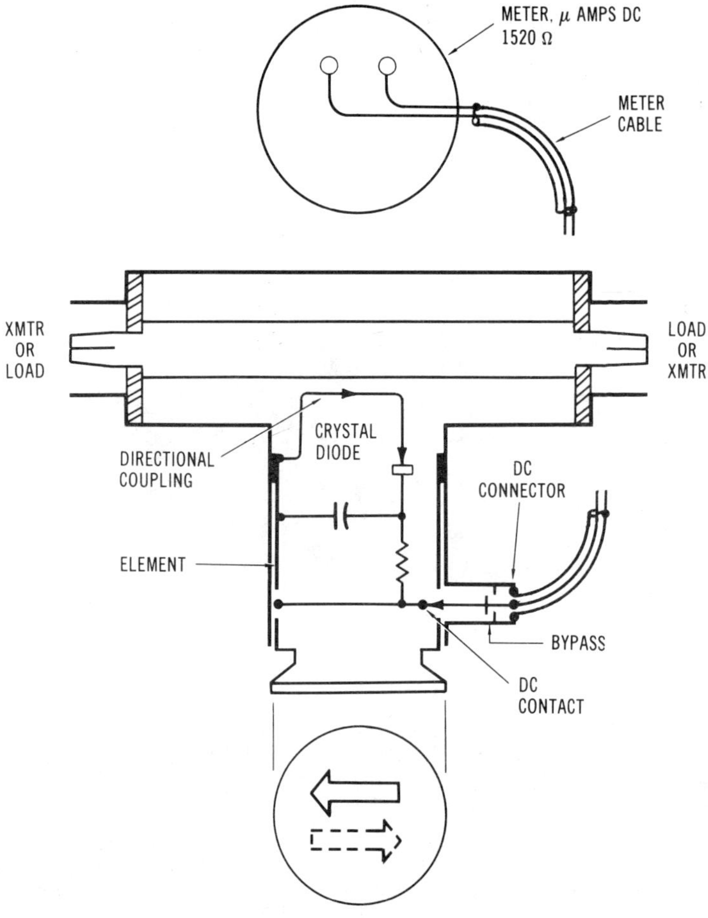

Fig. 6-11. Circuit of an rf wattmeter.

microamperes. To prevent meter and crystal damage, it is important that the proper size of plug-in element be used and that it be inserted correctly.

A complete coupling circuit which samples the traveling wave is mounted in each individual plug-in element. It consists of the pickup loop, crystal diode, and associated components. There are various plug-in elements corresponding to frequency and power rating. These are plugged into the front of the meter.

The pickup element of the rf wattmeter functions as a directional coupler. As shown in Fig. 6-11, the rf energy is conveyed to the crystal diode over two paths. One path is by way of capacitive coupling from the center conductor, which is part of the coaxial feedthrough arrangement. Radio-frequency energy is also inductively coupled into the pickup tube. By careful design, the two components are of equal amplitude and are either additive of subtractive.

The phase of the inductive component is a function of the direction of wave travel. In the arrangement of Fig. 6-12A, the energy picked up inductively from the wave traveling between transmitter and antenna adds to the capacitive component and produces a reading on the calibrated meter. Any reflected energy returning along the transmission line induces a voltage out of phase with the direct-coupled component from the reflected wave. Consequently, the influence of the reflected wave is canceled and its strength is not recorded.

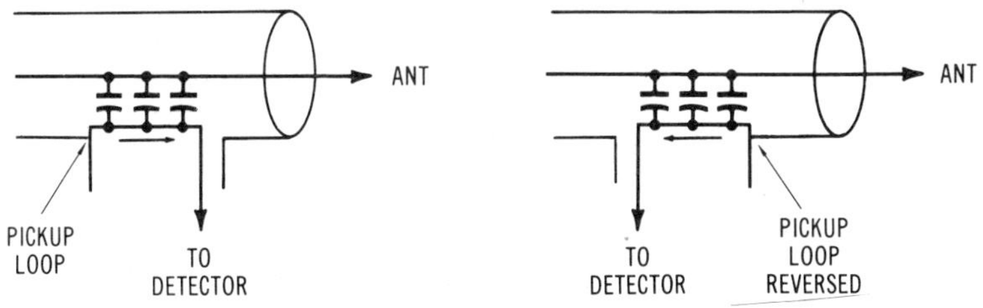

Fig. 6-12. Principle of a reflectometer.

To read the reflected power, the coupling plug-in element is inserted in the opposite direction, as shown in Fig. 6-12B. The reflected power induces into the coupling circuit a voltage in phase with the capacitively coupled component. Hence, the meter will read the reflected power. Insofar as the direct wave moving from the transmitter to antenna is concerned, the coupling loop is so connected that the induced component and the capacitive components are equal and opposite in polarity. The net voltage is zero; therefore, the direct power is not recorded on the meter.

This type of power meter, usually referred to as a *reflectometer*, has the ability to measure direct and reflected power. The power delivered to the antenna or other load can be calculated by subtracting the reflected power from the direct power.

In using an rf power meter of this type, it is very important to insert a plug-in element of the proper size and in the proper direction. For example, the reflected power of a closely matched system is much lower than the direct power. To make an accurate measurement of the reflected power therefore requires the use of a low-power plug-in element. Moreover, if this element is inserted in the improper direction, the strong direct power will be supplied to it, and the crystal and meter could be damaged.

The voltage standing-wave ratio (vswr) can be calculated if the direct and reflected powers are known. This equation is:

$$\text{swr} = \frac{1 + K}{1 - K}$$

where,

swr is the standing wave ratio,
K is the ratio of reflected power to direct power.

6-5. CB PERFORMANCE TESTER

The Sencore CB41 tester shown in Fig. 6-13 is designed specifically for CB radio servicing. Three key CB operating parameters can be checked with this instrument. These are rf power, swr, and modulation percentage. In the latter case, the tester allows checking both standard amplitude modulation and sideband transmission. The tester consists of the meter and control panel plus a sensor unit that is connected between the CB transmitter and its antenna system. The CB sensor should be positioned as close as possible to the transmitter.

For checking rf power and modulation, the unit provides a 50-ohm dummy load. When these measurements are taken, the antenna system is not connected to the sensor. When checking swr, the sensor is connected between the transmitter and the coaxial antenna line as shown in Fig. 6-14.

To set up the CB tester, first disconnect the antenna cable from the antenna jack of the CB radio. The CB transmitter lead from the

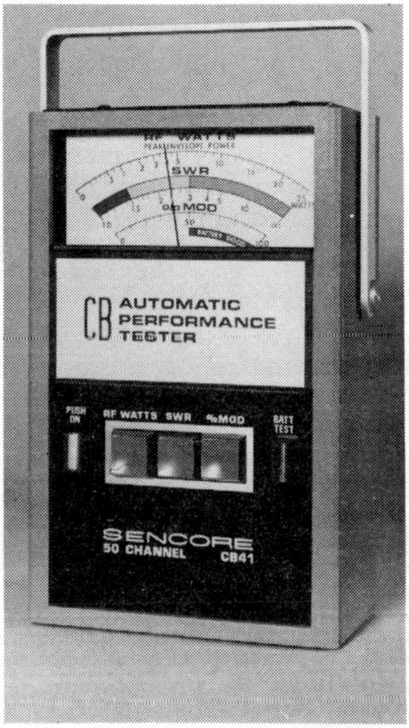

Courtesy Sencore, Inc.

Fig. 6-13. CB performance tester.

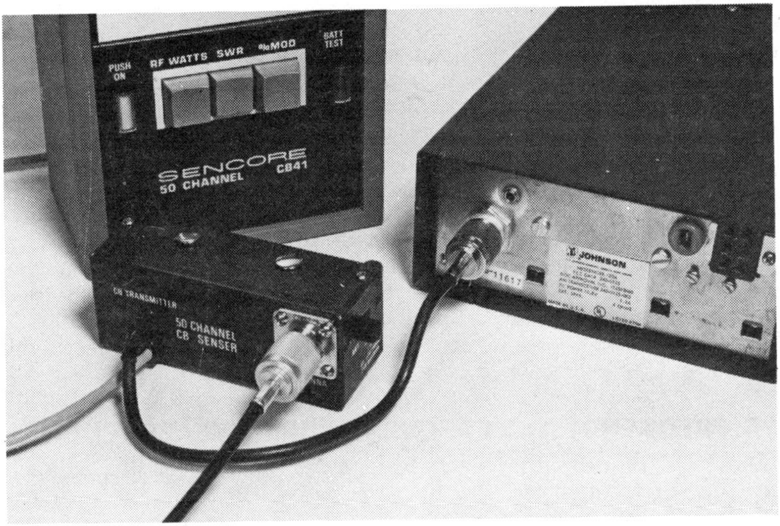

Courtesy Sencore, Inc.

Fig. 6-14. Meter and sensor connected to a CB radio.

tester is then connected to the antenna jack of the CB radio. Now it is possible to measure radio-frequency power and percentage modulation.

The sensor is set to the dummy load position, and the tester RF WATTS button is depressed. The transmitter now can be keyed to measure the power output. If it is a standard am transmitter and no modulation is applied the tester will read the carrier power. Typical readings will fall between 3 and 4 watts (the permissible maximum is 4 watts).

You can now modulate the transmitter by whistling or humming into the microphone. An increase in the power reading should be observed.

To check a sideband transmitter, the same interconnections are used with the sideband transmitter set for either upper- or lower-sideband operation. Inasmuch as the carrier is suppressed in sideband modulation, the meter on the tester should read zero rf watts when there is no modulation. If you modulate the transmitter by whistling or humming into the microphone, there will be an increase in the wattage reading up to a maximum of 12 watts pep.

An exact reading of the modulation percentage can be obtained by using the same interconnections. However, in this case the % MOD function button is depressed. The same procedure can be used to check standard amplitude or sideband modulation. When the transmitter is modulated (by whistling or humming at the normal speaking distance from the microphone), the meter needle deflection should read upward reaching 80% to a maximum of 100%.

In order to make an swr test, the same interconnections are made with the exception that the antenna cable should now be connected to the antenna receptacle of the sensor. The switch on the sensor is then set to the antenna position. By depressing the SWR function button, you should obtain an immediate reading of the standing-wave ratio on the transmission line. This particular tester circumvents the need for making a set adjustment, which is required with most swr meters.

If desired, the CB transmitter can be set to a channel near the middle of the band (approximately channel 20). Now the antenna length can be adjusted to obtain a minimum standing-wave ratio at this frequency. To assist you in this adjustment, an extension line for the tester can be obtained to allow the tester to be positioned where it can be seen as you make the antenna length adjustment. The sensor, of course, must always be connected near the transmitter.

6-6. FREQUENCY AND FREQUENCY-DRIFT MEASUREMENT

The FCC requires that each transmitter operate on its assigned frequency or frequencies. Furthermore, any drift from the assigned frequency must be maintained within strict tolerances. In most two-way radio services, the permissible percentage of drift is only 0.0005% at frequencies over 50 MHz. The very tight tolerances are necessary if an ample number of channels is to be provided for the many services using two-way radio. A transmitter should not drift significantly or it will interfere with similar two-way radios on adjacent channels. A minimum of drift is also very important in terms of reliability. Most installations use fixed-tuned receivers. Consequently, it is important that the transmitter maintain constant frequency in order to prevent distortion in the reproduced audio.

The frequency meter must be of the highest quality, and its tolerances should be even stricter than those of the transmitter (0.00025% or better) if exacting frequency measurements are to be made. A signal generator with a drift tolerance of 0.0001% permits accurate setting and measurement of transmitter frequency.

6-7. DIGITAL FREQUENCY COUNTER

The digital frequency counter, because of its compactness, accuracy, and speed with which a measurement can be made, has become popular as a frequency meter in the two-way radio services. The instrument of Fig. 6-15 permits frequency measurements between 10 hertz (Hz) and 1000 megahertz (1 GHz). Accuracies exceed the FCC specifications for communications work up to 1 GHz. A large 8-digit display permits direct reading with automatic placement of the decimal.

Three separate inputs are available. One is a fuse-protected 50-ohm termination for proper reading of high-frequency signals between 10 MHz and 1000 MHz. A second input is a high sensitivity one with 1 megohm loading to allow direct connections to lower-frequency circuits. It operates over a range of frequencies between 10 Hz and 100 MHz. A front-panel sensitivity control is

Fig. 6-15. Digital frequency counter.

included. A third input is a crystal-check facility which allows any crystal to be tested by simply inserting it into the front-panel test socket.

Two read-rate gate times of 1 second or 0.1 second are included. The fast rate provides a fast update and ability to follow a fast frequency change. The slower 1-second rate provides maximum resolution.

Four different frequency ranges provide the best resolution for any measurement according to the following:

10-500 Hz–0.01 Hz
500-20,000 Hz–0.1 Hz
20 kHz-100 MHz–1 Hz
10-1000 MHz–10 Hz

These figures indicate the ability of the instrument to make a precise frequency measurement.

A 50-ohm input jack provides proper termination for a 50 ohm coaxial cable. An rf protection fuse is located inside along with two spare fuses. In addition, an adjustable antenna is provided that permits a pickup of off-the-air signals. Thus a frequency check can be made by bringing the frequency counter close to a base or mobile transmitter.

The antenna, too, is useful because of its ability to test the output frequency of a walkie-talkie or a repeater transmitter that cannot be removed from service without causing down time. The frequency of a mobile transmitter can be measured by simply driving up next to the transmitter. There is no need to pull out the good transmitter from service.

There is a counter probe with a direct/isolate switch for use when direct connection to a signal source is desired. Also included is an rf pickup loop for making high-frequency or high-power frequency measurements when a direct connection is not desired. The pickup loop is especially useful because critical oscillator frequency measurements can be made without placing any load capacitance in the circuit. At the same time the very weak oscillator signal can be picked up and a reading obtained.

There is a 10-MHz output signal available at the rear of the counter. This signal component can be used as a reference source for calibrating less accurate counters or scopes. In addition it may be used in conjunction with the reception of a WWV signal for testing the accuracy of the counter frequency calibration.

6.8 FM DEVIATION MEASUREMENT

An oscilloscope and appropriate calibration system can be used to make an accurate measurement of fm deviation. Such a system is shown in Fig. 6-16. Fundamentally it is an fm receiver that develops a deviation-calibrated output signal. A double-conversion technique is used.

The tuner must be designed to operate over the frequency range of the transmitters to be checked. In the case of the commercial fm two-way radio bands one tuner might tune between 25–55 MHz, a second between 145–175 MHz, and a third and fourth tuner to cover the 450- and 800-MHz bands.

In a typical arrangement the high if frequency output of the tuner could be 10.7 MHz. A low intermediate frequency of 4.5 MHz could be used. In this case the carrier conversion oscillator frequency would be 6.2 MHz (10.7–4.5 MHz). At this frequency the dc output of the fm detector would be 0 volts in a typical example.

The conversion oscillator could be used to generate, depending on marker switch setting, a frequency 5000 Hz above 6.2 MHz or 5000

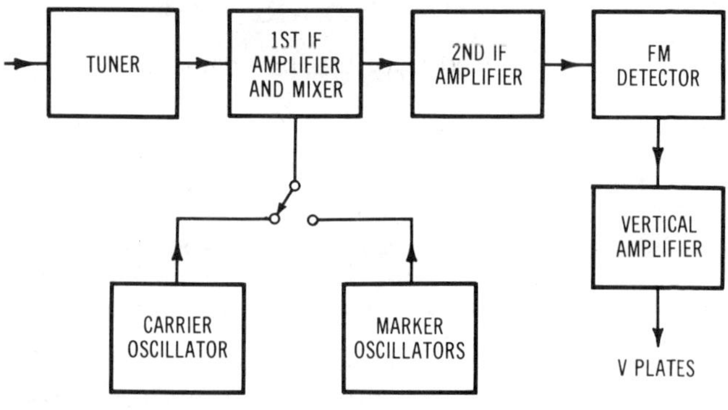

Fig. 6-16. Functional plan of a deviation meter using an oscilloscope.

Hz below 6.2 MHz. This means by proper setting of the marker switch it is possible to supply a 4.505-MHz or 4.495-MHz signal to the low if amplifier. By so doing, reference dc voltages are made available at the output of the detector; one voltage indicates a frequency of 5000 Hz above the center frequency and the other a frequency 5000 Hz below the center frequency.

The vertical deflection graticule of the oscilloscope can then be calibrated in terms of frequency deviation. This is shown in Fig. 6-17. In operation, the center carrier-oscillator marker line is made to coincide with the center line of the graticule. If the plus marker conversion oscillator is now turned on there will be an upward movement of the calibration line. This corresponds to an upward

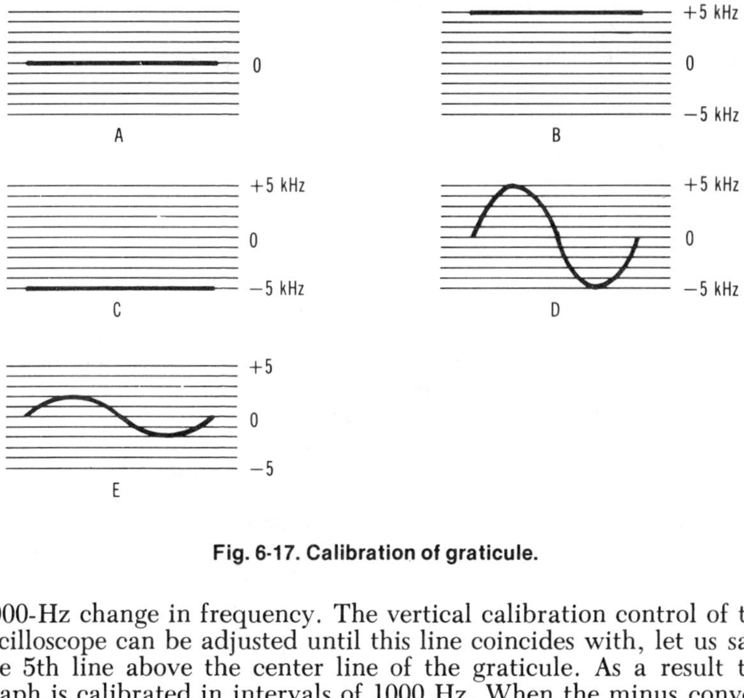

Fig. 6-17. Calibration of graticule.

5000-Hz change in frequency. The vertical calibration control of the oscilloscope can be adjusted until this line coincides with, let us say, the 5th line above the center line of the graticule. As a result the graph is calibrated in intervals of 1000 Hz. When the minus conversion oscillator is turned on there will be a similar calibration of the horizontal lines below the center line. This time it will be in steps of 1000 Hz below center frequency.

Of course additional marker oscillator frequencies can be made available to calibrate the oscilloscope scale for various other amounts of deviation. In so doing the scale can be calibrated for lower deviation values such as might be useful for measuring tone deviation and other special signal components.

If the incoming if frequency were now frequency modulated by a sine wave with a maximum deviation of ± 5000 hertz, it would be displayed as a sine wave on the oscilloscope screen with the positive peaks reaching up to +5000 hertz and −5000 hertz calibration lines as shown in Fig. 6-17. Were the deviation only ±2000 hertz, the sine-wave peaks would reach up only to the 2000-Hz calibration lines.

In making an actual fm modulation measurement it is necessary to supply a signal of appropriate magnitude to the input of the tuner. Off-air signals can actually be measured by picking them up on a receiving antenna. If transmitter deviation is to be measured right in the shop it would be necessary to apply the transmitter output through an appropriate attenuator to the input of the tuner. Usually the receiver section will include a meter that indicates when a signal of proper level is applied to the tuner input.

Such an fm deviation meter assembly not only measures the actual deviation of the transmitter but permits an observation of the actual received signal. Thus the effects of audio processing can be evaluated. Likewise any distortion in the modulated wave will show up on the demodulated signal displayed on the oscilloscope screen. You can also check the influence of deviation limiter operation. In such a test the audio input to the transmitter can be increased somewhat above the normal level and the influence of deviation limiting action observed. The function of the deviation limiter, of course, is to prevent deviation in excess of the FCC prescribed limit for the particular fm two-way radio service.

6.9 SPECTRUM MONITOR

The spectrum monitor has become increasingly popular in checking out the performance of two-way radio transmitters and receivers. Such a monitor (Fig. 6-18) permits you to measure the relative levels of carrier, sidebands, harmonics, and other spurious signals. Components are displayed in such a manner that individual frequencies can be determined as well.

The spectrum monitor or analyzer is referred to as a frequency-domain oscilloscope because it presents an amplitude vs frequency display rather than the amplitude vs time display or a more conventional time-domain oscilloscope. The horizontal base is calibrated in terms of frequency rather than time. For example, an am transmitter modulated by a 3000-Hz note is seen as a carrier, an upper sideband, and a lower sideband, each displayed separately along the base. The vertical deflections of the three are a true indication of their relative amplitudes.

Spurious signal components generated during the modulation process are also displayed and can be measured relative to the carrier level. As you know spurious signal components, according to FCC regulations, must be a specified number of dB below the actual carrier level. In fact, for Citizens Band operation, these spurious components and harmonics must be at least 60 dB down relative to the desired carrier.

A functional block diagram of the Cushman's spectrum monitor is given in Fig. 6-19. In effect, it is a triple-conversion receiver which

Courtesy Cushman Electronics, Inc.

Fig. 6-18. Spectrum monitor.

processes an incoming signal and applies it eventually to the vertical deflection plates of the oscilloscope. A sweep-frequency system makes the monitor sensitive to a selected range of frequencies on either side of the incoming signal, displaying desired sidebands as well as undesired components that appear on each side of the carrier frequency.

The monitor can also tune in harmonics and widely separated spurious signal components that may be emitted by the transmitter under test. Thus the monitor actually makes a spectrum analysis of the transmitted signal to make sure that it is in compliance with FCC technical standards.

The monitor also evaluates local-oscillator radiation from the unit when the transceiver is on either the transmit or receive mode of operation. Remember that spurious signal components can be radiated even though the transceiver is set to its receive mode.

The input to the monitor is made sensitive to a specific range of

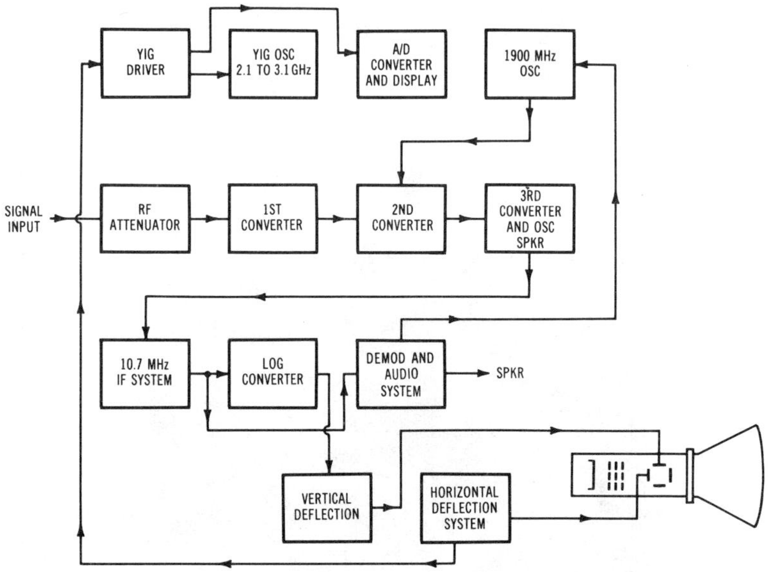

Fig. 6-19. Functional block diagram of the spectrum monitor.

frequencies because of the use of a sweep oscillator. The frequency of this local oscillator (a YIG oscillator in the case of the Cushman monitor) is swept in frequency by a ramp voltage generated in the horizontal deflection system and applied to the oscillator through the YIG driver, top left of Fig. 6-19. It is this tie-in between the horizontal deflection waveform and the local oscillator that permits accurate calibration of the horizontal sweep of the oscilloscope in terms of frequency.

The term YIG (yttrium, iron, garnet) denotes a special microwave oscillator that operates in the 2.1- to 3.1-GHz frequency range. Oscillator frequency is determined by the strength of the associated magnetic field in which it operates. This arrangement permits a smooth control of the oscillator frequency, varying the strength of the magnetic field by way of the YIG driver.

The YIG oscillator operates at this extremely high frequency to ensure that the instrument itself does not produce spurious responses and image frequencies within the measurement range of the instrument. This range extends between 1 and 1000 megahertz.

The output of the first converter is applied to a second converter where the signal is reduced to the 200-MHz second if range. Local oscillator injection is from a 1900-MHz oscillator. A third converter follows which brings the frequency of the signal down to the 10.7-MHz if range.

Output from the third if system is applied to the log converter and the fm demodulator. The output of the fm demodulator is applied to an audio amplifier and speaker. Additionally a component is applied back to the 1900-MHz oscillator in the form of an afc feedback loop.

The log converter changes the linear amplitude signals to a logarithmic output before the signal is applied to the vertical-deflection system. Thus the resultant vertical deflection is calibrated in decibels. This permits a dB comparison to be made between the various signal components displayed on the screen. The horizontal-deflection system, of course, generates the ramp voltage that must be applied to the horizontal-deflection plates as well as the component that is applied to the YIG driver and the followup swept YIG oscillator.

In summary, a frequency-domain display permits individual observations of carrier and sideband components of a transmitted signal. Also the display can be used to check the transmitter final amplifier for spurious output and breakup, identify intermodulation interference, check for receiver local-oscillator radiation, measure transmitter final amplifier for spurious output and breakup, identify intermodulation interference, check for receiver local-oscillator radiation, measure transmitter harmonics, monitor and identify carriers with either am or fm modulation, and do rf and if signal tracing through a transceiver.

A typical transmitter interconnection plan for evaluating the output of a transmitter is shown in Fig. 6-20. The transmitter signal is applied through an in-line wattmeter to a 50-ohm load. Thus, the

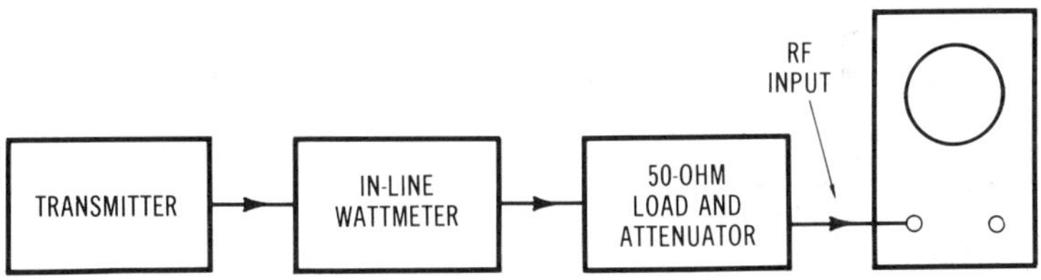

Fig. 6-20. Connection of the spectrum analyzer to transmitter output.

power output can be checked and preset for making the spectrum analysis. An included attenuator is needed to reduce the level of the signal for proper application to the radio-frequency input of the spectrum analyzer.

As an example of the type of troubles that can be identified refer to the patterns taken by a Cushman spectrum analyzer as shown in Fig. 6-21. The example in Fig 6-21A shows the spectral purity of a 25-watt fm radio transmitting at center frequency of 158.55 MHz. The vertical scale is calibrated in decibels while the horizontal scale is set for 10 MHz per division. Note that there are two spurious signals 40 and 46 dB down at about 12 MHz on each side of the carrier. According to FCC rules the spurious signals at these points should be down at least 57 dB.

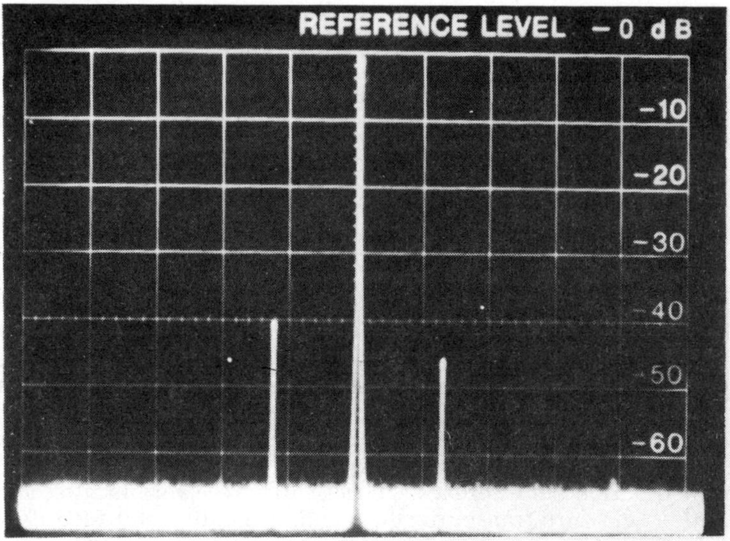

(A) Fm radio transmitting at a center frequency of 158.55 MHz.

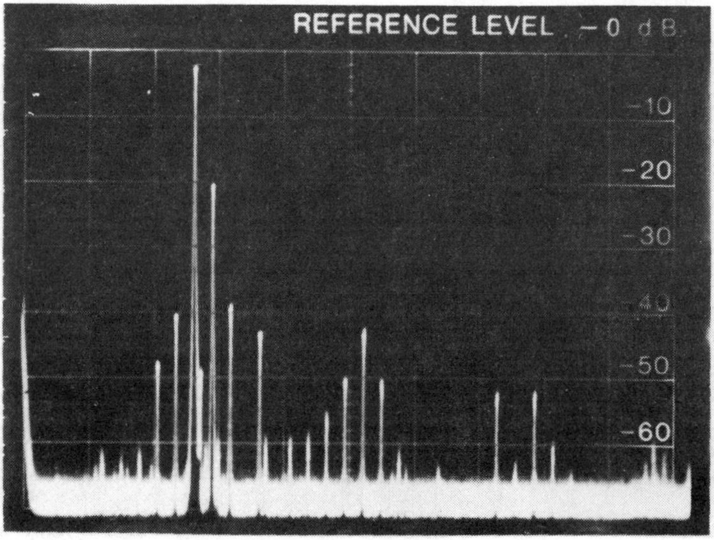

(B) CB radio transmitting on Channel 13.

Fig. 6-21. Typical trouble patterns as displayed by spectrum analyzer.

The pattern of a CB radio transmitting on channel 13 can be seen at the left of the illustration in Fig. 6-21B. In this case the center frequency of the analyzer is 52 MHz, and quite a bit of spurious radiation occurs at this second harmonic-frequency range. FCC rules require that second- and higher-order radiation must be suppressed by 60 dB in the Citizens Band (CB) Radio Service.

6-10. COMMUNICATIONS MONITOR

The communications monitor is an all-in-one test instrument that can be used to monitor, measure, and repair two-way radios. The *Cushman CE50A* (Fig. 6-22) is an example of a portable test instrument that operates in the vhf/uhf range to 1000 MHz. In the shop it can be powered by the 120-volt ac power supply or, at remote sites, by vehicular power or optional battery.

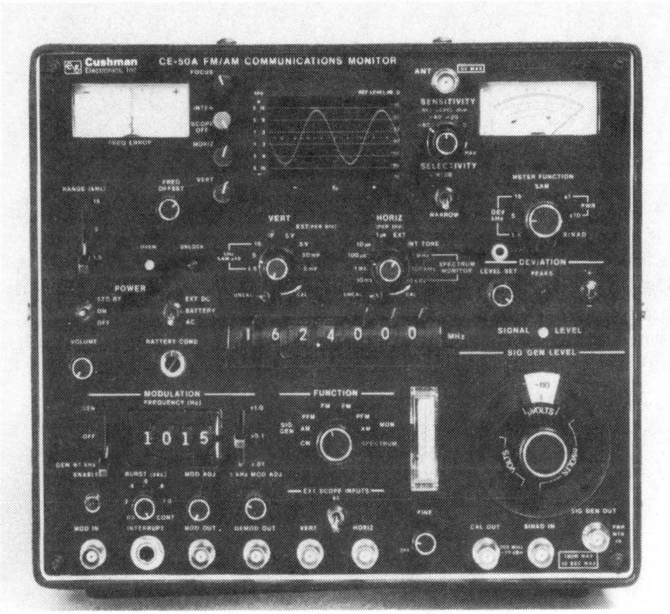

Courtesy Cushman Electronics, Inc.

Fig. 6-22. Communications monitor.

A wide variety of measurements can be made including carrier frequency and modulation characteristics of transmitted signals. Transmitted radio-frequency power measurements can be made. There is a capability for displaying demodulated audio frequencies on its internal oscilloscope for measurement or visual signal monitoring.

The receiver section of the monitor is a triple conversion superheterodyne with an adjustable sensitivity for monitoring signals of 2 microvolts to 500 millivolts in level. Thus transmitter testing can be made at separations from a few feet to several miles. Fm deviation as well as am modulation percentage can be measured.

The unit also generates accurate radio-frequency signals for testing sensitivity, alignment and performance of two-way radio receivers. The standard SINAD receiver sensitivity measurements can be made. (Refer to Section 6-13-4.) The generated radio-frequency signal can be a continuous-wave (cw) or it can be am or fm modulated by an audio frequency tone synthesizer.

The signal generator is an accurate frequency synthesizer with a frequency range of 100 kHz to 999.9999 MHz. Output can be connected directly to a transceiver and includes protection from accidental transmitter keying with an automatic electronic circuit breaker. Output level between 0.1 microvolt and 300 millivolts rms is set with precision attenuator switching.

In monitoring two-way radio signals they are picked up at the antenna input terminal at the top right (Fig 6-22). After demodulation the signal is sent to the demodulating section where the type of modulation selected is removed and displayed on the meter at the top right. The demodulated signal can also be applied to the oscilloscope for display and measurement.

A component is removed and sent to the frequency-error discriminator. The discriminator output is displayed on the meter at the top left indicating the departure of the received signal from the accurately dialed frequency of the communications monitor. The reading is the actual difference between the dialed frequency and the transmitter carrier frequency.

In summary, frequency of operation of the transmitter as well as its modulation characteristics can be observed simultaneously. In

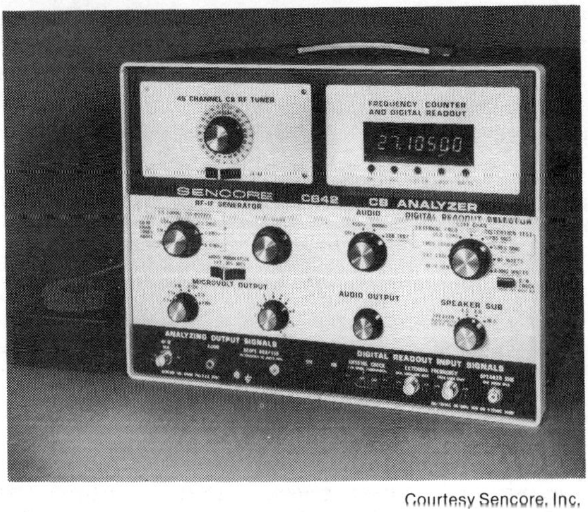

Courtesy Sencore, Inc.

Fig. 6-23. CB band analyzer.

addition the demodulated wave can be displayed on the scope screen.

To measure the transmitted power up to a maximum of 100 watts power output, the transmitter signal is applied to the power meter input jack at the bottom right. Measurements are made in two ranges, 0–10 watts and 10–100 watts. Using this input connection the same transmitter characteristics mentioned previously can be measured. Customarily this is the connection procedure followed when the transmitter and communications monitor are set up on the bench. In fact, only two connections need be made between the monitor and the transceiver to make almost a complete set of transmitter and receiver measurements including the SINAD sensitivity check. Such a check provides an accurate measurement of the useful sensitivity of a receiver. The measurement even includes the receiver output stage. A special version of the CE-50A communications monitor also includes an added spectrum monitor feature.

6-11. ANALYZER FOR CITIZENS BAND SERVICE

By using a unit such as that shown in Fig. 6-23, which is built by Sencore, Inc. for use in the 27-MHz Citizens Band Radio Service commonly known as the "CB" band, a bench can be equipped for CB service at a lower cost than with a general communications monitor.

This unit combines all functions needed for analyzing, repairing, and testing a transmitter or receiver operating in the CB band. The analyzer includes a high accuracy (1 ppm) rf generator. The output frequency is selected by a rotary switch that controls a digitally programmable phase-locked loop generator. The switch is preprogrammed to provide the 40 allocated CB channels with provisions for an additional 5 channels for later expansion of the band.

In addition to the rf output, the analyzer also provides an adjustable if generator and a selectable audio generator for alignment and troubleshooting purposes. Both the rf and if outputs pass through an internal attenuator allowing an output of .1 μV to 100,000 μV.

The analyzer uses a direct reading seven-digit digital frequency counter whose 1 ppm accuracy is well within the FCC specs. Transmitter frequency may also be measured as a percentage deviation from the proper channel frequency with a circuit that compares the internal rf generator with the transmitter output frequency, and displays the difference as a percentage readout.

The internal 50-ohm dummy load connects to a modulation detector that reads positive or negative modulation, and an rf power meter that reads peak envelope power (pep) to 25 watts for use on am or ssb transmitter checks. These readings, as well as audio output power, are indicated on the same digital display used for the frequency counter shown in Fig. 6-24.

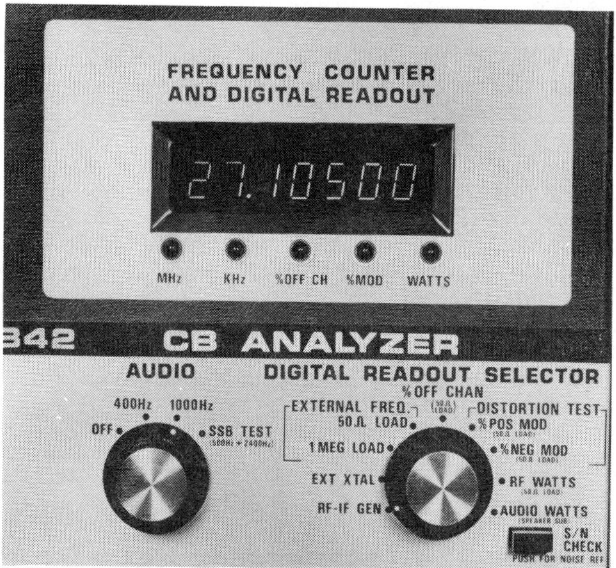

Courtesy Sencore, Inc.

Fig. 6-24. Digital display of analyzer shown in Fig. 6-23.

6-12. FCC TECHNICAL STANDARDS

Technical standards for the various radio services can be found in the appropriate volumes of the FCC Rules and Regulations. For example, technical standards for the private land-mobile services are found under Part 90 of Volume V. These volumes are available at FCC District Offices or can be purchased on a renewal basis. Such standards include requirements for standards of acceptability of equipment, frequency tolerance, modulation, emission, power, and bandwidth. The information that follows is a very abbreviated coverage of the standards as they apply to the Private Land-Mobile Radio Services. This will give you some idea of the importance of test equipment in evaluating the performance of two-way radio equipment.

6-12-1. Type Acceptance

With few exceptions only equipment which has been type-accepted by the Commission may be used in this particular radio service. Usually the manufacturer of the radio transmitting equipment requests type-acceptance. At the FCC offices there is a publication entitled "Radio Equipment List, Equipment Acceptable for Licensing." This is available for public reference at the FCC field offices.

6-12-2. Power

Applications for authorizations must specify no more power than the actual power necessary for satisfactory operation. In cases of harmful interference, the Commission may order a change in power or antenna height, or both. Except where otherwise specifically provided, the maximum power that will be authorized can be usually found in chart form in the FCC Rules and Regulations. A simplified chart for the private land-mobile radio services is given in Table 6-1.

Table 6-1. Authorized Maximum Powers

Frequency Range (MHz)	Maximum Output Power (watts)
1.3-3	1500
3-25	750
25-100	300
100-216	350
216-220	75
220-470	350
806-821	ERP*
851-866	ERP*

*Effective radiated power permissible depends upon transmitter power, antenna gain, and height.

Note in the 800-MHz band the maximum power is *effective radiated power* (erp) which takes into consideration both the transmitter power output and the antenna gain. Height also has an influence on the actual maximum erp value.

6-12-3. Types of Emission and Bandwidth Limitations

Types of emission were covered in Section 2-2 in Chapter 2 of this book. Type of emission is a part of the station authorization. Normally operations authorized in the private land-mobile services provide voice communications between stations. Although there are exceptions, the fundamental authorizations are for A3 or F3 emission. These authorizations also include the use of emissions for tone signals or signaling devices whose sole functions are to establish and maintain communications, to provide automatic station identification, and for operations in the public safety and special emergency radio services, to activate emergency warning devices used solely for the purpose of advising the general public or emergency personnel of an impending emergency situation.

F3 emission is authorized only on frequencies above 25 MHz. The bulk of the A3 emission is sideband type A3J authorized for audio telephony systems on frequencies below 25 MHz. Emission information for other types of services can be obtained from the appropriate FCC volume.

The maximum authorized bandwidth for standard A3 emission shall be 8 kHz. However, for type A3J operation below 10 MHz the bandwidth occupied by the emission shall not exceed 3500 Hz. The frequency coinciding with the center of the authorized band of emission shall be the assigned frequency. Both the authorized carrier frequency and assigned frequency shall be specified in the authorization. The authorized carrier frequency shall be 1400 Hz lower in frequency than the assigned frequency. Only upper-sideband emission shall be used.

For all F3 emissions on frequencies below 947 MHz, the maximum authorized bandwidth shall be 20 kHz and the maximum authorized frequency deviation shall be 5 kHz. There are exceptions to the above general technical requirements. These are given in detail in the appropriate FCC volume.

Emissions outside of the assigned bandwidth shall be held to a minimum. In fact, the mean power of emissions shall be attenuated below the mean output power of the transmitter in accordance with the following schedule:

1. On any frequency removed from the assigned frequency by more than 50 percent, but not more than 100 percent of the authorized bandwidth, at least 25 decibels;
2. On any frequency removed from the assigned frequency by more than 100 percent, but not more than 250 percent of the authorized bandwidth, at least 35 decibels;
3. On any frequency removed from the assigned frequency by more than 250 percent of the authorized bandwidth, at least 43 plus 10 $\log_{10}$ (mean output power in watts) decibels or 80 decibels, whichever is the lesser attenuation.

For single sideband operations (A3J) below 10 MHz, the carrier frequency power shall be at least 40 dB below the peak envelope power and the mean power of emissions shall be attenuated below the mean output power of the transmitter in accordance with the following schedule:

1. On any frequency removed from the assigned frequency by more than 1.75 kHz up to and including 5.25 kHz, at least 25 decibels;
2. On any frequency removed from the assigned frequency by more than 5.25 kHz and up to and including 8.75 kHz, at least 35 decibels;
3. On any frequency removed from the assigned frequency by more than 8.75 kHz, at least 43 plus 10 $\log_{10}$ (mean output power in watts) decibels.

6-12-4. Modulation Requirements

In general, the maximum audio frequency required for satisfactory radiotelephony intelligibility in the private land-mobile radio services is considered to be 3000 Hz (2800 Hz for single-sideband operation below 25 MHz).

Each transmitter shall be equipped with a device which automatically prevents modulation in excess of that specified which may be caused by greater than normal audio level. In most frequency ranges each transmitter must be equipped with a modulation limiter and a low-pass filter. Such a filter shall be installed between the modulation limiter and the modulated stage and shall meet specified bandwidth requirements.

When amplitude modulation is used for radiotelephony, the modulation percentage shall be sufficient to provide efficient communication and normally should be maintained above 70% on peaks, but shall not exceed 100% on negative peaks.

Regulations have been drawn up for the use of scrambling devices and digital voice modulation. Special authorization must be obtained.

6-12-5. Frequency Tolerance

A licensee in the private land-mobile radio services shall maintain carrier frequency of each authorized transmitter within a specified percentage of the assigned frequency. The abbreviated chart in Table 6-2 tabulates various key percentage values according to frequency, power and type of station.

Table 6-2. Frequency Tolerance In Percent

Frequency Range (MHz)	Fixed and Base Stations		Mobile Stations	
	Over 200 W	200 W or less	Over 2 W	2 W or less
Below 25	.005	.01	.01	.02
25 to 50	.002	.002	.002	.005
50 to 450	.0005	.0005	.0005	.005
450 to 470	.00025	.00025	.0005	.0005
470 to 512	.00025	.00025	.0005	.0005
806 to 821	.00015	.00015	.00025	.00025
851 to 866	.00015	.00015	.00025	.00025

Specific rules concerning transmitter measurements are as follows:

Sec. 90.215 Transmitter measurements.

(a) The licensee of each station shall employ a suitable procedure to determine that the carrier frequency of each transmitter authorized to operate with an output power in excess of two watts is maintained with the tolerance prescribed. This determination shall be made, and the results entered in the station records in accordance with the following:

(1) When the transmitter is initially installed;

(2) When any change is made in the transmitter which may affect the carrier frequency or its stability.

(b) The licensee of each station shall employ a suitable procedure to determine that each transmitter authorized to operate with an output power in excess of two watts does not exceed the maximum figure specified on the current station authorization. On authorizations stating only the input power to the final radio-frequency stage, the maximum permissible output power is 75 percent for frequencies below 25 MHz and 60 percent of the input power for frequencies above 25 MHz. If a non-dc final radio-frequency stage is utilized, then the output power shall not exceed 75 percent of the input power. This determination shall be made, and the results thereof entered into the station records, in accordance with the following:

(1) When the transmitter is initially installed;

(2) When any change is made in the transmitter which may increase the transmitter power input.

(c) The licensee of each station shall employ a suitable procedure to determine that the modulation of each transmitter, which is authorized to operate with an output power in excess of two watts, does not exceed the limits specified in this part. This determination shall be made and the following results entered in the station records, in accordance with the following:

(1) When the transmitter is initially installed;

(2) When any change is made in the transmitter which may affect the modulation characteristics.

(d) The determinations required by paragraphs (a), (b), and (c) of this section may, at the opinion of the licensee, be made by a qualified engineering measurement service, in which case the required record entries shall show the name and address of the engineering measurement service as well as the name of the person making the measurements.

(e) In the case of mobile transmitters, the determinations required by paragraphs (a) and (c) of this section may be made at a test or service bench: Provided, That the measurements are made under load conditions equivalent to actual operating conditions; and provided further that after installation in the mobile unit the transmitter is given a routine check to determine that it is capable of being received satisfactorily by an appropriate receiver.

6-12-6. Operator Maintenance and Tests

All transmitter adjustments or tests during the installation, servicing, or maintenance of a radio station, which may affect the proper operation of such stations, shall be made by or under the immediate supervision and responsibility of a person holding a general operator license, who shall be responsible for the proper functioning of the station equipment.

Additional Rules and Regulations are available for referencing at FCC Field Offices or they can be purchased from the Washington D.C. office.

Part 0	Commission Organization
Part 1	Practice and Procedure
Part 2	Frequency Allocations and Radio Treaty Matters; General Rules and Regulations
Part 5	Experimental Radio Services
Part 13	Commercial Radio Operators
Part 15	Radio Frequency Devices
Part 17	Construction, Marking, and Lighting of Antenna Structures
Part 21	Domestic Public Radio Services
Part 81	Stations on Land in the Maritime Services and Alaska-Public Fixed Stations
Part 83	Stations on Shipboard in the Maritime Services
Part 87	Aviation Services
Part 90	Private Land Radio Mobile Services
Part 94	Private Operational-Fixed Microwave Service

6-13. TEST SETUP SUMMARY

Modern two-way radio test equipment usually includes two or more test functions. As described in the preceding sections, a number of all-in-one units are available to provide multiple test functions in a single instrument. The transmitter and receiver functional diagrams (Figs. 6-25 and 6-26) given in this section emphasize the more important tests that are made. Remember that many of these tests can be made by a single test instrument.

6-13-1. RF Power Measurements

The rf power meter, whether it is a separate instrument or part of a composite instrument, provides a 50-ohm load at the transmitter output. A calibrated meter provides a watts-output

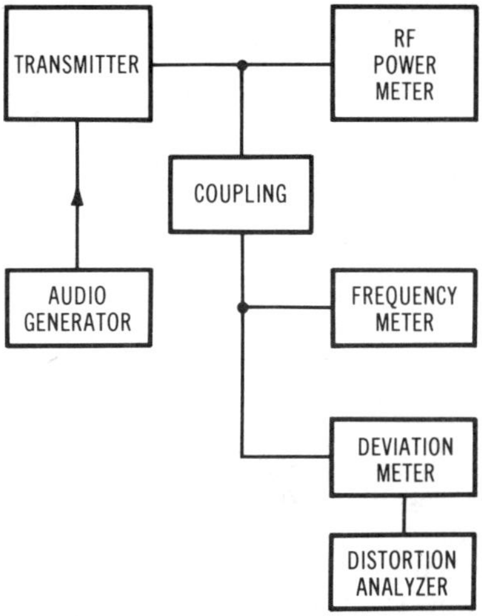

Fig. 6-25. Transmitter test setup.

reading. The resistive termination must be able to dissipate the power delivered by the transmitter.

If the measured output is below the value suggested by the transmitter manufacturer it may be an indication of a need for tuning or a transmitter defect. Transmitter tuning and loading can be adjusted. Some transmitters also include a power adjusting control to set the output level to the recommended value. Power output should not be adjusted higher than normal to minimize component wear.

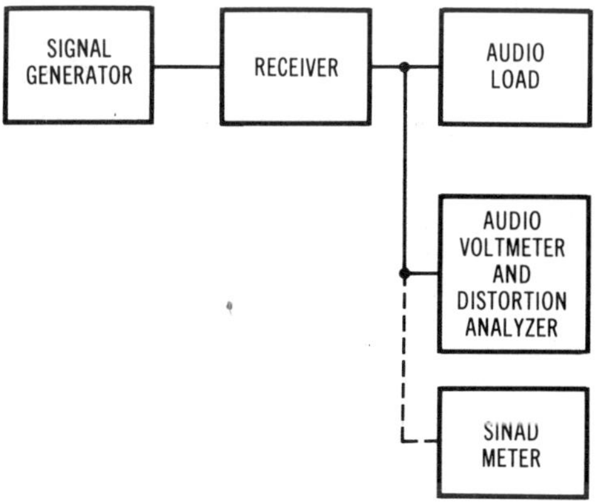

Fig. 6-26. Receiver test setup.

6-13-2. Transmitter Frequency

A counter or frequency meter can be used to make a direct measurement of the transmitter frequency. When a direct connection is made between transmitter and frequency meter an attenuator coupling system is employed to prevent the application of full transmitter power to the frequency meter. Modern counters and frequency meters are usually supplied with small antennas and a good separation between transmitter and meter can be made to prevent frequency-meter damage. Also some rf power meters have a built-in coupling system that provides a weak signal takeoff for connection to a frequency meter or counter.

Frequency readings are usually expressed in one of three ways. These are specific frequency, percentage difference between transmit frequency and actual assigned frequency, or the actual frequency difference between transmit frequency and assigned frequency.

If frequency is incorrect be certain to follow the specific instructions given in the service manual of the particular transmitter. In fact, if incorrect readings are obtained for any of the important transmitter or receiver tests the corrections you make should be those recommended in the instruction manual for the particular two-way radio unit.

The accuracy of the frequency meter is an extremely important consideration because of the required tight frequency tolerances of modern transmitters. Refer to the figures given in the frequency tolerance chart of Section 6-12. For best results the accuracy of the frequency meter should be 5 to 10 times better than the permissible tolerance on the particular assigned frequency. For example, if the frequency tolerance for the particular assigned frequency is ±0.0005%, the frequency meter used to make the measurements should have an accuracy of 0.00005%.

As mentioned previously some frequency meters give an actual reading in percent frequency tolerance. If the readings are given in actual frequency or frequency error the following equations can be used to calculate the actual frequency error in percentage:

$$\text{(1)}\quad \%\ \text{error} = \frac{\text{measured frequency} - \text{assigned frequency}}{\text{assigned frequency}} \times 100$$

$$\text{(2)}\quad \%\ \text{error} = \frac{\text{frequency difference}}{\text{assigned frequency}} \times 100$$

The accuracy of the frequency meter must be considered in deciding on an acceptable error percentage. Let us assume that a 150-MHz transmitter operates on a frequency with a frequency tolerance of 0.0005%. If the frequency meter itself has an accuracy of 0.00005%, an *acceptable percent error for a frequency measurement* must be less than 0.00045% (0.0005-0.00005).

Some figures will demonstrate the point. For example, according to the FCC frequency tolerance of 0.0005% on 150 MHz, a transmitter could be a theoretical 750 Hz off frequency and still be acceptable (150,000,000 × 0.000005). However, with the accuracy of the meter limiting the measurement a frequency error reading of more than 675 Hz would be *considered unacceptable* (150,000,000 × 0.0000045).

In summary then, it is important in making a frequency measurement that the actual accuracy of the frequency meter be taken into consideration. In our example any frequency error less than 675 hertz would mean that the transmitter is operating within the FCC frequency tolerance. If you had a perfect frequency meter any frequency error less than 750 Hz would be acceptable. The latter is an impossible situation.

6-13-3. Frequency Deviation

Deviation measurements are made following the procedure given in the instruction manual. Usually a 1000-Hz signal of a specific amplitude is applied to the transmitter from the audio generator. This is the value that should produce maximum transmitter deviation. In most applications this maximum reading is normally ± 5.0 kHz. A transmitter maintenance manual may recommend a lower value for a particular transmitter making certain that the transmitter will not be overmodulated on strong voice peaks. Consideration must also be given to the deviation produced by an included tone squelch system. Thus in some conditions final deviation setting may be as low as 3.75 kHz. Follow manufacturers' recommendations.

The symmetry of the deviation may be judged by comparing the positive and negative peaks. With perfect symmetry positive and negative deviation will be identical. If a distortion analyzer and audio voltmeter are included in the test setup the transmitter audio sensitivity as well as distortion can be measured.

6-13-4. Receiver Performance

A general interconnection plan for checking out the receiver section of a two-way radio is shown in Fig. 6-26. The signal generator must be highly accurate and capable of setting to the assigned received frequency very precisely. In fact, the frequency measuring unit for the signal generator is usually the same section of test equipment that is used to measure the transmit frequency. Thus the signal generator can be set on frequency to the same accuracy as a transmit frequency measurement can be made. It should be possible to frequency-modulate the signal generator output with an audio tone.

The signal generator must include an accurate attenuator to permit setting of the output to some exact output level. Often an isolation pad of 6 to 20 dB is located between the signal generator output and the receiver input.

An appropriate load must be placed on the speaker output of the receiver. Across this load an accurate audio voltmeter and/or distortion analyzer is connected.

A test setup of this type permits one to measure the audio distortion and sensitivity of a given receiver. Follow the instructions given in the maintenance manual. Arrangement also permits a measure of modulation bandwidth and squelch system sensitivity and performance.

A distortion analyzer or an optional SINAD meter can be used to make the important 12-dB SINAD sensitivity measurement. The SINAD ratio is:

$$\text{SINAD} = \frac{\text{signal} + \text{noise} + \text{distortion}}{\text{noise} + \text{distortion}}$$

The signal level at the receiver input which produces a 12-dB SINAD ratio is called the 12-dB SINAD sensitivity of the receiver. Such a level produces an intelligible and useful signal and is a measure of a receiver's useful sensitivity. In making a SINAD measurement the signal generator is set to the assigned receive frequency and its output to a standard signal level. Signal is deviated ±3 kilohertz with a 1-kHz modulating tone. The receiver is set to rated audio input. Now the signal generator output is decreased until the SINAD meter reads −12 dB. The signal level applied to the receiver input is the 12-dB sensitivity of the receiver.

CHAPTER 7

Transceiver Circuits and Systems

As in most electronic equipment, improper transceiver adjustment results in faulty operation. Circuits do drift, or their operating characteristics change with shifts in the supply voltage. Likewise when components must be replaced their characteristics sometimes differ from those originally used in the circuit. Environmental changes such as temperature or humidity can also alter operating conditions over a period of time. These variables require that a transceiver be retuned periodically if peak performance is to be maintained. Base station and mobile units must occasionally be brought into the shop and checked out and returned if necessary in accordance with a definite preventive-maintenance schedule. To do so it is important that you have a knowledge of circuits and systems.

7-1. BUSINESS RADIO TRANSCEIVER

An example of a modern solid-state mobile two-way fm radio is the *Motorola* model shown in Fig. 7-1. The unit can be purchased for low-band operation (25–50 MHz with 25-watt output), high-band (150.8–174 MHz with either 10 or 20 watts output), or uhf band (450–470 MHz with 10 watts output). Deviation is ±5 kHz and frequency stability for the high band and uhf band models is ±0.0005%.

A functional interconnect diagram of the transceiver is shown in Fig. 7-2. There are three major circuit boards with the main circuit board containing the receiver and the transmitter exciter. A pa section builds up the level of the signal at the output of the exciter to the final transmit power. There is a separate encoder/decoder which provides private-line operation. The private line arrangement

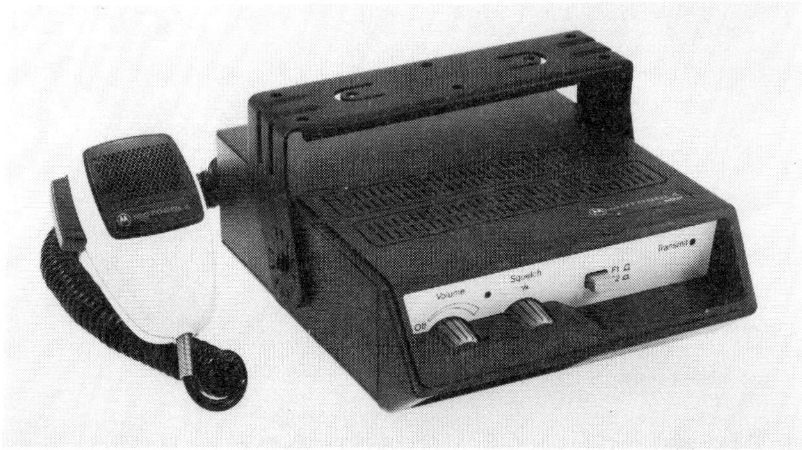

Courtesy Motorola, Inc.

Fig. 7-1. Business band two-way fm radio.

causes the radio receiver to reproduce only those signals with the correct incoming tone code. Thus undesirable messages on the same frequency are not heard, therefore reducing operator fatigue.

The audio from the microphone is applied to the exciter of the main circuit board. In addition a tone signal from the pl encoder is also applied to the exciter. Both signals modulate the transmitter. The private line CTCSS signal is a low-frequency audio tone that is transmitted along with the voice frequencies. Its frequency is subaudible and is not heard.

Output from the exciter is built up in level by the pa section and is passed to the antenna through an appropriate harmonic filter. Everything possible is done in modern two-way radio design to prevent the radiation of spurious signal components that can cause interference on other channels. In the uhf band model the rejection of spurious and harmonic emissions is greater than − 56 dB.

Also associated with the pa section is the electronic antenna switch that takes care of the automatic switching of the antenna between transmitter and receiver. When the push-to-talk switch on the microphone is depressed the entire unit switches over to the transmit mode. As soon as that switch is released there is an automatic change back to the receive mode.

If you will notice, at the left there is a hang-up button and clip usually associated with the mobile microphone. When the microphone is hung up, the private-tone operation is in effect. However, as soon as the microphone is removed from the hanger the private-tone activity is released and the operator can listen on the channel to make certain there is no other system using that channel before he begins to transmit. Hence he will not interfere with communications that are already in progress. He can wait until the communications end and then make his call.

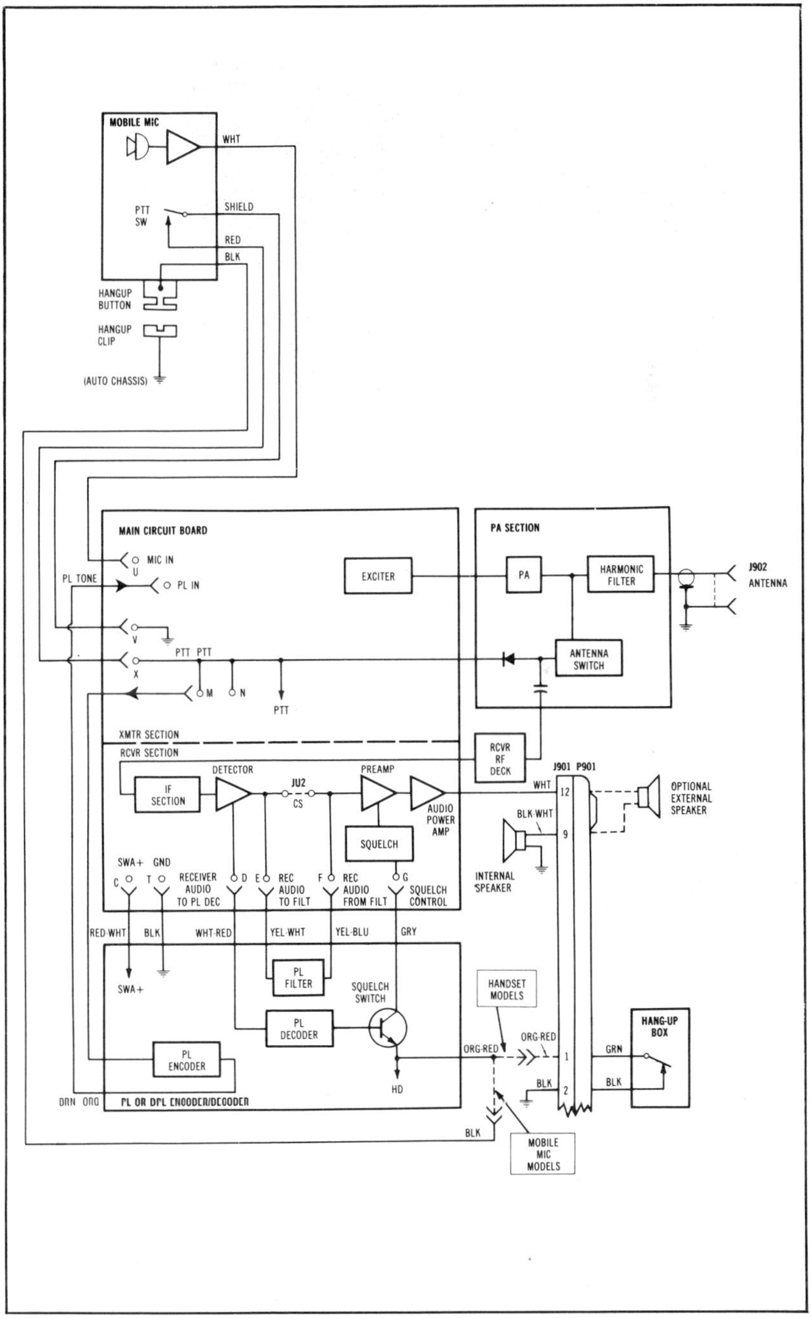

Fig. 7-2. Functional interconnect diagram.

On receive the incoming signal is first applied to the receiver rf deck where mixing action takes place and the signal is reduced to the intermediate frequency before application to the detector of the receiver section of the main circuit board. The detector not only demodulates the voice frequencies but it also must demodulate the incoming private-line tone signal. If one chooses not to incorporate private-line facility, units can be purchased with a jumper connected across the two terminals indicated at JU2. Otherwise the incoming signal passes through a pl filter to remove the tone component before the demodulated voice signal is applied to the preamplifier. An audio power amplifier follows, from whence the signal is applied to the speaker system.

The receiver also includes the usual squelch circuit which can be adjusted and used to block out signals that are below a decided power level. This is in the form of an external adjustment that can be adjusted by the operator to meet the given receiving requirements in the area in which he is active. The output of the detector is also applied to a private-line decoder which also operates in conjunction with the squelch switch. Its application is to keep the receiver squelched until that time that the desired tone signal is received.

7-2. CIRCUIT DESCRIPTION

A schematic diagram of the fm exciter is shown in Fig. 7-3. Audio signal from the microphone and its built-in amplifier is applied to the bases of transistors Q110 and Q111. This stage acts as an amplifier and a limiter of the microphone signal. Both positive and negative peaks are limited. Audio is transferred from the common-collector circuit through pre-emphasis network resistor R120 and capacitor C116 to the base of transistor Q112.

Output is removed from the emitter circuit and is applied through a splatter filter to the base of transistor Q113. The splatter filter attenuates harmonics when the voice-frequency signal is great enough to cause limiting action. Transistor Q113 itself acts as an isolator between the filter and the modulator. Audio level is set by deviation control potentiometer R126.

The transceiver is a two-channel affair using separate crystal oscillators and direct-frequency modulators. The variable-capacitance diodes are CR110 and CR111 which vary the capacitance associated with the crystals to produce frequency modulation. The crystals are mounted in ovens which heat the crystal whenever the temperature is below −10°C. Transistors Q114, Q115, and Q116 act as a temperature reference and control the current through the heater resistors of the two ovens.

The output resonant circuit of the crystal oscillator circuits is tuned to three times the crystal frequency, thus operating as a tripler and increasing the frequency up into the 50- to 56.9-MHz frequency range. This double-tuned resonant circuit is followed by a buffer amplifier (transistor Q119) which provides isolation and gain This stage ensures adequate drive to a succeeding tripler which multiplies the frequency to the 150- to 170.7-MHz frequency range. An rf amplifier stage is next (transistor Q121) followed by a power tripler (transistor Q122). Frequency has now been tripled into the uhf band and the power level has been raised to 300–400 milliwatts.

Spurious signal components are removed by the tripler resonant circuit and filter. Transistors Q801 and Q802 serve as power control and driver regulator. Transistor Q801 senses the pa current and then controls the power tripler and transmitter power-amplifier drive levels. Pontentiometer R804 serves as a manual adjustment of the power level by setting the operating point of transistor Q802 as well as transistor Q603 located in the power amplifier.

A two-stage uhf power amplifier is used (Fig 7-4). Fifty-ohm microstrip transmission line sections are used extensively as inductors in the power amplifier. They are indicated by the small lined blocks. The input resonant circuit is broadbanded and provides impedance matching and a low-pass filter characteristic. The power output of the pa driver is controlled automatically as well as manually from the exciter by way of transistor Q603, as mentioned previously.

The low-pass filter resonant circuit at the output of the pa driver is impedance matched to the input of the final power amplifier. Capacitor C608 is factory tuned for maximum power output at 470 MHz, the high frequency end of the band. Conversely capacity C617 in the collector output circuit of the final power amplifier is factory tuned for maximum power output at 450 MHz, the low end of the uhf band. The output resonant circuit is followed by a harmonic filter that attenuates the frequency components above the uhf band.

The antenna switch acts as an isolator for the receiver input during transmission. The 50-ohm microstrip line is a quarter-wavelength section. On receive CR601 is nonconducting and there is a signal transfer path through the microstrip line, there being a high-impedance load presented by the rf circuit. This permits the receive signal to pass to the receiver rf input. On transmit CR601 conducts and places a short at the output of the quarter-wave section of microstrip. As a result the transmitter rf output does not feed into the receiver rf input.

7-3. TRANSMITTER ADJUSTMENT

Transmitter adjustment has two major objectives. One of these is to ensure efficient and safe operation of the equipment along with convenient and effective operation of the two-way radio system. Another objective is to ensure compliance with FCC regulations. Three major considerations are frequency tolerance, modulation level, and power output. Essential pieces of test equipment are frequency meter, deviation meter, audio generator, dummy antenna load with wattmeter, and appropriate meters and alignment tools.

The transmitter alignment procedure is given in Chart 7-1. An understanding of techniques can be gained by going through the

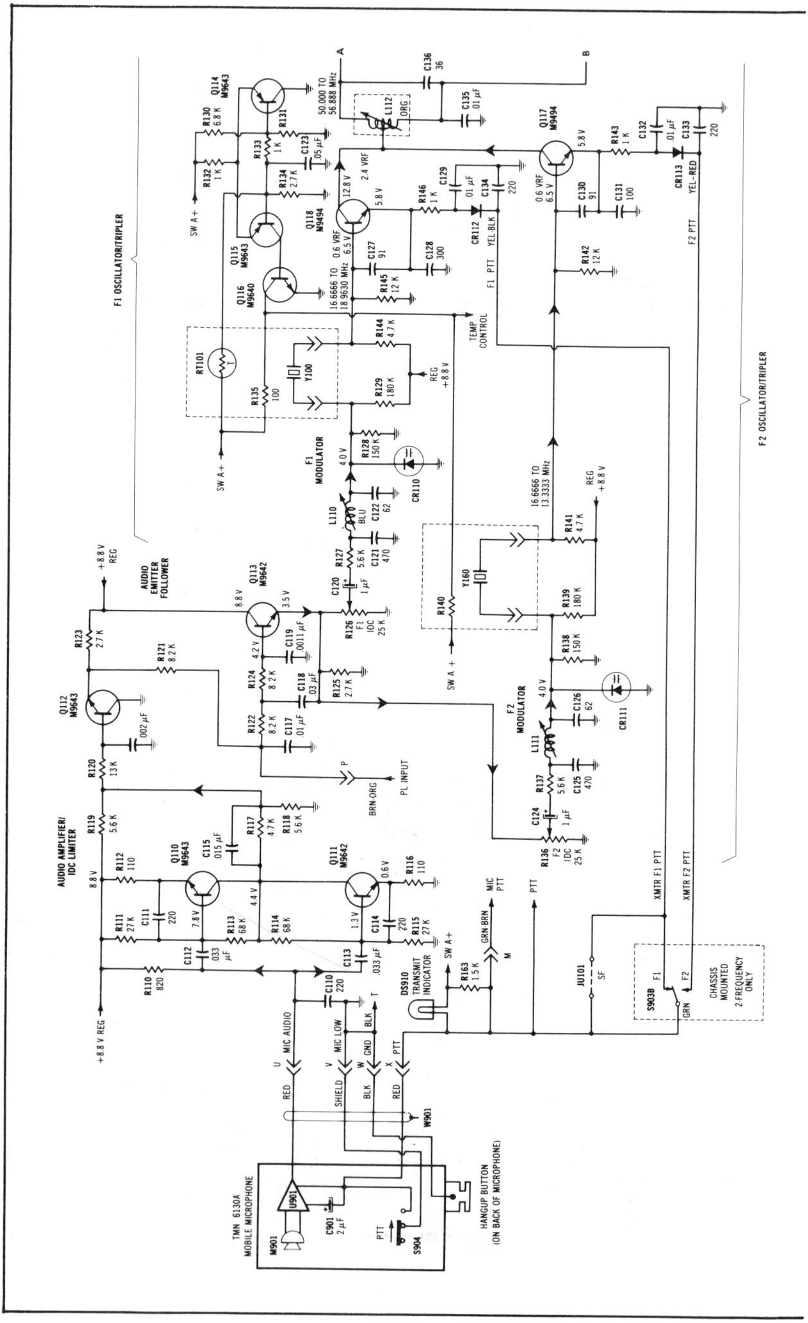
F1 OSCILLATOR/TRIPLER
F2 OSCILLATOR/TRIPLER
AUDIO AMPLIFIER/IDC LIMITER
AUDIO EMITTER FOLLOWER
F1 MODULATOR
F2 MODULATOR
TEMP CONTROL
PL INPUT
TRANSMIT INDICATOR
CHASSIS MOUNTED 2-FREQUENCY ONLY
TMN 6130A MOBILE MICROPHONE
HANGUP BUTTON (ON BACK OF MICROPHONE)

Fig. 7-3.

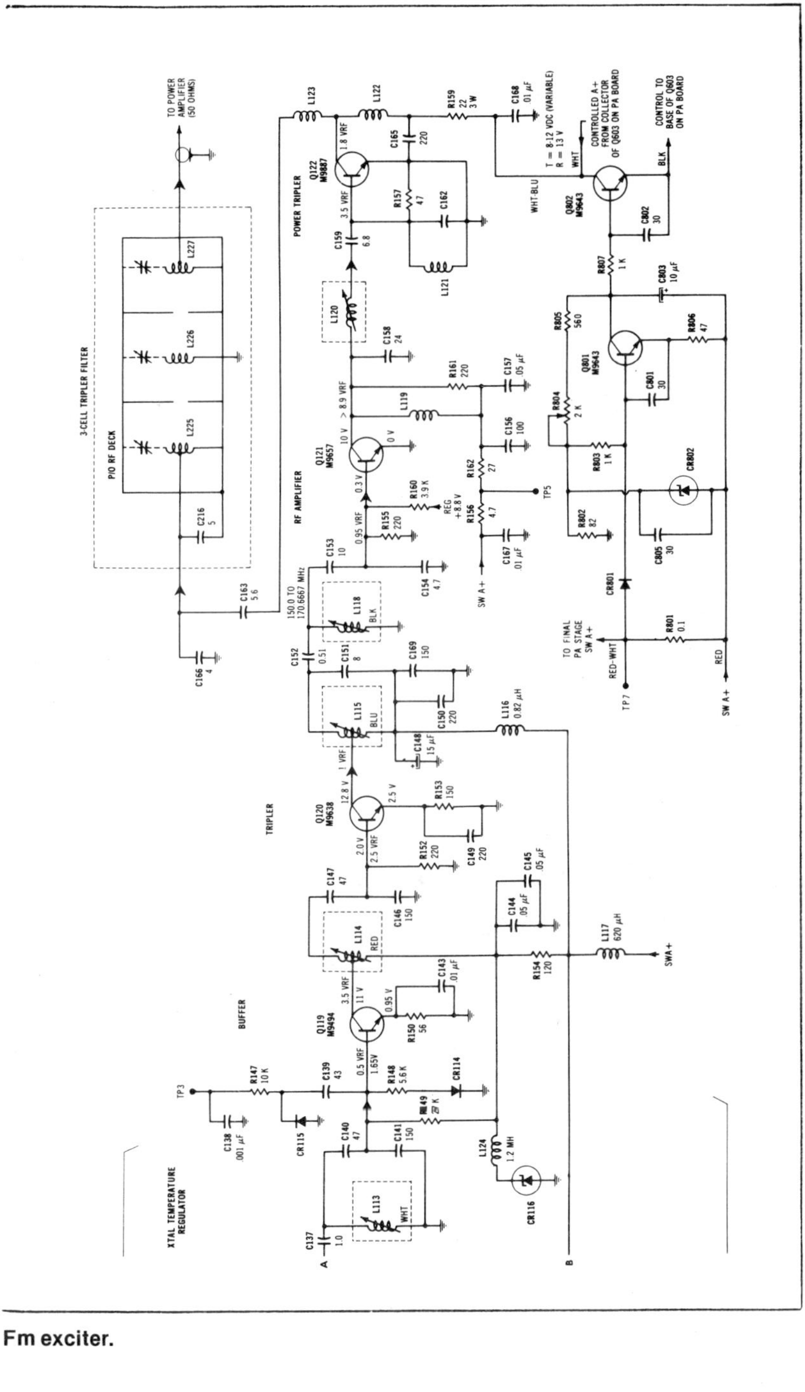

TO POWER AMPLIFIER (50 OHMS)
3-CELL TRIPLER FILTER
P/O RF DECK
POWER TRIPLER
RF AMPLIFIER
TRIPLER
BUFFER
XTAL TEMPERATURE REGULATOR
CONTROL TO BASE OF Q603 ON PA BOARD
CONTROLLED A+ FROM COLLECTOR OF Q603 ON PA BOARD
TO FINAL PA STAGE SW A+

Fm exciter.

Chart 7-1. Transmitter Alignment

Step	Test Point	Adjustment	Desired Reading
1 Oscillator Output	TP3	L112, then L113	Peak
2 Buffer Output	TP5	L114	Peak if TP5 has a reading.
	TP3		Dip if no TP5 reading.
3 First Tripler Output	TP5	L115, then L118	Peak
4 Second Tripler Input	TP5	L120	Peak on wattmeter or 24 μA max at TP5
5 Second Tripler Output Filter	Wattmeter	L225, 226, 227	Peak
6 Repeat all of the adjustments made to this point			
7 Oscillator Output	Wattmeter	L112	Detune for a reading of 5 watts
8 Second Tripler Input	Wattmeter	L120	Peak
9 Oscillator Output	TP3	L112	Peak
10 Repeat Step 5			
For widespaced operation (greater than 2 MHz separation between transmit frequencies on two-frequency radios), go to step 11A; otherwise, go to step 12.			

Chart 7-1. Transmitter Alignment (cont.)

NOTE: STEPS 11A, B, C, AND D ARE FOR WIDE-SPACED FREQUENCY OPERATION ONLY

Step	Test Point	Adjustment	Desired Reading
11A	TP3	L112	Peak on higher frequency
11B	TP5	L115	Peak on higher frequency
11C	TP3	L114	Peak on lower frequency
11D	Wattmeter	L225	Peak on higher frequency
12 Power Output	Wattmeter	R804	Adjust R804 to reduce power output to 11 watts. For widespaced operation, set R804 so that the channel with the lowest power output is 11 watts.

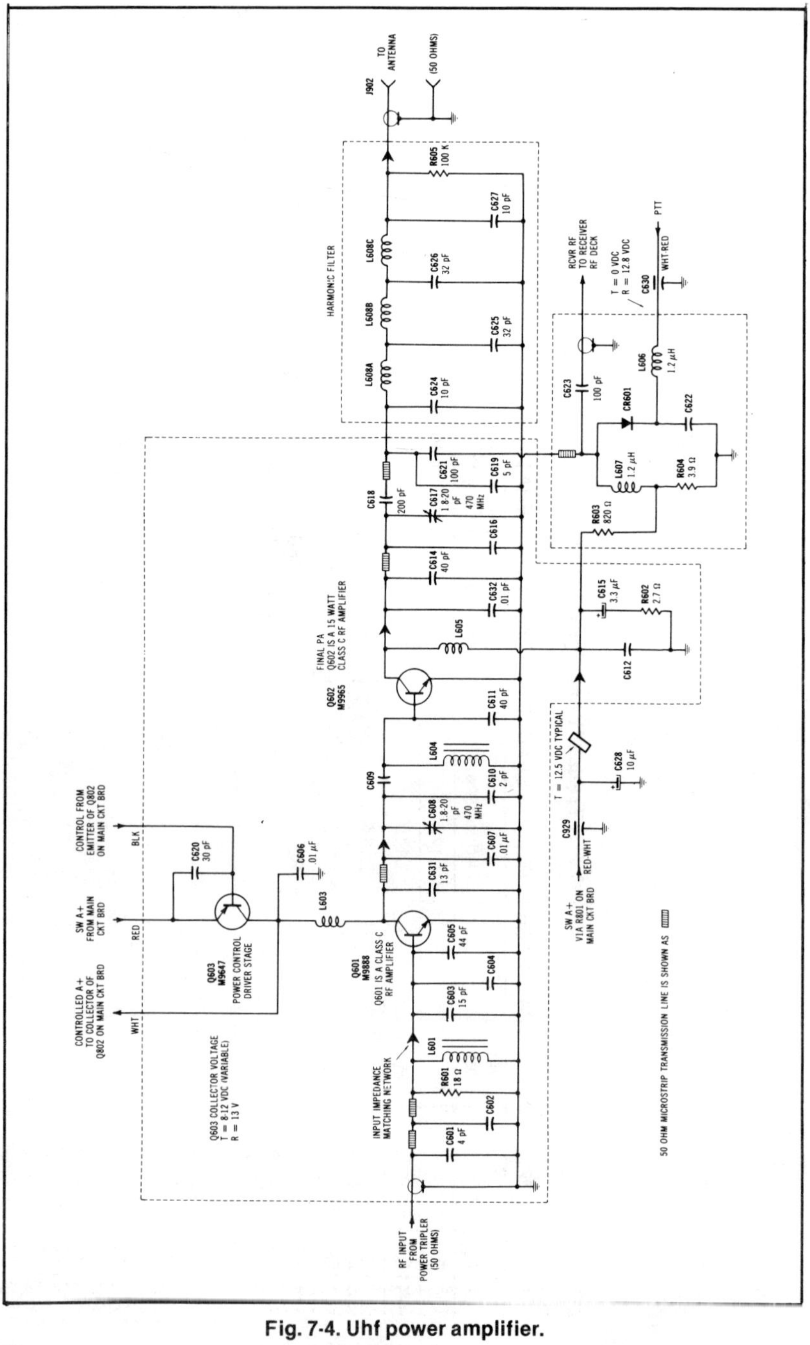

Fig. 7-4. Uhf power amplifier.

table step by step and referring to the schematic diagrams of Figs. 7-3 and 7-4. Locate the test points and the adjustment controls. A voltmeter is connected at the appropriate test point; a wattmeter is connected across the output.

Go through each of the twelve steps. When the maximum power output obtainable is less than 11 watts, it is suggested that the technician go back to step 2 of the exciter alignment and repeat the procedure. If the output remains below 11 watts, it is necessary to readjust potentiometer R804. When the output is less than 10 watts it is often an indication of a defect in the transmitter.

Chart 7-2 shows the procedure for adjusting frequency and fm deviation. In measuring the frequency, a frequency meter or counter must be connected across the output. In making the deviation adjustment, the microphone must be disconnected and an audio signal generator connected to the input. An ac FET or vacuum-tube voltmeter is used to measure the audio signal level.

Chart 7-2. Frequency and Deviation Adjustments

	Frequency
Step	**Procedure**
1	Select F1
2	Key transmitter with no modulation
3	Adjust L110 for F1 frequency ± 100 Hz
	Two-Frequency Models Only
4	Select F2
5	Adjust L111 for F2 frequency ± 100 Hz
	Deviation
Step	**Procedure**
1	Set audio signal generator output frequency to 1000 Hz and output level to 0.8 V.
2	Select F1 and key transmitter. Adjust F1 "IDC" control R126 for ±5 kHz deviation meter reading.
3	On 2-freq. models, select F2 and key transmitter. Adjust "IDC" control R136 for ±5 kHz deviation meter reading.
4	Reduce audio signal generator output level until deviation meter reads ±3.0 kHz. Audio oscillator output level should be 160 mV ±3 dB.

7-4. 800 MHz TRANSCEIVER

The Johnson 800-MHz transceiver (Fig. 7-5) is a completely solid-state fm transceiver operating from 806–821 MHz in transmit mode and from 851–866 MHz in the receive mode. Separation between transmit and receive frequencies is 45 MHz. The transceiver operates on 1 to 5 channels with one crystal-controlled oscillator used per transmit/receive channel. Arrangement is unique and only one oscillator is required since the injection signal into the first

Courtesy E. F. Johnson Co.

Fig. 7-5. Modern 800-MHz transceiver.

mixer of the receiver is the same frequency as the transmit frequency.

A functional block diagram of the receiver is given in Fig. 7-6. The receiver is a dual-conversion type with intermediate frequencies of 45 MHz and 10.7 MHz. Overall selectivity of the transceiver is established by helical and crystal filters.

Incoming signal from the transmit/receive switch is applied to a two-section helical filter at the top left. This signal is amplified by Q100 and is emphasized again using a three-section helical filter. Signal frequency is applied to the first mixer Q101. Here it is mixed with a signal component derived from a temperature-compensated crystal oscillator (tcxo) shown at the center right. Its oscillating frequency is the difference between the received frequency minus 45 MHz, divided by 48. The output of the crystal is amplified and applied to a fourth-harmonic filter which passes the fourth harmonic component of the tcxo crystal signal which is then doubled and tripled and passed on to a two-stage helical filter. Output is again doubled in frequency prior to its application to a two-stage helical filter that comes ahead of the first mixer. The overall frequency division between the tcxo crystal oscillator and the local-injection signal at the mixer is 48 $(4 \times 2 \times 3 \times 2)$. This places the injection signal on a frequency that is 45 MHz below the received signal.

The 45-MHz component at the output of the first mixer is applied to a 4-pole crystal filter which is very effective in passing a narrow band of frequencies in the vicinity of 45 MHz but has a high rejection to off-frequency components. Selectivity of the receiver is largely determined by its crystal filters. After amplification by Q200

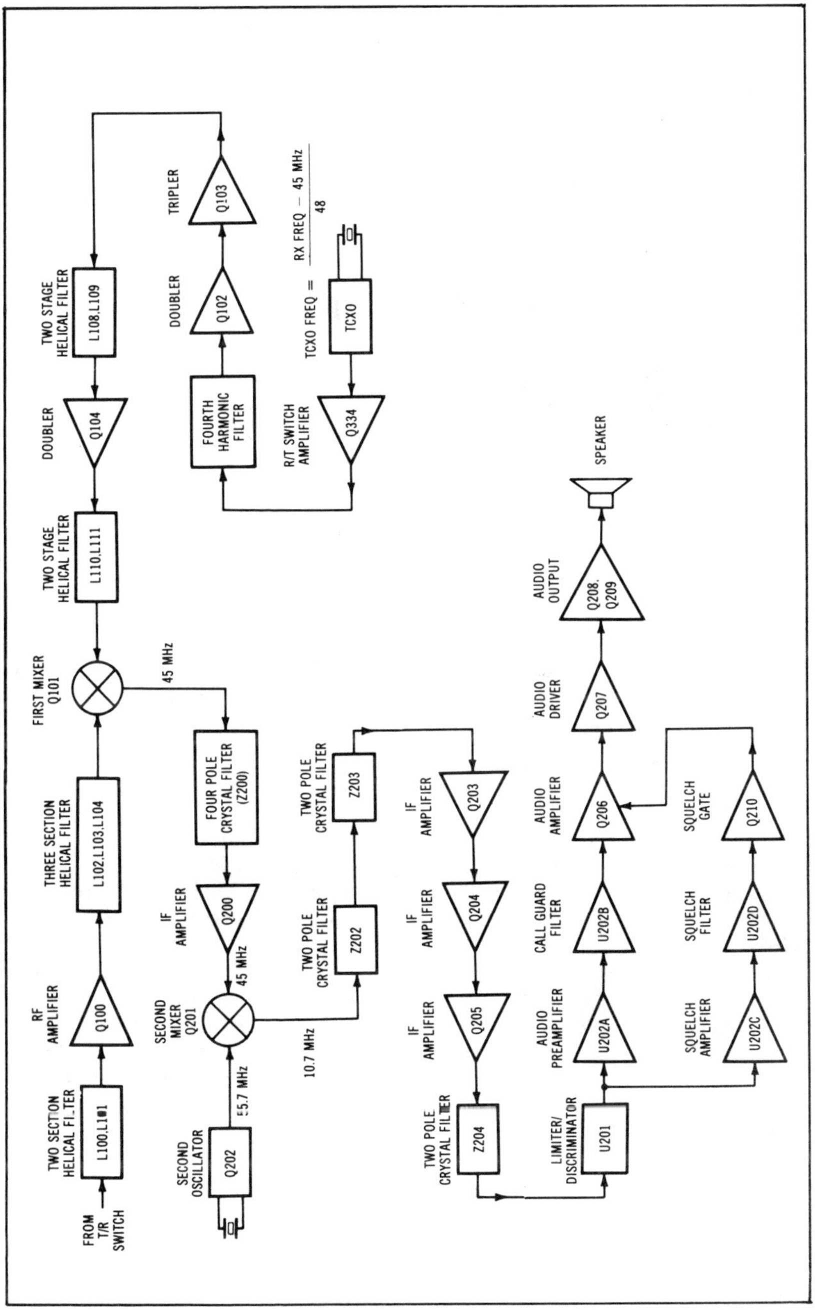

Fig. 7-6. Functional diagram of receiver.

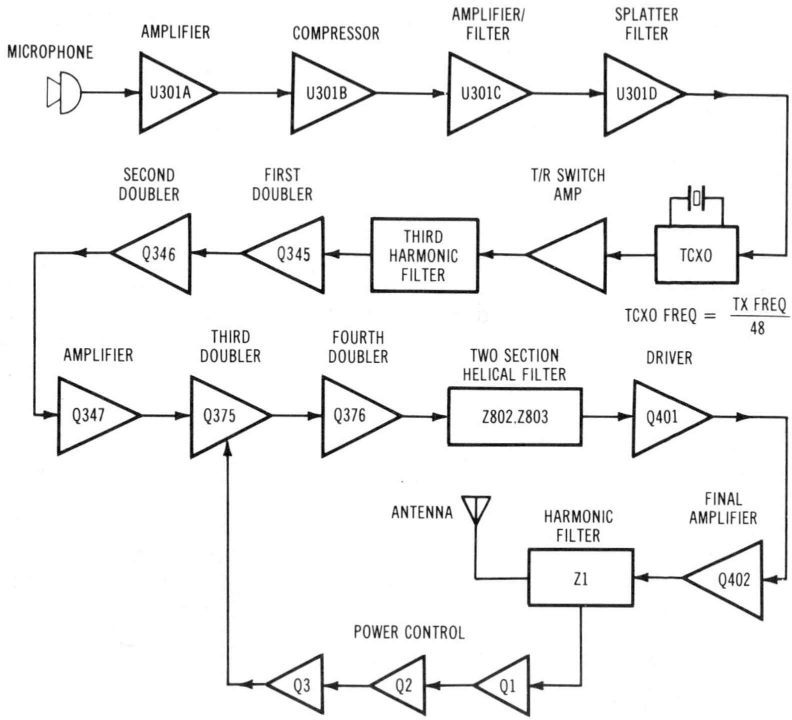

Fig. 7-7. Transmitter functional diagram.

the signal is applied to a second mixer. A crystal-controlled oscillator provides a 55.7-MHz injection component for the second mixer. The resultant difference frequency is 10.7 MHz. Two 2-pole crystal filters follow the second mixer before the signal is increased in level by 3 successive if amplifier stages. Additional filtering follows before the signal is applied to limiter discriminator U201.

Demodulated audio at the output of the discriminator is increased in magnitude by audio preamplifier U202A. A call-guard filter and audio amplifier follow. The call-guard circuit filters out any incoming tone that may be associated with the transmitted signal. Audio driver and output stage follow. A squelch system consisting of amplifier, filter, and gate is located between the discriminator output and the squelch audio amplifier (Q206).

The transmitter block diagram is shown in Fig. 7-7. Transmit audio processing is done by a quad operational amplifier (U301). The first quad provides impedance matching and amplification of the microphone audio. The second quad is a part of the compression and limiting circuits of the transmitter. Voice peaks are clipped and the overall signal is compressed to obtain a high average modulation percentage. The third quad provides amplification and some additional filtering to confine the audio frequency range for good intelligibility and low bandwidth. The final quad is a splatter filter that removes all frequencies over 3000 Hz.

Output is applied to the tcxo and that oscillator is frequency-modulated by the audio signal. The crystal-oscillator frequency is the transmit frequency divided by 48. This means the crystal frequency for transmit is the same as the crystal frequency for receive.

Output of the crystal oscillator, now frequency-modulated, is applied through the switch amplifier to a third-harmonic emphasis filter. The frequency-modulated third harmonic output is now applied to a group of four doubler stages and an intermediate amplifier. This builds up the signal to the transmit frequency range in the 806- to 821-MHz spectra.

A two-section helical filter follows the final doubler. The driver and final amplifier build up the power level of the frequency-modulated signal to an output of approximately 13 watts. Output is then applied to a harmonic filter.

An output is derived from this filter which is used to operate the power control circuits of the transmitter. This circuit senses if the rf power is too high or too low and makes a compensating adjustment in the third doubler to maintain a constant output power. A manual-adjust power control circuit is associated with the fourth doubler. Thus, the power output can be preset to the decided value by the manual adjustment, and it will be held at that value by the power feedback system.

7-5. BANDPASS HELICAL FILTERS

Helical filters and resonators are popular in modern vhf/uhf transceivers because of their very high values of unloaded Q and exceptional selectivity. In receiver application such resonant circuits offer protection from images, spurious signals, and intermodulation distortion components. In transmitter application they are able to select desired harmonics in a frequency-multiplier system and are a big assist in radiating a pure signal on the desired channel. In addition they are physically compact and sturdy.

Basically the helical resonant circuit is a quarter-wave section of transmission line housed in a circular or square container (Fig. 7-8). The container acts as a shield with a nearly lossless dielectric. The bottom of the helix is connected to the shield. The top of the helix is connected to the shield through a low-loss variable capacitor that can be used for tuning the resonator. Coupling can be made in various ways with inductive loops or capacitive probes. However, the direct tap on the coil, as shown in Fig. 7-8, is the most common.

A two-section or two-pole helical filter can be formed by using two sections in a single shield, setting up a double-tuned transformer arrangement. Input and output connections are made by taps on the input and output coils. Each section has its own

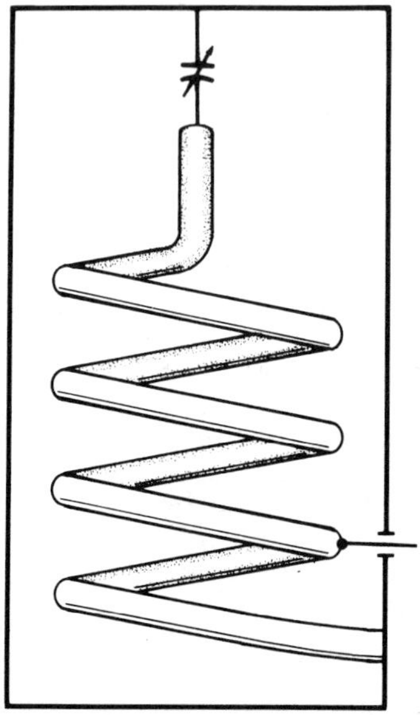

Fig. 7-8. Basic helical resonator circuit.

tuning capacitor. Coupling from one tuned circuit to the other is by way of an aperture; the location and size of the aperture has a decided influence on the characteristics of the two-pole filter.

The radio-frequency amplifier of the Johnson receiver uses the circuit arrangement of Fig. 7-10. A two-pole helical filter couples the antenna to the base input of the amplifier. A three-section helical filter couples the collector output of the rf amplifier to the mixer. Such an rf stage is highly selective and is effective in reducing intermodulation distortion components.

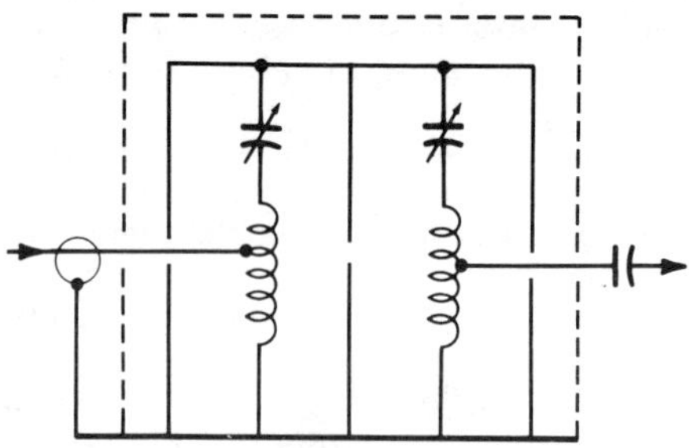

Fig. 7-9. Two-sectional helical filter.

7-6. RECEIVER SQUELCH SYSTEM

The squelch system of a two-way radio receiver disables the audio system when there is no receive signal. Thus the annoying and inherent receiver noise is not heard at the speaker output.

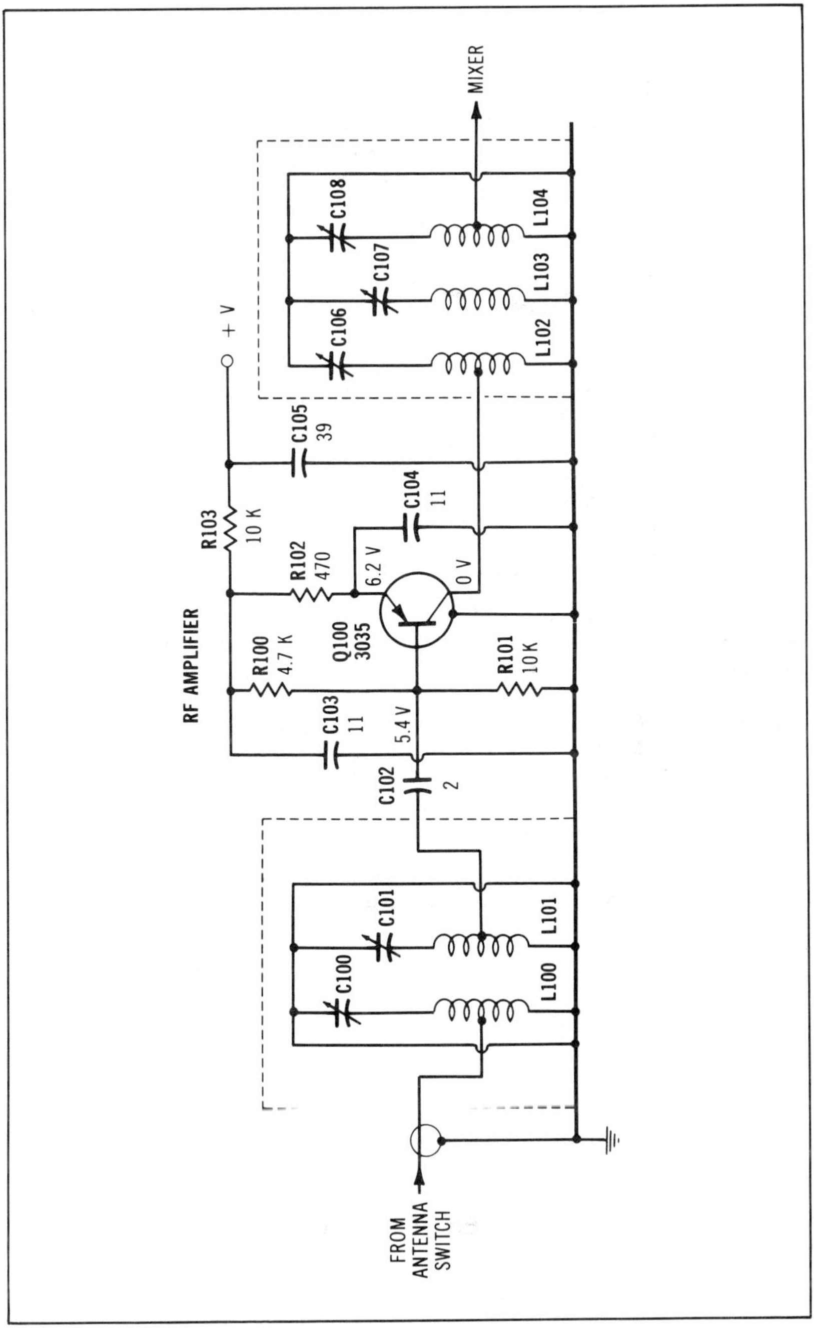

Fig. 7-10. Rf amplifier with input and output helical filters.

Squelch circuit operation is related to the noise present in the discriminator output. With no on-frequency signal being received, the receiver noise is amplified and present at high level at the discriminator output. When an on-frequency signal is being received, the signal will dominate the noise in the receiver, and there will be a decrease in the noise output. The limiter circuit ahead of the discriminator also becomes active and removes noise pulse, thus resulting in a further decrease in the noise. Actually the noise-level output is inversely proportional to the received signal strength. This condition makes the noise-signal component useful for squelch circuit operation.

The Johnson squelch system (Fig. 7-11), consists of a noise amplifier, a squelch filter, and a squelch gate. The noise signal is derived from the discriminator output and is applied to the input of an operational amplifier that increases the level of the noise. Amplification is determined by the ohmic value of resistor R1. Output is supplied to a voltage divider consisting of resistors R2 and R11 which set the level of the noise information applied to the squelch filter. The squelch filter is a high-pass type and removes frequencies below 3000 Hz, passing only the high-frequency noise.

Noise is supplied to a rectifier doubler circuit composed of capacitor C4 and diodes D3 and D4. The rectifier activity and the filtering of capacitors C5 and C7 develop a positive dc voltage at the base of the squelch gate transistor.

When no signal is being received there is a high noise level present which causes the dc level on the base of the squelch gate transistor to increase. As a result the gate conducts and grounds the supply voltage that is normally applied to the collector of an audio amplifier stage by way of resistor R9. This grounding of the supply voltage to the audio amplifier mutes the speaker audio. The actual squelch level can be controlled with the squelch control resistor that is also connected to the base of the squelch gate transistor.

When an on-frequency signal is being received there is little or no noise applied to the squelch amplifier input. As a result there is little rectified dc voltage applied to the base and the gate transistor does not conduct. Therefore the supply voltage is applied to the collector of the audio amplifier through resistors R8 and R9. The audio system then operates normally and the incoming signal is heard.

To prevent intermittent squelching when receiving a weak fading signal, a delay or hysteresis activity is caused by resistor R10 and diode D1. Diode D1 is forward-biased in the unsquelched mode providing a shunt through resistor R10 and capacitor C8 for the noise signal. This effectively reduces the sensitivity of the squelch circuit to noise. The result is that a greater noise change is required to change from an unsquelched to a squelch condition than vice-versa. Capacitor C9 at the base circuit also helps in preventing the gate from changing state during brief fading in or out of the received signal.

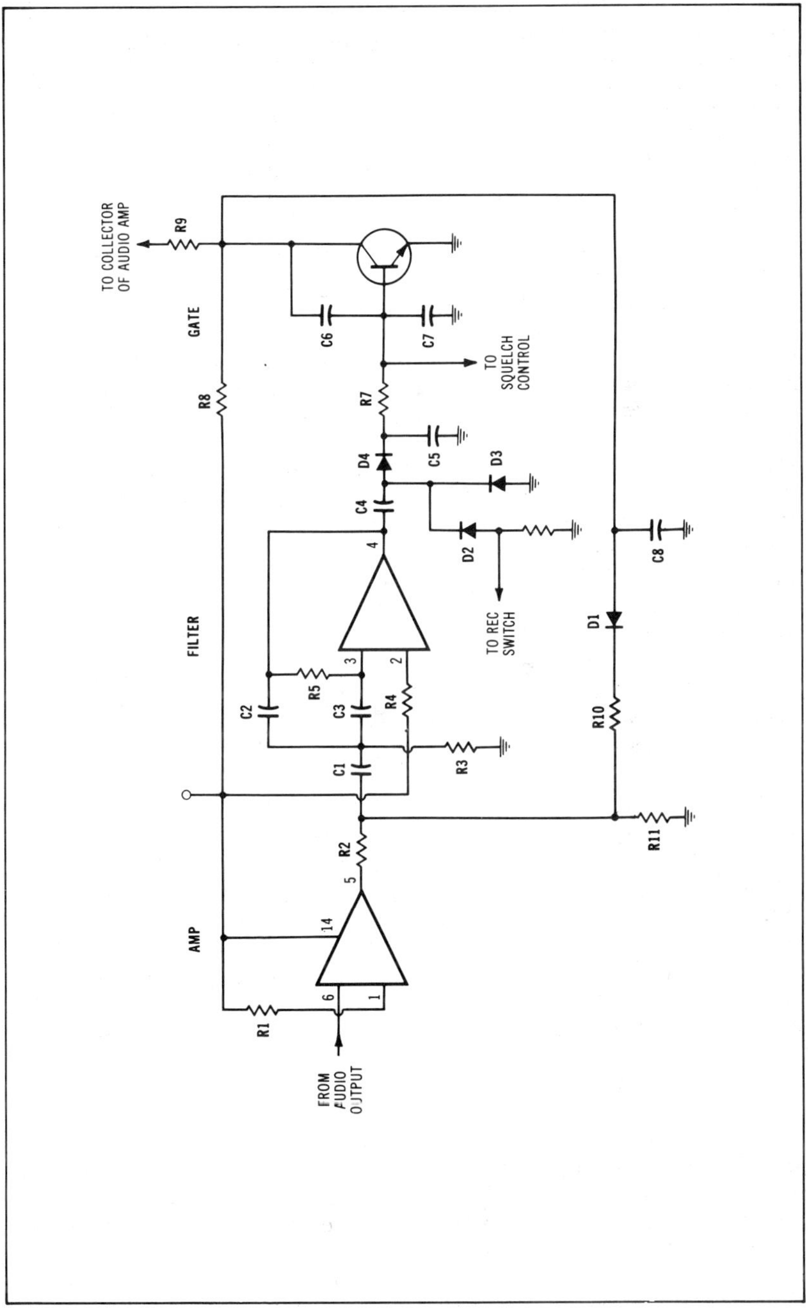

Fig. 7-11. Squelch system.

The combination of diode D2 and resistor R6 prevents the short bursts of noise from the receiver when the push-to-talk switch is released. This is often known as a squelch tail which results during the changeover between transmit and receive. The voltage taken from this point is used in the transmit received switching and maintains a positive voltage on the base of the squelch gate and a resultant squelched receiver until the regular circuit has time to function.

7-7. TRANSMIT AUDIO PROCESSING

The transmit audio processing is done by a quad operational amplifier IC (Fig. 7-12) which amplifies, compresses, and filters the audio signal. In the amplifier input stage, capacitor C301 shapes the response curve of the microphone circuit, while resistors R301 and R302 set the amplifier gain. Capacitor C303 and resistor R304 provide part of the 6-dB-per-octave pre-emphasis required in the modulation process.

From here the signal is applied to the compressor which cuts back on the amplification when the output reaches a certain level. Resistors R304 and R305 determine the normal amplification of the stage and, when the output level reaches above 0.5-volt peak, diodes CR301 and CR302 forward bias and provide additional feedback through capacitor C304 resulting in less amplification.

Output of the compressor is now applied to a limiter circuit which limits the peaks of the signal to prevent overmodulation of the transmitter (higher than normal deviation). Diodes CR303 and CR304 are forward-biased by resistors R307, R308 and R309. When the peak audio level exceeds the bias voltage, the diodes reverse-bias, clip, and limit the signal to that level.

The third quad serves as the amplifier/filter. Its gain is established by resistors R310 and R311 and feedback to attenuate high frequencies is provided by capacity C315. The remainder of the pre-emphasis is provided by capacitor C307 and resistor R313.

The final circuit is the splatter filter which filters out frequencies above 3000 Hz to prevent adjacent channel interference. The output of the splatter filter is coupled through capacitor C310 to the temperature-compensated crystal oscillator. The audio frequency modulates the tcxo frequency. Audio level can be adjusted by an associated deviation control associated with the oscillator.

7-8. STRIPLINE INDUCTORS AND RESONANT CIRCUITS

Sections of stripline serve as excellent resonant circuits and inductors in uhf transmitter amplifier circuits. Striplines are printed-circuit board pads especially shaped to provide a characteristic impedance similar to a transmission line. Electrically they are transmission lines with series inductance and parallel capacitances distributed along the line. Physical characteristics that determine the impedance of a stripline are width of the stripline, circuit-board thickness, and the dielectric or circuit-board material. Since the cir-

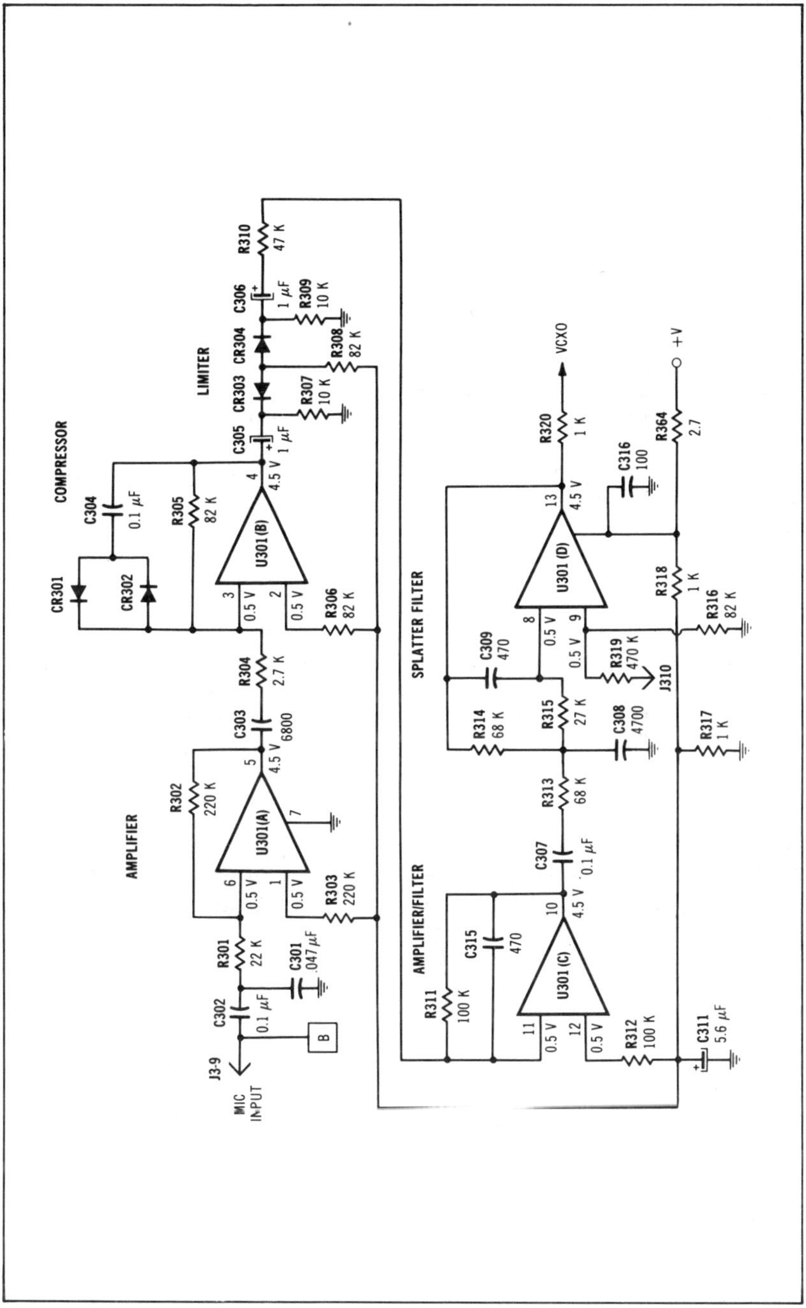

Fig. 7-12. Audio processing prior to fm oscillator.

cuit-board thickness and material cannot easily be changed, the method used to set the impedance is to alter the width of the pad. Generally, the wider the pad is, the lower is the characteristic impedance and vice-versa.

A simplified schematic of the Johnson final rf power amplifier is given in Fig. 7-13. Seven stripline sections, labeled T1 through T7, are shown. The output of the driver is connected by a 50-ohm coaxial line to the first stripline (T1) that provides impedance matching to the input circuit of the common-base rf amplifier. The wider section of the stripline, as designated by the horizontal portion of the T-shaped symbol on the schematic, along with capacitor C3 matches the input impedance of the transistor to that of the coaxial cable.

The thinner section of the stripline, as designated by the vertical portion of the T-shaped symbol, is a quarter-wavelength transmission line that is grounded by resistor R1, inductor L1, and capacitor C2. This grounded quarter-wavelength transmission line presents a high impedance to signals in the 800-MHz range and other frequencies are filtered to ground through resistor R1. The inductor provides a dc ground, while the capacitor provides an ac ground for 800-MHz signals.

The output of the transistor is fed to another stripline that functions in a manner similar to the one at the input. The supply voltage is fed through inductor L2 and the quarter-wavelength section of stripline. Since the stripline is thin and cannot carry large amounts of current, it is important always to have the power cable fused with a fast-blow fuse no larger than 15 amperes.

Output is coupled to another stripline (T3) which, in association with capacitor C6, adjusts the output impedance, permitting the signal to be fed through a 50-ohm coaxial cable to T4 which is a part of the antenna switch/filter circuits. Z1 is a harmonic filter designed from microstripline as represented by the horizontal portion of the symbol. The vertical portion of the Z1 symbol is a shorted quarter-wavelength section of transmission line which presents a high impedance to 800 MHz signals and provides a path for the application of dc to antenna switch diodes D1 and D2.

Diodes D1 and D2 are forward-biased in the transmit mode by the current passing through current-limiting resistor R3. The signal then has a low-impedance path through diode D1 and through another stripline (T5) that is used to isolate the transmit signal from the receiver. Diode D2 is forward-biased at the same time and it shorts one end of the quarter-wavelength section of the stripline to ground creating a high impedance for 800 MHz and, at the same time, placing the input side of capacitor C11 at ground potential. Signal passes through the final section of stripline (T6) to the coaxial cable that connects to the antenna jack of the transceiver.

In the receive mode diodes D1 and D2 are reverse-biased providing a high impedance path to the transmitter. However, there is now a low-impedance path from the antenna to stripline T7 and on to the receiver input. Capacitor C11 and inductor L3 are

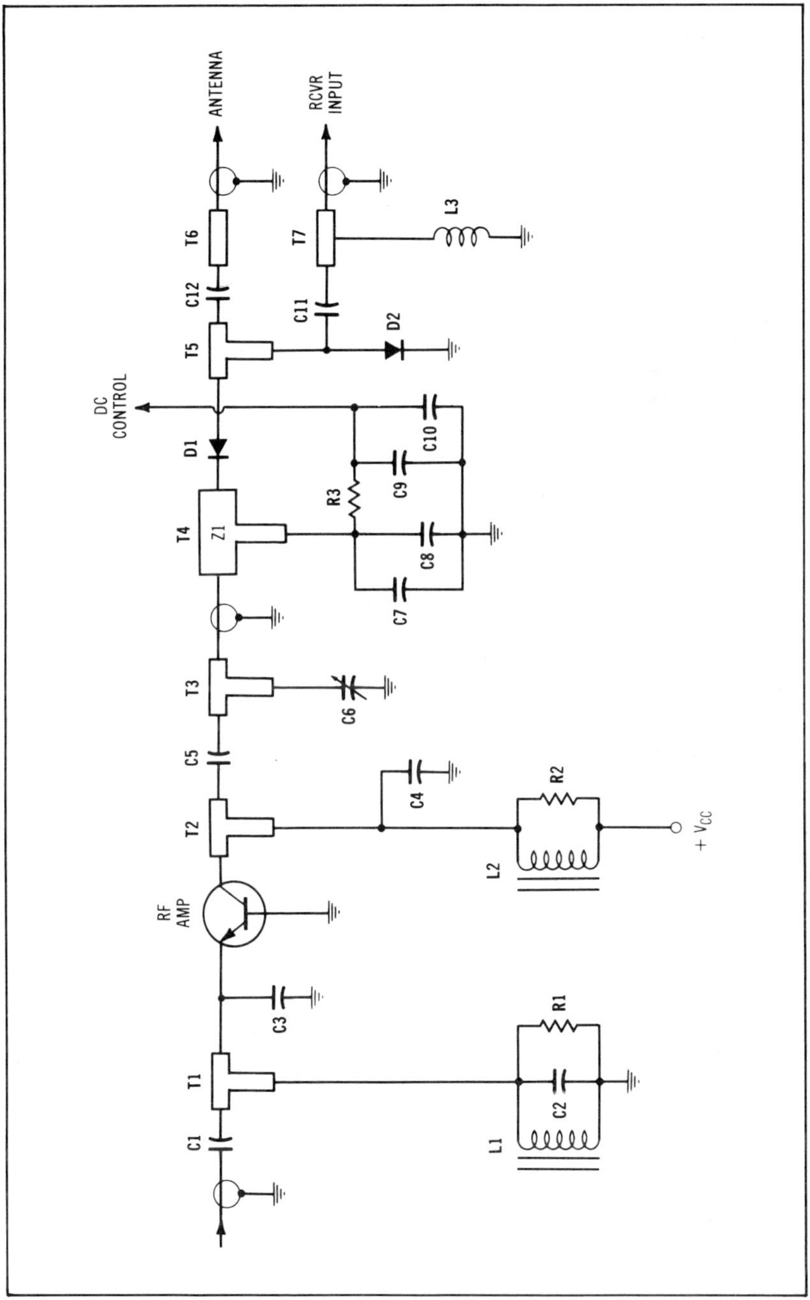

Fig. 7-13. Final rf amplifier using a number of striplines.

used to suppress low frequency signals that might be emitted from the receiver. Thus they are not radiated by the antenna system; spurious radiation and possible intermodulation distortion components are suppressed.

7-9. POWER SUPPLY AND CROWBAR OVERLOAD PROTECTION

A simplified schematic of the Johnson base-station power supply is shown in Fig. 7-14. It is a voltage-regulated supply using a bridge rectifier. The pass circuit consists of three transistors connected in parallel, although only one pass transistor is shown on the diagram. The regulated supply contains a reference voltage transistor and its 9.1-volt zener diode. Output can be adjusted precisely to 13.8 volts using potentiometer R6.

The collector output of the reference transistor is direct-coupled to the base of the driver transistor. Emitter output of the driver is connected to the base of the pass transistor. Thus the dc voltage on the base varies whenever there is an attempted change in the 13.8-volt output voltage value. This change in base bias regulates the resistance between collector and emitter of the pass transistors. Thus the output current will vary up or down to make a compensating change in the output voltage to maintain a constant value of 13.8 volts. Overload voltage protection is offered by the silicon-controlled rectifier (SCR) and the 16-volt zener diode of the output circuit.

The primary of the power transformer is connected to the 120-volt ac line through an appropriate 5-ampere fuse. Capacitors C1 and C2 filter out noises on the ac line. The secondary voltage output is about 20 volts rms and is rectified and filtered. Resistor R1 is a bleeder resistor that discharges capacitor C3 when the power input is removed.

The unregulated 25-volt dc output is supplied to the collector of the series regulator transistor through a 15-ampere fuse. A sense line connected to the output supplies a sample of the regulated voltage to resistor network R5 through R7. The emitter of the regulator transistor is maintained at 9.1 volts by the zener diode. The voltage on the base of the regulator transistor can be varied to raise or lower the regulated output.

The sense line also supplies an input to resistor R8, the 16-volt zener diode, and resistor R9. This combination forms a voltage divider to provide the SCR triggering voltage. When the regulated output reaches about 17.5 volts, the SCR will trigger and open the 15-ampere fuse at the output of the bridge rectifier. Capacitor C8 is a filter that prevents noise from triggering the silicon-controlled rectifier. This type of supply is called a *crowbar overload protection*. The crowbar circuit, by monitoring the output voltage, prevents any power supply defect from developing a high voltage output that could destroy the load to which the power supply is connected. Many integrated circuits are very critical of the supply voltage and

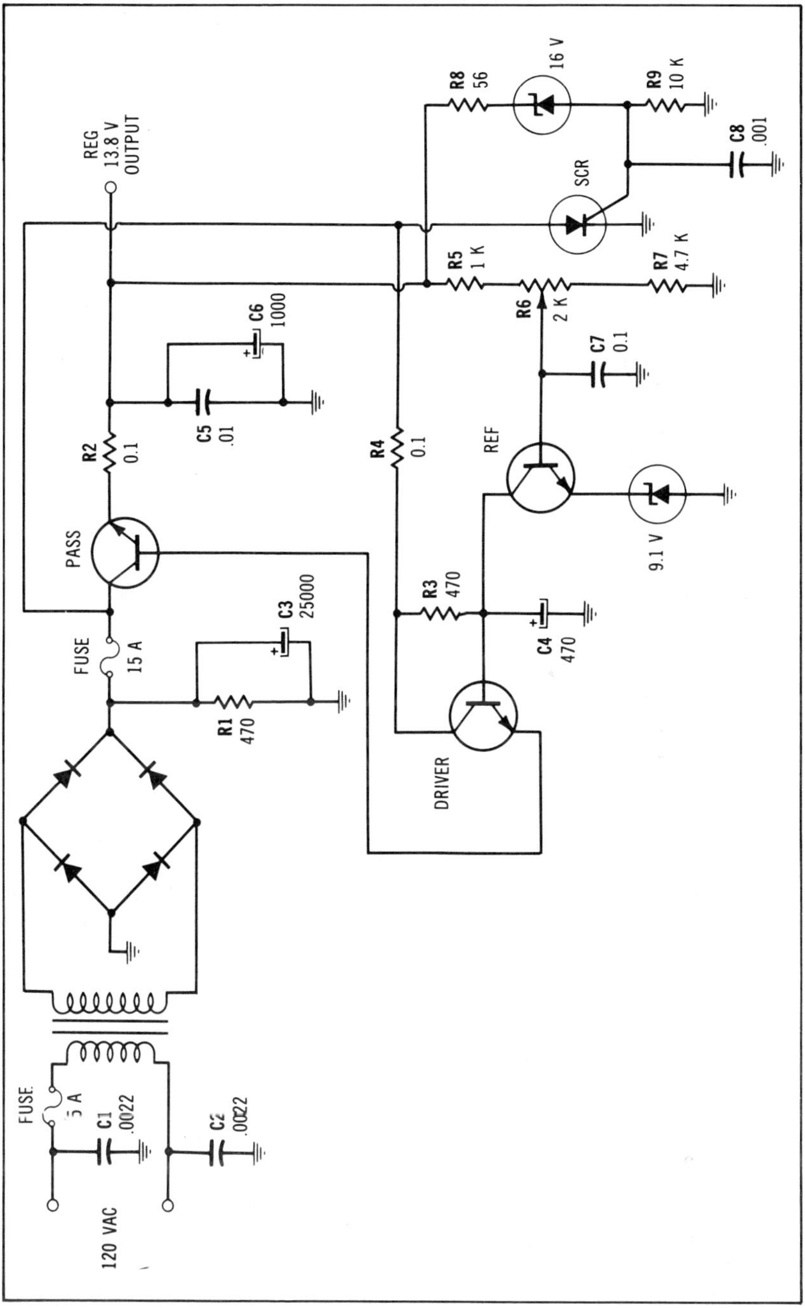

Fig. 7-14. Regulated power supply with crowbar overload protection.

the crowbar circuit provides a high order of protection that can prevent IC burn-outs.

7-10. CONTINUOUS-TONE-CONTROLLED SQUELCH SYSTEMS

CTCSS systems are used widely in the private land-mobile radio service. Appropriate tone transmissions minimize the annoyance of hearing signals from other systems having the same assigned carrier frequency. System definition and operating qualifications, as established by the Electronic Industry Association (EIA), are as follows:

> "The term 'Continuous-Tone-Controlled Squelch System' (herein referred to as CTCSS) shall define a system wherein the radio receiver(s) are equipped with tone-responsive devices which allow audio signals to appear at the receiver audio outputs only when a carrier modulated with a specific tone is received. Said tone must be continuously present for a continuous audio output. The transmitter emitting the carrier shall be modulated with a continuous tone, whose frequency is the same as the tone required to operate the tone-responsive device in the receiver, in order to permit audio at the receiver output. The purpose of the defined system is to minimize the annoyance of hearing communications directed to others sharing the same carrier frequency. By using a specific tone, each user may code his carrier to prevent his reception of audio signals from any uncoded or differently coded carriers.
>
> The CTCSS is considered primarily as a means for group calling as distinguished from individual calling.
>
> Since the CTCSS receiver is equipped to prevent hearing non-CTCSS or differently coded carriers, the transmitter operator may cause interference if he does not monitor the channel prior to his own transmission. To minimize this problem, all CTCSS receivers must be provided with a means for monitoring all signals on the receiver channel."

The standard tone frequencies are given in Table 7-1. Assignments in a given area shall be made from one of the groups to the maximum degree possible.

Table 7-1. Standard Tone Frequencies

Group A		Group B		Group C
67.0	151.4	71.9	146.2	74.4
77.0	162.2	83.5	156.7	79.7
88.5	173.8	94.8	167.9	85.4
100.0	186.2	103.5	179.9	91.5
107.2	203.5	110.9	192.8	
114.8	218.1	118.8	210.7	
123.0	233.6	127.3	225.7	
131.8	250.3	136.5	241.8	
141.3				

The transmitted tone frequency deviates the fm transmitter to a lesser degree than the audio. Nevertheless the sum of the two deviations shall not exceed the maximum rated system deviation for the particular radio service. When the rated system deviation is ±5 kHz, the CTCSS deviation must fall somewhere between 0.5 to 1.0 kHz. If a value of 0.5 CTCSS deviation is employed, the audio deviation may not exceed 4.5 kHz, maintaining a maximum total deviation of 5 kHz (4.5 + 0.5).

To prevent interference on a given channel a pretransmission monitoring system must be included. In one system it must be possible to monitor the receive channel prior to transmission without requiring the operator to perform any additional functions other than preparing to energize his transmitter by lifting the microphone from its hang-up mechanism. This arrangement can be considered to be as an automatic pretransmission monitoring. It is also acceptable to provide a switch that the operator can use to monitor the channel manually before each transmission or at any time deemed necessary in the course of operation.

An example of an *RCA* Encode/Decoder is given in Fig. 7-15. In the receive mode the tone gate holds the receiver in a squelch position until a carrier is received with a tone modulation of the correct frequency. Tone signal from the receiver discriminator is applied through a low-pass filter to the input amplifier (transistor Q1). The low-pass filter removes frequency components above the tone-modulation band. Output from transistor Q1 is fed through a limiter to the base of transistor Q2 which has a reed unit in its collector circuit. The reed unit has a bandwidth of less than ±0.5% and only the correct, or resonant-frequency, tone is passed to the base of transistor Q3.

The reed unit can work as an audio tone filter with sharp selectivity, or as a frequency source for a stable audio generator. It has isolated input and output coils which are coupled by a high-*Q* resonant reed. When the proper frequency is applied to the input or drive coil, the tuned reed resonates and generates a sine-wave signal of the same frequency in the pickup or output coil. The reed is an extremely sensitive device, requiring less than 0.1 mW input power. The input of the reed unit is isolated from the output and the output is proportional to the input. It is possible that the same device can serve as a decoder and as a tone-generating source.

The emitter output of transistor Q3 is coupled through capacitor C16 to a rectifier and filter composed of diode CR3 and CR4 along with capacitor C20. The resultant positive dc voltage is applied to the base of transistor Q6 causing it to conduct. When Q6 conducts the voltage at the junction of resistors R31 and R32 drops, turning off transistor Q7. Since the collector of this transistor is connected into the squelch circuit of the receiver, the turn-off will unsquelch the receiver and turn the audio module of the receiver on.

Transistor Q5 acts as a dc amplifier. If the drive to the reed unit is excessive, voltage from the output of the filter is applied to the base of transistor Q5 causing it to conduct. As a result some of the signal at the base of transistor Q2 is shunted to ground reducing the drive to the reed.

In the transmit mode the unit supplies a tone to modulate the transmitter. When the transmit bias voltage is applied to resistor R26 and on to the base of transistor Q4 (at the center right of Fig. 7-15) it conducts and establishes a feedback path from the collector

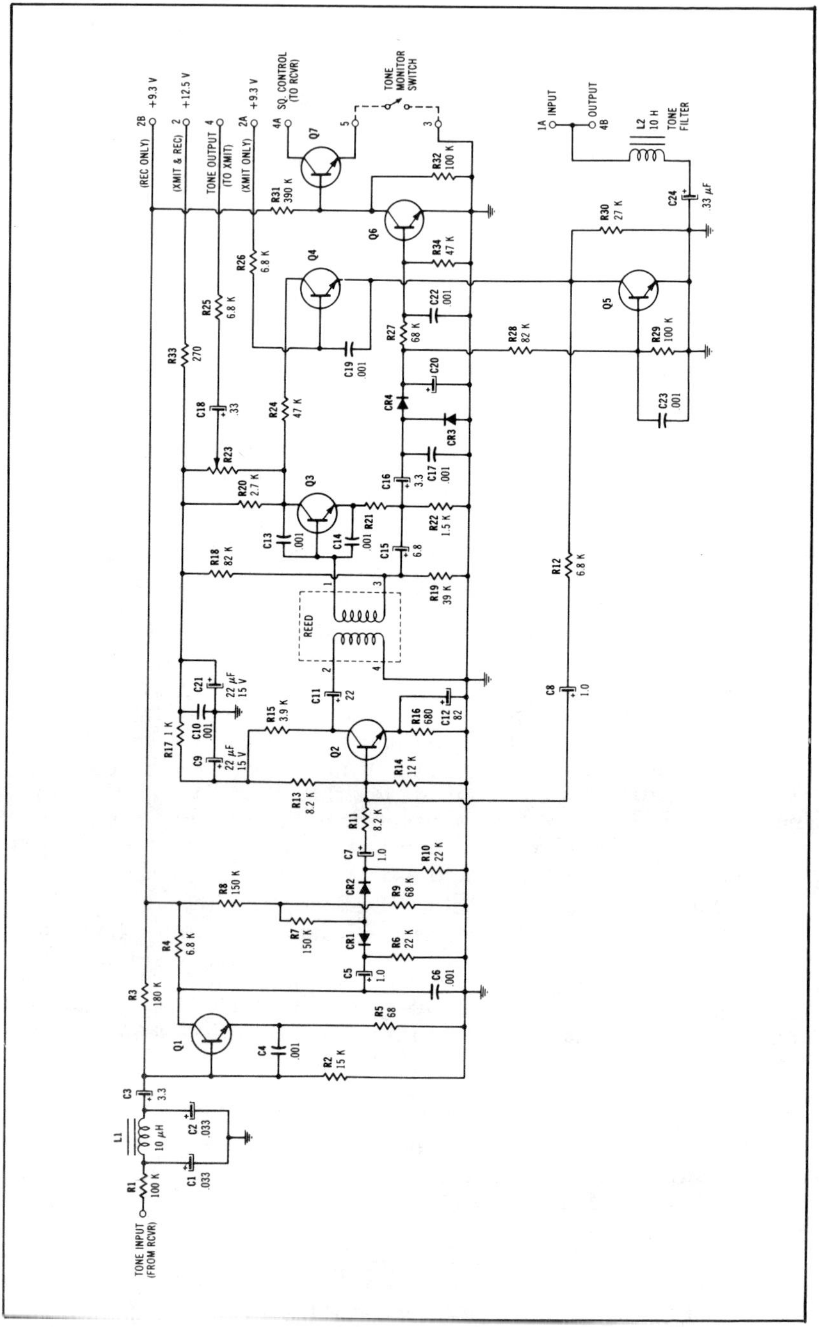

Fig. 7-15. RCA encoder/decoder.

of transistor Q3 back to the base of transistor Q2. This feedback path establishes oscillation, with the frequency of oscillation being determined by the resonant reed unit. A constant level of tone signal is maintained by transistor Q5 circuit which acts as an agc system just as it did in the receive mode.

The generated tone signal at the collector of transistor Q3 is supplied to the modulator module through variable resistor R23 and capacitor C18. The potentiometer permits adjustment of the tone deviation.

A tone monitor switch is located in the emitter circuit of transistor Q7. When this switch is open the receiver is unsquelched and the operator can monitor the channel before transmission. Normally the switch is closed and the receiver is unsquelched only when the appropriate tone signal is received.

7-11. SIDEBAND TRANSCEIVER

In a standard amplitude modulation system, a carrier and two sidebands are transmitted. The carrier remains fixed in amplitude and carries none of the modulation, while the two sidebands contain identical modulating information. As far as conveying information is concerned one could well abandon the carrier. If this is done and both sidebands are transmitted, the method of modulation is called double-sideband modulation (dsb). Since both sidebands carry identical modulation, it is also possible to dispose of one of the sidebands, Thus the carrier and one sideband are removed and all of the desired data can be conveyed by the single remaining sideband. This is single-sideband modulation (ssb).

The advantages of single-sideband modulation over standard amplitude modulation (am) are narrower bandwidth, better signal-to-noise ratio, better signal-to-interference ratio, and more economical use of available power. The last advantage facilitates the design of compact and lightweight gear for a given rf sideband output.

On the medium-frequency marine band (2 to 3 MHz) sideband transmission has become compulsory. The FCC divides single-sideband transmission into three categories. Emission A3H refers to the transmission of a full carrier and one sideband. Although this is a form of sideband emission, it too is to be phased out over an interval of time on specified high frequencies. A3A emission refers to the transmission of a single sideband at reduced carrier level. A3J emission refers to the transmission of a single sideband and the complete suppression of a carrier. It is to be anticipated that the A3J emission will be the most widely used. Sideband transmission is used extensively for commercial and amateur radio communications over a frequency range of 1.8 to 30 MHz.

Three steps are involved in the generation of a single-sideband signal. First the carrier itself must be suppressed at the same time the two sidebands are modulated. This activity is done in a circuit referred to as a balanced modulator. Second, the undesired sideband must be dropped. The actual sideband removed depends upon whether the upper or lower sideband is to be transmitted. Sideband removal is done with a sideband filter. (A phasing method of sideband removal is also possible but this method is not common in commercial sideband transceivers.) Third, the single sideband signal must be amplified and converted in frequency to the transmit channel. Linear mixers and amplifiers are needed if the sideband modulation is not to be distorted. Class-C amplifiers cannot be employed. Class-A, AB, and B linear amplifiers are required.

An example of a medium- and high-frequency sideband transceiver is the Sideband Associates (SBA) 2- to 23-MHz, 150-watts pep model of Fig. 7-16. This model can be used for either simplex or duplex operation with the capability of 12 duplex channels or 24 simplex channels. It is capable of all three types of emission—A3H, A3A and A3J. The transceiver is solid-state except for the vacuum-tube driver and power amplifier output stage. It is ideal for marine operation requiring only a 12-volt power source, marine whip antenna, and antenna coupler. A functional block diagram is given in Fig. 7-17.

Courtesy SBA, Inc.

Fig. 7-16. Marine and hf sideband transceiver.

The antenna relay is shown at the left center. Its purpose is to change the antenna between transmitter output and receiver input. The receiver signal path is shown by the solid line, and the transmitter path by the dashed line. The incoming signal is increased in level by the rf amplifier stage Q1. This stage is followed by the channel mixer. The source of local oscillator component for the mixer is oscillator Q15, followed by buffer Q14. The oscillator is crystal-controlled and determines the transmit and receive frequencies.

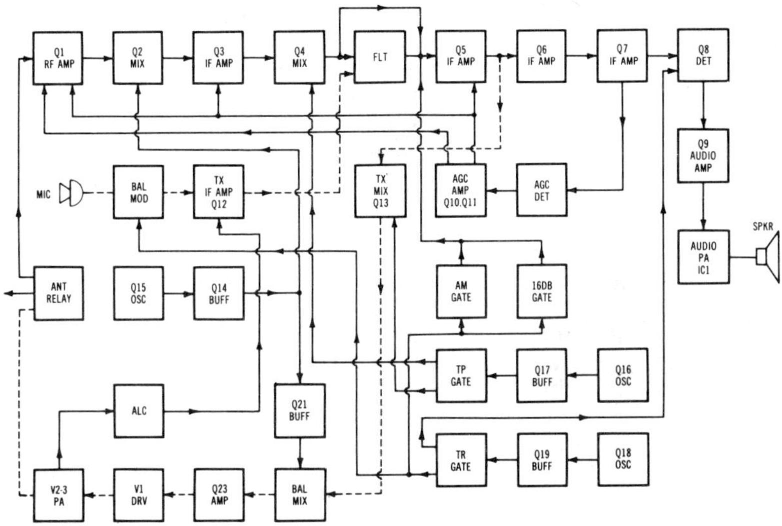

Fig. 7-17. Sideband transceiver block diagram.

The high-if amplifier builds up the level of the 1650-kHz difference component. It is followed by a second mixer (Q4) which reduces the signal frequency to 455 kHz.

A three-stage low-if amplifier follows. Its output is applied to product detector Q8. The beat-oscillator source is carrier oscillator Q18 at the lower right. Through a buffer and transmit-receive gate this component is applied to the product detector. The audio difference frequency at the output is built up in level by an audio amplifier and the audio output integrated circuit ICI.

One of the advantages of a sideband transceiver is that many of the circuits can be used interchangeably. All three oscillators and their associated circuits are used both on the receive and transmit modes. Follow now the signal path of the transmit signal, beginning with the microphone input directly above the antenna relay. Audio signal is applied to the balanced modulator along with a carrier frequency component generated by crystal oscillator Q18. When the mode of operation is to be upper sideband, the crystal frequency is 456.350 kHz; for lower sideband frequency it is 453.650 kHz. On receive, as mentioned previously, this very same signal is used for demodulation by the product detector. On transmit, the transmit-receive gate supplies the signal to the balanced modulator. The balanced modulator amplitude modulates the carrier at the same time the carrier is suppressed. However, the balanced modulator circuit also includes means for inserting full carrier or partial carrier when A3H or A3A is desired.

The balanced modulator is followed by an amplifier and then single-sideband filter FLT. This filter removes the undesired sideband, the one that is not to be transmitted. After amplification, the desired sideband is applied to transmit mixer Q13. Here a 1193.560-kHz component from oscillator Q16 is mixed with the sideband signal to produce an output in the high if frequency range of 1650 kHz. This latter component is applied to the final balanced mixer at bottom center (Fig. 7-17).

The channel oscillator component from Q15 is also applied to this mixer. The output of the mixer is tuned to the sum or up-conversion frequency which is the desired channel frequency of the transmitter. The balanced mixer configuration cancels the oscillator injection frequency so it does not appear in the output. A solid-state and two vacuum-tube amplifiers follow, increasing the sideband signal to the 150-watt pep level.

The automatic antenna coupler (Fig. 7-18) which is an accessory unit provides separate matching networks for each channel of the transceiver, providing optimum match between unit and whatever type of antenna is used for a specific channel. Each individual channel has an appropriate LC circuit which can be connected in either an L or pi configuration, depending on the frequency and antenna being used. Variable tuning and loading capacitors as well as tapped inductors can be adjusted for optimum antenna-matching conditions for each channel.

The appropriate channel is selected at the transceiver. When the transceiver bandswitch is turned to a selected channel, power is also applied to the 12-volt motor of the antenna coupler through a multiconductor cable. An associated switch with a five-deck arrangement selects the appropriate pretuned matching combination.

Fig. 7-18. Multifrequency antenna coupler.

7-12. SOME REPRESENTATIVE CIRCUITS

The carrier oscillator assembly (Fig. 7-19) uses two 40468A MOSFET transistors. The oscillator is basically a Pierce type followed by a buffer amplifier. Outputs are derived that can be applied to either the balanced modulator or the product detector.

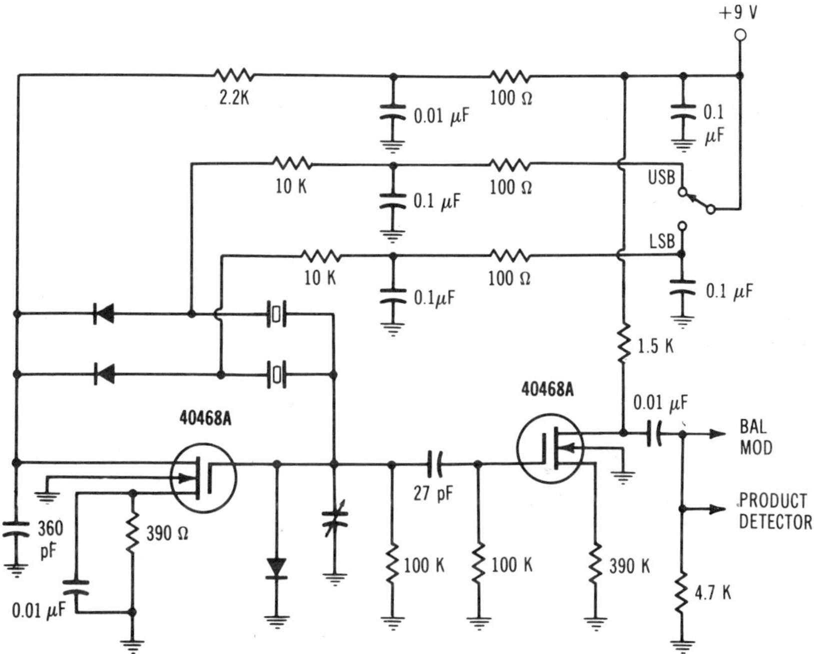

Fig. 7-19. Sideband carrier oscillator with usb-lsb diode switching.

Upper and lower sideband crystals are incorporated. A diode switch is used to close the feedback path between output and input depending on whether upper or lower sideband transmission is desired. Note that the switch applies +9 volts to the anode of the switching diode that connects the particular crystal into the circuit. cuit.

The channel oscillator (Fig. 7 20) operates as an untuned Colpitts arrangement using a 40673 MOSFET. A small variable capacitor is associated with the crystal so that the oscillator can be netted on the specific channel. A buffer amplifier follows the oscillator stage. The circuit is unique in that any crystal in the 2- to 24-MHz range will oscillate and supply adequate mixing and local oscillator injection signal. Output can be applied to either the first mixer of the receiver or the channel balanced mixer of the transmitter.

The microphone used with the transceiver incorporates a built-in amplifier, the output of the combination being applied directly to

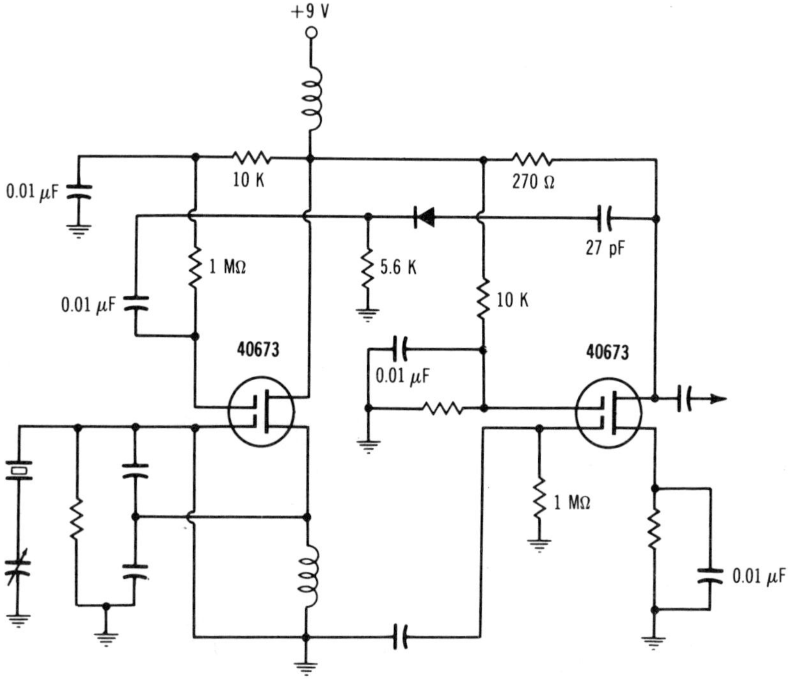

Fig. 7-20. Untuned wide-range hf oscillator.

the ring-type balanced modulator (Fig. 7-21). Output of the carrier oscillator assembly is applied to the same modulator. Within the ring modulator are two balance controls, a potentiometer, and variable capacitor, which are adjusted for complete suppression of the carrier component at the output. They do so by making certain the ring modulator operates in a completely balanced manner. Only two sideband components develop in the output. These are applied through a buffer amplifier stage to the sideband filter that removes the sideband component that is not to be transmitted. Alc voltage from the transmit rf power amplifier controls MOSFET bias and buffer gain. Thus the transmitter output does not distort on modulation peaks.

The diode switching circuit beneath the balanced modulator (Fig. 7-21) is used for the insertion of carrier when desired. This complete circuit is inoperative when A3J emission is desired. When single sideband and complete carrier is to be transmitted, +9 volts is applied to the am input. By so doing the top diode is turned on and there is a signal path between the carrier oscillator input and the output of the sideband filter. The 10K potentiometer can be used to control the level of this carrier which is now inserted into the transmitted signal.

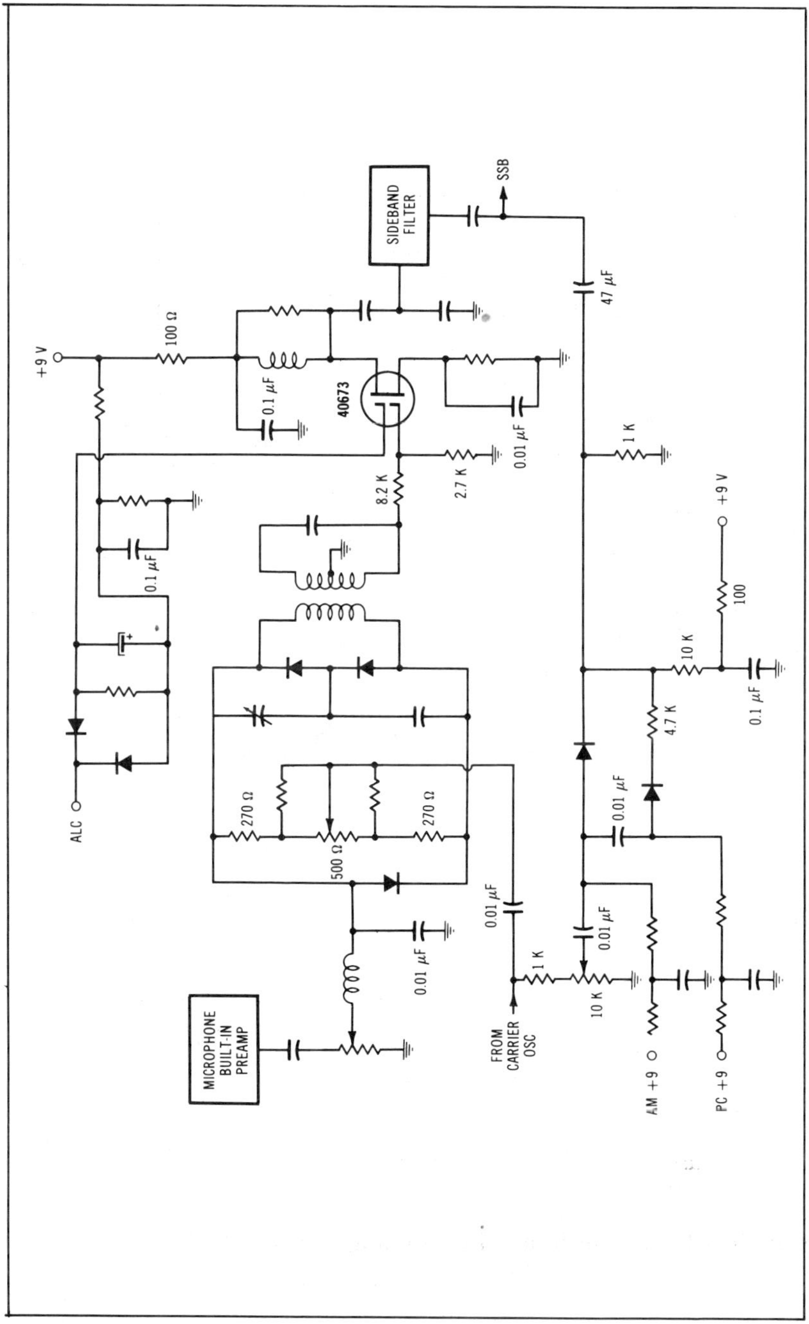

Fig. 7-21. Balanced modulator and sideband filter.

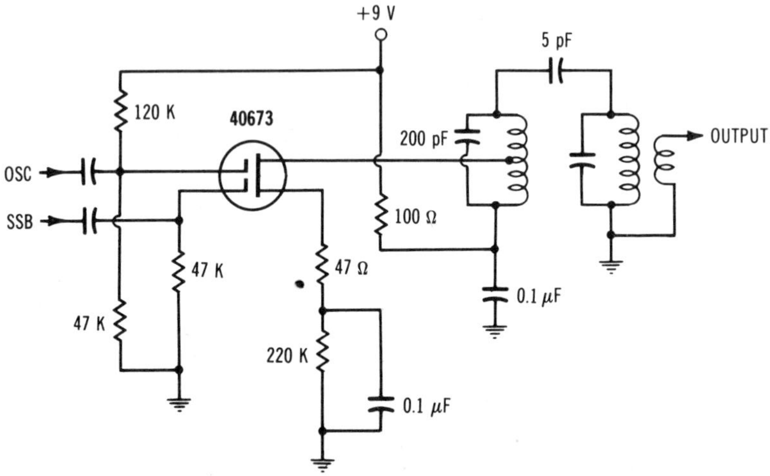

Fig. 7-22. Sideband mixer.

If only a partial carrier is to be transmitted, +9 volts is applied to the PC input. In this case the lower diode is switched on and the level of the inserted carrier is reduced sharply by the inserted 4.7K resistor. In this manner A3A type of emission is formed.

The first transmit mixer also uses a MOSFET (Fig. 7-22). Oscillator and sideband components are applied to the separate gates. the output is made resonant to the sum frequency.

The convenient manner in which a MOSFET stage can be used to drive the high impedance circuit of a vacuum tube is shown in Fig. 7-23. Channel-frequency sideband signal is applied to the one input

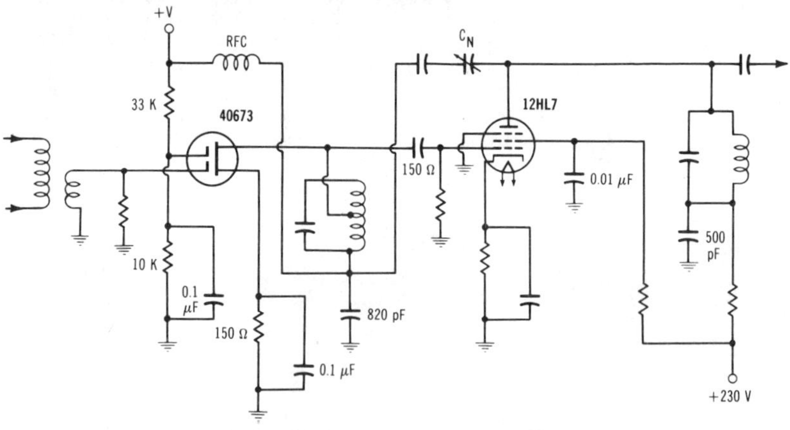

Fig. 7-23. MOSFET drive of vacuum-tube amplifier.

gate and bias is applied to the second gate. The amplified component developed across the drain resonant circuit is coupled to the grid of the 12HL7 linear rf amplifier. To prevent instability and possible self-oscillation a simple capacitive feedback link is used. This simple circuit will neutralize the amplifier over a 2- to 24-MHz frequency range. The small neutralizing capacitor C_N can be adjusted for maximum stability over the operating frequency range.

7-13. LOW-POWER MARINE SSB

Low-powered single-sideband operation is useful for many boats that travel off shore for pleasure or profit. Such units provide longer distance service than vhf fm marine installations. Furthermore telephone ship-to-shore contacts are possible as well as ship-to-ship communications beyond line-of-sight range. Typical installations would be found on off-shore fishing vessels, ocean racers, charter boats, work boats, research vessels, and rescue craft, as well as shore-side base stations that do not require long-distance radio communications. It must be stressed that single-sideband units are only licensed if the vessel already has a vhf/fm installation. An ssb unit is practical only if the vessel operates beyond the range of transmission of a vhf/fm unit.

A typical low-power 50-watt pep model is the Raytheon 1210 shown in Fig. 7-24. It operates in the marine bands of 2–2.85 and

Courtesy Raytheon Co.

Fig. 7-24. Low-power ssb marine transceiver.

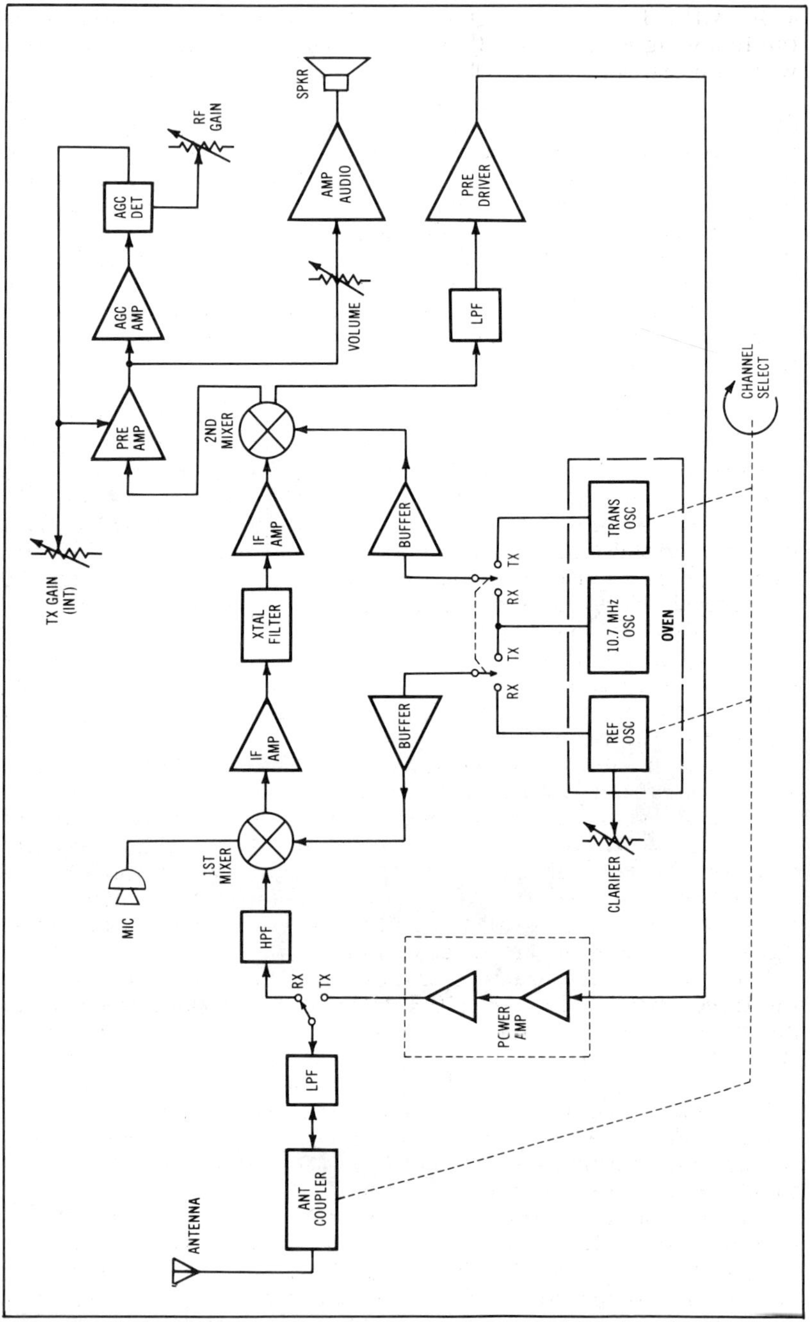

Fig. 7-25. Functional diagram of the transceiver.

4–4.6 MHz. It can be programmed to operate on 8 channels using the following emissions: A3J (ssb with suppressed carrier), A3A (ssb with pilot carrier), and A3H (ssb with compatible carrier).

A functional block diagram is given in Fig. 7-25; schematic diagram, in Fig. 7-26. Microphone output is applied directly to the first mixer which acts as a ring-type balanced modulator. Refer to the top center of the schematic diagram. Microphone signal arrives by way of capacitor C33 and resistor R21 and proceeds through capacitor C29 and the two radio-frequency chokes L11 and L12 to the ring modulator. The 10.7-MHz carrier component is generated by crystal oscillator, transistor Q10, and is supplied through the switching diodes CR17 and CR18 to the base of transistor Q13. Its output is transferred through capacitor C56 to the center tap of transformer T1 in the ring modulator.

The ring modulator is balanced by capacitor C25 and potentiometer R13. Output is removed from the secondary of transformer T2 and is applied through a resonant circuit composed of capacitors C26, C27, and inductor L13 which remove spurious modulation components. Signal level is increased by transistor Q4 prior to the application of the signal to sideband filter FL1. This filter is a crystal type and rejects the upper sideband. Low sideband signal only is applied to integrated circuit U1 which builds up the signal level prior to application to the second or transmit mixer.

This mixer sets the frequency of the transmit signal by combining sideband signal with the output of the transmit oscillator transistor Q12. This oscillator signal has a frequency above the if frequency and a sideband inversion occurs which generates an upper sideband signal at the output of the mixer.

The output across the secondary of transformer T5 is applied through a precise low-pass filter which has a cut-off frequency of 4.7 MHz. From here the signal goes to a predriver amplifier consisting of transistors Q15, Q16, and Q17. This amplifier increases the level of the signal to the point that it can drive a succeeding power amplifier. The final transistor (Q17) acts as an emitter follower to provide a low-impedance output for transferring signal to the input of the power amplifier.

Before discussing the power amplifier it is important to note that the three crystal oscillators (10.7 MHz, transmit, and receive) are mounted in a temperature-controlled oven. Three thermistors, RT1-RT3, positioned near the crystals, act as sensors connected to an operational amplifier which amplifies the bridge error signal and drives emitter-follower transistor Q23. The output of transistor Q23 controls the operation of transistor Q22 that sets the current through the heating resistors R112-R117. Potentiometer R103 is used to set the temperature at which the bridge balances.

The oscillators themselves are Colpitts types which include limiting diodes CR21–CR23. Switching action between transmit and receive is accomplished by diodes CR15–CR20 which function as a double-pole double-throw switch directing the correct oscillator signal for transmit and receive modes. Oscillator signals are amplified

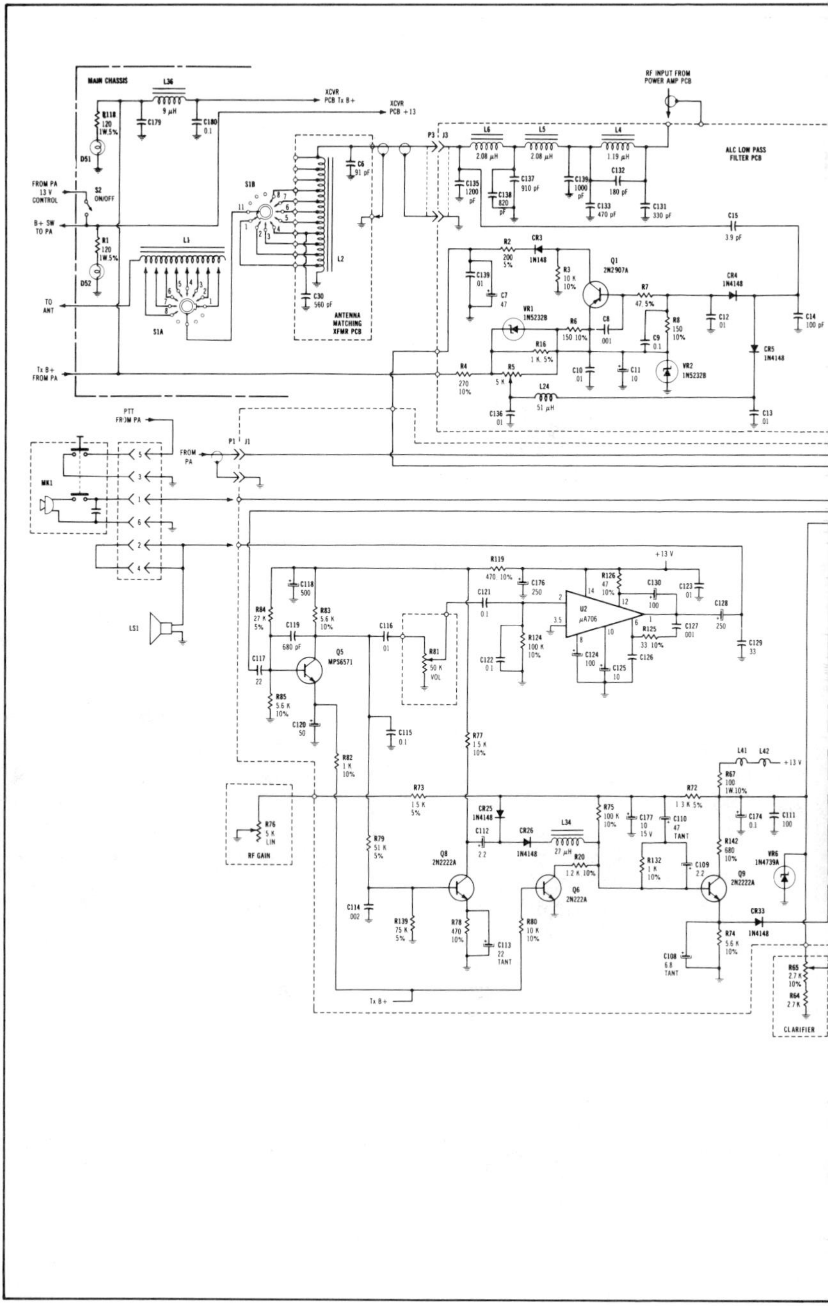

Fig. 7-26. Schematic of

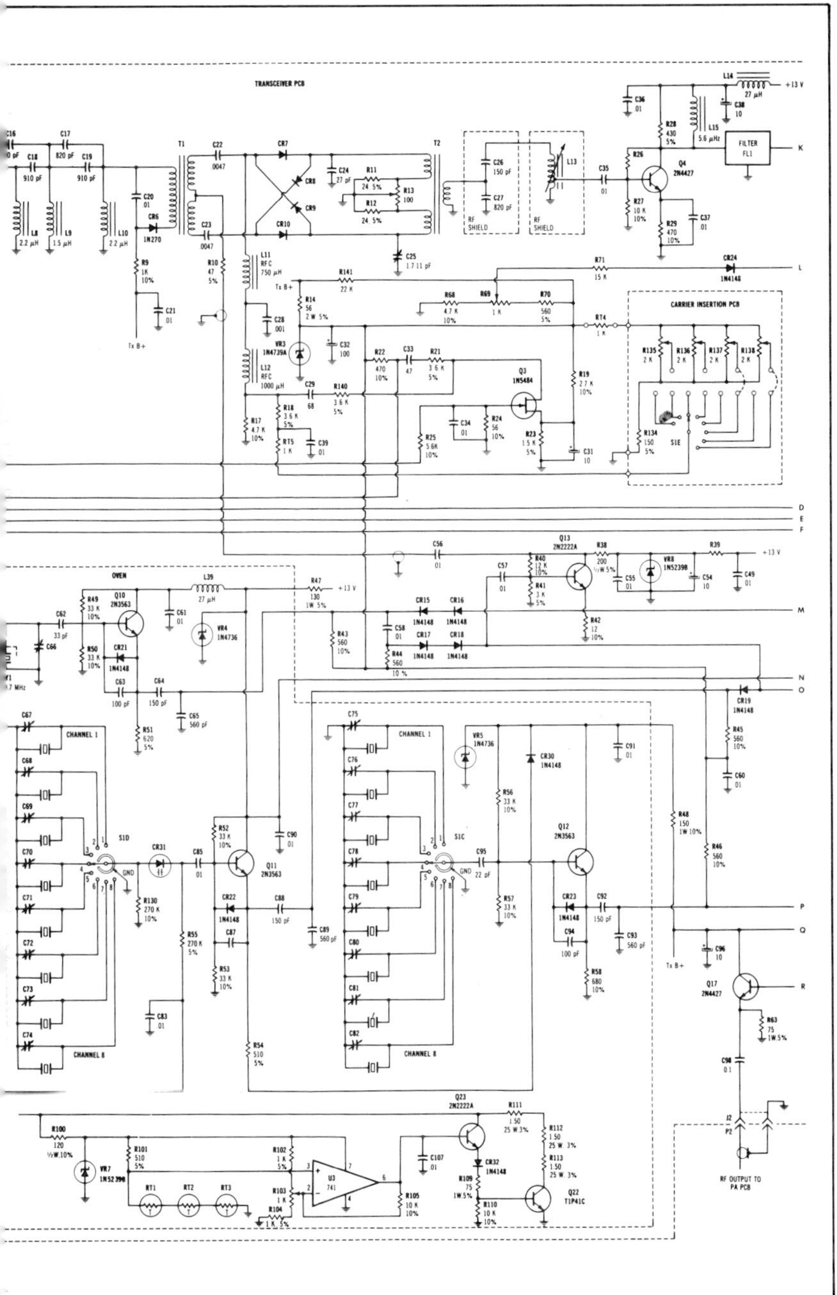

the Raytheon transceiver.

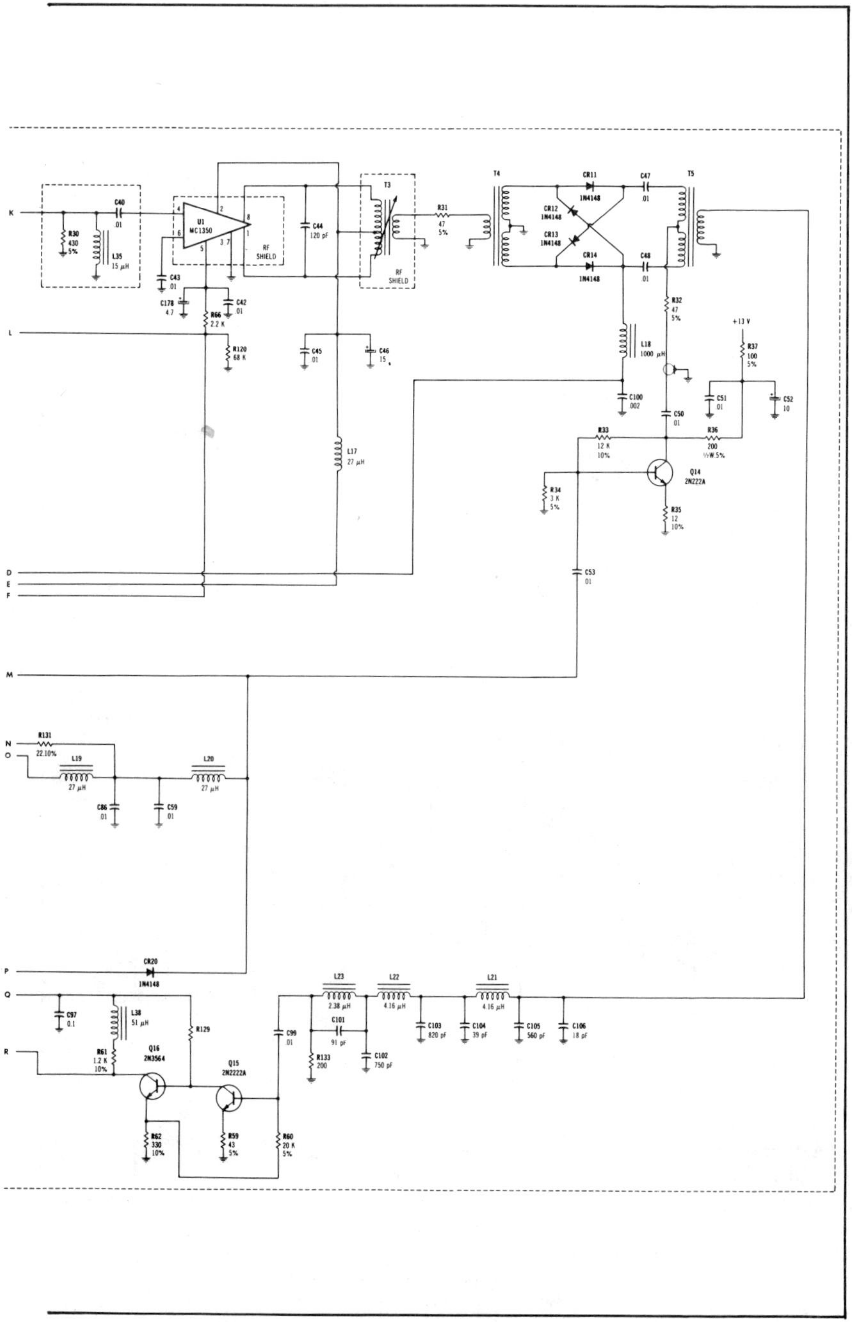

Fig. 7-26. Schematic of the Raytheon transceiver. (Cont.)

by the buffers (transistors Q13 and Q14) before they are applied to appropriate mixers.

The rf power amplifier schematic is shown in Fig. 7-27. This is a broadband amplifier that accommodates the range of frequency of operation of the transmitter. Tuning adjustments are not required. The input stage operates class A and has both collector-base feedback through resistor R89 and emitter degeneration to improve linearity. Output is coupled through transformer T7 to the input of the push-pull power amplifier which operates class B. Input impedance is stabilized by resistors R90 and R91. Transformer T8 is a shunt-fed arrangement that balances the collector output circuit well to reduce second harmonic output as well as providing a path for the dc collector current. A final output transformer is a broadband type that matches the collector impedance to 50 ohms. Signal is applied through the contacts of transmit-receive relay K1 to the antenna system. The relay is operated by the push-to-talk switch on the microphone.

Refer again to Fig. 7-26. Signal from the power amplifier is coupled through a low pass filter to the antenna matching network composed of inductors L1 and L2. These inductors are preset according to the desired frequencies of operation.

Capacitors C15 and C14 act as a voltage divider across the output of the low-pass filter and provide drive to the automatic-level-control (alc) circuit of the transmitter. Diodes CR4 and CR5 rectify this component and charge capacitor C9. The level at which these diodes conduct is set by the setting of the alc potentiometer. Transistor Q1 acts as a buffer, developing the alc bias across capacitor C7. This bias is applied to the field-effect transistor (Q3) located in the modulator circuit. In fact, this feedback path controls the level of the microphone signal applied to the balanced modulator preventing flattopping and the generation of spurious signal components.

When the transmit-receive relay (K1, Fig. 7-27) is on receive position, the output of the antenna coupler and low-pass filter is applied to the receiver input by way of P1-J1 (Fig. 7-26). Signal is applied directly to the ring circuit which in the receive position acts as a receive mixer. It arrives by way of a high-pass filter, the purpose of which is to block out high-powered broadcast stations. In the receive mode the output of the receive oscillator (transistor Q11) is applied to the first mixer (diode ring circuit). Output frequency of the first mixer is now 10.7 MHz which is built up in level by the if amplifier in the same manner as the transmit signal. Receiver gain is controlled by the voltage applied to pin 5 of integrated circuit U1. This control voltage is derived from an agc circuit.

In the receive mode the second mixer acts as a product detector with the output of the 10.7-MHz crystal oscillator being applied as the demodulated carrier. The demodulated audio is removed through the radio-frequency choke (inductor L18) and is applied through capacitor C117, bottom left of Fig. 7-26, to the base of

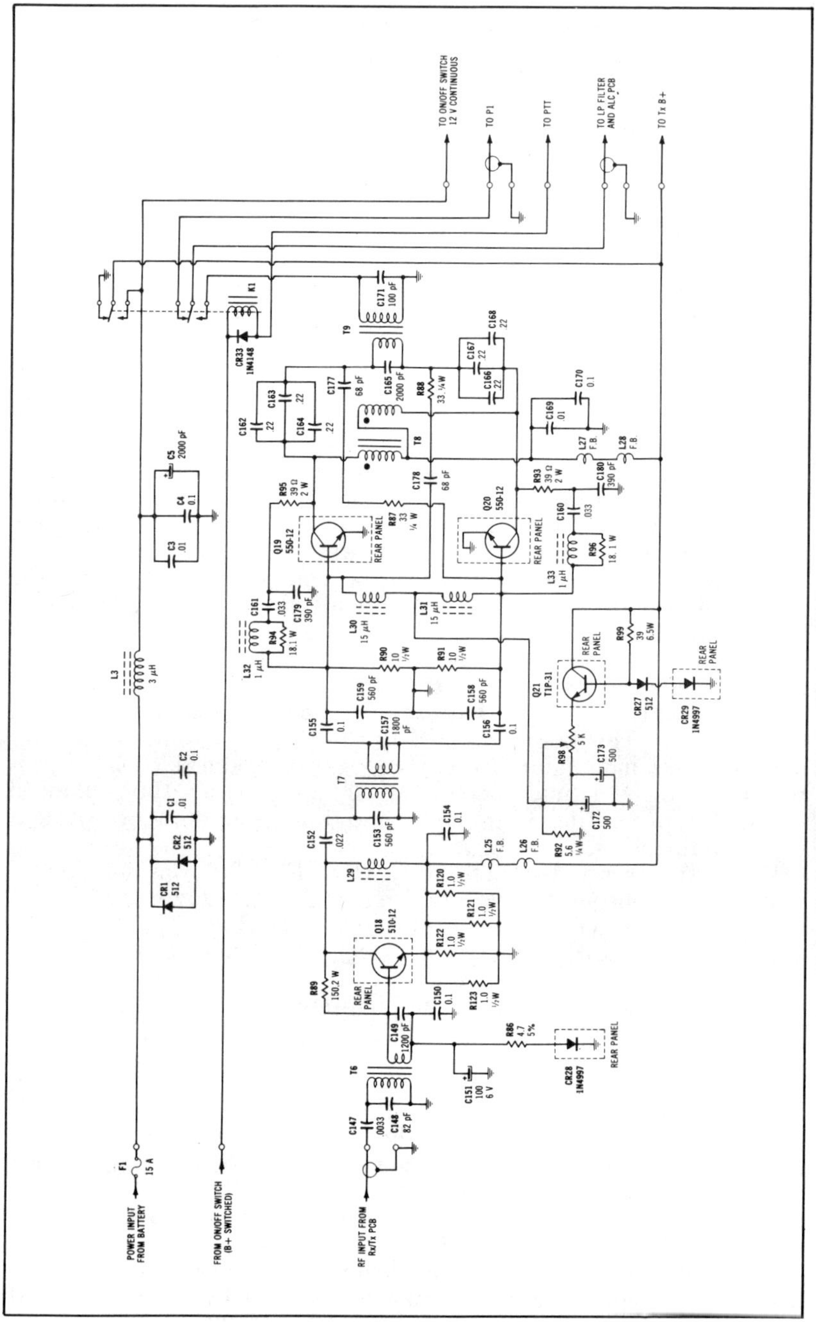

Fig. 7-27. Power amplifier module.

low-noise amplifier transistor Q5. Its output is applied through the volume control to the integrated circuit audio power amplifier U2. A portion of the signal is also amplified by transistor Q8 and is applied to the detector diodes (CR25 and CR26) of the agc circuit. When a signal is present the diodes rectify and add to the no-signal bias that reduces the gain of amplifier U1. Thus the agc activity is a function of the strength of the signal that is developed in the collector circuit of transistor Q5.

7-13. VHF-FM MARINE RADIO

An example of a modern vhf-fm marine radio is given in Fig. 7-28. It demonstrates the trend toward frequency synthesis and digital control. All USA and international marine channels can be used, which include 55 transmit channels and 72 receive channels. A specific channel can be activated by setting up the channel

Courtesy Raytheon Co.

Fig. 7-28. Compact vhf-fm marine radiotelephone.

according to number on the keyboard. The active channel is displayed on a digital readout. In addition the three USA and one Canadian weather channels can be received.

Emergency channel 16 can be activated by depressing a special quick-select button. Furthermore, channel 16 and one other channel can be receive-scanned to provide continuous monitoring (sea watch).

International marine channel activity can be activated by depressing a separate INT keyboard button. Thus the keyboard is programmed separately for USA and international frequency assignments.

The transmitter supplies 25-watt full power. In addition, by selecting the appropriate key the transmitter can be changed over to low-power, 1-watt operation, for harbor communications.

7-14. RADIO REPEATER SYSTEMS

Range and reliability or radiocommunications can be improved with the use of remotely controlled or repeater base stations that are located at a high transmitter site somewhere within the local area to be served. The repeater site can be a tall building or a mountain top. Often a shared facility is employed by a number of licensees. If a minimum of traffic is handled two or more users may even share the same remotely controlled transmitter and receiver. In most installations separate transmitters and receivers are employed and only the site is shared. Appropriate equipment must be installed to minimize interaction among the stations at the shared site.

Two basic repeater plans are given in Fig. 7-29. In example A the control unit and dispatch point is located at the business headquarters. There is a wire link between the control unit and the transmitter and receiver of the elevated base station. This link is

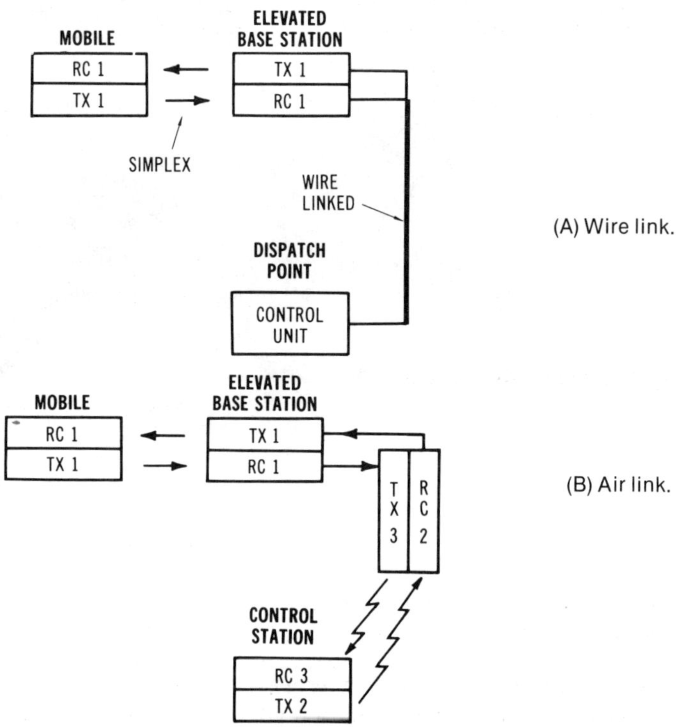

Fig. 7-29. Repeater systems.

usually supplied by the local telephone company. The elevated base station as well as all the mobile units operate on the the same frequency. Operation is similar to any two-way radio system except that the base station is located at an elevated site.

In example B there is an air link between the control station and the elevated base station. Three separate frequencies are employed. A call originating at the control station is transmitted on the No. 2 frequency and is picked up by the No. 2 frequency receiver at the elevated base station. This signal is applied directly to the No. 1 frequency transmitter at the elevated base station. The message is sent out and picked up by the mobile receiver on the No. 1 frequency.

The reply from the mobile station is sent to the elevated base station on the No. 1 frequency. The output of the No. 1 frequency receiver at the elevated base station is linked directly to the No. 3 frequency transmitter of the elevated base station. Signal is sent out and picked up by the control station receiver set to the No. 3 frequency. The same procedure occurs for any one of the mobile units that are a part of the system.

It should be pointed out that directional antennas can be used over the link that connects the control station to the elevated base station. Over a limited distance there can be direct communications among the mobile stations because they all operate on the same frequency.

A mobile-to-mobile radio relay system is shown in Fig. 7-30. This arrangement can extend the range over which reliable communications can be established among mobile stations. Mobile

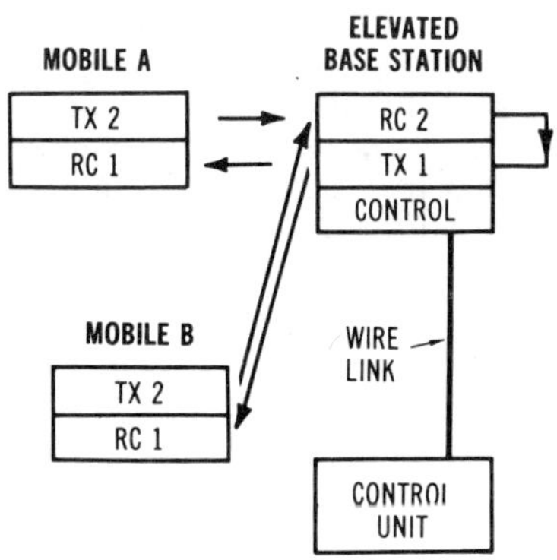

Fig. 7-30. Mobile-to-mobile relay.

units transmit on the No. 2 frequency and receive on the No. 1 frequency. The elevated base station is just the opposite, receiving on the No. 2 frequency and transmitting on the No. 1 frequency. A message initiated at mobile A is transmitted on the No. 2 frequency to the elevated base station where it is picked up by the No. 2 frequency receiver. There is a direct link between the receiver output and the transmitter input. Consequently the message is retransmitted on the No. 1 frequency and is picked up by mobile B on its No. 1 frequency receiver. The reply from mobile B is transmitted over the same path and is received by mobile A on its receiver which is tuned to the No. 1 frequency.

The control unit at the company headquarters is connected by a wire link to the elevated base station. It too can have a receiver tuned to the No. 2 frequency and a transmitter to the No. 1 frequency and therefore can communicate with the mobile stations. The wire link also provides control of the elevated base station and the operation of the system is the responsibility of the base station at the control unit.

There are quite a variety of repeater systems. However, some one point must be designated as a control point and the operation of the system is the control point responsibility. It should be possible to deactivate the system in case of unsatisfactory operating conditions or a system fault from this one location.

7-15. FAULT FINDING IN SOLID-STATE CIRCUITS

In Chapters 6 and 7 there is a significant coverage of test equipment, measurements, tuning and circuit arrangements. Some alignment data is given in Chapter 7. Since extensive test equipment is needed to ensure compliance with FCC technical requirements the very same test equipment is also available for pinpointing the general location of a fault.

If there is no loss of base-station or vehicular power and no erratic operation as a result of cabling, switching, or hardware breakdown, several basic test equipment measurements are advised. Is the power output and modulation normal on transmit? How are receiver sensitivity and squelch performance? If operation is nonexistent, poor, or suspect on a particular mode and the installation itself is apparently normal, it is customary to remove the unit for test-bench evaluation. Quite often a substitute unit is slipped into position. If the substitute unit performs properly there is added assurance that the installation is okay and the trouble lies within the unit removed.

The test bench is usually set up to provide a variety of important measurements and especially those that ensure compliance with FCC technical standards. In such a test setup the faulty section of module can soon be isolated. Actual steps taken to find the defective module vary depending on the design of the manufacturer's specific model. Sometimes a module substitute can be made immediately and the unit placed back in service. In other cases, individual circuits must be tested and faulty components located and replaced.

Because of the great variety of circuit types and functions it is important that you know the functions of the individual circuits that comprise the modules. This knowledge along with a good fundamental electronic background permits you to locate the faulty part with appropriate follow-up checks and measurements. The actual procedure can vary from model to model. You can save much time by following the instructions given in the manufacturer's manual for the specific model.

In commercial equipment there are often complex supply voltage and biasing systems. Thus schematics and voltage charts are of great value in locating faults. Always set the equipment to the operating conditions that the manufacturer suggests for making these voltage measurements.

Your fundamentals knowledge comes into its own when you have found the bad stage. Inspection, parts checks and voltage measurements will usually locate the failed component.

A high resistance voltmeter with an accurate low-voltage reading capability is of particular help in finding a bad part. Service manuals usually show typical voltage readings on schematic diagrams or include a voltage chart. In digital circuits the logic probe is essential. LEDs indicate if the logic is high or low. The typical probe accommodates both TTL and CMOS measurements. Even under signal conditions the probe can be used to indicate the presence or absence of a pulse waveform.

7-15-1. Bipolar and FET Device Voltages

Recall that for a bipolar transistor that is operating normally the base-emitter voltage is somewhere in the region of the barrier potential. Thus the voltage V_{BE} (Fig. 7-31) will usually be between 0.5 and 0.7 volt for low-power silicons. Voltage might range up to a volt or so for high-power types. If the bipolar is a germanium type the range may be between 0.1–0.3 for low power on up to 1 volt for high-power versions.

Collector voltage too is a measurable parameter. In a linear circuit the voltage V_{CE} should be less than the supply voltage depending on the voltage drop in the collector load. For a normally operating bipolar transistor the following relationship exists:

$$V_{CE} = V_{CB} + V_{BE}$$

Example B shows a bipolar transistor connected in a basic common-emitter circuit. Resistors R1 and R2 set the base bias

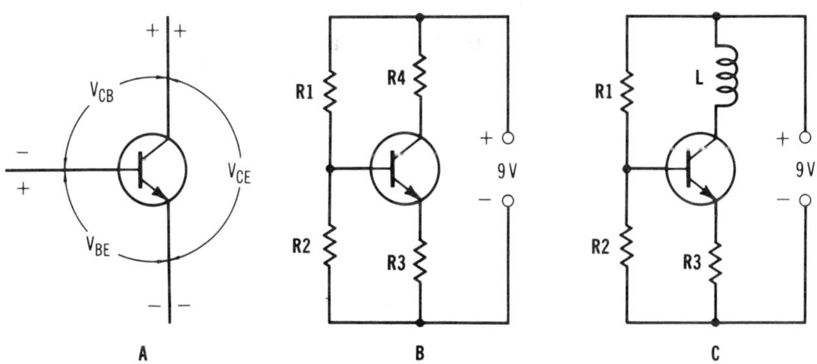

Fig. 7-31. Basic bipolar circuit-voltage evaluation.

current while resistor R3 supplies stabilization. Resistor R4 is the collector load. In a linear circuit the collector voltage as measured to common is less than the supply voltage because of the voltage drop in both resistors R3 and R4 (mainly R4). A typical value might fall between 4–4.5 volts.

If the collector voltage is 9 volts it indicates that the bipolar transistor is not conducting. The transistor is bad or the biasing system is not operating. If the collector voltage is near 0, the bipolar transistor may be drawing excessive current, indicating a transistor failure or a defect in the input circuit. In the resistor biasing circuit an open resistor R1 will result in no bias current. Therefore there will be no stage operation and the collector voltage will read 9 volts.

The circuit of Example C relates more to the operation of a bipolar transistor in a radio-frequency amplifier with the inductor L being part of a resonant circuit or, an audio power amplifier in which case inductor L would be the primary of the audio transformer. Measurements would be much the same as for example B except for the influence of this inductor. Especially in the case of a radio-frequency amplifier this inductor will have a very low resistance. Thus in normal operation the collector voltage will be quite high approaching the supply voltage of 9 volts. In audio power amplifiers inductor L would have some resistance. This resistance would have to be considered in evaluating the normal collector voltage for the stage. Nonetheless it would quite often be in the 6–8 volt range, assuming a 9-volt supply.

The polarities of the above voltages apply to an npn transistor. The very same conditions are appropriate for a pnp transistor except that a negative supply voltage would be used. Consequently the collector-to-common and base-to-common voltages would be negative rather than positive.

Similar voltage evaluations can be made for an FET transistor. Considerations would be a bit different from those of a bipolar because the FET is basically a voltage amplifier rather than a

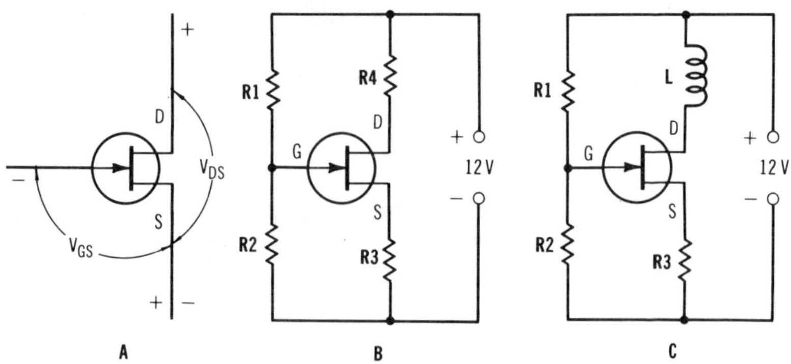

Fig. 7-32. Basic fet circuit-voltage evaluation.

current amplifier. Therefore the normal gate-to-source bias is a function of the characteristics of the particular FET type. In the case of the n-channel junction FET of Fig. 7-32 the base-to-common voltage would be negative with a value depending upon the characteristics of the particular type. Hence for a normally operating n-channel JFET the drain voltage would be positive and the gate bias negative. For a p-channel type the drain voltage is negative and the gate voltage is positive. The bias voltage for MOSFET types depends on whether they are the enhancement-mode or depletion-mode devices.

In the basic common-source FET stage as shown in Fig. 7-32B the biasing is largely determined by resistor R3. Augmenting bias is provided by resistor R1 and R2. For linear operation the drain-to-common voltage would be 6 volts or one or two volts higher or lower depending on characteristics. The voltage measured between gain and source will be of negative value. However, a positive voltage reading between gate and common would be normal because of the presence of resistors R1 and R2. Recall that it is the positive voltage across resistor R3 between source and common that has the main influence on the negative value of the gate bias.

A high value of drain voltage would indicate a failed FET or a defect in the biasing circuit. It indicates a nonconducting situation. A very low drain voltage reading approaching zero shows high conduction which could also be FET or biasing failure.

The influence of a drain circuit inductor as in example C would be about the same as for B circuit. There would be only a slight voltage drop across the inductor. Therefore it is normal when the drain voltage is the same, or not very much less, than the supply voltage. For a p-channel FET polarities would be the complement of those shown in Fig. 7-32.

7-15-2. Digital Circuit Voltage Evaluations

Voltage measurement techniques are different in checking out digital circuits. Recall that there are only two states representing logic 1 and logic 0. In positive-logic digital circuits (by far the more common), logic 1 equals or is near to the circuit supply voltage and

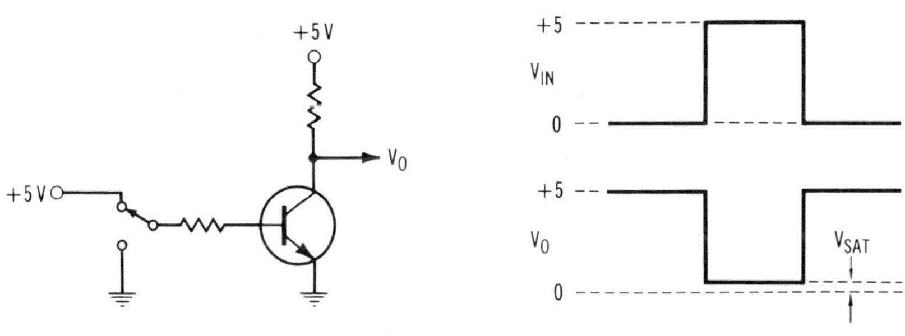

Fig. 7-33. Basic bipolar digital circuit.

logic 0 near the zero volts. Transistors in general operate between saturation and cutoff. In some high-speed devices this may not be so, but nevertheless there are only two steady-state voltage conditions, representing logic 1 and logic 0. Thus in most troubleshooting procedures a logic probe does as well or better than an actual voltmeter. The voltmeter however is handy in checking out power sources and supply voltage lines.

A basic bipolar digital circuit is given in Fig. 7-33. This is a very simple inverter that demonstrates the supply/saturation voltage concept basic to digital systems. Assume that the input switch is set to logic 1. The base-bias current will then be such that the transistor operates at saturation. Thus the output voltage will drop to near zero volts (the voltage saturation point). Were the output switch set to common, or logic 0, the bipolar transistor will not conduct. As a result, the output voltage will equal the supply voltage, or be logic 1.

A logic probe connected to the input (base) would read high. Transferred to the output (collector) it would read low. If now the input is connected to ground, the logic probe would read a low input and a high at the output. These readings represent normal operation. Measurements could also have been made using a voltmeter.

The waveforms show the operation with the application of 5-volt pulses. The output waveform is the complement of the input waveform. Also note that the output waveform swings between the supply voltage and the near-zero saturation voltage of the bipolar transistor.

If a low-frequency pulse is present the actual probe attached to input or output would show the pulse changing between high and low. A higher-frequency pulse would activate the pulse LED of the logic probe and would indicate the presence of the pulse at both the input and output if the bipolar transistor and its associated circuit are operating normally.

The same conditions and measurement techniques apply for FET and MOSFET circuit operation as shown in Fig. 7-34. Normal operation swings between two definite voltage levels. In the circuit shown it is between supply voltage and saturation voltage as indicated by output waveform V_O.

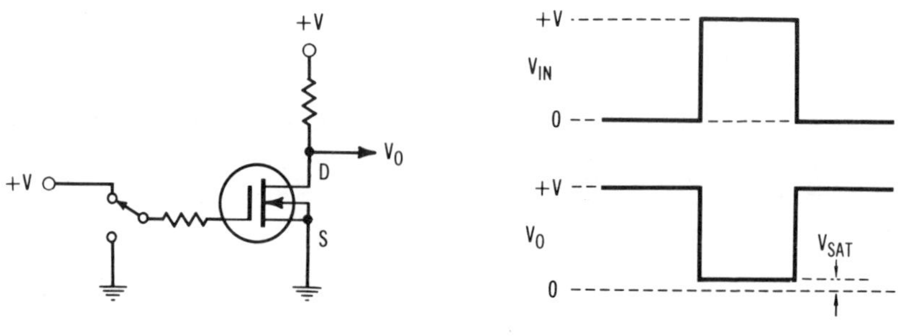

Fig. 7-34. Basic MOSfet digital operation.

Logic probe must be set in accordance with the supply voltage. Usually in the bipolar TTL digital circuits it is 5 volts. In CMOS digital circuits voltage is sometimes 5 volts but can be higher. Your logic probe will have the appropriate TTL/CMOS switching. If you use a voltmeter it must be set to the proper scale. In more specialized systems the logic levels for 1 and 0 may be somewhat different but will always be a two-level steady-state mode operation.

7-16. SCHEMATIC EVALUATIONS

In reviewing some of the basic techniques for locating circuit faults a number of representative circuits will be given in this section. Refresh your memory regarding the functions of various parts and try to anticipate how a specific part failure may have an adverse influence on an associated stage. In the sample schematics parts functions and the influence of part failures are emphasized. Component values are given in these representative circuits and in a school or home-study situation they might be built up and checked out to verify results.

Both transistor and vacuum-tube circuits are given. Although this handbook does not provide a detailed study of vacuum-tube circuits a useful review is present in the form of the questions and answers associated with later chapters as well as the descriptions associated with the vacuum-tube circuits of this section.

7-16-1. Transistor Circuits

Component faults in transistor circuits produce specific operational defects. Failures affect the operating voltages and currents of a transistor stage. Such changes, in turn, cause complete signal loss, attenuation, or distortion.

A common-emitter circuit is shown in Fig. 7-35. The source of signal is a 1000-Hz sine wave; its signal level is assumed to be that value which will produce maximum nondistorted output voltage. The functions of the components are:

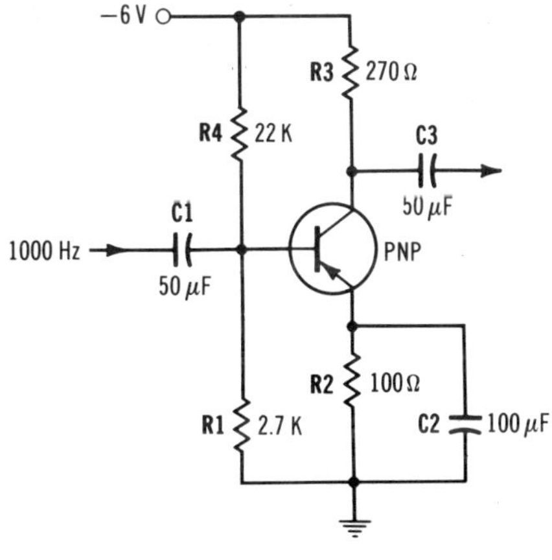

Fig. 7-35. Common-emitter stage.

C1. Input capacitor that serves as a dc blocking capacitor to isolate the dc components of the input stage from the source of signal.

R1-R4. Base-bias voltage-divider resistors.

R2. Emitter stabilization resistor.

C2. Emitter bypass capacitor.

R3. Collector load resistor.

C3. Output coupling capacitor. Isolates the dc components of the collector circuit from the next stage.

Component failures will have the following influences on circuit operation:

C1 open. No signal will be applied to the input of the amplifier.

C1 shorted. Exchange of dc components that could have an adverse influence on the biasing of the stage. If an audio generator with a low impedance and no output capacitor is used, the forward bias is removed from the stage. There is no output, the collector current goes to zero, and the collector voltage rises to the potential of the supply source.

R1 shorted. Same as for C1 shorted because the bias is shorted out.

R1 open. Output distorted, collector current rises, and collector voltage falls. An improper operating point is established by the high forward bias applied to the base.

R4 shorted. Output falls to zero, collector voltage falls, and collector current rises. Again an improper operating point has been established.

R4 open. Bias is again removed. Output falls, collector current goes to zero, and collector voltage rises to the value of the supply voltage.

R2 or C2 shorted. Output increases, collector current rises, and collector voltage falls. There will be an adverse influence on stability.

C2 open. There is ac degeneration. Dc operating conditions are unchanged.

R2 open. There is no forward biasing because of the interruption of the dc path. Output falls to zero, collector current goes to zero, and collector voltage rises to the supply value.

R3 open. Same as R2 open because of the interruption of the dc collector current path.

R3 shorted. Output falls to a low value, collector voltage and collector current both rise.

C3 open. No signal is transferred to the next stage.

C3 shorted. Interchange of dc components between stages, with possible distortion and shift of operating points.

In the common-base configuration of Fig. 7-36, the signal is applied to the emitter and removed at the collector. A 1½-volt emitter junction bias battery is needed. Recall that biasing polarity depends upon the type of transistor. For an npn type, both battery polarities would have to be reversed to those shown.

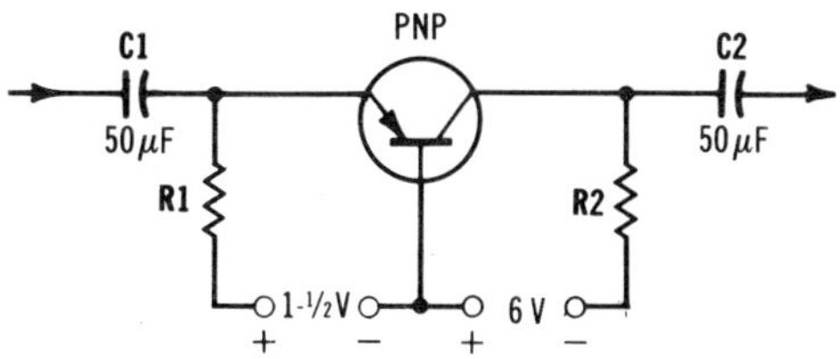

Fig. 7-36. Common-base amplifier.

The functions of the components are:

C1. Input coupling and dc blocking capacitor.

R1. Bias resistor.

R2. Collector load resistor.

C2. Coupling and dc blocking capacitor.

Component faults will have the following influence on circuit operation:

C1 open. No ac signal is applied to the input of the stage.

C1 shorted. Exchange of dc components between stages.

R1 shorted. A very high forward bias is applied. The collector current rises and the collector voltage falls. There will be no ac output.

R1 open. Stage is inoperative because of the removal of bias. There is no output, collector current falls to zero, and collector voltage rises to the supply value.

R2 open. Stage is inoperative because of the interruption of the dc path. Collector voltage and collector current fall to zero.

R2 shorted. Output decreases, while collector voltage and collector current rise.

C2 open. No ac signal is transferred to the next stage.

C2 shorted. Exchange of dc components between stages.

A Colpitts oscillator is shown in Fig. 7-37. A small forward bias is needed to start oscillations, and when oscillations stop in the transistor circuit the collector current falls to zero or declines substantially. The functions of the parts are as follows:

L1-C1. Oscillator tank circuit using a split capacitor arrangement to obtain feedback of suitable magnitude and polarity.

C2. Dc blocking capacitor and capacitor of the base time-constant RC combination.

R1-R2. Base-bias divider resistors.

R3. Collector-circuit resistor across which adequate rf voltage must develop for feedback use.

Component failures have the following influences on circuit operation:

L1-C1 open. When the tank circuit is disconnected, oscillations stop and the collector current falls to the low value set by the low forward bias.

L1 open. Oscillations again stop, with the collector current falling to the low forward-bias value.

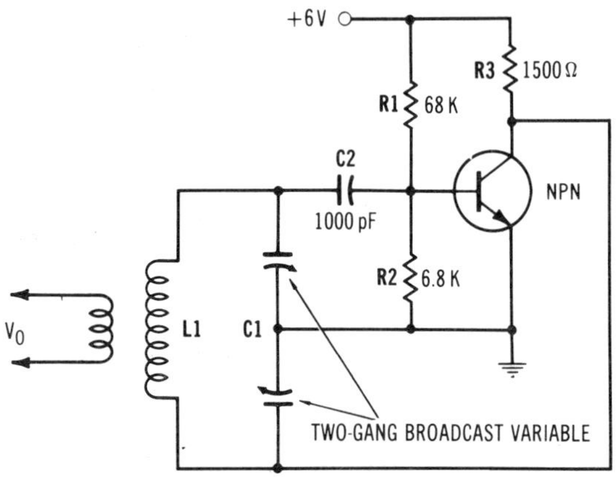

Fig. 7-37. Colpitts transistor oscillator.

L1 shorted. Same as L1 open.

C2 open. Same as L1 open.

C2 shorted. Forward biasing is disturbed causing an increase in the collector current, a decrease in the output voltage, and a shift in the oscillating frequency.

R1 open. No change in operating conditions once oscillations start. However, oscillator starting is very erratic.

R1 shorted. Collector current drops and there are no oscillations.

R2 open. Collector current falls slightly with approximately the same output voltage. However, there is poorer circuit stability.

R2 shorted. Forward biasing is removed and there is no collector current or output voltage.

R3 open. Circuit becomes inoperative because of the breaking of the dc current path. Collector current is zero and there is no rf output voltage.

R3 shorted. There is no feedback and, therefore, there are no oscillations or rf output. The collector current falls to the low forward-bias value.

Two basic transmitter rf stages are given in Fig. 7-38. These are a crystal oscillator and a follow-up class-C amplifier. The functions of the components are:

R1-R2. Base-bias resistors. Crystal-oscillator starting requires the use of a small forward bias on the emitter junction.

R3. Emitter stabilization resistor.

C1-C2. Feedback divider capacitors that establish an optimum amount of feedback.

C3. This capacitor and the primary of transformer L1 serve as the collector tank circuit of the oscillator. The secondary of L1 serves as a low-impedance coupling link into the base circuit of the class-C amplifier.

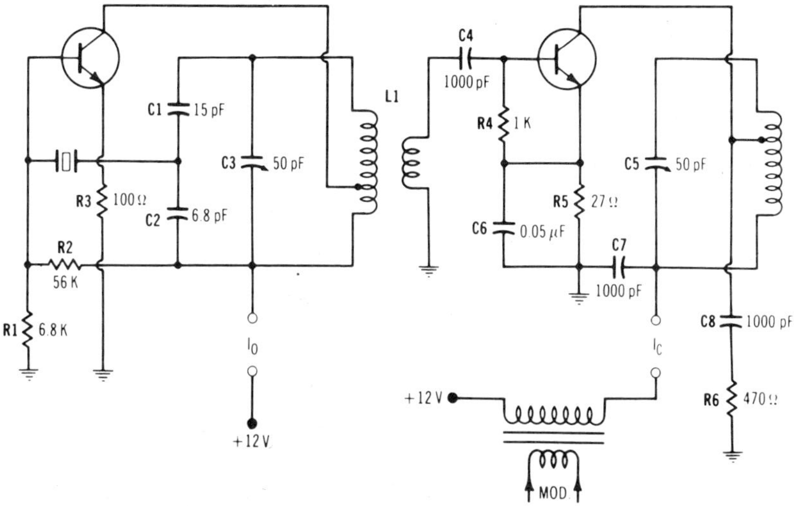

Fig. 7-38. Transistor crystal oscillator and class-C amplifier.

C4. Acts as the capacitance of the base-bias RC combination and is also a dc blocking capacitor. Class-C bias develops on this capacitor.

R4. Resistor of class-C base-bias RC combination.

R5. Emitter stabilization resistor.

C6. Emitter rf bypass capacitor.

C7. Collector rf bypass capacitor.

C5. Capacitor of collector tank circuit.

C8. Coupling and dc blocking capacitor.

R6. Load for transmitter output.

The transistor oscillator and class-C amplifier differ from their vacuum-tube counterparts in that a loss of oscillations and/or rf excitation results in a decline in the collector current (either to zero or low value set by the forward bias). If the crystal stops oscillating the collector current of the oscillator stage falls to the very low value established by forward-bias resistors R1 and R2. When the oscillator stops, no rf excitation reaches the next stage, and there can be no forward biasing. Consequently, the collector current of the class-C stage falls to zero.

Assume that the transmitter is adjusted properly and is being modulated near 100% by a 1000-Hz sine wave. Component failures will then have the following effects on the transmitter operation:

Crystal stops. There is no output. There is no class-C amplifier collector current and the crystal-oscillator current falls to the very low value established by the slight forward bias.

R1 shorted. This removes bias from the oscillator and it stops oscillating. The results are the same as for the previous condition.

R1 open. There is no great change in circuit operation. The oscillator current increases slightly and oscillator stability is impaired.

R2 shorted. There is no output and a high collector current is drawn by the oscillator stage because of the increase in the forward bias. However, it will not oscillate and therefore the collector current of the class-C amplifier will fall because of the absence of rf excitation from the oscillator.

R2 open. Again there is bias removal and no output. Collector currents fall. Oscillator does not start.

R3 shorted. Spurious oscillations result because of instability. Oscillator current rises.

R3 open. The circuit becomes inoperative because of the breaking of the dc path. Oscillations stop and the collector current falls to near zero. The class-C stage draws no current.

C1 open. Oscillations stop. There are no collector circuits.

C1 shorted. There is no great change in operations except that there is a loss of stability and a tendency to self-oscillations of a spurious nature.

C2 open. Same condition as for C1 shorted.

C2 shorted. Oscillations stop because of the lack of feedback. Collector current falls to near zero.

C3-L1 open. An open capacitor or inductor in the tank circuit stops oscillations and causes the collector current to drop to essentially zero.

C4 open. Oscillations continue but they are not supplied to the class-C amplifier. Therefore, the class-C current decreases to zero. However, the oscillator itself operates normally and there is collector current.

C4 shorted. The current rises and there are spurious oscillations generated in the class-C amplifier stage.

R4 open. Class-C stage is improperly biased. There is a decrease in the class-C amplifier collector current. Modulation is distorted because of the inability to follow the crest. The crystal oscillator operates in normal manner.

R4 shorted. Class-C stage is inoperative. Its collector current declines. The oscillator operates normally.

C6 open. There is rf degeneration and less gain in the class-C amplifier. Again the collector current of the class-C stage decreases. The modulation envelope is distorted because of the inability to follow modulation crests.

C6 shorted. There is a slightly greater output and the collector current rises. Stability is not good.

C5 or L2 open. Class-C stage does not function. It is possible that the class-C stage will oscillate in a self-excited manner and generate a spurious rf output.

C5 or L2 shorted. The class-C stage does not function efficiently. There is some direct feedthrough from the crystal oscillator. Output level is very low and modulation is distorted.

C7 open. No apparent change in operating conditions but the modulation system is less stable and there could be some feedthrough of rf energy into the audio section.

C7 shorted. The modulated amplifier becomes inoperative. There is no rf output. Crystal oscillator operates in normal manner.

C8 open. No rf output is transferred to the load.

C8 shorted. A dc component also appears across the load.

R6 open. No load is placed on the output of the transmitter. Class-C amplifier may self-oscillate and draw a high collector current.

R6 shorted. Output is shorted and no pattern is seen on the scope screen. However, a somewhat greater load is placed on the class-C stage and its collector current rises.

A radio-frequency power amplifier such as might be used at the output of a vhf or uhf fm transmitter is shown in Fig. 7-39. The output transistor is driven from the collector output of the preceding stage. Its output resonant circuit is inductor L1 and capacitor C1 which is series resonant and provides a low-impedance drive signal from across capacitor C1. This is applied to the base of the power-amplifier stage through capacitor C2 and additional input matching. Supply voltage is applied to the collector through a radio-frequency choke. Output pi-network and matching circuit supply output to a follow-up harmonic filter and antenna system. The various resonant-circuit inductors can be wire coils, microstrip lines, or a part of some other type of resonant circuit.

Inductor L1 or capacitor C2 open would prevent the transfer of drive signal to the base, resulting in no output or a very weak output. A shorted L1 would reduce the drive level to the output stage and a shorted C2 would result in the transfer of an improper bias to the base of the output transistor. This bias would be positive and result in a high collector current. Shorts of inductor L2 or capacitor C3 would result in a loss of RF excitation and no or very little output. There would be a low collector current.

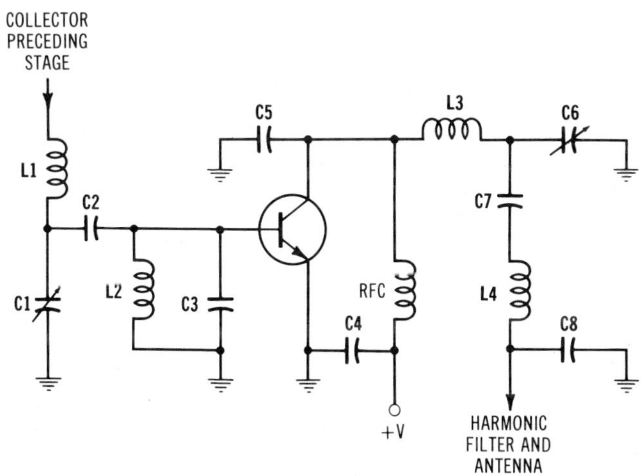

Fig. 7-39. Rf power output stage.

An open radio-frequency choke would shut off the output stage because there would be no collector current path. A shorted capacitor C4 would also cause an inoperative stage and possible overheating or blown fuse in connection with the +V supply voltage line. An open capacitor C4 would cause instability and possibly spurious oscillations. A shorted radio-frequency choke would result in no radio-frequency output, the collector being effectively at radio-frequency ground.

Any short or open of inductor L3 or an open capacitor C5 would upset resonant conditions in the output circuit reducing power level and causing instability. A shorted capacitor C5 would again place a very low resistance path on the supply voltage and the effects of such a short would be felt along the supply line. This would also be the case for a shorted capacitor C6. Opens and shorts associated with capacitor C7 and C8 as well as inductor L4 would disturb the output matching and radio-frequency power level would drop off. Furthermore a short of capacitor C7 would put the supply voltage potential on the line leading out to the harmonic filter and antenna system.

As mentioned earlier a radio-frequency power meter connected to the antenna or dummy load is of special help in determining what influence the power amplifier stage components have on the output. After individual components are checked and changed you can note the influence on the power meter reading after power is restored.

7-16-2. Vacuum-Tube Circuits

A triode grounded cathode amplifier is shown in Fig. 7-40. By using the component values shown, a voltage gain of approximately 50 can be obtained. That is, an input signal of one-half volt will result in a 25-volt output signal. The functions of the individual parts are:

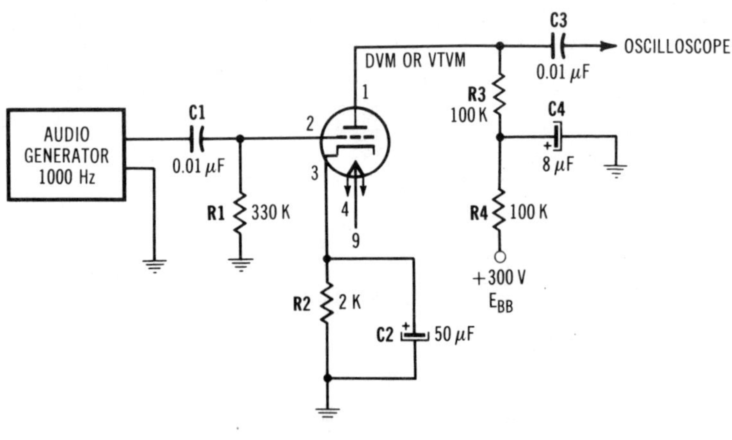

Fig. 7-40. Grounded-cathode amplifier.

C1. Coupling capacitor that provides dc isolation between input circuit and signal source.

R1. Grid resistor. The signal voltage is developed across this resistor. A secondary function of the resistor is to present a discharge path for electrons that accumulate on the control grid.

R2. Cathode resistor. The presence of plate current in this resistor develops a positive dc voltage between cathode and ground, thus developing the grid bias. The dc bias voltage acting between grid and cathode is negative.

C2. Cathode bypass filter capacitor. This capacitor prevents ac degeneration. If its filtering action were not present, an ac voltage would also develop across resistor R2 as a result of the changes in plate current with signal. The phase of such an ac voltage would be opposite to the phase of the grid signal and would subtract from the grid signal. It would reduce the effective grid signal and lower the overall gain of the stage.

R3. Plate load or plate coupling resistor. With applied signal, the changing plate current in this resistor causes a change in the plate voltage. The plate-voltage change is the useful output of the amplifier.

C3. Coupling capacitor. The coupling capacitor again isolates the dc of one stage from that of the next stage. However, such a capacitor offers a minimum reactance to the ac voltage change and, therefore, the ac output at the stage is coupled to the next stage.

R4. This is a decoupling resistor. It isolates the ac variations of a given stage from the power supply impedance. It also blocks any ac variation that might appear across the power supply from the sensitive input stages of an audio amplifier. Often a second function of a decoupling resistor is to lower the dc voltage of the supply line to the value required by a given vacuum-tube stage.

C4. This is a decoupling capacitor. Filtering requires the joint influence of both resistor and capacitor (R4-C4) in smoothing out undesired ac variations. It is the combination that does the filtering and makes available a steady dc voltage for the individual stage.

Next consider each individual part in terms of the trouble it might cause in a circuit if it became defective. The two extremes, open or shorted, are used. It is understood that similar defects, although not as severe, can be introduced when the component value changes. In general, it introduces the same sort of defect but the degree of influence will not be as great. In considering the ac signal performance of the stage, assume that the applied signal level approximates the desired maximum signal level of operation for the stage.

C1 shorted. If the course of signal also contains a dc component, this dc voltage is transferred to the grid and can cause improper biasing conditions.

C1 open. This results in no applied signal. The dc operating conditions of the stage are unchanged.

R1 open. The influence of the open grid resistor depends to a great extent on the circuit arrangement and the particular tube type. In some arrangements electrons accumulate on the control grid to such an extent that the tube is cut off. In other circuit arrangements, when too high a value grid resistor is used, a grid gas current could occur to bias the grid in the positive direction which would result in excessive dc currents in the tube.

R1 shorted. No ac input or no output because the applied signal is shorted between grid and ground.

C2 open. Ac degeneration results and there is less output signal for a given input signal. Dc operating conditions do not change.

C2 shorted. Stage is now improperly biased and the output may become distorted. There is a drop in the dc plate voltage and a rise in the dc plate current because of the bias removal.

R2 shorted. Same defect as that which occurs when capacitor C2 is shorted.

R2 open. Tube is cut off and becomes inoperative. There is no ac output, plate current is zero, and plate voltage is maximum.

R3 open. This defect is similar to the conditions that exist when R2 is open. The tube is cut off, there is no output and the plate current is zero. In this case the plate voltage is also zero. However there will be maximum supply voltage at the junction of capacitor C4 and resistor R4.

R3 shorted. There is no ac output voltage. The dc plate current is higher and so is the dc plate voltage.

C3 open. No ac signal voltage is transferred to the succeeding stage.

C4 open. Output voltage may rise because of the increase in the effective ac load resistance with the removal of the decoupling circuit and the addition of resistor R4 to the load. Output voltage may be distorted because of the nonlinear operating conditions. If the stage were associated with an amplifier chain, there would be a danger of self-oscillation, motorboating, or other disturbances that arise from improper decoupling.

C4 shorted. When C4 is shorted there is no supply voltage applied to the stage. Output is zero, plate current is zero, and plate voltage is zero. In many circuits there could be excessive current in the decoupling resistor R4 and it would overheat.

R4 open. The supply voltage would again be removed from the stage. Output voltage, plate current, and plate voltage would all be zero.

R4 shorted. Supply voltage increases as do the dc plate current and dc plate voltage. Output voltage may increase slightly. Again, if the stage were associated with a multistage amplifier, low-frequency feedback disturbances could arise.

A pentode grounded-cathode amplifier is shown in Fig. 7-41. Except that a screen-grid tube is used, note its similarity to the circuit of Fig. 7-40. A major advantage of the pentode stage, in addition to the better isolation between output and input, is the fact that the plate voltage has much less of an influence on the plate current. The screen grid takes over in the control of electron flow between cathode and plate. As a result the pentode has a greater voltage-amplification capability. That is to say, an output voltage of a given magnitude can be obtained with a much lower input voltage in comparison to a comparable triode circuit. The circuit is itself identical to that of a triode with the exception that a screen resistor R5 and a screen capacitor C5 are required. An input signal of one-half volt will result in a 40-volt output for the circuit of Fig. 7-41.

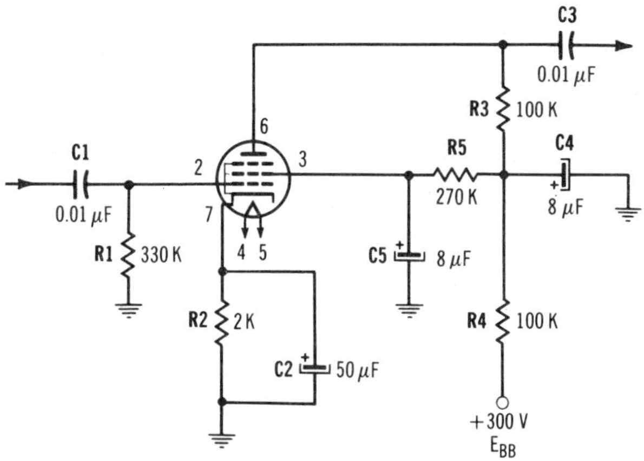

Fig. 7-41. Pentode grounded-cathode amplifier circuit.

The function of screen resistor R5 is to reduce the supply voltage to a value appropriate for the desired operating conditions. The screen capacitor C5 is a bypass or filter capacitor. It prevents degeneration in the screen-grid circuit. It does so by preventing the variations in the screen-grid current from causing a change in the screen-grid voltage. It holds the screen-grid voltage constant in the same manner that the cathode filter capacitor holds the cathode bias constant. The remainder of the components of the stage have functions identical to their like components in the triode circuit in Fig. 7-40.

Except for the screen resistor and screen capacitor combination, component defects produce the same faults as those covered in conjunction with Fig. 7-40. If screen-grid resistor R5 opens there is no screen-grid voltage and the stage reverts to triode operation. The output voltage is low and distorted. There is also a change in the dc operating conditions, plate current falls, and plate voltage rises. If resistor R5 were to short, there would be a higher than normal screen grid voltage. The screen and plate current would rise and there could be excessive screen-grid and/or plate dissipation.

If the screen-grid capacitor were to open, the screen-grid voltage would also vary with the applied signal. The stage would become degenerative and there would be a reduction in the output voltage. A shorted screen-grid capacitor would also remove screen grid voltage and the same circuit faults would develop that result when the screen-resistor is open.

In troubleshooting one should always remember that a shorted capacitor, more than likely, will also result in an overheated resistor. Thus what often appears to be a resistor defect is in fact being caused by a shorted or leaky capacitor. Thus, in the above example the shorted capacitor might also cause an overheated screen-grid resistor.

A single-tube phase splitter is shown in Fig. 7-42. Such a stage provides a means of obtaining opposite-polarity and nearly equal-amplitude output signal for driving the input of a push-pull stage. One output is removed across resistor R3; the other output is taken

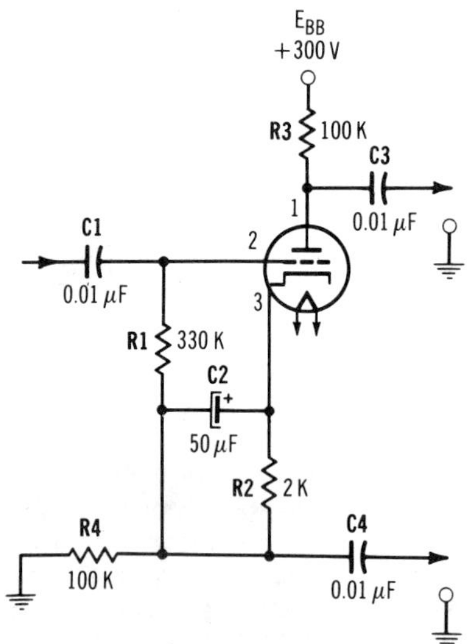

Fig. 7-42. Single-tube phase splitter.

across resistor R4. Inasmuch as the stage has both plate and cathode outputs, it functions in part as a cathode-follower, and, therefore, the gain is very low. For example with a 5-volt input signal there is approximately 4.9 volts output at the plate. Considering the two outputs as being connected in series, there would be a net overall gain approaching 2. It is most important to recognize that the two output voltages are of opposite polarity and almost equal in amplitude.

The functions of the parts are:

C1. Coupling capacitor. Function is the same as capacitor C1 of Fig. 7-40.

R1. Grid resistor. Function is the same as resistor R1 of Fig. 7-40. Note that the low end of resistor R1 is connected to the junction of resistors R2 and R4. This connection makes certain that the grid-cathode bias equals the dc charge placed on capacitor C2.

R2. Cathode bias resistor. The dc plate current in this resistor develops the dc bias voltage for the stage.

C2. The cathode bypass filter capacitor. The dc charge on this capacitor is the bias for the stage.

R3. Plate load resistor. The ac plate current variation in this resistor develops the output voltage change at the plate of the tube.

C3. Coupling capacitor. It has the same function as C3 of Fig. 7-40.

C4. Also a coupling capacitor for the opposite polarity output of the stage.

R4. This is the cathode output resistor. The cathode-current variation (same as plate-current variation) in this resistor develops the ac output voltage that is made available in the cathode circuit. The ac voltage across this resistor is less than the applied ac voltage because it is a cathode-circuit load. A similar but opposite polarity voltage is made at the plate.

The influence of component faults are as follows:

C1 open. Loss of grid signal.

C1 shorted. Possible change in operating conditions through the exchange of dc components. In this case the high dc voltage across R4 is transferred to the preceding stage or signal source.

R1 open. This open resistor results in an improper bias which causes a fall in the dc plate current and a rise in the dc plate voltage. Under certain circumstances, a distorted ac output also occurs.

R1 shorted. There is a reduction in the magnitude of the output signal that is made available by the basic design. The input signal is in effect applied across resistor R4.

R2 open. Circuit becomes inoperative because of the interruption of the dc current path. Plate current is zero and the plate voltage high.

R2 shorted. No bias is applied to the stage and improper operating conditions are established. There could be some distortion of the output voltage. Plate voltage falls and plate current rises.

C2 shorted. Same effects as R2 shorted.

C2 open. The influence would be limited although there would be some additional inequality in the output magnitudes.

R3 shorted. The plate voltage and plate current both increase. The stage itself reverts to a cathode-follower with no output made available at the plate.

R3 open. The stage becomes inoperative because of the interruption of the dc current path. The plate current and the plate voltage both fall to zero.

C3 open. No transfer of ac signal to next stage.

C3 shorted. Undesired exchange of dc components between stages.

C4 shorted. Likewise an undesired exchange of dc components between stages.

C4 open. No transfer of ac signal to next stage.

R4 open. Stage becomes inoperative because of the interruption of the dc current path. Plate current is zero and plate voltage is maximum.

A crystal oscillator is shown in Fig. 7-43. This oscillator crystal acts as the grid resonant circuit. The cathode and plate currents are one and the same, and therefore the plate current can be measured with a meter inserted in the cathode circuit.

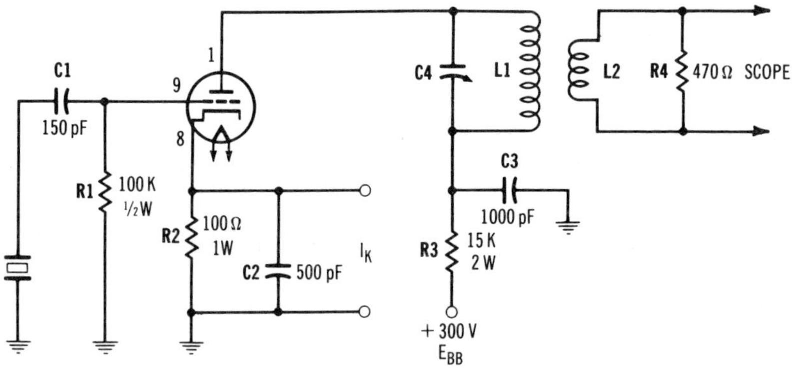

Fig. 7-43. Crystal oscillator.

A popular metering method is shown in Fig. 7-43. A low-value cathode resistor is used and the oscillator-circuit operation will remain essentially the same whether the meter is in or out of the circuit. This technique is adaptable to meter-switching arrangements because the meter can be switched in and out of the cathode circuit and among other circuits of the transmitters at will without affecting operation. The functions of the parts are as follows:

C1. Capacitor of grid RC combination. In some crystal circuits this capacitor is not used because the crystal and its holder can provide the capacitance of the grid RC combination.

R1. Resistor of grid RC combination.

R2. Metering resistor.

C2. Metering circuit bypass.

C3. Decoupling rf filter capacitor.

R3. Decoupling resistor and plate voltage-dropping resistor.

C4. Capacitor of plate tank circuit.

L1. Inductor of plate tank circuit.

L2. Low-impedance secondary coil.

R4. Output load of the oscillator.

It is important to know the tuning characteristics of a crystal oscillator. The plate current rises when the crystal stops oscillating or other defects develop. Likewise, detuning capacitor C4 from crystal resonance stops oscillations and causes a plate current rise. As the plate-tank capacitor is tuned to resonance, oscillations are indicated by a dip in the plate-current meter reading. The oscillator kicks out of resonance very quickly when the tank is tuned to the low side of the crystal frequency. Oscillations fall off more slowly when capacitor C4 is tuned to the high-frequency side of resonance. For best circuit stability it is customary to tune tank-circuit capacitor C4 slightly to the minimum-capacitance (high-frequency) side of maximum plate-current dip. Output can be displayed on a scope screen. Be certain scope can accommodate crystal frequency.

Component faults will have the following influence on stage operation:

C1 open. Oscillations cease, rf output falls to zero, and plate current rises.

C1 shorted. There will be no great change in operation because the crystal capacitance takes over.

R1 open. The rf output drops as does the dc plate current.

R1 shorted. No oscillations, no rf output, and plate current rises.

R2 or C2 shorted. No change in circuit conditions. However, metering circuit will not function.

R2 open. Stage becomes inoperative because of the breaking of the dc path to the supply line. There will be no oscillations and the plate current will fall to zero.

C2 open. No significant change in circuit operation except perhaps some slight detuning.

C4 or L1 shorted. No oscillations, no output, plate current rises.

C4 open. No oscillations, no output, and plate current rises.

L1 open. Circuit becomes inoperative because of the interruption of the supply voltage. There will be no oscillations, no output, and the plate current falls to zero.

R3 shorted. There is a somewhat greater output as well as a higher plate current. When the supply voltage is too high and/or the feedback current too great, the crystal can be fractured.

R3 open. Circuit again becomes inoperative because of the interruption of the supply voltage. There will be no oscillations and the plate current will fall to zero.

C3 open. No oscillations because of the interruption of the rf path. The plate-current meter reading rises.

C3 shorted. The stage becomes operative because of the removal of the plate voltage. However the current drawn through R3 can be excessive and the resistor overheats.

R4 shorted. A heavier load will be placed on the oscillator output and the plate-current meter reading rises. Too severe a load on a crystal stage may cause the circuit to stop oscillating completely.

R4 open. A very light load will be placed on the oscillator and the dip in the plate current will be very decided. Output voltage as displayed on the oscilloscope will increase because of the lighter load.

A basic, simple, and lowpower am transmitter circuit is shown in Fig. 7-44. The rf circuit consists of a triode Pierce crystal oscillator and follow-up class-C amplifier. The audio section consists of a triode voltage amplifier and a pentode class-A modulator. Only three receiving-type tubes are used.

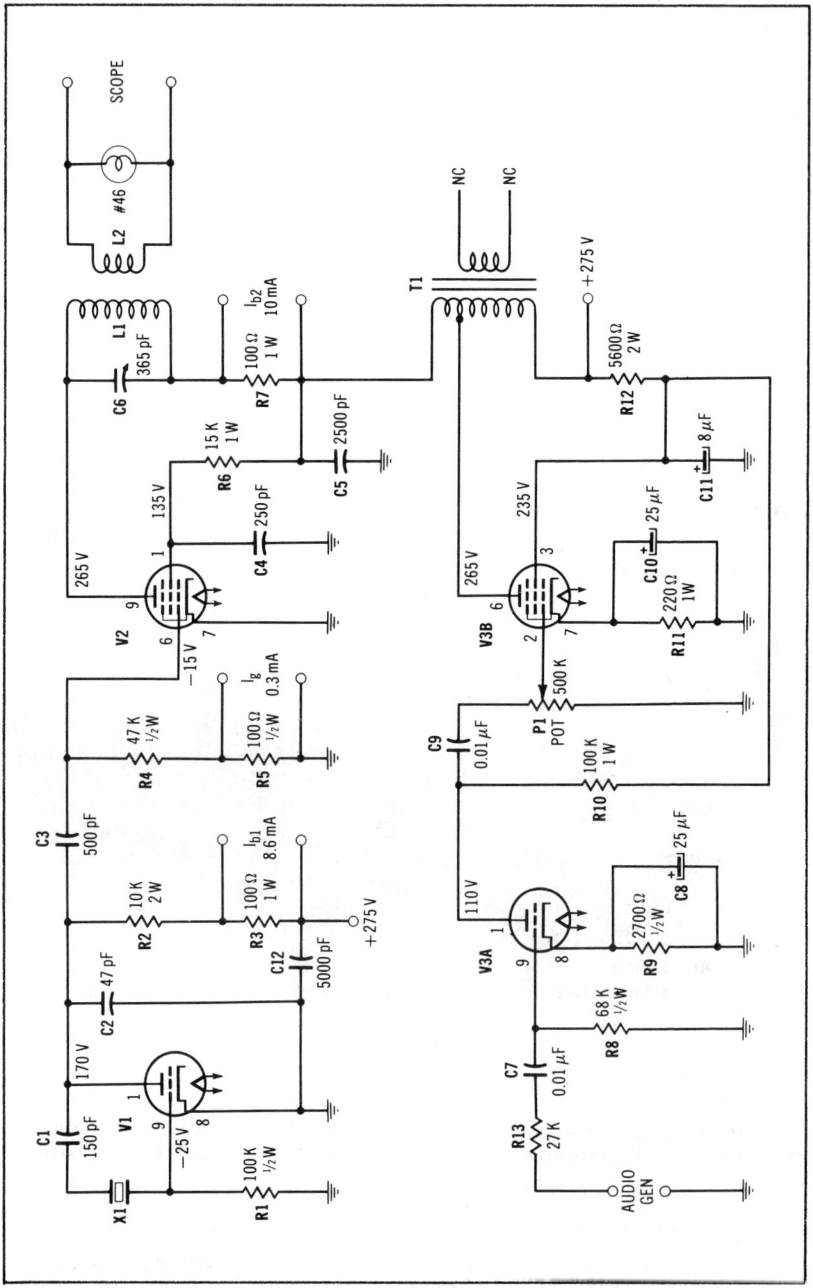

Fig. 7-44. An am transmitter.

The functions of the parts are as follows:

X1. Crystal.

C1. Dc blocking capacitor that keeps the dc plate voltage from the crystal. It is also a part of the grid RC combination.

R1. Resistor of grid RC combination.

C2. Capacitor that provides easier starting and a more favorable feedback division for the Pierce crystal oscillator.

R2. Plate-load resistor of the Pierce crystal oscillator.

R3. Metering resistor.

C3. Interstage coupling capacitor and capacitance of grid RC combination.

R4. Resistor of grid bias RC combination for the class-C amplifier.

R5. Metering resistor.

C4. Screen-grid rf bypass capacitor.

R6. Screen-grid voltage-dropping resistor.

C5. Rf plate-circuit filter capacitor. It is made high enough in value that it also filters out the high audio frequencies above 3000 Hz.

C6-L1. Plate tank circuit.

R7. Metering resistor.

B1. No. 46 pilot-bulb load.

C7. Input coupling capacitor of the audio amplifier.

R8. Grid resistor.

R9. Cathode-bias resistor.

C8. Cathode-filter capacitor.

R10. Plate-load resistor of the voltage amplifier.

C9. Interstage coupling capacitor.

P1. Audio gain control.

R11. Cathode-bias resistor.

C10. Cathode-filter capacitor.

C11. Decoupling and audio filter capacitor.

R12. Voltage dropping and decoupling resistor.

R13. Isolation resistor for supplying signal from an audio generator.

C12. Rf filter capacitor.

T1. Transceiver modulation transformer or audio choke.

The rf section of the transmitter has only a single control; this is the plate tank capacitor of the class-C amplifier. At resonance there will be a dip in plate current. The capacitor is adjusted for maximum glow of the bulb. The rf waveform can be observed by connecting an oscilloscope probe across the output.

The audio generator can now be connected to the transmitter and the gain control advanced until the modulation envelope appears. Almost 100% modulation can be obtained. The modulating-frequency range extends from several hundred hertz up to 3000 or 4000 Hz.

The effects of rf excitation loss on an am transmitter can be checked by removing the crystal from its holder. Both the oscillator and rf amplifier plate currents rise. The grid current falls to zero and there is no rf output.

The influence of modulation on the plate current and output can also be observed. The plate current of the class-C amplifier remains essentially constant as the percentage of modulation is increased slowly. The bulb also increases in brightness, indicating that with modulation more power is being delivered to the load. In effect the pilot bulb is functioning as an rf antenna meter which normally reads higher with modulation. At percentages near 100% the modulation becomes nonlinear and there is a slight drop in the plate-current reading (negative carrier shift).

Component failures influence circuit operation in the following manner:

X1 inoperative. Crystal stopping results in no output and no class-C amplifier grid current. Both dc plate currents rise.

R1 shorted. Oscillations stop, producing the same events as above.

C1 open. Same as above.

C1 shorted. No significant change.

C2 open. Oscillator is sluggish in starting.

C2 shorted. Oscillations stop and crystal-stage plate current rises. Class-C amplifier grid current falls, class-C plate current increases, and there is no rf output.

R2 shorted. Oscillations stop and a high plate current is drawn by the oscillator tube. Grid current falls to zero for the class-C stage, and its plate current increases. There is no rf output.

R2 open. Oscillator stage becomes inoperative. Grid current falls to zero and class-C plate current rises.

R3 open. Same conditions as with R2 open.

R3 shorted. No influence on circuit operation but metering circuit will not function.

C12 open. Possible rf feedback and instability.

C12 shorted. Complete transmitter becomes inoperative because of the short across the supply line.

C3 open. No rf excitation to class-C stage. Grid current goes to zero, plate current rises, and there is no rf output. Oscillator stage will operate in normal fashion.

C3 shorted. Output falls to zero and grid-meter current direction is opposite because of the plus dc grid voltage. Class-C plate current rises.

R4 shorted. Output falls and plate current rises. Oscillator operates normally.

R4 open. Grid current goes to zero, plate current falls, and output drops substantially. Modulation becomes nonlinear.

R5 open. Same conditions as with R4 open.

R5 shorted. No circuit effect. Metering position will be inoperative.

C4 open. Rf output declines.

C4 shorted. Output falls and plate current rises. Screen-grid voltage goes to zero.

R6 shorted. Plate current rises and output increases slightly. In some cases screen-grid dissipation would be exceeded.

R6 open. Screen-grid voltage goes to zero causing the output to drop and plate current to fall.

C5 open. Rf energy can feed into the modulator causing instability. Modulation frequency response can be affected adversely.

C6 or L1 shorted. Output falls to zero and plate current rises.

L1 open. Circuit becomes inoperative because of the removal of plate voltage.

C6 open. Tank circuit seriously mistuned and plate current rises as the rf output falls to zero.

R7 open. Circuit becomes inoperative because of removal of supply voltage.

R7 shorted. Circuit operations unaffected. Metering position will not function.

Failures in the audio section will result in the removal of modulation or the application of a distorted modulating wave to the class-C amplifier. The first stage is a triode grounded-cathode circuit. The influence of part failure on the operation of a voltage amplifier was covered in conjunction with Fig. 7-40. Failures in the audio-output stage have the following effects:

C9 open. No audio is conveyed to the output stage and the transmitter is not modulated.

C9 shorted. Modulating wave is distorted because of the application of a positive voltage to the grid of the output tube. Output-tube plate current rises.

P1 fault. Improper operation of potentiometer can result in modulation instability and the introduction of noise into the modulating system.

R11 open. Output stage becomes inoperative because of the interruption of the dc current path.

R11 shorted. Improper biasing and a distorted modulating wave.

C10 shorted. Same as R11 shorted.

C10 open. Decrease in modulation because the degenerative feedback reduces the amplitude of the modulating signal.

C11 shorted. No screen-grid voltage with a resultant reduction in plate current and audio power output.

C11 open. Screen-grid degeneration and a reduction in the magnitude of the modulating wave and degree of modulation.

R12 open. Removal of screen-grid voltage and a resultant decrease in plate current and audio power output.

R12 shorted. An increase in the screen-grid voltage and, in the circuit arrangement of Fig. 7-44, an increase in the plate-supply voltage of the triode voltage amplifier. Some small increase in audio output will result but safe dissipation powers may be exceeded.

T1 open. No modulation and no rf output because of the removal of the supply voltage from the class-C stage.

T1 shorted. No modulation, but there is rf output because the supply voltage to the class-C stage is not interrupted.

Some transmitters use a combination of solid-state and vacuum-tube devices. Typical are the radio-frequency power amplifiers previously shown in Fig. 7-23. Note that there are two separate supply voltage points: a low-voltage one for the MOSFET; and a high-voltage source for the vacuum tube. In this case a MOSFET stage is driving a vacuum-tube rf amplifier. The vacuum tube has a high-impedance input. Therefore a simple resonant circuit can be used to supply the drive signal.

Biasing of the input stage is handled by the two fixed resistors, while the rf signal is applied to the second gate. Any failures in the biasing or input stage would result in a loss of rf output and drive to the vacuum tube. An open radio-frequency choke would shut off the input stage. A shorted radio-frequency choke would result in instability and loss of neutralization because it would place a radio-frequency short across the 820 pF capacitor which is a part of the neutralization network along with capacitor C_N.

The bias for the vacuum-tube output stage is set by the grid resistor-capacitor combination and the safety cathode resistor. A loss of a component in the input system would result in a loss of drive and no, or very weak, radio-frequency signal at the vacuum-tube plate. However, the tube itself is protected from a loss of excitation by the cathode resistor-capacitor combination.

The screen-grid resistor and capacitor combination will influence the radio-frequency power output in the manner that was described previously for vacuum tube stages. Likewise the plate circuit components would have the same influence on operation as covered previously.

CHAPTER 8

Antenna Systems and Interference

The two antenna systems, base and mobile, are important essentials of a two-way radio system. They must be of good design and installed properly. The transmitter and its output system must be designed and tuned to transfer power efficiently from transmitter to antenna.

Since the two-way radio spectra is well used and crowded with signals, interference among stations as well as between transmitters and home-entertainment electronics is prevalent. Special consideration must be given to reducing interference. Such reductions involve antenna systems, the output systems of transmitters, the antenna input systems of receivers, and even the home entertainment devices.

8-1. ANTENNA BASICS

The prime responsibility of the antenna system is to make the most effective use of the radio-frequency power provided by the transmitter. The antenna system comprises the output system of the transmitter, the transmission line, and the antenna proper. The radio-frequency energy must be transferred to the transmission line, and the line must convey that energy to the antenna as shown in Fig. 8-1. The antenna, in turn, must radiate it. Infrequently there may be a tuning unit at the antenna in the form of matching hardware or LC tuner.

There is a great variety of basic antenna styles and modifications. However, most antennas stem from the fundamental Hertz and Marconi types shown in Fig. 8-2. These antennas are said to be "resonant" because their physical lengths are related to the wavelength of the rf energy they radiate. The Hertz antenna is usually called a half-wavelength or half-wave antenna because its physical length corresponds to that dimension of a radio wave. Thus the actual physical length depends on the frequency.

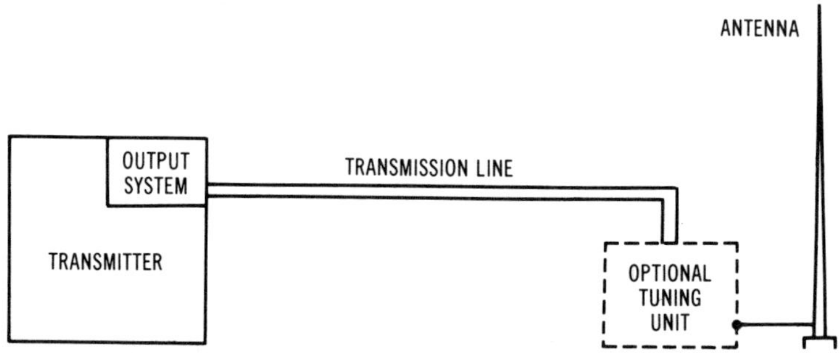

Fig. 8-1. Major segments of an antenna.

The radio-frequency energy is most often supplied to the center of the Hertz antenna. Such an antenna is referred to as a half-wave dipole. However, it can be supplied at the end as shown in Fig. 8-2A or at some other position along its length.

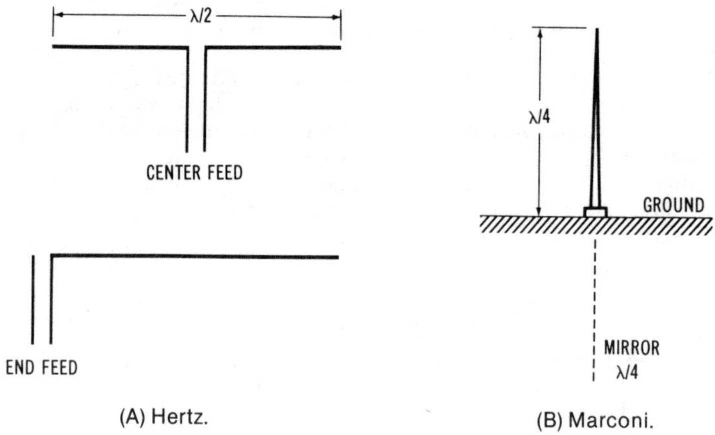

Fig. 8-2. Basic types of antennas.

Physically, a Marconi antenna is one-quarter wave long. Electrically, it acts like a half-wavelength antenna because the ground beneath it forms a mirror quarter-wavelength segment as shown in Fig. 8-2B. The characteristics and quality of the mirror segment are related to ground conditions. A network of conductors or large area metallic surface is often laid out beneath the antenna to provide a more uniform and lower resistance ground system.

The quarter-wave element can be mounted on a base insulator or grounded directly. In practice, the former is more common because the results are more uniform and matching is less of a problem.

A horizontal antenna radiates a horizontally polarized wave (electric vector is horizontal); a vertical antenna, a vertically polarized wave (electric vector is vertical). If a vertically polarized wave is desired, it is possible to mount a half-wave dipole vertically. However, the Marconi is not used as a horizontal antenna because it needs a mirror-ground.

A very practical formula for calculating a half-wavelength is:

$$\frac{\lambda}{2} = \frac{492}{f}$$

where,

$\lambda/2$ is the length of a half-wavelength in feet,
f is the frequency in MHz.

The free-space dimension calculated from this formula represents the half-wavelength separation of radio waves traveling in free space. Inasmuch as a radio wave travels slower in most other media than in air, the half-wavelength separation would be shorter than the free-space calculation. This means a radio wave travels slower in a transmission line or antenna than in free space; therefore the separation between adjacent waves is less than in free space.

A similar relationship exists for a radio wave on an antenna—a half-wave antenna is physically from 1% to 6% shorter than its free-space measurement. A point of discontinuity between the antenna conductor and space exists at the ends of the antenna. To a degree, space acts as a dielectric material and the capacitive effect links the two ends of the antenna. This end effect lowers the resonant frequency of the antenna. For this reason, it must be cut shorter than the free-space calculation to obtain resonance at a given frequency. In other words, an antenna must be cut to an *electrical half wavelength* which is physically shorter than a free-space half wavelength.

A thin antenna element must be cut only slightly shorter than its free-space value, whereas a large-diameter antenna must be made significantly shorter. Since a shorter physical length is required at the higher frequencies, the length is much more critical at these frequencies.

The formula most often used for calculating antenna resonant length for operation in the high-frequency spectrum is:

$$\frac{\lambda}{2} = \frac{468}{f}$$

where,

$\lambda/2$ is the half wavelength in feet,
f is the frequency in MHz.

The voltage and current distribution on a basic half-wave antenna is shown in Fig. 8-3A. The voltage maxima are at the ends and are of opposite polarity because the ends are one-half wavelength apart. The difference in polarity is comparable to that between the peaks of adjacent sine-wave alterations. The magnitude of the voltage is related to the strength of the radio wave supplied by the transmitter.

The current is maximum at the center of the antenna and falls away to zero at both ends. Since impedance is the ratio of voltage to current, the high-voltage ends display maximum impedance, and the high-current center, minimum impedance. Fig. 8-3B shows the impedance variation along the length of an antenna.

Impedance is a very important consideration in connecting an antenna to a transmission line. With a center-feed system, the line must match a low impedance. Conversely, an end-feeding arrangement calls for a high-impedance line.

A half-wave antenna fed at the center (Fig. 8-3C) has inductance, resistance, and capacitance; in this respect, it resembles a series-resonant circuit as shown in Fig. 8-3D. When the antenna is cut to the proper length for the applied frequency, the inductive and capacitive reactances are equal, and the antenna is said to be resonant. At resonance the impedance is resistive and minimum, just as it is in a series-resonant circuit. This resistive impedance is called the radiation or antenna resistance (R_A).

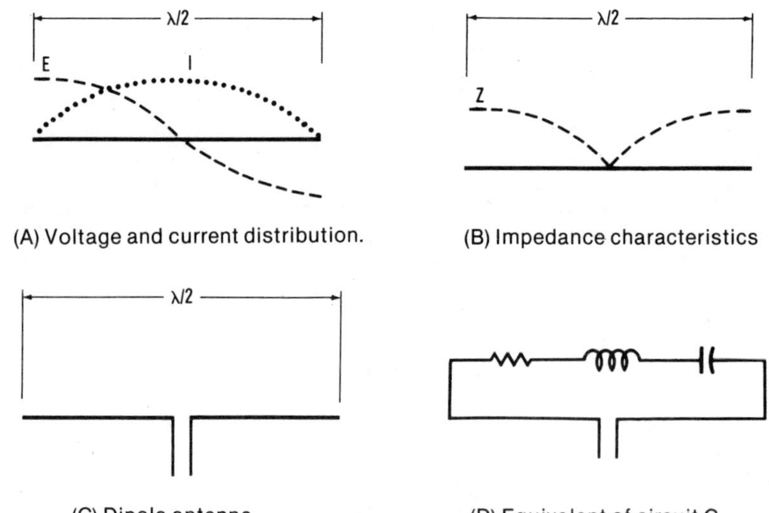

Fig. 8-3. Characteristics of a half-wave antenna.

The radiation resistance of a center-fed half-wave antenna is approximately 72 ohms; this value is influenced to some extent by the height of the antenna above ground. The radiation resistance of a base-fed quarter-wave Marconi is 36 ohms; this value is affected by ground conductivity beneath the antenna.

Antenna resistance is also affected by the shape and area of the antenna element, the location of the feedpoint, and the proximity of other elements. For example, as the element is made larger, the antenna resistance rises. This effect is similar to increasing the resistance of a series-resonant circuit—the antenna has a lower Q and a lower ratio of reactance to series resistance. Likewise its bandwidth increases. Thus larger-diameter and specially shaped elements permit the antenna to operate over a wider span of frequencies. Also the length to which it is cut is less critical. The capacitive end effect is somewhat greater; therefore such an antenna must usually be cut shorter than a thin dipole operating on the same center frequency.

8-2. RADIATION PATTERNS

If an antenna could be isolated in space and operated as a point source, its electromagnetic radiation pattern would be perfectly spherical. However, antennas are of a finite dimension and are operated near ground or other objects. Consequently the electromagnetic energy travels from the antenna at differing intensities—both horizontally and vertically.

Where a uniform horizontal radiation in all directions is desired, the pattern can be represented by a perfect circle, as shown in Fig. 8-4A. In other cases one may want to concentrate the radiation in a given direction in order to transmit the maximum signal between

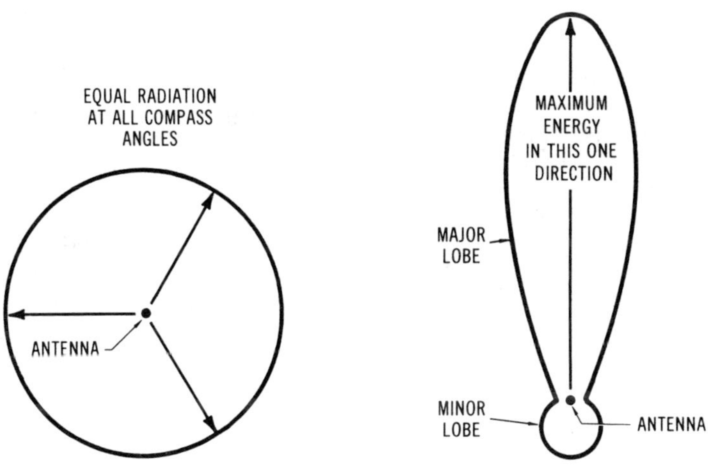

(A) Nondirectional. (B) Directional.

Fig. 8-4. Horizontal radiation patterns.

two fixed stations. A typical directional pattern is shown in Fig. 8-4B.

There are also reasons why one must be concerned with the vertical angle at which the electromagnetic energy leaves the antenna. Is the energy to be directed skyward or toward the horizon? For short-range communication one would prefer to direct the energy at a low vertical angle as shown in Fig. 8-5A.

In establishing a contact over a long-range path at a certain time and season of the year, it may be necessary to direct the signal skyward at a specific angle, as shown in Fig. 8-5B. The signal can be bounced off one of the ionized layers and returned to earth at the desired receiving point.

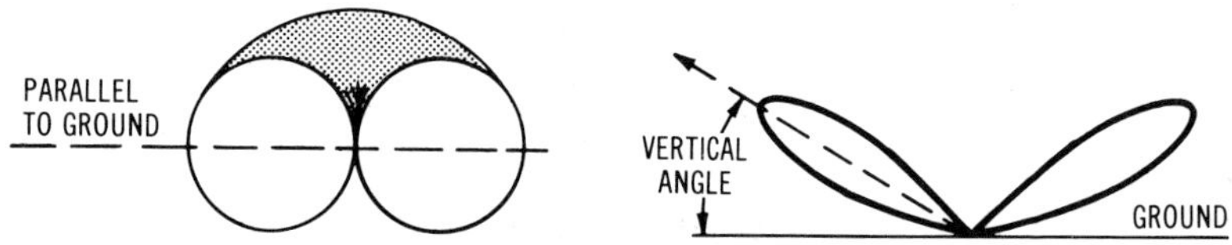

(A) Energy directed toward horizon.

(B) Energy directed skyward.

Fig. 8-5. Vertical radiation patterns.

The horizontal and vertical radiation patterns of the two basic antennas—the Marconi vertical quarter-wave and the Hertz horizontal half-wave—are given in Fig. 8-6. The basic vertical antenna has an omnidirectional (i.e., circular) horizontal pattern. Approximately the same level is sent out in all directions. The vertical antenna above a good conductive ground has a low-angle vertical radiation pattern. In other words, more radiated energy is directed over the surface of the earth, toward the horizon, instead of skyward. (Of course there is some skyward radiation.) Special antennas are used to further concentrate the vertical radiation at low angles.

The radiation pattern of a half-wave horizontal antenna is shown in Fig. 8-6B. The doughnut-like emission has been shifted through 90°. Now the vertical radiation pattern is circular and an equally strong signal is transmitted up, down, and to the sides. Even though the vertical free-space pattern is circular, the presence of ground does influence the vertical radiation for a horizontal antenna. In fact, the angular magnitudes vary in relation to height above ground. The actual pattern results from the addition and subtraction of radiation from the antenna itself and components reflected from ground.

The horizontal radiation pattern is a figure 8 with most of the energy moving out perpendicular or broadside to the direction of the antenna element. There is minimum signal radiation horizontally off the ends of the antenna. If such an antenna is to send out the strongest signal in a given direction, it must be positioned at right angles to the desired direction.

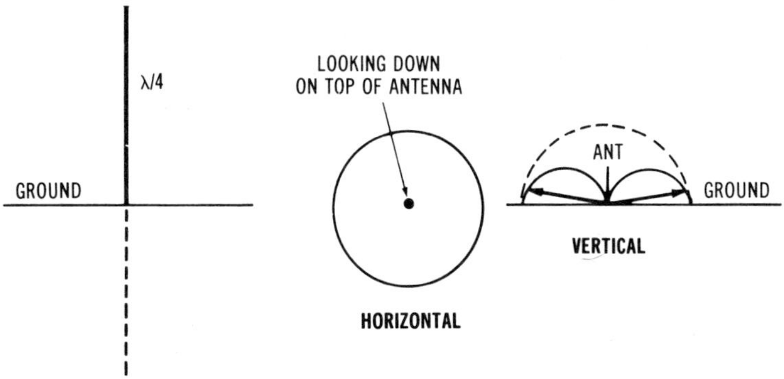

(A) Quarter-wave Marconi.

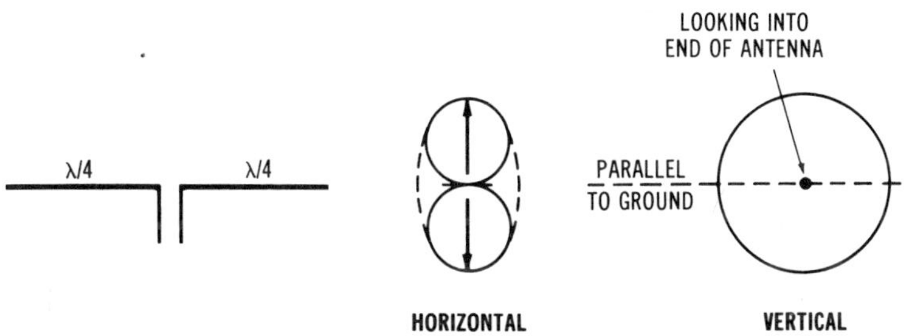

(B) Hertz.

Fig. 8-6. Patterns of the basic antennas.

8-3. THE VERTICAL ANTENNA

In most communication services the vertical antenna is king. Particularly in vehicular communications, the vertical antenna presents fewer installation problems. In general, vertical polarization provides more reliability when the antenna must be mounted close to ground (in mobile installations). Furthermore, the radiation patterns are more appropriate. In general the horizontal radiation pattern of a vertical antenna is circular. Consequently the same level of signal is sent out at all angles, a preferred condition in maintaining contact with cruising mobile units. The vertical antenna also has a low-angle vertical radiation pattern. Such a pattern is preferred in sending out signals over the countryside toward the horizon. In other words, the power supplied to the vertical antenna is used more effectively (with respect to the radio service to be rendered) than for a horizontal antenna.

Most dipole antennas, along with the various higher gain and directive combinations, can be made to radiate and receive vertically polarized signals. One of the most common radiocommunications antennas is the simple quarter-wave vertical, with or without modification. Basically it is a quarter-wave Marconi.

The most common antenna for two-way mobile radio systems is the simple quarter-wave whip shown in Fig. 8-7A. The antenna has a low enough resistance that it can be fed directly via a 50-ohm coaxial line. Its attractions are ease of mounting, limited length, plus a simple feed system that can take advantage of the shielding provided by a coaxial cable. Shielding is, of course, important in vehicular installations because of ignition and other electrical noise problems.

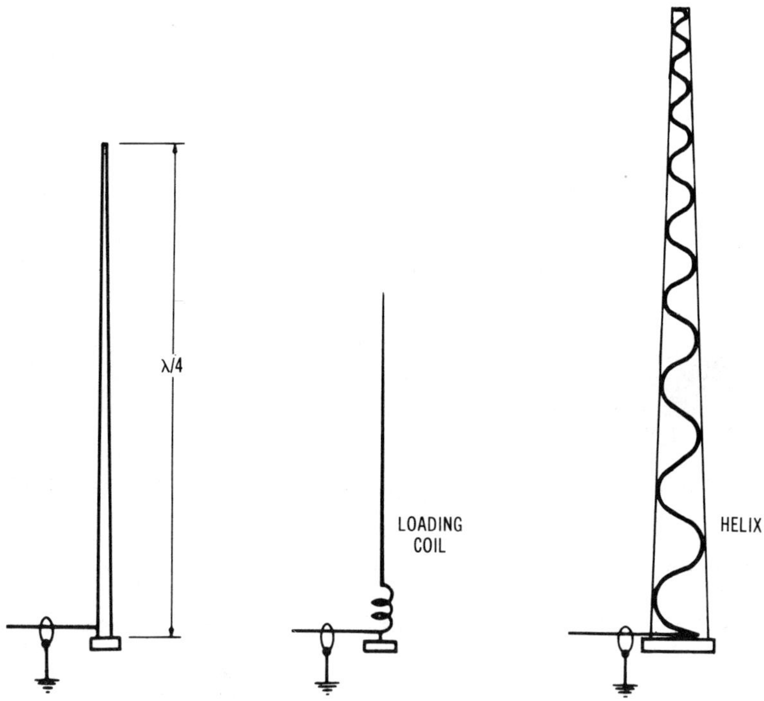

(A) Simple quarter-wave whip.

(B) Whip with loading coil.

(C) Helix antenna.

Fig. 8-7. Basic quarter-wave vertical antennas.

At some frequencies, even a quarter-wavelength antenna is too cumbersome for vehicular installation. There are ways of physically shortening quarter-wavelength antennas. However, there is some sacrifice in gain and performance. A popular method is to use a loading coil at the base of the antenna, as shown in Fig. 8-17B. When an antenna is cut shorter than a quarter wavelength at the desired frequency, the antenna becomes capacitive. If we now insert a reactance of opposite sign (inductive reactance of a coil), the antenna can still be brought into resonance, even though its overall length is shorter than a quarter wave.

An antenna has two important parameters—the overall length it occupies in space, and its resonant frequency. When the two factors are optimized, the antenna system gives peak performance. Use of the loading coil does permit the antenna to be tuned to resonance; therefore performance is improved over that of a short antenna that is not tuned. However, the performance does not attain the level possible with both proper overall length and resonant tuning.

Overall antenna length can also be reduced with the coil-like antenna structure as shown in Fig. 8-7C. This antenna usually takes the form of a helical conductor supported by an insulator such as fiberglass. In fact, the conductor is often mounted inside the insulator to protect it from the environment.

Although the simple quarter-wave whip (Fig. 8-8A) is used, more elaborate vertical antennas are usually employed at the base stations of two-way radio systems. Often associated with a quarter-wave whip is a ground plane such as that shown in Fig. 8-8B. The presence of a ground plane makes the performance of the antenna more uniform with relation to its mounting position above actual ground. Furthermore the ground plane is so positioned that the low angles of the vertical radiation pattern are emphasized. The signal coming off the antenna is radiated across the surface of the earth toward the horizon, an ideal condition for communicating with mobile units and their relatively low antennas. Because of this concentration, an antenna of this type has more gain than a simple quarter-wave whip.

Three other popular vertical antennas are shown in Fig. 8-8. The overall length of the vertical whip can be extended to more than a quarter wavelength (Fig. 8-8C). If such an antenna is suitably matched, it will also have gain because of its ability to concentrate the vertical radiation at low angles. The overall length of the coaxial antenna, (Fig. 8-8D) is approximately a half-wave. This arrangement also provides a means of matching when feeding the transmission line up through the center of the larger-diameter lower section. It has gain and an improved vertical radiation pattern.

In the uhf and vhf spectra, collinear verticals (Fig. 8-8E) have become increasingly popular. They provide a substantial gain and concentration of the vertical pattern to low angles as compared to a single quarter-wave element.

8-4. PHASED ARRAY

The use of additional antenna elements permits the energy transmitted or received to be concentrated at more narrow horizontal and/or vertical angles. Such an antenna is said to have gain because more of the power is directed into specific angles of radiation. Thus, antenna gain does not mean that the antenna operates like an amplifier; instead, it means that the antenna concentrates the power into preferred horizontal and/or vertical radiation patterns.

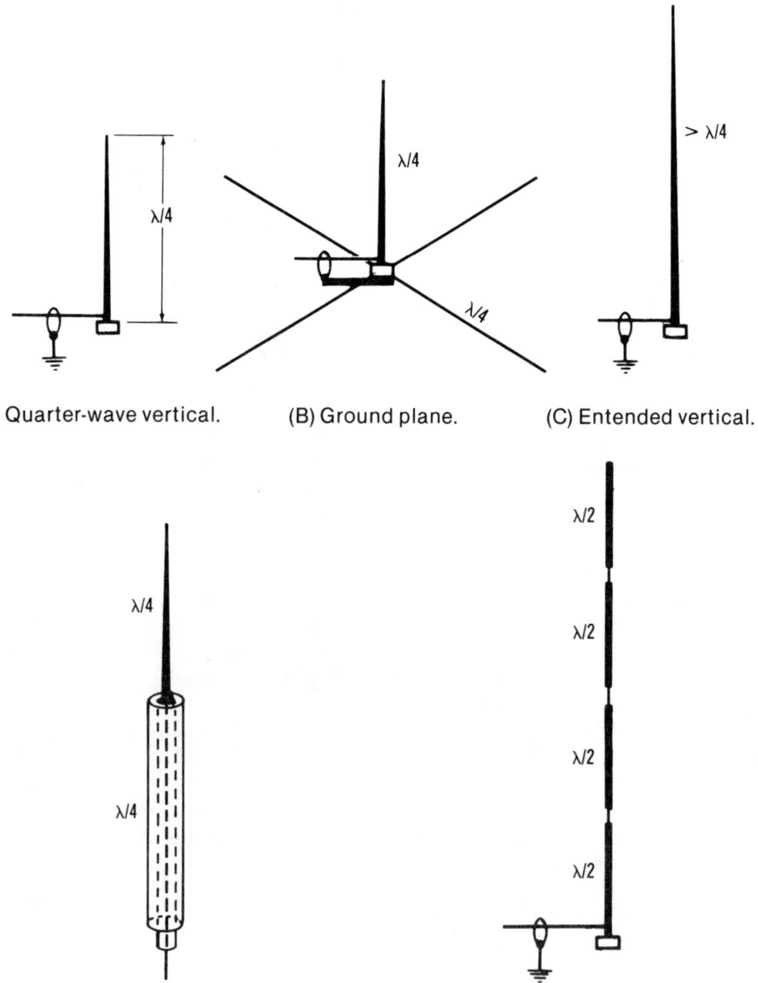

(A) Quarter-wave vertical. (B) Ground plane. (C) Entended vertical.

(C) Coaxial. (E) Collinear.

Fig. 8-8. Base station antennas.

In the measurement of antenna gain a half-wave dipole or a quarter-wave vertical can be used as a reference antenna and its gain is assigned a value of 0 dB. At other times the 0 dB reference is a theoretical point referred to as an *isotropic* source. If an antenna

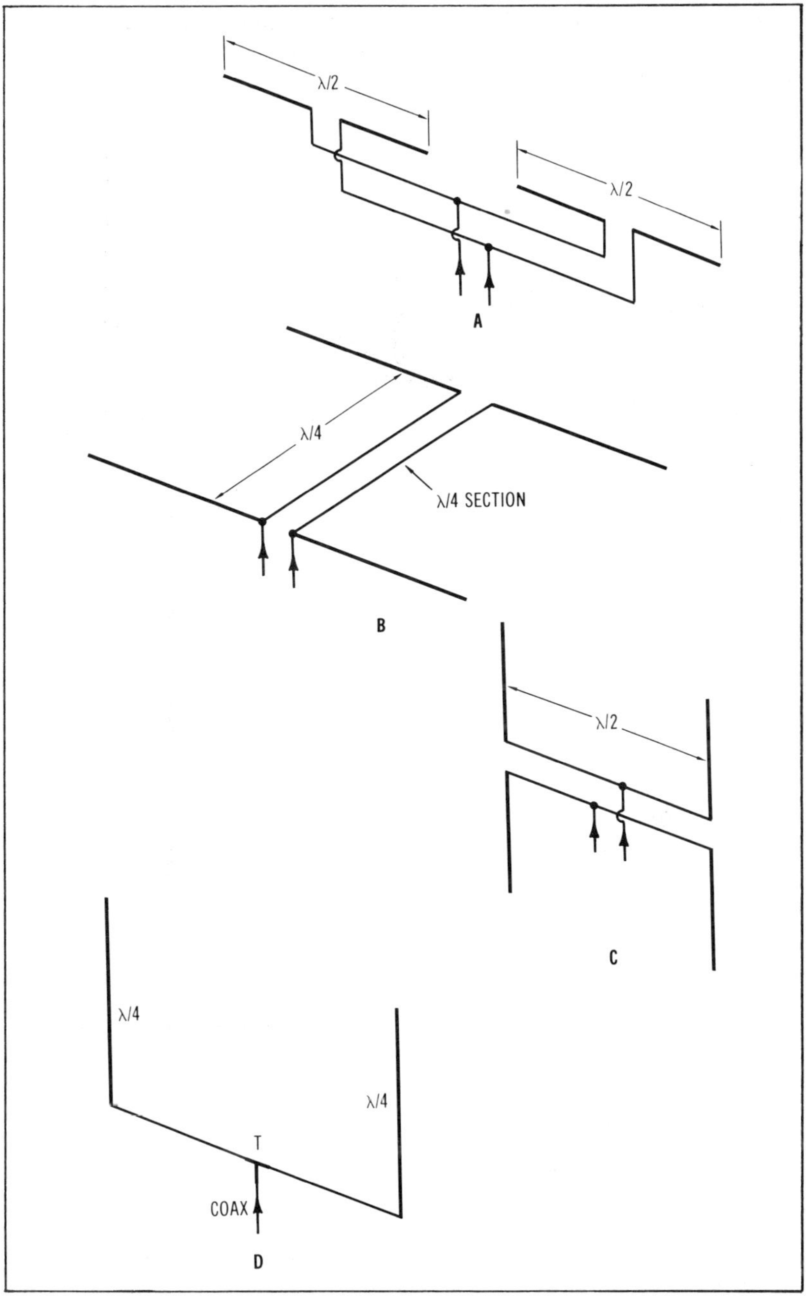

Fig. 8-9. Collinear, end-fire, and broadside phased arrays.

is said to have a gain of 3 dB, this means that if this antenna is directed toward a specific receiving location it will deliver 3 dB more signal than the reference antenna.

There are three ways to interconnect antenna elements into an array, Fig. 8-9. These are broadside, end-fire, and collinear. A collinear involves stacking one antenna immediately above the other as shown in Example A. Note that both antennas are fed from the transmission line. This antenna has a circular horizontal radiation pattern. Its gain comes from the fact that its vertical radiation pattern is concentrated at very low angles and the signals are sprayed out toward the horizon. Only a minimum amount of energy is sent skyward. Example B is an end-fire arrangement in which one element is placed in front of the other. Transmission line is connected to the first antenna and then through a second small piece of transmission line, only one-quarter wavelength long, is connected to the front antenna element. This style of antenna is directed horizontally and maximum signal is sent out along the line from the back antenna element toward the front antenna element.

The third type is known as a broadside array. The two antenna elements are positioned side by side approximately one-half wavelength apart. The feed system divides and over the same distance connects to the two antenna elements. Thus the elements are fed in phase. Maximum radiation from this antenna is broadside to the plane of the two elements. There is minimum radiation off the sides.

A common version of the broadside array is the popular phased vertical array shown in Example D. In a mobile installation these elements might be quarter-wavelength types fed by coaxial transmission lines using a centrally located T-joint.

Antenna gain can also be obtained by using antenna elements of various lengths that are not electrically connected to the driven element to which the transmission line is attached. These elements

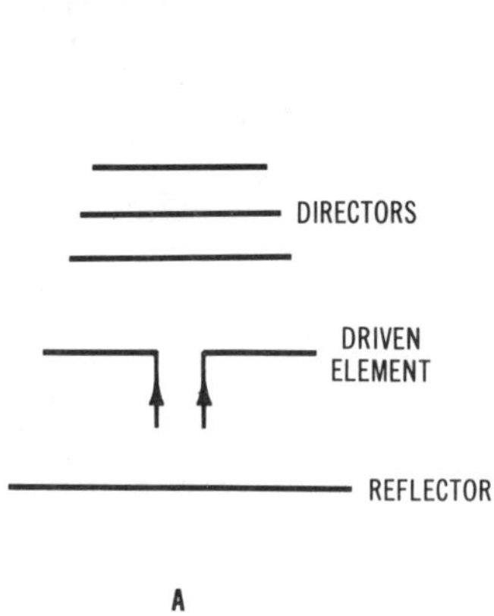

Fig. 8-10. Yagi antenna and radiation pattern.

are called parasitic elements. An element longer than the driven element acts as a reflector. A parasitic shorter than the driven element is called a director. A popular version of this antenna is the Yagi shown in Fig. 8-10. In the example there is one reflector and three directors. Maximum signal is sent out along the line of the elements. A typical radiation pattern is shown. Note how sharp it is, indicating that the antenna has a very high gain. However, the signal is radiated very strongly in a single direction. Very little radiation goes out at other angles. It is said to have a very small beamwidth. This beamwidth is usually given as the angle between the half-power positions on the directional pattern as shown. Yagi antennas are most often used at fixed locations from which a very strong signal is to be directed toward another fixed location.

8-5. THE TRANSMISSION LINE

The transmission line (Fig. 8-11) serves as the link between transmitter and antenna. When the transmission line is suitably matched at input and output, maximum energy is transferred. The equivalent circuit of a transmission line is shown in Fig. 8-11B. Even though the line is made up of continuous wires, these conductors have a very small inductance. The space between the wires, whether air or some other insulating material, inserts a small but definite capacitance between conductors. This distributed capacitance, as well as the inherent inductance, is uniform along the entire length of the line. In effect, the transmission line is an LC (inductance-capacitance) network. A radio wave requires a finite time to travel the length of the line, because capacitance opposes any voltage change, and inductance opposes any current change. There is also a small but significant opposition to current flow by the actual resistance (R) of the line.

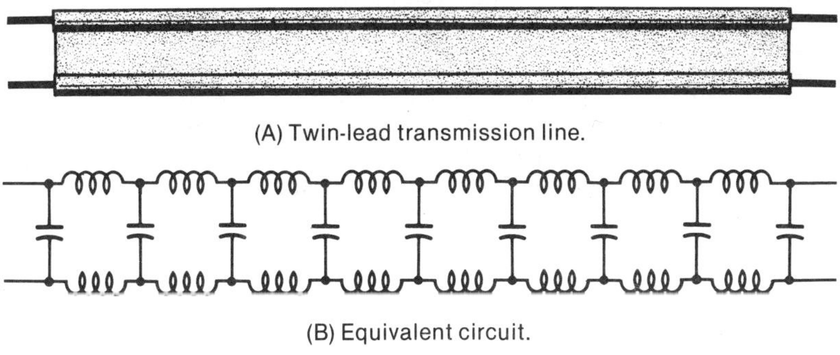

(A) Twin-lead transmission line.

(B) Equivalent circuit.

Fig. 8-11. A transmission line and its equivalent circuit.

Three important transmission-line parameters are associated with the LCR values of a practical line. The line displays a definite resistive impedance called its characteristic, or surge, impedance. It has an approximate value of:

$$Z_o = \sqrt{\frac{L}{C}}$$

where,
Z_o is the characteristic impedance,
L is the inductance per unit length,
C is the capacitance per unit length.

Usually L and C are stated as so much inductance and capacitance per foot.

The surge impedance is related directly to the diameter and spacing of the conductors. The larger the diameter is, the lower is the inductance per unit length. Likewise, the closer their spacing is, the greater is the capacitive effect per unit length.

The above factors apply to both parallel and coaxial lines, in accordance with the following formulas:

For parallel lines:

$$Z_o = \frac{276}{\sqrt{k}} \log \frac{b}{a}$$

where,
Z_o is the characteristic impedance,
k is the dielectric constant of the insulating material,
b is the center-to-center spacing between conductors,
a is the radius of the conductors.

For coaxial lines:

$$Z_o = \frac{138}{\sqrt{k}} \log \frac{b}{a}$$

where,
Z_o is the characteristic impedance,
k is the dielectric constant of the insulating material,
b is the inside diameter of the outer conductor,
a is the outside diameter of the inner conductor.

Of course, for open-wire line where air serves as the dielectric material, the k factors are omitted and the formulas become:

$$Z_o = 276 \log \frac{b}{a}$$

and

$$Z_o = 138 \log \frac{b}{a}$$

The characteristic-impedance values are the figures you state when you purchase transmission line. When you ask for 50-ohm line, you are stating the characteristic impedance. Again this is not a value you can measure with an ohmmeter; it is a resistive impedance.

Another important transmission-line parameter is its velocity factor. Because of the finite inductance and capacitance of a line, a radio wave travels more slowly along a transmission line than in air. The velocity factor is the ratio of the velocity along the line to the velocity in air or free space. The velocity of propagation along a line is computed by the formula:

$$\text{Velocity} = \frac{1}{\sqrt{LC}}$$

A third important parameter is the electrical length of a line, which is stated in wavelengths or fractions of a wavelength. For example, a one-wavelength section of transmission line has a physical length that corresponds to the separation between positive peaks of the radio wave on the line. A half-wavelength section corresponds to the separation between adjacent positive and negative peaks.

In determining physical length, we must consider the velocity factor. If the radio wave traveled along the transmission line at the same velocity it traveled in space, the regular half-wave free-space wavelength formula given previously could be used. However, the velocity of propagation along a practical transmission line is slower than the free-space velocity. Consequently the half-wave section of line will be shorter than the free-space dimension according to the velocity factor of the transmission line.

$$\frac{\lambda}{2} = \frac{492}{f} \times V$$

where,

$\lambda/2$ is the half-wavelength of the line in feet,
f is the frequency in MHz,
V is the velocity factor.

The electrical length of a transmission line is often given in electrical degrees, 360° being equal to one wavelength. Thus a half-wavelength section is often called a 180° line; a quarter-wavelength section, a 90° line.

Line attenuation refers to the losses a radio wave encounters in traveling from input to output. It is the result of I^2R (current squared times resistance) heat loss in the conductors and the heating of the dielectric material between them. The dielectric loss is insignificant for an air dielectric line and is much higher in lines with a dielectric material.

The lower the surge impedance of the line is, the higher the conductor loss becomes, because more current flows for a given output power. On the other hand, the dielectric loss increases as the surge impedance rises, because of the voltage stress placed across the conductors.

Conductor and dielectric losses both increase with frequency because of skin effect and decreasing dielectric reactance. Thus the attenuation of practical lines increases with frequency.

In the radiocommunications services, the coaxial line is the most common, even though it has a higher loss than open-wire and twin-lead types. A coaxial line has better shielding because of its grounded outer conductor. The performance of the line thus remains more uniform with environmental changes and is less affected by nearby metallic objects.

The two most common lines are RG-58/U and RG-8/U. The latter has less loss and is less flexible. Thus it is preferred for long runs of line, higher-powered transmitters, and higher operating frequencies. The small-diameter and more flexible RG-58/U is used in many vehicular installations because it is easier to route.

8-5-1. Lines in Operation

A theoretically perfect line has no power loss. In fact, if the surge impedance of the line matches the internal impedance of the generator (radio-frequency source), maximum power will be transferred to the line. This is a basic electrical law—maximum power is delivered to any load, when it has the same impedance as the source.

If the line were infinitely long, radio-wave energy would travel along it indefinitely with no loss and would not return to the source. Likewise, if the line were broken at some point, as shown in Fig. 8-12, and a pure resistance of the same ohmic value as the surge impedance were placed across the line, all the power fed to its input would be dissipated across the resistive termination. Here, too, none of the energy impressed at the input would ever return to the source because it would all be dissipated across the termination.

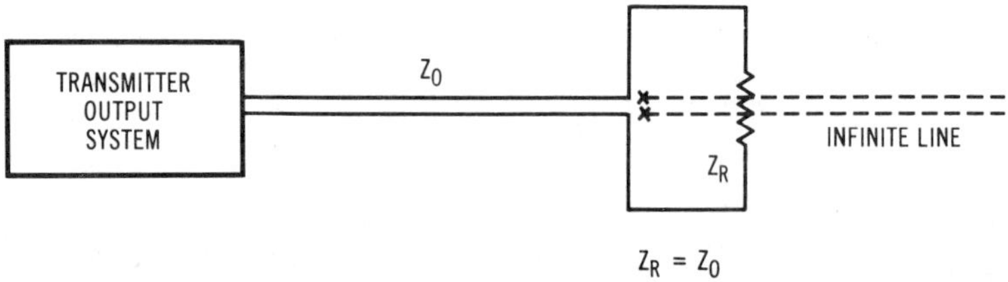

Fig. 8-12. The untuned transmission line.

This means that a properly terminated line displays the same characteristics as an infinitely long line. Furthermore, a properly terminated line will operate effectively regardless of its length.

No lines can be theoretically perfect. However, under reasonably ideal conditions the only signal loss will be due to line attenuation. In other words, when properly matched, the line will act as an efficient means of transferring the output of the rf power amplifier to its antenna termination.

A transmission line that is so matched is called an untuned line because its length is unimportant to the matching. The untuned line can be matched by designing the antenna with an antenna feedpoint impedance that matches the surge impedance of the line. Matching sections can be used when the antenna feedpoint impedance and the line impedance differ.

When an untuned line is matched exactly, the current is the same at any point along the conductor. The voltage will show only a gradual and usually insignificant decline (as a result of attenuation) from the input to the termination.

8-5-2. Influence of Termination

Let us now consider the characteristics of a line terminated in other than its surge impedance. Fig. 8-13A shows the far end of the line short-circuited. Here the energy will move down the line until it reaches the short. There is no load to absorb the energy and only one way for it to travel—so back it goes along the line, over the same path. The rf power traveling down the line to its termination is called the *incident power*; the power reflected from the termination and moving back toward the source is the *reflected power*.

Since the termination in Fig. 8-13A is a short circuit, there is maximum current flow and no voltage at this point. Inasmuch as the reflected current now moves back along the line at the same velocity it came down the line, the instantaneous magnitude and phase of the incident and reflected currents will differ everywhere along the line. At some points their phase will be the same and their amplitude almost double. At other points, the two currents will be out of phase, producing almost no current. Their amplitudes will be between minimum and maximum at still other points, as shown by the waveforms.

At the short, the two currents are in phase and there is a maximum reading. Exactly one-quarter wavelength back from here, the two currents will be out of phase, and the net current will be minimum. One-half wavelength back from the short, the two currents will be back in phase and there will be another maximum, and so on, back to the source.

A similar relationship exists for the voltage; it is zero at the short circuit and maximum one-quarter wavelength back (at the same point where the current is minimum). One-half wavelength back the voltage is back to minimum (at a maximum current point).

The variation of current and voltage along the line is referred to as a *standing wave*. Maximum current or voltage points are called loops or antinodes and minimum points are nodes. Thus a voltage loop occurs at the current node and vice-versa.

Similar but opposite relationships exist for an open-circuited line as shown in Fig. 8-13B. As before, the rf wave traveling down the line meets a termination that cannot dissipate the energy; again it turns back toward the source. Since the termination is an open circuit, the current flow is zero and the incident and reflected currents, although equal in value, are opposite in phase. However, the two voltage components are in phase and combine to form a maximum. Standing waves again form on the line, but the positioning of the loops and nodes is opposite that of the short-circuit distribution.

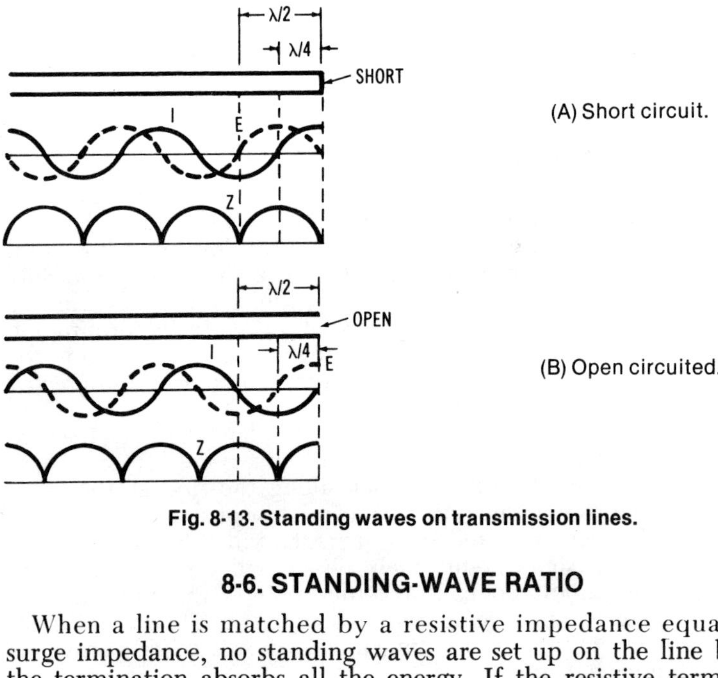

Fig. 8-13. Standing waves on transmission lines.

8-6. STANDING-WAVE RATIO

When a line is matched by a resistive impedance equal to its surge impedance, no standing waves are set up on the line because the termination absorbs all the energy. If the resistive termination has an ohmic value higher than the surge impedance, the standing-wave pattern set up on the line will resemble that of an open-circuit condition, as shown in Figs. 8-14A and B. The higher this ohmic value is, the more closely the standing-wave condition approaches that of an open-circuit condition. Moreover the voltage loop is always maximum at the termination, just as it is in an open circuit.

As the resistance of the termination is increased further and further from the surge impedance, less power is absorbed by the termination and more is reflected down the line. Hence, the standing waves are of higher magnitude in lines terminated in high resistance, as shown in Fig. 8-14B.

A similar relationship exists when the ohmic value of the termination is less than the surge impedance (Figs. 8-14C and D).

Now, however, the termination acts more like a short circuit—the standing-wave magnitude increases as the ohmic value of the termination decreases. As the value of resistive termination is decreased, more power is reflected. In a short-circuit condition (zero resistance), all the incident power is reflected. But when the ohmic value of the termination is only slightly less than the surge impedance, most of the incident power is absorbed, and what little power is reflected sets up a weak standing wave on the line as shown in Fig. 8-14C.

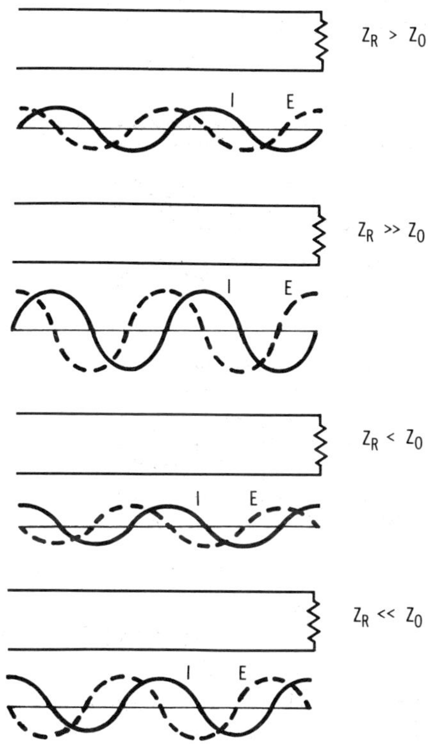

(A) Terminating resistance larger than surge impedance.

(B) Terminating resistance much greater than surge impedance.

(C) Terminating resistance less than surge impedance.

(D) Terminating resistance much less than surge impedance.

Fig. 8-14. Influence of termination on standing waves.

A measure of the magnitude of the standing wave on the line is called the standing-wave ratio, or swr. It can be determined from the following formulas:
where the current is known:

$$\text{swr} = \frac{I_{max}}{I_{min}}$$

where the voltage is known:

$$swr = \frac{E_{max}}{E_{min}}$$

Swr can be determined by measuring the voltage at a loop and a node with an rf voltmeter, or the current magnitudes at a loop and node with an rf current meter.

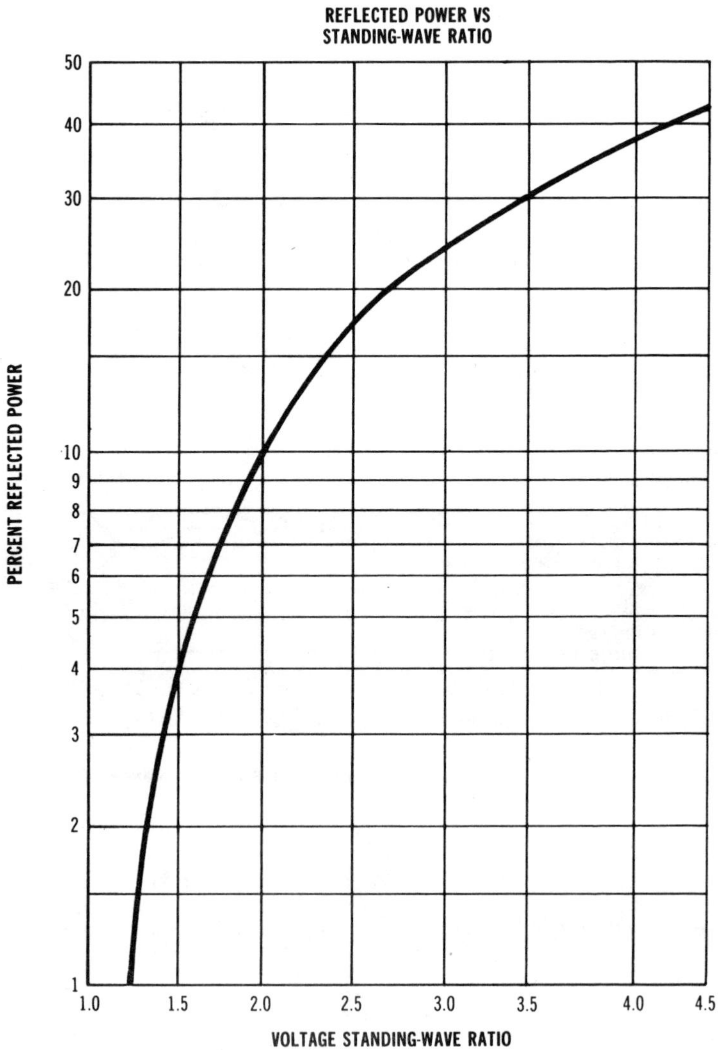

Fig. 8-15. Reflected power versus standing-wave ratio.

Impedance measurements can also be used. The larger quantity is always the numerator:

$$\text{swr} = \frac{Z_{\text{load}}}{Z_o} \text{ or } \frac{Z_o}{Z_{\text{load}}}$$

Commercial instruments are available that can directly measure standing-wave ratio in the transmission lines. These were considered in Chapter 6.

The swr reading tells much about the operation of an antenna system—the higher it is, the greater is the mismatch. Consequently, the less power reaches the termination.

A useful chart that depicts percent of reflected power in terms of vswr is given in Fig. 8-15. Note for an swr of 2.0 the reflected power is 10%. This indicates that with a power output capability of 20 watts only 18 of those watts would reach the antenna load $[20 - (0.1 \times 20)]$.

8-7. TYPICAL MOBILE AND BASE ANTENNAS

Three basic antenna types designed mainly for mobile use are shown in Fig. 8-16. Example A is a coaxial type that can be employed where there is an inadequate ground plane such as a large metallic surface or radials. The antenna is center fed with the upper quarter-wave section being a continuation of the inner conductor of the coaxial transmission line. The outer conductor of the coaxial line attaches to the quarter-wavelength sleeve that is immediately beneath the center insulator. This acts as a transformer and isolating section that permits an effective coaxial feed system up through the center of the sleeve. The section of tubing beneath the coaxial sleeve is a support mast for the antenna. Mounting options would be a snap-on bumper mount, vehicular side mount, or a trunk/body mount.

Example B is a 5/8-wavelength antenna which has a good low vertical angle of radiation. As a result the gain of the antenna is 2.5 dB at those angles which best favor vehicular radiocommunications. The coil at the bottom of the antenna serves as a combination base loading coil and spring. Thus an appropriate match can be made to a 50-ohm coaxial line. The coil provides loading and proper matching.

Example C is a collinear gain antenna comprised of two half-wave sections. Gain is 5.2 dB. There is a loading coil at the base of the antenna which permits a properly matched end-fed array. A phasing coil is located at the center and permits the proper in-phase feed of the two collinear elements.

Several base station antenna types are shown in Fig. 8-17. The driven element is a vertical folded dipole. The presence of the mounting mast provides some directivity in the direction away from the mast toward the driven element. Thus the antenna can be mounted in such a manner that a somewhat stronger signal is radiated in some favored direction. If an additional director rod is included as in example B there will be additional antenna gain in a favored direction.

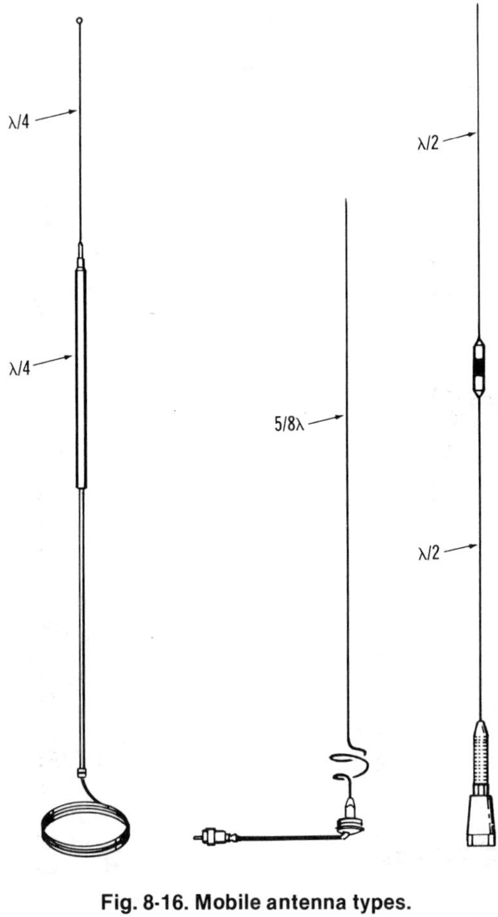

Fig. 8-16. Mobile antenna types.

Additional antenna gain can be obtained by stacking a group of driven elements on the mast as shown in example C. A gain of 9 dB is possible when each of the driven elements is mounted one beneath the other. A directional pattern results. If an omnidirectional pattern is desired, the elements can be staggered and not all facing the same direction. Possible gain is 6 dB.

Each of the elements is fed with a separate 50-ohm line. These four feedlines are cut to the proper length to provide equal power and in-phase energy excitation of the elements. The four lines are combined in a four-way matching assembly.

Antenna-mounting guidance as recommended by RCA for a

remotely controlled base station is given in Fig. 8-18. Proper construction and grounding is important in terms of lightning protection. Note that a lightning rod extends above the mast and it is suggested that the top of the antenna support be positioned below the top of the tower. To obtain a low resistance to ground a No. 4 copper conductor, or larger, extends from the top of the mast down through the center. This copper conductor serves as an appropriate ground for transmission line, bandpass cavity, and base station cabinet.

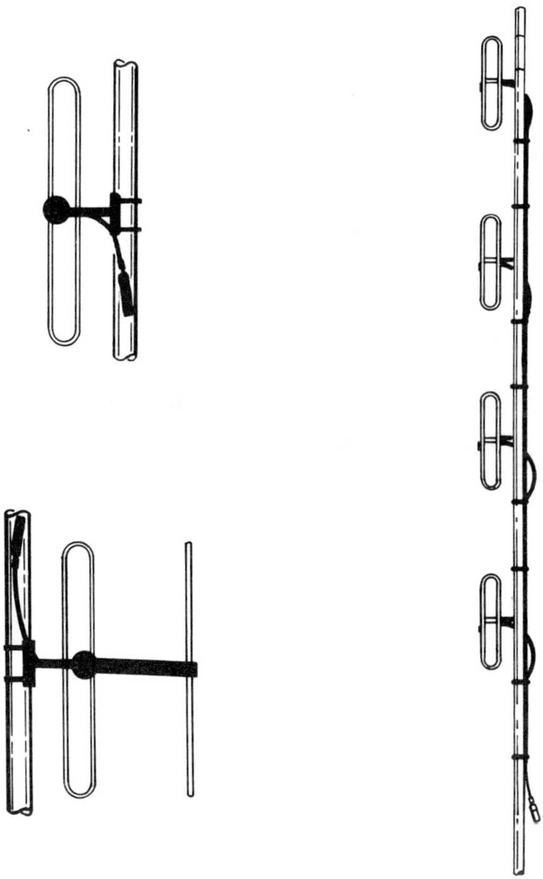

Fig. 8-17. Base station antenna types.

Guy wires, if used, are also bonded to the mast and when necessary the individual guys are broken up with insulators to minimize their influence on the loading and pattern characteristics of the antenna. A bandpass cavity blocks off-frequency signals from

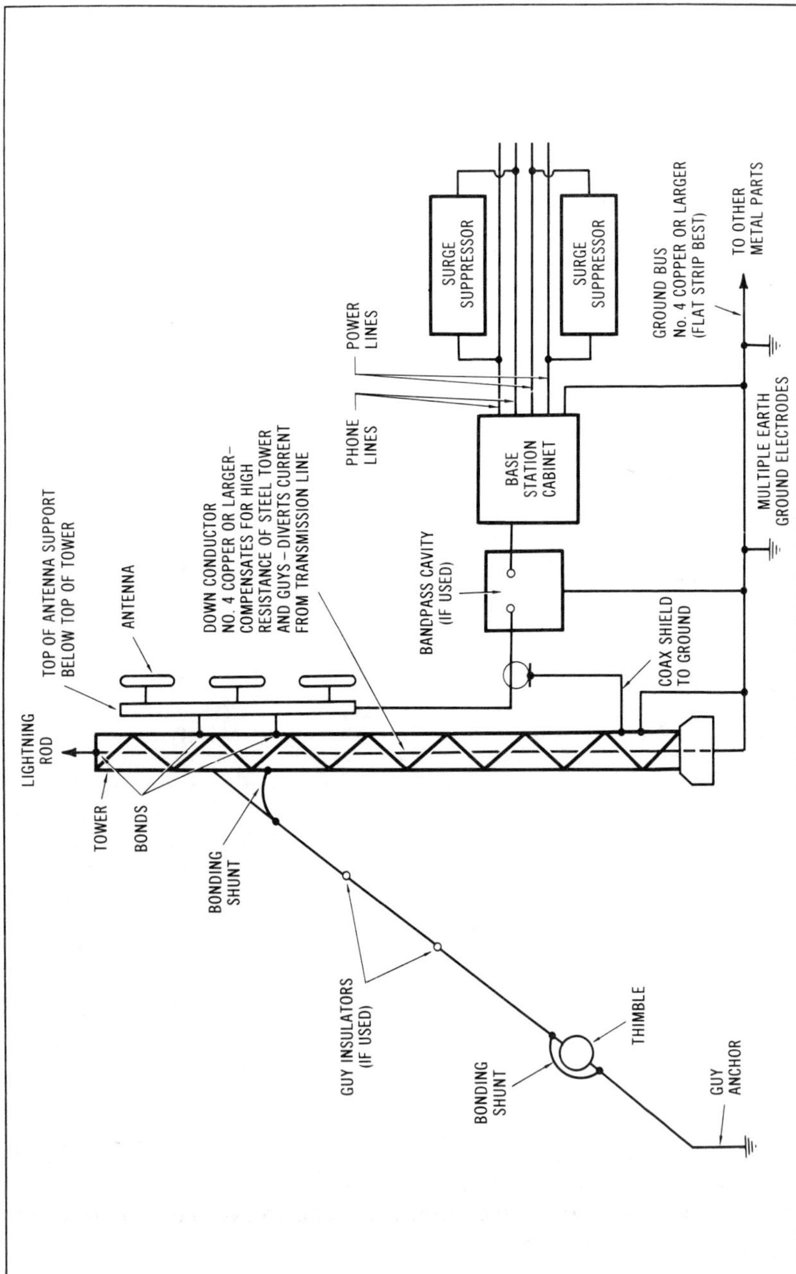

Fig. 8-18. RCA antenna system recommendations.

being radiated by the transmitter and also prevents off-frequency signals that are picked up from the antenna system from being supplied to the base station receiver. Surge suppressors are used across both the tower and phone line inputs to the base station.

8-8. RADIO INTERFERENCE AND HOME ELECTRONIC EQUIPMENT

In the last decade the growth of the two-way radio services has been phenomenal. Millions of transmitters are in operation. The Citizens Band Radio Service in particular has brought the radio transmitter into home and car, business, and truck. Significantly transmitters have been brought in close proximity to home entertainment devices such as television receivers, radios, audio amplifiers, tape recorders, electronic organs, etc. Radio interference problems have become more prevalent due to the very nearness of transmitter and home electronic equipment.

The FCC has technical standards for transmitters and compliance is mandatory. Despite this fact, the very closeness of a source of radiation to electronic equipment means the interference problem can prevail. The characteristics and performance of home electronic devices, other than transmitters, do not come under the jurisdiction of the FCC. Providing thorough shielding for such devices is also a cost consideration. Nevertheless many radio interference problems can be eliminated or reduced. Transmitter and its antenna system must be properly installed and adjusted. The same applies to the home electronic equipment. The use of simple low-cost filters, capacitors, and other components (most connected externally) can do much to reduce interference. Some examples follow.

Sources of radiation from a transmitter and/or antenna are the desired signal; undesired harmonics of the transmitted signal; and undesired spurious radiation from multipliers, oscillators, and mixers. Small as these components might be they can cause problems because of their very nearness to other electronic equipment. The undesired harmonics and spurious components are kept at very low levels by design and transmitter shielding. They must be kept below levels specified in the FCC technical standards for the particular radio service.

However, more ideal suppression of undesired components is attained with proper grounding and proper antenna match. Use a good earth ground for the transmitter via a large diameter wire or copper strap. This precaution minimizes radiation from the transmitter proper.

The typical transmitting antenna is mounted high and clear, and usually some distance away from the home electronic equipment. However, the transmission line itself originates at the transmitter and can be very near to television receivers, radios, etc. Therefore it is important to keep the line from radiating a strong signal by holding down the standing-wave ratio. Install the antenna system wisely and properly and keep the swr to a very minimum (Fig. 8-19).

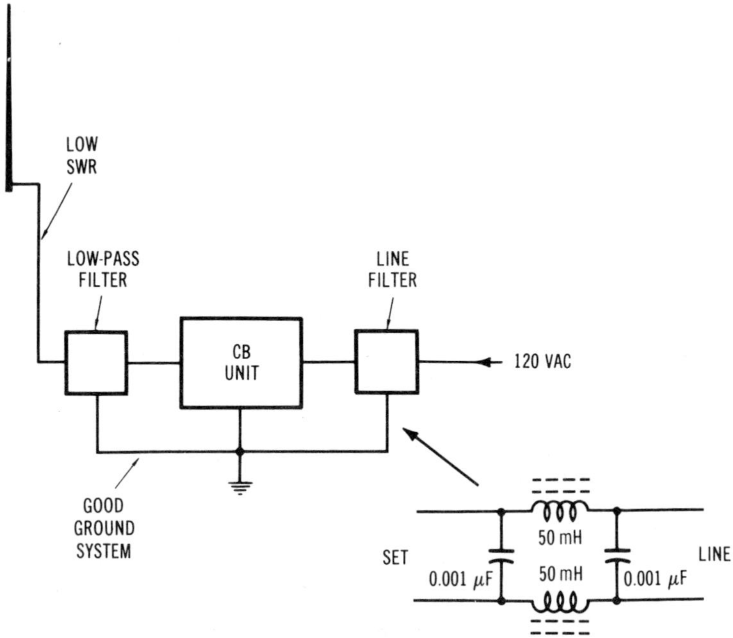

Fig. 8-19. Reduction of undesired radiation at transmitter.

The harmonic output of a CB transmitter can be troublesome. The second harmonic of the 27-MHz output of a CB transmitter falls on channel 2 of the television band. Third harmonic components fall on channels 5 and 6. To prevent this type of interference, it is important to keep the harmonic output of the transmitter low and, at the same time, reduce the pickup of the radio interference by the television receiver. Quite often the installation of a low-pass filter at the output of the CB transmitter will solve the problem. These filters are low cost and connect between the antenna transmission line and the transmitter. They have no adverse influence on the operation of the CB transmitter. However, they do offer high attenuation on frequencies above the 27-MHz band. Consequently the harmonics are suppressed and do not reach the transmission line and the antenna system.

On occasion harmonic and other spurious frequency components feed into the power line. An appropriate filter can be connected in the power line circuit (Fig. 8-19). Such a radio-frequency filter can also be added to the power line circuit of the receiver or other home electronic equipment that is troubled with radio interference.

Low-cost high-pass filters are available for receivers. These connect between the transmission line and the television receiver (Fig. 8-20). They are effective in blocking the fundamental frequency of a CB transmitter or any other low-frequency radio-frequency interference. At the same time there is little or no attenuation of the desired television signal components.

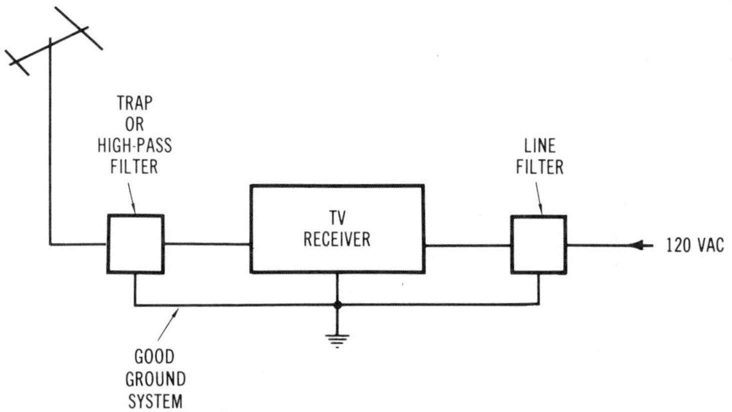

Fig. 8-20. Reduction of radio-frequency interference at receiver.

Resonant traps are also useful in blocking interference when the frequency of that interference is known. These are connected across or in series with sensitive inputs of television receiver, radio, or audio amplifier (Fig. 8-21). A parallel-resonant circuit would be placed in series with the signal path, while a series-resonant circuit would be connected directly across the input of the device. An adjustable trimmer capacitor can be used to tune the resonant circuit for maximum rejection of the undesired signal.

Radio-frequency signals can cause trouble even in an audio system. The rf energy can be rectified by any such nonlinear device as a transistor, integrated circuit, diode, etc. Proper grounding of transmitter as well as the electronic device being interfered with is important. Use large-diameter and as short as possible grounding leads. Power line filters can be of help.

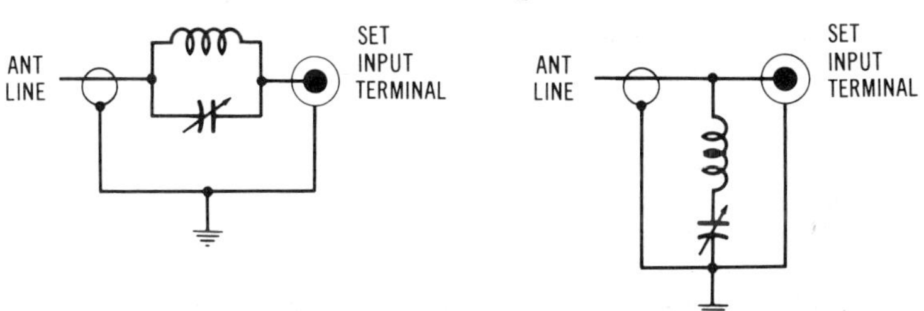

(A) Parallel resonant. (B) Series resonant.

Fig. 8-21. Use of resonant traps.

In the case of rectification in an audio amplifier a series inductor or a shunt capacitor can be of help in blocking the radio-frequency interference from a sensitive input (Fit. 8-22). A 0.001-μF capacitor would be a typical value. The value of the radio-frequency choke depends on the frequency of the interference. Typical values would be 1.5 mH up to 20 MHz, 500 μH for a range of frequencies between 20 and 50 MHz, and a value of 100 μH for a frequency range between 50 and 500 MHz.

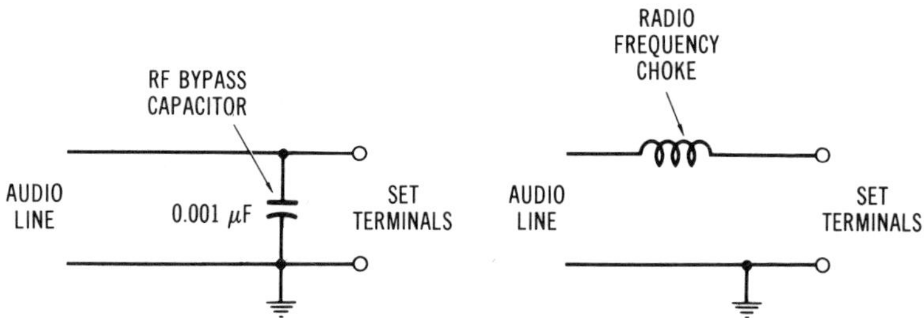

Fig. 8-22. Use of capacitor or rfc to block rf interference.

8-9. INTERMODULATION INTERFERENCE

Intermodulation interference has become a significant problem in areas crowded with two-way radio systems. These interference components can be generated by transmitter and receiver. Intermodulation components are generated when two or more signals are applied to a nonlinear device. These components are on differing frequencies and can cause interference with reception on other channels. If the interference is comparable or greater than the desired signal, the on-channel signal is interfered with severely. If the interference is weaker than the desired signal the interference will only be heard when the desired signal is not present.

When two or more transmitters are in close proximity to each other there is coupling between each transmitter and the others usually by way of the antenna system. Radiation from each transmitter enters the final amplifier and transmission system of the others. Intermodulation products are formed and those falling within the bandpass of the final amplifier and its output are radiated. This radiation can be on channels other than those to which the transmitters have been assigned.

Intermodulation products can also be formed at the receiver. Two or more strong off-channel signals may drive the rf amplifier into nonlinear operation or even intermodulate in the first mixer. These components can interfere with an incoming desired signal or can be heard in the output when no signal is arriving on the operating channel.

The example of Fig. 8 23 can be used to demonstrate the influence of intermodulation products. Normal operation exists between transmitter C and the receiver. Both are tuned to the desired channel of 160.35 MHz. Transmitters A and B are in close proximity to each other and transmit on the frequency assignments indicated. Even though their transmit frequencies are not the same as that of transmitter C, they can cause interference at the receiver, if they generate intermodulation components.

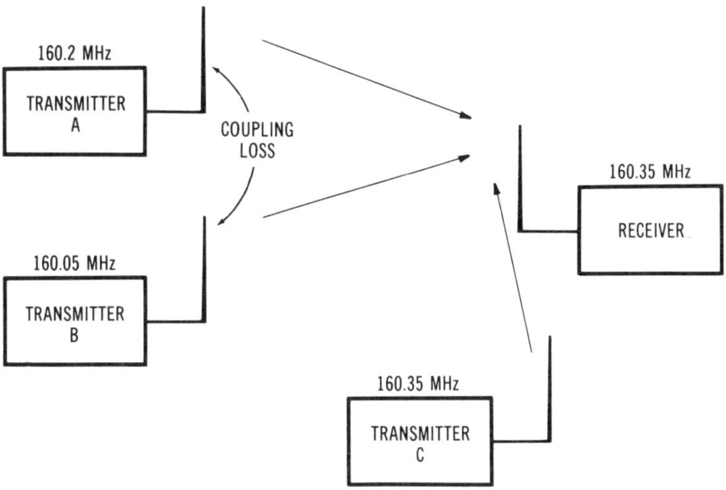

Fig. 8-23. Intermodulation basics.

Let us assume that there is a low coupling loss between the two transmitters. As a result some of the output from transmitter B enters the power output stage and antenna system of transmitter A. Likewise, some of the output of transmitter A enters transmitter B. Intermodulation components are generated in both transmitters. The difference frequency between the two transmitters is 0.150 MHz (160.200–160.050). This difference frequency subtracted from the carrier frequency of transmitter B produces an intermodulation frequency component of 159.900 MHz. This is radiated along with the desired 160.05 MHz signal whenever the two transmitters are operating simultaneously.

The difference frequency added to the frequency of transmitter A produces an output at 160.350 MHz. This component is transmitted along with the 160.2 MHz whenever the two transmitters are operating simultaneously. However, it should be noted that 160.35 MHz is the same frequency as transmitter C and is also the frequency to which the receiver is tuned. Thus it can be picked up by the receiving antenna and cause interference on the channel being used by transmitter C and the receiver.

It is important, if intermodulation components are to be eliminated, that there must be no significant coupling (high coupling loss) between the two transmitters in close proximity to each other. Thus, in the installation of transmitters that are near to each other, every attempt must be made to obtain as high as possible a coupling loss between the transmitters and antenna systems.

Intermodulation components can also be generated at the receiver. Assume that there is no exchange of signals between transmitters A and B. Therefore the intermodulation components are not generated. However, the transmit frequency of 160.05 and 160.02 MHz are transmitted and can possibly enter the receiver. If the two signals arrive simultaneously and reach the receiver in such strength that they can produce nonlinear operation of the receiver rf amplifier or reach the first mixer, they will interact with each other and generate the very same intermodulation frequency components. Thus there will be the same interference with the desired 160.35-MHz signal from transmitter C. This time the interference is the result of intermodulation products being formed at the receiver. This receiver intermodulation distortion can occur even though transmitters A and B are separated a great distance. If A and B come on the air simultaneously and are able to get into the receiver, the receiver will generate the intermodulation components.

It must be stressed that a variety of intermodulation frequency components can be generated. Most significant are the odd-order products. The even-order products (second, fourth, etc.) are so far removed in frequency that interfering effects do not occur. Most significant are the third- and fifth-order products. Although higher-order products can be generated they are usually of insignificant level.

The equations for the third- and fifth-order intermodulation products can be set down. Assume that A is the transmit frequency of transmitter A and B is the transmit frequency of transmitter B:

Second order $= 2A - B = 320.4 - 160.05 = 160.35$ MHz.
Second order $= 2B - A = 320.1 - 160.2 = 159.9$ MHz.
Third order $= 3A - 2B = 480.6 - 320.1 = 160.5$ MHz.
Third order $= 3B - 2A = 480.15 - 320.4 = 159.75$ MHz.

Of course if there are three or more transmitters in close proximity to each other there will be additional third- and fifth-order intermodulation products generated.

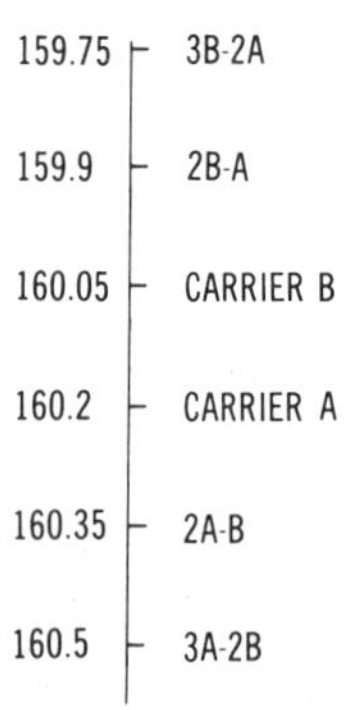

Fig. 8-24. Carriers and frequency location of 3rd- and 5th-order intermodulation products.

The relationship among the two carrier frequencies and the intermodulation products is shown in Fig. 8-24. Note that they are all separated from each other by 0.15 MHz which is the difference frequency between the carrier frequencies of the two transmitters. In other words the distortion products appear periodically above and below the frequencies of transmitter A and transmitter B.

8-10. INTERMODULATION REDUCTION

Transmitter intermodulation is reduced by increasing the coupling loss among transmitters. Refer to Fig. 8-25. A major step that can be taken is to separate the antenna systems as far as is practicable. The influence of antenna spacing is shown in Fig. 8-26. For example, if two antennas are vertically separated by 30 feet on the 160-MHz band there is an isolation of 55 dB. A horizontal separation of the same 30 feet results in an isolation of something more than 30 dB. This indicates that you can obtain a higher coupling loss by separating the antenna vertically rather than horizontally.

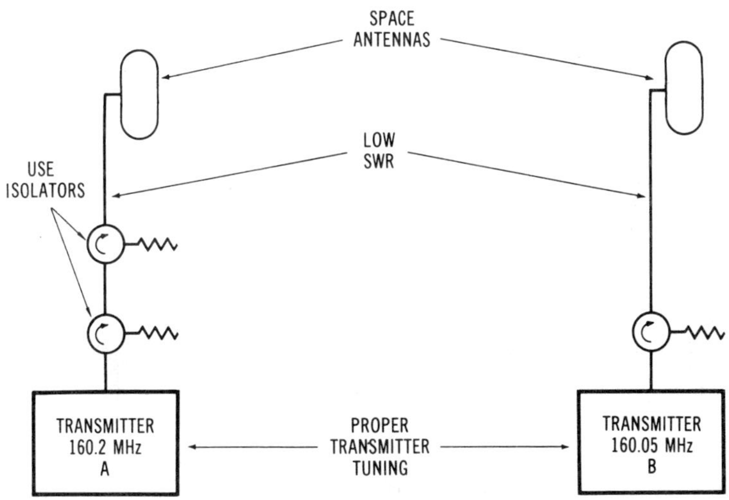

Fig. 8-25. Reduction of transmitter intermodulation.

Isolators are often used to increase the coupling loss as shown in Fig. 8-25. An isolator is usually a ferrite device that has a low impedance for the transfer of signal in one direction but an exceedingly high opposition to the transfer of signal in the opposite direction. Consequently the cross-coupling of frequencies is reduced without impeding the desired transmit signal on its way along the transmission line to the radiating antenna. Isolators usually include a terminating resistor across which the energy traveling in the wrong direction is dissipated. Thus the two isolators in the

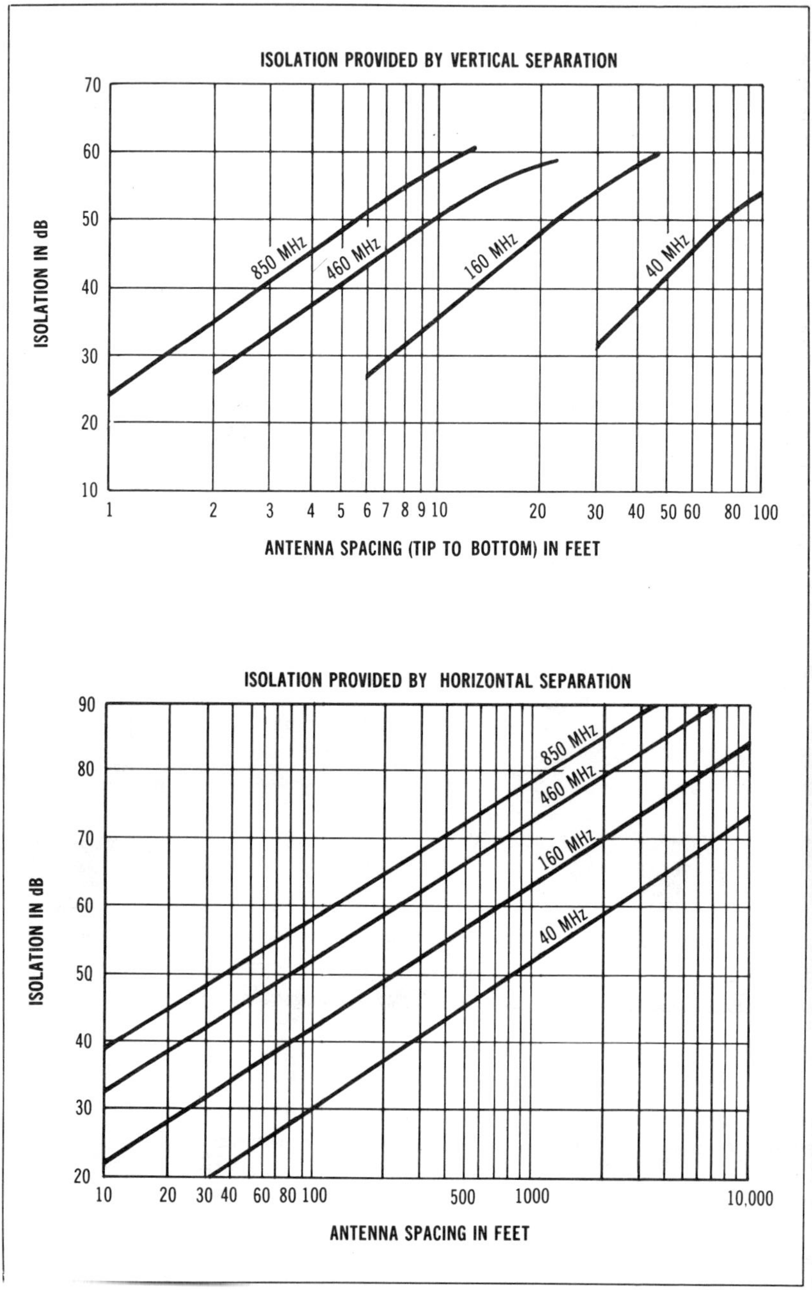

Fig. 8-26. Antenna isolation curves.

transmission-line system of transmitter A will allow a frequency of 160.2 MHz to reach the antenna while the coupling component of transmitter B on 160.05 MHz will not pass through the two isolators to transmitter A. In a similar manner the isolator in the antenna system of transmitter B will block any pickup of transmitter A signal and keep it from reaching transmitter B.

Intermodulation components can be reduced by proper tuning of both transmitters and antenna systems. Poor tuning and bad joints usually result in high swr readings. Thus matching and tuning are very important in maintaining a low swr on the transmission lines.

The generation of intermodulation products in the receiver is pretty much a matter of receiver design. Suppression filters connected between antenna system and receiver are a definite aid in blocking out those frequencies that are producing intermodulation components in the receiver. To set up such a filter arrangement it is of course necessary to first determine the incoming frequencies that are causing the receiver to generate intermodulation components. For example, two filters ahead of the receiver of Fig. 8-23, which are tuned to 160.05 MHz, would reduce the intermodulation components generated by the receiver. Of course, the filters must be tuned sharply and must not result in any serious attenuation of the desired 160.35-MHz incoming signal.

CHAPTER 9

Rules, Regulations, and Operating Practices

9-1. INTRODUCTION

Chapters 9 through 17 provide a question and answer coverage of some additional topics that are a part of the FCC syllabus. The subject matter also reviews previous chapters of the handbook as well. Additional coverage given includes basic electricity, vacuum tubes, basic and more advanced vacuum-tube circuits. These latter questions and the follow-up answers will help you judge your understanding of the topics listed under Section A and C of the syllabus. In studying these chapters try to answer each question before you read the follow-up discussion.

Chapter 9 is a thorough coverage of the topics presented in sections H and I of the syllabus. Stressed are key FCC rules, operating practice, radiotelephone distress procedures, and important FCC Rules and Regulations. The chapter concludes with the first set of review questions.

9-2. THE FEDERAL COMMUNICATIONS COMMISSION

The FCC was created by the Communications Act of 1934 for the purpose of regulating communications. Among the general powers given to the Commission is the authority to classify radio stations, prescribe the qualifications of operators, and to issue radio station and radio operator licenses.

9-3. RADIO STATION LICENSES

Radio stations come in many shapes and sizes. When the FCC talks about a "radio station" they don't just mean the local broadcast stations that transmit entertainment and news. A radio station from the FCC's point of view can be anything from a million-watt television station to the CB set next door. One thing most radio stations have in common is a station license issued by the FCC. With this station license the FCC gives the station authority to transmit.

In many ways a station license is like an automobile registration. Almost anyone may own a car but it may be used on a public road only after a registration has been issued by the state. This registration assigns a unique combination of letters and numbers to the vehicle, defines the purpose for which the vehicle is to be used (personal transportation, farming, commercial, etc.), and imposes a requirement that the vehicle be used only in accordance with a set of regulations, called a motor vehicle code.

Similarly, a radio station license assigns a unique station identification, states the purpose for which the station is being used (broadcasting, business, aviation, etc.) and requires that the licensee comply with the Federal Communications Commission Rules and Regulations. Except for certain low power stations which are clearly labeled as not requiring a license, all radio stations must be licensed before they are operated. Operators should check to be sure that the station they wish to operate has a valid station license before making any transmissions.

9-4. DISTRESS, URGENCY, AND SAFETY MESSAGES

A distress message indicates that the sending station is in grave and imminent danger and requests immediate assistance. Distress messages are preceded by the international distress call which is the word *MAYDAY* spoken three times.

An urgency message indicates that the sending station has a very urgent message concerning the safety of life or property. Urgency messages are preceded by the international urgency signal which is the words *PAN PAN* spoken three times.

A safety message indicates that the sending station is about to transmit a message concerning safety of navigation or give an important meteorological warning. Safety messages are preceded by the international safety signal which is the word *SECURITY* spoken three times.

9-5. PRIORITY OF COMMUNICATIONS

As a radio operator, you must give priority to communications in the following order:

First—Distress Messages.
Second—Urgency Messages.
Third—Safety Messages.
Fourth—Other messages.

You must give absolute priority to any distress messages or any communications relating to a distress. Control of distress messages is the responsibility of the persons in distress. If you hear a radio station in distress but are unable to help, and it appears that no one else has heard the distress message, the station licensee or the person in charge of your radio station (such as a ship's captain) may give you permission to retransmit the distress message.

9-6. STATION RECORDS

The FCC requires some stations to post the licenses of their operators at the station location. Many stations are also required to keep station logs as a record of the station's operation. These logs may not be erased or destroyed. Corrections to them may only be made by the operator who made the original entry.

Some radio stations are required to paint and light their transmitting tower for safety purposes. Requirements of this nature may be found on the station license.

9-7. RADIOTELEPHONE OPERATING PRACTICE

Each radiotelephone operator should know and abide by the rules and accepted procedures of operation. Courtesy and consideration are two very important factors in minimizing unnecessary interference and maintaining reliable communications on active channels. The radiotelephone license holder should be an example to operators of lower grade, and he should do everything possible to encourage proper operations in the service or services with which he is concerned.

A fine summary of radiotelephone operating practice, recommended by the FCC, follows:

A licensed radio operator should remember that the station he desires to operate should be licensed by the Federal Communications Commission. In order to prevent interference and to give others an opportunity to use the airways, he should avoid unnecessary calls and communications by radio. He should remember that radio signals normally travel outward from the transmitting station in many directions, and, therefore, these transmitted signals could be intercepted by unauthorized persons.

Before making a radio call, the operator should listen on the communications channel to ensure that interference will not be caused to communications which may already be in progress. At all times in radiocommunications the operator should be courteous. If your radio has a squelch capability, it should be defeated and the volume turned up to make certain the channel is not occupied before you transmit. If your radio has subaudible tone control, it should be defeated before making the radio call to make certain you cause no interference to communications already in progress. Defeat can be done with a switch or automatically when the microphone is removed from its hanger.

Station identification should be made clearly and distinctly so as to avoid unnecessary repetition of call letters and enable other stations to clearly identify all calls.

An operator normally exhibits his authority to operate a station by posting a valid operator license or permit at the transmitter control point.

While a radio transmitter is in a public place, it should at all times be either attended by, or supervised by, a licensed operator; or the transmitter should be made inaccessible to unauthorized persons.

A radio transmitter should not be on the air except when signals are being transmitted. The operator of a radiotelephone station should not press the push-to-talk button except when he intends to speak into the microphone. Radiation from a transmitter may cause interference even when voice is not transmitted.

When radiocommunications at a station are unreliable or disrupted due to static or fading, it is not a good practice for the operator to continuously call other stations in attempting to make contact, because his calls may cause interference to other stations that are not experiencing static or fading.

A radiotelephone operator should make an effort to train his voice for most effective radiocommunication. His voice should be loud enough to be distinctly heard by the receiving operator, and it should not be too loud, since it may become distorted and difficult to understand at the receiving station. He should articulate his words and avoid speaking in a monotone as much as possible. The working range of the transmitter is affected to some extent by the loudness of the speaker's voice. If the voice is too low, the maximum range of the transmitter cannot be attained; if the voice is too loud, the range may be reduced to zero due to the signals becoming distorted beyond intelligibility. In noisy locations the operator sometimes cups his hands over the microphone to exclude the extraneous noise. Normally, the microphone is held from 2 to 6 inches from the operator's lips.

It is important in radiotelephone communications that operators use familiar and well-known words and phrases in order to ensure accuracy and eliminate undue repetition of words. Some radio operating companies, services, networks, associations, etc., select and adopt standard procedure words and phrases for expediting and clarifying radiotelephone conversations. For example, in some services, *Roger* means "I have received all of your last transmission." *Wilco* means "Your last message received, understood, and will be complied with." *Out* or *Clear* means "This conversation is ended and no response is expected." *Over* means "My transmission is ended, and I expect a response from you." *Speak slower* means "Speak slowly." *Say again* means "Repeat;" and *Words twice* means "Give every phrase twice."

Often in radiotelephone communications a phonetic alphabet or word list is useful in identifying letters or words that may sound like other letters or words of different meanings. For example "group" may sound like "scoop" or "bridge" may sound like "ridge." A phonetic alphabet or word list consists of a list of 26 words, each word beginning with a different letter for identifying that particular letter. If the letters in "group" are represented in a phonetic word by *George*, *Roger*, *Oboe*, *Uncle*, and *Peter*. "Group" is transmitted as: "Group"; G as in George, R as in Roger, O as in Oboe, U as in Uncle, P as in Peter."

In making a call by radio, the call sign or name of the called station is generally given 3 times followed by the call letters of the calling station given 3 times.

In testing a radiotelephone transmitter, the operator should clearly indicate that he is testing, and the station call sign or name of the station, as required by the rules, should be clearly given. Tests should be as brief as possible.

If a radio station is used only for occasional calls, it is a good practice to test the station regularly. Regular tests may reveal defects or faults which, if corrected immediately, may prevent delays when communications are necessary. Caution should be observed by persons testing a station to make certain their test message will not interfere with other communications in progress on the same channel. Technical repairs or adjustments to radiotelephone communication stations are made only by or under the immediate supervision and responsibility of operators holding a general class radiotelephone license.

When a licensed operator in charge of a radiotelephone station permits another person to use the microphone and talk over the facilities of the station, he should remember that he continues to bear responsibility for the proper operation of the station.

If an operator wishes to determine the specification for obstruction marking and lighting of antenna towers he should look in Part 17 of the Rules and Regulations of the FCC. If he wishes to determine the specification for a particular station, he should examine the station authorization issued by the Commission.

9-8 RADIOTELEPHONE DISTRESS PROCEDURE

The international radiotelephone distress signal consists of the spoken expression **MAYDAY**. This signal shall be used to announce that the ship, aircraft, or other vehicle that sends the distress signal is threatened by serious and imminent danger and requests immediate assistance. The distress signal shall be followed by the distress messages containing the identity of the station in distress, its position, the nature of the distress, and the nature of the assistance requested.

The international radiotelephone urgency signal consists of the word **PAN**, spoken three times, and is to be used when the calling station has a very urgent message to transmit concerning the safety of the ship, aircraft, or other vehicle, or concerning the safety of some person on board or sighted from on board.

The international radiotelephone safety signal consists of the word **SECURITY** spoken three times, and announces that the station is about to transmit a message concerning the safety of navigation, or giving important meteorological warnings.

The distress *sent* by *radiotelephony* comprises:

1. The distress signal **MAYDAY** spoken three times;
2. The words **THIS IS**, followed by the identification of the mobile station in distress, the whole repeated three times.

Chart 9-1. Recommended Radiotelephone Distress Procedure

DISTRESS COMMUNICATIONS FORM

Instructions: Complete this form now (except for items 6 through 9) and post near your radiotelephone.

Speak **SLOWLY — CLEARLY — CALMLY**

1. Make sure your radiotelephone is on.
2. Select either *VHF Channel 16* (156.8 MHz) or *2182* kHz.
3. Press microphone button and say: **"MAYDAY — MAYDAY — MAYDAY."**
4. Say: **"THIS IS** ________, ________, ________,
   (your boat name, your boat name, your boat name)
   ________."
   (your call letters)
5. Say: **"MAYDAY:**________."
   (your boat name)
6. **TELL WHERE YOU ARE** (What navigational aids or landmarks are near?).
7. **STATE THE NATURE OF YOUR DISTRESS.**
8. GIVE NUMBER OF ADULTS AND CHILDREN ABOARD, **AND CONDITIONS OF ANY INJURED.**
9. **ESTIMATE PRESENT SEAWORTHINESS OF YOUR BOAT.**
10. **BRIEFLY DESCRIBE YOUR BOAT:**
    ________; ________ **FEET;**
    (State Registration No.; Length)
    ________**FEET;** ________; ________**HULL;** ________**TRIM;**
    (Draft; Type; Color; Color)
    ________**MASTS;** ________**POWER;** ________.
    (Number; Type; Horsepower; Construction Material)
    ____
    (Anything else you think will help rescuers to find you.)
11. Say: **"I WILL BE LISTENING ON *CHANNEL 16/2182.*"**
    (Cross out channel no. or frequency that does not apply)
12. End Message by saying: "THIS IS ________ OVER."
    (your boat name and call sign)
13. Release microphone button and listen: Someone should answer. **IF THEY DO NOT, REPEAT CALL, BEGINNING AT ITEM 3.** If there is still no answer, switch to another channel and begin again.

The distress call should be followed as soon as possible by the distress message, which comprises:

1. The distress signal **MAYDAY**;
2. The name of the ship, aircraft, or vehicle in distress;
3. Particulars of its position, the nature of the distress and the kind of assistance desired;

4. State number of adults and children aboard and conditions of any injured.
5. Any other information which might facilitate the rescue.

As a general rule, a ship shall signal its position in latitude and longitude, using figures for the degrees and minutes, together with one of the words **NORTH** or **SOUTH** and one of the words **EAST** or **WEST**.

As a general rule, and if time permits, an aircraft shall transmit in its distress message the following information: estimated position and time of the estimate; heading in degrees (state whether magnetic or true); indicated air speed; altitude; type of aircraft; nature of distress; type of assistance desired, and any other information which might facilitate the rescue (including the intention of the person in command, such as forced alighting on the sea or crash landing).

After the transmission by radiotelephony of its distress message, the mobile station may be requested to transmit suitable signals followed by its call sign or other identification, to permit direction-finding stations to determine its position. This request may be repeated at frequent intervals if necessary. In radiotelephony, the signal is made by holding the transmitter on the air for the specified period of time, without speaking into the microphone other than to identify the station.

Immediately before a crash landing or forced landing (on land or sea) of an aircraft, as well as before total abandonment of a ship or aircraft, the radio apparatus should be set for continuous emission, if considered necessary and circumstances permit.

A radiotelephone distress procedure as recommended by the Radio Technical Commission for Marine Services is given in Chart 9-1.

9-9. COMMUNICATIONS ACT

Sec. 301. It is the purpose of this Act, among other things, to maintain the control of the United States over all the channels of interstate and foreign public transmission; and to provide for the use of such channels, but not the ownership thereof, by persons for limited periods of time, under licenses granted by Federal authority, and no such license shall be construed to create any right, beyond the terms, condition, and periods of the license. No person shall use or operate any apparatus for the transmission of energy or communications or signals by radio (a) from one place in any Territory or possession of the United States or in the District of Columbia to another place in the same Territory, possession, or district; or (b) from any State, Territory, or possession of the United States, or from the District of Columbia to any other State, Territory, or possession of the United States; or (c) from any place in any State, Territory, or possession of the United States, or in the District of Columbia, to any place in any foreign country or to any vessel; or (d) within any State when the effects of such use extend beyond the borders of said State, or when interference is caused by such use or operation with the transmission of such energy, communications or signals from within said State to any place beyond its borders, or from any place beyond its borders to any place within said State, or with the transmission or reception of such energy, communications, or signals from and/or to places beyond the borders of said State; or (e) upon any vessel or aircraft of the United States; or (f) upon any other mobile stations within the jurisdiction of the United States, except under and in accordance with this Act and with a license in that behalf granted under the provisions of this Act.

9-10. SECRECY OF COMMUNICATIONS

Sec. 605. No person receiving or transmitting, or assisting in transmitting, any interstate or foreign communication by wire or radio shall divulge or publish the existence, contents, substance, purport, effect, or meaning thereof, except through authorized channels of transmission or reception, to any person other than the addressee, his agent, or attorney, or to a person employed or authorized to forward such communication to its destination, or to proper accounting or distributing officers of the various communicating centers over which the communications may be passed, or to the master of a ship under whom he is serving, or in response to a subpoena issued by a court of competent jurisdiction, or on demand of other lawful authority; and no person not being authorized by the sender shall intercept any communication and divulge or publish the existence, contents, substance, purport, effect, or meaning of such intercepted communication to any person; and no person not being entitled thereto shall receive or assist in receiving any interstate or foreign communication by wire or radio and use the same or any information therein contained for his benefit or for the benefit of another not entitled thereto; and no person having received such intercepted communication or having become acquainted with the contents, substance, purport, effect, or meaning of the same or any part thereof, knowing that such information was so obtained, shall divulge or publish the existence, contents, substance, purport, effect, or meaning of the same or any part thereof, or use the same or any information therein contained for his own benefit or for the benefit of another not entitled thereto: Provided, That this section shall not apply to the receiving, divulging, publishing or utilizing the contents of any radio communication broadcast, or transmitted by amateurs or others for the use of the general public, or relating to ships in distress.

9-11. RULES AND REGULATIONS

Commercial Radio Operators

13.64. Obedience to lawful orders.

All licensed radio operators shall obey and carry out the lawful orders of the master or person lawfully in charge of the ship or aircraft on which they are employed.

13.65 Damage to apparatus.

No licensed radio operator shall willfully damage, or cause, or permit to be damaged any radio apparatus or installation in any licensed radio station.

13.66 Unnecessary, unidentified, or superfluous communications.

No licensed radio operator shall transmit unnecessary, unidentified, or superfluous radio communications or signals.

13.67 Obscenity, indecency, profanity.

No licensed radio operator or other person shall transmit communications containing obscene, indecent, or profane words, language or meaning.

13.68 False Signals.

No licensed radio operator shall transmit false or deceptive signals or communications by radio, or any call letter or signal which has not been assigned by proper authority to the radio station he is operating.

13.69 Interference.

No licensed radio operator shall willfully or maliciously interfere with or cause interference to any radio communication or signal.

13.72 Exhibiting signed copy of application.

When a duplicate or replacement operator license or permit has been requested, or request has been made for renewal, or a request has been made for an endorsement, higher class license or permit, or verification card, the operator shall exhibit in lieu of the original document a signed copy of the application which has been submitted to the Commission.

Stations on Shipboard in the Maritime Services

83.352 Frequencies for use in distress and search and rescue operations.

(a) The frequency 2182 kHz is the international distress frequency for radiotelephony: it shall be used for this purpose by ship, aircraft, and survival craft stations operating in the authorized bands between 1605 and 4000 kHz when requesting assistance from the maritime services. The frequency 156.8 MHz is the international distress, safety and calling frequency for radiotelephony for stations of the maritime mobile service when using frequencies in the authorized bands between 156 and 174 MHz.

(b) The frequency 121.5 MHz (using class A9 emission) is available for assignment to survival craft stations for radio-beacon purposes. The frequency 121.5 MHz (using class A9 emission) is available for assignment to stations for facilitating search and rescue operations. The frequencies 121.5 MHz and 123.1 MHz (class A3 emission) are available for assignment to ship stations for scene of action search and rescue operations between ships and aircraft. The frequency 121.5 MHz will be assigned to ship stations for search and rescue operations only if the frequency 123.1 MHz is also authorized. Communication in support of search and rescue operations is permitted on the frequency 121.5 MHz only when communications on the frequency 123.1 MHz or other vhf frequencies is not practicable. Ships and aircraft engaged in such communications on 121.5 MHz should shift to 123.1 MHz as soon as possible.

(c) The frequency 243 MHz (class A9 emission only) is available to EPIRB stations for facilitating search and rescue operations.

83.353 Frequencies for calling.

The frequency 2182 kHz is the international general calling frequency for radiotelephony: it may be used for this purpose by ship and aircraft stations operating in the maritime mobile service in the bands between 1605 and 4000 kHz. The frequency 156.8 MHz is the international distress, safety and calling frequency for radiotelephony for stations of the maritime mobile service when using frequencies in the authorized bands between 156 and 174 MHz. In addition, these frequencies may be used for transmission of:

(a) The international urgency signal, and very urgent messages (preceded by this signal) concerning the safety of a ship, aircraft, or other vehicle, or the safety of some person on board or within sight of such ship, aircraft, or vehicle.

(b) The international safety signal, and messages (preceded by this signal) concerning the safety of navigation or giving important meteorological warnings; however, safety messages shall be transmitted, when practicable, on a working frequency after a preliminary announcement on 2182 kHz.

(c) Brief radio operating signals.

(d) Brief test signals in accordance with the provisions of 83.365, as may be necessary to determine whether the radio transmitting equipment of the station is in good working condition on this frequency.

AVIATION SERVICES

87.183 Frequencies available.

(f) 121.5 Megahertz: This is a universal simplex clear-channel frequency for use by aircraft in distress or condition of emergency. Except for transmissions of signals by an aircraft station operated with only an emergency locator transmitter using an A9 emission, it will not be assigned to aircraft unless other frequencies are assigned and available for normal communications. The channel is available, as follows:

(1) For emergency communications when circumstances beyond the control of the pilot prevent communication between the aircraft and ground stations on regularly assigned channels.

(2) For emergency direction finding purposes.

(3) For establishing air-to-ground contact by aircraft in distress, emergency or when lost.

(4) In connection with search and rescue operations, to provide a common channel for aircraft (either civil or military) not equipped to transmit on 123.1 MHz. This includes communications between aircraft, and between aircraft and ground stations. Stations having the capability should change to 123.1 MHz as soon as practicable.

(5) To provide a common frequency for survival communications and for survival radio beacons (emission A2).

(6) For air/ground communications between aircraft and ocean vessels for safety purposes when service on other vhf channels is not available.

(g) 3023 kHz, 5680 kHz, 122.9 MHz, and 123.1 MHz: When engaged in search and rescue activities in accordance with subpart K of this part, these frequencies may be used by aircraft stations for:

(1) Air-to-air communications with other aircraft engaged in search and rescue activities.

(2) Air-to-ground communications with aeronautical search and rescue stations.

(3) The frequencies 3023 and 5680 kHz may be used by aircraft and ship stations for search and rescue scene-of-action coordination purposes, including communications between stations and participating land stations. Ship stations communicating with aircraft stations shall employ 2.83J emission.

Private Land Mobile Radio Services

90.113 Station authorization required.

No radio transmitter shall be operated in the services governed by this part except under and in accordance with a proper authorization granted by the Commission.

90.135 Modification of license.

(a) The following changes in authorized stations require an application for modification of license:

(1) Any change in frequency.

(2) Any change in type of emission.

(3) Any increase in output power beyond that authorized.

(4) Any increase in antenna height beyond that authorized.

(5) Any increase in the number of base transmitters beyond that authorized.

(6) Any change in the authorized location of the base or fixed transmitter or the area of mobile operation.

(7) Any change in ownership, control, or corporate structure.

(8) Any change in class of station.

(b) The following changes do not require an application for modification of license:

(1) Change in mailing address of licensee.

(2) Change in name only of licensee, without changes in the ownership, control, or corporate structure.

(3) Change in the number and location of station control points, or of control stations meeting the requirement of 90.119(a) (2) (ii).

(4) Change in the number of mobile units for stations operating below 470 MHz; or in the number of mobile units operated by Radiolocation Service licensees.

(5) Any other change not listed in paragraph (a) of this section.

(6) In case of a change listed in subparagraph (1) or (2) of this paragraph, the licensee shall promptly notify the Commission of such change. The notice, which may be in letter form, shall contain the name and address of the licensee as they appear in the Commission's records, the new name or address, the call signs and classes of all radio stations authorized to the licensee under this part and the radio service in which each station is authorized. The notice shall be sent to the secretary, Federal Communications Commission, Washington, DC 20554, and to the Engineer in Charge of the Radio District in which the station is located, and a copy shall be maintained with the license of each station until a new license is issued.

(7) In the case of a change listed in subparagraph (3), (4) or (5) of this paragraph the licensee shall promptly notify the Commission within 30 days of the change. The notice shall be filed on the appropriate application form (FCC Form 400 or Form 425), and shall be sent to the Secretary, Federal Communications Commission, Washington, D.C. 20554.

90.135 (b) (3) amended eff. 8-1-79; V(79)-1

90.145 Special temporary authority.

(a) In circumstances requiring the temporary use of radio facilities, the Commission may issue special temporary authority for new or modified operations. A request for special temporary authority may be made in letter form signed in accordance with 90.125 of this part. It should be submitted, in duplicate, at least 10 days prior to the date of the proposed operation. However, in cases of emergency involving danger to life or property, or due to damage to equipment, the request may be made by telephone or telegraph under the condition that a letter request is submitted within the following ten days. All requests for special temporary authority shall be clear and complete within themselves and shall not rely on any pending application.

(b) Every request for special temporary authority should contain the following information:

(1) Name and address of the applicant;

(2) Need for special action, including a description of any emergency or damage of equipment;

(3) Type of operation to be conducted (such as field test, dispatching etc.);

(4) Purpose of operation;

(5) Times and dates of operation;

(6) Class of station and name of radio service;

(7) Location, including geographical coordinates if known, of transmitter and/or mobile area of operations;

(8) Number of fixed transmitters and number of mobile units;

(9) Operating frequency;

(10) Output power of the transmitter;

(11) Type of emission;

(12) Description of antenna, including height above ground, and power gain;

(13) Statement of eligibility for a radio service under this part.

90.155 Time in which station must be placed in operation.

(a) All stations authorized under this part must be placed in operation within 8 months from the date of grant or the authorization shall be invalid and must be returned to the Commission for cancellation.

(b) For local government entities only, a period longer than eight months for placing a station in operation may be authorized by the Commission on a case-by-case basis, where the applicant submits a specific schedule for the completion of each portion of the entire system, along with a showing that the system has been approved and funded for implementation in accordance with that schedule.

90.157 Discontinuance of station operation.

If a station licensed under this part discontinues operation on a permanent basis, the licensee shall forward the station license to the Commission for cancellation. A copy of the request for cancellation shall be sent to the Engineer in Charge of the district in which the station is located. For the purposes of this section, any station which has not operated for 1 year or more is considered to have been permanently discontinued.

90.203 Type acceptance required.

(a) Except as specified in paragraph (b) of this section, each transmitter utilized for operation under this part and each transmitter marketed as set forth in section 2.803 (of part 2) must be of a type which is included in the Commission's current Radio Equipment List as type accepted for use under this part; or, be of a type which has been type accepted by the Commission for use under this part in accordance with the procedures in subparagraph (2) of this paragraph.

(1) The Commission periodically publishes a list of equipment entitled "Radio Equipment List, Equipment Acceptable for Licensing." Copies of this list are available for public reference at the Commission's offices in Washington, D.C., and at each of its field offices. This list includes type accepted equipment and, also, until such time as it may be removed by Commission action, other equipment which appeared in this list on May 16, 1955.

(2) Any manufacturer of radio transmitting equipment (including signal boosters) to be used in these services may request type acceptance for such equipment following the procedures set forth in subpart J of part 2 of this chapter. Type acceptance for an individual transmitter or signal booster also may be requested by an applicant for a station authorization by following the procedure set forth in part 2 of this chapter. Such equipment if type accepted will not normally be included in the Commission's "Radio Equipment List" but will be individually enumerated on the station authorization.

90.215 Transmitter requirements.

(a) The licensee of each station shall employ a suitable procedure to determine that the carrier frequency of each transmitter authorized to operate with an output power is excess of two watts is maintained within the tolerance prescribed in 91.213. This determination shall be made, and the results entered in the station records in accordance with the following:

(1) When the transmitter is initially installed;

(2) When any change is made in the transmitter which may affect the carrier frequency or its stability.

(b) The licensee of each station shall employ a suitable procedure to determine that each transmitter authorized to operate with an output power in excess of two watts does not exceed the maximum figure specified on the current station authorization. On authorizations stating only the input power to the final radio-frequency stage, the maximum permissible output power is 75 percent for frequencies below 25 MHz and 60 percent of the input power for frequencies above 25 MHz. If a non-DC final radio-frequency stage is utilized, then the output power shall not exceed 75 percent of the input power. This determination shall be made, and the results thereof entered into the station records, in accordance with the following:

(1) When the transmitter is initially installed;

(2) When any change is made in the transmitter which may increase the transmitter power input.

(c) The licensee of each station shall employ a suitable procedure to determine that the modulation of each transmitter, which is authorized to operate with an output power in excess of two watts, does not exceed the limits specified in this part. This determination shall be made and the following results entered in the station records, in accordance with the following:

(1) When the transmitter is initially installed;

(2) When any change is made in the transmitter which may affect the modulation characteristics.

(d) The determinations required by paragraphs (a), (b), and (c) of this section may, at the opinion of the licensee, be made by a qualified engineering measurement service, in which case the required record entries shall show the name and address of the engineering measurement service as well as the name of the person making the measurements.

(e) In the case of mobile transmitters, the determinations required by paragraphs (a) and (c) of this section may be made at a test or service bench: Provided, That the measurements are made under load conditions equivalent to actual operating conditions; and provided further, that after installation in the mobile unit the transmitter is given a routine check to determine that it is capable of being received satisfactorily by an appropriate receiver.

90.443 Content of station records.

Each licensee of a station in these services shall maintain records in accordance with the following:

(a) For all stations, the results and dates of the transmitting measurements required by 90.215 of this part and the name of the person or persons making the measurements.

(b) When service or maintenance duties are performed on stations, the responsible operator shall sign and date an entry in the station record giving:

(1) Pertinent details of all duties performed by and under the supervision of the operator;

(2) The operator's name and address, and;

(3) The class, serial number and expiration date of the operator's license. Information stipulated in subparagraphs (2) and (3) of this paragraph need be entered only once in the station record at any station where the responsible operator is employed on a full time basis and an operator's license is properly posted.

(c) For stations whose antenna or antenna supporting structure must be illuminated, the licensee will record:

(1) The time the tower lights are turned on and off each day, if manually controlled.

(2) The time the daily check of proper operation of the tower lights was made, if an automatic alarm system is not provided.

(3) Any observed or otherwise known failure of a tower light by indicating:
(i) The nature of such failure.
(ii) The date and time the failure was observed or otherwise noted.
(iii) The date, time and nature of adjustments, repairs or replacements made.
(iv) The identification of the Flight Service Station (Federal Aviation Administration) notified of the failure or any code or rotating beacon light not corrected within 30 minutes, and the date and time such notice was given.

(v) The date and time notice was given to the Flight Service Station of the Federal Aviation Administration that the required illumination was resumed.

(4) The completion of the 3-month periodic inspection required by 90.441 of this part by indicating:

(i) The date for the inspection and the condition of all tower lights and associated tower lighting control devices, indicators and alarm systems.

(ii) Any adjustments, replacements, or repairs made to insure compliance with the lighting requirements and the date such adjustments, replacements, or repairs were made.

90.445 Form of station records.

(a) Station records shall be kept in an orderly manner, and in such detail that the data required are readily available. Key letters or abbreviations may be used if proper meaning or explanation is set forth in the record.

(b) Each entry in the records of each station shall be signed by a person qualified to do so, having actual knowledge of the facts to be recorded.

(c) No record or portion thereof shall be erased, obliterated, or wilfully destroyed within the required retention period. Any necessary correction may be made only by the person originating the entry, who shall strike out the erroneous portion, initial the correction made, and indicate the date of correction.

90.447 Retention of station records.

Records required by this part shall be retained by the licensee for at least one year.

9-12. REVIEW QUESTIONS

9-1. **What should an operator do when he leaves a transmitter unattended?**—The transmitter should be made inaccessible to unauthorized persons.

9-2. **What are the meanings of clear, out, over, roger, words twice, repeat, and break?**—Refer to the appropriate words in the Section 9-7. "Repeat" means to transmit message or section of message again to make certain it has been copied correctly. "Break" means that the transmitting station will pause briefly for an acknowledgment from the station copying his message, or that the receiving station wishes to break in on the transmitting station to request a repeat.

9-3. **How should a microphone be treated when used in noisy locations?**—The operator can cup his hand over the microphone to exclude extraneous noise.

9-4. **What may happen to the received signal when an operator has shouted into a microphone?**—The range of the transmission may be decreased, because the signal may become distorted beyond intelligibility.

9-5. **Why should radio transmitters be off when signals are not being transmitted?**—Radiation from the transmitter may cause interference even though voice or other information is not being transmitted.

9-6. **Why should an operator use well-known words and phrases?**—It is advisable to ensure accuracy and save time from undue repetition of words.

9-7. **Why is the station call sign transmitted?**—So that other stations may clearly identify all calls. Call signs must also be given at prescribed times according to the FCC Rules related to the particular radio service.

9-8. **Where does an operator find specifications for obstruction marking and lighting where required for the antenna towers of a particular radio station?**—Data may be found in Part 17 of the FCC Rules and Regulations.

9-9. **What should an operator do if he hears profanity being used at his station?**—He should take steps to prevent the profanity from going out on the air.

9-10. **Who bears the responsibility if an operator permits an unlicensed person to speak over his station?**—The licensed operator continues to bear responsibility for the proper operation of the station.

9-11. What is meant by a phonetic alphabet in radiotelephone communications?—It is a list of words helpful in identifying letters or words that may sound like other letters or words that have different meanings.

9-12. How does the licensed operator of a station normally exhibit his authority to operate the station?—A valid operator license must be displayed at the transmitter control point.

9-13. What precautions should be observed when you are testing a station on the air?—Be certain testing does not interfere with other communications in progress on the same channel. Clearly indicate that you are testing.

9-14. What information must be contained in distress messages? What procedure should be followed by a radio operator in sending a distress message? What is a good choice of words to be used in sending a distress message?—Distress message should contain the following information:

a. Distress call.
b. Name of ship.
c. Geographical position.
d. Nature of distress.
e. What kind of assistance is needed.
f. Any additional information that will help in the rescue, such as type of ship, color, length, etc.

The international radiotelephone distress signal is the word "MAYDAY." Words should be simple and to the point. On advisement, it may be necessary to transmit suitable signals followed by call sign or name to permit direction-finding stations to determine position.

The distress message can be repeated at intervals until an answer is received. If there are no answers, the message may be repeated on any other available frequency on which attention might be attracted.

9-15. What do distress, urgency, and safety signals indicate? What are the international urgency, safety, and distress signals? In the case of a mobile radio station in distress, what station is responsible for the control of distress-message traffic?—A distress signal indicates that a mobile station is threatened by grave and eminent danger and requests immediate assistance. The urgency signal indicates that the calling station has a very urgent message to transmit concerning the safety of a ship, aircraft, or other vehicle, or the safety of a person. The safety signal indicates that the station is about to transmit a message concerning the safety of navigation or giving important meteorological warning. These signals are as follows:

	Radiotelegraph	Radiotelephone
Distress	S O S	MAYDAY
Urgency	X X X	PAN
Safety	T T T	SECURITY

Whenever possible the mobile station in distress is responsible for the control of distress-message traffic.

9-16. In regions of heavy traffic why should an interval be left between radiotelephone calls? Why should a radio operator listen before transmitting on a shared channel? How long may a radio operator in the mobile unit service continue attempting to contact a station which does not answer?—An interval should be left between radiotelephone calls to permit the shared use of the channel. An operator should listen before transmitting on a shared channel to make certain that he does not interfere with communications in progress. A call shall not continue for more than 30 seconds. If the called station does not reply, a second call should not be made until after an interval of two minutes. If there is no reply to a call sent three times at intervals of two minutes, the calling shall cease and shall not be renewed until after an interval of fifteen minutes. However, if there is no reason to believe that harmful interference will be caused, the call sent three times at intervals of two minutes may be repeated after a pause of not less than three minutes.

9-17. Why are test transmissions sent? How often should they be sent? What is the proper way to send a test message? How often should the station call sign be sent?—If the station is used at infrequent intervals, test transmission can be sent out to ascertain if the transmitter continues to operate in a normal manner. If the station has not been in operation, the transmitter can be checked with a test transmission according to an adopted schedule (once each 24 hours is often considered appropriate). When tests are required, precautions should be taken to prevent any emission that will cause harmful interference. Radiation must be reduced to the lowest practicable value and entirely suppressed if possible. The licensed operator shall listen carefully before test emission so as not to interfere with transmission in progress.

The official call sign of the testing station followed by the word "test" shall be announced as a warning that test emissions are to be made. If any other station transmits the word "wait," testing shall be suspended for an appropriate interval of time.

During tests the operator then announces the word "testing" followed by a voice transmission test by numerical count. Test signals should be transmitted that do not conflict with normal operating signals or that will actuate automatic alarms. The test signal shall have a duration not exceeding 10 seconds.

At the conclusion of the test a voice announcement of the official call sign, name of the ship or station, and general location is to be made. It is customary to give the call sign three times. The test transmission shall not be repeated until at least one minute has elapsed; on the frequency 2182 kHz or 156.8 MHz a period of at least five minutes shall elapse before another test transmission is made.

9-18. **In the mobile service, why should radiotelephone messages be as brief as possible?**—Transmissions should be brief to give other stations a chance to use the channel and to minimize interference.

9-19. **How does the licensed operator of a ship station exhibit his authority to operate a station?**—The original license should be exhibited at a conspicuous place at the principal location on board ship. For certain stations of a portable nature and when the operator holds a restricted radiotelephone operator permit, the person may have on his person either the operator license or a verification card.

9-20. **What is the difference between calling and working frequencies?**—Calling frequencies are used for transmissions from a station solely to secure the attention of another station for a particular purpose. A working frequency is a frequency that is used strictly for carrying on radio-communications for a purpose other than calling by any station or stations using telegraphy, telephony, or facsimile.

CHAPTER 10

Alternating and Direct Current

10-1. DEFINITIONS AND UNITS

10-1-1. By what other expression may a difference of potential or electromotive force be described?—It may be described as voltage or voltage drop. The unit of electromotive force is the volt.

10-1-2. By what other expression may an electric-current flow be described?—It is a motion of electrons. The unit of electric-current flow is the ampere.

10-1-3. Explain the relationship between the physical structure of the atom and electric-current flow.—An atom consists of a nucleus (composed of protons and neutrons) and one or more encircling electrons in one or more shells. With the application of sufficient external energy, electrons can be removed from the outer shell and made to move from atom to atom. *The controlled drift of electrons is referred to as electric-current flow.*

10-1-4. With respect to other electrons, what is the difference between conductors and nonconductors?—In a conductor, *the electron or electrons of the outer shell (valence electrons)* of the atomic structure can, under the proper stress, be made to *move freely from atom to atom.* The outer electrons of a nonconductor (insulator) are bonded more closely to each other; electrons are not exchanged freely with neighboring atoms, and there is no, or a minimum, current flow.

An ion is a charged atom or grouping of atoms. If an electron has been removed from the basic atomic structure it is called a positive ion; if an electron has been added, it is called a negative ion.

10-1-5. What is the difference between electrical power and electrical energy? In what unit is each expressed?—Electrical energy is the latent or stored ability to do work electrically. It is potential energy. Electrical power represents this stored energy in action. This is kinetic energy. In other words, power is energy in motion.

The unit of electrical power is the watt; the unit of electrical energy is the joule. One watt equals one-joule-per-second. The practical unit of electrical energy is the kilowatt-hour or 3600 joules.

10-1-6. What is the relationship between impedance and admittance? Between resistance and conductance?—Conductance is the reciprocal of resistance; admittance is the reciprocal of impedance. Impedance is measured in ohms; admittance, in ohms. Resistance is measured in ohms; conductance, in mhos.

10-1-7. Define the term reluctance.—**Reluctance** is the opposition that a magnetic path offers to the flow of magnetic lines of force. Reluctance is also sometimes explained as magnetic resistance.

10-1-8. What is permeance?—Permeance is the reciprocal of reluctance and is a measure of the ease with which an inductance allows the passage of magnetic lines of force.

10-1-9. Define the term permeability.—Permeability refers to how well one type of material passes magnetic lines as referred to another. Usually the permeability of air is assigned a value of 1.

10-1-10. Define hysteresis.—Hysteresis is a flux lag in magnetic material that does not follow exactly in step with the increase and decrease of the applied magnetizing force.

10-1-11. What is a joule?—The joule is the unit of energy. In electricity it represents 1 watt-second.

10-1-12. What is a coulomb?—The coulomb is a unit of electric charge and is defined as the charge that passes a point in one second as a result of a steady current of 1 ampere.

10-2. ELECTRONIC COMPONENTS

10-2-1. What is the relationship between wire size and resistance of the wire?—The resistance of a wire of a given length and composition varies inversely with the square of its diameter. This is to say, if the diameter of the wire is doubled, its resistance will be ¼ of the original value. Resistance increases directly with the length of a wire of given diameter.

The circular mil is often used to designate wire size. (Circular mil is a unit of area corresponding to the area of a circle with a diameter of 1 mil or 0.001 inch.) Since it is a unit of area, one can state: Wire resistance increases directly with length and inversely with its diameter in circular mils.

The lower the wire gauge size of a wire, the greater is its circular mil area. A No. 20 wire has a greater circular mil area than does a No. 22 wire.

10-2-2. What is skin effect? How does it affect the resistance of conductors at the higher frequencies?—The higher the frequency the more electrical energy tends to flow on or near the surface of a conductor. Hence, there is no uniform flow of rf energy throughout the cross section of the conductor. This is called skin effect and it causes a rise in the rf resistance of a conductor with frequency.

Since the high-frequency energy travels mainly at the surface, copper and aluminum tubing are popular. Conductor weight can be kept down by the tubular construction and the surface is of a large effective area and low resistance. The large conducting surface areas presented by thin metal strips of metal sheets, plates, or surfaces coated with silver and other low-resistance metals are attractive as conductors, resonant circuits, and filters at vhf and uhf frequencies.

10-2-3. Name four materials that make good insulators at low frequencies but not at uhf or above.—Insulators such as glass, porcelain, bakelite, rubber, cloth, paper, and wood are often found inadequate for good insulation at uhf and higher frequencies. Ceramics and various plastics are good insulators for vhf and uhf.

10-2-4. What is the value and tolerance of a resistor which is color coded (left-to-right) red, black, orange, gold?—The resistance is 20,000 ohms and the tolerance is 5%. Refer to color codes.

X **10-2-5. What would be the value, tolerance, and voltage rates of an EIA mica capacitor whose first row colors were (from left to right) white, red, green; second row: green, silver, red?**—The capacitance is 2500 pF (0.0025 μF), the tolerance is 10%, and the dc working voltage is 500 volts.

10-2-6. Explain the operation of a break-contact relay; a make-contact relay.—When a break-contact relay is energized the movable contact moves away from the fixed contact and an associated circuit is opened. When a make-contact relay is energized the movable contact is pulled against the fixed contact and an associated circuit is closed. A relay provides a means of opening and closing electronic circuits.

10-2-7. List three precautions which should be taken in soldering electrical connections to ensure a permanent junction.—(1) The connections to be joined together should be clean and mechanically firmed (2) the two surfaces to be joined should be properly heated (heat to connection and then solder to connection); and (3) a rosin-core solder should be used to ensure the making of a low-resistance, noncorrosive joint.

10-2-8. Explain the theory of molecular alignment as it affects the magnetic properties of materials.—Materials according to their magnetic properties are influenced differently by an external magnetic field. The encircling and spinning motions of the atomic electrons are influenced by the field in such a manner that the atom itself then possesses magnetic qualities. According to the spacings among atoms and the molecular make-up of the material, these atomic magnetic fields become additive in small sections called domains. If the domains are so widely spaced that their fields are not additive, the material is classified as a nonmagnetic material.

In other materials there are many domains, and they line up in a definite orderly fashion. The number of aligned domains increases with the strength of the external magnetic field. Such a material is classified as a magnetic material.

Some magnetic materials lose their magnetism as soon as the external magnetic field is removed. Others may retain the alignment of the domain a length of time after the external magnetic field is removed. *This is called residual magnetism.* The latter magnetic materials are said to have a high retentivity. If the retentivity of the magnetic material is low, the domains return to a random orientation as soon as the field is removed. With the application of an external field, the domains will again align themselves.

The stronger the magnetic field, the greater is the number of aligned domains. The material is said to be saturated when the strength of the external field is such that practically all of the domains are aligned.

The force that produces a magnetic field is called *magnetomotive force.* The unit of magnetomotive force is the gilbert.

10-2-9. What is the relationship between the inductance of a coil and the number of turns of wire in the coil; the permeability of the core material used?—The inductance increases with the number of turns. The inductance also increases with the permeability of the core material. In the common iron-core inductor, the inductance increases directly with the permeability of the core and with the number of turns squared.

For a given number of turns and dimensions, a coil with an iron core will have a higher inductance than one with an air core because the permeability of iron is greater than that of air.

10-2-10. What factors influence the direction of magnetic lines of force produced by an electromagnet?—The two principal factors are the direction of the winding and the direction of the current through the wires of the coil.

10-2-11. Which factors determine the amplitude of the emf induced in a conductor that is cutting magnetic lines of force?—These factors are the number of lines being cut, the strength of the magnetic field, the speed with which the conductor cuts the lines, and the angle of the conductor relative to the magnetic lines of force.

10-2-12. In what way does an inductance affect the voltage-current phase relationship of a circuit? Why is the phase of a circuit important?—The presence of inductance causes the current to lag the applied voltage. The higher the series inductive reactance is in comparison to the resistance of the circuit, the greater is the lag of the current (up to a maximum of near 90° when the inductive reactance is much greater than the resistance of the circuit). The phase relation between voltage and current, or *phase angle,* is important because it determines the power factor of the circuit (ratio of true power to apparent power).

$$\text{Power Factor} = \frac{\text{True Power}}{\text{Apparent Power}}$$

$$\text{True Power} = \text{Apparent Power} \times \cos\theta$$

(θ is the phase angle).

10-2-13. Explain how self- and mutual inductance produce transformer action.—The current in the coil produces a surrounding magnetic field. When there is any change in this current, there is a resultant change in the magnetic field. The changing magnetic field, in turn, induces a voltage (counter emf) across the coil which is opposite in polarity to the applied voltage and, therefore, opposes the change in current. The opposition the coil offers to the change in current is called *self-inductance.* Electrical energy is contained in the associated magnet field.

If the changing magnetic field cuts the turns of an adjacent coil (positioned near to the coil to which the electrical energy is applied) a voltage will be induced into the second coil by any mutual flux lines. If a load is placed on the output of the second coil there will be a transfer of power between the primary and secondary via the mutual flux lines. This so-called *mutual inductance* is related to the extent of the flux linkages between the two coils.

In summary, self-inductance has to do with the flux associated with a coil to which energy is applied, while mutual inductance is concerned with the flux of one coil as it is linked to a second coil. It is this linkage that produces transformer action.

The equation for mutual inductance is

$$M = K\sqrt{L1 \cdot L2}$$

where,
M is the mutual inductance in henrys,
K is the coefficient of coupling,
L1 is the self-inductance in henrys,
L2 is the self-inductance in henrys.

10-2-14. What does coefficient of coupling mean?—The coefficient of coupling refers to the extent of coupling between the primary and secondary. It indicates that fraction or percentage of the total flux of one coil that is linked to a second coil.

The coefficient of coupling is high (slightly less than unity) for cores of high permeability. With air-core coupling the coefficient of coupling is usually substantially less than 10%.

$$K = \frac{M}{\sqrt{L1 \cdot L2}}$$

where,
K is the coefficient of coupling,
M is the mutual inductance in henrys,
L1 is the self-inductance in henrys,
L2 is the self-inductance in henrys.

10-2-15. How is power lost in an iron-core transformer? In an air-core transformer?—The two dominant iron-core transformer losses are hysteresis and eddy-current losses. Hysteresis is a heat loss, with power being dissipated in altering the orientations of the magnetic domains in the magnetic material. The eddy-current loss is a resistive loss that is a result of circulating currents within the core. Such currents are induced by the changing magnetic flux.

In an air-core transformer, power is lost as a result of the IR drop in the windings. The coefficient of coupling in the air-core transformer is low, and power is sacrificed because of the low percentage of the primary flux that links primary and secondary. There is much so-called leakage flux having no primary-secondary linkage. Power also is lost by radiation.

10-2-16. In an iron-core transformer, what is the relationship between the transformer turns ratio and primary to secondary current ratio; between turns ratio and primary to secondary voltage ratio (assume no losses)?—The primary to secondary current ratio varies inversely with the turns ratio. The primary to secondary voltage ratio varies directly with the turns ratio. This is to say, assuming a step-up transformer, that the secondary voltage will be greater than the primary voltage, and the secondary current will be less than the primary current by the amount of the step-up turns ratio.

$$\frac{N_s}{N_p} = \frac{E_s}{E_p} = \frac{I_p}{I_s}$$

10-2-17. What prevents high currents from flowing in the primary of an unloaded transformer?—The transformer is not loaded and makes no power demand on the primary. There is no energy being transferred by way of the flux linkage between primary and secondary. As a result, there is no current demand being made from the source of power being applied to the primary except that which the primary requires. Since this current is limited by the self-inductance of the primary, the primary current is minimum.

Were a load placed on the secondary, there would be power extracted from the primary via the flux linkage. This transferred power would have to be replaced by electrical energy from the power source connected to the primary. There would be a corresponding increase in the primary current.

10-2-18. An audio transformer has a resistive load connected across its secondary terminal. What is the relationship between this resistance, the turns ratio, and the input impedance at the primary terminals? How is this principle useful in matching impedances?—The impedance of the load reflected to the primary by the load resistance connected across the secondary varies as the square of the turns ratio:

$$Z_p = Z_s \left(\frac{N_p}{N_s}\right)^2$$

For example, if the turns ratio were 2 to 1, primary to secondary, the input impedance would be four times the ohmic value of the secondary load resistance.

$$Z_p = Z_s \times \left(\frac{2}{1}\right)^2 = Z_s \times 4$$

The principle can be used to match a generator of a specific output-impedance requirement to a load which is of a different impedance. For example, a specific vacuum tube or transistor stage may operate in an optimum fashion when a load of a specific value is placed on its output. Perhaps this load value is not the same as that to which the stage must deliver its power. In this case a transformer can be used to establish a suitable match between the preferred impedance at the output of the stage and the actual resistance of the load.

10-2-19. In what ways are electrical properties of common circuit elements affected by electromagnetic fields? Are interstage connecting leads susceptible to these fields?—Electromagnetic fields have numerous useful purposes. However, stray electromagnetic fields can also be troublesome. Coils, inductors, and transformers are frequently shielded to prevent undesired electromagnetic coupling into sensitive circuits. For example, in an audio system electromagnetic fields of power-line frequency from transformers can induce hum into the low signal level stages of the amplifier which can produce an audible hum in the audio system. Electromagnetic radiation from radio-frequency coils and coupling systems of a transmitter can induce signals into the low-level audio or radio-frequency circuits of a transmitter and cause feedback howl, instability, and other defects.

It is true that interstage connecting leads of sensitive weak-signal stages are susceptible to stray electromagnetic fields. In high-impedance, low-signal stages interconnecting leads are often short and/or shielded to reduce susceptibility. Stray and undesired electromagnetic fields can have an adverse influence on the operation of television and other types of cathode-ray tubes.

In general, high-impedance circuits and coupling networks are more susceptible to the influence of electromagnetic fields than lower-impedance arrangements. The more sensitive and the lower the level of the desired signal is, the more adverse is the influence of the stray electromagnetic field.

10-2-20. Compare some properties of electrostatic and electromagnetic fields.—The electromagnetic field is established by a permanent magnet or an electromagnet. An electromagnetic field is established by the flow of current through the conductors of a coil. The flux lines of an electromagnetic field enter and leave the magnet at the so-called poles. They leave the magnet at its north pole and follow the easiest path as they make complete loops and return to the magnet's south pole.

Electromagnetic fields can repel or attract each other according to the orientation of their poles.

The total amount of flux that flows varies directly with the applied magnetomotive force and inversely with the reluctance (magnetic resistance) of the flux path. The flux density is the quotient of the total flux and the area.

A very similar field, called an electrostatic field, is set up between two charged points, particles or bodies. This occurs when there is no closed circuit to equalize the two charges. A stress or force field is set up between the two charged particles. The electrostatic flux lines move in a direction from positive particle to negative particle. The intensity of the field is related directly to the difference of potential between the two charged particles and inversely with the distance separating the two charges. Field density is again the quotient of total flux and area.

The electrostatic flux lines, if visible, would show attraction between unlike charged bodies and repulsion between bodies of like charge, similar to the electromagnetic lines between like and unlike poles of a magnet. It is such an electrostatic field that extends between the charged plates of a capacitor.

10-2-21. How does the capacitance of a capacitor vary with the area of the plates; spacing between the plates; and dielectric material between the plates?—The capacitance increases with plate area and the dielectric constant of the dielectric material; it decreases with an increase in the spacing between plates.

10-2-22. Assuming the voltage on a capacitor is set at or below the maximum

allowable value, does the value of the capacitor have any relationship to the amount of charge it can store? What relationship does this storage of charges have to the total capacitance of two or more capacitors in series? In parallel?—The charge that a capacitor can store is the product of the voltage and capacitance. When two or more capacitors are connected in parallel the total charge that can be stored is increased and is greater than the storage capacity of any one capacitor in the group. The capacitance adds when capacitors are connected in parallel.

When capacitors are connected in series, the storage capability decreases because of the lower net capacitance. When capacitors are connected in series, the net capacitance is always less than the capacitance of any individual capacitor in the series group. There is no additional movement of charges once the smallest capacitor has been charged.

Capacitor charge is the product of voltage and capacitance ($Q = CE$). Assuming there is no loss, the voltage across a group of capacitors is the quotient of charge divided by the capacitance. For a given charge, therefore, doubling the capacitance will halve the voltage.

The charge placed on a capacitor stays there for a period of time determined by the resistance shunted across the capacitor. The higher the resistance, the longer the charge remains. It would remain indefinitely if there were no shunting resistance.

10-2-23. How should electrolytic capacitors be connected in a circuit in relation to polarity? Which type of low-leakage capacitor is used most often in transmitters?—Electrolytic capacitors must be connected with proper polarity. They should be connected so that the more positive voltage of the circuit is connected to the positive terminal of the capacitor, and the less positive point to the negative terminal of the capacitor. Low-leakage mica and ceramic capacitors are common in transmitters.

10-2-24. What is the rms value of an ac sine-wave voltage? The peak value? The peak-to-peak value?—The peak value of a sine-wave voltage is that voltage measured from the zero axis to the peak of the positive or negative alternation (Fig. 10-1). The peak-to-peak value is twice this value and equals the voltage measured between the positive and negative crests of the sine wave. The rms, or root-mean-square, value equals 0.707 times the peak value. This is also known as the effective value and corresponds to a direct-current value that would produce the same heating effect.

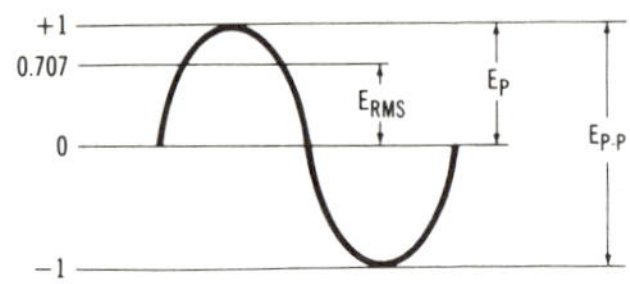

Fig. 10-1. Key sine-wave voltages.

10-2-25. Draw a vector diagram showing the voltages present in a series circuit composed of an inductive reactance, a capacitive reactance, and a resistance.—Refer to Fig. 10-2. In such a series circuit the current and resistive voltage are in phase as shown. The capacitive voltage lags both the current and the resistive voltage by 90 degrees. The inductive voltage leads both the series current and the resistive voltage by 90 degrees. The actual lengths of the vectors depend upon the relative ohmic values of the resistance and the reactances.

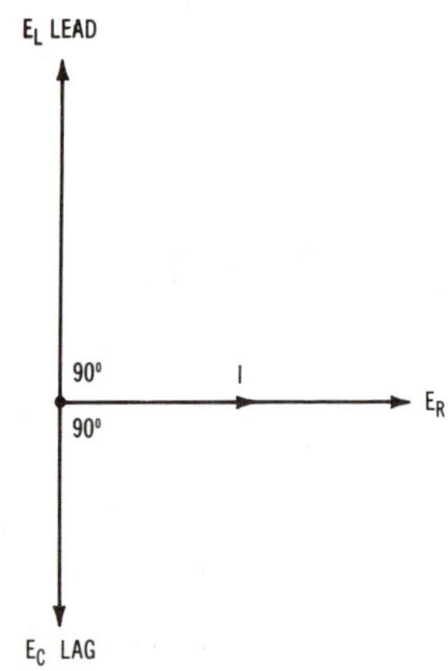

Fig. 10-2. Voltage relations in a series RLC circuit.

10-2-26. What component has the highest rate of breakdown? (a) Resistor, (b) capa-

citor, (c) inductor or (e) crystal.—The answer is capacitor.

10-2-27. How do you calculate the impedance of a resistor, inductor and capacitor combination in series? In parallel?—The equation for the series combination is:

$$Z = \sqrt{R^2 + (X_L - X_C)^2}$$

The equation for the parallel combination is:

$$Z = \frac{1}{\sqrt{\left(\frac{1}{R}\right)^2 + \left(\frac{1}{X_L} - \frac{1}{X_C}\right)^2}}$$

10-2-28. What is the formula for calculating the resonant frequency of an inductor and a capacitor in parallel?—This is the classic resonant-frequency formula or:

$$fr = \frac{1}{2\pi\sqrt{LC}}$$

10-2-29. What are harmonics?—A harmonic frequency is an integral multiple of a fundamental frequency. The second harmonic of a 1-MHz wave is 2 MHz. The fifth harmonic of the same wave would be 5 MHz.

10-2-30. What frequency sine waves combine together to form a square wave?—A square wave consists of a fundamental wave plus a series of harmonic waves. The greater the number of harmonics in the wave the squarer is the resultant wave (steep sides and a flat top).

10-2-31. What are the characteristics of ferrite? What is a ferrite bead?—Ferrite is a powdered and complex material with a high resistance. As a result there are very low eddy-current losses and excellent performance results at high frequencies up into the uhf spectrum and higher. Ferrite can be shaped into a tiny bead that can be slipped over a wire conductor. In position it functions as a radio-frequency choke and is especially effective in reducing vhf/uhf parasitics.

10-2-32. What is a toroid coil?—A toroid is a coil or transformer that is wound on a doughnut-shaped ferrite core. At high frequencies, up into the vhf/uhf range, such a device has minimum flux leakage (radiation) and a low-reluctance magnetic path within.

10-2-33. What is strip line?—Strip line is a thin narrow conductor that is deposited on dielectric material in a printed-circuit fashion. Such lines can be made to display the characteristics of transmission lines in the uhf/microwave spectra. They can be used to form low-loss efficient resonant circuits and wave filters.

10-2-34. What are the characteristics of tantalum, ceramic, and polystyrene capacitors?—The tantalum capacitor is an electrolytic type that uses tantalum and an associated oxide as the dielectric. For a given capacitance and voltage rating the tantalum type weighs less and it is smaller than the standard aluminum electrolytic capacitor. Ceramic capacitors use a ceramic material as a dielectric. Such capacitors perform well in radio-frequency applications up into the vhf/uhf frequency range and higher. Ceramic materials are also the dielectric of conventional disc capacitors which can be manufactured with a wide range of temperature characteristics. The polystyrene capacitor uses a plastic material as a dielectric. It is in the form of a thin film that is laid between strips of metal foil. In radio-frequency applications such a capacitor has a limited and known change in capacitance with temperature change. In high-stability oscillator circuits it has replaced the mica capacitor.

10-3. CIRCUITS

10-3-1. What is the impedance of a parallel circuit composed of a pure inductance and a pure capacitance at resonance? Of a series circuit at resonance?—The impedance of a parallel circuit at resonance is maximum. Power factor is 1. Theoretically, if there were no resistance in the circuit the impedance would be infinite. The impedance of a series circuit at resonance is minimum. Power factor is 1. Theoretically, if there were no resistance in the circuit, the impedance would be zero.

10-3-2. What is the Q of a circuit?—The *Q* is the figure of merit of a resonant circuit. This quality factor of a resonant circuit is the quotient of circuit reactance divided by the circuit resistance, or:

$$Q = \frac{X}{R}$$

The higher the circuit resistance the lower is the *Q*. The lower the *Q* of a resonant circuit, the greater is the bandwidth of the circuit.

A resonant circuit with a high Q has a sharp high-peaked response. However, the circuit of lower Q has a greater band of frequencies over which its response is essentially uniform.

10-3-3. Explain how the values of resistance and capacitance in an RC network affect its time contact.—Time constant is the product of resistance and capacitance—the time needed for the capacitor to charge to 63.2% of its peak value. Note that increasing either the resistance or capacitance increases the time constant.

$$T_{seconds} = R_{ohms}\, C_{farads}$$

The magnitude of the output of an RC network is very much dependent on the frequency of the input wave. The higher the frequency of the input wave, the higher the percentage of the input wave amplitude that appears across the resistor and the lower the percentage of the input wave that appears across the capacitor.

An RC network will not change the form or "shape" of an applied sine wave; it will only influence its amplitude.

10-3-4. Explain how the values of resistance and inductance in an RL network affect its time constant.—The time constant of an RL network is the quotient of the inductance and resistance:

$$T_{seconds} = \frac{L_{henrys}}{R_{ohms}}$$

The higher the value of the inductance, the greater is the time constant. The higher the value of the resistance, the lower is the time constant.

In an RL circuit an applied sine wave is not distorted, but its amplitude depends on the value of the resistance and inductance. The higher the frequency, the higher is the percentage of the input wave that appears across the inductance, and the lower is the percentage that appears across the resistor.

10-3-5. Draw a circuit diagram of a low-pass filter composed of a constant-K and an M-derived section.—Refer to Fig. 10-3. M-derived sections can be used to alter the frequency-response characteristics of low-pass, high-pass, bandpass, and band-reject filters. In the case of the former two filters, the addition of an M-derived section will introduce a very high

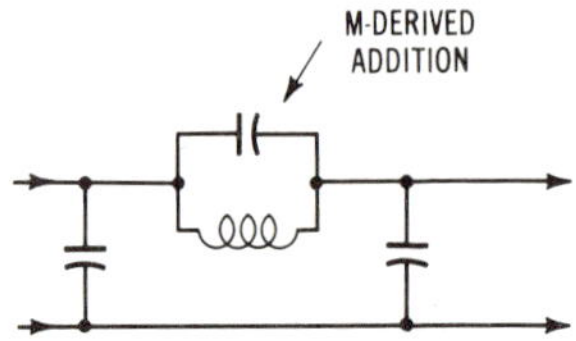

(A) Parallel combination.

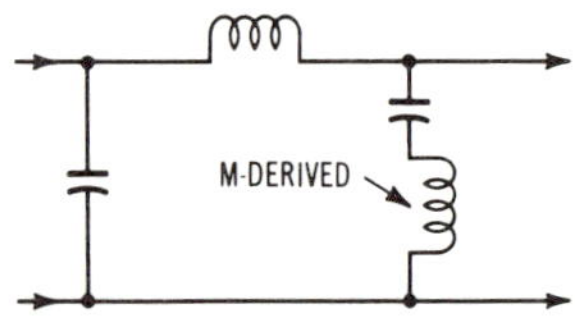

(B) Series combination.

Fig. 10-3. Constant-K low-pass filter with M-derived section.

attenuation at some specific frequency within the bandpass. For the latter two filters, the addition of an M-derived section will introduce very high attenuation at two specific frequencies in the attenuation band of the filter. For example, such an M-derived section can be used in establishing an essentially flat response for a desired bandpass along with sharply attenuated response skirts.

Sharper cutoff characteristics and flatter bandpasses can be obtained by adding additional M-derived sections. An M-derived section can be a series combination LC placed across a filter to attenuate a specific frequency, as in Fig. 10-3B.

10-3-6. In general, why are filters used? Why are band-stop, high-pass, and low-pass filters used? Draw schematic diagrams of the most commonly used filters.—A filter is a frequency-selective network that can emphasize one range of frequencies and de-emphasize other frequencies, or it can be arranged to attenuate a specific range of frequencies and pass, largely unattenuated, other frequencies.

A band-stop filter is one that attenuates or blocks the transfer of a specific band of frequencies. In communications such a filter can be used to block a source of interference from a receiver or prevent the transmission of a spurious frequency by a transmitter

A high-pass filter will pass frequencies, essentially unattenuated, above a desired cutoff frequency. Frequencies below this cutoff frequency will be attenuated. A low-pass filter will pass essentially unattenuated all frequencies below a desired cutoff frequency. All frequencies above the cutoff frequency will be attenuated. In establishing a certain voice-frequency range in a communications system, a low-pass filter can be used to attenuate all voice frequency components above a desired high-frequency cutoff. A high-pass filter can be used to attenuate all voice frequencies below a desired low cutoff frequency. In this way the audio bandwidth of a communications system can be confined, let us say, within a range extending from 250 to 2500 Hz. Refer to Fig. 10-4 for typical filters.

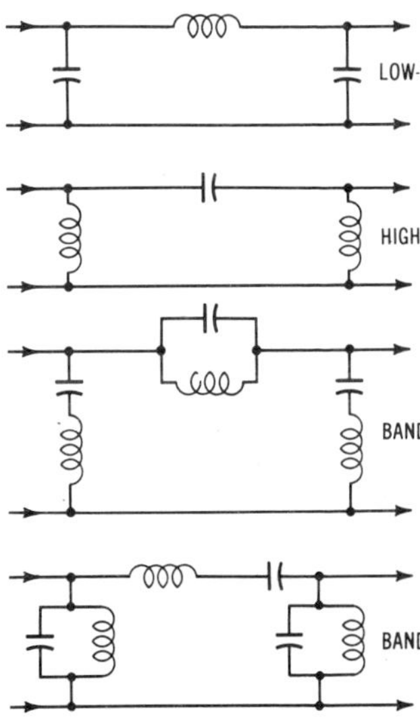

Fig. 10-4. Basic filter types.

10-3-7. Why is impedance matching between electrical devices an important factor? Is it always to be desired? Can it always be attained in practice?—*Exact impedance matching permits the maximum transfer of power from a source of electrical energy to a load.* Exact match cannot always be attained because of the variables and the circuit constants involved.

Quite often it is neither practical nor economical to establish an exact match. In other cases an exact match may produce distortion or some other undesired side effect. Sometimes an exact match may not be desirable because of the need to obtain maximum voltage gain or a specified current gain. In some instances an exact match may place too great a load on a given source.

10-3-8. A speaker with an impedance of 4 ohms is working in a plate circuit which has an impedance of 4000 ohms. What is the impedance ratio of an output transformer used to match the plate circuit to the speaker? What is the turns ratio?—The impedance ratio primary-to-secondary is:

$$\text{Impedance Ratio} = \frac{Z_p}{Z_s} = \frac{4000}{4} = 1000 \text{ to } 1$$

Since the impedance ratio is a function of the turns ratio squared, the actual required turns ratio becomes:

$$\left(\frac{N_p}{N_s}\right)^2 = \frac{Z_p}{Z_s}$$

$$\frac{N_p}{N_s} = \sqrt{\frac{Z_p}{Z_s}} = \sqrt{\frac{1000}{1}} = 31.6 \text{ to } 1$$

10-4. CALCULATIONS

10-4-1. A certain power company charges seven cents per kilowatt hour. How much would it cost to operate for 24 hours, three 120-volt lamp bulbs connected in parallel, each having an internal resistance of 100 ohms?—The first step is to determine the number of watts consumed each hour by the three bulbs.

$$P = \frac{E^2}{R} = \frac{(120)^2}{100} = 144 \text{ watts}$$

$$P_{total} = 3 \times 144 = 432 \text{ watts, or } .432 \text{ kW}$$

Next, determine the total number of kilowatts consumed over the 24-hour period.

$$\text{Total kWh in 24 hours} = 24 \times .432 = 10.37 \text{ kWh}$$

Finally it is possible to determine the total cost on the basis of the kilowatt-hours consumed:

$$\text{Cost} = 10.37 \times 7¢ = \$0.73$$

10-4-2. The output of an amplifier stage having a voltage gain of 30 dB is 25

volts. What is the input voltage level?—A 30-dB rise corresponds to a voltage gain of 31.6. The calculation is as follows:

$$E_{Ratio} = \log^{-1}\frac{30}{20}$$

$$E_{Ratio} = \log^{-1}1.5 = 31.6$$

$$E_{Input} = \frac{Output}{31.6} = \frac{25}{31.6} = 0.791 \text{ volt}$$

It is advisable to commit to memory certain key dB voltage and power gain figures. It can help you avoid a more time consuming calculation.

dB	V&I Ratio	Power Ratio
3	1.41	2
6	2	4
10	3.16	10
20	10	100
30	31.6	1000
40	100	10,000
50	316	100,000
60	1000	1,000,000

10-4-3. Draw two cycles of a sine wave on a graph of amplitude versus time. Assume a frequency of 5 MHz.—Refer to Fig. 10-5.

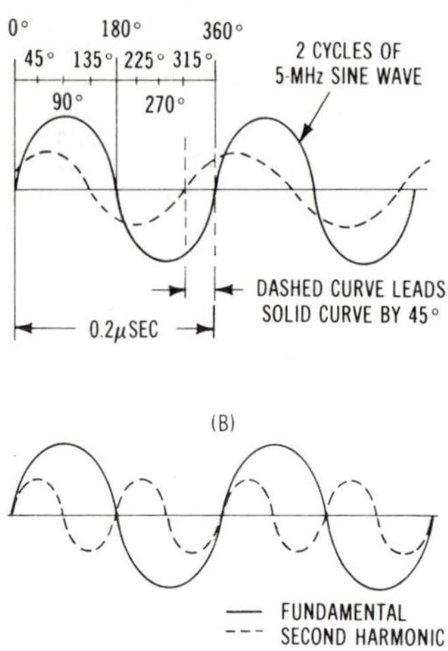

Fig. 10-5. Waveform drawings used for Question 10-4-3.

A. What would be the wavelength of one cycle in meters? In centimeters?—The wavelength is 60 meters as per the following wavelength calculation:

$$\lambda = \frac{Velocity}{Frequency} = \frac{300 \times 10^6}{5 \times 10^6} = 60 \text{ meters}$$

Velocity is a constant of 300,000,000 meters per second. Inasmuch as one meter is equivalent to 100 centimeters, the wavelength in centimeters is 6000 (60 × 100).

B. How many degrees does one cycle represent?—360°.

C. How much time would it take for the wave to rotate 45°; 90°, 280°?—First it is necessary to determine the period of a 5-MHz wave. One cycle of a 5-MHz frequency occurs in 0.2 microsecond:

$$\text{Period} = \frac{1}{Frequency} = \frac{1}{5 \times 10^6} = 0.2\ \mu s$$

Inasmuch as 45° corresponds to one-eighth of a cycle, the time required for a wave to rotate 45° becomes 0.025 microsecond. The calculations are as follows:

Time

$$45° = \frac{45°}{360°} \times 0.2 = 0.025\ \mu s.$$

$$90° = \frac{90°}{360°} \times 0.2 = 0.05\ \mu s.$$

$$280° = \frac{280°}{360°} \times 0.2 = 0.155\ \mu s.$$

D. If there were a second harmonic of this frequency, how many cycles thereof would be represented on this graph?—The second harmonic frequency is twice the fundamental frequency, and there would be two harmonic cycles to each fundamental cycle. Four cycles of the harmonic are shown as dashed-line curves in Fig. 10-5B.

E. On the same graph draw two cycles of another sine wave leading the first by 45°.—Refer to Fig. 10-5A.

F. What would be the velocity of this wave or any other electromagnetic wave in free space?—The velocity of wave propagation is the product of frequency and wavelength. For the above example the answer is:

$$\text{Velocity} = f\lambda = 5 \times 10^6 \times 60 = 300 \times 10^6 \text{ meters/second}$$

This velocity applies to any electromagnetic wave regardless of frequency. The velocity of propagation is often stated as 186,000 miles/second.

10-4-4. Explain how to determine the sum of two equal vector quantities having the same reference point, but whose directions are 90° apart; 0° apart; 180° apart. How does this pertain to electrical currents or voltages?—When two equal vector quantities are in phase, they are additive, and the resultant is twice the value of either vector. When two equal vectors are 180° related, they cancel and produce a resultant of zero. The two equal 90° vectors have a right triangle relationship, and the resultant is the vector sum of the two quantities or 1.414 times either vector. The actual vector relations are shown in Fig. 10-6.

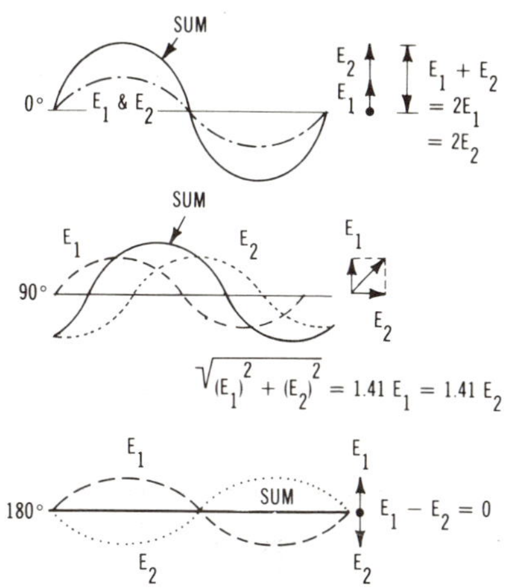

Fig. 10-6. Equal vector relations.

More than one voltage of a given frequency (*or* current of a given frequency) is often present in a circuit. They combine to form a net value, the magnitude of which depends on their individual magnitudes and phase relation.

10-4-5. A relay with a coil resistance of 500 ohms is designed to operate when a 0.2-ampere current flows through the coil. What value of resistance must be connected in series with the coil if it is to be energized by a 110-volt dc source?—The first step is to determine the total resistance that must be placed in the circuit to limit the current to 0.2 ampere:

$$R_T = \frac{E_s}{I_s} = \frac{110}{0.2} = 550 \text{ ohms}$$

Since the coil resistance is given as 500 ohms, the additional resistance needed in the circuit becomes:

$$R = R_T - R_{coil}$$
$$R = 550 - 500 = 50 \text{ ohms}$$

10-4-6. Draw a circuit composed of a 12-volt battery with three resistors (10, 120, and 300 ohms, respectively) arranged in a pi-network.—Refer to Fig. 10-7.

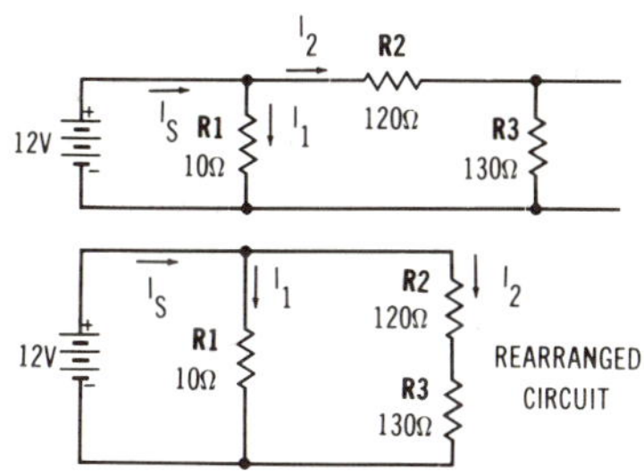

Fig. 10-7. Resistance pi-network.

A. What is the total current; the current through each resistor?—It is to be noted that there are two current components I_1 and I_2. The current I_1 is the function of the resistance of R1 and the applied voltage. The current through resistors R2 and R3 is a function of their total series resistance and the applied voltage. Calculations are as follows:

$$I_1 = \frac{E}{R1} = \frac{12}{10} = 1.2 \text{ amps}$$

$$I_2 = \frac{E}{R2 + R3} = \frac{12}{120 + 300}$$
$$= 0.0286 \text{ amp}$$

The total current is the sum of I_1 and I_2 or:

$$I_T = I_1 + I_2$$
$$I_T = 1.2 + 0.0286 = 1.2286 \text{ amps}$$

B. What is the voltage across each resistor?—It is apparent that the full voltage appears across resistor R1. The voltage drop across resistors R2 and R3 can be determined using Ohm's law.

$$E_{R1} = I_1R1 = 1.2 \times 10 = 12 \text{ volts}$$
$$E_{R2} = I_2R2 = .0286 \times 120 = 3.432 \text{ volts}$$
$$E_{R3} = I_2R3 = .0286 \times 300 = 8.58 \text{ volts}$$

C. What power is dissipated in each resistor? The total power dissipated by the circuit?—The power dissipated by each resistor is the product of the voltage across it and the current through that resistor. Calculations are as follows:

$$P_{R1} = E_{R1}I_1 = 12 \times 1.2 = 14.4 \text{ watts}$$
$$P_{R2} = E_{R2}I_2 = 3.432 \times 0.0286 = 0.098 \text{ watt}$$
$$P_{R3} = E_{R3}I_2 = 8.58 \times 0.0286 = 0.245 \text{ watt}$$

The total power consumed is the sum of the individual powers or:

$$P_T = 14.4 + 0.098 + 0.245 = 14.743 \text{ watts}$$

The total power is also the product of the applied voltage and the total current drawn by the network. This calculation will verify your previous work:

$$P_T = E_sI_s = 12 \times 1.2286 = 14.743 \text{ watts}$$

10-4-7. What is power factor? Give an example of how it is calculated.—Power factor is the ratio of true power to apparent power. A wattmeter, for example. reads the true power in an ac circuit. The apparent power is the voltage times the current. When the current and voltage are in exact phase, the power factor is 1, and true power and apparent power are the same. The greater the phase angle between voltage and current, the lower the power factor; hence the higher the apparent power to the true power in the circuit. Stated in an equation:

$$PF = \frac{\text{True Power}}{\text{Apparent Power}} = \frac{I^2R}{VA} = \frac{R}{Z} = \cos\theta$$

If a wattmeter reads 100 watts supplied to an electrical device but voltmeter and ammeter read 120 volts and 1 ampere, respectively, the power factor would be:

$$PF = \frac{100}{120 \times 1} = 0.833$$

10-4-8. Draw a circuit composed of a voltage source of 100 volts and 1000 Hz, a 1-microfarad capacitor in series with the source, followed by a T-network composed of a 2-millihenry inductor, a 100-ohm resistor and a 4-millihenry inductor. The load resistor is 200 ohms.—Refer to Fig. 10-8.

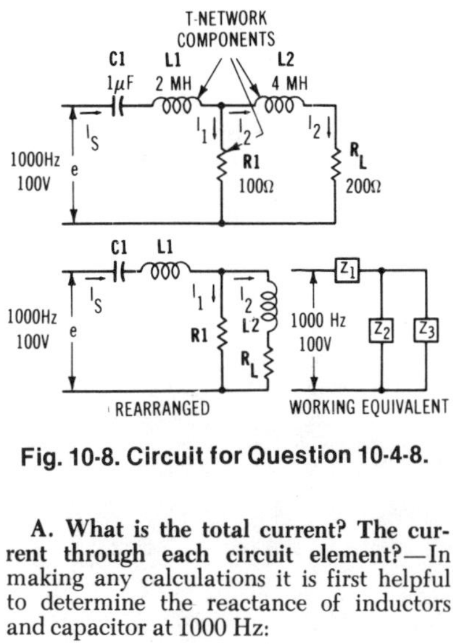

Fig. 10-8. Circuit for Question 10-4-8.

A. What is the total current? The current through each circuit element?—In making any calculations it is first helpful to determine the reactance of inductors and capacitor at 1000 Hz:

$$X_C = \frac{1}{6.28 \times 1000 \times 1 \times 10^{-6}} = 159.2 \text{ ohms}$$
$$X_{L1} = 6.28 \times 1000 \times 2 \times 10^{-3} = 12.56 \text{ ohms}$$
$$X_{L2} = 6.28 \times 1000 \times 4 \times 10^{-3} = 25.12 \text{ ohms}$$

Setting up three impedance sections, the net impedance of the circuit is:

$$Z = Z_1 + \frac{Z_2Z_3}{Z_2 + Z_3}$$

Individual impedance values are:

$Z_1 = jX_{L1} - jX_C = -j(159.2 - 12.56)$
$= -j147$ ohms

$Z_2 = R1 = 100$ ohms

$Z_3 = R_L + jX_{L2} = 200 + j25.12$
$= \sqrt{(200)^2 + (25.12)^2}$

$202\angle 7°$

$Z_2 + Z_3 = R1 + (R_L + jX_{L2})$
$= 100 + 200 + j25.12$
$= 300 + j25.12$
$\sqrt{(300)^2 + (25.12)^2}$
$= 301\angle 5°$

Impedance of parallel Z_2 and Z_3:

$$Z_e = \frac{Z_2 Z_3}{Z_2 + Z_3} = \frac{100 \times 202/7°}{301/5°} = 67.1\angle 2°$$

In rectangular form:

$R_e = Z_e \cos 2° = 67.1 \times 0.9994$
$= 67.06$ ohms

$X_e = Z_e \sin 2° = 67.1 \times 0.0349$
$= 2.34$ ohms

$Z_e = 67.06 + j2.34$ ohms

The total impedance is now:

$Z = Z_1 + Z_e$
$Z = -j147 + (67.06 + j2.34)$
$Z = 67.06 - j145$
$Z = \sqrt{(67.06)^2 + (-145)^2}$
$= 160\angle -65°$

Now it is possible to determine currents and voltages:

$$I_s = \frac{E_s}{Z} = \frac{100}{160} = 0.625 \text{ amp}$$

Voltage drop across Z_e (Z_2 and Z_3 in parallel) is:

$E_{Ze} = I_s Z_e = 0.625 \times 67.1 =$
41.9 volts

Currents I_1 and I_2 are:

$$I_1 = \frac{E_{Ze}}{R1} = \frac{41.9}{100} = 0.419 \text{ amp}$$

$$I_2 = \frac{E_{Ze}}{Z_3} = \frac{41.9}{202} = 0.207 \text{ amp}$$

In summary:

$I_s = 0.625$ amp

Current in C and L_1 $= I_s = 0.625$ amp

Current in R1 $= I_1 = 0.419$ amp

Current in L_2 and R_L $= I_2 = 0.207$ amp

B. What are the individual voltage drops?—Individual voltage drops become:

$E_c = I_s X_c = 0.625 \times 159.2$
$= 99.5$ volts

$E_{L1} = I_s X_{L1} = 0.625 \times 12.56$
$= 7.85$ volts

$E_{R1} = I_1 R1 = 0.419 \times 100$
$= 41.9$ volts

$E_{L2} = I_2 X_{L2} = 0.207 \times 25.12$
$= 5.2$ volts

$E_{RL} = I_2 R_L = 0.207 \times 200$
$= 41.4$ volts

C. What is the apparent power of the circuit?—Apparent power is:

$P_A = E_s I_s = 100 \times .625$
$= 62.5$ watts

D. What is the true power?—True power is:

$P_{true} = E_s I_s \cos\theta$
$= E_s I_s \cos(-65°)$

$P_{true} = 62.5 \times 0.4226$
$= 26.4$ watts

E. What are the powers consumed by R1 and R_L?—Powers consumed by R1 and R_L are:

$P_{R1} = E_{R1} I_1 = 41.9 \times .419$
$= 17.6$ watts

$P_{RL} = E_{RL} I_2 = 4.14 \times .207$
$= 8.57$ watts

Note that the sum of the powers consumed by the resistors equals the previous true power calculation.

10-4-9. In the previous example what is the total reactive power?—Reactive power is the difference between apparent power and true power. Thus in the previous example the reactive power is 36.1 (62.5 – 26.4) var. The var is the unit of reactive power. Another equation for reactive power is:

$$\text{Var} = I_{rms} \times E_{rms} \times \sin\theta$$

θ equals the phase angle between current and voltage.

10-4-10. A series-parallel circuit is composed of a 5-ohm resistor in series with the parallel combination of a capacitor having a pure reactance of 20 ohms and an inductance having a pure reactance of 8 ohms. What is the total

impedance of the circuit? Is the total reactance capacitive or inductive?—In the parallel circuit of inductor and capacitor the lowest reactance sets the direction of the net current. Inasmuch as this is the inductive reactance, the net reactance is inductive and has a value of:

$$X_N = \frac{-X_C X_L}{-X_C + X_L} = \frac{-20 \times 8}{-20 + 8}$$
$$= 13\tfrac{1}{3} \text{ ohms}$$

The total impedance of the circuit becomes:

$$Z = \sqrt{R + (X_N)^2}$$
$$= \sqrt{(5)^2 + (13\tfrac{1}{3})^2} = 14.26 \text{ ohms}$$

10-4-11. A 10-amp alternating current is flowing in a series circuit composed of 5-ohms resistance, 25-ohms capacitive reactance, and 12-ohms inductive reactance. What is the voltage across each component? What is the total voltage? Why is the total voltage not simply the sum of the individual voltages?—The current magnitude is the same for each component of a series circuit. Therefore, the voltage drops across the capacitor, inductor, and resistor are as follows:

$$E_R = I \times R = 10 \times 5 = 50 \text{ volts}$$
$$E_C = I \times X_C = 10 \times 25 = 250 \text{ volts}$$
$$E_L = I \times L_L = 10 \times 12 = 120 \text{ volts}$$

The total voltage is the vector sum of the individual voltages. This is not a simple arithmetical addition, because the voltages are not in phase. The voltage across the resistor is in phase with the current, the voltage across the capacitor lags the current by 90°, and the voltage across the inductor leads the current by 90°, as in the vector diagram Fig. 10-9. Note that the inductor and capacitor voltages are out of phase by 180°. They are opposite in polarity, and therefore one subtracts from the other. Thus the net voltage across the capacitor and inductor in series becomes:

$$E_N = E_C + E_L = -250 + 120$$
$$= -130 \text{ volts}$$

This net voltage lags the voltage across the resistor by 90° and a simple right-triangle solution can be used to determine the total voltage.

$$E_T = \sqrt{(E_R)^2 + (E_N)^2}$$
$$= \sqrt{(50)^2 + (-130)^2} = 139.3 \text{ volts}$$

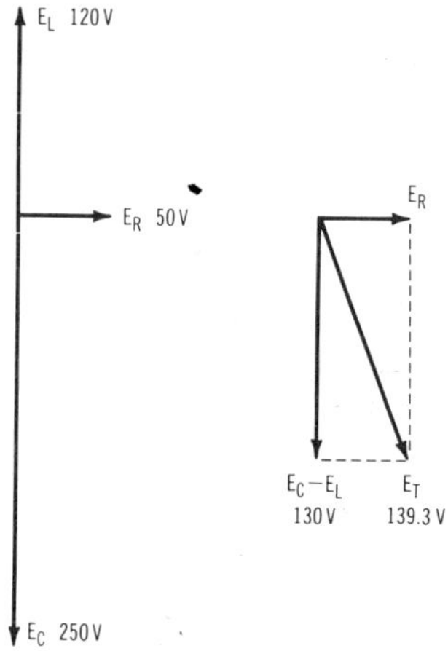

Fig. 10-9. Vector relations for Question 10-4-11.

In summary the total voltage is not the simple sum of the individual voltages, because of the phase relationships among the three voltages.

10-4-12. If a capacitor is charged to 10 volts, then placed in parallel with a resistor, what will the voltage across it be after one RC time constant?—The voltage across the capacitor will be approximately 3.67 volts. The capacitor is discharging exponentially and it will discharge to 63.3% of its maximum value in one time constant.

Table 10-1 shows the charge and discharge percentages for time-constant values of 1 through 5.

Table 10-1. Charge and Discharge Percentages in Terms of Time Constant

RC Time Constant	% Charge	% Discharge
1	63.3	36.7
2	86.5	13.5
3	95	5
4	98.3	1.7
5	99.4	0.6

10-4-13. What is Ohm's law? Show an example to demonstrate the law.—Ohm's law states: The current in a circuit is directly proportional to the voltage across that circuit and inversely proportional to the impedance of the circuit. Stated as an equation:

$$I = E/Z$$

By mathematical rearrangement the equation can be rewritten

$$Z = E/I \text{ or } E = IZ$$

If the impedance is resistive the equations become:

$$I = E/R$$
$$R = E/I$$
$$E = IR$$

What is the current in a resistor if the resistance has a value of 100 ohms and the voltage across the resistance is 200 volts?

$$I = E/R = 200/100 = 2 \text{ amperes}$$

10-4-14. What is a voltage divider circuit? Show a schematic diagram that demonstrates its operation.—A voltage divider is formed by connecting two or more resistors in series across a supply voltage. The voltage across each resistor depends on the relative values of the other resistors in the series chain. Any number of resistors can be used. The examples of Fig. 10-10 are for a three-resistor divider. For each resistor a simple

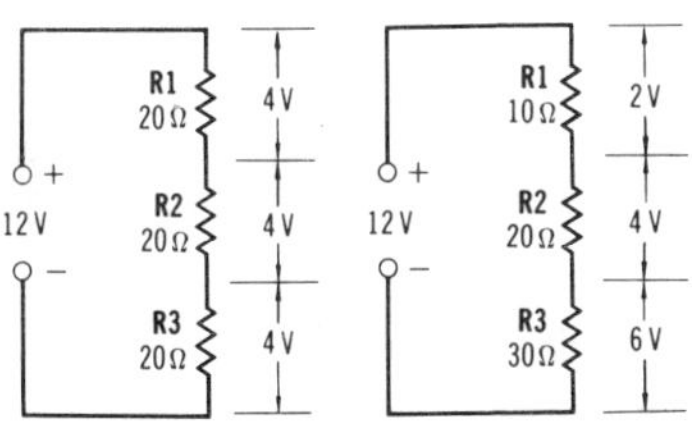

Fig. 10-10. Voltage divider circuits.

proportion can be set up to determine the voltage across a specific resistor. The voltage across resistor R1 would be expressed as:

$$\frac{E_{R1}}{E_T} = \frac{R1}{R1 + R2 + R3}$$

$$E_{R1} = E_T\left(\frac{R1}{R1 + R2 + R3}\right)$$

Set down this relationship for each of the resistors of Fig. 10-10. In example A the voltages divide equally because each resistor is of the same value. In example B they divide according to the relative values of the divider resistors.

The general voltage divider equation is:

$$E_{RX} = E_T \frac{R_X}{R1 + R2 + \ldots R_N}$$

A second example is shown in Fig. 10-11. A voltage of 5 volts is to be made available from a 12-volt source using a two-resistor voltage divider. The current in the divider is to be 100 mA. Find the values of resistors R1 and R2.

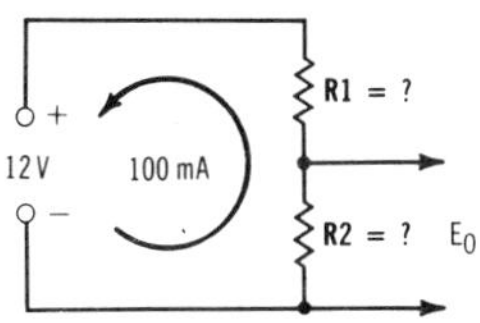

Fig. 10-11. Problem for determining resistor values.

Since the current is to be set at 100 milliamperes, the sum of the two resistances must be:

$$R_T = \frac{E_T}{I} = \frac{12}{0.1} = 120 \text{ ohms}$$

The next step is to find the value of resistor R2 that will produce a 5 volt output when the current is 100 mA or:

$$R2 = \frac{E_o}{I} = \frac{5}{0.1} = 50 \text{ ohms}$$

The value of resistor R1 is:

$$R1 = 120 - 50 = 70 \text{ ohms}$$

10-4-15. What is a current divider? Show a sample circuit.—Current dividers are constructed by connecting components in parallel. Two two-resistor current dividers are shown in Fig. 10-12. The voltage across each resistor equals the supply voltage. However, the current in each resistor depends on its ohmic value. Again a proportion can be set up as follows:

$$I_T = \frac{E}{\left(\frac{R1 \times R2}{R1 + R2}\right)} = \frac{E(R1 + R2)}{R1 \times R2}$$

By proportion:

$$\frac{I_{R1}}{I_T} = \frac{\frac{E}{R1}}{\frac{E(R1 + R2)}{R1 \times R2}}$$

Relationship can be inverted and simplified:

$$I_{R1} = I_T \frac{R2}{R1 + R2}$$

A similar equation can be set up for the current in resistor R2:

$$I_{R2} = I_T \frac{R1}{R1 + R2}$$

Use these equations to verify the currents shown in each of the resistors of Fig. 10-12.

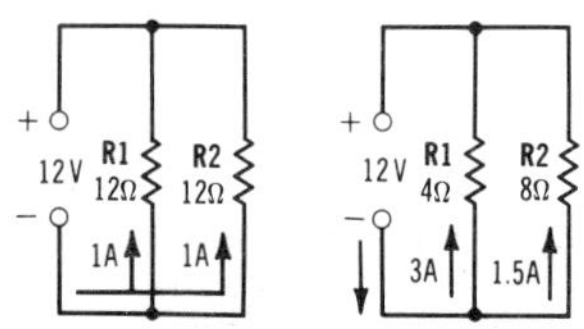

Fig. 10-12. Basic current divider.

In the circuit of Fig. 10-13 no more than 200 mA is to be drawn from the source. What must be the values of the resistors needed to obtain 5/16 of the total current in resistor R1 and 11/16 of the total current in resistor R2? Ohm's law can be used to determine the total resistance:

$$R_T = 16/0.2 = 80 \text{ ohms}$$

Simple proportions can be set up to determine the values for resistor R1 and R2:

$$\frac{I_{R1}}{I_T} = \frac{R_T}{R1}$$

$$\frac{5/16\ I_T}{I_T} = \frac{R_T}{R1}$$

$$\frac{5}{16} = \frac{80}{R1}$$

$$R1 = 256 \text{ ohms}$$

Value for R2 is:

$$\frac{11}{16} = \frac{80}{R2}$$

$$R2 = 116 \text{ ohms}$$

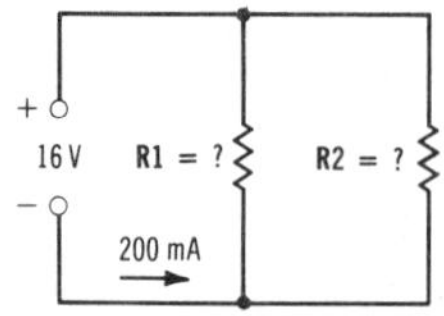

Fig. 10-13. Problem for determining resistor values.

10-4-16. What is Kirchhoff's law? Demonstrate with an example.— Kirchhoff's law of circuit analysis states: The algebraic sum of the voltage drop around a circuit is equal to 0. The circuit of Fig. 10-14 can be used to demonstrate the law. The total resistance connected across the 12-volt source is 300 ohms (100 + 200) and, therefore, by Ohm's law is:

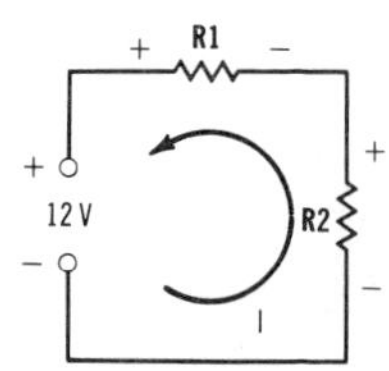

Fig. 10-14. Kirchhoff's law example.

$$I = E/R1 + R2 = 12/100 + 200 = 0.04 \text{ A}$$

The voltage drops across the two resistors are:

$$E_{R1} = IR_1 = 0.04 \times 100 = 4 \text{ volts}$$
$$E_{R2} = IR_2 = 0.04 \times 200 = 8 \text{ volts}$$

We can now prove Kirchhoff's law. Start at the negative terminal of the battery and follow the arrow. The first negative sign is found at the bottom of resistor R2 and the next negative sign is encountered by the arrow on the right side of resistor R1. The arrow meets a + sign at the top of the battery. Stated as an equation:

$$-8 \text{ volts} - 4 \text{ volts} + 12 \text{ volts} = 0$$

It is apparent from the above that in many simple and more complex circuits we often use Kirchhoff's law to solve a problem although we are not really aware of that fact. Perhaps the most common application is one we use often when we assume that the sum of the voltage drops around a circuit is equal to the supply voltage. A simple rearrangement of the previous equation representing Kirchhoff's law proves that fact.

$$12 = 4 + 8$$

10-4-17. What is Thevenin's theorem? Use an example to demonstrate the theorem.—Thevenin's theorem states: The current in any load impedance connected across a two-terminal network is the same as if it were connected across a simple constant-voltage generator that has a voltage equal to the open-circuit voltage across these two terminals and a series impedance equal to the impedance of the network looking back from the terminals. What is the load current when an 80-ohm load is connected across the two terminals of the representative network of Fig. 10-15A? The first step in solving the problem is to determine the open-circuit voltage across the two terminals. Resistors R1 and R2 act as a voltage divider across the 10-volt source.

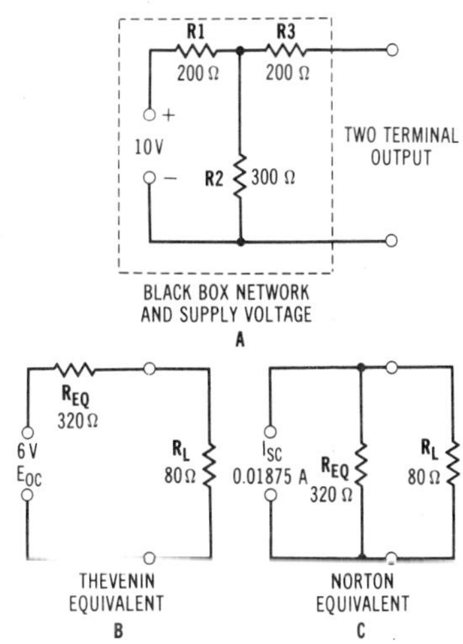

Fig. 10-15. Thevenin and Norton theorems concepts.

Therefore using the two-resistor voltage-divider concept the voltage across resistor R2 is 6 volts $(3/5 \times 10)$. Inasmuch as there is no load placed on the output there is no voltage drop across resistor R3 and the open circuit output voltage also becomes 6 volts.

To determine the equivalent series resistance of the network it is necessary to know the short-circuit output current. In effect, resistor R3 then appears in parallel with resistor R2. By Ohm's law the network current becomes:

$$I_T = \frac{10}{200 + \dfrac{300 \times 200}{300 + 200}} = \frac{10}{200 + 120} = 0.03125\text{A}$$

The voltage across resistors R2 and R3 in parallel has a value of:

$$E_{R2R3} = 0.03125 \times 120 = 3.75\,\text{V}$$

The output short-circuit current will be this voltage divided by the ohmic value of resistor R3 or:

$$I_{SC} = \frac{E_{R2R3}}{R3} = \frac{3.75}{200} = 0.01875\text{A}$$

The equivalent resistance is the open-circuit output voltage divided by the short-circuit output current or:

$$R_{EQ} = \frac{E_{OC}}{I_{SC}} = \frac{6}{0.01875} = 320 \text{ ohms}$$

From this information you can draw the equivalent Thevenin's circuit as shown in Fig. 10-15B. Ohm's law can be used to calculate the load current as follows:

$$I_L = \frac{E_{OC}}{R_{EQ} + R_L} = \frac{6}{320 + 80} = 0.015\text{A}$$

A very important fact to be disclosed about Thevenin's theorem is that a simple high-resistance voltmeter and a low-resistance current meter connected across the two terminals would permit you to obtain the open-circuited output voltage and the short-circuited output current with two simple measurements. A voltmeter connected across the two terminals would give you a reading for the E_{OC}. Replace the voltmeter with a current meter and the reading will be the short-circuit current (I_{SC}).

(If you wish to verify the previous calculations you can complete the network of Fig. 10-15A by connecting an 80-ohm resistor between the two terminals and using strictly Ohm's law to calculate the current that will be present in the 80-ohm load resistor.)

In fact, using these measurements it would not have been necessary that you know the ohmic values of the resistors in the network or even the supply voltage. How the network performs could be determined simply by measuring the open circuit output voltage and the short-circuit output current. Knowing this you would know how the network performs when you connect any value of load resistance to the output terminals. In this case not having any knowledge of the network or supply voltage it would be called a two-terminal *black box*. How the black box performs is made known just by making the two measurements.

10-4-18. What is Norton's theorem? Demonstrate its capability.—Norton's theorem states: The current in any load impedance connected across a two-terminal network is the same as if it were connected across a simple constant-current generator that has a current equal to the short-circuit current between these two terminals and with a parallel impedance equal to the impedance of the network, looking back from the two terminals.

This concept is very similar to Thevenin's theorem. The difference is that the black box is to be considered a current generator rather than a voltage generator. The equivalent impedance then becomes parallel-connected rather than series connected.

In demonstrating the concept it is again necessary to determine the open-circuit output voltage and the short-circuit output current. From these two calculations you can determine the equivalent resistance of the network. If the circuit were identical to that of Fig. 10-15A, the values would be the same as those calculated previously. However, this time the concept is a constant-current source and a shunt equivalent resistance as shown in Fig. 10-15C.

Note that the circuit is a current divider and therefore the load current becomes:

$$I_L = 0.01875 \times \frac{320}{80 + 320} = 0.015\text{A}$$

In summary, the same two measurements (E_{OC} and I_{SC}) at the output of the two-terminal network can be used to establish either a constant current (Norton's theorem) or a constant voltage (Thevenin's theorem) interpretation of the black box circuit.

CHAPTER 11

Tubes and Transistors

11-1. ELECTRON TUBES

11-1-1. Discuss the physical characteristics and a common usage for each of the following tube types: (a) diode, (b) triode, (c) tetrode, (d) pentode, (e) beam power, (f) remote cutoff, (g) duo-triode, (h) cold-cathode, and (i) thyratron.—(a) The diode is a two-element tube consisting of a cathode and an anode, often referred to as a plate (Fig. 11-1). The cathode is a thermionic generator of electrons. When the cathode is heated, the electrons are emitted from the surface of the cathode and move through the interelement space to the anode, which is operated at a positive potential relative to the cathode. Electrons can only flow in this one direction. If the anode is made negative with respect to the cathode, there is no motion of electrons from cathode to anode.

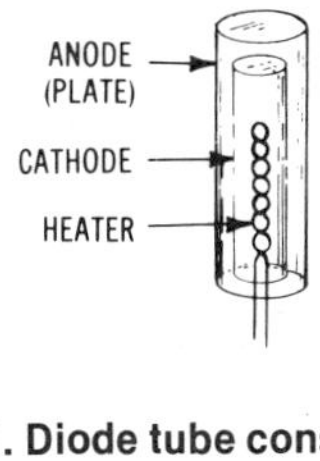

Fig. 11-1. Diode tube construction.

The cathode itself can be a directly heated filament, the filament wire emitting the electrons. The cathode surface can be heated indirectly, the heat from the filament wire raising the cathode surface to a temperature that causes the special coating on the cathode surface to emit electrons.

Three common uses for a diode in electron circuits are as a rectifier, a detector, or a clipper-limiter.

(b) A triode is a three-element tube consisting of a heater-cathode, control grid, and plate, all mounted within an evacuated glass, metal, or ceramic case (Fig. 11-2). At the center of the structure is the cathode, which acts as a source of electrons when heated. The cathode electron-emitting surface can be heated directly or indirectly.

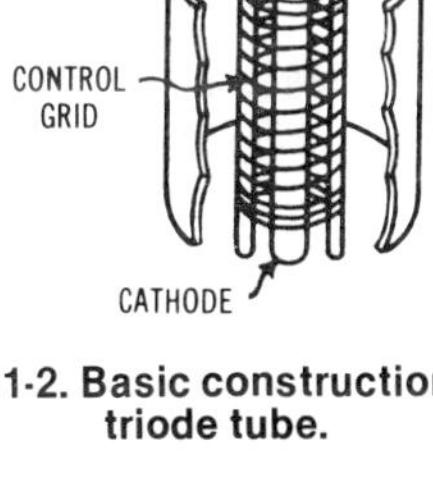

Fig. 11-2. Basic construction of a triode tube.

The cathode is surrounded by a control grid made of a fine-wire mesh. Its gridlike construction is such that the electrons can readily pass through the wires and move on toward the more positive plate. The grid function is to control the number of electrons flowing from cathode to plate.

The plate is a solid electrode that surrounds the grid. A positive potential (with respect to the cathode) is applied to the plate, which then attracts the electrons emitted from the cathode. The control grid is able to control the number of these emitted electrons that reach the plate.

The triode can be used as an amplifier of audio, video, and radio frequencies. It can serve as either a voltage or power amplifier. It can also be used as an oscillator, limiter-clipper, and in other special services.

(c) A tetrode is similar to a triode except for an additional electrode between the grid and the plate (Fig. 11-3). This additional electrode, called a screen grid, makes the plate current much more independent of the plate voltage. In effect, it acts as a shield between the plate and the grid circuit, minimizing feedback between the input and output circuits by way of the internal tube capacities. The screen grid, also being a fine wire mesh, permits the easy passage of electrons enroute to the plate. Because it operates at a positive potential, the screen grid has a substantial influence on the amount of plate current.

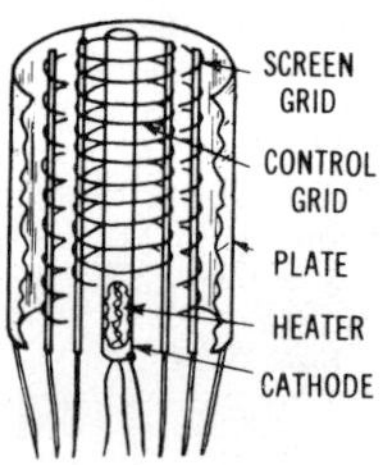

Fig. 11-3. Basic construction of a tetrode tube.

In a tetrode, the plate voltage may not be driven too far in the negative direction. If the plate voltage becomes lower than the screen voltage during a portion of the input cycle, electrons strike the plate and dislodge secondary electrons which are attracted to the screen (the screen grid now has a higher potential than the plate). This (dynatron effect) limits the linear range over which the plate voltage can be made to change by an applied signal.

The tetrode can be used in the same services as a triode, and it is common as an amplifier of radio frequencies. Although not as popular as the pentode, it sees service as a high-powered radio-frequency amplifier.

(d) The pentode is a five-element tube. It is similar to the tetrode in construction (Fig. 11-4), except that it has an additional electrode (suppressor grid) between the screen grid and the plate. The primary function of this electrode is to counteract secondary emission at the plate.

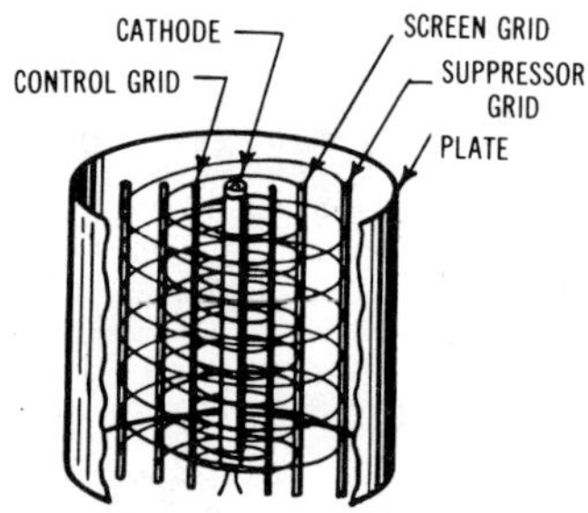

Fig. 11-4. Basic construction of a pentode power tube.

The inserted suppressor grid, held at or near ground potential, retards the emission of secondary electrons, causing them to fall back on the plate. This retarding action can be anticipated, because the suppressor grid is essentially negative with regard to the plate. As a result, the positive plate voltage can swing to a value that is substantially lower than the actual screen-grid potential.

The pentode can be put to the same uses as the triode and tetrode. However, the pentode is especially effective as a radio-frequency small-signal amplifier. They are also used as high-gain voltage amplifiers in audio and video circuits. Other pentodes, because of their good power sensitivity, are also employed as audio power amplifiers.

The curves of Fig. 11-5 provide a summarizing comparison among the triode, pentode, and tetrode. Triode tubes are characterized by low plate resistances

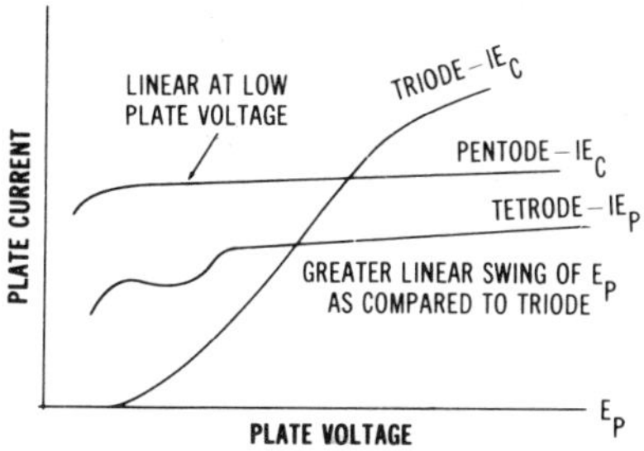

Fig. 11-5. Characteristics of triode, tetrode and pentode tubes.

(R_p) and low to moderate values of amplification factor (μ). A triode is often referred to as a *constant-voltage generator* because its plate voltage is rather independent of plate current for any constant value of grid bias (refer to the triode curve). The triode has good fidelity as an audio amplifier when operated on the linear portion of its transfer characteristic. It has a significant second-harmonic component which limits its efficiency if the second-harmonic distortion is not to exceed a tolerable percentage. Such second harmonics can be suppressed by using two triodes in a push-pull arrangement. In class-C rf amplifier operation, a triode requires neutralization because of the low-impedance feedback path from plate to grid. Moreover, the triode requires more drive signal than a comparable tetrode or pentode audio or class-C amplifier.

The tetrode has a high plate resistance, and it can have a high g_m. In addition, there is more effective shielding between the plate and grid circuits. Tetrode and pentode tubes are referred to as *constant-current generators*, because their plate currents are rather constant regardless of changes in plate voltage for a given grid bias. A given change in the grid voltage of a pentode, compared with that of a triode, causes a much greater change in plate current and a greater output. In class-C operation a good pentode or tetrode normally requires no neutralization, except when operated at very high frequencies. Even at high frequencies, only a limited amount is required.

The pentode has a high plate resistance and can be made with a very high g_m. Thus, a substantial change can occur in the plate current and plate voltage with only a small change in the applied grid signal. Note from the pentode curve that its performance is linear to a much lower value of plate voltage than a comparable tetrode.

In general, vacuum tubes have a high grid-input resistance. For most practical designs, it can be neglected and considered infinite in comparison to the impedance of the source of input signal. The plate-output resistance of a vacuum-tube is also relatively high but much lower than its input resistance.

Pentode and tetrode tubes have a substantially higher output resistance than a triode. Again, this verifies the logic of considering a triode as a low-impedance constant-voltage source of signal, while the tetrode and pentode are considered more appropriately a high-impedance constant-current source of signal.

Since a triode has a low output resistance, a certain level of output voltage can only be obtained with both a greater plate-current variation and high dc component of plate current as compared to tetrode or pentode operation. This does not mean that tetrodes and pentodes do not operate with high dc plate currents and ac current variations. With comparable levels of current variation they incorporate an even higher output capability plus a so-called greater "power sensitivity." This means that a greater power output can be obtained for a given level of applied signal. Stated another way, more driving power is required for a triode than a pentode for a given level of power output.

(e) The beam-power tube (Fig. 11-6) is a special form of tetrode-pentode with characteristics more like those of a pentode. It provides the ultimate in power sensitivity and absolute value of output

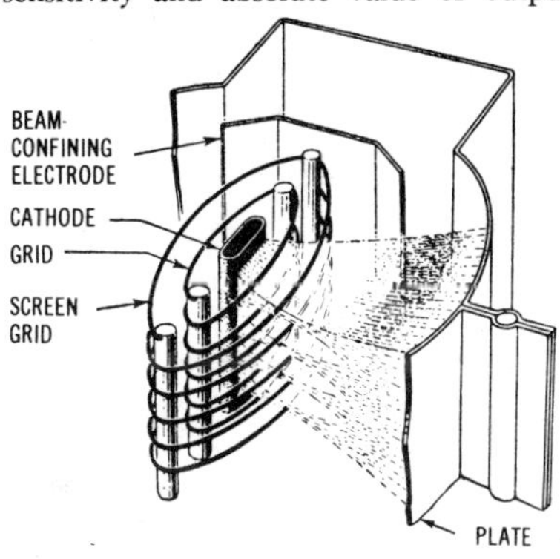

Fig. 11-6. Basic construction of a beam-power tube.

power. Beam-forming plates between the screen grid and the plate guide the electrons in an efficient beam-like manner between the cathode and plate. Because these plates are at cathode potential, their focusing action sets up a simulated suppressor grid between the screen grid and the plate. Hence, even at low plate voltages there is no flow of secondary electrons from the plate to the screen grid. Such a tube has both a high plate efficiency and power sensitivity. It is used widely in audio power stages and throughout the radio-frequency sections of transmitters. It has the same uses as triode and pentode with special emphasis on those applications that require the generation of considerable power output.

(f) Over the years a variety of special tubes has been designed for special services. One example is the remote-cutoff tube. Such a tube uses a variable spacing in the control-grid construction, with the wires of the mesh being closely wound at the ends and with a progressively wider spacing toward the center. This differs from the essentially uniform control-grid construction of a regular tube. The influence of uniform and variable spacing on the grid-voltage and plate-current characteristic of a tube is shown in Fig. 11-7. For a conventional

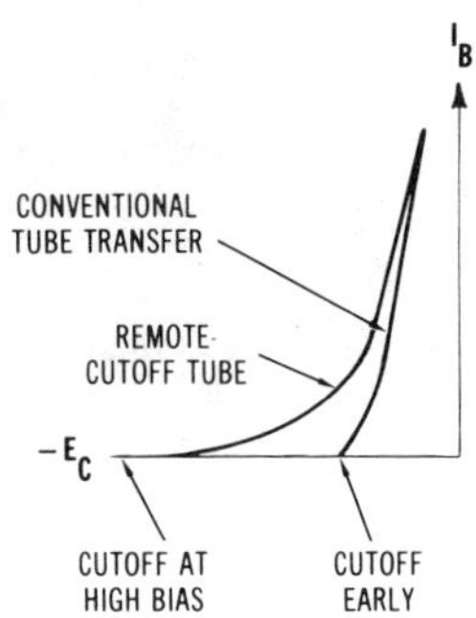

Fig. 11-7. Remote-cutoff operation.

control-grid construction, the tube cuts off very sharply at the grid cutoff bias. With the changing control-grid spacing, the cutoff of the plate current is approached more gradually with rising negative grid voltage. Over this range the μ of the tube falls gradually. When the tube is driven hard, the gradual swing to cutoff produces less distortion than a sharp cutoff.

The above feature provides an excellent means of changing the gain of an amplifier using either manual or automatic means. As the dc component of grid bias is increased, the gain of the stage will decrease in a gradual manner. The remote-control or variable-μ pentode is used widely in radio receivers as an rf and if amplifier. Using the remote cutoff characteristic, it is possible to use avc-agc systems to regulate amplifier gain in accordance with the magnitude of an incoming signal. Remote-cutoff tubes are also used in other services for which a smooth control of amplifier gain is a prerequisite.

(g) A duo-triode is a tube that contains two individual triode sections within the same envelope. Usually the triodes have similar characteristics except for interelectrode capacitances. Such a tube can be used in any service appropriate to triode operation, affording a space-saving mount and convenient circuit arrangement when more than one identical triode is required in a piece of electronic gear.

(h) The cold-cathode diode or triode requires no heater. Instead, it contains a gas which ionizes upon application of the proper voltages between anode and cathode or between a starting electrode and the cathode (Fig. 11-8). Once the gas is ionized and current flow starts between cathode and anode, it does not stop until the anode-cathode voltage drops to a level that causes the gas to deionize. In a diode cold-cathode tube, a starter extension, attached to the main anode, is positioned relatively near to the cathode. The gas first ionizes across this short distance, and then spreads to produce

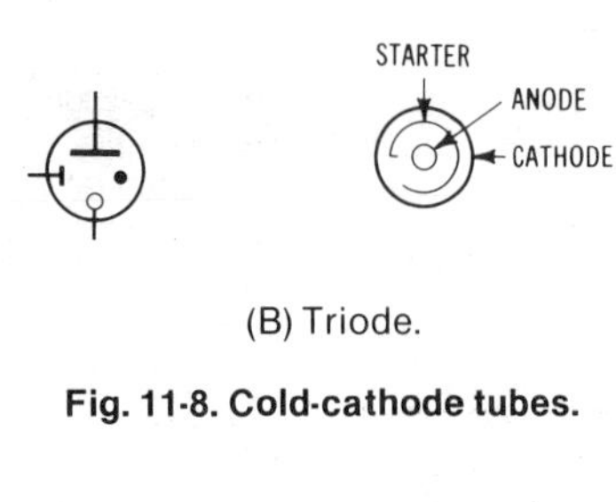

(A) Diode.

(B) Triode.

Fig. 11-8. Cold-cathode tubes.

complete ionization over the entire anode surface.

In a triode cold-cathode tube, a separate starting electrode is included. It acts as a control element and when there is a proper difference of potential between it and the cathode, the gas begins to ionize and then spreads over to the plate surface. Cold-cathode tubes can be used as voltage regulators and light-producing strobotron devices.

(i) The thyratron is a hot-cathode gas-filled control tube consisting of anode, cathode, and control or "firing" grid. When an adequate difference of potential exists between plate and cathode, the gas within the tube will ionize and present a low-resistance path that permits a very high current flow.

A negative potential applied to the control grid (Fig. 11-9) prevents gas ionization despite a high anode potential. When the grid voltage is made less negative or positive (depending upon thyratron design), the positive voltage on

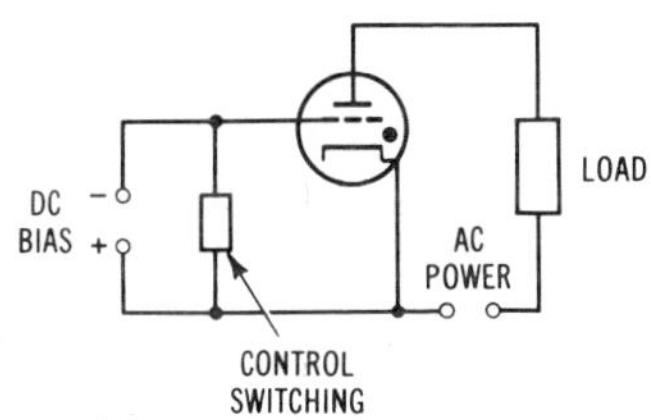

Fig. 11-9. Thyratron control circuit.

the plate will be adequate to start ionization which will spread rapidly and permit a high current flow. The grid then loses control of thyratron operation until a negative or significantly lower anode voltage is applied to deionize the gas. The thyratron is used as a high power switch. A control-grid signal of insignificant power level can be used to switch a very high-powered circuit that contains the thyratron tube.

11-1-2. Draw a simple circuit diagram of each of the following and describe the operation. Show a signal source, and include coupling and bypass capacitors, power-supply connections, and plate load: (a) af "grounded-cathode" amplifier with cathode resistor biasing, as for class-A operation; (b) af "grounded-cathode" pentode amplifier with battery biasing, for class-A operation; (c) rf "grounded-grid" triode amplifier with LC tank plate-load, for class-B operation; (d) af "cathode-follower" triode amplifier; and (e) af push-pull pentode amplifier operated class B, with transformer coupling to a speaker.—(a) In a vacuum-tube amplifier as shown in Fig. 11-10, the dc voltage (E_b) supplied to the plate attracts the electrons emitted from the cathode. The number of electrons

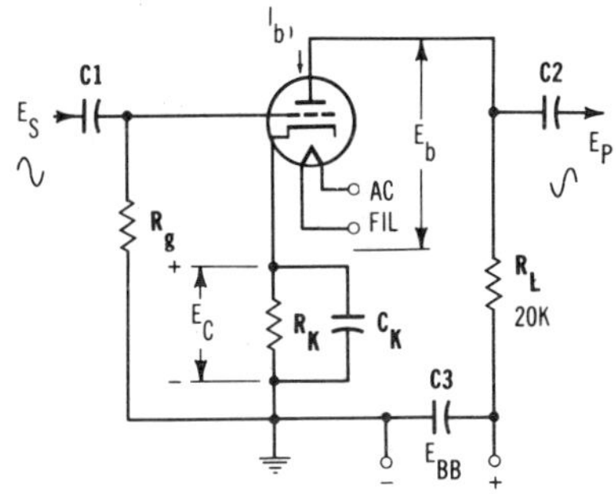

(A) Circuit.

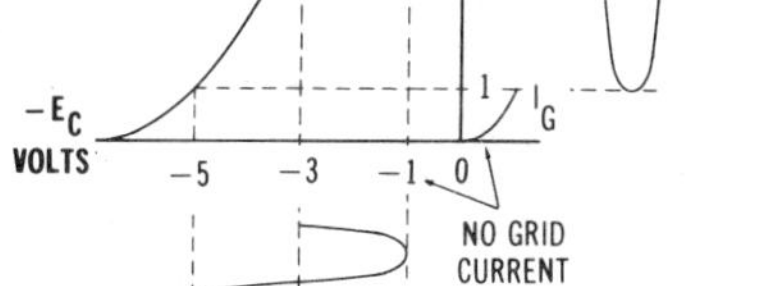

(B) Transfer Curve.

Fig. 11-10. Typical grounded-cathode stage.

reaching the plate and flowing through the plate-load resistor (R_L) is determined by the negative voltage supplied to the control grid. In this circuit the plate current (I_b) in flowing through the cathode resistor (R_k) develops a potential of 3-volts dc across the cathode resistor-capacitor combination. The direction of the plate current through the cathode

resistor is such that the grid is made negative, with respect to the cathode, by 3 volts.

The so-called transfer curve of Fig. 11-10B shows that the plate current is a function of the negative grid voltage. Notice that over the most linear portion of the curve (between -1 and -5 grid volts), the plate current changes linearly with respect to the grid voltage. The negative dc bias is usually at the center of the linear portion of the transfer characteristic (-3 volts in the example). With this amount of negative grid bias supplied to the vacuum-tube stage, the dc component of plate current I_b would be 2.5 milliamperes. This is referred to as the *operating-point plate current.*

To understand the operation of the triode as an amplifier, assume that an ac sine wave (E_s) of 2-volts peak is being supplied to the control grid. As the sine wave rises on its positive crest, the negative grid voltage decreases and the plate current increases. At the crest of the positive alternation, the instantaneous grid voltage is —1 volt and the plate current is 4 milliamperes. The less negative the grid is driven, the weaker the space charge is and the more freely the electrons flow between cathode and plate.

On the negative alternation of the grid-voltage signal, the grid swings in the negative direction. Consequently, the plate current decreases, because the negative grid has a greater holding power on the space-charge electrons. At the crest of the negative alternation, the instantaneous grid voltage is —5 volts and the plate current is only 1 milliampere.

When the plate current increases with a rise in grid voltage, the plate voltage drops because of the greater voltage drop across the load resistor. Conversely, when plate current falls during a negative swing of the grid, the plate voltage rises. This is to say that input and output voltages are 180° *out of phase.*

A very important fact to recognize is that the plate-current change follows the grid-voltage change. As the grid voltage swings between peaks, it changes between —1 and —5 volts. Instantaneously, the plate current also changes between the limit of 4 milliamperes and 1 milliampere. Thus the peak ac plate current is 1.5 milliamperes, which flows in the 20,000-ohm plate load resistor (R_L).

The plate-voltage change can be determined by multiplying the plate current times the ohmic value of the plate-load resistor or:

$$E_p = I_pR_L = 0.0015 \times 20{,}000 = 30 \text{ volts}$$

The plate-voltage change is 30 volts peak. Inasmuch as the initial grid-voltage change was 2-volts peak, the amplifier has a gain of 15 (30/2).

This is how a triode vacuum tube functions as a voltage amplifier. By correct selection of tube and operating conditions, a vacuum-tube stage can also be made to operate as a power amplifier.

In the example of Fig. 11-10, the stage is said to operate class A, because there is plate current during the entire cycle of the grid-voltage input signal. Likewise, the grid is always negative, and, hence, does not draw any current from the cathode. The plate-current and plate-voltage changes are essentially replicas of the grid-voltage change.

The dc component of plate current I_b, in flowing through the cathode resistor, develops the required dc bias for the tube. This bias is such that the operating grid bias is at the center of the most linear portion of the transfer characteristic. To prevent negative feedback that could result from any ac voltage variation across the cathode resistor, it is necessary to use a cathode filter capacitor (C_k) that is large enough to filter out any ac voltage variations across the cathode resistor, keeping the voltage across the cathode combination constant at -3 volts. A capacitor (C3) is also connected across the plate supply. It filters out ac variations and blocks the transfer of undesired components in either direction between amplifier and power circuits. Capacitors C1 and C2 provide a means of transferring ac variations between stages.

The stage is referred to as grounded-cathode amplifier, because the cathode is at ac ground potential. If the cathode capacitor were not present, the cathode would not be at exactly ac ground potential, because of the ac voltage drop that would appear across the cathode resistor. Such a stage would still be considered as a grounded-cathode amplifier. The small ac variation across an unbypassed cathode resistor provides a form of negative feedback.

(b) A pentode grounded-cathode amplifier is shown in Fig. 11-11. Instead of using cathode bias, a grid-bias battery is used. No cathode resistor-capacitor combination is employed. However, it is

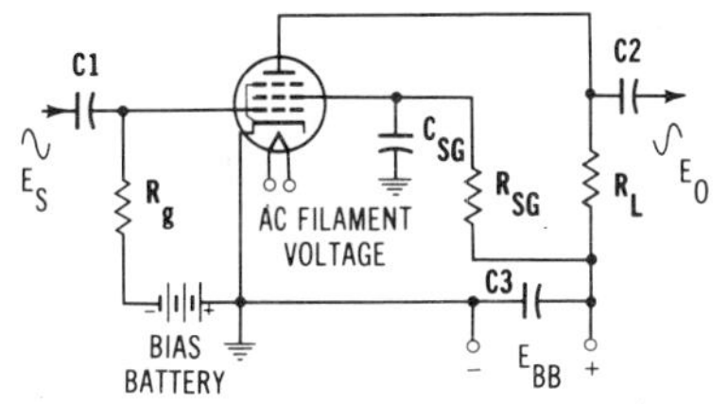

Fig. 11-11. Pentode grounded-cathode amplifier stage.

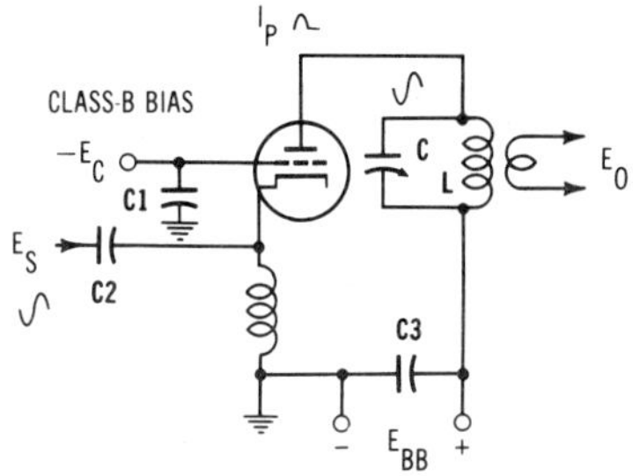

Fig. 11-12. Grounded-grid amplifier.

necessary to supply screen-grid voltage to the pentode. Quite often, the screen-grid voltage is lower than the operating plate-supply voltage. To obtain the desired amount of screen-grid voltage, a screen resistor (R_{SG}) provides the necessary voltage drop from the supply voltage. A screen-grid filter capacitor (C_{SG}) is used between the screen grid and ground to keep the screen-grid voltage constant despite the variations in the screen-grid current with applied signal. The suppressor grid is internally tied to the cathode.

The operation of the pentode grounded-cathode amplifier is similar to a triode grounded-cathode stage. The ac voltage variation on the grid causes a similar change in the pentode plate current. The variations in the plate current through load resistor R_L, in turn, produce an ac output-voltage variation. Again the output voltage change is 180° out of phase (opposite polarity) with the input voltage change.

In the usual pentode voltage amplifier, a substantially higher value of plate-load resistance (R_L) can be used as compared to that of a triode. Thus a smaller change in plate current, under the control of an applied grid-signal voltage, produces a higher output voltage.

The influence of the screen voltage is such that it exerts more control over the plate current than the plate voltage does. Therefore, the plate voltage can be made to swing over a substantially greater voltage range, and its variation along with that of the plate current will result in a high output that is a good replica of the input signal. The suppressor is operated at ground or cathode potential.

(c) A grounded-grid triode rf amplifier is shown in Fig. 11-12. In this type of circuit the control grid rather than the cathode is at ac ground potential (via capacitor C1). The input signal, instead of being applied to the control grid, is applied to the cathode. Regardless of the point of application of the input signal, grid or cathode, there is an ac voltage variation between grid and cathode. However, the impedance is low with cathode input and a low-impedance load is placed on the preceding stage.

The particular stage of Fig. 11-12 is biased class B from a source of negative voltage of the proper value to bias the tube at, or very near, cutoff. A filter capacitor (C1) is used to hold this bias voltage constant, and in so doing, it maintains the control grid at rf ground potential.

When a signal is applied between cathode and ground, its positive alternation swings in the cutoff direction (application of a positive voltage to the cathode is felt as a negative grid-cathode voltage). Inasmuch as the tube is biased to cutoff, the positive alternation will swing beyond the cutoff bias voltage.

The negative alternation of the applied cathode signal will act as though a positive-going signal were applied to the control grid. Consequently, the plate current will rise and follow the negative alternation of the applied signal. As the plate current increases, the plate voltage drops, and there is a negative swing across the plate resonant-tank circuit. The plate current will, of course, cut off as soon as the cathode signal swings positive. However, a resonant LC circuit is an energy-storing combination, and the high negative charge placed on the resonant-circuit capacitor during the plate current flow will discharge. Thus there will be a flow of electrons in the resonant-tank circuit, even during the positive alternation of the applied cathode signal. This circulating current in the resonant circuit produces a positive alternation of voltage across the resonant circuit. As a result, a sine wave of voltage will be developed across the output resonant circuit despite the fact that the plate current appears for only the negative alternation of the input wave.

If a resistive load were used instead of the resonant load, the reconstruction of the sine wave would not occur. The output voltage in this case would vary only during one alternation of the input wave, that alternation of the wave which causes the change in plate current.

(d) An audio-frequency cathode follower is shown in Fig. 11-13. A cathode

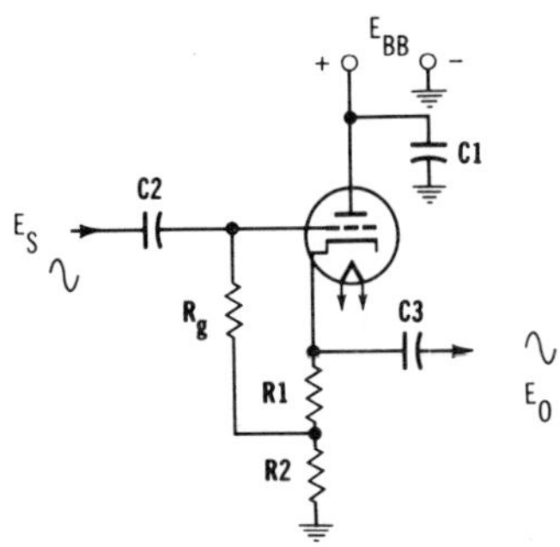

Fig. 11-13. Cathode-follower stage.

follower operates as a grounded-plate stage. The normal plate voltage is applied to the tube. However, notice that a filter capacitor (C1) connects directly from the plate to ground. This connection holds the plate at ac ground potential.

In this question we have now covered the three basic types of vacuum-tube connections. These are grounded-cathode, grounded-grid and grounded-plate (cathode follower). The cathode-follower stage has a very high input impedance and therefore places a very light load on any preceding stage. The dc component of voltage developed across the R1 portion of the total cathode resistance serves as the bias for the stage. External battery or supply-line bias can be used if desired.

The cathode follower has a low output impedance, and unlike the grounded-cathode or grounded-grid configurations, the voltage gain of a cathode follower is always less than 1. The input signal is applied between grid and ground, and the output signal is removed from between cathode and ground. The positive alternation of the input signal causes an increase in plate current. This plate current, in flowing through the cathode resistor as cathode current, causes an increase in voltage across the cathode resistor. A negative alternation of the input signal causes a decrease in plate and cathode currents, and therefore a decrease or negative swing across the cathode resistor. It is to be noted that with the cathode-follower connection, the output and input voltages are in phase. Input and output voltages are also in phase for a grounded-grid amplifier. Only with the grounded-cathode configuration are input and output voltage exactly 180° out of phase.

(e) A push-pull pentode audio amplifier for class-B operation is shown in Fig. 11-14. In a push-pull audio class-B amplifier, it is possible to obtain an out-

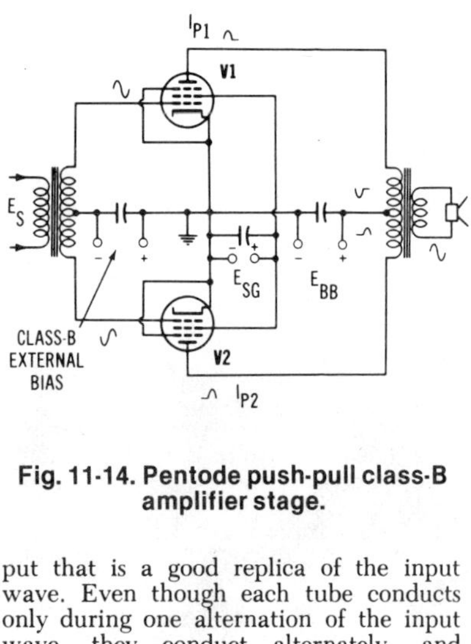

Fig. 11-14. Pentode push-pull class-B amplifier stage.

put that is a good replica of the input wave. Even though each tube conducts only during one alternation of the input wave, they conduct alternately, and therefore the complete input wave is reconstructed in the output. Both tubes are biased class B, with the use of an external bias. The circuit arrangement is such that an ac ground potential exists at the center of the input-transformer secondary and the center of the output-transformer primary. Plate-supply voltage is inserted at the center point of the output-transformer primary.

The turns ratio of the input transformer is such that a proper impedance match and voltage step-down exists between the low-impedance input of the class B grids and the higher-impedance output of the preceding stage. Input impedance is low because the class-B tubes draw grid current during the peaks of the positive alternations. The turns ratio of the output transformer is such that the very low impedance of the speaker is matched to the substantially higher output impedance of the class-B tubes.

The positive alternation of the input wave, as it appears across the secondary of the input transformer, drives the grid of V1 positive and the grid of V2 negative. Inasmuch as the tubes are biased class B, V2 swings below cutoff and has no output, while the grid of V1 swings in the positive direction, and there is a like change in its plate current. As a result there is a negative swing of the plate voltage of V1 and an ac variation across one-half of the primary of the output transformer. During the negative alternation of the input wave, the grid of V1 swings negative into cutoff, while the grid of V2 swings in the positive direction. Therefore the plate current rises in V2 and follows the negative alternation of the input wave. This results in a decrease in the plate voltage of V2, and, consequently, a negative alternation of the input wave appears as a positive alternation across the other half of the output-transformer primary.

In summary, with a class-B push-pull connection, the complete input wave is reconstructed in the output circuit. Its phase, of course, is of opposite polarity to that of the input wave. Although amplifiers of this type, according to their biasing, are referred to as class-B amplifiers, they are not always biased exactly at cutoff. By biasing them just very slightly above cutoff, there can be a noticeable reduction in distortion.

11-1-3. What is meant by "space charge"? By "secondary emission"?—In the normal operation of a vacuum tube, the cathode is able to emit more electrons than the plate can accept. These excess electrons collect in a cloud between the cathode and control grid. This grouping of electrons is referred to as the space charge. Inasmuch as it is a grouping of electrons, the net charge is negative and it offers opposition to the further emission of electrons from the cathode.

The control grid, however, is able to influence the electron density of this space charge; the more negative the control grid, the higher is the density. The control that the grid exerts is used to regulate the number of electrons that break away from the cloud and move on to the plate. The less negative the grid is made, the more electrons are pulled out of the space charge and pass on through the grid wires to the plate.

The secondary emission of electrons refers to those electrons dislodged from a surface by the bombardment of arriving electrons. In a vacuum tube, electrons arrive at the plate with such impact that they dislodge other electrons from it. These latter electrons either fall back onto the plate or are attracted by another electrode momentarily operating at a higher potential. The release of these secondary electrons from the plate is called secondary emission.

11-1-4. What is meant by "amplification factor" (μ) of a triode vacuum tube (amplifier)? Under what condition would the amplifier gain approach the value of μ?—The μ is a measure of the effectiveness of a vacuum tube in producing a plate-voltage change from an applied grid-voltage change. More exactly, it expresses the quotient of a change in plate voltage that results from a very small change in grid voltage, or:

$$\mu = \frac{\Delta E_p}{\Delta E_g} \quad (I_{p\ \text{constant}})$$

The gain of a vacuum-tube amplifier can be made to approach, but never reach, the value of the μ of a given tube. The higher the impedance of the load placed on the output of the tube (providing the appropriate dc operating voltages can be maintained), the more nearly the gain of the associated amplifier approaches the value of μ. It can be said then that over the practical range of operation, the higher the value of the plate load resistance (R_L), provided the proper plate voltage and grid bias are maintained and the tube is operated only over the linear portion of its characteristic, the higher is the possible voltage gain and the nearer the stage gain approaches the value of μ.

11-1-5. What is meant by "plate resistance" of a vacuum tube? On what does its value depend?—The plate resistance is a measure of the influence that a plate-voltage change has on a plate-current change. It can be expressed mathematically as the quotient of a small change in plate voltage over the resultant small change in plate current, or:

$$R_p = \frac{\Delta E_p}{\Delta I_p}$$
$(E_{g\ constant})$

Its value depends on the influence that a changing plate voltage has on the plate current. As compared to a triode, a tetrode or pentode tube with its screen grid affords better isolation, and, therefore, a change in plate voltage has much less influence on the plate current. This means that the plate resistance of a tetrode or pentode is much higher than that of a triode, as discussed previously in connection with Question 11-1-1. The more effective the screen-grid voltage is in controlling the plate current, the higher the plate resistance of a tube.

11-1-6. What is meant by the voltage gain of a vacuum-tube amplifier? How does it achieve this gain?—The voltage gain of an amplifier is the quotient of the ac output voltage over the ac input voltage. Expressed as an equation:

$$\text{Gain} = \frac{E_o}{E_{in}}$$

Voltage gain is the result of the amplification ability of the tube. A small grid-voltage change results in a substantial plate-current change. This plate-current change in a moderate to high value of load resistance results in a plate-voltage change that is of greater magnitude than the input-signal change. The activity of the tube in amplifying an applied signal was covered in detail in part (a) of Question 2.

11-1-7. Draw a rough graph of plate current versus plate-supply voltage for three different bias voltages on a typical triode vacuum tube. (a) Explain, in a general way, how the value of the plate-load resistance affects the portion of the curve over which the tube is operating. How is this related to distortion? (b) Operation over which portion of the curve produces the least distortion?—(a) Such a set of curves is shown in Fig. 11-15. On these curves several load lines corresponding to different values of load resistance have been drawn. It can be noted that for a given grid-voltage (peak-to-peak swing between 0 and −2 volts), a higher output voltage with a lower operating dc plate current can be obtained with the highest value of load resistance.

However, distortion is also a factor

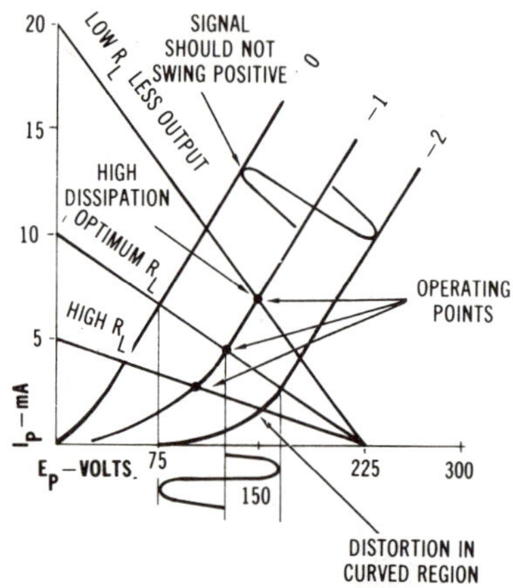

Fig. 11-15. Load lines on triode characteristics curve.

that is influenced by the value of the load resistance. Thus an optimum value of load resistance is a function of both gain and tolerable distortion. The highest resistance load line crosses over some of the curved portions of each of the bias curves. Therefore, the response of the amplifier is not linear in these segments, and distortion can result. With too low a value of load resistance the gain is limited, and the operating point may be of some value that will cause the safe plate dissipation of the tube to be exceeded.

The load line must also be positioned at such a location that the applied signal does not swing the grid any further positive than zero, or any further negative than the curved portion of the bias curve near cutoff. Any positive swing of the grid voltage results in grid-current flow and distortion of the input wave. A swing into cutoff clips off a portion of the output wave.

(b) An operating point and load-line value must be selected in accordance with the supply voltage and the maximum amplitude of the input wave. It should be positioned so that the signal variation occurs over the linear portions of the bias curves. Operations should be concentrated mainly in the section of the characteristics where there is uniform spacing among the individual bias curves. Distortion is caused by a positive swing beyond zero and a negative swing into the curved and closely spaced regions of the bias curves near cutoff.

11-1-8. Draw a rough graph of plate-current versus grid-voltage (I_o vs E_g) for various plate voltages on a typical triode vacuum tube. (a) How would output current vary with input voltage in class-A amplifier operation? class-B operation? class-C operation? (b) Does the amplitude of the input signal determine the class of operation? (c) What is meant by current-cutoff bias voltage? (d) What is meant by plate current saturation? (e) What is the relationship between distortion in the output waveform and: (1) the class of operation; (2) the portion of the transfer characteristic over which the signal is operating; and (3) amplitude of input signal? (f) What occurs in the grid circuit when the grid is "driven" positive? Would this have any effect on biasing? (g) In what way is the output current related to the output voltage?—A typical set of I_p vs E_g triode curves are shown in Fig. 11-16. The transfer-characteristics drawing of Fig. 11-17 can also be used in understanding the various classes of amplifier operation.

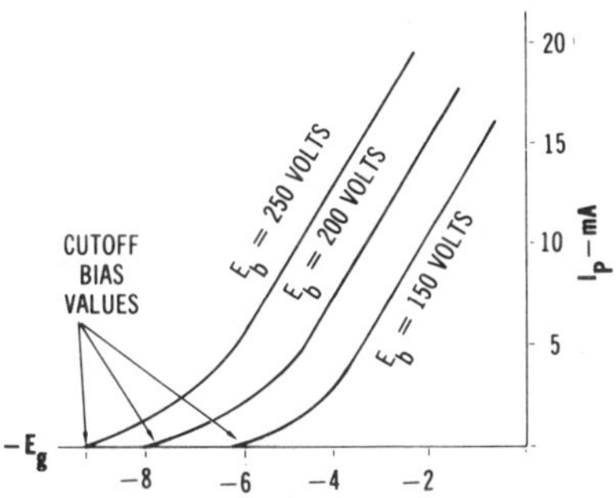

Fig. 11-16. I_p—E_g characteristics of a triode.

(a) In a class-A amplifier, the output plate current is a replica of the input voltage (Fig. 11-17). Plate current appears during the entire cycle of the input-voltage wave. A class-B amplifier as shown is biased at cutoff. Thus, with an applied input sine wave the tube conducts only during the positive alternation of the input wave. This is to say, plate current occurs for only 180° of the input wave.

Class-AB bias refers to some level of bias between strictly class A and strictly class B. In this class the plate current occurs for something more than 180° but less than 360° of the input cycle, and

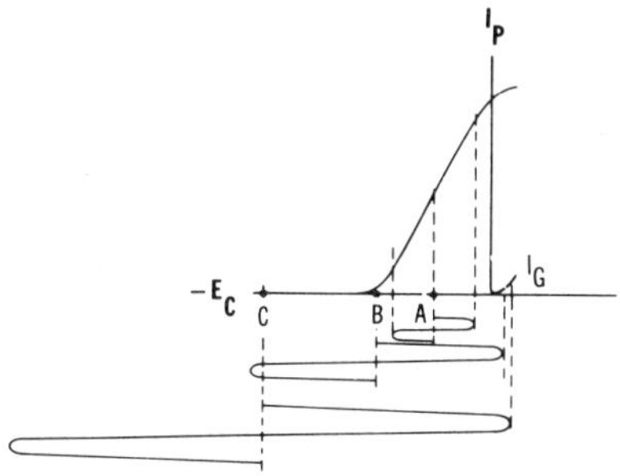

Fig. 11-17. Class-A, -B, and -C bias.

plate current occurs for the entire positive alternation of the input wave and for a portion of the negative alternation. In class-C amplifier operation the tube is biased beyond cutoff. As a result, plate current occurs for only a portion of the positive alternation of the input wave. In a typical class-C amplifier the plate current may occur for only 90° to perhaps 150° of the input cycle.

(b) The class of operation of an amplifier is mainly determined by the dc biasing of that amplifier and can be influenced by the magnitude of the alternating grid voltage. It must be stressed, depending on the type of circuit, that the amplitude of the input signal can influence the type of bias if it in some way changes the bias or, in the case of class-A operation, is great enough to swing beyond cutoff. In the latter case, plate current will not be present during the entire input wave. In signal-biased circuits the type of bias established is very much a function of the magnitude of the input signal.

(c) Current-cutoff bias voltage is the negative grid voltage at which the plate current falls to zero. It should be noted from Fig. 11-16 that the higher the operating plate voltage, the higher the negative cutoff bias needed to reduce the plate current to zero.

(d) The plate is said to be saturated when it accepts all the electrons emitted from the cathode, and an increase in the plate voltage causes no significant increase in the plate current.

With plate saturation there is no space charge, because all of the electrons are attracted to the plate as fast as they are emitted from the cathode.

(e) Related distortions in the output waveform are dependent on:

(1) In a typical single-ended audio amplifier an essentially nondistorted output is obtained only when the amplifier is operated linear class A. For push-pull operation an essentially nondistorted output can be obtained using class A, class AB, or class B.

The above statements are also correct for radio-frequency amplifiers. In addition, with an appropriate resonant load (LC circuit) radio-frequency single-ended and push-pull amplifiers can also be biased class B or class C, and the output wave becomes a good replica of the input wave. However this latter statement does not always apply for a class-C amplifier when the radio frequency wave is modulated.

(2) Distortion also depends on the portion of the transfer characteristic over which the tube is operated. Nondistorted output is only obtained when the tube is operated over the linear portion of its characteristic or, as in the case of class B, when a push-pull connection is employed. Driving the grid beyond zero or into the saturated region of the tube characteristic produces a distorted output. Likewise, a signal swing into the crowded portion of the curves near cutoff results in a distorted output.

(3) The amplitude also influences distortion. Even though a tube is biased on the linear part of its characteristic, a high-amplitude signal can swing the grid positive or into the crowded region near cutoff. Distortion will result despite the attempted linear biasing of the stage.

(f) When the grid is driven positive, the control grid attracts some of the electrons emitted from the cathode. Thus it removes electrons from the stream moving between the cathode and plate. The higher the positive swing of grid voltage, the more electrons it removes from the electron stream.

The grid current reduces the input resistance of the stage and places a greater load on the preceding source of signal. When grid current occurs, additional power must be supplied from the source of signal if the desired rise in grid voltage is to be attained. If that required power is not available, the input wave will be limited or clipped by the flow of grid current.

If there is an RC time contact in the grid circuit, the grid current will build up a charge on the capacitor. The average charge on the capacitor will be a dc component that, in effect, will change the normal grid bias of the stage. With a time constant present, the grid current will also affect the dc operating bias of the stage in addition to its possible influence on the instantaneous grid-voltage rise.

(g) In the typical audio amplifier and in other amplifier types, there is a linear relationship between output (plate) current change and output voltage (plate) change. The two components, however, are 180° out of phase. When the plate current increases, the output voltage decreases, and vice versa. An increase in the plate current results in a greater voltage drop across the load and a lower instantaneous plate voltage. Since the output is removed between the plate and cathode, this plate-voltage change is the actual output voltage of the amplifier. Conversely, with a decrease in plate current, there is a lower voltage drop across the load and a higher plate and output voltage.

11-1-9. Why is the efficiency of an amplifier operated class C higher than one operated class A or class B?—The efficiency of a vacuum tube, as related to its plate circuit, has to do with the efficiency of the conversion from dc plate-power input to ac power output. In a class-A amplifier the plate current occurs for the entire input cycle. The dc component of plate current is high and at the level of the operating point.

In a class-B stage, plate current occurs only during one alternation of the input wave, and therefore the average dc component of plate current is less and the dc power input is lower.

In a class-C amplifier the plate current occurs for an even smaller percentage of the total input-wave cycle. The dc input current is the average of these bursts of plate current. Therefore the average or dc component of current and power input are substantially less for a given swing in plate current and plate voltage, as compared to class-A or class-B operation. The plate must only dissipate power when current is drawn and as a result the class-C stage operates at an even higher efficiency; that is, a greater power output for a given dc power input can be obtained from class-C operation as compared to class A and class B.

11-1-10. Compare tetrode tubes to triode tubes in reference to high plate current and interelectrode capacitance.—In a tetrode tube the significant interelectrode grid-plate capacitance (C_{gp}) is much less than that of a triode. A triode has a substantially higher input capacitance than a tetrode. The screen grid functions as an electrostatic shield between the control grid and the plate. Therefore the effective or net capacitance between grid and plate is very much less.

The plate resistance of a tetrode is much higher than that of a triode, and, therefore, it can operate into a high-impedance load in a more optimum manner. Therefore, for a comparable output, a smaller output plate-current variation and dc plate current is required.

This does not mean that tetrode tubes cannot be made with high dc or ac current capabilities. High-powered tetrode and pentode tubes are available. When they are used, a high-current and high-power output can be obtained with less input driving power than that needed for a comparable output with a triode tube. It is said that tetrodes and pentodes have a higher power sensitivity than a triode.

11-1-11. What is the principal advantage of a tetrode tube over a triode tube as a radio-frequency amplifier?—The principal advantage of a tetrode tube as a radio-frequency amplifier is the fact that it requires no neutralization, because of the isolation between input and output provided by the screen grid. Almost as important an advantage is the fact that it has a higher power sensitivity. Thus, in higher-power rf amplifiers it provides a greater power output for a given rf driving power.

11-1-12. A triode, "grounded-cathode" audio amplifier has a μ (amplification factor) of 30, a plate impedance of 5000 ohms, load impedance of 10,000 ohms, plate voltage of 300 volts, plate current of 10 milliamperes, cathode-resistor bias is used. (a) What is the stage gain of this amplifier? (b) What is the cutoff-bias, E_{co}? (c) Assuming the bias voltage is one-half the value of E_{co}, what value cathode resistor would be used to produce the required bias? (d) What size capacitor should be used to sufficiently bypass the cathode resistor if the lowest approximate frequency desired is 500 hertz?— (a) The gain of the amplifier is determined with the basic voltage gain equation of a triode audio amplifier:

$$\text{Gain} = \frac{\mu R_L}{r_p + R_L}$$

$$\text{Gain} = \frac{30 \times 10{,}000}{5000 + 10{,}000} = 20$$

(b) The approximate cutoff bias or projected cutoff can be calculated as follows:

$$-E_{co} = \frac{E_b}{\mu} = \frac{300}{30} = -10 \text{ volts}$$

(c) If it is assumed that the operating bias voltage is one-half the cutoff value, the ohmic value of the cathode resistor must be as follows:

$$R_k = \frac{\frac{E_{co}}{2}}{I_b} = \frac{5}{0.01} = 500 \text{ ohms}$$

(d) Quite often the low-cutoff frequency refers to that frequency at which the response is down 29.3% (−3 dB) from the stage gain at mid-frequency. In this case the reactance of the capacitor at 500 hertz would have to be equal to the ohmic value of the resistor. It is also a common practice to assume that the response would be down no more than several percent when the reactance of the capacitor is made equal to 1/5 or even 1/10 of the ohmic value of the resistance at a desired low frequency. Based on this latter condition the value of the cathode bypass capacitor would have to be:

$$X_c = \frac{R_k}{10} = \frac{500}{10} = 50 \text{ ohms}$$

$$C_k = \frac{1}{2\pi f X_c} = \frac{1}{6.28 \times 500 \times 50} = 6.37\ \mu\text{F}$$

11-1-13. Are there any advantages or disadvantages of filament-type vacuum tubes when compared to indirectly heated types?—Disadvantages of the filament-type tube are that they are more subject to hum than corresponding heater-type tubes, lower emission efficiency, and high operating temperatures. Their advantages are the ability to generate high currents and to operate under adverse operating conditions including very high anode voltages.

A filament type is also subject to uneven loss of filament emission because one end of the filament is at a higher potential than the other end. This can be avoided by reversing filament voltage polarity on a scheduled basis.

11-1-14. What kind of vacuum tube responds to filament reactivation, and how is reactivation accomplished?—Tubes using thoriated filaments can be reactivated. Low emission results when the thorium layer has partially dissipated. However, there is some thorium still dissolved in the filament, and it can be brought to the surface by a temporary increase in the filament temperature. It is accomplished by first using a high flashing temperature with a filament voltage several times higher than normal for a short interval, and then holding the temperature for a lengthier period of time while operating at a temperature set by a filament voltage 50% above normal.

11-1-15. The following are excerpts from a tube-manual rating of a beam pentode. Explain the significance of each item. Control grid-to-plate capacitance, 1.1 pF; input capacitance, 2.2 pF; output capacitance, 8.5 pF; heater voltage, 6.3 volts; maximum dc plate-supply voltage, 700 volts; maximum peak-positive pulse voltage, 7000 volts; maximum negative-pulse plate voltage, 1500 volts; maximum screen-grid voltage, 175 volts; maximum peak-negative control-grid voltage, 200 volts; maximum plate dissipation, 200 watts; maximum screen-grid dissipation, 30 watts; maximum dc cathode current, 200 mA; maximum peak-cathode current, 700 mA; maximum control-grid circuit resistance, 0.47 megohm.—The control grid-to-plate capacitance is the interelectrode capacitance that exists between the control grid and the plate electrodes of the tube. This is the important interelectrode capacitance (C_{gp}).

The input capacitance has to do with the capacitance that exists mainly between the grid and the cathode (C_{gk}) plus the influence that C_{gp} and the output circuit might have on the total net capacitance seen across the input. The output capacitance is that capacitance that exists between the plate and the cathode (C_{pk}).

The heater voltage is that normal potential that must be applied across the filament pins for the proper heating of the cathode surface.

Maximum dc plate-supply voltage is the safe limit of the voltage that can be present from plate to cathode of the tube. The maximum peak-positive pulse voltage is that safe limit of positive peak voltage that may appear instantaneously from plate to cathode. Maximum-negative plate pulse voltage is the safe instantaneous peak voltage that may appear as a negative potential between plate and cathode.

Maximum plate dissipation is the maximum power that can be dissipated by the plate of the tube. The maximum screen-grid dissipation is the maximum power that can be safely dissipated by the screen grid.

Maximum dc cathode current is the maximum average or dc component of current that can be safely drawn by the tube. The maximum peak-cathode current is the maximum instantaneous peak current that may be safely drawn. Maximum control-grid circuit resistance is the maximum resistance that may be inserted between grid and cathode without causing unfavorable operation as a result of the accumulation of electrons at the grid and a resultant flow of an undesired reverse current.

11-1-16. Name at least three abnormal conditions which would tend to shorten the life of a vacuum tube; also name one or more probable causes of each condition.—Three abnormal conditions would be excessive plate or screen-grid voltage, excessive plate or screen-grid current, or excessive heater voltage. Proper heater voltage is important to the optimum and efficient operation of a vacuum tube. *If the heater voltage is too low, it is not possible to obtain optimum operating conditions. If it is too high (such as might occur with the improper adjustment of a variac, use of an incorrect tap on a heater power transformer, or frequency line-voltage surges and conditions that fail to keep the filament voltage reasonably constant), early burnout or early loss of emission capability may result.*

Excessive plate or screen-grid current can overheat tubes and result in a higher than safe screen-grid or plate dissipation. Loss of excitation is a frequent cause of excessive tube current in a class-C rf power amplifier. Loss of bias and improper tuning are two other causes. Power circuit and supply-voltage line defects, loss of supply-voltage regulation, power surges, or changes in loading can result in improper voltages. Loss of plate voltage can result in the flow of excessive screen-grid current.

11-1-17. Describe a pentagrid converter and its common usage.—A pentagrid is a five-grid tube employed as a combined oscillator, mixer, and first detector. The first two grids and the cathode form the oscillator circuit. The incoming radio-frequency signal feeds the third grid. The fourth grid is the screen grid; fifth grid, the suppressor grid. Such converters are common in vacuum-tube superheterodyne receivers.

11-1-18. What are some possible causes of insufficient plate current in a pentode circuit?—Since the screen grid controls the plate current of a pentode tube, a low screen-grid voltage would result in insufficient plate current. Too high grid bias could also be a cause of insufficient plate current. Plate current is also influenced by plate voltage and a significant drop in plate voltage would also result in a lower plate current.

11-1-19. What is the difference between a series-fed and a shunt-fed plate circuit? Draw a diagram of each.—In a series fed arrangement the supply voltage is, in effect, in series with the plate load and the plate-cathode circuit of the vacuum tube as shown in Fig. 11-18. In the

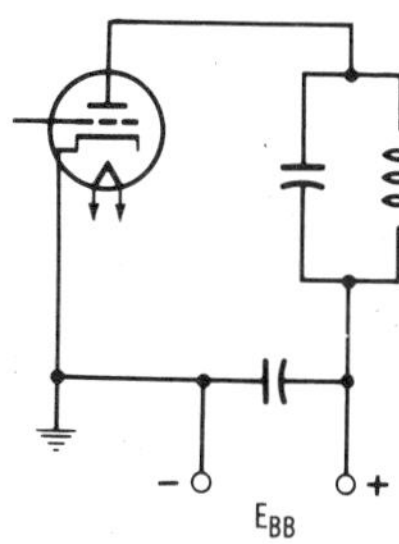

(A) Series feed.

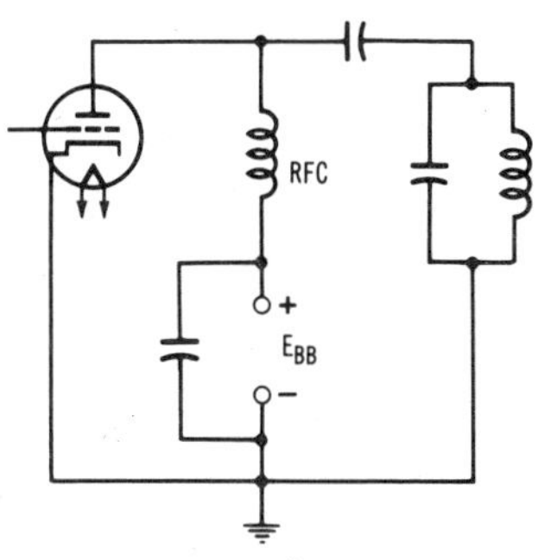

(B) Shunt feed.

Fig. 11-18. Plate-voltage circuits.

shunt-fed arrangement of example B note that the supply voltage is connected through a radio-frequency choke directly to the plate of the tube. In effect, the supply voltage is fed in shunt with the plate load and the plate-cathode circuit of the vacuum tube.

11-1-20. What is thermionic emission? How is it used in electron tubes?—Thermionic emission is the freeing of electrons from a material as a result of heat. In a vacuum tube a chemically treated metallic surface is heated by the vacuum-tube heater circuit and, as a result, electrons are freed and can be attracted by other elements of the tube. In the case of a triode it is the plate that attracts the electrons freed from the cathode surface that releases electrons by way of thermionic emission. The control grid regulates the flow of these electrons to the plate.

11-2. TRANSISTORS

11-2-1. Describe the difference between positive (p-type) and negative (n-type) semiconductors with respect to: the direction of current when an external emf is applied; the internal resistance when an external emf is applied.—An n-type semiconductor contains electron carriers. As shown in Fig. 11-19A the electrons in the n-type semiconductor will flow toward the positive side of an applied external emf. The direction of external current is determined by the polarity in which the emf is connected. As the arrows indicate, current (electron flow) is from the negative to the positive side of the applied emf source. Reversing the polarity of the applied emf causes a change in the direction of current in the external circuit and through the semiconductor material.

In p-type semiconductor material the current carriers are holes, or positive charges. Their direction of motion is opposite to that of electrons. Therefore, the hole current is toward the negative side of the applied emf source, as shown in Fig. 11-19B. The direction of current is again a function of the polarity of the external emf.

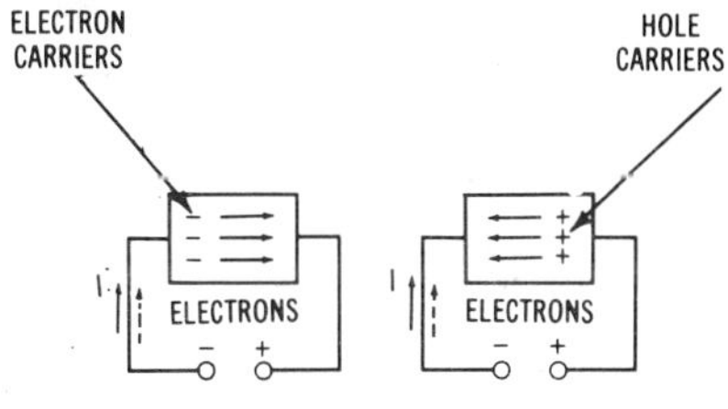

(A) N-type. (B) P-type.

Fig. 11-19. P- and n-type semiconductors.

The resistance of semiconductor material falls between that of a conductor and a nonconductor (insulator). The actual resistance depends very much on the number of free carriers (electrons or holes) in the semiconductor material. The number of these free carriers depends on the percentage of specified impurities that have been added to the pure crystalline semiconductor material. This is referred to as doping the pure semiconductor material—the greater the doping, the more current carriers and the lower the resistance of the semiconductor material with an applied emf.

When two such portions of semiconductor material are combined to form a junction, they provide rectifier action and a unidirectional current. When the junction is said to be forward biased (Fig. 11-20A), the positive terminal of the external source of emf is connected to the p-side; the negative terminal to the n-side. The electron carriers of the n-side flow through the junction to the positive terminal, and the hole carriers of the p-side flow in the opposite direction through the junction to the negative terminal. The junction now has a low resistance and a significant external current is present in the circuit.

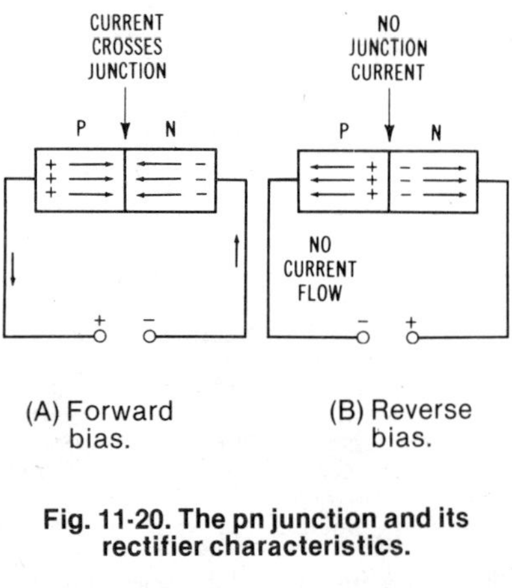

(A) Forward bias.

(B) Reverse bias.

Fig. 11-20. The pn junction and its rectifier characteristics.

If the junction is back or reverse biased, as in Fig. 11-20B, the electron carriers of the n-side move to the positive terminal and the hole carriers of the p-side to the negative terminal. There is no motion of carriers across the junction, and the resistance of the junction is high. Very little or no external current flows.

11-2-2. What is the difference between forward and reverse biasing of transistors?—When a transistor junction is forward biased, it is so biased that the junction resistance is low, and there is a free motion of carriers through that junction. The reverse and forward biasing of a junction is shown in Fig. 11-20. With forward bias, the junction resistance is low. With reverse or back biasing the junction resistance is high.

In a normal transistor circuit (Fig. 11-21) the emitter-base junction is forward biased, and the collector-base junction is reverse biased. Note that the manner of connection of the bias voltages depends on whether the transistor is a pnp or npn type.

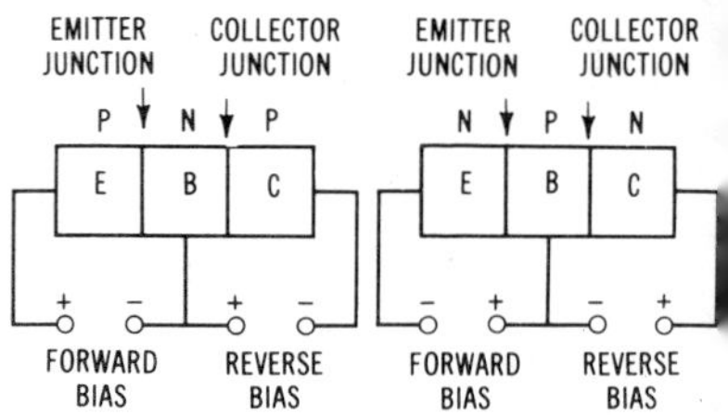

(A) Pnp type.

(B) Npn type.

Fig. 11-21. Normal biasing of transistors.

Notice that the emitter-base junction or just *emitter junction* is forward biased. Since the emitter junction is forward biased, there is, in the case of an npn transistor, a motion of electrons across the junction into the base. Some of these electrons are neutralized by holes in the p-type base. However, the emitter is more highly doped than the base (fewer holes than injected electrons), and therefore a great many of the electrons are not neutralized and move through the base to the collector junction. Here they are attracted to the positive voltage of the collector and are pulled across the collector junction (even though it is reverse biased in terms of the collector-base bias) and out into the external circuit. The transistor amplifies because this current is greater than the much smaller base current that occurs as a result of charge neutralization in the base. Also, gain is possible because the transistor output resistance is high (collector junction is reverse biased) and its input resistance is low (emitter junction is forward biased), and even an identical current would produce a greater output voltage (IR) and power (I^2R).

11-2-3. Show connections of external batteries, resistance load, and signal source as would appear in a properly (fixed) biased common-emitter transistor amplifier.—Basic circuits for pnp and npn transistors are shown in Fig. 11-22. Note that for normal operation the polarity of the bias sources is opposite for pnp and npn transistors. The polarity, of course, must be selected to forward bias the emitter junction and reverse bias the collector junction.

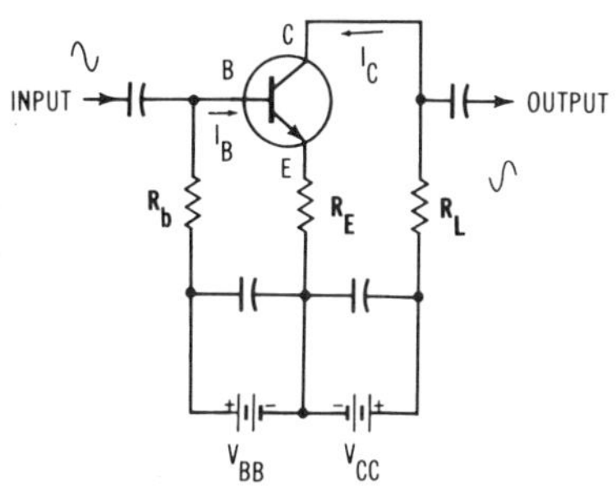

(A) Npn type.

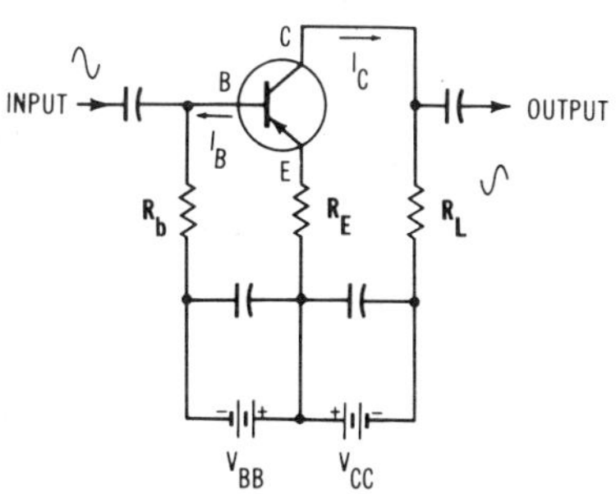

(B) Pnp type.

Fig. 11-22. Properly biased common-emitter stage.

In the operation of the common-emitter stage the emitter junction is forward biased from the battery (V_{BB}). Resistor R_b has a value that results in a proper bias current through the emitter junction. The direction of current is such that the emitter junction is biased in the forward direction. Some bias appears across R_E but it is not of a polarity to forward bias the emitter junction. However, it is useful in stabilizing the transistor operating point with relation to temperature change. Resistor R_L is the output load resistor. The amplified current (I_c) through R_L develops the output voltage.

When an ac signal is applied to the base circuit, it causes a variation in the base-emitter voltage and base current. There is a like variation in the number of carriers injected into the base region and a follow-up change in the collector current. The variation of the collector current is greater than the signal variation of the base current, and amplification results. The change in collector current through the load resistor develops the amplified signal across the output of the common-emitter amplifier. A more detailed explanation follows:

In the common-emitter amplifier of Fig. 11-22A, the base-to-emitter junction is forward biased and the base-to-collector junction is reverse biased. The signal is applied between the base and emitter. Since the base is forward biased, electrons move from the emitter, across the emitter junction, and into the base. It is said that carriers are injected into the base through the emitter junction. The magnitude of the carrier injection is a function of the signal variation applied to the base. This is similar to the action of a conventional vacuum tube, where the control grid controls the variation of electron flow through the grid wires as they travel from cathode to plate.

Many more electron carriers are injected into the base than are required to sustain the base-emitter current. These carriers diffuse across the base to the collector junction. At the collector junction they are attracted by the positive potential of the collector and they are propelled through the collector junction into the collector element.

Even though the collector-based junction is reverse biased, there has been a free motion of electrons across the base, through the collector junction and the collector element, into the collector output circuit. As a result, there is a collector-emitter current through the output load resistor (R_L). This current is a function of the signal voltage applied to the base. Inasmuch as the collector-emitter circuit has a high resistance, an amplified voltage is developed across the collector load resistor.

Input voltage and output voltage are

out of phase for the common-emitter circuit. When the input signal swings positive the base-emitter junction swings in the direction of more forward bias. Hence, the base current increases and so does the collector current. The higher collector current produces a drop in the collector voltage and a negative swing of the output voltage. Oppositely, when the input signal swings negative, the forward bias is reduced on the emitter junction. As a result the base current and the collector current decline correspondingly. There is a resultant rise in the collector voltage and a positive swing of the output voltage.

11-2-4. The following are excerpts from a transistor handbook describing the characteristics of a pnp alloy-type transistor as used in a common-emitter configuration. Explain the significance of each item.—

Maximum and Minimum Ratings

Collector-to-Base Voltage (emitter-open)	−40 Max. Volts
Collector-to-Emitter Voltage (base-to-emitter volts = −0.5)	−40 Max. Volts
Emitter-to-Base Voltage	−5 Max. Volts
Collector Current	10 Max. mA
Transistor Dissipation	
At Ambient Temperature of 25° for Operation in Free Air	120 mW
At Case Temperature of 25° for Operation with Heat Sink	140mW
Ambient-Temperature Range (operating and storage)	−65 to +100 °C

The maximum value given in each of the various items must not be exceeded under any circumstances. The ambient-temperature range lists both a minimum and maximum voltage range is the maximum voltage that can be applied safely. If it is exceeded, there will be a sharp damaging increase in the current between the collector and base as a result of the break-up of the crystal lattice structure.

The collector-to-emitter voltage is the maximum voltage that can be safely applied between these two elements before the transistor will become inoperative or subject to damage. There is also a safe limit to the emitter-to-base voltage.

The particular transistor should not be operated in any manner that causes the maximum collector current to be exceeded. Excessive current can overheat and damage the collector junction.

A transistor, according to its construction, is capable of dissipating safely a specific amount of heat before its junction will be damaged. Of course, if a heat sink is employed to conduct some of the heat away from the junction and case, a greater amount of heat can be dissipated safely. Maximum safe dissipation is given in milliwatts or watts.

Minimum- and maximum-temperatures are usually stated. When the transistor is operated at high temperature, the safe transistor dissipation wattage must be derated. Note that in the table the transistor dissipation is based on a case temperature of 25° C. If the operating temperature is higher than this value, the transistor dissipation wattages must be made correspondingly less according to specified derating information.

11-2-5. Draw a circuit diagram of a method of obtaining self-bias, with one battery, without current feedback, in a common-emitter amplifier. Explain the voltage drop in the resistors.—The circuit of Fig. 11-23A derives its bias from the same supply-voltage source as the collector-emitter circuit. The value of R_b is chosen with relation to the supply voltage (V_{CC}) so that the proper series current goes through R_b and the base-emitter circuit to correctly bias the emitter junction. The voltage drop across R_b is the difference between supply voltage V_{CC} and the bias required between base and emitter; or, the bias is the difference between V_{CC} and the voltage drop across R_b or:

$$\text{Junction Bias} = V_{CC} - V_{Rb}$$

The self-bias circuit shown in Fig. 11-23B uses local feedback to stabilize the operating point. In this bias arrangement the base bias is derived from the collector voltage rather than the collector-supply voltage. The collector voltage, of course, is the difference between collector-supply voltage V_{CC} and the dc voltage drop across collector load resistor R_L. The value of base-bias resistor R_b must be such that the proper base current will flow through the emitter junction of the transistor.

The foregoing being so, the required value of base-bias resistor R_b is lower as

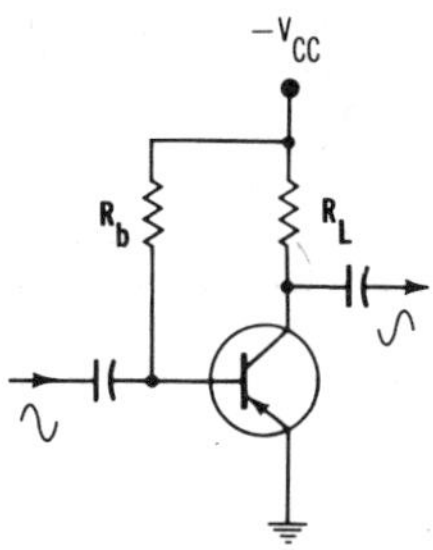

(A) Single-battery bias, no feedback.

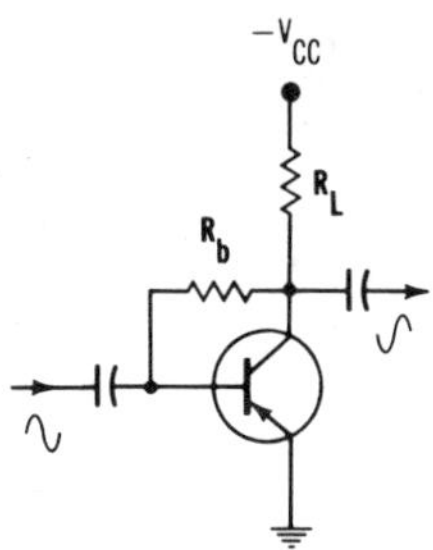

(B) True self-bias with feedback.

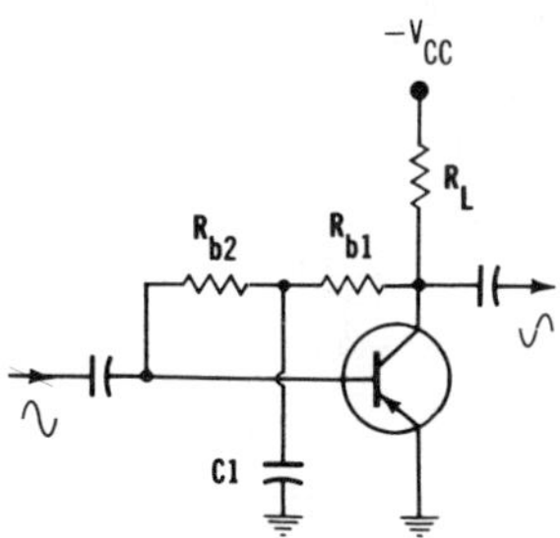

(C) Self-bias with no ac feedback.

Fig. 11-23. Single battery bias and self-bias arrangements.

compared to that for the fixed-bias arrangement of Fig. 11-23A. If there is any output drift with temperature there will be a corresponding change in the collector voltage. The attempted change in collector voltage will cause a corrective change in the base-bias current. The direction of the base-bias current change will be such that it will oppose the direction of drift of the collector current.

The circuit arrangement of Fig. 11-23C is similar to Fig. 11-23B, except that there are split bias resistors. The filter combination R_b1 and C1 prevent ac feedback.

11-2-6. Draw a circuit diagram of a common-emitter amplifier with emitter bias. Explain its operation.—The common-emitter circuit of Fig. 11-22 uses some

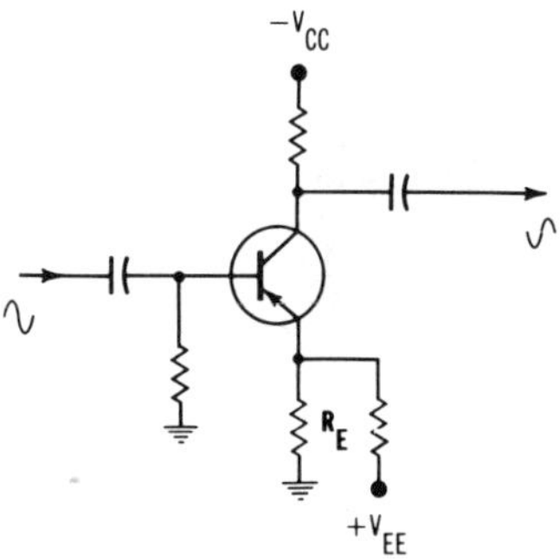

Fig. 11-24. Emitter-bias arrangement.

emitter bias as developed by the flow of emitter current. Because the connection does not provide forward bias, it is also necessary to bias the base circuit. Its function is one of stabilization. However, external emitter bias can be used.

In a true emitter-bias arrangement (Fig. 11-24), the emitter resistor connects to a separate supply voltage. The actual voltage that determines the base bias current is established by the difference of potential between this supply voltage and the voltage drop across emitter resistor R_E. This voltage sets the emitter-base bias voltage.

The circuit arrangement also stabilizes the operating point because any drift in the collector current will be felt as a change in the emitter current and in the voltage drop across emitter resistor R_E. The resultant bias change produces a compensating change in the base-bias current in a direction that will oppose any tendency for the collector current to drift.

The operation of the common-emitter amplifier, except for the biasing arrangement, is similar to that discussed in conjunction with Question 11-2-3.

11-2-7. The value of the alpha cutoff frequency of a transistor is primarily

dependent on what factor? Does the value of alpha cutoff frequency normally have any relationship to the collector-to-base voltage?—The alpha cutoff frequency is that frequency at which the current amplification of the transistor based on the common-gate configuration is 3 dB down from the alpha rating of the transistor. This frequency is dependent on the chemistry and physical structure of the transistor particularly the base. Collector-to-base voltage has no significant influence on the alpha cutoff frequency.

11-2-8. What is the gain factor of a transistor?—The fundamental gain factor is a term referred to as *alpha*. It is the ratio (less than unity) of a small change in collector current as caused by a small change in emitter current in a common base configuration:

$$\text{Alpha } (\alpha) = \frac{\Delta I_c}{\Delta I_e} \text{(with } V_c \text{ held constant)}$$

Equally important and appropriate for the common-emitter type of transistor circuit is the factor of *beta*. This represents the current gain of a common-emitter configuration and is the ratio of a small change in collector current to a small change in base current:

$$\text{Beta } (\beta) = \frac{\Delta I_c}{\Delta I_b} \text{(with } V_c \text{ held constant)}$$

11-2-9. Why is stabilization of a transistor amplifier usually necessary? How would a thermistor be used in this respect?—The amplifier circuit must usually be stabilized to prevent a drift in the operating point. This drift in the operating point is a result of the collector leakage current which varies with the collector-junction temperature. The transistor is a temperature-conscious device, and the operating point can drift substantially with temperature change. Bias-stabilizing circuits are used to minimize any such drift.

A thermistor-stabilized circuit is shown in Fig. 11-25. A thermistor has a negative temperature coefficient (resistance decreases with an increase in temperature). In the example this change in resistance compensates for any increase in emitter-junction current and voltage with temperature rise. In a transistor circuit the leakage current increases with temperature. This in turn results in an increase in emitter current. An increase in emitter-junction voltage follows. Such a tendency would produce a change in the operating point. However, the forward-bias voltage of the base depends on the resistance of the thermistor. Its resistance will decrease with an increase in temperature. As a result there will be a compensating decline in the base-bias voltage. Thus the base current stabilizes and compensation is made for the attempted drift of the operating point with a rise in temperature.

11-2-10. Draw simple schematic diagrams of the following transistor circuits, and explain their principles of operation. Use only one voltage source; state typical component values for low-power 10-MHz operation: (a) Colpitts-type oscillator, (b) class-B push-pull amplifier, (c) common-emitter amplifier, (d) a pnp transistor directly coupled to an npn type.—(a) A Colpitts radio-frequency oscillator is shown in Fig. 11-26. In an oscillator, a portion of the radio-frequency output of

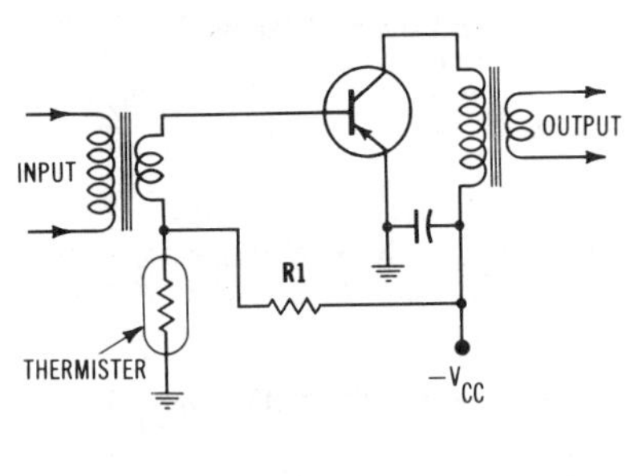

Fig. 11-25. Thermistor-stabilizer bias system.

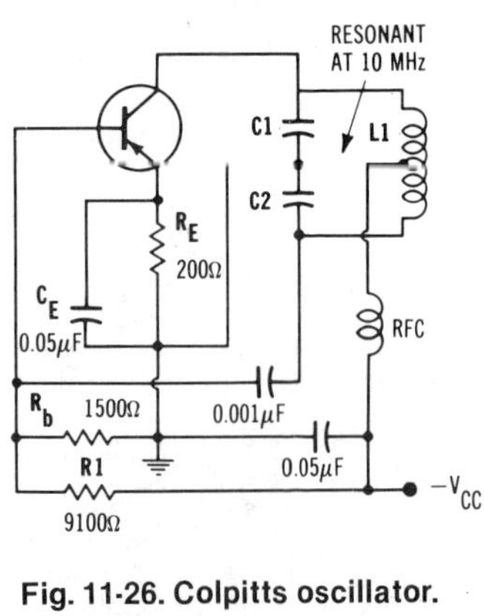

Fig. 11-26. Colpitts oscillator.

an amplifier is fed back, in phase, to its input. In this manner a transistor or vacuum-tube amplifier can be self-excited. This means that the stage itself supplies the exciting input signal which, if the stage were being used as a normal amplifier instead of an oscillator, would come from a preceding stage. If this exciting feedback signal is of the proper amplitude, the amplifier will operate continuously as a self-generating oscillator. Oscillators differ mainly in the manner that is used to obtain the in-phase feedback.

In the case of the Colpitts oscillator, the output is developed across the collector resonant circuit composed of coil L1 and capacitors C1 and C2. The split capacitors are used to obtain the proper feedback amplitude. The portion of the output voltage present in the resonant circuit, as developed across capacitor C2, appears between the base and the emitter of the transistor stage. This is the in-phase feedback voltage. Actually, the split-capacitor feedback arrangement provides a 180° phase shift. This shift in conjunction with the 180° polarity shift of the common-emitter transistor connection provides a total shift of 360° and, therefore, the base-emitter circuit is excited in phase. Were the base connected to the junction of the two capacitors and the emitter to the bottom of the resonant circuit, oscillations would not occur because the voltage fed back from the collector circuit would appear out of phase instead of in phase with the base-emitter variation.

The oscillator is stabilized with emitter resistor R_E and its rf filter capacitor C_E. A small amount of emitter-junction forward bias is needed to initiate the cycle that builds up to a continuous oscillation. If this bias were not provided, the oscillator would not start. The base-bias resistors are resistors R1 and R_b.

The Colpitts is a widely used transistor oscillator operating from a very low frequency to an exceptionally high radio frequency. If high-value capacitors were used along with an iron-core choke, the basic oscillator, as shown in Fig. 11-26, could be made to operate at an audio frequency.

(b) A class-B push-pull rf amplifier is shown in Fig. 11-27. Input and output resonant circuits are employed. On the input side the resonant circuit is in the primary and is excited with input from the preceding stage. The secondary is one of low impedance and provides a match to the low-impedance input of the transistor.

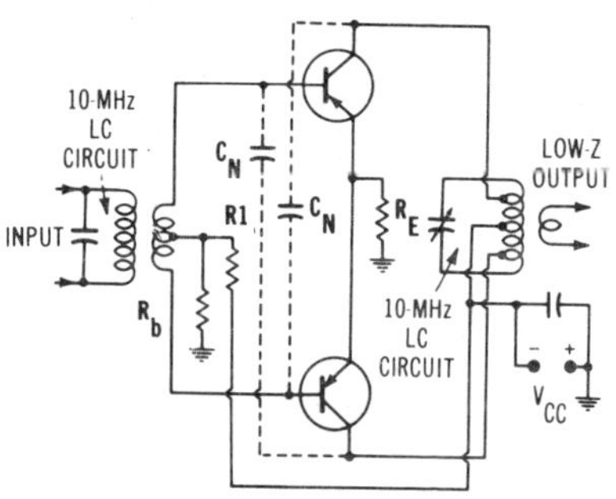

Fig. 11-27. Class-B rf amplifier.

Inasmuch as the transistors are biased at class B, they conduct during opposite alternations of the input wave. On alternate alternations, each draws a burst of collector current into the output resonant circuit. The energy-storing ability of the resonant circuit permits the development of a radio-frequency wave of good sine-wave form across the output resonant circuit. This rf energy can be transferred to an antenna or to a succeeding stage by way of the low-impedance secondary winding. A small emitter resistor can be used for stabilization.

Optimum class-B biasing requires the use of base bias (resistors R1 and R_b). This bias is usually just enough to overcome the emitter junction barrier potential. Sometimes no external biasing is used and the transistors are biased near class B (transistors are essentially non-conducting with no applied forward bias).

The class-B stage is to operate as an amplifier and not as an oscillator. Because of the high input conductance, neutralization is often not required. For some low-power transistors (lower input conductance) steps must sometimes be taken to prevent feedback between the output and input resonant circuits. To prevent oscillations and reduce this tendency, crisscross neutralizing capacitors are employed. Note that they connect from the collector of one transistor to the base of the second transistor. In so doing, they provide a certain amount of negative feedback which opposes any tendency to positive feedback. With appropriate values the negative feedback cancels the positive feedback exactly, and the stage will operate as a stable amplifier.

(c) A common-emitter rf amplifier is

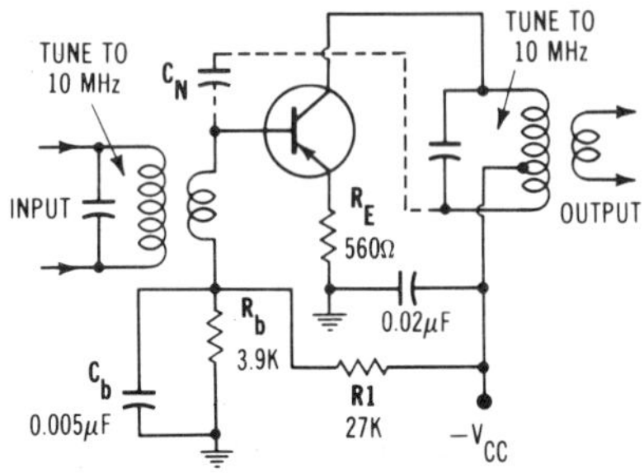

Fig. 11-28. Common-emitter rf amplifier.

shown in Fig. 11-28. This is typical of the circuit arrangements used in receiver rf and if amplifiers. Most are biased for class-A operation, using a base divider combination composed of resistors R1 and R_b and bias filter capacitor C_b. An emitter resistor is used for operating-point stabilization. This resistor can be unbypassed if some ac negative feedback is required for stabilizing a high-gain radio-frequency amplifier.

Again the primary side of the input transformer is resonant, while the secondary provides a low-impedance match to the transistor. In the more sensitive and higher frequency-amplifier circuits, neutralization is used to improve stability and avoid self-oscillations. In the arrangement of Fig. 11-28 it would be possible to connect a neutralizing capacitor (C_N) from the bottom of the collector resonant circuit to the base of the transistor.

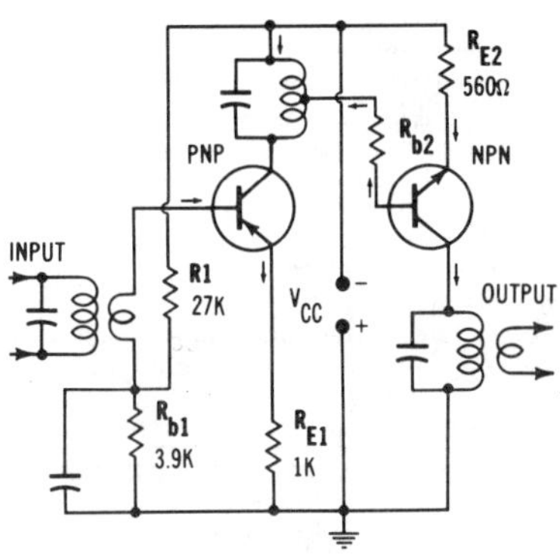

Fig. 11-29. Direct-coupled amplifier.

(d) Pnp and npn transistors, because of the opposite biasing polarity, can be conveniently direct-coupled. Two such amplifiers are shown in Fig. 11-29; a pnp transistor is direct-coupled to a succeeding npn transistor stage. The amplifiers shown are tuned 10-MHz stages. With proper choice of components and transistors a video amplifier can also be made to operate up to a frequency of 10 MHz and higher.

Note that there is no coupling capacitor, and a direct connection is made from collector of the first transistor to the base of the second stage. The biasing of the first stage is handled by resistors R1, R_{b1}, and emitter-stabilization resistor R_{E1}. Insofar as dc is concerned, there is a direct path from the collector of the first stage to the base of the second stage. Therefore, the bias voltage (V_{BE}) on the second stage is set by the current in the series circuit composed of the collector circuit of the input stage, R_{b2}, emitter junction, and R_{E2}. Its polarity is such that the base is made positive with respect to the emitter (the potential of the base of the second transistor is less negative than the potential at its emitter). The collector of the second transistor is, of course, at a positive potential as is required for proper biasing of the npn type.

The arrows on the illustration represent electron current. For the pnp transistor the current is out of the emitter and into the collector and base. Current direction is opposite for the pnp type. Observe that the first transistor has two output current paths from ground. One path is through the first transistor via the emitter junction, collector junction, and output load. The second path finds its way back to the negative supply through the emitter junction of the second transistor and emitter resistor R_{E2}. The current is in a direction that will forward bias the second transistor.

11-2-10. Discuss etched-wiring printed circuits with respect to the following: (a) determination of wiring breaks, (b) excessive heating, (c) removal and installation of components.—In a printed-circuit wiring chassis, soldered connecting wires have been replaced by conductive surfaces which have been printed or etched on an insulated panel. A thin copper-plated plastic laminate is used as a base. A light-sensitive chemical is first coated on the copper when employing the photo-etching process. A mask of the desired circuit-wiring arrangement is then positioned over the treated laminate and exposed to light. In the developing process the exposed photosensitive chemical is removed. However the coating remains on the unexposed lines and areas. An etching bath is next used to remove the copper from the exposed portion. Finally, the photo chemical is washed away and the etched copper design remains on the laminate.

The repair of printed-circuit boards requires a low-wattage pencil-tip soldering iron which should be handled carefully so as to avoid excessive heating of the board. A small wire brush or desoldering tool can be used to clean away solder. Rosin-core solder with a low melting point is preferred in making any solder connections.

Overheating or other physical damage may cause the copper wire strips to break away from the laminate or breaks may occur in the actual interconnection strips. Some of these breaks are of the hair-line type and are difficult to locate. A magnifying lens is helpful in locating very thin breaks. An ohmmeter can also be used to advantage in finding such breaks. If a break is found between two principal connection points, one of the ohmmeter prods can then be moved in short steps toward the other until the area of the break is localized more definitely.

Excessive heating is to be avoided. Soldering irons with ratings in excess of 25 to 35 watts are not recommended. When a copper wiring strip has broken away from the laminate or has a crack, it may be repaired by cutting away the damaged section. A small piece of wire can then be soldered across the break. In all such repairs, to avoid further damage to the connecting copper strips do not overheat connections. Excessive solder and overheating can also damage or short out adjacent copper connections if the solder drips or spreads. The wire brush can come in handy in removing solder drippings, although a small probe and/or heat may be necessary to free some of the drippings.

Gentle handling is essential in the repair and removal of component parts and wiring from a printed-wiring board. Prying or excessive force is to be avoided to prevent damage to connections and the thin-skinned laminate. Defective parts can be removed or replaced in one of two ways. The component pigtails can be clipped as close as possible to the defective part. The pigtails that remain connected to the board and those of the replacement part can then be soldered together.

If the removal of the component is necessary, its leads should be clipped as near as possible to the connection points. The pigtail sections remaining can then be removed with the careful application of heat. The connecting areas should then be cleaned carefully, and the pigtails from the replacement part inserted, clipped, bent into position, and soldered. Finally, the area should be inspected very carefully and cleaned of excess solder. A protective coat of lacquer can then be applied to provide long-term protection.

11-2-11. What are the properties of varactor diodes? of zener diodes? of field-effect transistors? Describe some uses of each.—The varactor diode has a capacitance that varies with the applied voltage. The varactor diode can be used to frequency-modulate an oscillator or part of an automatic-frequency-control system where a dc voltage is used to set the frequency of the oscillator. The varactor diode can be used as a frequency multiplier.

A zener diode is a constant voltage device and despite a changing current through the device the voltage across its terminals will remain constant. It is used as a voltage regulator or reference voltage in voltage-regulated power supplies. It can be used to set a constant voltage at various positions along supply-voltage lines of an electronic system. For more information on varactor and zener diodes read Section 3-2.

The field-effect transistor is a voltage-controlled electronic device that can be used as an oscillator, amplifier, or other type of electronic circuit. It has a high input impedance and a high output impedance. It is used frequently as a radio-frequency amplifier and also as an input audio amplifier. However, it has a great variety of uses in all types of electronic circuits. Refer to Sections 3-12 through 3-14 of Chapter 3.

11-2-12. What is a light-emitting diode (LED)? Why does it use a series resistor for normal operation?—A light-emitting diode is a semiconductor diode that emits light when it conducts a current. The LED when forward-biased and connected in series with a resistance to set its current at a proper value will have a long life. The series resistance limits the current to a safe value and avoids burnout. Refer to Section 5-4.

11-2-13. Discuss the use of LED digital

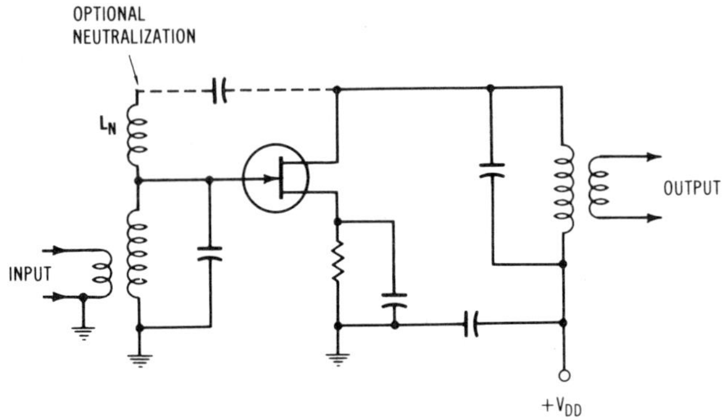

Fig. 11-30. Fet rf amplifier.

readouts in displays.—In a LED digital readout, segmented light-emitting diodes can be switched on and off to display numerals, letters or other symbols. In a digital system these numbers or letters can be switched on and off at a prescribed rate to establish the digital readouts used so commonly in computer, calculator, digital display meters (particularly digital voltmeters), frequency counters, etc. Refer to Sections 5-5, 6-7, 6-10, and 6-11.

11-2-14. Draw a schematic diagram of a field-effect transistor (fet) radio-frequency amplifier. Are its input and output impedances high or low?—Refer to Fig. 11-30. Input and output impedances are high. Study Sections 3-12 through 3-14.

11-2-15. Draw a schematic diagram of a grounded-base rf amplifier and explain its operation.—Refer to Fig. 11-31. In the grounded-base configuration the base itself is operated at rf ground potential. The signal is applied between emitter and ground. Inasmuch as this is a very low-impedance input circuit a step-down transformer (T1) is used to match the incoming signal to the low-impedance input of the grounded-base stage. Resistors R1 and R2 and filter capacitor C1 provide base bias for the stage. Amplified output is removed in the collector circuit. A step-down transformer matches this output to the low-impedance input of a succeeding stage. The resonant circuit associated with transformers T1 and T2 is tuned to the signal frequency. For more details refer to Chapter 3. In particular check out Sections 3-4, 3-5, 3-9, and 3-10.

11-2-16. Draw a schematic of a grounded-emitter rf amplifier and explain

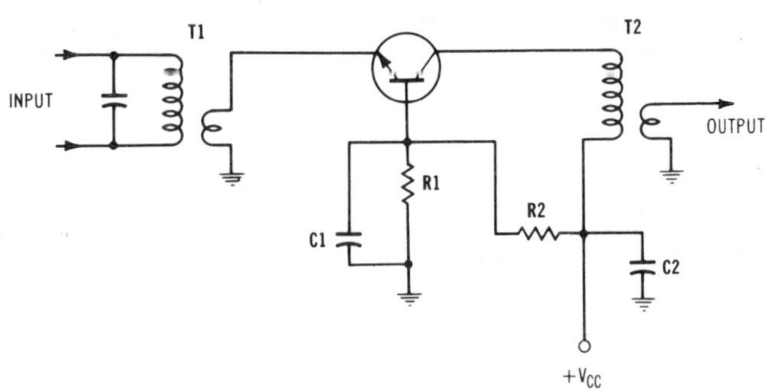

Fig. 11-31. Grounded-base rf amplifier.

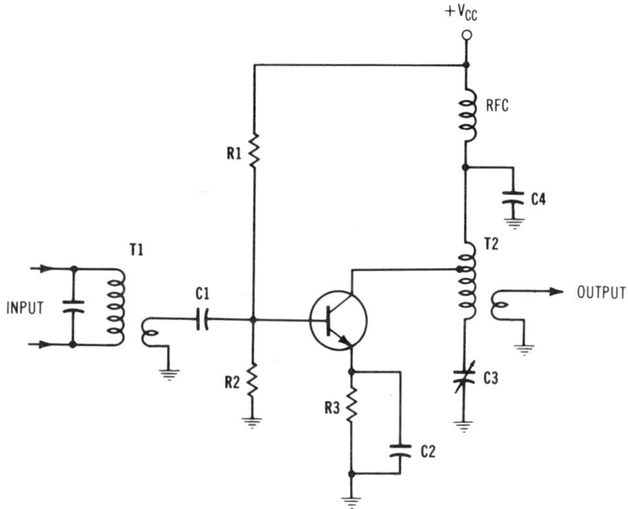

Fig. 11-32. Grounded-emitter rf amplifier.

its operation.—Refer to Fig. 11-32. In the grounded-emitter circuit the emitter is operated at rf ground potential. The signal is applied to the low impedance input provided by the base-emitter circuit. Consequently transformer T1 is a step-down type to match the incoming signal to the low-impedance input. In many rf power amplifier applications such a stage is operated at near class-B bias. Resistors R1 and R2 provide the necessary base-bias current. Emitter stabilization is provided by the combination of resistor R3 and capacitor C2. Resistor R3 is often a very low value and its particular application is to prevent thermal runaway. The collector output is tapped in at a low-impedance point of the output resonant circuit consisting of the primary of transformer T2 and tuning capacitor C3. Again a step-down transformer is used to match into the low impedance of a succeeding stage or antenna system. The supply voltage is fed through a radio-frequency choke that blocks the rf energy from the supply voltage line. Study Chapter 3 and in particular Sections 3-3, 3-4, 3-9, and 3-10.

11-2-17. How do transistors and vacuum tubes compare?—The triode vacuum tube, bipolar transistor, and junction field-effect transistor are basically three-element devices. Input elements for the tube, bipolar, and fet are respectively grid, base, and gate. Output elements are plate, collector, and drain. The common elements are cathode, emitter, and source. In addition the vacuum-tube triode requires a filament or heater that is necessary in obtaining thermionic emission. The vacuum-tube basically requires a high supply voltage, while the two transistor types operate at low supply voltages. Usually the bipolar transistor operates at a somewhat lower voltage than the field-effect transistor. The vacuum tube and the field-effect transistors are basically voltage amplifiers and have a high input impedance. They respond to an input voltage change. The bipolar transistor is basically a current amplifier and has a low input impedance. It responds to an input current change. Output impedances of the vacuum tube and the junction field-effect transistor are generally high but are lower than their input impedances. The bipolar transistor has a very low input impedance and a low but somewhat higher output impedance than its input impedance. Basic theory of operation differs substantially. Refer to Sections 3-3, 3-4, 3-5, 3-12, and 3-13. Basic vacuum-tube operation is covered in this chapter.

11-2-18. Discuss the differences among common-base, common-emitter, and

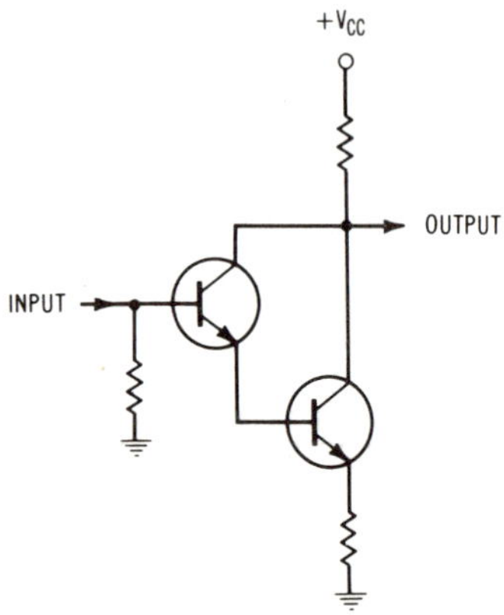

Fig. 11-33. Basic Darlington pair.

common-collector circuits and explain the operation of each.—Refer to Sections 3-3 through 3-6.

11-2-19. Draw a diagram of two transistors connected in a Darlington configuration. Refer to Fig. 11-33. Why is this done?—The Darlington configuration has two basic advantages. The circuit permits a high input impedance. A second advantage is that it permits a high output current. Consequently a very small input current is able to develop a significantly higher output current.

11-2-20. Why do high-power transistors sometimes require heat sinks?—Heat is a major consideration in power-transistor application. The heat must be conducted away from the collector junction. If it is not, the transistor can be damaged permanently or it can destroy itself through thermal runaway. There can be a shift in the operating point, distortion, poor efficiency, relatively short life, and a high power demand when heat is excessive. The lower-powered power transistors are often cooled by metallic heat sinks that fit over the transistor case. For higher-powered types the transistor is made to establish thermal contact with a chassis or a metallic radiating surface called a heat sink. Heat sinks are thin structures made of copper or aluminum which function well as heat radiators. They are painted black to increase the efficiency of heat radiation. Their size depends on the amount of heat that must be conducted away from the transistor.

11-2-21. How do silicon and germanium transistors differ?—Both silicon and germanium have semiconductor properties with four outer valence electrons. The atomic number of germanium is 32; for silicon, 14. The emitter-base voltage drop for germanium is about 0.25 volt while for silicon it is 0.65 volt. Generally low-cost germanium transistors are used in general purpose applications and for low levels. There are some power germaniums too. Silicon transistors run the gamut of applications and are more common in high-frequency applications than germanium types. However special germanium types are used in microwave applications.

11-2-22. Give the voltage gain equations for the three bipolar transistor and FET configurations.—Bipolar equations were given in Fig. 3-21. Fet equations are as follow:

Common source

$$VG = g_{fs}\frac{r_{ds}R_D}{r_{ds}+R_D}$$

Common drain

$$VG = \frac{g_{fd}R_s}{1+g_{fd}R_s}$$

Common gate

$$VG = \frac{(g_{fg}r_{ds}+1)R_D}{r_{ds}+R_D+(g_{fg}r_{ds}+1)R_s}$$

11-2-23. Give the voltage gain equations for triode and pentode.—The voltage gain of a triode amplifier is:

$$VG = \frac{\mu R_L}{r_p+R_L}$$

The voltage gain of a pentode amplifier is:

$$VG = gmR_L$$

11-2-24. What is the logic symbol for an AND gate? an OR gate? a NAND gate? a NOR gate? an inverter?—Symbols are given in Fig. 11-34. Study Section 5-3.

11-2-25. Why is it important that digital logic circuitry be shielded from radio-frequency sources?—Digital circuits use pulses and fast switching. Radio-frequency components exist in the makeup of these pulses and switching activities.

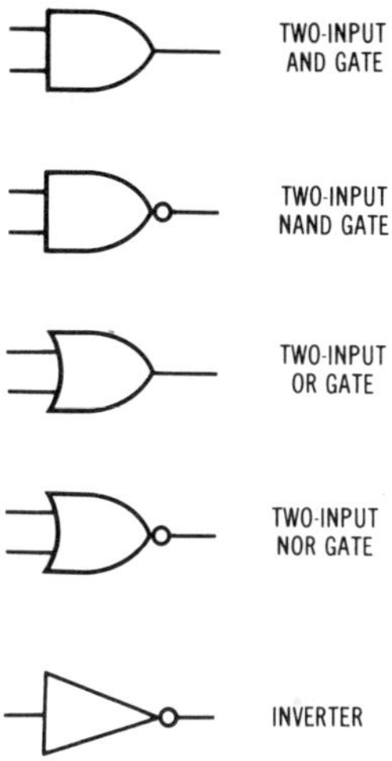

Fig. 11-34. Logic symbols.

Consequently they must be shielded to prevent possible interference with radio-frequency circuits and even possible modulation of radio-frequency sources.

11-2-26. What are the characteristics of an operational amplifier (Op Amp)?—An operational circuit amplifier is a feedback-controlled circuit and its gain and other operational characteristics are set by its external circuit components. Refer to the operational amplifier discussion in Section 3-16.

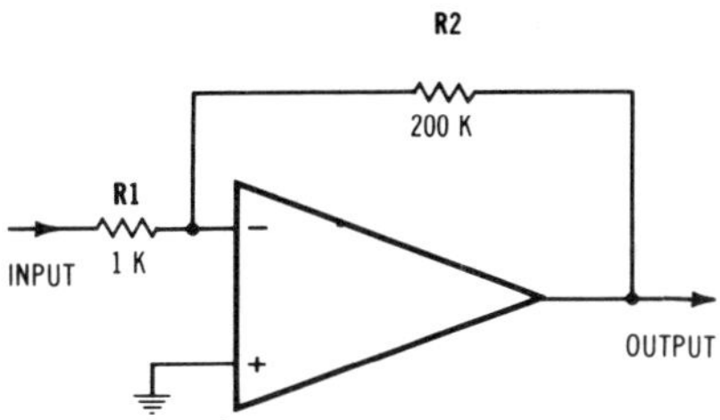

Fig. 11-35. Inverting op-amp with gain of 200.

11-2-27. Name some ways in which an operational amplifier can be used.—An operational amplifier can be used as an amplifier at audio, video, and rf frequencies, or as a low-pass, bandpass, high-pass, and band-reject active filter. It can be used as a signal conditioner, a signal processor, or a signal generator. There are also many measurement and control circuits that employ operational amplifiers. Refer to Section 3-16.

11-2-28. Draw a circuit diagram of an inverting amplifier with a gain of 200 using an operational amplifier and two resistors.—Refer to Fig. 11-35.

CHAPTER 12

Power Sources

12-1. RECTIFIERS AND FILTERS

12-1-1. Discuss the relative merits and limitations as used in power supplies of the following types of rectifiers: (a) mercury-vapor diode, (b) high-vacuum diode, (c) copper oxide, (d) selenium, (e) silicon.—(a) A mercury-vapor rectifier tube permits a high current flow, because of the low resistance path presented by the ionized gas molecules. Such a tube has a low and fixed internal voltage drop and excellent regulation. The current capability is high, and the tube is efficient in terms of required filament power and anode dissipation. Such a tube is critical of operating temperature, and some delay is involved in bringing the cathode up to optimum temperature. It has a lower peak-inverse voltage (piv) than a high-vacuum type and includes output transients (gas ionization process) that can be a source of rf interference.

(b) The high-vacuum rectifier can withstand a higher peak-inverse voltage and is less critical of operating temperature as compared to the mercury vapor. It is not as economical and presents some voltage-regulation problems for power demands that change substantially. Rf interference and transients are not a problem.

(c) The copper-oxide diode is a metallic rectifier. It is a small, low-current, low-voltage rectifier principally used in ac meters. It has a long life and constant characteristics and can be made economically.

(d) The selenium diode is also a metallic rectifier. Low- and high-current versions can be fabricated and made to have a high peak-inverse voltage (approximately three times that of a copper oxide type). They are somewhat more costly than the copper oxide type but are less subject to temperature changes and are more rugged.

(e) The silicon solid-state rectifier is taking over many of the duties of both high-vacuum and mercury-vapor rectifiers. Its regulation is good, and it can be fabricated to handle very high currents. Furthermore, no filament power is required. Peak-inverse voltages are lower than high-vacuum types, but this disadvantage is readily overcome by connecting a number of silicon rectifiers in series. They are subject to damage by current surges, and a protective series resistance is usually required.

12-1-2. Explain the action of a voltage regulator (vr) tube.—The voltage regulator is a cold-cathode type tube and maintains a constant output voltage over its operating range.

As shown in Fig. 12-1, the supply voltage for the load is taken off directly across the tube. A series resistor confines the current within the limits of operation of the vr tube. An adequate voltage must

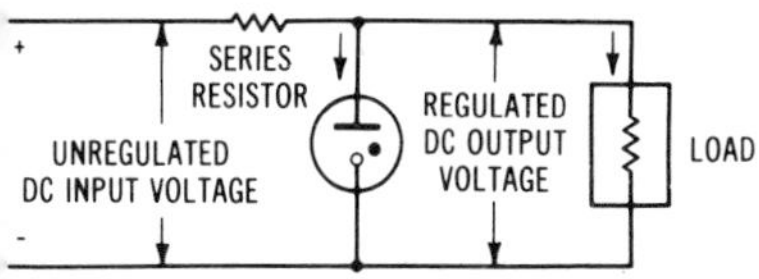

Fig. 12-1. Voltage-regulator circuit.

be made available so that the gas in the vr tube ionizes. As the load current changes, more or less current is shuttled through the vr tube and a constant voltage drop is maintained. When the load current increases, there is less current through the vr tube. Conversely, when the load current decreases, more current is shuttled through the vr tube. The resistance of the conducting path (offered by the ionized gas molecules) varies with output-current demand in such a manner that a constant voltage drop is maintained across the tube.

12-1-3. Draw a diagram of each of the following power-supply circuits. Explain the operation of each, including the relative input- and output-voltage amplitudes, waveshapes and current waveforms: (a) vacuum-tube diode, half-wave rectifier with a capacitive-input "pi-section" filter; (b) vacuum-tube diode, full-wave rectifier with choke-input LC filter; (c) silicon diode, doubler-circuit rectifier with a resistive load.—(a) A half-wave rectifier with a capacitive-input filter is shown in Fig. 12-2. Appropriate waveforms are shown. To understand its operation let's first assume that the filter is not connected (broken at point X). During the positive alternation of the input sine wave as it appears across the secondary of the transformer, a positive voltage is applied to the plate of the diode rectifier. As a result, electrons are attracted from the cathode emitting surface. The more positive the anode is with respect to the cathode, the more electrons there are that flow and the higher the current. This current through resistance R_L develops a load voltage that corresponds to the positive alternation of the input sine wave. During the negative alternation of the input wave, the plate is made negative with respect to the cathode. Consequently, no electrons are attracted over to the plate, and the circuit resistance is essentially infinite. No current occurs, and there is zero voltage drop across resistor R_L. Current has been in one direction, and it only occurs when the plate is made positive with respect to the cathode. A pulsating voltage, corresponding to the positive alternations of the input wave, is developed across the load.

Next consider how the filter influences the operation of the power circuit. During the conducting alternation, the input capacitor charges toward the peak amplitude of the positive alternation. When little power is being demanded by the load, the capacitor charges to the very peak of the positive alternation. If a considerable amount of current is drawn from the supply by the load, the capacitor does not charge to the peak magnitude.

The load continues to demand current even after the peak of the positive alternation has been passed. As a result, charge is withdrawn from the capacitor to supply current to the load, and the magnitude of that capacitor charge (volt-

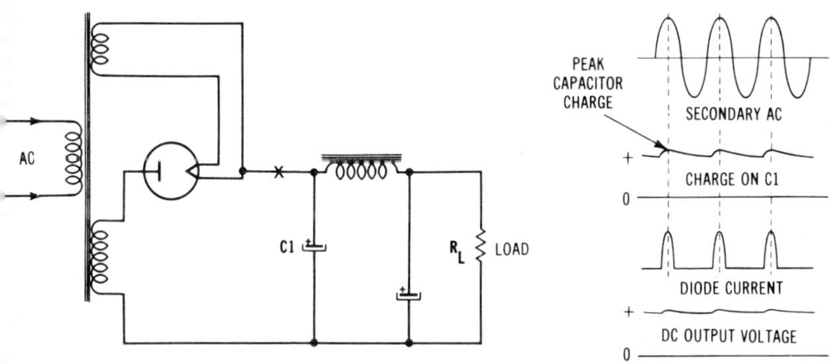

Fig. 12-2. Half-wave rectifier with capacitive-input filter.

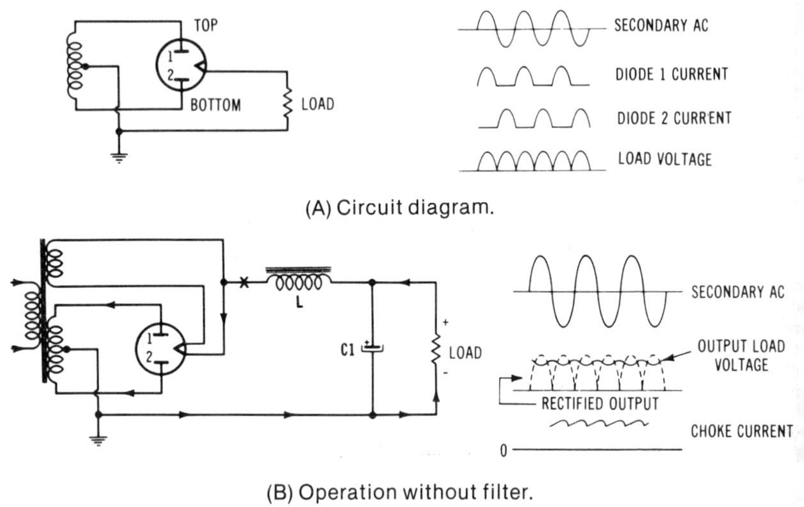

Fig. 12-3. Full-wave rectifier with choke-input filter.

age) decreases. However, during each positive alternation the charge is replenished by the conducting interval of the rectifier tube. Note the capacitor and output voltages.

As shown in the waveforms, the tube conducts for even less than the time interval of the positive alternation of the input wave. It conducts just long enough to replenish the charge on the input capacitor—the greater the power demand of the load for a given set of filter components, the longer the period of the rectifier current. It follows from this that there is a higher average current and a higher dc output current as demanded.

Current through the load is in only one direction. The purpose of the rectifier is to cause this unidirectional current. The purpose of the filter is to smooth out the pulsations so that the dc output voltage remains essentially constant, and a constant dc current goes through the load. Heater power is derived from the transformer filament winding.

The larger the input capacitor and the greater the energy storing capability of the pi-section filter, the more effectively the pulsations are filtered out into a smooth dc voltage and current. It is permissible to state that the capacitor opposes a change in voltage and therefore holds the voltage constant to the best of its ability. The inductor opposes a change in current and helps smooth out the current in the rectifier-filter circuit. The time constant of the filter and load is long in comparison to the period of the ac sine wave.

(b) A full-wave rectifier with a choke-input filter is shown in Fig. 12-3. Consider its operation first without the filter (Fig. 12-3B). The top rectifier plate is connected to the top of the secondary winding of the transformer. Its operation is identical to that of the single half-wave rectifier of Fig. 12-2. It produces positive pulsations across the load during the positive alternations of the secondary input wave. The magnitude of the pulsations depends on the amplitude of the sine wave as it appears between the top of the secondary and the center tap.

The bottom half of the rectifier tube has its plate connected to the bottom side of the transformer secondary. As a result, it is the negative alternation of the secondary sine wave that causes the plate of the bottom rectifier to be positive with respect to its cathode. During this negative alternation, electrons flow from its cathode to the anode. Note that the current, as a result of the conduction of the bottom rectifier, is in the same direction as that of the top rectifier. Therefore both rectifiers cause current alternately in the same direction through the load. In this case, positive alternations are developed across the load which correspond to the time intervals of the negative alternations of the input sine wave. Hence, these pulsations fit in between the pulsations that result from the conduction of the top rectifier.

It is apparent that with a full-wave rectifier there will be a higher average dc current, because current occurs for both alternations of the input wave. There will be less variation in the output voltage as compared to the much greater voltage variation for a half-wave rectifier circuit. Furthermore, the variation that does exist at the output of a full-wave rectifier occurs at twice the frequency of the variation of the output of a half-wave rectifier. This imposes less critical operating conditions on the associated filter, because for a given set of component values, it is easier to filter a high frequency than a lower one.

Again, it is the function of the associated filter to smooth out the current and voltage variations (ripple components) at the output of the rectifier so that a reasonably constant load voltage and current results. A choke opposes a change in current. Thus, during the times of rectifier current, some of the electrical energy is stored in the magnetic field of the inductor. In a practical choke-input filter circuit, the output filter capacitor does not charge to the peak amplitude of the pulsating waveform. Usually the capacitor charge is somewhat less than the rms value of the secondary voltage (one side to center tap). However, when the load demands current, it is contributed by the inductor, and the load voltage and dc current are sustained.

Of course, the output capacitor also stores energy, but its purpose is mainly that of a voltage-smoothing component to remove any ripple voltage present. In the case of the capacitor-input filter, the bulk of the energy is stored by the input capacitor, and its voltage drops as the load withdraws current from its charge. Since the capacitor charges up to the peak value of the sine-wave alternation, there can be a considerable voltage variation when there is a heavy or changing load demand. The choke-input filter, although the available dc voltage is less than that of a capacitor-input filter, provides better voltage regulation with changing load demands and has a lower ripple-voltage component.

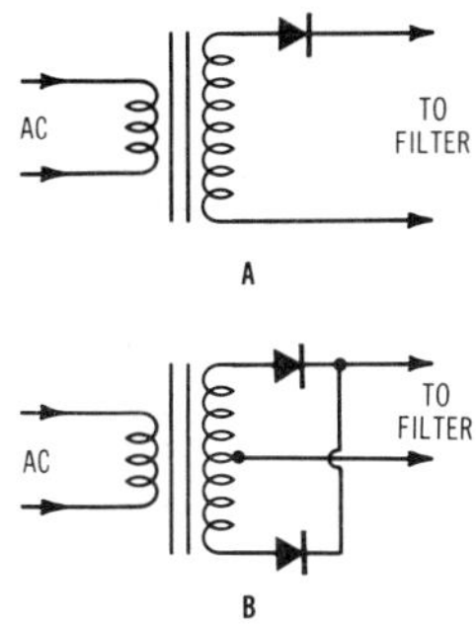

Fig. 12-4. Silicon rectifier half-wave and full-wave circuits.

Silicon diode rectifier half-wave and full-wave rectifier circuits are shown in Fig. 12-4. Note the circuit simplicity as compared to the vacuum-tube counterparts. There is no filament power needed thus reducing transformer size and weight. Theory of operation and waveforms are the same as in the vacuum-tube discussions.

(c) A silicon rectifier voltage-doubler

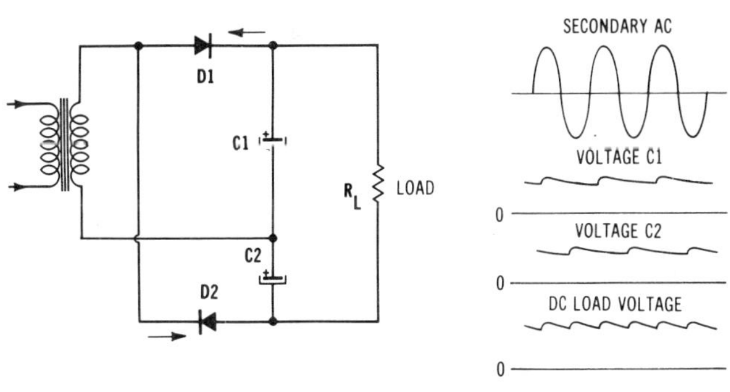

Fig. 12-5. Silicon-rectifier voltage-doubler circuit.

circuit is shown in Fig. 12-5. Filament power and a filament winding on the transformer are not required. In a voltage-doubler circuit one rectifier can be made to conduct on one alternation of the input wave and the second rectifier on the next alternation, just as in a full-wave rectifier arrangement. However, their currents are made to charge two different capacitors. In the circuit, diode D1 conducts on the positive alternation it places a positive charge on capacitor C1. During the negative alternation of the input wave. During this positive alternation of the input wave, diode D2 conducts. Its direction of current is such that capacitor C2 is charged as shown. The circuit arrangement charges two capacitors that are connected in series across the load. Therefore, in terms of the load, the total voltage approaches a level that is twice the peak amplitude of the sine wave as it appears across the secondary of the power transformer.

Again, as in the case of any capacitor-input filter, whether or not the capacitor charge actually reaches the peak value depends on the current demand, voltage doublers are popular for loads of low and constant power demand. The voltage regulation is not good, and a considerable amount of ripple must be filtered out by the filter circuit. In fact, the ripple is greater and the regulation is poorer than a comparable full-wave rectifier. However, the dc load voltage for low-power demands is almost four times that attainable with a full-wave rectifier circuit. A two-fold increase is made possible with the voltage-doubler circuit. Another two-fold increase is possible because the full secondary voltage can be used, while a full-wave rectifier load voltage is related to the voltage across one-half of the secondary winding. The higher the resistance of the load (R_2L_3), the higher is the dc load voltage, and the lower is the output ripple voltage.

12-1-4. What is meant by the "peak-inverse voltage" rating of a diode, and how can it be computed for a full-wave power supply?—This is the maximum value of inverse voltage that can be safely applied between anode and cathode (negative to the anode and positive to the cathode).

The full-wave arrangement of Fig. 12-6 shows the high negative inverse voltage being supplied to rectifier 1, when rectifier 2 is on its conducting alternation. In the example, it has been assumed that 1000 volts rms appears

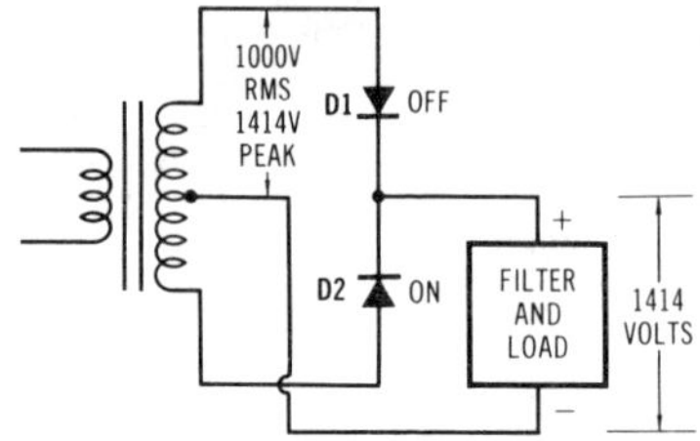

Fig. 12-6. Peak inverse voltage across nonconducting diode of full-wave circuit.

across each half of the secondary. This corresponds to a peak amplitude of 1414 volts. On the peak of the negative alternation of the secondary sine wave, this amount of negative voltage is being applied t the anode of D1. At the same time D2 is conducting, and, therefore, approximately the same voltage, except for the small voltage drop across the conducting diode, appears at the input to the filter. Thus, approximately 1414 volts positive is applied to the cathode of the top diode. The net voltage drop between anode and cathode is twice this value, or 2828 volts (minus the small voltage drop across D2). This corresponds to twice the peak amplitude of the rms voltage that appears across one half of the secondary.

In a practical full-wave rectifier with a choke-input filter, the dc output or load voltage approaches 90% of the rms voltage across one-half the secondary. In our example, this would be a dc output of 900 volts. Inasmuch as the calculated voltage of the peak-inverse voltage across D1 was 2828 volts, it can be noted that the rectifier must be able to withstand 3.14 (2828/900) times the dc load voltage of 900 volts.

12-1-5. What advantages may a bridge-rectifier circuit have over a conventional full-wave rectifier?—The bridge rectifier is able to develop a dc output that corresponds to the full secondary voltage of a given transformer, thus utilizing the power capability of the transformer more efficiently. Each rectifier is subjected to a lower inverse voltage as compared to the rectifiers in a full-wave circuit.

12-1-6. What are the characteristics of a capacitor-input system as compared to a choke-input system? What is the effect on a filter choke of a large value of direct current?—The capacitor input-filter permits a higher dc load voltage, but at a sacrifice in voltage regulation. The peak voltage across the input-filter capacitor is higher than that of the first capacitor of a choke-input type. In using a capacitor-input filter, the ratio of the peak rectifier current to the dc output current is higher than for a choke-input type.

An excessive current through the filter choke may result in overheating or a decrease in inductance, because of core saturation.

12-1-7. What effect does the amount of current required by the load have upon the voltage regulation of the power supply? Why is voltage regulation an important factor?—Voltage regulation refers to how constant the dc output voltage remains with changes in load. Stated as an equation:

$$\% \text{ Regulation} = \frac{E_{\text{No Load}} - E_{\text{Full Load}}}{E_{\text{Full Load}}} \times 100$$

If voltage regulation is good, changes in loading (current demand) have a minimum influence on the dc voltage made available by the power supply. Changes in supply voltage to the stages of an electronic unit can result in improper or erratic operation of the equipment.

Usually with an increase in current demand (greater load) there is a drop in the dc output voltage. This can be the result of an additional voltage drop across the dc resistance of a filter inductor, an increase in the dc drop across the rectifier, or a decrease in the ac voltage made available by the transformer. The output dc voltage remains more constant when there is little or no change in current demand.

12-1-8. What is the purpose of a "bleeder" resistor as used in connection with power supplies?—A bleeder resistor maintains a more constant load on the power supply by maintaining some current drain on the supply even though the load proper is demanding little or no current from the power supply. It is connected across the power-supply output and maintains a specific minimum current drain through the filter choke and improves the power-supply regulation. A second function of a bleeder resistor is to discharge the filter capacitors when the power is switched off.

12-1-9. Would varying the value of the bleeder resistor in the power supply have any effect on the ripple voltage?—Yes. A bleeder resistor of an appropriate value reduces the ripple, because it does not permit the filter capacitors to charge up to peak value under light loading of the power supply. Of course, a bleeder resistor of an impractical low value could place such a load on the output of the power supply that the ripple voltage would increase. Thus, there is a certain optimum value which will provide a minimum ripple voltage.

12-1-10. What are swinging chokes? Where are they normally used?—A swinging choke improves regulation when the power supply operates into a widely changing load. It is designed in a manner that causes the inductance to change inversely with the current. By so doing a high inductance is obtainable with low current; yet the inductance is sufficient for proper operation under heavy loads. If a more conventional choke were used, its inductance would drop at low currents and cause the output voltage to rise, causing poor power-supply regulation under changing loads. Such a swinging choke is common in transmitter power supplies, especially class-B modulators, because the current demand often varies with the modulation.

12-1-11. Show a method of obtaining two voltages from one power supply.—Refer to Fig. 12-7. A bridge rectifier permits a half voltage output via the center tap of the transformer. In this case diodes D1 and D4 operate in a full-wave rectifier circuit for the 300-volt output.

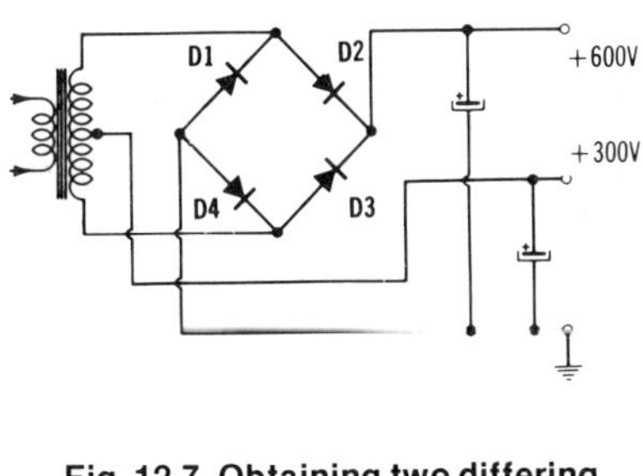

Fig. 12-7. Obtaining two differing voltages from a single supply.

12-1-12. What does a blue haze in the space between the filament and plate of a high-vacuum rectifier tube indicate?—A blue haze indicates the presence of gas in the high-vacuum tube. It is an indication that it is losing its vacuum and should be replaced to avoid failure or unstable operation.

12-1-13. If the plate or plates of a rectifier tube suddenly become red-hot, what could be the cause and how could remedies be effected?—The glow of red indicates high current caused by a short circuit or a low resistance shunt across the rectifier output. A common defect is a shorted filter capacitor. The cause of severe loading should be reduced, and a shorted filter capacitor should be replaced (or at least removed from the circuit until a replacement can be made).

12-1-14. If a high-vacuum–high-voltage rectifier tube should suddenly show severe internal sparking and then fail to operate, what elements of the rectifier-filter system should be checked for possible failures before installing a new rectifier tube?—The trouble could be a bad tube or a severe load on that tube. Check the power-supply capacitors, choke, and transformer for a short circuit, or a filter capacitor that breaks down under load. An ohmmeter can be used to localize the defect between the power supply and the external load by making readings with the load connected and then with it disconnected.

12-2. CIRCUITS

12-2-1. What is meant by the term *power factor*?—Power factor is the ratio of true power to apparent power. It is often given in percentage relating resistance to impedance. Equations are:

$$PF = \frac{\text{True Power}}{\text{Apparent Power}}$$

$$\% \; PF = \frac{\text{Resistance}}{\text{Impedance}} \times 100$$

12-2-2. What is the purpose of a bleeder resistor as used in connection with power supplies? What type of resistor is preferable, carbon or wirewound? Why?—The bleeder resistor prevents the output voltage from rising to a high value when there is no load, especially when using a swinging choke. The bleeder resistor also provides a means of discharging filter capacitors when the power is removed and there is no other discharge path. This eliminates the hazard involved when it is possible for a capacitor to hold a charge for an extended period of time. Carbon bleeder resistors are preferred because on occasion a wirewound type might open up and filter capacitors may not be discharged.

12-2-3. How may a series circuit of silicon diodes be protected against voltage transients? Unequal peak reverse voltages?—The ill effects of reverse voltages can be avoided by equal-value equalizing resistors shunted across the individual diodes as shown in Fig. 12.8. Equal value capacitors shunted across the individual diodes afford protection from voltage transients.

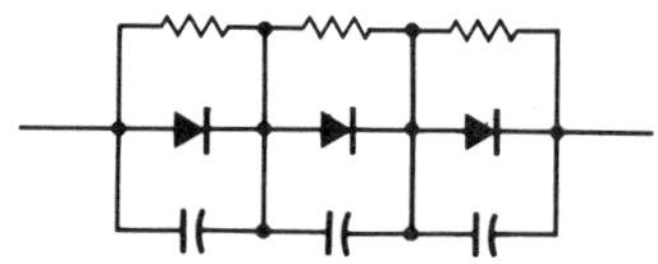

Fig. 12-8. Protection of series-connected diodes.

12-2-4. How do you calculate power supply regulation percentage?—The equation for voltage regulation is:

$$\% \text{ Voltage Regulation} = \frac{E_{\text{No Load}} - E_{\text{Full Load}}}{E_{\text{Full Load}}} \times 100$$

12-2-5. How is the minimum acceptable voltage rating for power-supply filter capacitors determined?—The value of the filter capacitor is set by the rms voltage output of the rectifier. In the case of a capacitor-input filter the capacitor working voltage should be 1.41 times this rate. This is the very minimum acceptable voltage rating. In practice it is customary to have a minimum working voltage at least twice the maximum voltage that will be present across the capacitor.

12-2-6. Explain the use of three-terminal integrated circuit regulators.—The three-terminal integrated circuit regulator is a self-contained voltage regulator containing a pass element, reference, sensing circuit, and error amplifier. Only three terminals are necessary as shown in Fig. 12-9. These are for the unregulated input voltage, common, and regulated dc output voltage. These devices are availa-

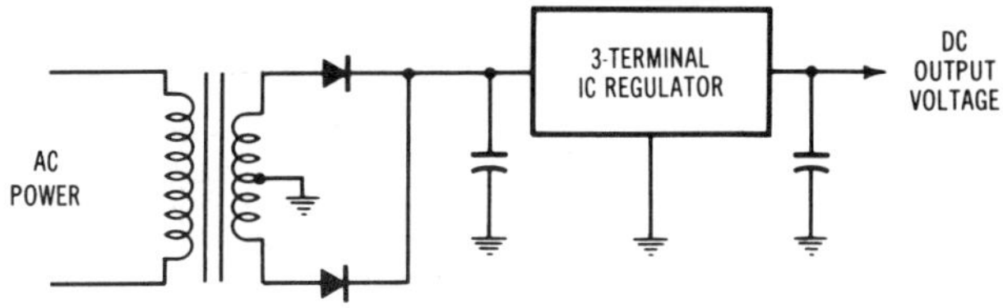

Fig. 12-9. Voltage-regulated supply using 3-terminal IC regulator.

ble for regulation of a positive voltage or a negative voltage. Certain types, with the use of an appropriate external circuit, can be used to obtain an adjustable and regulated dc output voltage. They have become a very popular form of power supply in all types of electronic systems. Their main use is in low-voltage and high-current power supplies.

12-2-7. Draw a schematic diagram of a transistor voltage regulator using discrete components such as a transistor and a zener diode. Explain its operation.—A transistor voltage regulator is shown in Fig. 12-10. Transistor Q1 is the pass transistor and its conductivity is made to vary with changes in load current, maintaining the output voltage at a constant value. This is accomplished with a zener diode (Z1) which determines the base bias current of the pass transistor. Resistor R_S sets the safe zener diode current. The zener diode current varies about this value as it changes the base current and consequently the conductivity of the pass transistor. The regulated dc output voltage is determined by the rated voltage of the zener diode. Other regulated dc output voltages can be obtained by changing over to a zener diode of a different voltage value.

An increase in load on the output of the power supply will result in a greater current demand and an attempt for the output voltage to decrease. However, there will be a compensating change in the zener diode current. This current is also the base current of the pass transistor. As a result there will be an increase in the conductivity of the pass transistor and, as a result, the new current demand will be met at the same time output voltage is held at a constant value.

12-2-8. What would happen if the pass transistor of a transistor regulator short-circuited? Draw a schematic of the circuit to protect against damage to the load in the event of a pass transistor short circuit.—If the pass transistor short-circuits there will be no restraining of the output voltage. Consequently the output voltage will rise and possibly cause damage to the load circuit to which the power supply is connected. In the usual protective circuit the rise in the output voltage and current with a short-circuit condition will blow a fuse and prevent the supply from delivering a damaging current to the load circuit. Refer to the crowbar overload protection arrangement shown in the power-supply circuit given in Section 7-9 of Chapter 7.

12-2-9. Draw a voltage-regulating circuit using a zener diode. Label the input and output voltages and the diode voltage rating.—Refer to Fig. 12-11. In a typical

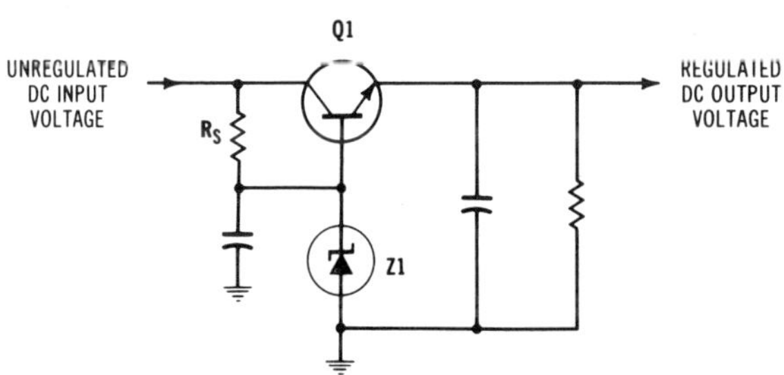

Fig. 12-10. Transistor voltage regulator.

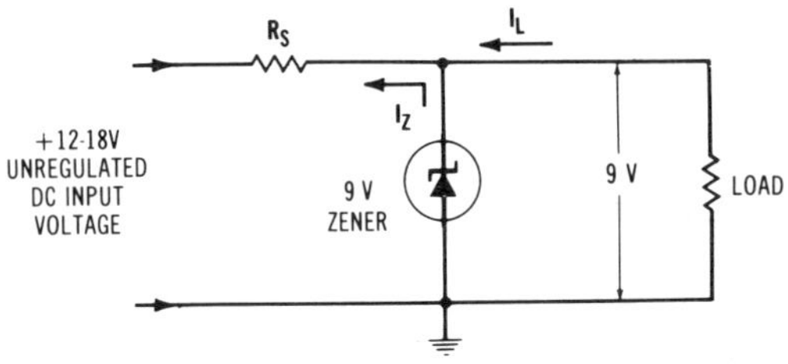

Fig. 12-11. Zener diode voltage regulator circuit.

9-volt voltage-regulated supply using a 9-volt zener diode, the unregulated dc input voltage will probably fall between 12 and 18 volts. A 9-volt regulated voltage is made available for the load. In operation any attempted change in the output voltage with a change in load will shuttle more or less current through the zener diode. Consequently the voltage drop across resistor R_S will be such that a constant 9 volts appears across the output. The zener maintains a constant voltage across its terminals despite the change of current through the zener over its normal operating range. Thus, if the load demands more current through resistor R_s, there will be less current in the zener and the voltage drop across R_S will be such that the voltage applied to the load will remain at 9 volts.

12-2-10. Compare the general solid-state structure and operation of a silicon-controlled rectifier (SCR) and a Triac. Draw representative circuits and describe operation.—SCRs and Triacs are used as solid-state switches and power controls. The SCR is a single-direction current device and is used in dc and ac applications in power control and various pulse-switching circuits. The Triac is a bidirectional device with current controlled in either direction. It is used mainly for ac power control applications. Usually its purpose is to switch and control power derived from an ac power source.

An SCR has three junctions instead of the single junction of an ordinary silicon-diode rectifier or the two junctions of a bipolar transistor. Its three terminals, Fig. 12-12, are anode, cathode, and gate. When there is an appropriate gate signal the SCR conducts just like a vacuum-tube thyratron. The SCR will continue to conduct even though the gate signal is removed or even reverse-biased. To stop operation the supply voltage must be switched off or decreased to a near zero value. The SCR permits a very weak gate signal to switch on a much higher-powered circuit.

The basic four-layer construction and its three junctions are illustrated in Fig. 12-12. The top junction is forward-biased and the next junction lower is reverse-biased. The junction between the gate and cathode connections must be forward-biased to turn on the SCR. When the gate is made positive (such as with the application of a positive bias or a positive pulse), electrons from the cathode layer pass into the gate layer and holes from the gate layer into the cathode layer. As a result there is also an increase in the leakage current through the middle junction. Since the top junction is forward-biased the transistor action results in an output current. There is now a rapid current rise, the level of which is set by the external load. Nevertheless a small gate current has resulted in a much greater load current. The cathode-anode current has now been established and cannot be broken by the events that take

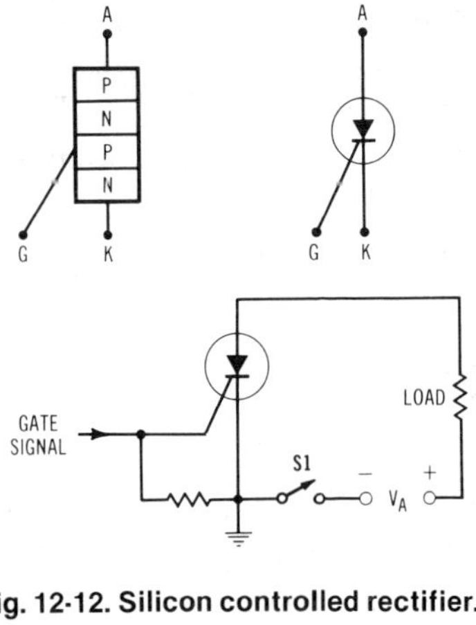

Fig. 12-12. Silicon controlled rectifier.

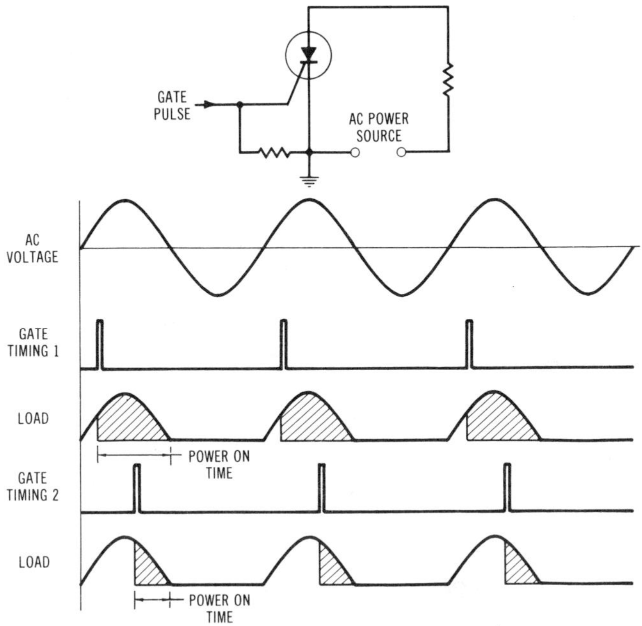

Fig. 12-13. Operation of SCR in an ac circuit.

place at the gate terminal. Load current can be stopped by a break in the load circuit or the reduction of the anode voltage to near zero.

A basic circuit for dc operation is given in Fig. 12-12. The anode-cathode circuit consists of the load, the supply voltage V_A and switch S1. When a positive gate pulse arrives the SCR is switched on and current is present in the load. This current continues until switch S1 is opened momentarily. If you now close switch S1 the current remains shut off until another positive gate pulse arrives.

The operation of an SCR in an ac circuit is demonstrated in Fig. 12-13. The anode-cathode voltage in this case is supplied from an ac power source. An arriving gate pulse turns on the SCR if the anode voltage is on its positive ac alternation. Since the ac voltage swings next toward 0 and its negative alternation, the SCR is shut down. It will not turn on again until an applied positive gate pulse again occurs when the anode voltage is on a positive alternation. This is demonstrated in the waveforms of Fig. 12-13. Note for example -1 that soon after the ac voltage swings positive there is an arriving gate pulse. Consequently the load current corresponds to the shaded area of the positive alternation of the ac voltage. It is during this interval of time that current is delivered to the load. This load current switches off during the negative alternation of the ac power cycle. Load current will resume again upon arrival of the next gate pulse.

This timing of the load current relative to the arrival of the gate pulse can be used to regulate the actual amount of power delivered to the load. Assume that the gate pulse is delayed relative to the positive alternation of the ac power as shown in the second example. As a result the load current is present for a shorter interval of time as compared to the conduction time of example #1. This tells us that a much lower average power is being delivered to the load. It is apparent then that the timing of the gate pulse cannot only be used to turn the load current on and off but can also be used to regulate the average power delivered to the load.

In the ac application of the SCR there is no need for any switching arrangement in the load circuit to shut off the power because of the negative sweep of the ac power source. Removal of the gate pulse

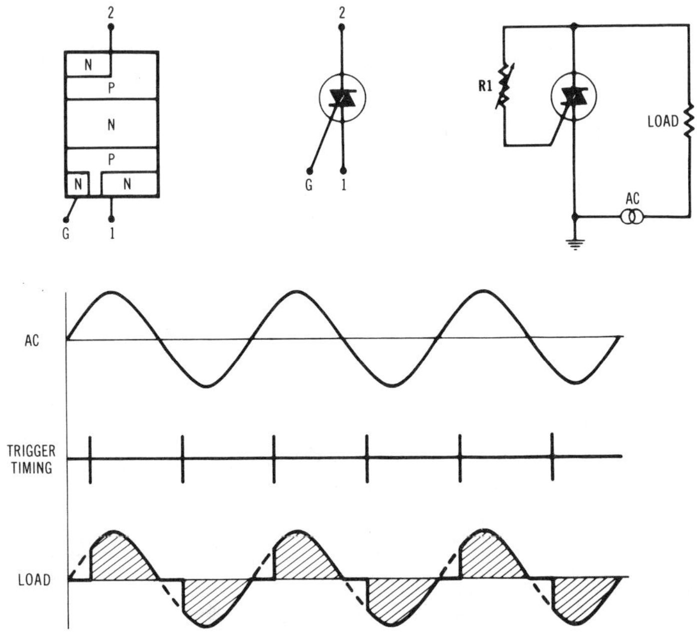

Fig. 12-14. Triac arrangement and circuit.

itself shuts off the load current completely.

A Triac basic construction, symbol and circuit are given in Fig. 12.14. The Triac differs from the SCR in that the current can flow in either direction in the output load. Therefore the device is ideal for ac applications. In effect the Triac acts as two SCRs connected in reverse direction with but a single gate. A positive gate arriving when the ac voltage is on its positive alternation will initiate current in the direction from terminal 1 to terminal 2. Operation is like that of an SCR. The central n-region is lightly doped and is able to store a charge. This hold-over charge and the Triac layering permits conduction on the negative alternation of the power cycle. This action too is initiated by gate triggering. Thus the Triac responds to both positive and negative gates.

In the sample circuit of Fig. 12-14 the Triac is being triggered by the ac voltage. The ohmic value of resistor R1 sets the trigger timing relative to the ac cycles. Thus the power delivered to the load varies with the setting of the resistor R1. In the waveform sample, power is delivered to the load during the filled-in portion of the load waveform. Shifting the trigger timing to the left would increase the power delivered to the load. A shift in the timing to the right would decrease this power.

12-2-11. What is a power dc-to-ac inverter? Draw a schematic diagram and briefly explain its operation.—In some radio-communications installations there is need for 120 volt ac power when the primary power source is a low dc voltage. Solid-state power inverter circuits can be used to make the changeover.

The most modern inverters are solid-state and employ variations of the basic circuit shown in Fig. 12-15. Fundamentally it is an oscillator switching circuit that generates 60-hertz squarewaves. There are two power transistors and a saturable transformer (T1). The characteristics of the saturable transformer provide the necessary feedback and set the frequency of operation. Inasmuch as the dc supply voltage also can have an influence on operating frequency, a voltage regulator is often included. The collector to collector voltage variation appears across primary power transformer T2. The 120-volt ac power is de-

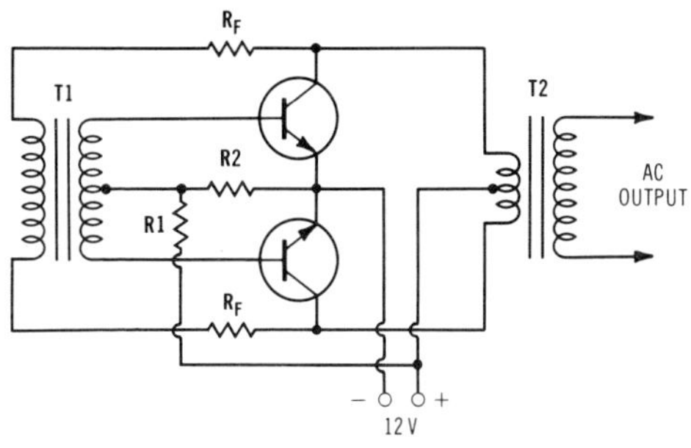

Fig. 12-15. Basic inverter circuit.

veloped across its secondary. Transformers can be designed for operation at a different frequency or to develop a different ac voltage.

Upon application of power one of the transistors, despite their similarity, will conduct more than the other. The feedback path drives this transistor toward a saturation level set by the saturable transformer. In this drive, transistor Q2 may reach saturation, while transistor Q1 is switched off. Upon transformer saturation the feedback cycle reverses and transistor Q1 goes to saturation and Q2 shuts down. The feedback path between the collector output and base input is by way of the two resistors that link the primary of transformer T2 to the primary of transformer T1. Proper forward-biasing that permits the power transistors to conduct is set by resistors R1 and R2.

12-2-12. What is a mobile dc-to-ac converter? Draw a schematic diagram and give a brief description of its operation.—In mobile two-way radio installations a power supply must often be capable of stepping up the 12-volts dc of the vehicle battery to a much higher dc voltage for use by a high-power solid-state transmitter/receiver combination. Such a voltage step-up would also be necessary if the transmitter includes one or several high-powered vacuum tubes.

The circuit of Fig. 12-16 consists of switching transistors and an appropriate feedback winding on transformer T1 to produce oscillations. Switching rate might be as low as 50 hertz on up to several thousand hertz. The high switching rate is preferred because less output filtering is needed and smaller transformer cores of less weight can be used.

Supply voltage from the vehicular

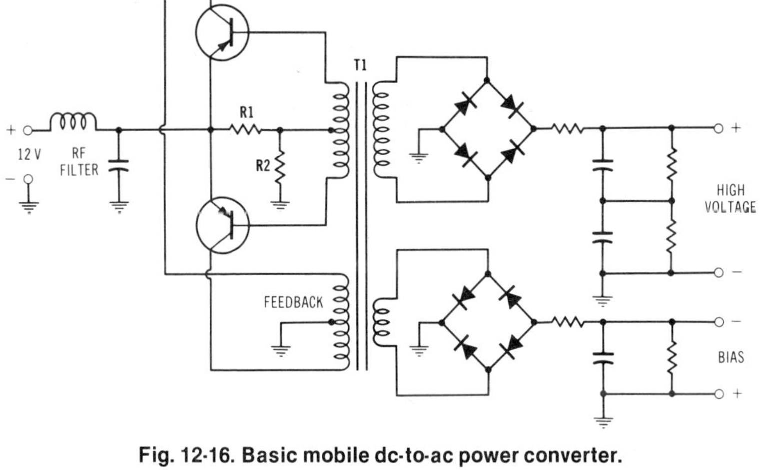

Fig. 12-16. Basic mobile dc-to-ac power converter.

battery is supplied to the switching transistors through a radio-frequency filter. Resistors R1 and R2 provide forward-biasing. Two stepup secondary windings develop the high-voltage ac supplied to bridge rectifiers. The output of the high-voltage winding is rectified and filtered to develop a high dc output voltage. A second winding and associated bridge rectifier can be used to develop any required negative bias.

12-3. BATTERIES

12-3-1. How does a primary cell differ from a secondary cell?—When a primary cell discharges, the electrochemical activity destroys one or both electrodes, and the cell is not rechargeable. In a secondary, or storage cell, the battery can be discharged and recharged many times. The chemical activities are reversible; therefore, the battery is recharged by applying a reverse current.

12-3-2. What is the chemical composition of the electrolyte of a lead-acid storage cell?—The electrolyte is a solution of sulfuric acid and distilled water having a specific gravity of approximately 1.275. Specific gravity is the ratio of the weight of a solution to that of an equal volume of pure water.

When the battery is in operation (discharging), the electrolyte breaks down and deposits sulfuric acid on both plates of the battery. The electrolyte gradually loses its acid content, and its specific gravity drops; the battery makes available less and less electrical energy. When a reverse current is applied, the battery is under charge; the acid is returned to the water; and the electrolyte is restored to its original specific gravity.

12-3-3. What may cause sulfation of a lead-acid storage cell?—Sulfation can be caused by frequent undercharging, improper specific gravity, leakage or internal defect, and standing idle in a discharged condition.

12-3-4. What will be the result of discharging a lead-acid storage cell at an excessively high current rate?—The heat generated by an excessively long and high-current discharge can buckle the plates. If the discharge is too prolonged, the specific gravity will drop below a safe level, and excessive sulfation of the plates will occur. This sulfation cannot be broken down successfully when the battery is recharged. As a result, the internal resistance of the battery will increase.

12-3-5. If the charging current through a storage battery is maintained at the normal rate, but its polarity is reversed, what will result?—The current will discharge instead of charging the battery. If this discharge is excessive and for too long an interval, heat and sulfation problems will develop.

12-3-6. What is the approximate fully charged voltage of a lead-acid cell?—An individual cell is 2.1 volts. Thus the fully charged voltage of a so-called 6-volt storage battery is 6.3 volts; for the 12-volt storage battery, it is 12.6 volts.

12-3-7. Describe the care which should be given a group of storage cells to maintain them in good operating condition.—Maintain the normal specific-gravity level for full charge. The condition of such a battery is best checked with a hydrometer which measures specific gravity. Do not discharge below the minimum safe specific-gravity level. Keep the electrolyte at the proper level. Keep the terminals clean, dry, and coated. Never charge at an excessive rate. Do not permit uncharged batteries to stand. Keep batteries well-ventilated and do not store in an excessively hot, cold, or moist place.

For long-term inactive storage, drain the acid, rinse out the battery, and then refill with distilled water.

12-3-8. What steps may be taken to prevent corrosion of lead-acid storage cell terminals?—Keep the terminals clean and the connections firm. Preparations are available for putting a protective layer on the terminal. Vasoline petroleum jelly will provide a thin protective coating over the terminals.

12-3-9. How is the capacity of a battery rated?—Batteries are rated in ampere-hours. For example, a 60-ampere-hour battery can deliver 7.5 amperes of current for 8 hours before recharging or the same battery could deliver 5 amperes for almost 12 hours.

Usually the rating is based on an 8-hour operation. Actual ampere-hours available are greater for a slow discharge than for a fast one. Longer life is obtained when a battery is not permitted to discharge to too low a level before it is recharged.

12-3-10. What is meant by the specific gravity of an electrolyte, and how can it be used to determine the state of charge of a lead-acid cell?—Specific gravity is the weight of a solution relative to that of an equal volume of pure water. The electrolyte of a lead-acid storage cell is a solution of sulfuric acid and distilled water having a specific gravity of approximately 1.26. This is the approximate reading of specific gravity at full charge. When the battery is in operation (discharging), the electrolyte breaks down and deposits sulfuric acid on both plates of the battery. The electrolyte gradually loses its acid content and its specific gravity drops. When a reverse or charging current is applied to the battery, the acid is returned to the water and the electrolyte is restored to its original specific gravity. Thus a hydrometer which is an instrument that can be used to measure specific gravity can be used to check out the charge of a lead-acid cell.

12-3-11. How do the characteristics of nickel-cadmium and lead-acid cells differ?—Lead-acid cells form a low-cost, reliable and high-powered battery. Such a battery is capable of delivering a high current, both peak and average. Both lead-acid and nickel-cadmium cells can be recharged. Nickel-cadmium cells come in various shapes and sizes. Some are similar to dry-cell batteries and, in fact can be used as rechargeable replacements in many electronic circuits that use dry cells. Higher-powered nickel-cadmium cells are also available but as yet do not approach the power capability of lead-acid cells. Nickel-cadmium cells provide a constant voltage output over the normal discharge cycle. In the charge and discharge activities of a nickel-cadmium cell there is oxidation change but no change in the physical state. A lead-acid cell, however, deteriorates over a period of time because of the changes in the physical state with charge and discharge. As a result a nickel-cadmium cell has a longer operating life than a lead-acid cell and also has a very much longer shelf life in either a charged or discharged condition. The lead-acid battery employs a liquid electrolyte while most nickel-cadmium batteries use a dry electrolyte. Some of the higher-powered nickel-cadmium types do use a wet electrolyte.

12-3-12. Discuss the differences in charging methods of nickel-cadmium and lead-acid cells.—A very simple charger can be used for a lead-acid battery requiring only that it have a voltage output that can be maintained somewhat higher than the battery voltage and also the current capability to supply the necessary charging current. As the lead-acid battery charges the charging current declines. A nickel-cadmium cell battery requires a more constant-current charger. The charging voltage is significantly higher than the voltage of the cell and the charger is designed to supply a constant charging current. The nickel-cadmium battery supplies a reasonably constant voltage output and, when that voltage falls below a specified value, it is an indication that the battery should be recharged. It is difficult to use voltage as an indicator of the state of charge of a lead-acid cell. The preferable method is to use a hydrometer to measure the specific gravity and when that reading falls below a specific value it is necessary to recharge the lead-acid battery.

12-3-13. What are the characteristics of alkaline batteries?—Alkaline batteries are available both as primary and secondary types. They comprise a zinc anode, manganese-dioxide cathode, and a potassium-hydroxide electrolyte. This battery differs from the conventional zinc-carbon primary battery because of its alkaline electrolyte. However, the alkaline battery has a lower internal resistance and impedance as well as a high service capacity. Thus, when fabricated into the standard battery sizes, it will provide a higher current capability and a longer service life for given operating conditions, as compared to the zinc-carbon type. The overall efficiency is attractive for applications that require higher currents and longer drain intervals.

The alkaline secondary battery has some of the same characteristics as a nickel-cadmium type. Initial cost is less. However, the alkaline battery cannot be discharged (cycled) as often. The alkaline battery should not be discharged as deeply because its life will be shortened. When used with care it provides an economical advantage in numerous applications.

CHAPTER 13

Amplifiers and Oscillators

13-1. AUDIO AMPLIFIERS

13-1-1. What are the factors that determine the correct bias voltage for the grid of a vacuum tube?—One factor is whether the amplifier is to be biased on the linear portion of its transfer as in class A, or whether it is to be biased for class-B or class-C operation. The bias voltage must be set in accordance with the desired output, input voltage, permissible distortion, supply voltage, optimum load impedance, and so that the performance of the tube is held within safe operating limits.

13-1-2. Draw schematic diagrams illustrating the following types of grid biasing, and explain the operation of each type: (a) battery, (b) power supply, (c) voltage divider, (d) cathode resistor.—The various bias arrangements are shown in Fig. 13-1.

(a) In the simple battery-biasing arrangement, the negative side of the battery connects to the grid, and the positive side connects to the cathode. In operation, other than class C, there is no significant current through grid resistor R_g, and thus this resistor does not influence grid biasing.

(b) In more elaborate electronic equipment, a separate bias winding on the power transformer or an entirely separate bias-voltage supply may be used. Often the bias-supply bleeder arrangement is tapped, which permits various values of bias voltage to be distributed throughout the electronic gear. Taps are selected according to the bias of the various tubes.

(c) A voltage-divider arrangement composed of two or more resistors is often used to obtain a specific bias. In this arrangement the bias-supply voltage is at some definite and constant voltage. A two-resistor voltage divider is then used to divide this distributed voltage down to the specific value needed by the given tube in the electronic system.

Occasionally instead of using negative voltage applied to the control grid, a voltage divider connected accross a positive supply-voltage source will divide that voltage down to the proper bias level and supply it as a positive voltage to the cathode. Of course, this is the same as supplying a negative voltage to the control grid, making it negative with respect to the cathode.

(d) The cathode-resister method of biasing was discussed previously in Chapter 11, Questions 11-1-2 and 11-1-12. The dc component of plate current in the cathode circuit develops the necessary bias voltage across the cathode resistor with the proper choice of cathode resistor value. In a tetrode or pentode stage it must be remembered that the screen-grid current also is present in the cathode circuit, and the bias value becomes a function of the total cathode current through the cathode resistor.

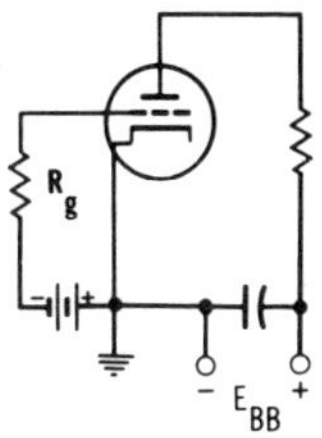

(A) Battery biasing.

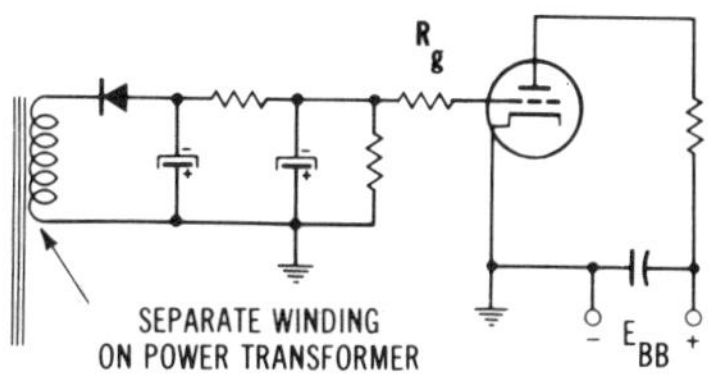

(B) Bias power supply.

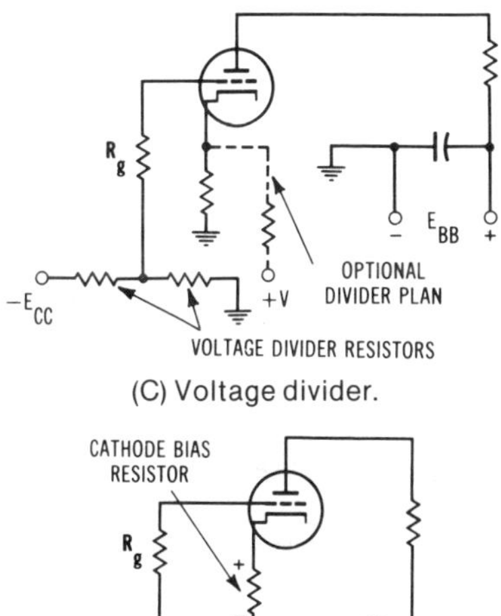

(C) Voltage divider.

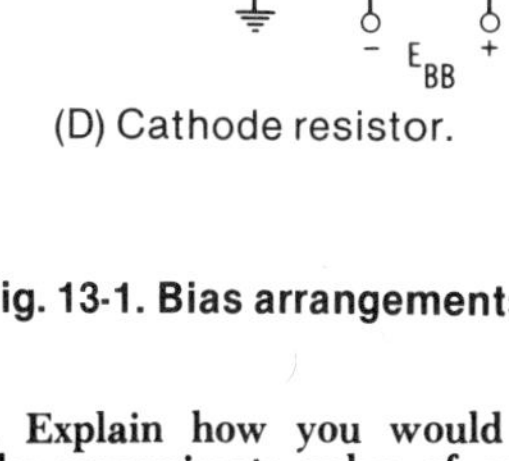

(D) Cathode resistor.

Fig. 13-1. Bias arrangements.

13-1-3. Explain how you would determine the approximate value of cathode-bias resistance to provide correct grid bias for any particular amplifier.—In selecting the proper value of cathode resistor it is first necessary to know the exact value of bias that is to be required and the total cathode current when this amount of bias is established. Remember again that for a triode vacuum tube the cathode current is equal to the plate current. In tetrode and pentode tubes the cathode current is equal to the sum of the plate current and the screen-grid current.

When the desired bias and cathode current have been selected according to the manufacturer's ratings or characteristic curves and the desired operating conditions, Ohm's law can be used to determine the ohmic value of the cathode resistor. For example, if the desired grid bias is −3 volts and the total dc cathode current is 2.5 milliamperes, the ohmic value of the cathode resistor becomes:

$$R_k = \frac{-E_c}{I_k} = \frac{-3}{0.0025} = 1200 \text{ ohms}$$

The power that must be dissipated by this resistor can be obtained with a second simple calculation:

$$P = E \times I = 3 \times 0.0025$$
$$= 0.0075 \text{ watt}$$

13-1-4. Is grid-leak biasing practical in audio-amplifier stages?—No. This type of biasing is not used because in the normal linear operation of vacuum tubes there is no grid current. Thus no significant voltage can be developed across the grid resistor. If an operating point and signal amplitude were selected so that there would be grid current, it would result in distortion and, in most cases, improper loading of the preceding audio stage. Grid-leak biasing, however, is a common method of obtaining negative grid voltage for class-C rf amplifiers and oscillators.

13-1-5. Draw a diagram showing a method of obtaining grid bias for a filament-type vacuum tube by the use of resistance in the plate circuit of the tube.—This technique is shown in Fig. 13-2. In such an arrangement the negative side of the plate-voltage supply returns to ground through a bias-resistor combination. The control grid is returned to the junction of the resistor and the negative side of the plate voltage supply.

13-1-6. Draw simple schematic diagrams illustrating the following types of coupling between audio-amplifier stages, and between a stage and a load: (a) triode vacuum tube inductively coupled to a speaker, (b) resistive coupling between two pentode vacuum tubes, (c) impedance coupling between two tetrode

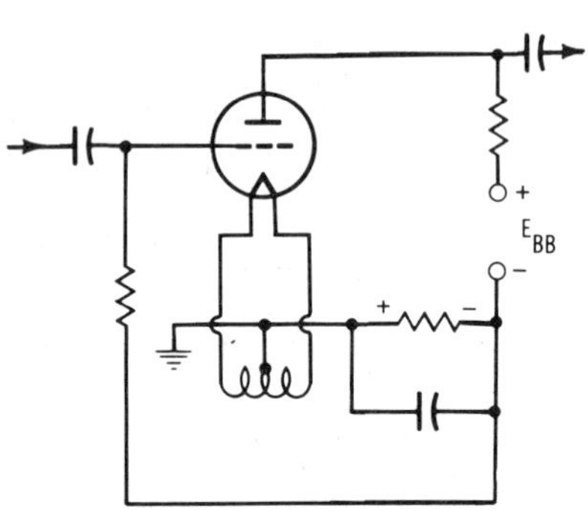

Fig. 13-2. Biasing a tube from the plate-circuit resistor.

vacuum tubes, (d) a method of coupling a high-frequency speaker to an audio-frequency audio-amplifier tube without plate current going through the speaker winding, and without the use of a transformer.—Refer to Fig. 13-3.

13-1-7. Draw circuit diagrams and explain the operation (include input-output phase relationships, approximate practical voltage gain, approximate stage efficiency, uses, advantages, and limitations) of each of the following types of audio circuits: (a) class-A amplifier with cathode-resistor biasing, (b) cathode-follower amplifier, (c) at least two types of phase inverter for feeding push-pull amplifiers, (d) cascaded class-A stages with a form of current feedback, (e) two class-A amplifiers operated in parallel, (f) class-A push-pull amplifier.—A class-A amplifier with cathode-resistor biasing was discussed in Chapter 11, Question 11-1-2 and Fig. 11-10. Input and output voltages are 180° out of phase. practical voltage gains extend from below 10 up to in excess of 100. Practical efficiencies fall between 10% and 30% with a theoretical maximum of 50%. Such stages can be used as audio-amplifiers and, with appropriate compensating circuits, as video amplifiers.

In conjunction with resonanat circuits instead of resistive loads, they can function as radio-frequency amplifiers. When operated correctly they serve as linear amplifiers.

Maximum voltage gain can be obtained when such stages are employed as small signal amplifiers. As the signal level becomes higher and higher, there is a corresponding reduction in the practical value of voltage gain that can be obtained.

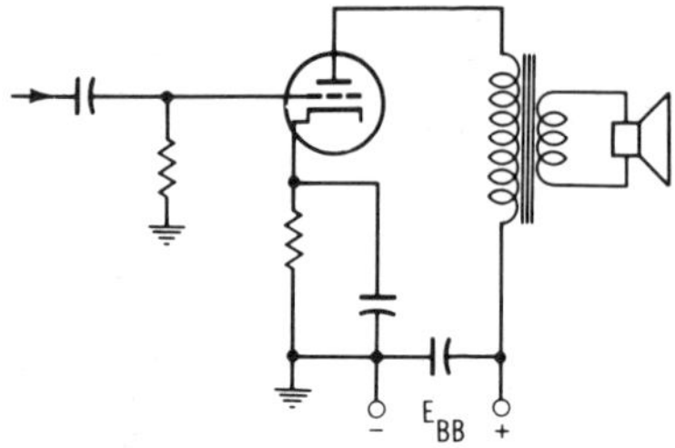

(A) Inductive coupling to a speaker.

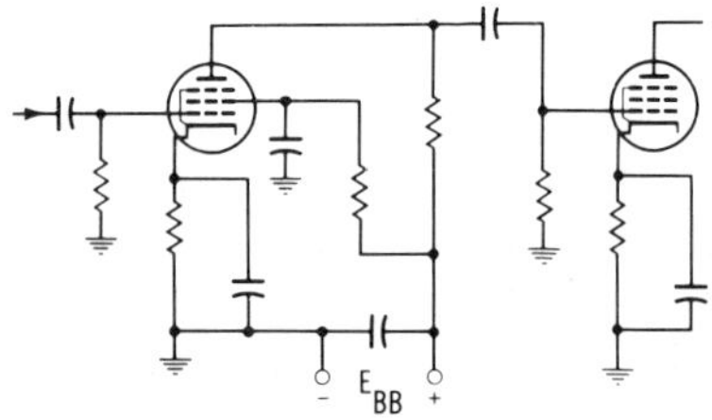

(B) Resistance-coupled stages.

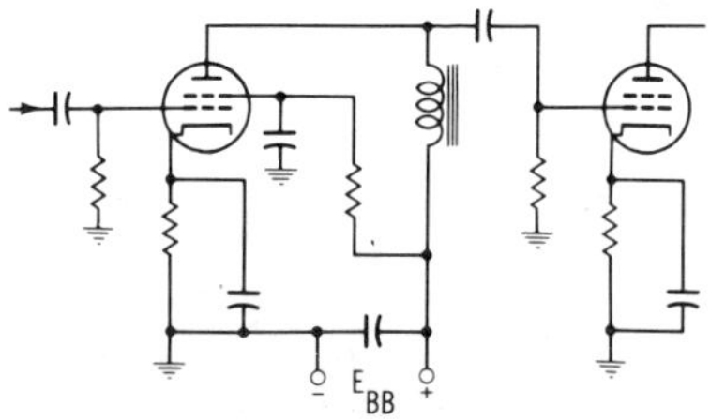

(C) Impedance coupling.

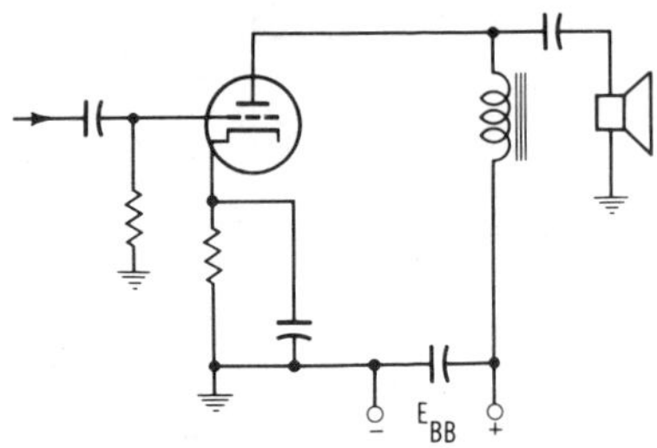

(D) Coupling to a high-impedance speaker.

Fig. 13-3. Coupling systems.

(b) The circuit and general operation of a cathode-follower was covered in Chapter 11, Question 11-1-2 and Fig. 11-13. The cathode-follower is limited to a voltage gain of less than 1. Input and output voltages are in phase. A major use for a cathode follower is to match a high impedance to a low impedance; it serves as an effective impedance transformer with a wideband frequency response. The cathode-follower has a high input impedance and a low input capacitance. Consequently a cathode follower can be used where a very light load is to be placed on the preceding stage.

(c) The two most popular forms of phase inverters are shown in Fig. 13-4. The purpose of a phase inverter is to obtain as nearly as possible equal-amplitude and opposite-polarity signals for application to a push-pull amplifier. The split-load inverter (Fig. 13-4A) uses both plate and cathode output. The plate output is of course 180° out of phase with the input wave, while the cathode output is in phase with the input voltage. As a result, the opposite polarity condition is met. By equalizing the ohmic values of

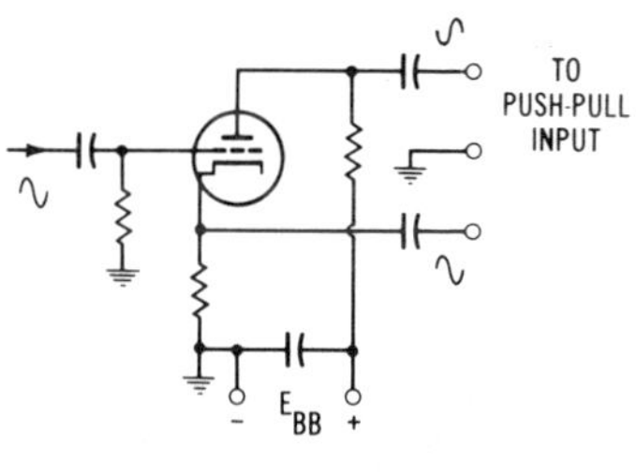

(A) Split-load type.

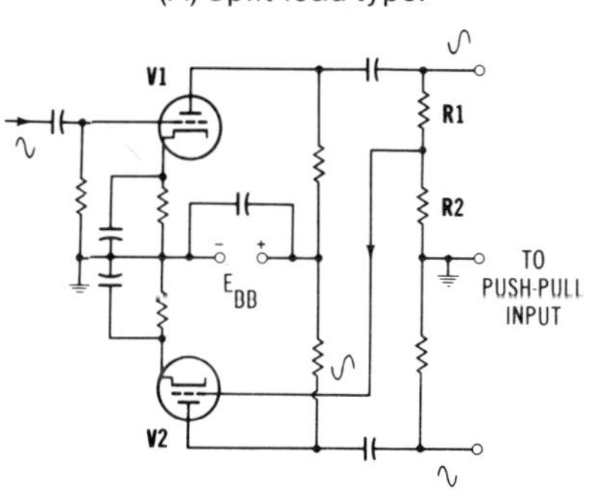

(B) Two-tube type.

Fig. 13-4. Phase inverters.

plate-load resistor and cathode resistor, the amplitude can also be made essentially the same. Amplifier gain is limited to less than unity. The two output impedances differ, and, therefore, the coupling system must be designed properly to make certain that essentially equal-amplitude components over the desired frequency range are delivered to the next stage.

The second phase-inverter arrangement (Fig. 13-4B) uses two triode sections. It provides gain along with required equal-amplitude and opposite-polarity components. Output impedances, too, are essentially the same for both sources of signal. The top tube (V1) is connected as a class-A resistance-coupled amplifier. It develops a signal component which is out of phase with the input wave. A portion of this output signal is taken off across the output voltage-divider resistor R1 and R2. The weaker voltage variation is connected to the grid-cathode circuit of the lower class-A amplifier tube, V2. The voltage divider reduces its level to approximately the same magnitude as the original input wave. An amplified component appears at the plate circuit of V2. Because of the additional polarity shift of V2, the output signal is brought back in phase with the input wave. By proper choice of load divider, its magnitude at the output at V2 can be made the same as that present at the plate circuit of V1. In this case, amplified equal-amplitude and opposite-polarity signals are made available for driving a following push-pull stage. Both of the phase inverters are fundamentally voltage-amplifier stages. They are designed to operate into a high impedance. They are not usable when the following push-pull stage requires a significant amount of driving power.

(d) Two resistance-coupled stages with current feedback are shown in Fig. 13-5. The total gain of the two-stage amplifier is the product of the individual stage gains. Less gain is possible when feedback is used. However, feedback stabilizes the amplifier operation, maintaining a constant and uniform gain over a desired audio-frequency range. Negative feedback also reduces nonlinear distortion. In feedback systems, the feedback may be proportional to either the output voltage or the output current. When proportional to the output voltage, the feedback is called *voltage feedback;* when proportional to the output current, it is called *current feedback*.

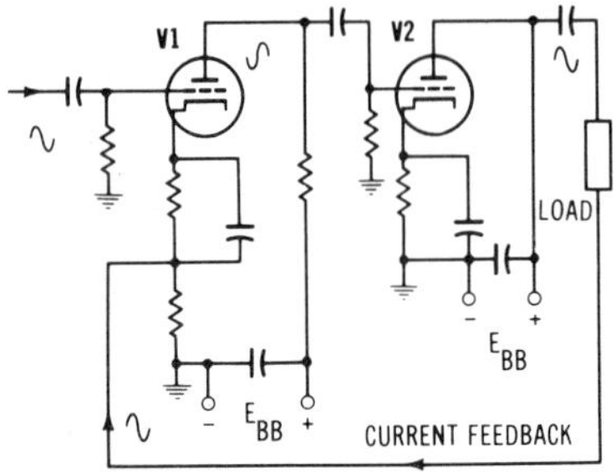

Fig. 13-5. Two-stage amplifier with current feedback.

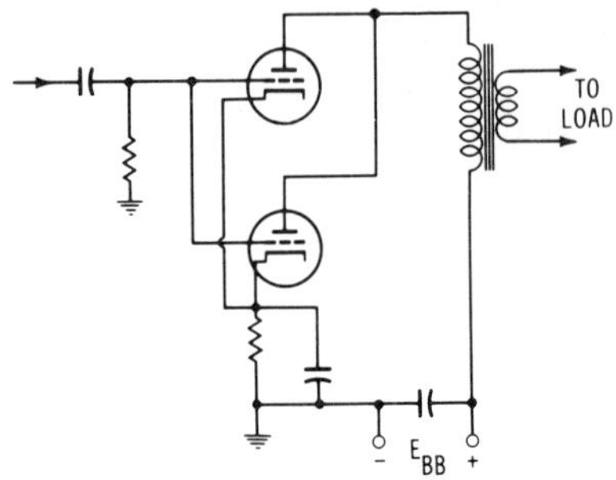

Fig. 13-6. Two triode tubes connected in a parallel arrangement.

The output load is the source of current feedback which is proportional to the current in that load. To obtain negative feedback, it is of course necessary that the feedback voltage be out of phase (opposite polarity) relative to the signal voltage at the point where the feedback is inserted, as shown in Fig. 13-5.

The phasing relationship can be verified when one considers that plate output is of opposite polarity from the input signal, while the cathode output is of the same polarity as the input signal. Note that the plate output of V1 is of opposite polarity with respect to the input to V1. However, the plate output of V2 has the same phase as the input of V1. As a result, when it is reintroduced into the cathode circuit of the first tube it is of the same polarity as the grid-input wave. Therefore it is degenerative and provides negative feedback.

It should be remembered that use of negative feedback corrects for distortion and instability which arise within those amplifiers that are a part of the feedback loop. Also the use of negative feedback makes no corrections for the condition of the input wave. If the input wave is distorted, it will appear distorted at the output of the feedback loop.

(e) Two triode *power amplifiers* connected in parallel are shown in Fig. 13-6. Two similar tubes so connected can approximately double the power output. The output impedance is approximately halved, and it is necessary that the transformer primary carry twice the current normal for a single-tube stage. The cathode current also is higher, and to obtain the same operating bias it is necessary to halve the value of the cathode resistor. In other respects its operation is quite similar to a single-tube circuit; efficiency is approximately the same.

No significant improvement can be expected when paralleling tubes in a class-A *voltage amplifier.* In this case, the plate resistance is halved, and to establish the same operating-point plate voltage it is necessary to halve the value of the load resistor because of the doubled current flow.

(f) A class-A push-pull amplifier is shown in Fig. 13-7. The push-pull connection permits power outputs, for a given distortion percentage, of more than twice that which can be obtained from a single-ended stage using the same tube. In a class-A push-pull amplifier, both tubes draw plate current during the entire period of the input cycle. This differs from the class-B mode of operation discussed in Chapter 11, Question 11-1-2 and Fig. 11-4. The class-A push-pull amplifier sees both tubes biased on the linear portion of their characteristic curves. Hence, with or without input signal, both tubes draw approximately the same dc component of plate current. A strictly class-A push-pull amplifier requires no driving power, and

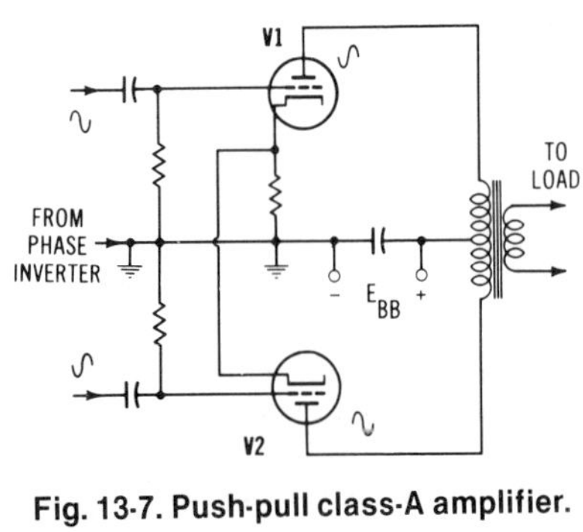

Fig. 13-7. Push-pull class-A amplifier.

an input transformer is not necessary. It can be driven from a phase-inverter circuit which provides the proper opposite-polarity and equal-amplitude input waves.

In operation, the plate current of V1 rises and the plate current of V2 falls during the positive alternation of the input wave. During this alternation the plate of V1 swings negative, and the plate of V2 swings positive. The output components therefore appear additive across the two sections of the primary winding. Then, during the negative alternation of the input wave, the plate current of V1 falls, and that of V2 rises. Input and output phase relations are shown in Fig. 13-7.

The push-pull connection has a number of advantages over a single-ended or a parallel pair of tubes. No cathode capacitor is required, because the plate current of one tube is increasing at the same time the plate current of the other is decreasing. Therefore the combined cathode current is constant and develops the required dc grid bias for both tubes. There is no trouble with core magnetization, because the plate current occurs in opposite directions through the two sides of the primary winding. If there is any ac hum in the supply voltage, the push-pull connection provides cancellation. Any change in the supply voltage as a result of hum is cancelled across the primary of the transformer because the two plate currents are equal and occur in opposite directions.

When the amplifier is operating on the nonlinear portion of a tube transfer characteristic, odd and/or even harmonic distortion components are generated. An advantage of the push-pull connection as compared to the single-ended type is the fact that the generated even-order harmonics are reduced. As a result the push-pull connection provides an inherently lower distortion percentage. Thus two tubes in class-A push-pull have a substantially lower distortion percentage than the same type of tube operated either as single-ended or as a parallel pair.

If the stage is so designed that it is not limited to strictly class-A operation, an even higher output can be obtained. In class-AB or class-B operation the plate current and voltage are made to swing over a greater segment of the transfer characteristic into some of the nonlinear region. However, the push-pull connection aids in the reduction of any introduced even-harmonic distortion. Therefore an even greater power output can be obtained relative to a distortion percentage of a given value. In push-pull class-A operation, efficiency can be made to approach nearer to the 50% limit.

13-1-8 What would probably be the effect on the output amplitude and waveform if the cathode-resistor bypass capacitor in an audio stage were removed?—An unbypassed cathode resistor introduces negative current feedback. As a result there will be a reduction in the amplitude of the output wave. If nonlinear distortion is introduced by the stage, it will be reduced, and therefore the output waveform would appear less distorted. Otherwise, there would be no change in waveform.

13-1-9 Show by use of circuit diagrams two ways of using single-ended stages to drive a push-pull output stage.—A transformer drive of a push-pull output stage is shown in Fig. 13-8. The secondary of

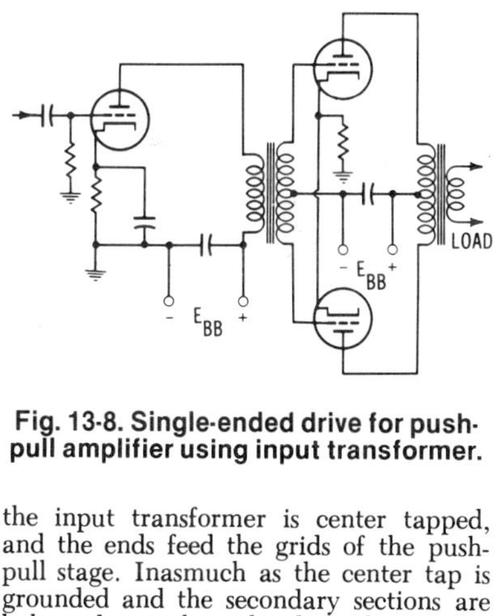

Fig. 13-8. Single-ended drive for push-pull amplifier using input transformer.

the input transformer is center tapped, and the ends feed the grids of the push-pull stage. Inasmuch as the center tap is grounded and the secondary sections are balanced, equal-amplitude and opposite-polarity signals are made available for driving the push-pull input. An advantage of the transformer method of driving a push-pull stage is that a reasonable impedance match can be made and power can be transferred from the output of the driver stage to the input of the push-pull stage. In class-B and in *certain* class-AB push-pull connections, a certain amount of driving power is required from the preceding stage.

The phase-inverter type of drive for push-pull stages was covered in connection with Question 13-1-7 (c) and

Figs. 13-4 and 13-7. Phase inverters can be used when no significant driving power is required at the push-pull input, such as in class-A and certain class-AB modes of operation.

13-1-10. Why does a class-B audio-frequency amplifier require considerably greater driving power than a class-A amplifier?—The input to a class-A amplifier requires only that a required voltage variation be present between grid and ground. This changing voltage through the action of a control grid has an influence on the plate current but does not draw electrons out of the stream. Therefore, the input impedance to the class-A stage is very high, and the driving power is insignificant. The usual class-B stage draws grid current, during which time the input impedance drops to a low value. Furthermore, the tube is driven over a greater grid-voltage range because the plate current must swing from near zero to a high maximum value. As a result, a considerable amount of grid driving power is required. The power must be such that grid losses resulting from the grid current are overcome.

13-1-11. What factors should be taken into consideration when ordering a class-A output transformer? A class-B audio-output transformer feeding a speaker of known ohmic value?—The important class-A output-transformer considerations of concern here are the *turns ratio*, which establishes a proper impedance match for efficient transfer of power; the *primary inductance*, which must be adequate to pass the lowest frequency desired; the *distributed capacity* of the windings must be such that the highest frequency to be passed is not attenuated; the *wire size* must be capable of handling the tube current; and the *transformer construction* must be such that the core does not saturate with normal levels of output signal. Magnetic-core saturation is less of a problem when a class-A push-pull amplifier is used. For class-A push-pull triodes, the impedance seen by each tube is one-half of the reflected plate-to-plate impedance.

The same considerations apply to the characteristics of a class-B audio-output transformer. Power rating, frequency response, turns ratio, etc., must correspond to the tubes being used in the push-pull circuit and the desired operating characteristics. The turns ratio is a significant factor in the operation of single-ended and push-pull amplifiers.

In class-B push-pull the impedance seen by each tube is one-fourth of the reflected plate-to-plate impedance.

13-1-12. Why are decoupling resistors and capacitors used in stages having a common power supply?—They are used to prevent amplifier oscillations, motorboating, and other low-frequency instability. A power supply has an impedance that is common to all circuits with which it is associated. Any ac variation across the impedance from the higher-powered or higher voltage-level stages can be fed back to earlier, more sensitive stages by means of this common impedance and the supply-voltage lines. The decoupling resistors and capacitors are used in the supply-voltage lines to filter out any such voltage variations so that they are not inserted into the earlier stages in the process of supplying the proper dc electrode voltages.

13-1-13. Draw circuit diagrams and explain the operation of two commonly used tone-control circuits.—Two typical tone-control circuits are shown in Fig 13-9. In the simple tone-control circuit of Fig. 13-9A, a series capacitor and

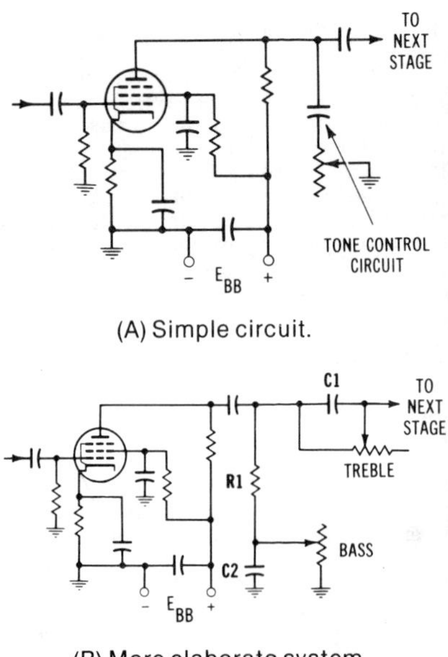

(A) Simple circuit.

(B) More elaborate system.

Fig. 13-9. Tone-control arrangements.

potentiometer are connected across grid-input or plate-output circuits. When the arm of the potentiometer is located at the capacitor side, the capacitor appears directly across the load resistor. At the high frequencies it places a low reactance across this load, and, therefore, the highs are attenuated more than the middle and low audio ranges. As more and more resistance is added, the influence of the capacitor is reduced due to the rising impedance of the series combination of resistor and capacitor. Thus the highs are attenuated to a much lesser degree and the high-frequency performance of the amplifier is emphasized.

The more elaborate tone-control system of Fig. 13-9B includes separate high- and low-frequency tone controls. Capacitor C1 has a low reactance at high audio frequencies and an increasing reactance over the middle- and low-frequency ranges. Thus there is relatively much less attenuation of the highs. The attenuation of the middle- and low-frequency ranges relative to the highs depend on the setting of the potentiometer. The lower the resistance inserted across the capacitor, the more uniformly the highs and lows are passed to the next stage. In effect, the highs are emphasized by using a controlled amount of attenuation of the middle- and low-frequency ranges.

Low-frequency emphasis is obtained by regulating the amount that the effective gain is stepped up at low frequencies. A resistor-capacitor combination is used across the normal load resistor R1. The reactance of the capacitor is high at low frequencies, and therefore the series impedance of resistor R1 and capacitor C2 is high. Over the middle- and high-frequency ranges, the reactance of the capacitor is low, and the impedance is determined mainly by the resistance of R1. Therefore the stage gain is lower at the mid- and high-frequency ranges. The extent to which the lows are boosted depends on the setting of the potentiometer. The lower the resistance that it shunts across the capacitor, the more uniform is the gain over the audio-frequency range. The lows are emphasized by placing maximum resistance across C2.

13-1-14. Why do vacuum tubes produce random noises?— Although there are several types of noises generated within a vacuum tube, the most pronounced are the so-called shot-effect noise components. These are generated by the non-uniformity of the emission from a cathode surface, resulting in minute fluctuations in plate and other electrode currents. This particular random fluctuation along with other much lower-level tube noises is referred to as vacuum-tube random noise.

13-1-15. Name some causes of hum and self-oscillation in audio amplifiers. Give the methods for reducing them.—Self-oscillation is produced by positive feedback which can occur between the input and output of an individual stage, or between the output of a higher-voltage or power-level stage and the more sensitive lower-level input stage feedback of an amplifier. Individual stage feedback is reduced by proper circuit design and adequate isolation between input and output circuits. The use of tetrode and pentode tubes reduces the possibility of individual stage oscillations. Neutralizing capacitors are other types of neutralizing circuits can be used to cancel feedback by introducing a controlled amount of feedback of the opposite polarity, minimizing the tendency of a stage to self-oscillate. In an amplifier it is important that there be a proper isolation between high-level output and sensitive input stages. Shielding and the in-line construction of stages as well as the complete isolation of output and input leads remove those conditions which encourage oscillations. Decoupling capacitors and resistors and special filter circuits also reduce the oscillation tendency. Low-impedance circuits at certain critical points in an amplifier can do much to overcome any tendency to oscillation.

Hum in an amplifier can be a result of inadequate filtering of the dc supply voltage, or hum that leaks from ac filament lines into the high-gain sensitive-input stages of an amplifier. Hum can be reduced by providing adequate power-circuit filtering and proper decoupling of supply-voltage lines. Feed lines should be routed carefully and properly balanced and bypassed. In some cases the early stages of a very high-gain amplifier may employ dc filament voltage. Push-pull circuits are less subject to the influence of hum.

Hum can result from undesired capacitive or inductive coupling. Transformers should be well constructed and shielded and should be mounted in a manner that keeps them isolated from signal leads and sensitive stages. Signal-input leads and any leads that carry signal over other than a short path between one stage and the next should be of a shielded low-impedance type which is much less subject to hum and noise pick-up. Possible sources of hum should be kept well isolated from input lines and high-gain input stages so as to minimize both inductive and capacitive coupling of hum into the signal path.

13-1-16. How would saturation of an output transformer create distortion?—Output-transformer saturation causes the transformer to respond in a nonlinear manner to the amplitude changes of an applied signal. As a result there is an introduction of harmonic distortion. Transformer-core saturation causes flattening of the high-amplitude portions of an applied strong signal.

13-1-17. Why is noise often produced when an audio signal is distorted?—Any nonlinearity in the tubes or other components of an amplifier will distort an audio signal. In the process, both harmonic- and intermodulation-distortion components are generated. The former are in the form of harmonics, second, third, fourth, fifth, etc., of the original wave. The latter components result from mixing activities among two or more of the audio frequencies that comprise the original audio signal, and possibly their harmonics. These spurious signals of various frequencies are carried along with the desired signals. Inasmuch as they are frequency components that did not exist in the original signal (noise), they tend to raise the background noise level of the signal.

13-1-18. What is a *decibel?*—The decibel unit is an expression of the magnitude of a change in sound level. The human ear responds logarithmically to an increase in sound level; one decibel being a change in sine-wave sound level that is barely distinguishable by the ear.

The decibel unit is also used to express a change in signal level. In this case it must be based on some reference level. Quite often an audio signal power of 1 milliwatt across a 600-ohm impedance is a standard corresponding to zero decibel (O dB).

13-1-19. What is the formula for determining the power loss or gain in a circuit in decibels?—The basic equation is:

$$\text{dB} = 10 \log \frac{P_{out}}{P_{in}}$$

There are specific reference levels used in electronics and in broadcast audio systems. Reference level is usually stated in terms of dBm which corresponds to 1 milliwatt across 600 ohms.

If the impedances across which signals are being measured are the same, voltage and currents can also be stated in terms of dB as follows:

$$\text{dB} = 20 \log \frac{E_{out}}{E_{in}}$$

$$\text{db} = 20 \log \frac{I_{out}}{I_{in}}$$

13-1-20. What is the purpose of deliberately introduced degenerative feedback in audio amplifiers?—Negative or degenerative feedback stabilizes amplifier gain and reduces distortion. The degenerative feedback reduces the distortion introduced by the amplifier but it does not correct for any distortion that may already exist in the signal.

13-1-21. Given the gain of an amplifier employing feedback and the overall voltage gain of the circuit, how is it possible to determine the amount of feedback used? State the formula used, and solve a sample problem.—The feedback is the fractional part (β) of the output voltage fed back to the input of the amplifier. The greater the negative feedback ($-\beta$) is, the greater is the reduction in the amplifier gain. The basic equation for a feedback amplifier is:

$$A' = \frac{A}{1 - A\beta}$$

This can be rearranged to solve for the negative feedback ($-\beta$):

$$-\beta = \frac{A - A'}{AA'}$$

In a typical example if the voltage gain (A) without feedback is 100 and the voltage gain with feedback is 10, the feedback β is:

$$-\beta = \frac{100 - 10}{100 \times 10} = 0.09 = 9\%$$

13-1-22. What is a volume unit (VU)? How does it respond to audio signals?—A volume unit or VU corresponds to the number of decibels a sound signal is above a standard reference level. A common standard sets O VU as a level of 1 milliwatt across 600 ohms. Many amplifier input and output terminals have a 600-ohm impedance and 600-ohm lines interconnect audio equipment. The typical VU meter is designed to follow average speech levels realistically with relation to human hearing. It is an underdamped meter so the pointer will pause momentarily in its upward swing and then start down slowly. Erratic vibrations and jumps are smoothed out.

13-1-23. Draw a diagram of a single-button carbon microphone circuit, including the microphone transformer and source of power.—Such a microphone and associated circuit are shown in Fig. 13-10.

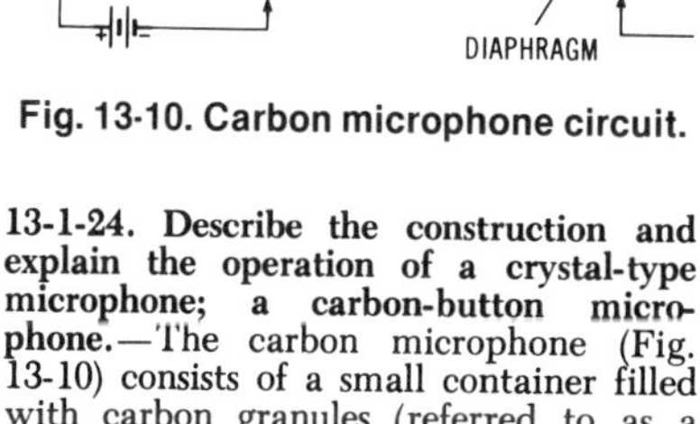

Fig. 13-10. Carbon microphone circuit.

13-1-24. Describe the construction and explain the operation of a crystal-type microphone; a carbon-button microphone.—The carbon microphone (Fig. 13-10) consists of a small container filled with carbon granules (referred to as a carbon button) attached to a diaphragm. As the sound vibrations strike the diaphragm they cause a varying pressure on the button. The carbon granules compress and expand in step with this pressure, resulting in a resistive change which influences the current in the external circuit. This external circuit consists of a source of dc voltage and the primary of a microphone transformer. The current variation in the primary of the microphone transformer induces a similar changing voltage into the secondary.

In summary, the changing sound pressure causes a variation in the microphone current. These variations are stepped up by the transformer and appear as an audio voltage across the secondary.

Sometimes the carbon granules pack tightly, cutting down on the output and increasing the noise. Usually they can be shaken loose.

The crystal microphone operates on the piezoelectric principle, in which certain crystals, such as Rochelle salt, develop an output voltage when subjected to a mechanical pressure. When sound vibrations are applied to a diaphragm linked mechanically to such a crystal (Fig. 13-11), the variations of mechanical pressure cause the crystal to generate an output voltage that follows the sound change. Another version of a crystal microphone uses a number of very small crystal elements which in themselves serve as a diaphragm. The individual outputs are combined to develop the total output of the microphone.

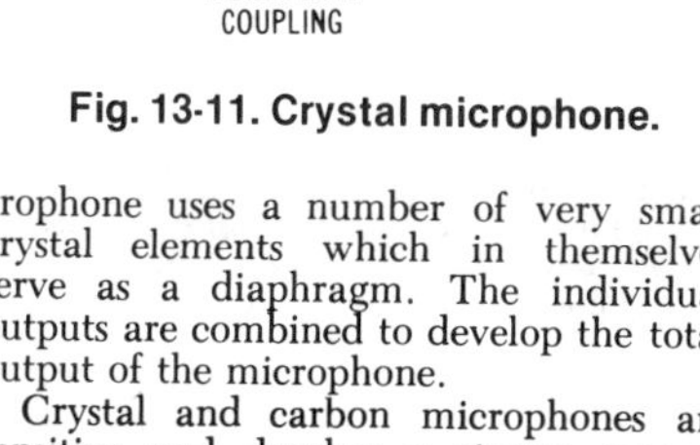

Fig. 13-11. Crystal microphone.

Crystal and carbon microphones are sensitive and develop a strong output. The crystal microphone has a high-output impedance compared to a carbon type, but it has a better frequency response.

13-1-25. What precautions should be observed when using and storing crystal microphones?—It is advisable to store a crystal microphone in a cool, dry location because of its susceptibility to environmental extremes, in particular, to extremes of high temperature and high humidity.

13-1-26. If low-impedance headphones on the order of 75 ohms are to be connected to the output of a vacuum-tube

amplifier, how may this be done to permit most satisfactory operation?—The preferred method is to use an audio transformer with a step-down impedance ratio. In this manner the rather low impedance of the headphones can be matched to the higher impedance of the amplifier, providing a proper load for the operation of the vacuum tube and the optimum transfer of signal from its output to the headphone.

13-1-27. Describe the construction and explain the operation of a dynamic microphone.—The dynamic microphone (Fig. 13-12) employs a moving-coil principle. A wire coil attached to a diaphragm is positioned in a strong magnetic field. Whenever sound waves strike the diaphragm and cause it to to move, the coil also moves, and it cuts the lines of force of the magnetic field. This motion induces a voltage in the coil, the ends of which are connected to a transformer in the microphone. These moving-coil movements can be made very rugged and can deliver a high output signal level. The output voltage varies with, and depends upon, the sound pressure changes on the diaphragm. Microphones in which the output voltage varies with the sound intensity and pressure are referred to as *pressure types*.

13-2. RADIO-FREQUENCY AMPLIFIERS

13-2-1. What is the difference between rf voltage amplifiers and rf power amplifiers with regard to applied bias? What kind of tube is generally employed in rf voltage amplifiers?—Rf voltage amplifiers are used extensively in receivers and less frequently in transmitters. They are to be found in single-sideband transmitters. Rf voltage amplifiers are usually biased class A and operate over the linear segment of their transfer characteristics. The pentode is the most common tube used in rf voltage amplifiers although, on occasion, in the earlier stages of a receiver, advantage is taken of the low-noise characteristics of a triode.

Rf power amplifiers are usually operated class B or class C. A typical class-C power amplifier is shown in Fig. 13-13. It includes input and output resonant circuits, which, in the case of a straight-through or fundamental-frequency rf amplifier, are tuned to the same frequency. The beam-power tube is the most common rf power amplifier, although a number of pentodes and some triodes are used in this function. In most instances, with appropriate circuits and tubes, there is no need for neutralizing a beam-power or pentode tube.

In a class-C stage the tube draws plate current only during the most positive portion of the positive alternation of the input wave. Thus the plate current is drawn in bursts of high-peak value. Such a burst supplies a lot of electrical energy into the plate resonant-tank circuit, and the capacitor is charged to a high negative value (sharp drop in plate voltage). However, a resonant circuit has smoothing and energy-storing ability. When the plate current burst ceases, the charge on the capacitor begins to fall off and energy is transferred to the coil. After the charge reaches zero, the collapsing field of the coil releases its stored energy back into the tank circuit. An opposite flow of current results and charges the capacitor in the opposite

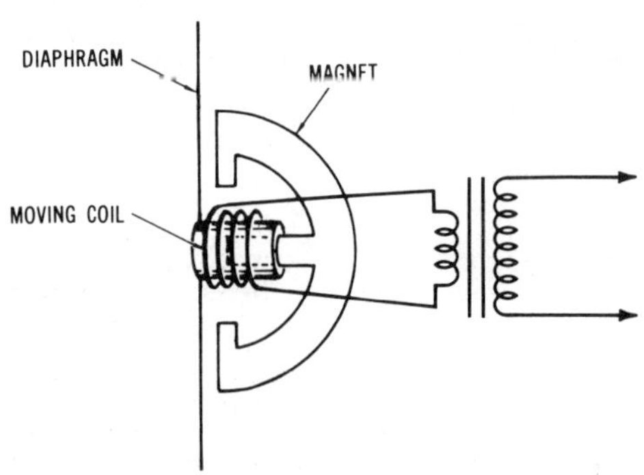

Fig. 13-12. Basic construction of a dynamic microphone.

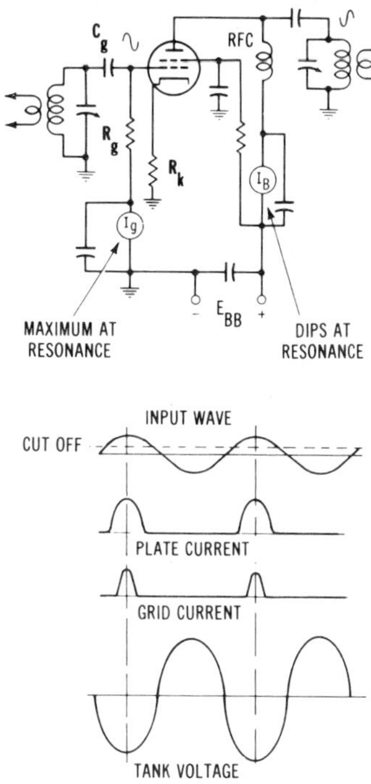

Fig. 13-13. Basic class-C amplifier.

polarity even though the plate current has stopped.

The magnetic field about the coil is now of opposite direction. When the capacitor loses its peak charge, the magnetic field again collapses, and a new cycle of original polarity begins.

Soon afterwards, a new burst of plate current is introduced because of the positive swing of the grid waveform, and the plate tank-circuit capacitor is again charged to a maximum negative value, initiating a new cycle of operation. Although the rf amplifier operates class C and plate current does not occur for the full cycle of the input wave, a good sine wave is developed across the plate tank circuit, with the proper choice of LC constants and load.

In most rf power amplifiers, the class-C bias is developed by the grid current in the grid circuit. On the positive peak of the input wave there is a substantial amount of grid current, which develops a negative charge on grid capacitor C_g. The current is such that the charge placed on the capacitor biases the tube beyond cutoff. Between intervals of grid current, the charge leaks off the capacitor through resistor R_g. If the time constant of R_gC_g is high enough, a reasonably constant bias will be retained during the entire cycle of the input wave. This is equivalent to a dc bias voltage of a value that is several times beyond cutoff. Of course, the capacitor is restored to full charge by the grid current at the crest of each positive input wave.

The grid current is often measured with a dc meter. Such a meter is useful in tuning a class-C stage. There is maximum dc grid current when the applied input wave is peaked at its maximum amplitude. A dc meter is often used in the plate circuit (or in the cathode circuit) to measure the dc plate current. This reading is also useful in tuning a class-C stage. When a plate resonant circuit is tuned to resonance, it displays a maximum impedance, and the dc plate current "dips" to a minimum. As energy is withdrawn to a load, more power must be supplied to the tank circuit, and there will be a rise in the dc plate current reading.

One of the hazards of using self-bias in operation of a class-C amplifier results from any loss in the rf input wave or exciting signal. If the rf excitation is lost, there will be no input to draw the burst of grid current which develops the class-C bias for the stage. As a result, the stage will operate with no bias, and there will be excessive plate current, which can damage the tube or other circuit components.

One method of avoiding trouble from the loss of rf excitation is the use of some cathode bias. The bias developed across the cathode resistor is unimportant in terms of the class-C operation of the stage. However, if the rf excitation fails, the attempted rise in the cathode-plate current will develop a high enough bias across the cathode resistor to prevent the plate current from rising to a harmful level.

13-2-2. What are the advantages of using a resistor in series with a cathode of a class-C radio-frequency amplifier tube to provide bias?—The cathode resistor is a safety resistor, and it develops a controlling bias when there is a loss of rf excitation to the stage. With no rf excitation the plate current attempts to rise to a harmful level. However, in flowing through the cathode resistor, a counteracting bias is developed which holds the plate current to a safe, reasonable level even though rf excitation has been lost.

13-2-3. What is an rfc? Why is it used?—Rfc refers to a radio-frequency choke. A radio-frequency choke has a high reactive impedance and is used to confine rf energy to the appropriate parts of a radio-frequency stage. Such chokes block the rf energy from supply-voltage lines and components, such as dc meters, which can be damaged with the presence of rf energy. Thus they permit the connection of dc voltages and other signal components without placing a severe load on the rf signal path.

For example, the radio-frequency choke (rfc) shown in Fig. 13-13 keeps the radio-frequency energy away from the dc plate-current meter and out of the supply-voltage lines. The fact that it has a very high reactance at the resonant frequency of the plate tank circuit means that it does not shunt any of the radio-frequency energy to ground, permitting the stage to operate with better efficiency. At the same time it provides a means of applying the dc plate voltage to the rf power-amplifier tube.

13-2-4. Draw a schematic diagram of a grounded-grid amplifier, and explain its operation.—A grounded-grid amplifier is shown in Fig. 13-14. In the grounded-grid stage the exciting signal is applied to the cathode circuit. Suitable chokes or a filter circuit must be a part of the filament leads to permit a means of supplying the filament current without placing a short circuit on the incoming rf wave. Inasmuch as the stage is operated class C, plate current occurs during the most negative portion of the negative alternation of the input wave (same as applying a positive signal to the control grid). Again, a burst of plate current occurs which shock-excites the plate-tank circuit as per the discussion given in association with Question 13-2-1.

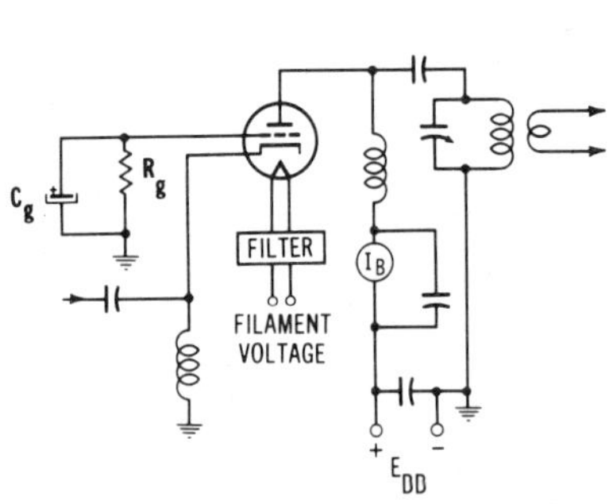

Fig. 13-14. Grounded-grid amplifier.

It should be noted that in a grounded-grid stage both the input and output rf energy are a part of the plate-cathode circuit. Consequently the total rf power output is the sum of the rf energy developed by the grounded-grid tube, plus the smaller amount of rf energy supplied from the preceding stage.

Even though a triode is used in a grounded-grid stage, no neutralization is needed, because the control grid is operated at rf ground potential and acts as a shield between input and output circuits. The stage can be operated with external class-C bias or from self-biasing, using the grid resistor and grid capacitor shown.

13-2-5. Draw a circuit diagram of a push-pull (triode) final power amplifier with transmission-line feed to a shunt-fed quarter-wave antenna, and indicate a method of plate neutralization.—Refer to Fig. 13-15.

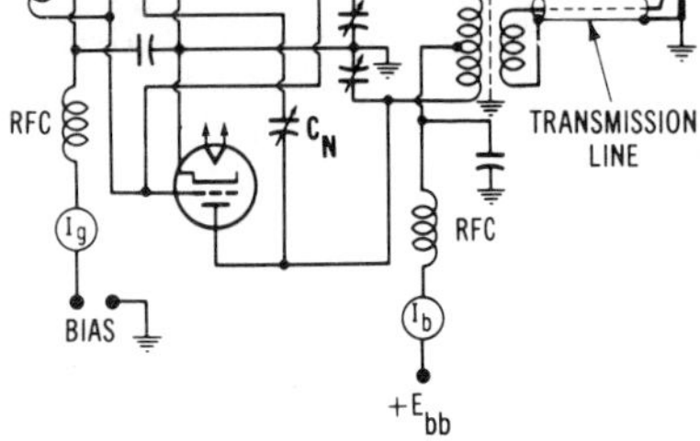

Fig. 13-15. Push-pull amplifier with plate cross-neutralization grounded quarter-wave antenna.

13-2-6. Explain the principle involved in neutralizing an rf stage.—A class-C amplifier self-oscillates when enough in-phase rf energy is fed back from the output circuit to the input circuit. The path that the feedback energy takes is through the interelectrode capacitances of the tube, mainly the plate-grid capacitance (C_{gp}). One method of preventing self-oscillation is to feed back from the plate circuit another component which will cancel out the feedback component that causes self-oscillation. Such an arrangement is shown in Fig. 13-16. The component fed back through neutralizing capacitor C_n is of opposite polarity from that fed back to the grid circuit through the interelectrode capacitance. By regulating carefully the size of capacitor C_n, the neutralizing component can be made of equal amplitude and opposite polarity from that which is fed back through the tube capacitance.

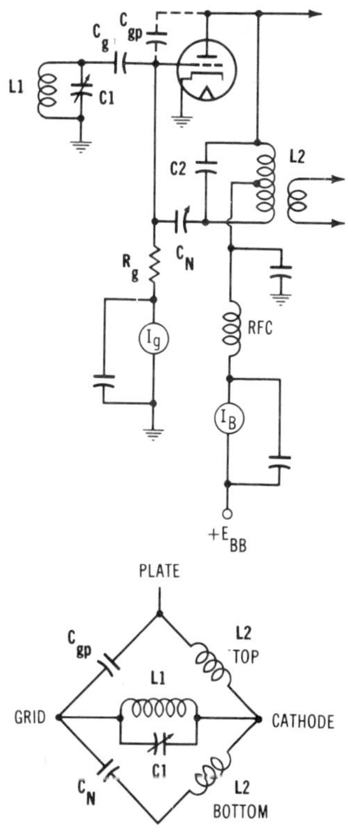

Fig. 13-16. Plate neutralization system.

One can also consider a neutralization system as a balanced bridge. When the bridge is adjusted, the rf currents in the inductive capacitive legs are balanced. Hence there is no difference of rf potential between grid and cathode as a result of the feedback currents.

13-2-7. Explain step by step at least one procedure for neutralizing an rf amplifier stage.—In the neutralization process the tuning objective is to minimize the net transfer of energy from the plate circuit back to the grid circuit of the stage. In most neutralization systems a converse plan is used in performing the actual neutralization procedure. The neutralization process is mainly accomplished by minimizing the amount of input signal that is transferred to the output tank when no plate supply voltage is supplied to the stage. Any such input signal must travel over the same path to the output as output signal would travel to input without neutralization. Thus, if we make the necessary adjustments to minimize the transfer of signal from input to output, we have in effect also minimized the possibility of the transfer of signal from output to input. Thus in neutralizing a stage, the normal rf excitation is applied to the grid of the stage to be neutralized. However, note that no power except filament power is supplied to that stage.

A step-by-step procedure, as applied to the circuit of Fig. 13-16, is as follows:

1. Place the filament circuit in operation, but turn off the plate voltage (screen voltage also if it is a multigrid tube).
2. Supply the normal rf signal to the control-grid circuit of the stage. If the stage is being neutralized for the first time, set the neutralizing capacitor to one extreme of its range. Let us assume that it has been set to maximum capacitance.
3. Resonate the grid tuned circuit for maximum rf excitation as indicated by a maximum reading on the dc grid-current meter (I_g).
4. Tune the plate tank capacitor over its range. If the stage is not neutralized, the grid-current meter reading will jump whenever the plate tank circuit is tuned through resonance. Likewise, there will be an indication of output when a sensitive rf indicator is placed close to the plate tank coil.
5. Decrease slightly the capacitance of the neutralizing capacitor. Tune the plate tank circuit through resonance again, and note the kick of the needle. Continue to decrease the setting of the neutralizing capacitor until there is a minimum kick of the needle and a minimum rf output when the plate

tank capacitor is tuned through resonance.

6. If the neutralizing capacitance is decreased too far, the rf output and grid kick will increase. There is just one optimum setting for the neutralizing capacitor, and it must not exert too little or too much capacitance into the path.
7. The stage is properly neutralized if the plate tank capacitor can be tuned through resonance without changing the grid-current reading. The actual procedure is an attempt to find the setting of the neutralizing capacitor which will give no or very little grid-current meter deflection.
8. Place some load on the output. Apply the power to the stage at reduced voltage. Tune the plate tank circuit for a resonant dip, using meter I_B. Finally, the full plate voltage can be applied, and the plate tank circuit and output coupling are adjusted for optimum operation.

13-2-8. State some indications of the presence of parasitic oscillations in a transmitter. Give some methods of suppressing them.—Parasitic oscillations are undesired oscillations in an rf amplifier stage. They are self-excited, usually at some frequency quite removed from the desired operating frequency of the amplifier. They are often indicated by the presence of rf energy in a stage, even though the rf excitation from the preceding stage has been removed. They are indicated when a signal can be picked up on a receiver tuned to some remote frequency when the transmitter is in operation. Meter readings are high or erratic.

Usually the presence of parasitics makes it difficult to tune a stage for optimum performance, or the tuning and operation of the stage seems to be erratic. The stability and efficiency of the stage are harmed with the presence of parasitics. Their presence in a modulated amplifier can have an adverse effect on the modulation quality or make it extremely difficult to tune for an optimum modulation characteristic. Parasitics in a stage that amplifies a modulated rf signal can introduce distortion components into the modulation.

In summary, parasitic oscillations can make themselves evident as self-oscillation when rf excitation is removed, instability, difficult tuning, poor efficiency, and the introduction of distortion components.

A sensitive rf indicator can be used to track down parasitic oscillations after the removal of the normal rf excitation from the stage. A grid-dip meter can even be used to make an approximate frequency measurement of the parasitic oscillations. Usually they occur at a very low frequency (several hundreds of kilohertz) as caused by radio-frequency chokes, capacitors, and other circuit components, or at a vhf frequency where they can be caused by lead lengths, stray capacitances, and circuit components. Changes in circuit wiring and rearrangement of components are often necessary to stop these parasitics. Parasitic chokes and resistors are also available and are very effective in the suppression of the undesired self-oscillations.

13-2-9. Draw a schematic diagram and explain the operation of a harmonic generator stage.—A typical harmonic generator is shown in Fig. 13-17. Triode

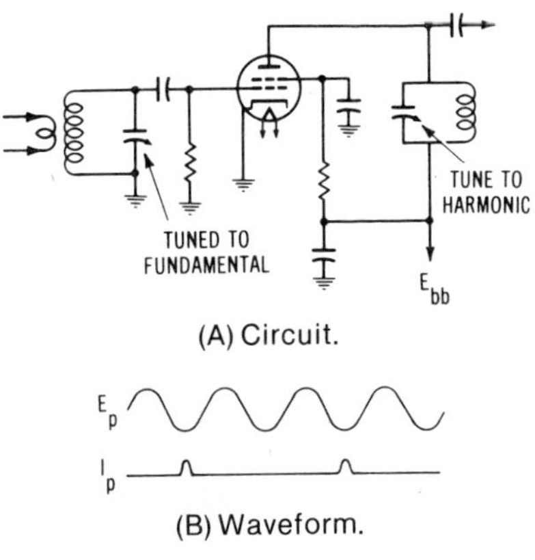

(A) Circuit.

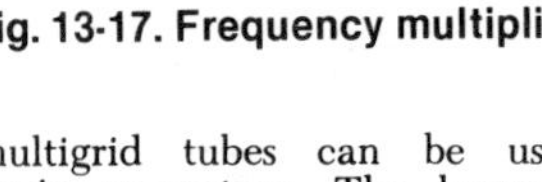

(B) Waveform.

Fig. 13-17. Frequency multiplier.

or multigrid tubes can be used as harmonic generators. The beam-power tube is very common because of its low driving power and capability of generating a strong harmonic output. No neutralization is necessary, because the output plate tank circuit is tuned to a frequency different from the input signal. It is tuned to the second, third, or higher harmonics of the input frequency, depending on the desired output frequency.

The harmonic generator is operated class C. In fact, to obtain a strong harmonic output the generator is often operated even further beyond cutoff than a normal straight-through class-C amplifier. To some extent, the desired harmonic, second, third, fourth, etc., determines the most favorable operating angle for the stage. (This refers to the number of degrees or portion of the input wave during which there is plate current.) *In general, the sharper the plate current burst, the stronger is its harmonic content.* However, efficiency is also a factor in selecting the proper beyond-cutoff bias.

If the harmonic generator is to operate as a doubler, the plate tank circuit is tuned to twice the frequency of the grid resonant circuit. A burst of plate current is drawn through the plate tank circuit each time the grid waveform rises to its positive crest. However, the amount of energy drawn into the plate tank circuit is great enough that the oscillations within the plate tank circuit can go through two cycles before energy must be added. Of course, this means that a harmonic generator operates at a lower efficiency than a straight-through amplifier. Furthermore, the higher the desired harmonic is, the lower is the operating efficiency. Thus the output capability drops off rather quickly as the order of the desired harmonic is increased.

13-2-10. What class of amplifier is appropriate in a radio-frequency doubler stage?—A class-C rf amplifier is appropriate to this application. In class-C operation the plate current occurs in bursts or pulses which have a high-harmonic content. Thus it is possible to develop a stronger harmonic wave in the plate tank circuit which is tuned to twice the frequency of the incoming signal.

13-2-11. Draw a circuit diagram of a push-push frequency multiplier, and explain its principle of operation.—In a push-push circuit (Fig. 13-18) the input is driven in push-pull, while the output is connected in a parallel arrangement. This type of circuit is an efficient even-harmonic generator. It is used frequently as a doubler.

The discussion associated with Question 13-2-9 covered the fact that with a conventional single-tube doubler there was a burst of current only on alternate input sine waves, a fact which lowers the efficiency. This reason for loss is avoided in the push-push arrangement.

In a doubler version the plate tank circuit is tuned to the second harmonic of the incoming wave. During the peak of

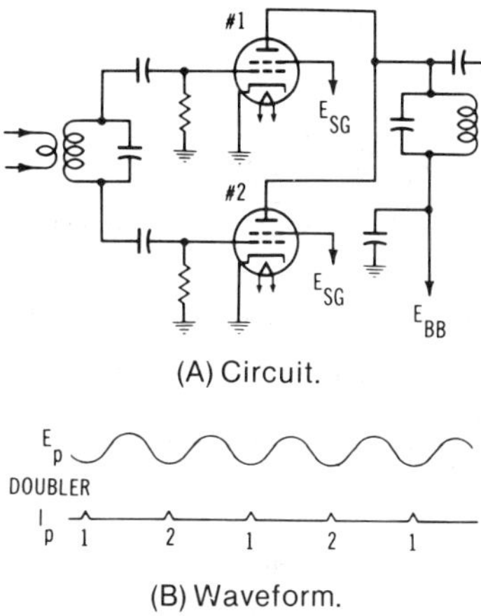

Fig. 13-18. Push-push frequency doubler.

the positive alternation, V1 conducts and draws a burst of current into the plate tank circuit. During the negative alternation of the input wave, V2 conducts because its grid is made positive with respect to its cathode. As a result, another burst of plate current is drawn into the tank circuit. Thus each alternation of the grid signal (two per cycle) results in plate current. Inasmuch as the plate tank circuit is oscillating at twice the frequency of the incoming signal, there is plate current in the tank circuit once during each harmonic cycle. Thus in the push-push arrangement plate current coincides with the negative peak of each output cycle, increasing the output and efficiency.

13-2-12. Push-pull frequency multipliers normally produce what order of harmonics; even or odd?—A push-pull frequency multiplier is useful in the generation of odd-order harmonics (third, fifth, etc.). It is a poor even-harmonic generator because the even components cancel in the push-pull output circuit.

13-2-13. Draw schematic diagrams of the following circuits, and give some possible reasons for their use: (a) link coupling between a final rf stage and an antenna (include a low-pass filter); (b) capacitive coupling between an oscillator stage and a buffer amplifier; (c) a method of coupling a final stage to a quarter-wave

Marconi antenna other than link or transmission line.—(a) A link-coupled arrangement is shown in Fig. 13-19. Such an arrangement is used to transform the high-impedance output of the rf power amplifier to the low impedance of the antenna and transmission-line system.

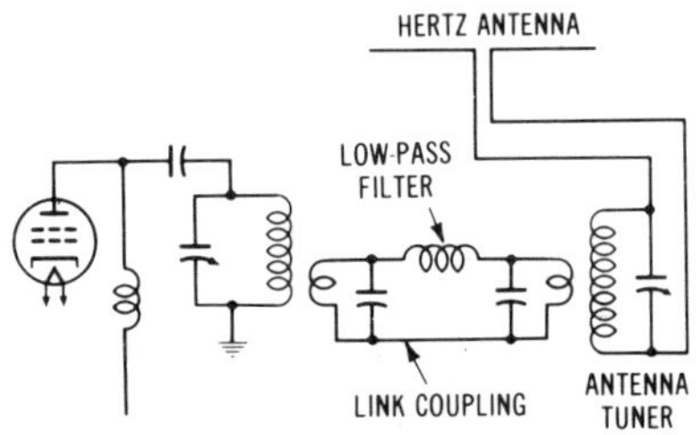

Fig. 13-19. Link-coupling arrangement to a antenna system.

The link pickup coil has a very low impedance and minimizes loss and undesired radiation. It also blocks the transfer of undesired signal components that may be present in the rf amplifier output. To further reduce the transfer of harmonic components, a low-pass filter is inserted in the signal transfer path. Such a filter does not contribute any serious attenuation to the desired frequency, but it offers great opposition to harmonics of that frequency. Thus harmonic components are not radiated, and possible interference to other services is avoided. Link coupling can be used whenever rf energy must be transferred between two widely separated resonant circuits.

(b) Capacitive coupling of the type shown in Fig. 13-20 is effective and economical because only a few components are required. It is a rather high-impedance method of interstage coupling, but it affords an excellent means of transferring a signal from the high-impedance stage of an amplifier to the even higher input impedance of the next stage, provided the two stages are mounted very near to each other. Usually the oscillator and the next stage are positioned very closely together.

(c) The simple pi network arrangement of Fig. 13-21 can be used to couple the transmitter output to a Marconi antenna when the antenna itself is very near to the transmitter. In such an arrangement a single wire, often copper tubing, is connected to the proper impedance-matching point on the Marconi antenna.

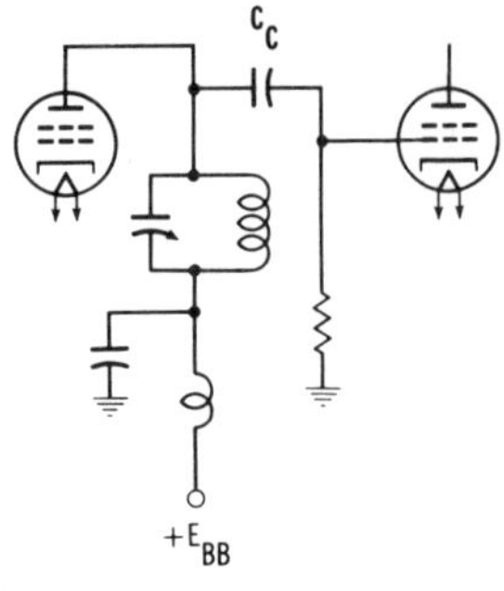

Fig. 13-20. Interstage capacitive coupling.

13-2-14. Draw a schematic of a grounded-emitter frequency doubler and explain its operation.—Refer to Fig. 13-22. The appearance of the frequency doubler does not differ from that of a conventional grounded-emitter tuned rf amplifier. The exception is that the output resonant circuit is tuned to twice the frequency of the incoming signal. Thus a second-harmonic component is emphasized and developed across the collector output transformer (T2). Again it is a step-down transformer that permits the transfer of the second harmonic energy to the low-impedance input of a succeeding stage. Biasing of the transistor is a function of the level of the incoming signal and is adjusted to emphasize the generation of harmonic components.

13-2-15. Explain what problems might be caused by a loss of drive in an rf amplifier.—Many rf power amplifiers are signal biased. Consequently the absence of signal removes that bias and faulty operation of the stage might occur. In the case of a vacuum-tube rf amplifier excessive plate current might be drawn. In

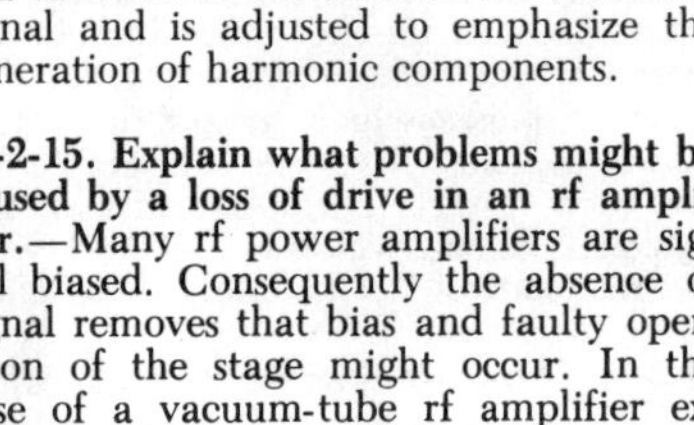

Fig. 13-21. Single line direct coupling to a Marconi antenna.

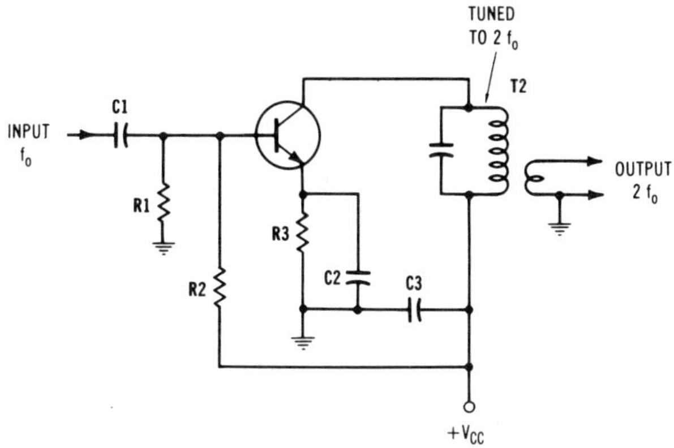

Fig. 13-22. Grounded-emitter frequency doubler.

some cases such a loss of drive may result in the blowing of fuses or set off a warning that the transmitter is not functioning properly.

13-2-16. Describe some factors in connection with the following items which should be considered in vhf and above, but would not be of particular concern at mf or below: (a) wire diameter and lead length, (b) wiring configuration (placement and bending), (c) coaxial cables and transmission lines, (d) capacitor types.— (a) In vhf operation even short lengths of wire can become an appreciable segment of the wavelength. Likewise, stray capacitances which are inconsequential at low frequencies now have a low enough reactance to become of concern in the operation of vhf circuits.

Furthermore, at vhf frequencies there is more pronounced skin effect, and conduction is on the surface of a wire. Therefore the surface area of a wire carrying a given rf current must be greater than for a similar current at low frequencies. In fact, tubing is an economical means of transferring vhf signals because it affords a large conducting outer surface and is hollow where there would normally be no vhf current. Lead lengths must be kept to a minimum to avoid adverse resonant effects. Likewise, wire diameters must be capable of carrying the vhf current but must not be so large as to introduce stray capacitive effects.

(b) Wiring must be as straight and direct as possible. Sharp bends often increase required lead length and introduce discontinuity losses. Leads that carry vhf currents at high impedance must be kept short and clear of other components to avoid capacitive losses.

(c) Coaxial cables and flat-type transmission lines have greater losses at vhf frequencies than at low frequencies. High-quality lines of minimum loss must be selected, and routing must be made as direct as possible to minimize attenuation and other losses. Proper matching is very important at vhf frequencies because mismatch accentuates the attenuation losses of a transmission line.

In many high-powered vhf circuits, sections of transmission line are often used as resonant circuits or resonant filters. Such transmission-line sections are of a practical length for high-frequency application and can be made to have more favorable resonance characteristics than a so-called lumped-constant capacitor-and-inductor tuned circuit.

(d) Capacitors may be small in terms of actual capacitance, however, a little capacitance affords a very low reactance at the very high frequencies. Not only must they be designed physically to present a low capacitance, but they must also be planned to display a minimum of inductance (made to have an exceedingly high self-resonant frequency).

Some variable vhf capacitors look like conventional lower-frequency capacitors except that they are smaller in physical size, have fewer plates, and may have a wider spacing. Other capacitors combine capacitance and inductance in one physical structure such as the so-called butterfly resonant circuit which, in effect, is a variable capacitor and a surrounding single turn inductance, forming a complete vhf resonant circuit. Another popular vhf capacitor is simply two flat circular discs, one of which can be moved in its relative spacing to the other.

Fixed capacitors are small and use high-quality, high-frequency dielectric material. Capacitors are often coaxial in construction, with their capacity determined by the capacitance between an inner conductor and an outer shield. Small ceramic feed-through capacitors of this type are popular in high-frequency circuits. They permit a means of feeding supply-voltage lines into a shielded high-frequency circuit, and at the same time act as an rf filter capacitor between the supply-line lead and the shielding to which they are mounted.

13-2-17. Name at least three circuit factors (not including tube types and component values) in a one-stage amplifier circuit which would not be of particular concern at vlf, but should be considered at vhf.—These are lead lengths which could become an important part of a wavelength in the vhf spectrum; stray capacitances (parts and wiring), although of very low value, would have a low significant reactance in the vhf spectrum; and careful close-spaced arrangement of components so as to obtain efficient stage operation. Shielding the output from input and proper grounding is a particular concern in vhf circuits; leads should run in a short, direct path to a common ground. Keep lead lengths short to minimize interaction among stages.

13-2-18. Why are special tubes sometimes required at uhf frequencies and above?—The three problems of uhf and higher-frequency operation are stray capacitances, lead inductances, and transit-time effects. These factors limit the uhf and higher-frequency performance of conventional tubes. Uhf and higher-frequency tubes are constructed with minimum interelectrode capacitances and minimum lead lengths and inductance. Conventional tubes can be made very small and made to have small electrodes, such as the subminiature types. However, these are hampered in terms of power capability.

Transit time is a particular problem in conventional tubes if they are to be employed in uhf and higher-frequency services. It requires a finite time for an electron to move from the cathode through the control grid to the plate. At these exceptionally high frequencies the transit time of the electron is significant in terms of the period of the input rf wave. As a result, losses and inefficient operation result. Special tubes, such as the klystron, traveling-wave tube, and magnetron use the electron transit time to advantage in the design of tubes that operate in a very different manner from the conventional types.

13-2-19. What is a lighthouse triode? An acorn tube? These tubes are designed for operation in which frequency range?—Acorn and lighthouse tubes were designed for operation in the vhf-uhf frequency ranges. Special versions of the lighthouse or planar tube construction are used well up into the microwave region. These tubes are of a special construction (Fig. 13-23) that minimizes interelectrode capacitances and lead inductance. The acorn tube is very small, the electrodes are closely spaced, and leads are brought out via a glass seal instead of through a base. Some special acorn types employ multiple leads for versatility in mounting arrangements that permit very short spacings between tube electrodes and external circuit components.

In a lighthouse tube the cathode, control grid, and plate are flat cylindrical surfaces. They are mounted in parallel planes with their extensions serving as metal disc leads. Such a tube has minimum interelectrode capacitances and, at the same time, a high mutual conductance because of the close proximity of the electrodes. Such a planar-disc construction is ideally arranged for use in a coaxial cavity. Grid-to-cathode and grid-to-plate resonant coaxial-tank circuits can be used. The cylindrical structure can be made to fit snugly against the cathode, grid, and plate-electrode discs of the lighthouse tube. The high-Q and well-shielded resonant circuits almost become a part of the actual tube.

13-2-20. Draw a schematic diagram of a varactor diode tripler and describe its operation.—A basic varactor diode tripler circuit is given in Fig. 13-24. The L1-C1 combination at the input is used for impedance matching. Capacitor C2 and inductor L2 along with the varactor diode tune to the fundamental frequency of the incoming signal. Thus a resonant current exists in the input portion of the varactor circuit. However, the voltage variation across the varactor itself is rich

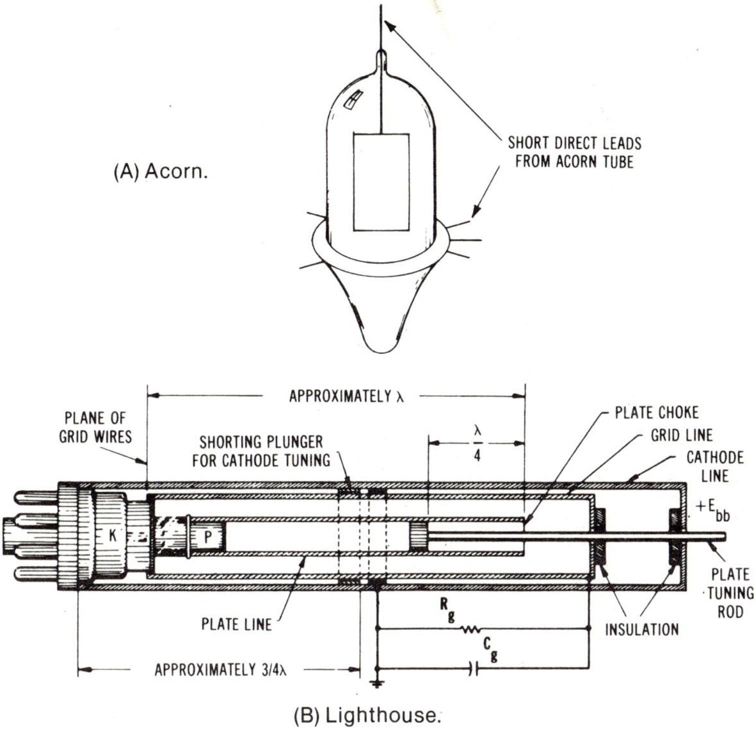

Fig. 13-23. Acorn and lighthouse triodes.

n harmonics. The third harmonic can be brought out and permits operation as a tripler by tuning capacitor C4 and inductor L4 to the third harmonic. Even greater emphasis of the third harmonic output can be obtained with another resonant circuit connected directly across the varactor. This added resonant circuit composed of capacitor C3 and inductor L3 is tuned to the second harmonic. Therefore a fundamental component and a strong second-harmonic component are present in the varactor. Inasmuch as the varactor is a nonlinear device, these two

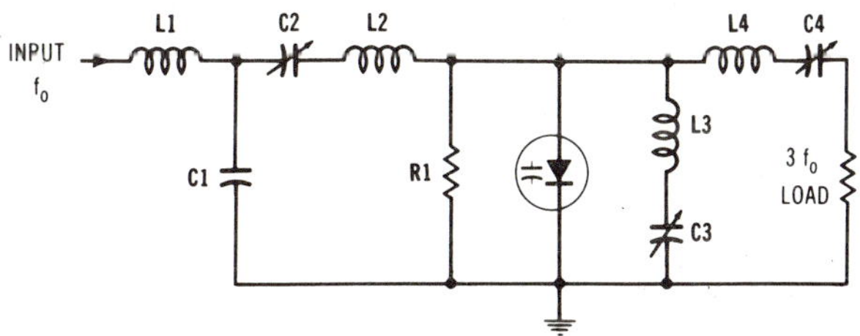

Fig. 13-24. Varactor tripler.

current components produce a sum component which is, in effect, a third-harmonic current. Thus the presence of the second harmonic or idler tank circuit takes advantage of the mixing activity of the varactor and a much higher third-harmonic component results than would be present if no idler tank circuit were used.

A varactor multiplier circuit provides an efficient means of obtaining a strong harmonic output. The source of power is the signal itself and no external dc power demand is made. A biasing resistor (R1) is added. The small current in this resistor develops the required dc voltage drop to reverse-bias the varactor.

13-3. OSCILLATORS

13-3-1. Draw circuit diagrams of each of the following types of oscillators (include any commonly associated components). Explain the principles of operation of each. (a) Armstrong, (b) tuned-plate tuned-grid (series fed and shunt fed, crystal and LC controlled), (c) Hartley, (d) Colpitts, (e) electron coupled, (f) multivibrator, (g) Pierce (crystal controlled).—(a) An Armstrong circuit is shown in Fig. 13-25. In a vacuum-tube rf oscillator, the tube and its associated components (which usually include a resonant circuit), are combined into a feedback-amplifier arrangement capable of making a continuous conversion of dc power to radio-frequency variations. As shown in Fig. 13-25A, an oscillator is an amplifier with a portion of its output fed back to its input. Oscillations will occur when this feedback is of the proper amplitude and phase. Because of a 180° phase shift in the common-cathode amplifier, the feedback arrangement must provide an additional 180° phase change to bring the output voltage back in phase with the input variation. This in-phase condition, called *positive feedback* or *regeneration*, permits the feedback signal to add to any input change, and self-sustained, continuous oscillations can be generated. There are many types of oscillators for generating rf energy; all use this fundamental feedback principle. In general, oscillators differ only in the manner in which they obtain the required feedback.

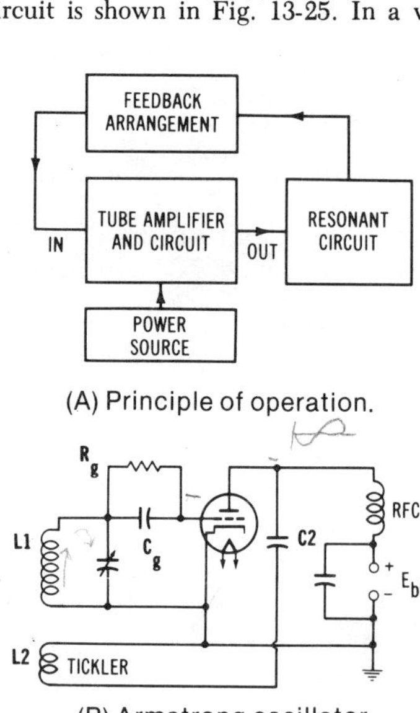

(A) Principle of operation.

(B) Armstrong oscillator.

Fig. 13-25. Oscillator principle and circuit.

The Armstrong oscillator (Fig. 13-25B) is a classic example, and it will be used to explain the operation of oscillators in general. In the Armstrong oscillator the necessary phase reversal of the feedback is determined by the proper connection of grid, cathode, and plate to the resonant circuit coil and the small so-called tickler coil that feeds back a portion of the plate variation to the control-grid circuit. This fed-back variation is induced in the coil of the grid resonant circuit.

In a vacuum-tube circuit, some tiny voltage or current input variation is inevitable. Such a change is amplified, and in an oscillator arrangement, linked back to the grid circuit through the feedback arrangement from the plate circuit. In the case of the Armstrong oscillator, it is via the small tickler coil. The feedback amplitude, little as it might be initially, will cause an amplified plate-voltage variation which is again fed back to the grid. Hence the original grid-voltage change is reinforced, and an even greater grid-voltage and follow-up plate-voltage change occurs. These variations build up until they are self-supporting, and the stage goes into continuous oscillation. The final magnitude of the oscillations is confined by grid-limiting and saturation, and by the dc supply voltage and power capability of the tube.

To build up to an oscillating state, the gain of the amplifier must be greater than unity, and in the initial build-up period it is substantially greater than unity. In the case of the Armstrong oscillator, the magnitude of the feedback is determined by the relative turns ratio between the two coils, L1 and L2, and their proximity. As the build-up continues, the gain of the amplifier falls to the point just a bit greater than unity which overcomes all the circuit losses and sustains continuous oscillation. The polarity of the feedback is determined by the relative winding direction of the two coils, and the connection of the tube electrodes to these windings. If the plate (Fig. 13-25B) were connected to the top of the tickler winding and the bottom of the tickler winding were grounded, the stage would not oscillate, because the feedback voltage would be out of phase with the control-grid variation, resulting in negative feedback instead of positive feedback. As it is wired in Fig. 13-25B, the feedback is positive, and the voltage induced into the grid resonant circuit from the tickler coil is in phase with the initial control-grid change.

The radio-frequency oscillator is self-biased by the grid current which occurs during the positive alternation of the grid cycle. This grid current charges grid capacitor C_g to a negative level. When the grid current ceases, the charge remains on the capacitor because of the high ohmic value of grid resistor R_g. The time constant of R_gC_g is high in comparison to the period of the generated wave.

A power radio-frequency oscillator is usually biased class C by the grid current. Therefore the plate current occurs in bursts which shock-excite the resonant circuit. The flywheel energy-storing ability of the resonant circuit causes a sine wave to be developed. Lower-powered oscillators and receiver oscillators are, in general, biased to a much lower level than that for class-C operation.

Capacitor C2 is a blocking capacitor which prevents the dc plate voltage from being shorted to ground. In some cases its reactance, along with the reactance of the tickler coil, can be used to control the amount of feedback. The radio-frequency choke (rfc) prevents the power supply from placing an ac short on the plate output of the oscillator, and it also blocks the radio-frequency energy from the power supply.

(b) Various types of tuned-plate tuned-grid oscillators are shown in Fig. 13-26. Fig. 13-26A shows a shunt-fed tuned-plate tuned-grid oscillator. In this type of oscillator there are two resonant circuits, one in the plate and one in the grid. These resonant circuits can be isolated from each other, and there is no need for any mutual linkage between coils as in the case of the Armstrong oscillator. Feedback between the two resonant circuits is by way of the interelectrode capacitance of the tube, largely the grid-plate capacitance (C_{gp}). In previous rf amplifier questions we were concerned with this capacitance because in a straight-through amplifier it is objectionable and encourages self-oscillation. In the case of a tuned-plate tuned-grid oscillator we use this capacitance to advantage, and we can design circuit arrangements that permit the capacitance to be the means of supplying feedback from output to input.

In the tuned-plate tuned-grid oscillator the feedback is capacitively coupled from the plate resonant circuit to the grid resonant circuit. When such an oscillator is first turned on, the plate current shock-excites the plate tank circuit into oscillation. Some of this energy is capacitively coupled back to the grid resonant circuit

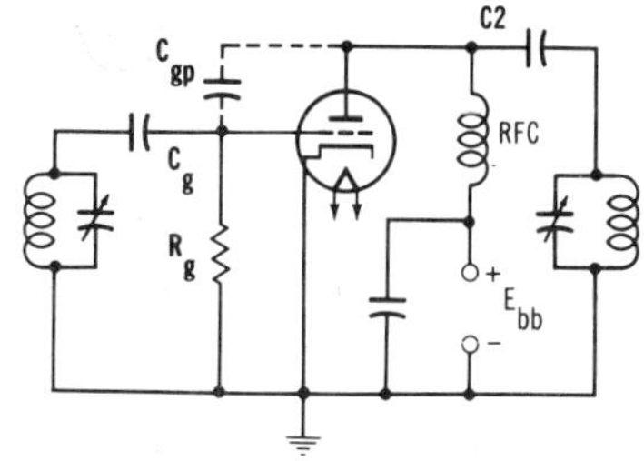

(A) Shunt fed.

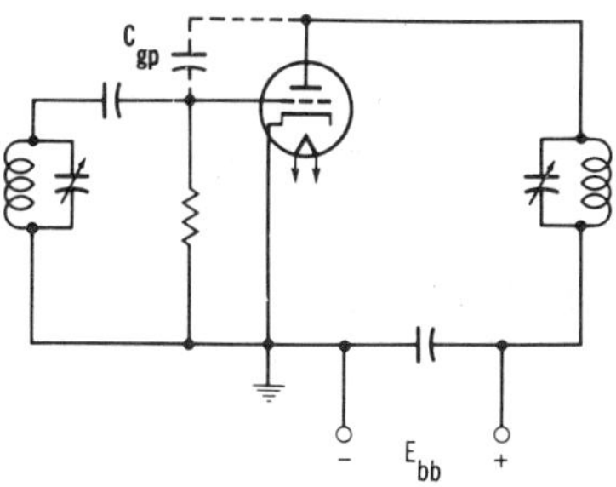

(B) Series fed.

Fig. 13-26. Tuned-grid tuned-plate oscillators.

and causes a grid voltage variation that appears amplified in the plate resonant circuit. A portion of the amplified oscillation is fed back into the grid circuit to reinforce the original grid-voltage change. In this manner the oscillations build up to the limits set by the tube and circuit. After this level is attained, oscillations are sustained by the continuous feedback activity.

As in the case of the Armstrong oscillator, the circuit is self-biased with the R_gC_g combination. Capacitor C2 acts as a dc block, and the radio-frequency choke has the same function as discussed previously.

The oscillator is said to be a shunt-fed type because of the direction of the dc plate current. Notice that the path taken by the dc current is through the radio-frequency choke and the tube. The plate resonant circuit is in parallel, or in shunt, with the power supply and the tube.

The series-fed tuned-plate tuned-grid circuit is shown in Fig. 13-26B. Its operation is identical to that of the shunt-fed tuned-plate tuned-grid oscillator. It differs only in the direction of the dc plate current with relation to the resonant plate circuit. In this case, the path of the dc plate current is through the coil of the resonant circuit and the tube. In effect, the tube, power supply, and resonant circuit are in series.

Examples of series- and shunt-fed crystal oscillators are shown in Figs. 13-27A and B. Certain crystalline materials produce a voltage when placed under mechanical stress. One such material is quartz, used in the manufacture of transmitting crystals. If the pressure on such a crystal is varied and an alternating voltage is applied across the crystal, a vibration is set up. The vibration is at a particular frequency determined by the crystal size and structure. Such a crystal has the characteristics of a resonant circuit. When inserted into an oscillator stage which is tuned to the natural frequency of the crystal, it will vibrate strongly. In fact, a potential difference much like a resonant voltage across a tank circuit is developed across the crystal. This is the variation that is supplied as feedback to the oscillator grid. A similar variation set up in the plate tank circuit is now fed back to the crystal to keep it in mechanical vibration. Inasmuch as the feedback is strongest at the mechanical resonant frequency of the crystal, the frequency of the oscillator is determined by the natural frequency of the crystal. Thus a crystal oscillator is a highly stable oscillator dependent on the mechanical-vibration frequency of the crystal.

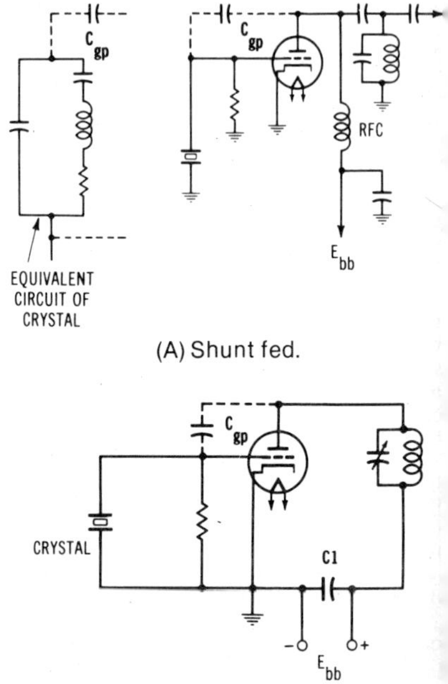

(A) Shunt fed.

(B) Series fed.

Fig. 13-27. Crystal oscillators.

As shown in Fig. 13-27A, the crystal is itself a series-resonant circuit having a high inductance and a low capacitance. The capacitance of the crystal and its holder, along with the distributed external capacitance shunted across the crystal, forms an effective parallel-resonant circuit. Thus the crystal oscillator is basically a tuned-plate tuned-grid oscillator, the crystal itself forming an equivalent grid resonant circuit. Feedback is by way of interelectrode capacitance C_{gp}, as in the case of the tuned-plate tuned-grid oscillator. The shunt-fed arrangement requires the appropriate dc blocking capacitor and radio-frequency choke. In the series-fed arrangement, capacitor C1 is the plate circuit rf bypass.

(c) A Hartley oscillator is shown in Fig. 13-28. The popular Hartley oscillator employs a single resonant circuit, and its inductor is tapped to obtain the required feedback. There are two sections

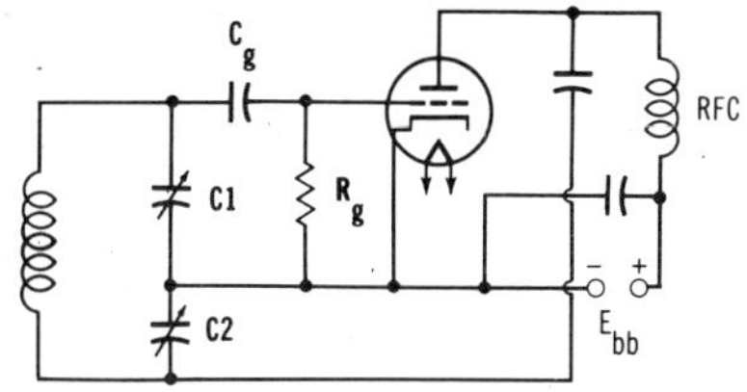

Fig. 13-28. Hartley oscillator.

to the resonant circuit: the portion between the plate and ground, and the smaller segment between the control grid and ground. In operation, an initial change in plate current will shock-excite the resonant circuit into oscillation.

Let us assume that the plate is swinging negative. Under this condition the bottom of the resonant circuit is positive with respect to the plate side. It is this low side which is connected to the control grid, supplying in-phase feedback to the grid (remember that there was an additional 180° phase shift by the tube). Such in-phase feedback permits the oscillations to build up to the operating limits of the tube and stage. The amount of feedback is controlled by the position of the tap on the inductor. This tap can be moved, within limits, to obtain a desired oscillating characteristic and output. The fixed resistors, capacitors, and rfc serve the same function as in the previously discussed oscillators.

(d) A Colpitts oscillator is shown in Fig. 13-29. Its operation is identical to that of the Hartley oscillator, except for the manner in which feedback is obtained. In the case of the Colpitts oscillator, the feedback is obtained by using split capacitors rather than a tapped coil. The tank-circuit connections are the same as for the Hartley. The amount of feedback is controlled by the relative reactances of C1 and C2. The remaining

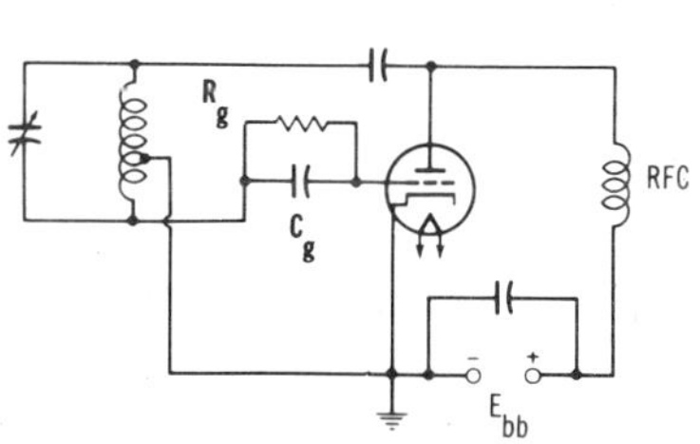

Fig. 13-29. Colpitts oscillator.

components have the same function as covered for previous oscillators.

(e) An electron-coupled (eco) or variable-frequency (vfo) oscillator is shown in Fig. 13-30. Electron-coupled oscillators use pentodes or beam-powered tubes. The plate circuit serves as a means of transferring the oscillations to a load in a manner that minimizes the influence of the load on the oscillator frequency. The oscillator section of the tube consists of cathode, control grid, and screen grid.

In the operation of the oscillator, the screen grid has the same function as the plate in the oscillators described previously. In fact, in the circuit of Fig. 13-30, the screen grid, control grid, and cathode are connected into a Hartley oscillator circuit which is identical to that of Fig. 13-28. The oscillations are coupled to the output circuit by way of the electron stream between cathode and plate. The rather loose coupling provided by this arrangement provides high stability, high gain, and is much less subject to load variation.

As shown, the plate output circuit can be a resonant circuit, a choke, or a resistor. Although a resonant circuit, when used, can be tuned to the fundamental frequency of the oscillator, it is usually tuned to the second-harmonic frequency. The electron-coupled circuit serves then as both an oscillator and frequency doubler. When a choke or resistor is used in the plate circuit, the output frequency is the same as the fundamental frequency.

(f) A multivibrator circuit is shown in Fig. 13-31. In a multivibrator circuit one stage is resistance-capacitance coupled (RC coupled) to the second stage, and the second stage, in turn, is RC coupled back to the input of the first stage. In this manner in-phase feedback can be obtained because the output of the second stage will be in phase with the input to the first stage (two 180° polarity shifts by the tubes). This is a necessary condition for feedback. However, in a multivibrator the feedback is of such a magnitude that the tube conduction is switched on and off alternately by the charges placed on the two grid networks R1C1 and R2C2. Thus the multivibrator does not generate a sine wave, but a squared wave of the type shown in Fig. 13-31. This type of circuit is called a *relaxation oscillator*.

In explaining its operation, let us assume that tube V1 is beginning to conduct, and its grid voltage is rising. Consequently there is a corresponding increase in its plate current and a fall in its plate voltage. This change, in turn, drives the grid of V2 negative and the plate of V2 positive. Since the plate of tube V2 is RC coupled back to V1, there results an amplified increase in V1 grid voltage. The net effect of this feedback activity is to drive V1 quickly positive to a limiting level and V2 quickly negative beyond cutoff. At this point the feedback activity stops; however, still another change has occurred. The amplified feedback voltage has placed a high negative charge on capacitor C2, and before V2 can conduct again, this charge must leak off through resistor R2. In fact, V2 remains cut off, and V1 conducts until the negative voltage being developed across R2 by the declining discharge current falls to the conduction bias of V2. At this moment V2 plate current begins to rise, and its plate voltage falls. In turn, V1 grid voltage swings negative, and its plate voltage swings positive. There results a further increase in V2 grid voltage, and the feedback activity

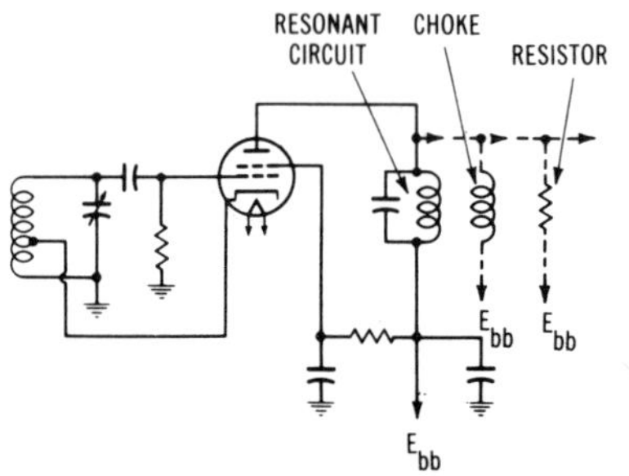

Fig. 13-30. Electron-coupled oscillator.

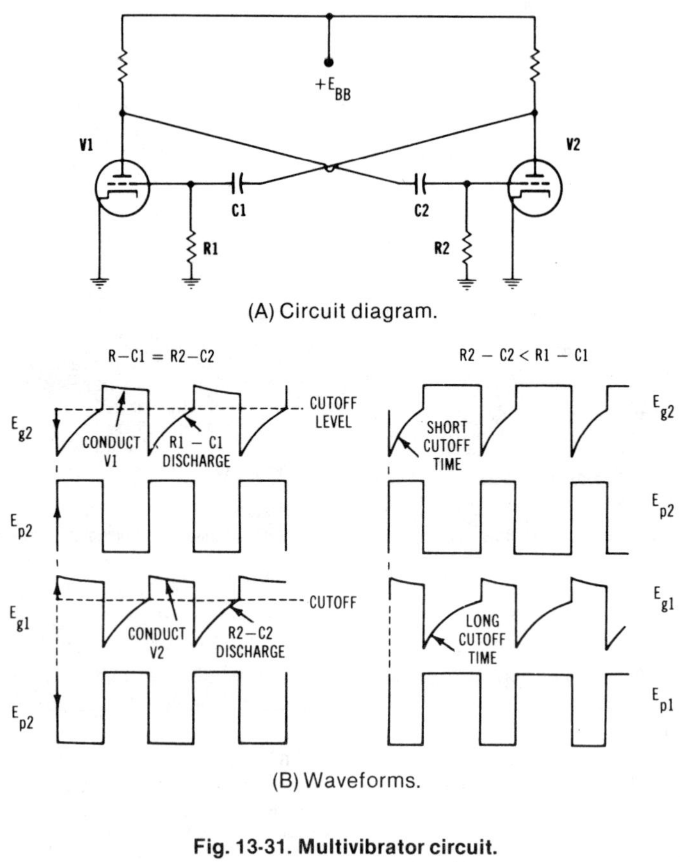

Fig. 13-31. Multivibrator circuit.

once again takes over. This time, feedback is of opposite polarity with V1 grid being driven quickly to cutoff and V2 to the limiting level at which point a further increase in grid voltage will cause no additional increase in plate current.

Tube V1 is now cut off and V2 remains conducting, as capacitor C1 discharges through resistor R1. After an interval of time set by the time constant of R1C1, the grid of V1 once again reaches the conduction level, and the entire cycle of activity begins to repeat.

The waveforms show that the actual period of the generated pulselike waveform corresponds to the sum of the V1 and V2 cutoff periods. *Thus the frequency of operation is determined largely by the two time constants R1C1 and R2C2.* If the time constants are equal, both tubes are cut off for the same time interval, and a symmetrical waveform (square wave) is developed in the plate circuit. If a more asymmetrical and pulselike output is desired, it can be obtained by making the two time constants unequal. In so doing, one tube cuts off for a longer time interval than the other.

The frequency of a multivibrator can be increased by decreasing one or both of its time constants; the frequency can be decreased by increasing one or both of its time constants.

(g) A Pierce crystal-controlled oscillator is shown in Fig. 13-32. In a Pierce oscillator, the crystal is connected between plate and grid, usually in series with a blocking capacitor. The crystal itself serves as the only resonant circuit of the oscillator. Feedback is obtained in a Colpitts like manner, the split capacitor consisting of interelectrode capacitance C_{gp} and a small capacitor C2 connected between control grid and ground (in effect, between the bottom of the crystal resonant circuit and ground). The Pierce crystal oscillator has the advantage that no tunable resonant circuit is associated with the oscillator. Thus crystals can be switched in and out to change frequency

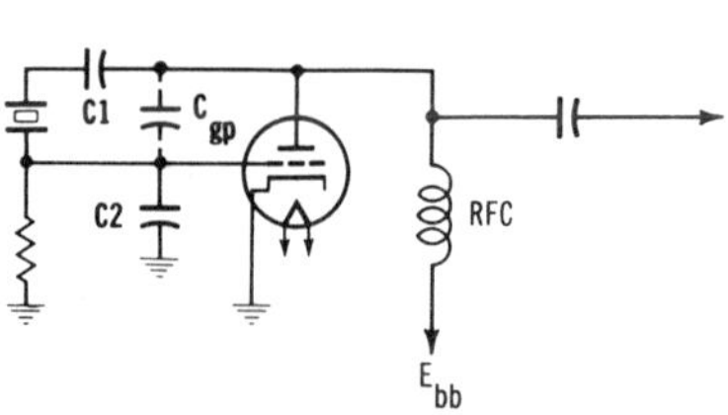

Fig. 13-32. Pierce crystal oscillator.

without having to retune the oscillator. However, the oscillator output is weak and usually operates at a low plate voltage so as not to damage the crystal.

13-3-2. What are the principal advantages of crystal-controlled oscillators over tuned-circuit oscillators?—Frequency stability is the major advantage. A crystal oscillator operates on a frequency determined by the highly stable characteristics of a crystal. Once a crystal oscillator has been tuned carefully to a precise frequency it will hold that frequency without frequent retuning. A tuned-circuit oscillator must be readjusted and checked more often. The frequency of a tuned-circuit oscillator is more subject to change as a result of the proximity of other objects and changes in voltage and tube characteristics.

13-3-3. Why should excessive feedback be avoided in a crystal oscillator?—Too much feedback and crystal current can damage the crystal by cracking or shattering it.

13-3-4. Why is a separate source of plate power desirable for a crystal oscillator stage in a radio transmitter?—A separate or well-regulated source of plate power is desirable when the crystal oscillator must operate on a precise frequency with a very strict tolerance in terms of permissible frequency drift. Small as it might be, the supply voltage does have some influence on the frequency of a crystal, and this influence can be quite significant when the frequency of operation must be held within a very close tolerance. The modulator and/or other stages of a transmitter have changing power demands, which may reflect back as a change in the supply voltage to the crystal oscillator and result in a consequent change in oscillator frequency.

13-3-5. What may result if a high degree of coupling exists between the plate and grid circuits of a crystal-controlled oscillator?—If the coupling is too great, the feedback may be excessive and result in frequency instability (crystal heating) and possible fracture of the crystal.

13-3-6. What determines the fundamental frequency of a quartz crystal?—The fundamental frequency of the crystal is determined by its physical dimensions (mechanical resonance) and the type of crystal-cut.

13-3-7. What is meant by the temperature coefficient of a crystal?—The frequency of a crystal is temperature dependent. The amount of frequency change with temperature change is referred to as the temperature coefficient. The temperature coefficient is usually stated in terms of frequency change per degree centigrade. The temperature coefficient can be either positive or negative, depending on the direction of frequency change as a result of an increase in temperature. When a crystal has a negative temperature coefficient, the crystal frequency decreases as temperature increases. Conversely, when it has a positive temperature coefficient, the crystal frequency increases as the temperature increases.

13-3-8. Is it necessary or desirable that the surfaces of a quartz crystal be clean? If so, what cleaning agent may be used that will not adversely affect the operation of the crystal?—Yes. Reliable starting and frequency stability require that the crystal be clean. Carbon tetrachloride or even soap and water and a soft cleaning tissue or cloth can be used to clean the crystal surfaces. A dusty or oily crystal may not oscillate.

13-3-9. What are the characteristics and uses of an overtone crystal? A third-mode crystal?—At high frequencies, above 20 MHz, crystals must be extremely thin for a fundamental frequency of operation. As a result they are very fragile and subject to instability and fracture. However, a crystal can be made to operate on a mechanical-overtone frequency. Crystals can be cut to emphasize odd overtone frequencies, such as third, fifth, seventh, etc. In so doing, a larger and thicker crystal with a lower fundamental frequency can be employed. Overtone crystals can be placed in conventional crystal-oscillator circuits, if such circuits are tuned to the appropriate overtone frequency. However, improved performance can also be obtained by using oscillator circuits that are designed to take the best advantage of overtone crystals.

A third-mode type is an overtone crystal that has been processed to emphasize its third overtone frequency. A third-overtone frequency is close to but not exactly the third harmonic of the fundamental frequency.

13-3-10. What is the purpose of a buffer-amplifier stage in a transmitter? A buffer amplifier is usually placed between a crystal oscillator and a following rf amplifier or multiplier stage. A buffer is operated in a conservative manner and is rather loosely coupled to the crystal oscillator. Its purpose is to serve as an isolation stage between the crystal oscillator and the transmitter stages that follow. As a result, changes in operating conditions will have much less influence on the frequency and output of the oscillator. Some buffers are operated class A to further reduce loading.

13-3-11. Explain some methods of determining if oscillation is occurring in an oscillator circuit.—Oscillations can be ascertained with some form of rf indicator such as an absorption wavemeter, dip meter, neon bulb, single turn of wire with a series pilot bulb, etc. A receiver tuned to the oscillator frequency can also be used to indicate the presence of oscillations.

In the case of a vacuum-tube oscillator, the presence of negative self-bias as measured on a vtvm will indicate the presence of oscillations. Certain types of transistor-oscillator circuits will not draw any collector current unless they are in an oscillating condition.

13-3-12. Explain some of the factors involved in the stability of an oscillator (both crystal and LC controlled).—The frequency of oscillation of a crystal is temperature dependent. For this reason, some crystals are mounted in thermostatically controlled ovens when it is necessary to maintain operation on some precise frequency. In an LC type of oscillator, frequency stability is very much dependent on the characteristics of the frequency-controlling resonant circuit. Such resonant circuits must be designed so that there is a minimum drift in component values with changes in ambient temperature.

Crystals and tuned circuits that minimize capacitive changes in associated components (such as a tube or transistor) with variations in supply voltage and/or loading must be used. In other words, design emphasis is given to those factors which may change the vibration frequency of a crystal, or the resonant frequency of the frequency-determining LC resonant circuit. In general, for both types of oscillators the objective is to minimize changes in component values with change in temperature, and reduce and minimize the influence of capacitive changes with shifts in supply voltages and/or loading conditions.

13-3-13. What is meant by parasitic oscillations? How may they be detected and prevented?—Occasionally oscillators may operate at a frequency other than that of the crystal or the frequency-determining LC circuit. Rf amplifiers too, as covered in Question 13-2-8, may self-oscillate at some remote frequency. Such oscillations are called parasitic. They can be detected with some form of rf indicator or a receiver. Abnormally high or erratic grid- or plate-current readings often indicate the presence of parasitics.

Parasitics can be avoided with proper layout and selection of parts. Similar type and value chokes and capacitors should be avoided in grid and plate circuits to avoid parasitic oscillation of the tuned-plate tuned-grid type. Parasitic resistors and chokes used in grid and plate circuits insert rf resistance in a manner that can prevent the build-up of vhf parasitic oscillations. Lead length and wiring arrangements, as well as components, should be dissimilar enough that grid and plate tuned-line type of resonant conditions are not set up that can cause vhf parasitics.

13-3-14. Describe the physical structure of a klystron tube, and explain how it operates as an oscillator.—A typical klystron oscillator is shown in Fig. 13-33.

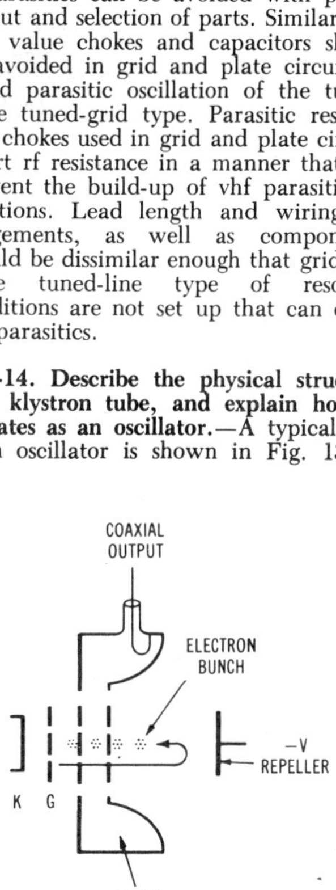

Fig. 13-33. Klystron oscillator.

This tube takes advantage of electron transit time. The tube consists of a cathode, focusing electrode, a cavity resonator which serves as the anode, and a repeller electrode. Electrons emitted from the cathode are focused into a fine beam which passes a gap in the cavity resonator in its motion toward the repeller plate. The repeller has a negative potential which retards the electrons, causing them to return and pass across the cavity gap a second time. The cavity functions as a resonant circuit which determines the oscillation frequency.

There is an rf variation that builds up across the cavity gap which changes the velocity of the electrons as they move across the gap toward the repeller. The gap field causes the electrons in transit to proceed at differing velocities as they leave the gap. They are said to be *velocity-modulated* and begin to bunch, the faster ones catching up with the slower ones. This bunching continues in the travel toward the repeller plate and back to the gap. Consequently the electrons arrive back at the gap in definite bunches at spaced intervals. If they are made to pass the gap in proper phase relationship, they will shock the cavity into oscillation. There is a similarity between this activity and the bursts of plate current that shock-excite a more conventional resonant circuit into oscillation.

The oscillations across the cavity gap then velocity-modulate additional electrons flowing past the cavity toward the repeller. As a result, there is continuous bunching of electrons, and continuous oscillation is established. The returning bunched electrons supply the necessary energy to keep the cavity in sustained oscillation. Rf energy can be removed from the klystron oscillator by a small pickup loop which is inserted into the klystron cavity.

13-3-15. Describe the physical structure of a multianode magnetron, and explain how it operates.—A typical magnetron oscillator is shown in Fig. 13-34. It is used to generate high-power microwave signals. The magnetron tube consists of a cylindrical cathode surrounded by a mul-

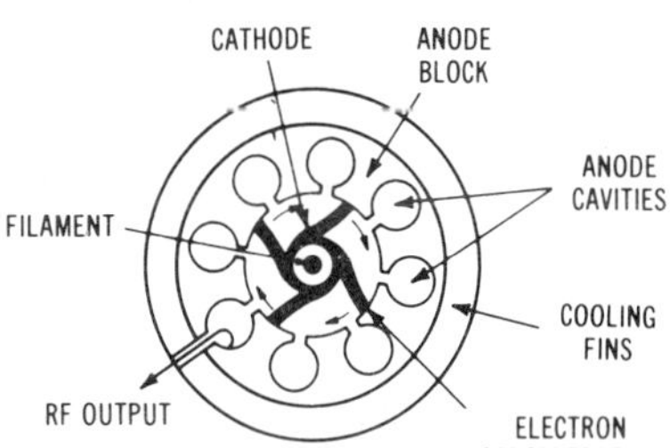

Fig. 13-34. Magnetron oscillator.

ticavity anode. The individual cavities are linked to the cathode-anode space by slots. A powerful magnet surrounds the anode structure.

Interaction between the magnetic field, the anode-cathode dc voltage, the fields surrounding the electrons in transit from cathode to anode, and the rf fields in the vicinity of the slots that result from cavity oscillations cause the electrons to trace an angular path toward the anode. Because the complex interaction results in differing velocities of electron motion, the electrons bunch in an almost spokelike manner. Here we have another microwave tube that is taking advantage of electron transit time in a manner that causes electrons to be velocity modulated.

As the electrons move in rotating spokelike fashion (bursts) across the cavity slots, energy is delivered to the resonant cavities causing them to be shock-excited into oscillation. The cavities are joined together to produce a net output which can again be extracted by inserting a probe into one of the cavities. The oscillations of each cavity influence the bunching of the electrons at each cavity slot, and in this manner oscillation is sustained.

13-3-16. Draw a diagram showing the construction of a traveling-wave tube. Explain its principle of operation.—A typical traveling wave tube (twt) construction is shown in Fig. 13-35. It is

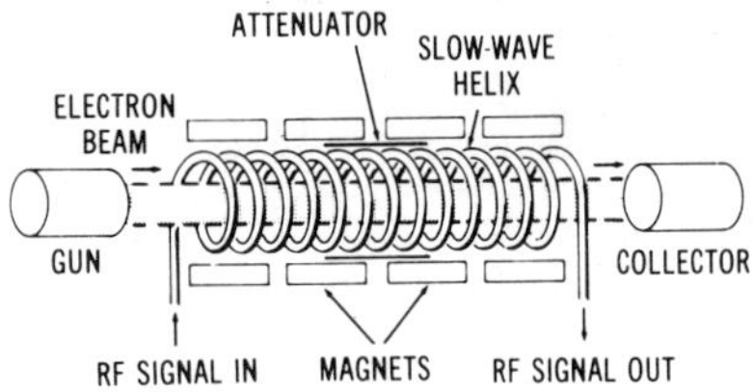

Fig. 13-35. Traveling-wave tube.

usually employed as a microwave amplifier. In a traveling-wave tube, velocity modulation is again employed.

The electrons are generated by an electron gun and made to move in a fine beam toward a collector by the focusing action of permanent magnets. The path of the beam is through the center of a helical conductor. It is the interaction between the magnetic field surrounding the helical conductor and the electrons which causes the electron bunching.

The microwave input signal is applied to the helix at the electron-gun side. The signal moves along the conducting helix at the velocity of light. However, the helical construction is such that the actual electric field that surrounds the helix travels at a slower rate, corresponding more closely to the velocity of the electrons of the beam as they travel toward the collector. Consequently, there is an interaction between the actual electric field of the helix and the electrons of the beam. In fact, as the helix electric field changes with the microwave variation, the electrons are alternately increased and decreased in velocity. Consequently they begin to bunch together just as in a klystron tube.

As electrons progress down the traveling-wave tube, they bunch together with increasing density. The dense electron bunches in motion, in turn, induce additional energy into the helix and cause the amplitude of the microwave signal to increase progressively as it winds around from the input to the output end of the helix. Thus, in the traveling-wave tube the amplification rises progressively as the bunching becomes more dense, and more energy is induced into the helix from the beam.

Coaxial couplings or waveguide transducers are used to couple signal to the input (gun side) of the helix and to remove the amplified signal from the output (collector side) of the helix.

Self-oscillation in a traveling-wave amplifier is prevented by an attenuator which is inserted into the helix to prevent the feedback component from building up to a level that would cause self-oscillation. In a suitable construction or with the use of a feedback path, the traveling-wave tube can be made to operate as a microwave oscillator.

13-3-17. What is the function of a Gunn diode?—The Gunn diode is a solid-state diode that in an appropriate circuit will function as a microwave oscillator. They are capable of generating microwave powers between 100 mW and approximately 1 watt. Efficiency is very low and in the order of 2 to 5 percent.

13-3-18. How may the oscillating frequency of a crystal be adjusted?—The frequency of a crystal itself is determined by the cut of the crystal and the size. The mounting assembly, including the pressure plates, also contributes capacitance that also influences the oscillating frequency. External shunt capacitance and series capacitance in the crystal circuit influence the oscillating frequency of the crystal oscillator. Such external capacitance if made adjustable can be used to set the frequency of the crystal oscillator to some precise frequency.

13-3-19. What is flywheel effect and how does it relate to operation of rf oscillators and rf amplifiers?—Flywheel effect refers to a characteristic of a resonant circuit that provides smoothing and energy-storing ability. Even though the current introduced into the resonant circuit is in the form of a burst of current or energy, the resonant circuit will smooth it out, and produce a sine wave of voltage. Thus an output resonant circuit of proper Q when present in an amplifier or oscillator circuit will produce a sine-wave output even though the output current is nonsinusoidal and the stage is operating class B or class C. Refer to Sections 3-10, 3-11, 3-14 and Questions 13-2-1 and 13-3-1 of Chapter 13.

13-3-20. Explain what is meant by the cut of crystal. How do fundamental cuts differ from overtone cuts?—The raw crystal quartz is basically a 6-sided prism. Such a crystal has three axes and a crystal can be cut such that it is normal to one of these axes. Temperature coefficient and other operating characteristics are influenced by the cut of the crystal. An overtone crystal is cut in such a manner that it vibrates strongly at a particular harmonic. The third-overtone cut is the most popular; others are processed for strong fifth- or seventh-overtone vibrations.

13-3-21. What is meant by locking two oscillators together?—If two oscillators capable of operating on the same frequency are coupled together, one of the oscillators will dominate and both oscillators will operate on the same frequency. It is said that the two oscillators are locked together. In a phase-locked loop system, a reference oscillator through a feedback system can be used to set the frequency of a voltage-controlled oscillator. This can be done at the fundamental frequency or through a frequency-divider arrangement, in which case the voltage-controlled oscillator may operate on some precise multiple of the reference oscillator. Again, the two oscillators are locked, but on widely separate frequencies. Refer to the phase-locked loop discussion in Section 3-16.

13-3-22. Describe how one crystal in a synthesizer can be used to generate several different frequencies.—Refer to the phase-locked loop discussion in

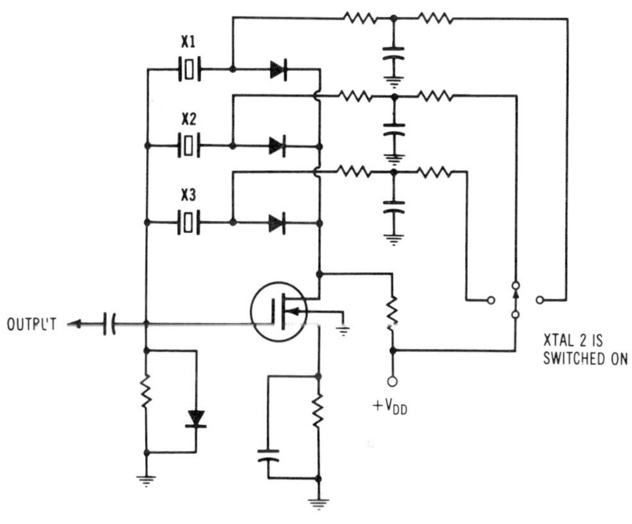

Fig. 13-36. Crystal selection by diode switching.

Section 3-16 and the discussion of frequency sysnthesizers in Section 3-17.

13-3-23. Draw a block diagram of a phase-locked loop frequency synthesizer using a frequency divider, a frequency/phase comparator, a low-pass filter, a reference oscillator, and a voltage-controlled oscillator.—Refer again to Sections 3-16 and 3-17.

13-3-24. Draw a simple schematic diagram of a crystal-controlled radio-frequency oscillator in which one of several frequencies may be selected by diode switching of the different crystals.—Refer to Fig. 13-36. Crystal is selected by supplying a bias that causes the diode associated with the particular crystal to conduct. Therefore the crystal-oscillator circuit is completed for that particular crystal and oscillations are initiated.

CHAPTER 14

Amplitude-Modulation Systems

14-1. TRANSMITTERS

14-1-1. Discuss the following items with respect to their harmonic attenuation properties as possibly used in transmitter or receiver: (a) link coupling, (b) tuned circuits, (c) degree of coupling, (d) bias voltage, (e) decoupling circuit, (f) shielding.—The main reason for suppressing harmonics in a transmitter is to block the radiation of harmonic signals that can interfere with other radio services. For efficient operation of the transmitter, many of the stages operate class C, and therefore the harmonic content of output signals can be quite high. These harmonics must be suppressed. In receivers, the presence of harmonic components can cause annoying heterodynes (birdies) and the introduction of distortion components. Harmonic components in a mixer are generated by the required nonlinearity of the mixing process, and by the strong signal components in the if amplifier that stem from operation on the nonlinear portion of the tube or transistor transfer characteristics (particularly with the reception of a strong signal). Such harmonic components leak back into the early rf stages of the receiver where they mix with off-frequency signals and produce an output in the if range. Such harmonic components in the receiver can also result in the production of intermodulation distortion components.

(a) A link-coupled arrangement (Fig. 14-1A) minimizes the transfer of harmonic components. Such a link presents a low-impedance coupling path for only that frequency to which the two resonant circuits are tuned. Very little harmonic energy is induced into the link and from the link into the secondary resonant circuit. In a link arrangement the two resonant circuits are usually not mutually coupled; that is, there is no direct linkage of magnetic lines between the primary and secondary. Because of this, a link can be used to couple two widely separated resonant circuits.

(b) Of course, the greater the number of resonant circuits in a chain which are tuned to the desired resonant frequency, the more effectively are the harmonic components rejected. This applies in both the transmitter and receiver amplifier chains. Two inductively coupled circuits, such as shown in Fig. 14-1B, are preferred to a single interstage resonant circuit in terms of harmonic rejection. However, such a coupling arrangement is not as effective as link coupling in the minimization of harmonics. When stages operate on the nonlinear portions of their transfer curve, the use of high-Q resonant circuits helps in the reduction of harmonic levels.

(c) The degree of coupling also influences the transfer of harmonic components. In general, the looser the coupling

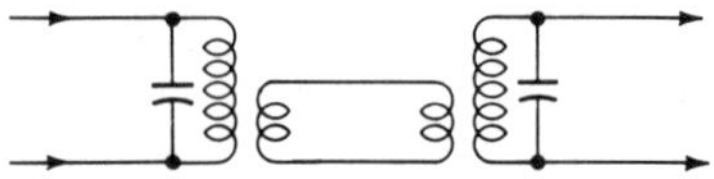

(A) Link coupling.

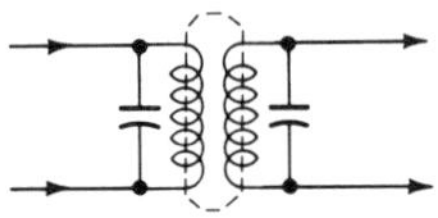

(B) Mutual coupling.

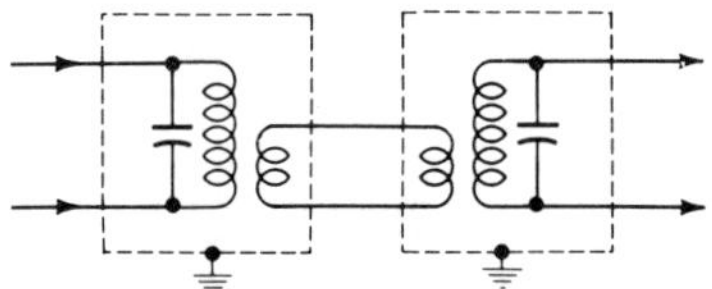

(C) Shielded resonant circuits.

Fig. 14-1. Coupling arrangments.

is, the lower is the harmonic transfer. For example, in the circuit of Fig. 14-1B, the looser the coupling is between the primary and secondary resonant inductors, the better is the harmonic suppression. In terms of transmitter operation, this means the degree of coupling between stages, and between the final stage and the load, must often be optimized in terms of the most effective transfer of power and the most effective harmonic suppression.

(d) The bias voltage has an influence on the harmonic level in class-C stages. A bias should be selected in accordance with the level of the rf excitation, so that the positive peaks of the input wave do not drive the stage into plate saturation. An angle of flow should be selected which is optimum for fundamental operation and maintains harmonic generation within reasonable levels.

Receiver rf stages operate in class A. However, incoming strong signals may cause a swing into saturation and cutoff, and result in the formation of harmonics. In a receiver, then, a good avc-agc system is helpful because signal levels can be better confined within the linear portions of the transfer characteristics, particularly in those stages immediately ahead of the detector.

(e) In a receiver, decoupling circuits are of particular advantage because they can prevent harmonic components, generated in the latter stages of the if amplifier, from reaching the much more sensitive rf input stages over voltage supply lines. Decoupling circuits are useful in transmitters because they also prevent the transfer of ac signal components (harmonics included) among the various stages.

Receiver selectivity is aided with the use of multiple resonant circuits. In general, loose coupling provides better selectivity than does tight coupling.

(f) Shielding is effective in minimizing the direct radiation of harmonic components. Even in a receiver, the shielding of if and detector transformers minimizes the direct radiation of harmonic components back to the sensitive rf stages. When link coupling is used between transmitter stages, the shielding of resonant tank circuits can prevent direct flux-linked harmonic paths from being established between two coupled resonant circuits (Fig. 14-1C).

14-1-2. List several frequently used methods of attenuating harmonics in transmitters, and explain how each works.—Low-impedance coupling arrangements such as link coupling (Fig. 14-1) block the transfer of harmonics because of the high impedance offered by the links to any frequency other than the resonant frequency of the coupled tank circuits.

The transfer of harmonic components between two coupled resonant circuits is often largely due to capacitive linkage rather than flux linkage. Therefore, the use of a so-called Faraday shield (Fig. 14-2) can minimize the transfer of harmonics by acting as a grounded capacitive shield. At the same time, the gridlike construction of the shield offers no impedance to the magnetic flux lines that link the primary and secondary at the resonant frequency.

Inasmuch as harmonic radiation must be held below certain tight FCC values, the low-pass filter (Fig. 14-2B) is used extensively. A low-pass filter is usually placed in the signal-transfer path between the output transmitter and the antenna, and, on occasion, even between stages of the transmitter. The low-pass filter passes frequency components up to a specific limit (that limit is not too far above the fundamental frequency which is to be transferred along the signal path). As a result, the filter offers high attenuation at harmonic frequencies.

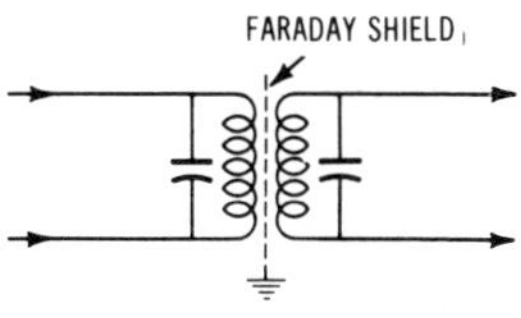

(A) Using a Faraday shield.

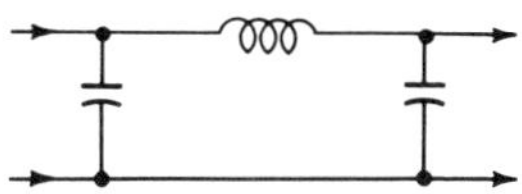

(B) Low-pass filter.

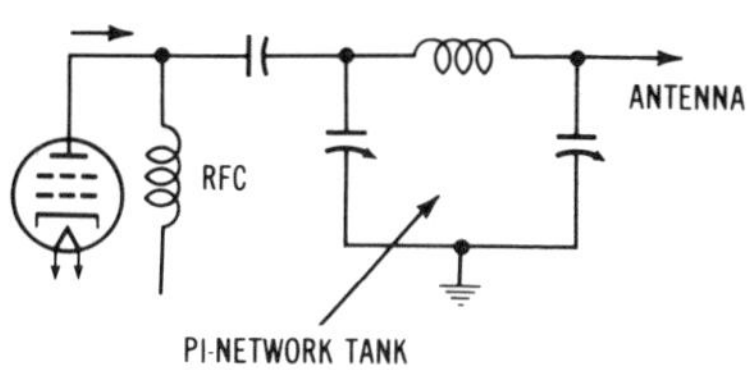

(C) Pi-network.

Fig. 14-2. Coupling systems used to reduce harmonics.

Sometimes additional resonant sections are added (M-derived section added to a constant-K filter) to offer maximum rejection at the specific harmonic frequency.

Another popular circuit is the pi-network resonant circuit (Fig. 14-2C). This arrangement includes an appropriate resonant circuit of the proper Q and impedance for the transmitter stage. It also provides an impedance transformation from the high-impedance output of the stage to the low impedance of an antenna system or feed system to a following stage. At the same time, it also acts as a low-pass filter and rejects the transfer of harmonic components from the resonant circuit to the load.

Of course, a variety of other operating factors, such as proper shielding, correct operating bias, tank circuits of an adequate Q, etc., also aid in the minimization of harmonics. Refer to Question 14-1-1.

14-1-3. State a probable cause and methods of reducing transmitter spurious emissions (other than harmonics).—Such spurious emissions can be in the form of self-excited oscillations. These result when an amplifier stage self-oscillates. These can be at or near to the desired frequency of amplification. In this case the spurious emission can be eliminated by the proper adjustment of an adequate neutralization system. Refer to Questions 13-2-6 and 13-2-7 of Chapter 13.

Lead inductances, stray capacitances, component parts and their arrangements can also produce parasitic oscillations at very high and very low frequencies. Proper selection of components can reduce low-frequency parasitics, and the use of parasitic resistors and chokes can block the start of vhf parasitics. Refer to Questions 13-2-8 and 13-3-13 of Chapter 13.

Another spurious emission is subharmonics. In a transmitter using frequency multipliers, there can be a leak-through of frequency components that are submultiples of the carrier frequency. This can be a particular problem in transistor transmitters because the fundamental-frequency output of a transistor frequency multiplier is often at an objectionably high level. Properly designed multiplier resonant circuits, and sometimes special filters, are used to reduce these submultiples.

14-1-4. Define transmitter intermodulation, its possible cause (or causes), its effect, and steps that should be taken to reduce it.—Transmitter intermodulation is the interaction between two or more signals in a given transmitter (particularly the rf output stage) because of nonlinear characteristics. Such nonlinearity results in output frequencies that are equal to the sums and differences of the interacting frequencies and multiples thereof. Transmitter rf intermodulation is the result of the close proximity of transmitters such as at a communications center or a multiple repeater site. Energy components coupled back from antennas to the transmitter final amplifiers can produce intermodulation components and spurious emissions. (Transmitter class-B and class-C power amplifiers operate in a nonlinear fashion.) Proper diplexers, filters, and isolators are used to minimize such interaction. Refer to Section 8-6 of Chapter 8.

Inasmuch as transmitters employ a variety of nonlinear stages, intermodulation can develop at various places in a transmitter. Mixers, frequency changers, modulators, and even the conventional class-C amplifier are nonlinear circuits. As a result, sum and difference components can be generated. Such spurious intermodulation components are not only radiated, but they can result in erratic operation of a transmitter and/or the introduction of distortion components. Circuits should be used to minimize their generation, and appropriate resonant circuits and filters must be employed to emphasize desired signal components far above the intermodulation component. Proper stage decoupling of voltage supply lines is important. Proper transmitter shielding and the use of appropriate filters in the transmission-line path between the transmitter and antenna system do much to prevent the radiation of such intermodulation components.

14-1-5. List several frequently used methods of attenuating harmonics in transmitters and explain how each works.—A common method of attenuating harmonics is to use a pi-network output resonant circuit. Such a resonant circuit can be designed to match impedance as well as attenuate frequencies above its bandpass. Inasmuch as the second harmonic and other higher order of harmonics are above the desired frequency these are attenuated by the output circuit. Another popular way of reducing harmonics is to use a pi-network filter, particularly between the transmitter output and the antenna system. This can be a multisection filter which has a very high order of attenuation to harmonic frequencies. At the same time the desired signal encounters only a minimal loss. The high-pass filter network is almost an essential in solid-state transmitter designs. Of course, harmonics can be attenuated by proper tuning of the transmitter. Follow the manufacturer's recommended procedure for optimum tuning with minimum generation of harmonics. In some instances resonant circuits are used to de-emphasize certain harmonics. A series-resonant circuit connected between the signal output and ground presents a low impedance to the harmonic frequency to which it is tuned and that component is not developed in the output. Sometimes a parallel-resonant circuit is placed in the series circuit over which the desired signal passes. It is tuned to a harmonic frequency and therefore offers maximum attenuation and does not permit harmonics to be transferred from its input side to its output side. In modern two-way radio equipment, resonant cavities and helical filters are also used for the rejection of undesired components, and the best transfer of a desired signal. Refer to some of the discussions in Chapters 3 and 7.

14-1-6. Why can lack of neutralization cause spurious emissions from a transmitter?—When a transmitter is not neutralized properly rf amplifiers can self-oscillate on a frequency or frequencies that are not the same as the desired frequency. Also intermodulation can occur which will generate additional spurious emissions.

14-2. AMPLITUDE MODULATION

14-2-1. Draw a block diagram of an am transmitter.—Refer to Fig. 14-3.

14-2-2. Draw a simplified circuit diagram of the final stages (modulator, modulated amplifier, push-pull linear amplifier) of a type of low-level plate-modulated transmitter utilizing a pentode tube in the modulated stage. Explain the principles of operation. Repeat the diagram, using a tetrode to provide high-level modulation.—Refer to Figs. 14-4 and 14-5.

In a low-level plate-modulated transmitter, some stage preceding the power amplifier is modulated. The rf excitation for the modulated pentode amplifier (V1) comes from a so-called rf exciter. The rf exciter includes the frequency-control oscillator and the amplifiers that are required to increase the rf signal to the

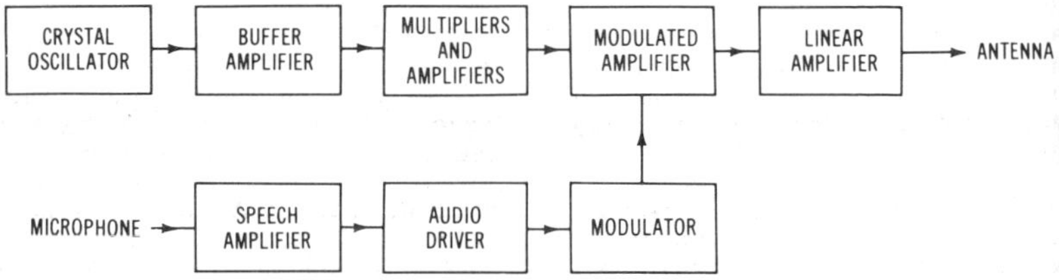

Fig. 14-3. Block diagram of an am transmitter.

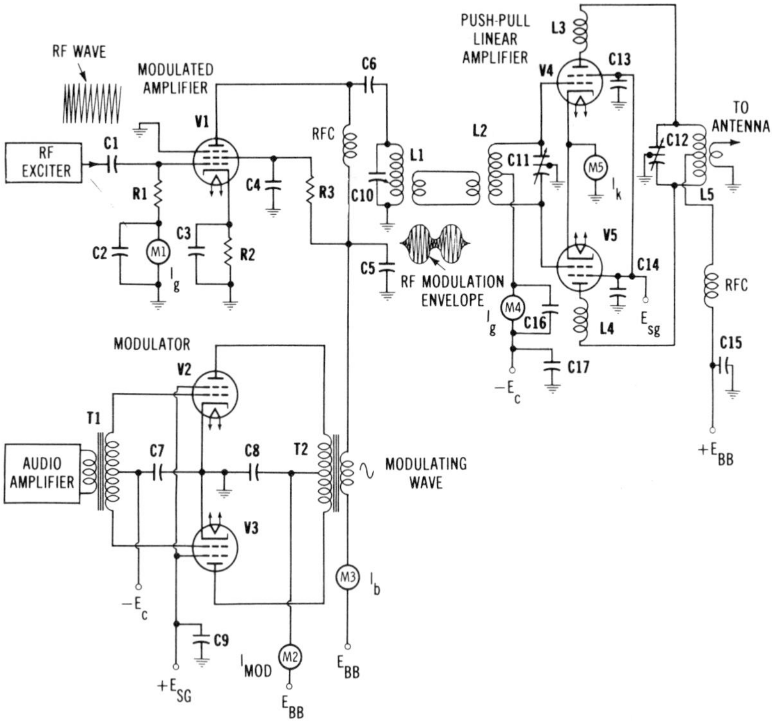

Fig. 14-4. Low-level plate-modulated amplifier and linear output amplifier.

magnitude needed at the input of the modulated amplifier. The modulated amplifier is biased class C, using the self-bias grid combination, R1C1. Audio-modulator tubes V2 and V3 are driven by a signal from an audio amplifier. The function of the audio amplifier is to build up the microphone or other audio variations to the proper level for driving the push-pull modulator. Transformer T1 provides a suitable match into the modulator. The audio variation across the secondary of output transformer T2 (usually called a *modulation transformer*) places an audio variation on the plate-supply voltage to the modulated amplifier.

For example, the supply voltage to the modulated amplifier may be 600 volts. The voltage variation across the secondary transformer T2 may have a peak amplitude to 600 volts. Therefore, the supply voltage for the modulated amplifier would vary between 0 and 1200 volts, following the audio variation at the output of the modulator. Let us consider how this influences the operation of the modulated amplifier.

The supply voltage is connected to the plate of the modulated amplifier through the radio-frequency choke (rfc). The same supply voltage is also supplied to the screen grid of the modulated amplifier through screen-grid dropping resistor R3. Thus, any variation across the secondary of the modulation transformer varies both the plate and the screen-grid voltages of the modulated amplifier.

In a class-C amplifier the plate current occurs in bursts, as shown in Fig. 14-6. The peak amplitude of these bursts of plate current depends on the screen-grid and plate voltages. When the screen-grid and plate voltages are maximum (positive peak of modulating wave), the plate-current bursts also have a maximum peak amplitude. When the screen-grid and plate voltages dip to a low value

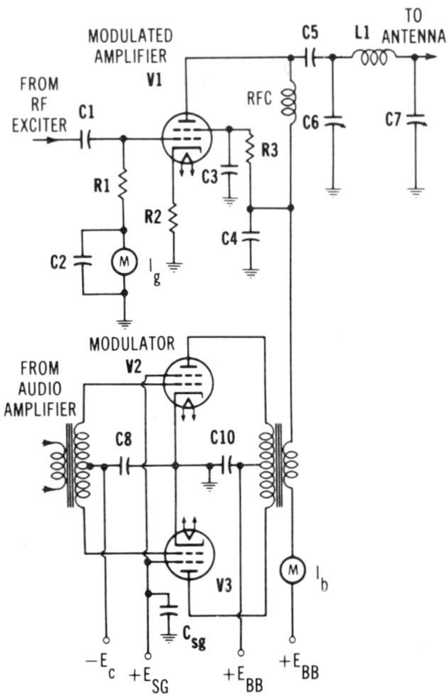

Fig. 14-5. High-level modulator.

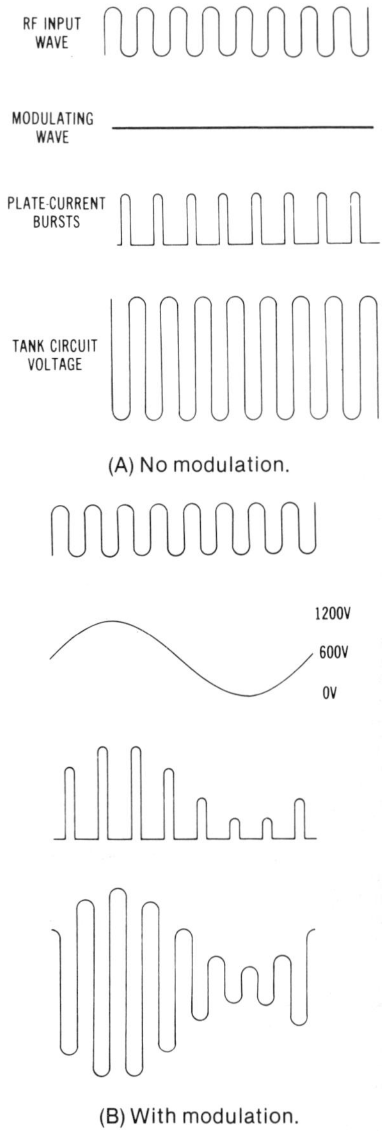

Fig. 14-6. The amplitude-modulation process in waveforms.

(negative peak of modulating wave across the secondary of the modulation transformer), the plate-current bursts are minimum in amplitude. It is apparent that the peak amplitudes of the plate-current bursts vary with the modulation.

It is these bursts of plate current that charge the resonant tank capacitor to its maximum negative value. The energy-storing ability (flywheel effect) of the resonant tank circuit reconstructs an rf sine wave.

When there is modulation across the secondary of the output transformer, there is a like variation in the peak amplitude of the plate current. This, in turn, determines just how far negative tank-circuit capacitor C10 is charged. Of course, the charge placed on capacitor C10 determines the amplitude of the sine wave that is reconstructed in the tank circuit. It follows, then, that the amplitude of the rf sine waves as they develop in the tank circuit follows the variations in the peak plate current and, in turn, the audio variations present across the secondary of the modulation transformer. Refer again to the waveforms of Fig. 14-6.

When the audio variations at the output of the modulator are of a lower amplitude, there is a reduced peak plate-current variation. Consequently, there is a smaller variation in the peak amplitude of the rf cycles developed across the output tank circuit. In this manner, the rf variations developed across the output of the modulated amplifier follow the changes in the modulating wave.

The output of the modulated amplifier is no longer an rf sine wave of a constant amplitude as it appears at the input of the modulated amplifier. Rather, the sine wave now varies in magnitude in accordance with the modulation. We have formed a so-called modulation envelope. If this envelope is to be increased in amplitude and power level without causing distortion, any following amplifier must operate over the linear portion of its characteristic curve. Thus, the stage that follows the modulated amplifier is not a class-C amplifier; rather, it is operated either in class AB or class B. It is said to be a push-pull "linear rf amplifier."

To minimize distortion, the operation of the two tubes and associated circuits should be balanced. The split-stator capacitors, C11 and C12, help to maintain a good rf balance. The split-stator type of capacitor has two separate stator-plate sections, while a single rotor on a single axle rotates in and out of the stator plates in a uniform manner. This rotor can then be placed at rf ground potential, as shown, to obtain a balanced circuit. In the linear amplifier, the modulation envelope, as it appears at the input circuit, is reproduced with fidelity in the output tank circuit.

Although not included specifically in the question, let us consider the basic function of each component part in order to give you a better understanding of the operation of the modulation circuit and, therefore, a better understanding of the specific answers to many of the questions that follow.

Capacitors

C1. This capacitor has a low reactance at the rf frequency, and it serves as a means of coupling the rf output of the exciter to the input of the modulated amplifier. It is also the capacitive portion of the combination, R1C1.

C2. An rf bypass capacitor that keeps radio-frequency energy out of dc meter M1.

C3. An rf filter capacitor that prevents rf degeneration.

C4. This is an rf filter capacitor that grounds the screen-grid in terms of the rf signal. However, its reactance is high enough at audio frequencies so as not to interfere with the variation of the modulated screen-grid voltage.

C5. This is a plate rf filter capacitor with a function similar to that of C4. It presents a short circuit to the rf signal, but at the same time it does not prevent the modulating wave from varying the plate supply voltage.

C6. A dc blocking capacitor that keeps the dc plate voltage from the tank circuit and prevents coil L1 from shorting out the dc plate voltage.

C7. This capacitor has low reactance in the audio frequency range, and it acts as a bias-source filter.

C8. This capacitor acts as a plate supply-voltage filter, keeping audio variations out of the power supply.

C9. This capacitor has a similar function to C8, but in terms of the screen-grid voltage.

C10. Resonant-circuit capacitor.

C11. Split-stator capacitor of the grid resonant circuit.

C12. Split-stator capacitor of the plate resonant circuit.

C13, C14. Screen-grid filter.

C15. Plate-voltage filter.

C16. Rf bypass around dc meter.

C17. Bias-supply filter capacitor.

Resistors

R1. Grid-bias resistor and part of R1C1 time constant.

R2. Safety cathode resistor that protects the tube with a loss of rf.

R3. Screen voltage dropping resistor.

Inductors

Rfc. Radio-frequency chokes that keep radio-frequency energy out of the supply-voltage lines and permit the connection of dc voltages without shunting the radio-frequency signals.

L1. Plate tank coil of the modulated amplifier.

L2. Grid tank coil of the linear amplifier.

L3, L4. Parasitic chokes.

L5. Coil of the linear-amplifier plate resonant circuit.

T1. Audio-input transformer.

T2. Modulation transformer.

Meters

M1. DC grid-current meter.

M2. Dc plate-current meter for the modulator. For class-AB and class-B operation the indication of this meter will vary with the audio modulation.

M3. Dc plate current meter for the modulated amplifier. With a properly designed and tuned am modulation system, this meter will not change with modulation.

M4. Dc grid-current meter for the linear amplifier.

M5. Dc cathode-current meter for the linear amplifier. This current reading approximates the total plate current drawn by the two linear-amplifier tubes.

The circuit of Fig. 14-4 is that of a "low-level modulation system." It is called low-level modulation because the modulation occurs ahead of the final rf power amplifier which, in this case, is in the form of a push-pull linear amplifier.

The circuit of Fig. 14-5 is called a "high-level modulation system" because the final radio-frequency amplifier (the one feeding the antenna) is plate modulated. The modulated amplifier is identical to that of Fig. 14-4, except that a tetrode tube instead of a pentode tube is being used, and the plate resonant-tank circuit is a pi-network instead of a simple parallel LC combination. The combination of C6, C7, and L1 is a parallel circuit that is resonant at the output rf frequency. However, the split-capacitor arrangement also functions as an impedance transformer. Looking into the pi-network from the tube, a high impedance is seen, matching the high-impedance output of the modulated amplifier. Looking into the network from the antenna side, one sees a low impedance, and a match is provided to a low-impedance antenna system. In this high-to-low impedance transformation, capacitor C7 is several times larger than capacitor C6. The pi-network arrangement is also fundamentally a low-pass filter. When the values are chosen properly, this type of tank circuit has a maximum rejection of harmonic and other higher-frequency components and minimizes their transfer to the antenna system.

The modulator is a push-pull type, similar to that of Fig. 14-4. The modulation process occurs in exactly the same manner, with the modulating wave varying the screen-grid and plate-supply voltages to the modulating amplifier. In this manner, the rf peak plate-current bursts of the modulated amplifier also vary in amplitude and a modulation envelope (amplitude modulated) is developed in the pi-network tank circuit.

14-2-3. Draw a simple schematic diagram showing a method of coupling the modulator tube to a radio-frequency power-amplifier tube to produce grid modulation of the amplified rf energy. Compare some advantages or disadvantages of grid modulation with those of plate modulation.—Refer to Fig. 14-7. In the example, the final rf power amplifier is grid-modulated instead of plate-modulated. Note the manner in which the circuit arrangements of Figs. 14-7 and 14-5 differ. The screen-grid and plate-output circuits are identical, with the exception that the screen-grid and plate voltages are not modulated. Rather, the modulation transformer is located in the grid circuit of the modulated amplifier.

The modulator can be either a single-ended or a push-pull amplifier. Inasmuch as the amount of audio power required to grid-modulate an amplifier is much less than that required to plate-modulate the same amplifier, a single-ended audio amplifier-modulator is often employed. Such a single-ended modulator stage is shown

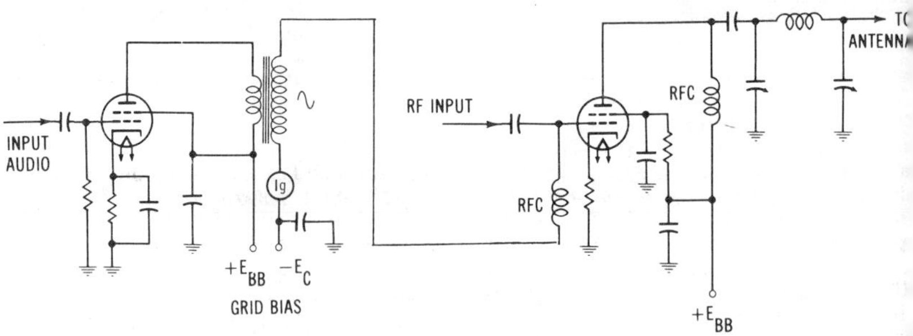

Fig. 14-7. Grid-modulation system.

in Fig. 14-7. The turns ratio of the modulation transformer is such that a match is made between the output of the class-A audio power amplifier and the input impedance, looking into the grid circuit of the modulated amplifier from the secondary.

In the circuits of Figs. 14-4 and 14-5, the modulation was handled by varying the screen-grid and plate voltages at an audio rate. In a grid-modulation system, similar results are obtained by modulating the dc grid-bias voltage at an audio rate. Note that the grid bias for the modulated amplifier in Fig. 14-7 is obtained from a separate dc grid-bias supply. If there is an audio variation across the secondary of the modulation transformer, there will be a similar variation in the grid bias.

As the rf waves come into the modulated amplifier, they are, of course, of constant amplitude. The maximum positive peaks of these waves determine the peak plate current drawn by the modulated amplifier. However, if the grid bias is modulated, the peak grid voltage will be the algebraic sum of the rf peak voltage and the instantaneous amplitude of the modulating wave. As a result, the peak grid voltage itself will vary with the modulation and, in like fashion, so will the peak plate current drawn by the modulated amplifier. Thus, as in the case of plate modulation, the grid-modulated circuit eventually results in the variation in the peak plate current and in the magnitude of the rf cycles as they develop across the output tank circuit.

A particular advantage of the grid-modulation system is that the audio power required to modulate the grid voltage over the necessary range is much less than that needed in plate modulating the same tube. This lowers the cost of the modulator system, since it requires fewer expensive modulation components. Likewise, the power demand made by the modulator is much less, and a simpler and less costly power supply is required by the audio system.

Grid modulation has several disadvantages. A grid-modulation system is more difficult to adjust for optimum operation, and the modulated amplifier cannot be operated with as high an efficiency as with plate modulation. Linear and full 100% modulation of the envelope is difficult to obtain with a grid modulation system because of the nonlinearity of the transfer characteristic.

14-2-4. What is the meaning of the term carrier frequency?—The carrier is the generated rf wave. It is a single-frequency rf wave that is transmitted whether modulation is or is not applied to a conventional am transmitter.

14-2-5. If a carrier is amplitude modulated, what causes the sideband frequencies?—In an am modulation system, there is a mixing process that occurs through the nonlinear characteristic of the modulated amplifier; this mixing process produces the sideband frequencies. In this mixing action, the single frequency carrier wave is combined with the modulating wave to produce sum and difference frequencies called *sideband* or *side* frequencies. Thus if a 1000-kHz rf carrier wave is modulated by a 1000-Hz note, sideband frequencies are produced at 1001 kHz (1000 kHz + 1000 Hz) and at 999 kHz (1000 kHz – 1000 Hz).

The 1000-Hz modulating wave never appears in the output tank circuit of an am modulated amplifier because the tank presents a short circuit to this frequency. Thus, the low-frequency fundamental used in the mixing process is lost. However, three radio-frequency waves do appear in the tank circuit. These are the original carrier frequency and the two sideband frequencies. The three rf waves combined in the output tank circuit form a resultant, amplitude-modulated rf signal. This resultant rf signal is the characteristic am modulation envelope.

14-2-6. What determines the bandwidth of emission for an am transmission?—It is determined by the highest frequency in the modulating wave. In the preceding example there were two side frequencies, appearing 1000 Hz to each side of the carrier frequency. Thus, the total signal occupies a bandwidth of 2000 Hz (± 1000 Hz). Perhaps the highest frequency transmitted by the station is 5000 Hz. Now the total bandwidth of emission that would be required is 10,000 Hz (± 5000 Hz) because the side frequencies extend as far as 5000 Hz on each side of the carrier frequency.

14-2-7. What is the relationship between the percent of modulation and the shape of the waveform envelope, relative to carrier amplitude?—The higher the modulation percentage is, the greater is the magnitude of the crest-to-trough variation of the modulation envelope, relative to the amplitude of the unmodulated carrier. Stated as a simple equation this is:

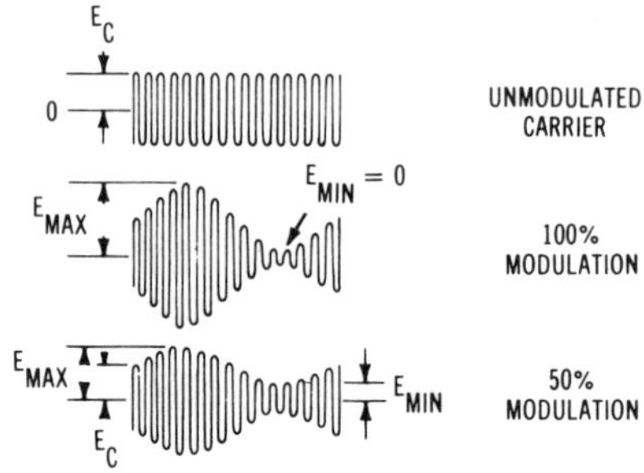

Fig. 14-8. Modulation percentages.

$$\% \text{ modulation} = \frac{E_{max} - E_{min}}{2E_{carrier}} \times 100$$

Refer to Fig. 14-8. When the modulation envelope is 100% modulated, the crest of the envelope rises to twice the peak amplitude of the unmodulated carrier, and in the trough the unmodulated envelope falls away to zero during the negative crest of the modulating wave.

14-2-8. Why does exceeding 100% modulation in an am transmission cause excessive bandwidth of emission?—When the modulation exceeds 100%, side-frequency components are usually harmonics of the modulating wave and are further displaced from the carrier than the normal sideband pair. As a result, the transmitter signal spreads out and occupies a much wider band of frequencies. In fact, the side frequencies can be so far removed from the carrier that they appear in the channels assigned to another station, causing interference.

The splatter takes place because the actual modulating sine wave is distorted when 100% modulation is exceeded. Such distortion introduces harmonic-signal components that are not normally present in the modulating wave, and thus a wider band of signal components is transmitted.

14-2-9. What is the relationship between the average power output of the modulator and the plate circuit input of the modulated amplifier, under 100% sinusoidal modulation? How does this differ when normal-voice modulation is employed?—To attain 100% modulation it is necessary that the average power output of the modulator be one-half of the dc power input to the plate-modulated amplifier. However, because of the high-peaked and irregular variations of speech waveform, the average power output needed to sustain 100% modulation on voice peaks is much less than that required for 100% sinusoidal modulation. It can be estimated that the average audio-power output in terms of voice transmission need only be about one-quarter of the dc power input to the dc plate-modulated amplifier.

14-2-10. What is the relationship between the amount of power in the sidebands and the intelligibility of the signal at the receiver?—The greater the power in the sidebands is, the higher is the amplitude of the recovered signal at the receiver. Thus, the higher the average sideband power is, the louder is the signal at the receiver. Greater intelligibility results in the presence of noise and interference, assuming there is no distortion caused by overmodulation, severe clipping, etc.

14-2-11. How does a linear power amplifier differ from other types?—It differs in that it is biased to operate in a linear manner in terms of the amplitude variations of an applied radio-frequency modulation envelope. Thus, the am envelope can be applied to its input, and an amplified replica will be developed across its output. If such a wave were applied to the input of a conventional class-C amplifier, the output would be distorted.

14-2-12. What would cause a dip in the antenna current when am is applied? What are the causes of poor carrier amplitude regulation or carrier shift?—Such a dip may be the result of the inability of the modulator or modulated amplifier to follow the modulation crest. Improper operating conditions in the grid circuit, such as incorrect bias or insufficient rf excitation, could prevent generation of the high peak power necessary to convey the modulation crest. A defective modulated amplifier, a power-supply component failure, or poor regulation can also kick the antenna current down. Too heavy loading of the modulated-amplifier output by the antenna system can cause the same disturbance.

Poor carrier amplitude regulation in an am modulated transmitter refers to a fluctuation in the dc plate current with modulation. Some of the causes of negative shift (plate current dip with modulation) are improper impedance match between modulator and modulated stage or between modulated stage and its load, a failing modulated

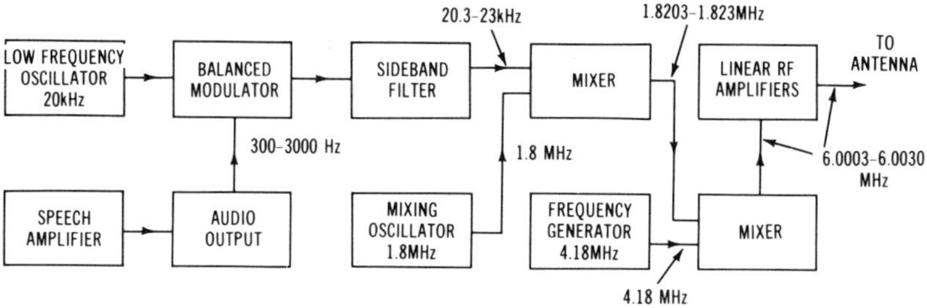

(A) Transmitter.

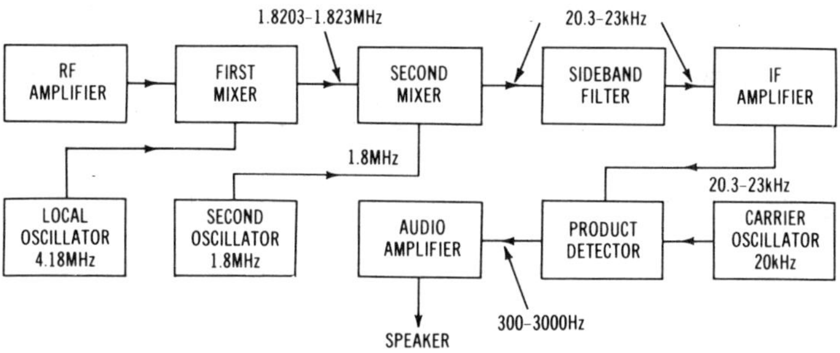

(B) Receiver.

Fig. 14-9. Functional plan of an SSSC system.

amplifier tube, poor power-supply regulation, and insufficient radio-frequency excitation at the grid of the modulated amplifier. An upward kick of the plate meter (positive shift) can be caused by improper neutralization or overmodulation.

14-2-13. What is meant by dynamic instability or frequency shift with reference to a modulated rf emission?—This refers to any change in the carrier frequency with modulation.

14-2-14. What might cause frequency modulation in an am radiotelephone transmitter?—The two most likely causes are changes in loading of the oscillator with modulation and changes in the oscillator supply voltage with modulation. The frequency of an oscillator, even though it might be crystal controlled, is influenced to some extent by its electrode supply voltages and by the load applied to its output. Thus if the oscillator supply voltage changes as a result of the modulator or modulated amplifier, there can be an actual shift in frequency of the oscillator with modulation, thus producing frequency modulation of the carrier. Likewise, if a changing load is reflected to the output circuit of the oscillator from a following modulated amplifier, a similar fm modulation of the am frequency can result.

14-3. SINGLE SIDEBAND

14-3-1. Draw a block diagram of an sssc transmitter (filter type) with a 20-kilohertz oscillator and an emission frequency in the range of 6 MHz. Explain the function of each stage.—Such a block diagram is shown in Fig. 14-9. Single sideband or single-sideband suppressed-carrier (sssc) transmission systems use an am modulation process. However, in the modulator and/or following phasing or filtering system, both the carrier and one sideband are removed. Only a single sideband is transmitted.

The process starts with a carrier oscillator; in the example, this oscillator operates at 20 kHz. Its output is applied to a balanced modulator, along with the audio signal. Let us assume that the audio bandwidth extends no higher than 3 kHz. The function of the balanced modulator is to amplitude modulate an rf wave at the same time that the carrier frequency itself is removed. The output of the modulator then consists of two sideband spectrums, one on each side of the carrier frequency. Both sidebands are present, but the carrier has been removed.

The next step is to remove one sideband. This is done with a so-called sideband filter. It offers maximum rejection of the carrier frequency and one of the sidebands. However, the other sideband frequency spectrum is passed to a follow-up mixer. In this manner, the single-sideband signal has been formed. In the example this signal contains those side frequencies present between 20.3 and 23 kHz. The low-frequency sideband and the carrier are no longer present.

The next step is to convert the sideband signal to the transmit frequency and provide the necessary power amplification. The frequency step-up occurs in one or several mixers. Finally, there is a channel oscillator or transmit frequency generator which sets the transmit carrier frequency. In the mixer, this component is combined with the single sideband signal. In the mixing process the sideband frequencies are raised to the 6-MHz level, with the sum frequency appearing in the output. The mixer, too, is a balanced arrangement, so that the carrier frequency is removed. No 6-MHz component is transmitted. The single-sideband frequency is then increased in level with one or more linear amplifiers to the final transmitting power. Linear amplifiers are used so as not to distort the single-sideband modulation envelope applied to the input. Conventional class-C amplifiers cannot be used.

14-3-2. Explain the principles involved in a single-sideband suppressed-carrier (sssc) emission. How does the bandwidth of emission and required power compare with that of full carrier and sidebands?—A single sideband is formed using an amplitude-modulation process. However, in the actual modulation process the carrier is removed. Furthermore, follow-up circuits also remove one of the sidebands, and only a single sideband remains for transmission. The required bandwidth is therefore slightly less than one-half of that needed to transmit the conventional full carrier and two sidebands.

A single-sideband system conserves a considerable amount of power. With 100% modulation, a standard am modulated transmitter, with a rated carrier power of 1000 watts, radiates 250 watts per sideband. The carrier power is 1000 watts and the total sideband power is 500 watts. Hence, the average transmitter-power rating is 1500 watts. To generate a 250-watt single-sideband signal, the power of the transmitter would only have to be 250 watts, instead of the much higher power requirements of a transmitter which transmits full carrier and sidebands with each sideband having a power level of 250 watts.

Comparisons are often made in terms of signal-to-noise ratio at the receiver. It can be proved that a single-sideband transmitter rated at 500 watts delivers a signal of the same ratio as a standard am transmitter with a 1000-watt carrier output (1500-watt total rating).

14-3-4. Explain briefly how an sssc emission is detected.—The single-sideband receiver is usually a superheterodyne type. Quite often double conversion is used, and the low-frequency if amplifier also includes a sideband filter. The filter has a narrow response and passes only the desired sideband to the detector. Furthermore, it offers maximum rejection of the carrier frequency and the unwanted sideband. Therefore, any interference components in this frequency range are prevented from reaching the detector.

The single-sideband receiver must have a stable detection system. A so-called product detector is most often used. It is again a mixing process which recovers the original audio variation. The oscillator associated with the product detector (Fig. 14-9) generates a very stable carrier-frequency component. In the example it is a 20-kHz oscillator. In the product detector, this 20-kHz component mixes with the incoming sideband signal. The output circuit of the product detector responds to the difference frequency. In this case the difference frequency is the original audio variation. Thus, the original audio is recovered (detected) using a mixing process.

14-4. RECEIVERS

14-4-1. Draw a block diagram of a single conversion am superhetrodyne receiver. Assume an incoming signal, and explain

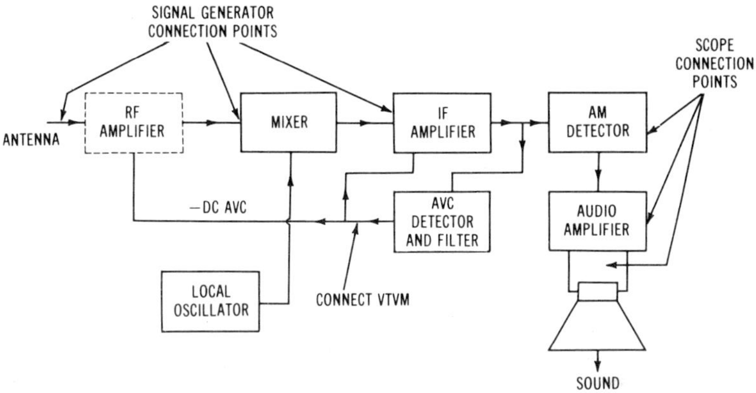

Fig. 14-10. Block diagrams of an am receiver showing test points.

briefly what occurs in each stage. Why are the various stages shielded from each other?—Refer to Fig. 14-10. The incoming signal picked up by the antenna system is first increased in level by a radio-frequency amplifier. (Those receivers that do not use a radio-frequency amplifier apply the incoming signal directly to the mixer.) The function of the mixer is to beat the incoming signal down to the if frequency range. A local oscillator generates the beating signal used in this process. In the example the local oscillator operates on a frequency higher than that of the incoming signal. For a 1000-kHz incoming signal, the local oscillator signal must be at 1455 kHz to produce an if output of 455 kHz. In the mixing arrangement the difference frequency is emphasized in the output, while the sum and two fundamental frequencies are rejected. In receiving a signal on a different frequency, say 1200 kHz, the local oscillator operates on 1655 kHz. Again a difference frequency signal of 455 kHz appears at the mixer output.

The if amplifier is fixed-tuned to 455 kHz, and it builds up the magnitude of the if signal taken off at the output of the mixer. Finally, the signal is increased to a level which permits linear operation of an am detector. Usually this am detector is a diode, rectifying either the positive or the negative alternation of the incoming modulation envelope. A pulsating, unidirectional diode current results. The detector also includes an output-filter arrangement. This part of the detector smooths out the pulsations and forms a replica of the original audio variation. The recovered audio signal is now increased in amplitude to the level necessary for driving a speaker. This is the task of the audio amplifier.

Shielding is appropriate because the stages and systems operate at high gain. Shielding prevents circuit instability and self-oscillation. Also undesired mixing activities among circuits are avoided.

14-4-2. Explain the relation between the signal frequency, the oscillator frequency, and the image frequency in a superheterodyne receiver.—The signal frequency is that of the desired incoming signal. It is mixed with the local-oscillator frequency to produce an if frequency component. An image frequency is an undesired signal frequency that is positioned on the other frequency side of the local-oscillator frequency. In the example in Question 14-4-1 and Fig. 14-10 it would be the high-frequency side of the local-oscillator frequency. Its position is exactly 455 kHz above the local oscillator frequency, or 1910 kHz. If the receiver picks up a 1910-kHz signal and permits it to pass the mixer, there will also be a difference frequency at 455 kHz (1910 − 1455) produced in the output. This signal would now interfere with the desired signal.

The frequency of 1910 kHz is called the *image frequency*. Receivers are designed to offer maximum rejection over the image-frequency spectrum so as to minimize the possibility of interference in the reception of stations operating over the desired frequency range.

14-4-3. Draw a circuit diagram of an am second detector and af amplifier (in one envelope) showing avc circuitry. Also show coupling to, and identification of, all adjacent stages. (a) Explain the principles of operation. (b) State some conditions under which reading of avc voltage would be helpful in troubleshooting a receiver. (c) Show how this circuit would be modified to give davc.—Refer to Figs. 14-11A and 14-11B. The output of the if amplifier is applied to the am detector. In Fig. 14-11, the plate of the diode is connected to the top of the detector resonant circuit. When the rf cycles in the modulation envelope swing positive, the diode conducts. The peak diode current depends on the peak amplitude of the if cycles. If the if signal is unmodulated, the peak diode current is the same for each cycle. During the negative alternation of the rf cycle, the plate is driven negative, and there is no diode conduction.

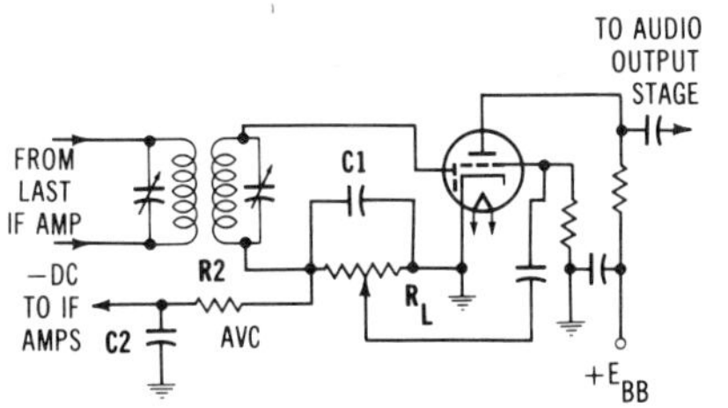

(A) Simple avc circuit.

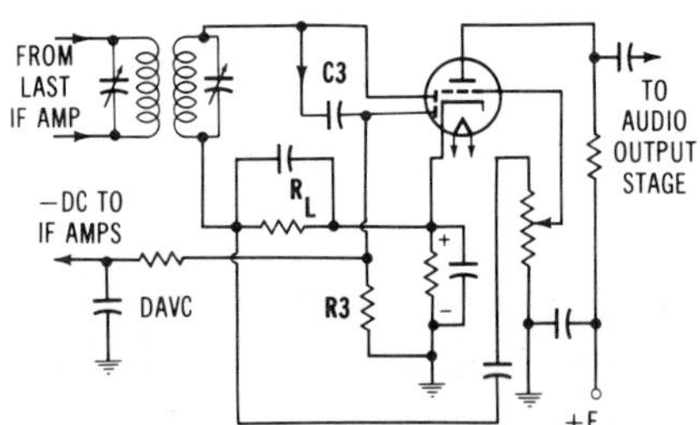

(B) Delayed-avc circuit.

Fig. 14-11. Detector, avc, and audio amplifier circuit.

(a) With modulation, the peaks of the if cycles which form the modulation envelope vary in amplitude according to the modulating wave. As a result, the peak diode current also varies in amplitude. In effect, the diode has been operating as a rectifier, drawing pulses of unidirectional current whose amplitudes vary in accordance with the modulation envelope.

These bursts of current go through the diode load resistor (R_L), and if no capacitor is across the load resistor, the current then develops pulses of voltage. However, capacitor C1 is present across the diode resistor, and it is charged to a peak value corresponding to the instantaneous peak amplitude of the input voltage (modulation envelope). Between pulses of diode current the charge remains on the capacitor, held there by the time constant of the load resistor and capacitor. The time constant is long in comparison to the period of the if cycle. However, each burst of diode current restores the charge on the capacitor, building up a peak value of the input wave at that instant. As a result, the charge across the capacitor varies with the modulation envelope; however, it is smoothed out and is not in bursts. In fact, the time-constant circuit has made a conversion from pulsations of diode current to a changing voltage which corresponds to the original modulating wave. The time constant of the output resistor-capacitor combination ($C1R_L$) must not be too great, or the actual audio variations would be filtered out as well. Thus, the capacitor is actually an rf filter capacitor, filtering out the rf variations, while the envelope changes are unaffected.

The output of the detector is then resistor-capacitor coupled to a triode audio amplifier. A potentiometer is inserted at this point to serve as a volume control. As the arm of the potentiometer is varied toward the left side of the potentiometer, more of the signal from the output of the diode detector is applied to the audio amplifier. The audio amplifier is a voltage amplifier which increases the magnitude of the relatively weak detected signal to a proper level for driving the audio power-output stage. This stage supplies the necessary power for driving the speaker.

Coming off the diode load is still another signal path, through a very high value of resistor (R2) and a follow-up large-value capacitor (C2). The time constant of this combination is very long, and it is capable of filtering even the audio variations at the output of the diode detector. Thus the R2C2 combination is both an rf and an audio filter, and no rf or audio variations appear across avc filter capacitor C2. Instead, a dc voltage which corresponds to the average level of the received signal is present across the capacitor. With no incoming signal there would be no diode current, and, therefore, no charge would be placed on the avc capacitor. The charge on the avc capacitor is a function of signal strength. The stronger the incoming signal is, the higher is the negative voltage developed across the capacitor.

A problem introduced by a strong incoming signal is that it drives one or several of the rf/if stages of the receiver into the distorted high-bias or saturated regions. However, if the negative dc voltage developed across the avc filter capacitor is fed back as an additional bias to certain earlier stages, it is able to reduce the gain of the stage. Therefore, a strong incoming signal is not amplified as much and is not permitted to swing into the distorted regions. By so doing, a receiver can be made much more sensitive to weak incoming signals because the hazard of distortion on strong incoming signals has been reduced.

When a weaker signal is being received, less negative voltage appears across the avc circuit, and the amplifier gain is not reduced as much, permitting the weaker incoming signal to be amplified a greater amount than a stronger incoming signal. The net result is to maintain the strength of the detected output signal more uniform than if no avc were employed.

(b) The avc voltage is a dc voltage which corresponds to the strength of an incoming signal. Inasmuch as it varies with signal level, it is a very useful alignment indicator. As the various stages of a receiver are peaked for best performance, it will be so indicated by a peak voltage reading on the avc line, as measured on a dc vtvm. When no signal is reaching the detector, there is of course no avc voltage.

A signal generator can be moved back through the if and rf amplifiers of the receiver to find a spot at which the signal has dropped out. This will be indicated by no reading, or a very sharp drop in the reading of a vtvm placed across the avc line. As the signal generator is moved from stage to stage from the detector back toward the antenna-output stage, there will be additional amplification, with less input signal needed to produce the same readings on the avc line.

(c) For the very best reception of a very weak signal, it is best that no avc voltage be applied. This can be accomplished by using a delayed automatic volume control (davc) system. In this arrangement the avc circuit does not function until the incoming signal has exceeded a certain level. Thus the amplifier is able to operate wide open with the reception of a weak signal. At the same time a strong signal will put the avc into operation and reduce the amplifier gain.

Delayed avc can be accomplished with the use of a second diode in the detector. This diode is connected so that a negative dc voltage appears between the diode plate and cathode, In Fig. 14-11B this negative voltage is a result of the dc cathode-bias voltage developed across the cathode resistor for the audio-amplifier section of the tube. Thus an incoming rf signal, arriving via capacitor C3, cannot cause the avc diode to conduct until it first overcomes the reverse diode dc bias. If the signal is strong enough to do so, diode current will go through resistor R3, and a negative dc voltage corresponding to the average strength of the incoming strong signal will appear on the avc line.

It is to be noted that the operation of the detector diode itself is not delayed, because diode load resistor R_L is connected directly to the cathode rather than to ground.

14-4-4. Draw a bfo circuit diagram, and explain its use in detection.—A beat-frequency oscillator (bfo) circuit is shown in Fig. 14-12. It is an electron-coupled oscillator (Hartley oscillating section) that operates at the if frequency. Such a circuit is used in the reception of cw (code) signals. In continuous-wave code transmission, the actual rf carrier at the transmitter is interrupted and turned on by the spaces and marks (dots and dashes) of

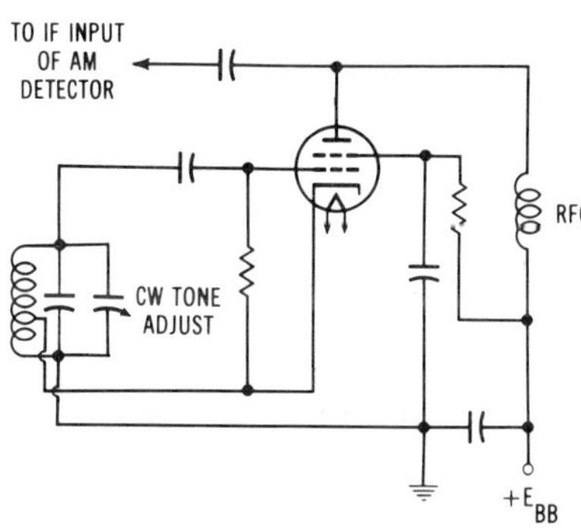

Fig. 14-12. Bfo circuit.

the code message. A dot turns on the carrier for a short interval of time; a dash turns on the carrier for a longer interval. During the spacings between marks, the carrier is turned off. The bfo at the receiver converts the incoming interrupted cw signal to an interrupted audible tone.

In operation the output of the beat-frequency oscillator (bfo) is applied to the same am detector as the incoming interrupted if carrier code signal. The frequency of the bfo signal is very near to the frequency of the incoming if carrier—so near, in fact, that the difference between the two frequencies is an audible tone (400 Hz, 1000 Hz, or whatever the adjusted frequency difference between the two signals is). During the space between marks, there is no incoming if carrier and no signal against which the bfo signal can beat. Therefore, during spaces, there is no output from the detector. During the dots and dashes, however, there is both an incoming carrier and a bfo component which beat together. The difference frequency is developed in the output and is amplified and reproduced as an audible tone by the headset or speaker. Thus the dot and the dash marks can be heard, and the message can be translated from the code to corresponding letters and words. The variable capacitor adjusts the cw tone.

14-4-5. What would be the advantages and disadvantages of utilizing a bandpass switch on a receiver?—A bandpass switch can be used to set the receiver bandwidth in accordance with the emission bandwidth of an incoming signal. A code signal has a rather narrow bandwidth, and therefore a bandpass switch can be set for minimum bandwidth. In so doing, the signal-to-noise and signal-to-interference ratios are reduced because of the much narrower range of frequencies that will be received.

However, if the bandwidth is made too narrow, a modulated signal which occupies an appreciable bandwidth will not be fully detected, and distortion results. A narrow bandwidth is a problem when the receiver is subject to drift; it may be necessary to frequently retune the receiver to be sure that the signal to be received comes through on the narrow bandpass established by the bandpass switch. The bandpass switch is an additional receiver adjustment, and it must be set in accordance with the emission bandwidth to be received. Some additional component parts and receiver complexity are involved, and therefore there is an additional cost factor.

14-4-6. Explain what is meant by the sensitivity of a receiver; selectivity of a receiver. Why are these important? In what units are they usually expressed?—Sensitivity has to do with the ability of a receiver to produce an intelligible output from an incoming weak signal. Thus, sensitivity is important in terms of the capability of a receiver to handle very weak incoming signals. The ultimate sensitivity of a receiver is influenced by noise. Exceptional gains are possible, and very weak signals can be amplified tremendously. However, a useful output can be obtained only when the signal can be made to dominate the noise.

The selectivity of a receiver has to do with the ability of a receiver to cope with interfering signals that are present on nearby frequencies. The narrower the bandwidth of the receiver is, the greater is its selectivity, and the more free it is from interference from signal components on nearby frequencies. However, receiver bandwidth must be adequate to pass the emission band of the incoming signal. Thus, for good selectivity, the response of the receiver should be very sensitive to the required emission bandwidth and then drop off very sharply on each side of the desired bandpass.

When the bandpass receiver is narrow, not only is the selectivity improved, but there can also be an improvement in sensitivity. The noise inherent in the receiver is a function of the receiver bandwidth. Thus, the noise content too is reduced, and there is a direct relationship between selectivity and the ultimate sensitivity for a given receiver design.

Receiver sensitivity is usually expressed in microvolts. If a receiver has a sensitivity of 2 microvolts, it means that this level of input signal will produce a usable, standard test output. For high-frequency operation where noise generated within the receiver is of significance, such noise becomes a limitation on the ultimate sensitivity. At lower frequencies, external noises are usually the limiting factor in just how weak a signal can be recovered.

Selectivity is usually stated in frequency bandwidth. Usually this bandwidth is the frequency difference between the high-frequency point at which the response is 6 dB down from maximum and the low-frequency point at which the response is also 6 dB down.

14-4-7. Explain, step by step, how to align an am receiver, using the following instruments: signal generator and speaker; signal generator and oscilloscope; signal generator and vtvm or dvm. Also explain briefly what is occurring during each step.—A block diagram for illustrating alignment in shown in Fig. 14-10. The speaker, oscilloscope, and vtvm or dvm are all forms of output indicators that are useful in the alignment of a receiver. The speaker is an aural indicator; one makes the alignment by listening to the tone emitted by the speakers. When this type of indicator is used, the signal applied from the signal generator must be modulated (usually by a 400-Hz tone) so that the output of the speaker can be heard. When an oscilloscope is used as an alignment indicator, it is often attached across the voice coil of the speaker, although it can be connected at various other points in the receiver between the speaker and the output of the detector. The oscilloscope provides a visual display of the 400-Hz tone which is being used to modulate the signal-generator output.

The vtvm or dvm is also a visual type of indicator, and it is connected across the avc line. When a vtvm or dvm is used as an alignment indicator, the signal-generator output need not be modulated. Recall that the avc line carries a dc voltage which corresponds to the average level of the applied rf signal. Many published alignment procedures for commercial models are based on the use of a vtvm as an alignment indicator. Often specific levels of avc voltage for suggested values of signal-generator input voltage are given in the alignment information.

The alignment procedure progresses from the diode detector toward the antenna input of the receiver. Usually the detector transformer is peaked first by applying an if frequency from the signal generator to the input of the last if stage (or the only if stage in some small am receivers). The secondary and then the primary resonant circuit is tuned for a maximum output. Such a maximum output is indicated by the loudest tone heard via the speaker, a maximum deflection of the modulating waveform on the oscilloscope screen, and a maximum reading of the dc voltage on the avc line by the vtvm or dvm. In many alignment procedures a suggested level of input signal is given, and/or the signal level is adjusted for a specific reading on the avc line. Inasmuch as alignment controls often interact, it may be necessary to jockey back and forth between secondary tuning and primary tuning in locating the optimum settings.

The signal generator is now moved to the grid of the next if stage in the direction of the antenna input, or if only a single if stage is employed, to the grid circuit of the mixer. Inasmuch as an additional stage of amplification is now added, it is necessary to reduce the output of the signal generator. It is important not to overload the receiver in performing the alignment, and in most cases it is advisable that alignment be made with a relatively weak input signal. The input signal should be comparable to the signal values that will be present in the reception of stations with medium-to-weak signal strengths.

The signal generator is now moved to the input of the mixer, and the if transformer between the mixer and the input of the first if amplifier is peaked. In the preceding steps we have applied a signal that corresponds to the if frequency of the receiver. The individual if transformers and their associated resonant circuits have been aligned for resonance, as indicated by maximum readings on the indicator.

The next general procedure for the alignment of the receiver is to tune the mixer-input circuit, local oscillator resonant circuit, and, if included, the resonant circuits of an rf amplifier. In this alignment procedure the signal generator is usually connected to the antenna input. Depending on the recommended alignment procedure, its output should be set to a specific level, or it should be adjusted until there is a particular voltage reading as measured on the avc line.

The next step is to align the antenna input and rf circuits. Usually this is done at the high-frequency end of the received band. If the antenna is supplied directly to the mixer, it is only necessary to peak the antenna-input resonant circuit. If the receiver uses an rf amplifier, both the antenna input and the resonant circuit between the output of the rf amplifier and the input of the mixer must be peaked.

Most often the mixer-oscillator is aligned with the use of a signal set to some frequency at the high-frequency end of the band. In fact, if at all possible, the generator should be set precisely to some particular frequency. When this frequency is tuned in on the receiver, the reading on the receiver dial should match the frequency to which the signal generator has been set. If this is not so, the dial of the receiver should be set to the exact frequency of the generator. Then the trimmer capacitor which is in the local-oscillator circuit should be adjusted until the incoming generator signal peaks at this setting of the receiver. This step has permitted you to calibrate the receiver dial at the high-frequency end of its range.

In some receivers this adjustment need be made only at the high-frequency end. However, in many receivers it is also necessary to make similar adjustment at the low-frequency end. The signal generator is set to some precise frequency at the low-frequency end. When the receiver is tuned to the low-frequency end, it should be able to pick up this signal. If the alignment of the local oscillator is correct, the dial reading will match the frequency of the signal generator. If it does not, reset the receiver until this dial reading is obtained. Now adjust the padder of the receiver local oscillator until maximum output is received at the frequency of the applied signal. In this step you have calibrated the low-frequency end of the band.

Now you must go back to the high-frequency end and reset the calibration. It may be necessary to jockey back and forth several times until the receiver "tracks" over the entire band. By "tracking" we mean that the local-oscillator frequency changes in a manner that provides an exact difference if frequency as the receiver dial is varied from one end of the band to the other.

In the two-way radio services, operations are on one or several fixed frequencies and the rf stages, mixers, and local oscillators are peaked for reception of the desired channel(s).

14-4-8. What is an image response in a receiver? How would a double-conversion receiver help?—The image response of a receiver is its sensitivity to a signal that reaches its input on the image frequency. An image frequency is an undesired signal frequency that is positioned on the opposite-frequency side of the local-oscillator frequency as compared to the frequency of the desired signal. Refer to Question 14-4-2. The desired signal frequency and the image frequency are separated by twice the intermediate frequency. In the case of a 455-kHz if frequency this separation would only be 910 kHz. In a double-conversion receiver the first if frequency is at a high frequency. If this high if frequency were 5 MHz, it would mean that the separation between the desired signal frequency and the image frequency would be 10 MHz. With this great frequency separation it would mean that the image response of the receiver would be very much less than if the signal and image frequencies were closer together.

CHAPTER 15

Frequency Modulation

15-1. MODULATION AND MODULATORS

15-1-1. Draw a schematic diagram of a frequency-modulated oscillator using a reactance-tube modulator. Explain its principle of operation.—Such a circuit is shown in Fig. 15-1. *A reactance tube can be made to operate as either a variable inductance or a variable capacitance across the frequency-controlling resonant circuit of an oscillator.* In this example the reactance tube is operating as a variable inductor. As the inductance of the reactance tube varies with the modulating wave, so does the frequency of the oscillator.

In operation, the reactance tube causes an rf current (I_x) in the oscillator tank circuit. This current lags the oscillator rf voltage (E_p) across the tank circuit by 90°. It introduces an inductive current component which is in phase with the normal inductive current (I_L) and out of phase with the normal capacitive current (I_c) circulating in the tank circuit. When this inductive current is made to vary in magnitude, the effect is the same as changing the reactance of the inductive leg of the tank circuit and, therefore, its frequency.

The lagging current of the reactance tube is formed by applying a part (E_g) of the tank voltage of the oscillator to the control grid of the reactance tube, through a phase-shifting network, R1C1. This network provides the phase shift that causes the rf plate current of the reactance tube to lag the rf voltage across the oscillator tank.

The next step in understanding the operation of a reactance oscillator is to understand how the amplitude of the rf current component is made to vary. The amplitude of the rf current component at the output of the reactance tube is a function of the mutual conductance of the reactance tube $\left(g_m = \dfrac{\Delta I_p}{\Delta E_g}\right)$. The g_m can be varied by changing the reactance tube control-grid bias. Consequently, the amplitude of the rf component (I_x) in the plate circuit of the reactance tube varies in amplitude in accordance with the modulating wave.

Any change in the amplitude of the rf current that is introduced into the tank circuit by the reactance tube causes a change in the inductive current (I_L) in the oscillator tank circuit. This is the same as changing the inductance of the resonant circuit, and thus the frequency of the oscillator. The oscillator frequency is made to vary (deviate) up and down with respect to the oscillating frequency which is present when no modulation is applied. This latter frequency is called the *center* or *resting frequency* of the frequency modulated oscillator.

The higher that the amplitude of the modulating wave is at the input of a re-

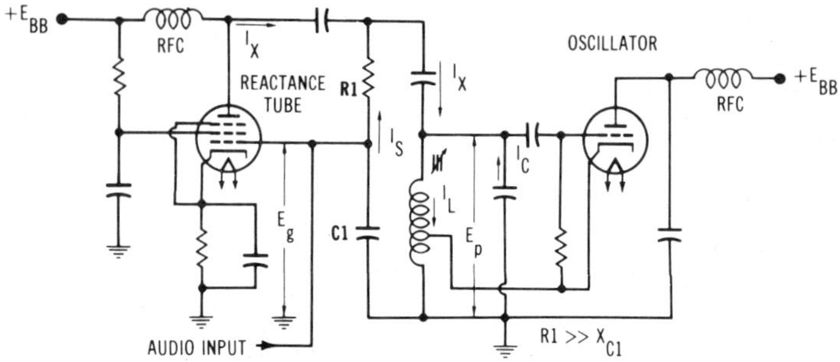

Fig. 15-1. Reactance-tube modulator.

actance tube, the greater is the change in the rf current (I_X) supplied to the tank circuit by the reactance tube—therefore, the greater is the frequency change (deviation) of the oscillator frequency. Thus, a strong voice component will cause a greater frequency deviation than a weaker one.

The higher the frequency of the modulating wave is, the faster is the rate of rf current change at the output of the reactance tube. Therefore there will be a faster change in the magnitude of the rf current introduced into the tank circuit, and the rate of change of frequency of the oscillator will be higher.

15-1-2. Discuss the following in reference to frequency modulation: (a) the production of sidebands, (b) the relationship between the number of sidebands and the modulating frequency, (c) the relationship between the number of sidebands and the amplitude of the modulating voltage, (d) the relationship between percent modulation and the number of sidebands, (e) the relationship between modulation index, or deviation ratio, and the number of sidebands, (f) the relationship between the spacing of the sidebands and the modulating frequency, (g) the relationship between the number of sidebands and the bandwidth of emission, (h) the criteria for determining the bandwidth of emission, (i) reasons for pre-emphasis.—(a) In an am modulation system only one pair of sidebands is generated. To effect the frequency deviation of a resultant fm modulated rf wave, it is necessary to generate more than one pair of sidebands. The actual number of sideband pairs that make up the resultant fm wave depends on a factor called the *modulation index*. The modulation index stated as an equation is

$$\text{Mod. Index} = \frac{\text{Deviation in Frequency}}{\text{Modulating Frequency}}$$

The higher the modulation index, the more generated sidebands there are.

Just as the rf modulation envelope of an am wave is the resultant of a carrier and a single pair of sidebands, the resultant rf fm wave is the combined carrier and multiple sidebands.

(b) The number of sidebands is influenced by the modulating frequency. The higher the modulating frequency is for a given deviation (amplitude of modulating voltage), the fewer are the number of sidebands generated, but the greater is their spacing.

(c) The higher the amplitude of a modulating voltage of a given frequency the greater is the deviation of the fm wave and the greater the number of sidebands.

(d) In an fm system the deviation is given in terms of percent modulation, with the maximum permissible deviation being represented as 100% modulation. Therefore, the higher the percent modulation (greater frequency deviation), the greater are the number of generated sidebands.

(e) As covered in (a), the higher the modulation index, the greater are the number of sidebands.

(f) The higher the modulating frequency, the greater is the spacing among the generated sidebands. For a 1000-Hz modulating wave, there would be a sideband one-thousand hertz on each side of the center frequency ($f_c \pm 1000$ Hz). Other sideband pairs are spaced in harmonic relation, that is, if the modulating frequency were 1000 Hz, the second sideband pair would be located

± 2000 Hz from the center frequency, while the third sideband pair would be located ± 3000 Hz from the center frequency. It is apparent that the higher the modulating frequency, the greater is the frequency separation that exists among the generated sideband pairs.

(g) The greater the number of sideband pairs, the wider is the bandwidth of emission. For example, if the modulating wave is 3000 Hz and three sideband pairs are transmitted, the total bandwidth is 18 kHz (± 9000 Hz). If the deviation were such that four sideband pairs were generated, the bandwidth of emission would be 24 kHz (± 12,000 Hz).

(h) The criteria for the bandwidth of emission are the maximum permissible deviation and the highest-frequency modulating wave that is to be conveyed. This stems from the fact that the greater the deviation, the greater is the number of generated sidebands, and the higher the frequency of the modulating wave, the greater is the frequency separation among sidebands. Thus, a station which is assigned a maximum-permissible deviation of ± 15 kHz (100% modulation) will occupy a greater span of frequencies (channel width) than a station which is assigned a maximum-permissible deviation of ± 5 kHz (100% modulation). A station which is permitted to transmit a maximum-modulation frequency of 15,000 Hz will occupy a greater span of frequencies than one which is assigned maximum audio frequency of 3000 Hz per a given frequency deviation.

The above criteria are often expressed in the form of a *deviation ratio:*

$$\text{Deviation Ratio} = \frac{\text{Maximum Permissible Deviation}}{\text{Highest Modulating Frequency}}$$

(i) *A frequency-modulation system is less susceptible noise.* However, its noise-rejection characteristics are more favorable at low modulating frequencies than they are at high modulating frequencies. For this reason, a system of pre-emphasis is used to obtain better overall signal-to-noise ratio. This is accomplished by permitting the higher-frequency components (per a given initial amplitude of modulating wave) to cause a greater deviation of the fm wave as compared to a lower-frequency component of the same magnitude. Before the audio is reproduced at the receiver, it is passed through a de-emphasis network which restores the frequency components of the modulating signal to their initial magnitude as existed prior to pre-emphasis.

15-1-3. Discuss wideband and narrowband reception in fm voice-communication systems with respect to deviation and bandwidth.—*The greater the frequency deviation of an fm wave per a given audio bandwidth, the better are the interference and noise-rejection characteristics. However, the resultant bandwidth of emission is also greater.*

Bandwidth is of concern when so many radio services require a frequency spectrum. However, the benefits of fm transmission can be made available by reducing the maximum permissible deviation and, at the same time, limiting the maximum frequency of the modulating wave. This technique is particularly adaptable to voice communications, in which case the intelligibility is not affected adversely by limiting the highest audio frequency to 2500 to 3000 Hz. Thus, with a maximum permissible deviation of ± 5 kHz, and in some cases ± 15 kHz, the desirable characteristics of frequency modulation can still be obtained.

Such a technique would not be applicable to fm broadcasting where the required highest modulating frequency is 15,000 Hz and higher. Under this circumstance a much greater deviation is appropriate if the good characteristics of fm transmission (dynamic range and frequency response) are to be obtained.

In narrowband fm transmission, the receiver need only have a narrowband response, and a very favorable sensitivity and selectivity can be obtained.

15-1-4. How is good stability of a reactance modulator achieved?—In assigning fm channels to radio stations, the FCC requires that the carrier or center frequency of the fm wave be maintained within a close tolerance. When a reactance modulator is used, it is customary to employ a self-excited oscillator that is frequency modulated. Such an oscillator is subject to frequency drift, and any drift in the operating characteristics of the reactance tube can also produce an undesired center-frequency shift of the oscillator. A feedback afc system or phase-lock loop (Fig. 15-2) is used to maintain both the center-frequency stability of the oscillator and the operating stability of the reactance modulator. The output of the frequency-modulated oscillator is supplied to a divider chain and a follow-up phase detector. A second signal for the

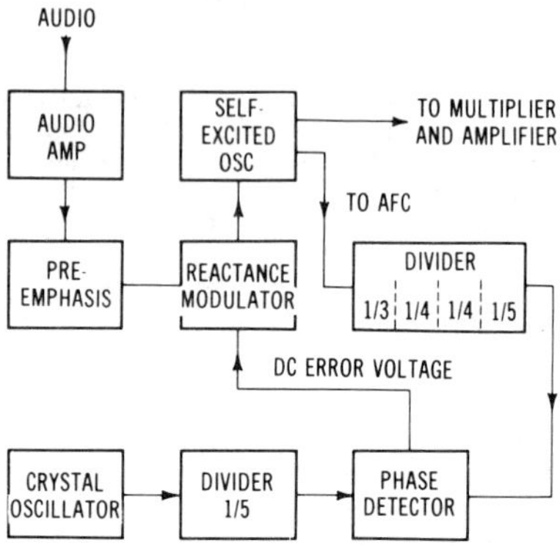

Fig. 15-2. Afc system used to stabilize the reactance-tube modulator.

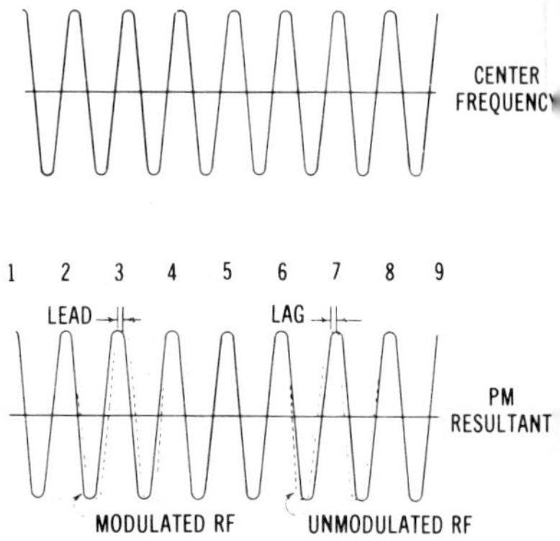

Fig. 15-3. The phase-modulated wave.

phase detector is derived from a highly stable crystal oscillator through a second divider chain. The two frequency components are compared in the phase detector. If there has been any drift in the center-frequency output of the frequency modulated oscillator, a dc error voltage is developed, and it is applied as a controlling dc bias to the reactance circuit. This will change the center-frequency reactive component placed by the reactance tube across the tank circuit of the oscillator, which will make a suitable correction in its center frequency.

15-1-5. Explain, briefly, what occurs in a waveform if it is phase modulated.—Phase modulation is an indirect means of generating an fm wave. In the process, the phase of the center frequency is varied and made to lead and lag its initial phase in accordance with the modulating wave. The frequency of the resultant wave also changes because a phase lead or lag changes the period of the rf wave, which results in a frequency change as well.

The waveform in Fig. 15-3 demonstrates the influence of phase lead or lag on the resultant frequency. The illustration shows a resultant phase-modulated wave. A phase lead is indicated between the second and third peaks. At the third peak the resultant wave is leading that which would be the phase of the unmodulated center frequency (dashed curve). It is to be noted that the period (as represented between peaks) is now shorter than for the unmodulated condition. This means that the frequency of the resultant wave has been increased.

The influence of a phase lag is shown between peaks six and seven. Now the peak of the seventh wave comes late with respect to the peak of an unmodulated condition. Therefore there has been an increase in the wave period and a resultant lower frequency. In the actual modulation process, the shift in phase is made to be gradual, and therefore there is also gradual change in frequency, just as in the case of the direct fm process.

15-1-6. Draw a circuit diagram of a phase modulator. Explain its operation. Label adjacent stages.—Such a modulator is shown in Fig. 15-4. The output voltage (E_p) of the rf oscillator follows two paths—one path is through capacitor C1 to the control grid of the phase modulator as rf voltage E_g, and the other path is directly to the plate circuit of the modulator through capacitor C2, as rf voltage E_1. Two rf components, therefore, appear in the plate circuit of the modulator; one component is E_1, and the other is E_2. The latter has been developed through the operation of the modulator tube with the application of the rf component (E_g) to its control grid. Ordinarily, because of the polarity shift of the tube, the two components would be out of phase. However, the reactances of C1 and C2 have been selected to establish a proper phase relationship between the rf voltage components E_1 and E_2, as shown in the vector diagram. These two components combine vectorially in the plate circuit to produce a resultant output voltage (E_0).

The important thing to recognize in

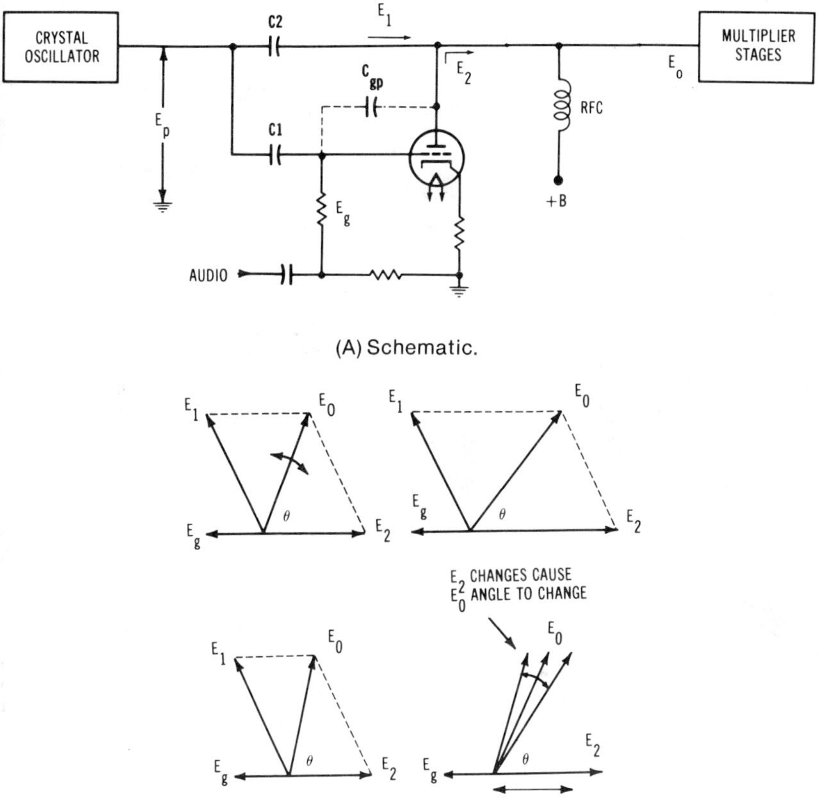

(A) Schematic.

(B) Vectors.

Fig. 15-4. Phase modulator.

the vector diagram is that the *phase angle of E_0* depends on the *magnitude of the E_2 component* that is present in the plate circuit. The magnitude of the E_2 rf component is a function of the grid bias on the phase modulator. As in the case of the reactance-tube modulator, this grid bias varies with the modulating wave. As a result, the amplitude of rf component E_2 also varies with the slow-changing modulating wave. In this activity the phase angle of the phase-modulator output voltage E_0 varies with the modulating wave.

The higher the amplitude of the modulating wave, the greater is the phase-angle deviation. The higher the frequency of the modulating wave, the greater is the rate of phase deviation.

As described in the previous question, the phase-modulation process produces a resultant frequency deviation. The resultant fm wave is then applied to succeeding multipliers and amplifiers. It should be noted from the vectors that the amplitude of the resultant wave (E_0) may also vary. However, these amplitude variations are smoothed out by the following class-C multiplier and amplifier stages.

15-1-7. Explain, in a general way, why an fm deviation meter (modulation meter) would show an indication if coupled to the output of a transmitter which is phase modulated by a constant-amplitude, constant-audio frequency. To what would this deviation be proportional?—As explained in Question 15-1-5,

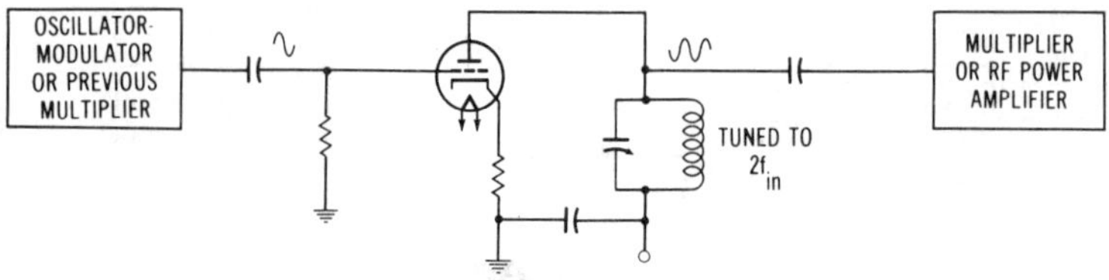

(A) Frequency doubler.

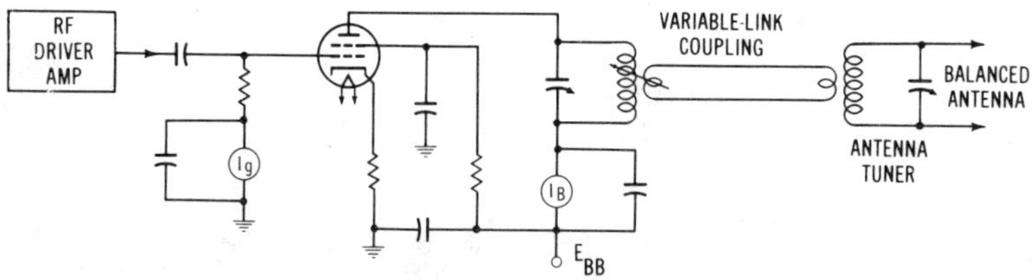

(B) Power amplifier with variable-link coupling.

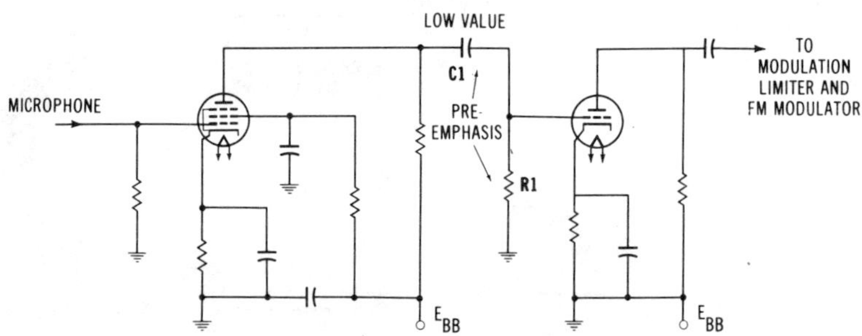

(C) Speech amplifier.

Fig. 15-5. Stages of a phase-modulated fm transmitter.

the phase shift in the resultant waveform produces frequency modulation by an indirect process. Thus, an fm deviation meter connected to the output of a phase-modulated transmitter will indicate the resultant frequency modulation of the wave. The greater the phase change, the greater is the frequency deviation of the resultant wave, and the higher is the meter reading.

It should be mentioned that in a pure phase-modulation process, the frequency deviation is also proportional to the frequency of the modulating wave. This would mean that the higher the modulating frequency of a given amplitude, the greater is the frequency deviation of the resultant wave. However, in an actual fm transmitter using a phase-modulation process, a suitable audio-frequency correction network is used ahead of the phase modulator. This network (often called a *predistorter*) regulates the amplitude of the modulating wave, with regard to frequency, in such a manner that the fm deviation at the output of the phase modulator for a given audio amplitude is the same, regardless of the modulating frequency.

15-1-8. Draw a circuit diagram of each of the following stages of a phase-modulated fm transmitter. Explain their operation. Label the adjacent stages: (a) frequency multiplier (doubler) with capacitive coupling on input and output; (b) power amplifier with variable-link coupling to antenna (include a circuit for

metering grid and plate currents); (c) speech amplifier with an associated pre-emphasis circuit.—A difference in the amplification and multiplication of an fm wave as compared to an am wave should be mentioned first. When an am modulation envelope is supplied to a frequency multiplier or class-C amplifier, it is distorted because the amplifier does not respond in a linear manner to the amplitude variations of the envelope. An fm wave, however, has a constant amplitude; it changes only in frequency. Thus, it is possible to amplify and step up in frequency an fm wave, with a class-C amplifier or multiplier, without introducing distortion.

(a) A frequency doubler with capacitive input and output coupling is shown in Fig. 15-5A. The input circuit is driven by a signal of the fundamental frequency; the output tank circuit is tuned to the second harmonic. The operation of frequency multipliers and doublers was covered in Questions 13-2-9 and 13-2-11 in Chapter 13.

(b) A class-C power amplifier with variable link coupling to an antenna system is shown in Fig. 15-5B. The grid and plate-current meters show the dc component of grid and plate current; the meters are bypassed with rf filter capacitors to keep radio-frequency energy out of the meter movements. The operation of such an amplifier was covered in detail in Question 13-2-1. The individual cycles of constant-amplitude fm draw bursts of plate current which shock-excite the output tank circuit. Since the spacing among the plate-current bursts varies as the result of the fm wave applied to the input circuit, there will be a similar frequency change in the rf waves developed across the output tank circuit. Inasmuch as the peak values of the plate-current bursts are of constant amplitude, the rf wave developed across the plate tank circuit will also be of constant amplitude, but of varying frequency. The resonant circuits of the fm transmitter must, of course, have the appropriate bandwidth (Q) to accommodate the emission band width of the generated fm signal.

The variable-link coupling permits the tank circuit to be loaded properly. By changing the coupling between the link and the tank-circuit inductor, a proper load can be reflected for optimum operation of the stage. The tighter the coupling is, the greater is the reflected load, as will be indicated by an increase in the reading of the dc plate-current meter. The loading is adjusted until the plate-current reading corresponds to optimum operation of the stage.

(c) A typical speech-amplifier system for an fm transmitter is shown in Fig. 15-5C. The pre-emphasis network is a simple series resistor-capacitor combination (R1C1). The reactance of the low-value capacitor decreases with frequency. As a result, with an increase in the audio frequency a higher percentage of the output voltage that is developed in the plate circuit of the first speech amplifier is transferred to the grid of the second stage. A 2000-Hz note as it appears in the plate circuit of the first stage would appear at higher amplitude at the grid of the second stage than a 500-Hz note of the same initial amplitude. Thus the 2000-Hz note would cause a greater frequency deviation of the transmitter, as is required in a pre-emphasis system. The two audio components will be restored to a like amplitude by a de-emphasis network located at the output of the fm detector of the fm receiver.

15-1-9. Do the harmonics of an fm transmission contain intelligible modulation?—Yes, for the same reason that the frequency of an fm wave can be multiplied without encountering distortion (in frequency multiplying we emphasize a harmonic component of a wave), the harmonics from an fm transmitter could also contain intelligible modulation. In the harmonic-generating process, both the center frequency and the deviation are multiplied. For example, a 150-MHz wave with a deviation of ± 5 kHz would have a second harmonic component of 300 MHz, with a ± 10-kHz deviation.

15-1-10. Explain the operation of a Foster Seeley discriminator. A phase-lock loop detector. A ratio detector.—Discussions of the Foster-Seeley and ratio detector are given in Questions 15-3-1 and 15-3-2. A phase-locked loop can be used as an fm detector because of its ability to respond to the frequency deviation of the incoming signal applied to its input (Fig. 15-6). The phase-locked loop has three main sections: phase comparator, low-pass filter, and voltage-controlled oscillator (vco) connected in a closed feedback loop. In operation the incoming signal and the locally generated signal from the vco are compared in the phase detector. The output of the phase detector will be a difference or beat frequency. Since it is low in frequency the output is able to pass through the filter

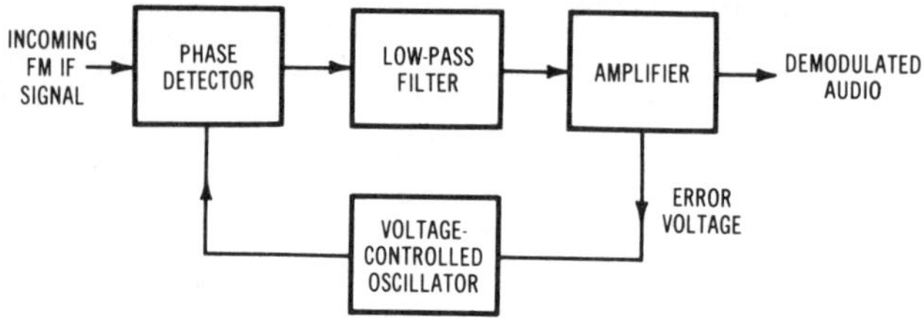

Fig. 15-6. Pll fm detector.

and on to the voltage-controlled oscillator. The passband of the filter is such that the incoming and vco components do not pass. The feedback component in a phase-locked loop detector keeps the frequency of the oscillator on the incoming center frequency. Consequently the error voltage at the output of the phase detector will be a copy of the audio variations that originally frequency-modulated the incoming signal. This low-frequency component can be removed at the output of the error amplifier and applied to the input of a succeeding audio amplifier. It should be noted that no tuned circuits are necessary with this type of a detector. More details on phase-locked loops are given in Section 3-16 in Chapter 3.

15-1-11. What are the characteristics of crystal if filters in receivers? Why would they be preferred over LC circuits?—A crystal is stable in its frequency and in association with the proper components can be used to form a highly selective and stable filter. Thus the desired frequency components are emphasized while off-frequency signals encounter maximum rejection and are not transferred from input to output. The crystal if filter can have an ideally shaped response that passes the desired band of frequencies with minimum attenuation. At the same time its sharp skirts offer maximum rejection to off-frequency signals. This ideal selectivity and off-band rejection are better than those that can be obtained with LC circuits of higher cost. In addition an LC circuit can drift and is influenced by its surroundings. Stray inductances and capacitances can influence the response of an LC circuit. The use of crystal filters is demonstrated in a

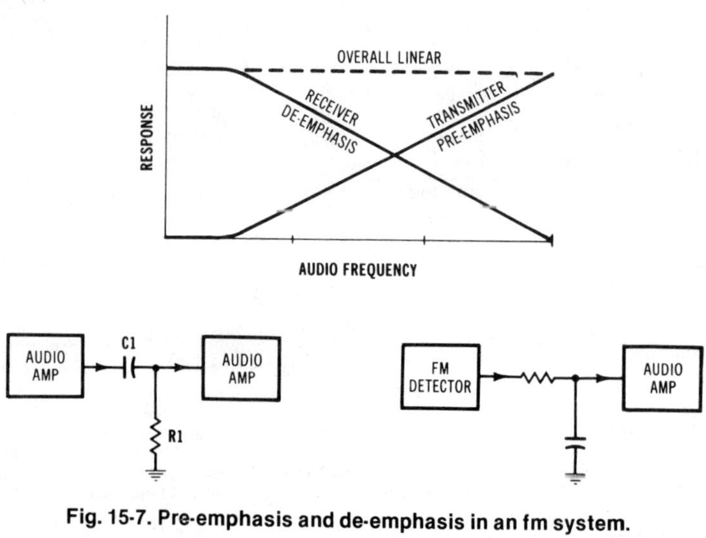

Fig. 15-7. Pre-emphasis and de-emphasis in an fm system.

number of the circuits presented in Chapter 7.

15-1-12. What is pre-emphasis? de-emphasis? How and why are they used?—A major advantage of fm transmission is its ability to reject the amplitude-varying noise and interference components. However, an fm system is not completely noise free because there are fm noise components to contend with. The fact that interference has less effect at low modulating frequencies than at higher frequencies is characteristic of an fm system. An improved signal-to-signal noise ratio is obtained when the fm deviation is greater for a higher-frequency modulating wave as compared with a low-frequency modulating wave that has the same initial amplitude. The application of this principle is called *pre-emphasis*. A simple series resistor-capacitor combination can be used to obtain pre-emphasis. Refer to Fig. 15-7. The pre-emphasis network can be inserted between audio amplifier stages. Since the capacitance of C1 is low, its capacitive reactance is quite high at low- and mid-range audio frequencies, decreasing to a rather low value at high modulating frequencies. Hence for a comparable input to the pre-emphasis network the higher-frequency components will be present at the input of the next stage with approximately the same amplitude as existed at the input to the pre-emphasis network. However, with a decrease in frequency the reactance of the capacitor increases and, as a result, the amplitude of the signal delivered to the next stage is decreased. The response of a typical pre-emphasis combination is shown by the pre-emphasis curve in Fig. 15-7.

Because pre-emphasis is used, the high-frequency components of the audio signal at the receiver output will appear at a grater relative level than the middle- and low-frequency components (as compared to their initial relative magnitudes at the audio input of the transmitter). The audio-frequency components are restored to their initial relative value with the use of a *de-emphasis* network at the output of the receiver detector. The frequency response curve of the de-emphasis network is of opposite slope to that of the pre-emphasis network. That is, the higher frequencies are attenuated more than the lower ones but in exactly the same proportion as the high frequencies were initially pre-emphasized at the transmitter. As a result the overall frequency response is made more linear and, at the same time, the overall signal-to-noise and signal-to-interference ratios have been improved.

15-1-13. How can a varactor be used to generate direct fm in a crystal oscillator?—A varactor can be placed in series or parallel with the crystal. The audio wave is then used to change the capacitance presented by the varactor to the crystal circuit. As a result there will be a change in the crystal-oscillator frequency as the capacitance of the varactor is made to vary.

The frequency of oscillation of a crystal-oscillator circuit can be made to vary over a limited range by changing the capacitance that is placed in parallel or in series with the crystal proper.

A circuit example is given in Fig. 15-8. This is a bipolar crystal oscillator circuit with the varactor diode placed in series with the crystal. The audio signal is applied through a transformer and an appropriate resistor. When an audio signal is applied the capacitance of the varactor varies, changing the effective crystal-circuit capacitance, and causing the frequency of oscillation to follow the audio variations. The varactor is biased suitably from the supply line and through resistor R1.

15-1-14. How are bandwidth, modulating frequency, and the number of significant sidebands related?—Bandwidth increases with an increase in the modulating frequency. Bandwidth also increases with an increase in the number of significant sidebands. The number of significant sidebands increases with the amplitude of the modulating frequency. Refer to Sections 4-1, 4-2, and 4-3. Also refer to Question 15-1-2.

15-1-15. What is the difference between frequency swing and frequency deviation?—Frequency swing is the peak difference between the maximum and minimum values of the instantaneous frequency of the carrier wave during modulation. Frequency deviation is the peak difference between the instantaneous frequency of the modulated wave and the carrier. Refer to Fig. 15-9.

15-1-16. Discuss some methods of modulation of the microwave baseband which allow a number of channels to be transmitted.—At microwave frequencies the actual bandwidth of the channel can be substantially greater than in the lower-frequency ranges. Consequently many

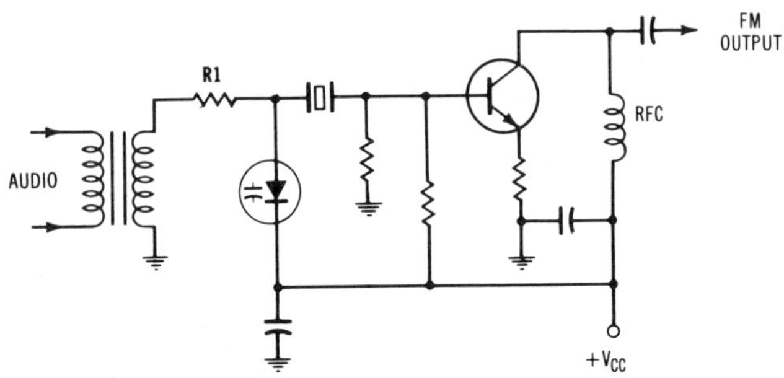

Fig. 15-8. Varactor diode used for direct fm generation.

voice or data channels can be combined in a baseband modulation arrangement to modulate a single microwave signal. A number of ways are used to assemble or multiplex these individual channels in a baseband modulation arrangement in such a manner that the entire group of individual voice or data channels can be used to modulate a single microwave carrier having a suitable bandwidth.

One system is known as *frequency division multiplex* (FDM). Each individual voice channel, frequency or phase, modulates its own low-frequency subcarrier. Groups of these modulated subcarriers are spaced at regular intervals along a span of frequencies known as a *baseband* and applied to a baseband modulator. Spacing is such that there is no interaction among the modulated subcarriers. In the baseband modulator this entire multiplexed group of signals is used to modulate (generally frequency- or phase-modulate) the single microwave carrier. Pilot and alarm signals are also added. The entire signal might occupy a bandwidth of several megahertz or more depending on the number of channels to be conveyed. However, this represents no bandwidth problem in the extremely high-frequency microwave spectra.

At the receiver the entire spectra of subcarrier channels is demodulated from the incoming microwave carrier. Then each subcarrier itself is demultiplexed and then demodulated individually to recover each original voice or data signal. The multiplex sequencing at the transmitter and the demultiplexing process at the receiver are synchronized with the use of a pilot signal. An alarm signal is included to indicate possible troubles in the baseband modulation system.

Another system of baseband modulation is referred to as time division multiplex (TDM). Usually this system is employed in conjunction with a digital pulse code modulation (PCM) process. In a typical system the level of audio frequency variations are divided into 128 relative voltage levels or quanta. Each quantum is represented by a specific 7-digit binary number. An additional digit or eighth bit is added for control and signaling purposes.

The actual sampling for the proper level of the voice-frequency variation is done once each 125 microseconds. This corresponds to a sampling rate of 8000 per second. In a typical system eight bits can be transmitted in a little more than 5 microseconds and, consequently, there is an exceedingly long time during which nothing is being transmitted if only a single voice channel were to be conveyed.

In a bandwidth of 750 kHz to 1 MHz as many as twenty-four simultaneous conversations can be conveyed using the

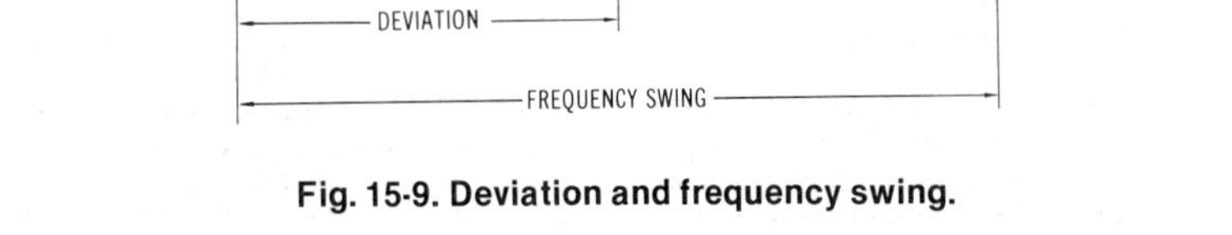

Fig. 15-9. Deviation and frequency swing.

time-division multiplex system. The quantum value for each channel is taken in sequence 1 through 24. In one sweep 192 (24 × 8) bits take a quantum measure of each of the 24 channels. One additional framing control bit is transmitted for each scan. Consequently there are a total of 193 bits for each sequence of twenty-four 8-bit numbers. This is repeated 8000 times per second (known as the frame rate). Hence more than 1.5 million bits are transmitted per second. This is the signal that is assembled by the baseband-modulation system. Order, alarm, and continuity pilot signals are also added in the baseband modulator. The composite baseband signal frequency modulates the carrier of the microwave transmitter.

All of this is decoded at the microwave receiver and individual signals are properly channeled. The individual signals are demultiplexed and reconstructed into the bit sequence that can permit the individual recovery of each of the voice-frequency variations of the 24 channels. Appropriate pulses and clocks keep the multiplex and demultiplex operations in synchronism.

Special compression techniques are often employed to crowd this transmitted data into a narrower bandwidth. For example, in the 2-GHz band the maximum permissible bandwidth is 3.5 MHz.

15-2. FM MAINTENANCE

15-2-1. Assume that you have the following instruments available: ac-dc vtvm, ammeter, heterodyne-frequency meter (0.0002% accuracy), absorption wavemeter and fm modulation meter. Draw and label a block diagram of a voice-modulated (press-to-talk microphone), indirect (phase-modulated) fm transmitter having a crystal multiplication of 12. (a) If the desired output frequency were 155.460 MHz, what would be the proper crystal frequency? (b) Assume transmitter strip completely detuned, ammeter jacks in multiplier control-grid circuits and final amplifier control-grid and cathode circuits. Explain, in detail, step by step, a proper procedure for tuning and aligning all stages except the plate circuit of final power amplifier (pa). (c) Assume a tunable antenna with adjustable coupling to the plate circuit of the final pa. With the ammeter in the cathode circuit of the pa, and with the aid of a tube manual, describe a step-by-step method of obtaining maximum power output, without damage to the tube. (d) If the pa in (c) were a pentode, how would you determine the power input of the stage? (e) In (c), how would you determine if the pa stage is self-oscillating? If so, what adjustment can be made? (f) Assume that the transmitter's assigned frequency is 155.460 MHz with a required tolerance of ±0.0005%. What would be the minimum and maximum frequencies, as read on the frequency meter, which would ensure the transmitter being within tolerance? (g) Assume that the 1-megahertz crystal oscillator of the frequency meter has been calibrated with WWV, and that the meter is tunable to any frequency between each 1-MHz interval over a range of 20 to 40 MHz, with usable harmonics up to 640 MHz. Explain in detail what connections and adjustments would be made to measure the signal directly from the transmitter, also by means of a receiver. (h) In checking the frequency deviation with the modulation meter, would you expect a greater deviation to be caused by whistling or by speaking in a low voice into the microphone? (i) If the transmitter contains a means of limiting, and is overmodulating, what measurements and adjustments could be made to determine and remedy the fault?—A functional block diagram of such an fm transmitter is shown in Fig. 15-10. *A particular advantage of phase modulation (pm) over frequency modulation (fm) is that a crystal-controlled oscillator can be used. This helps in maintaining the center-frequency stability required by the Federal Communications Commission.* If direct frequency modulation of the self-excited oscillator (fm) were used, an elaborate afc system would be required to maintain the carrier (center frequency) stability needed.

(a) The proper crystal frequency would be 12.955 MHz (155.460/12).

(b) The ammeter jacks provide points of insertion for an ammeter to measure the control-grid current in the multipliers and the power amplifier. Inasmuch as the meter is to be inserted in control-grid circuits, it will indicate maximum when there is maximum rf excitation of the stage with which the meter is associated. Hence, as the various resonant circuits are tuned, the resonant setting that permits the maximum transfer of power to the next stage will be indicated by a maximum meter reading in the grid circuit of that stage.

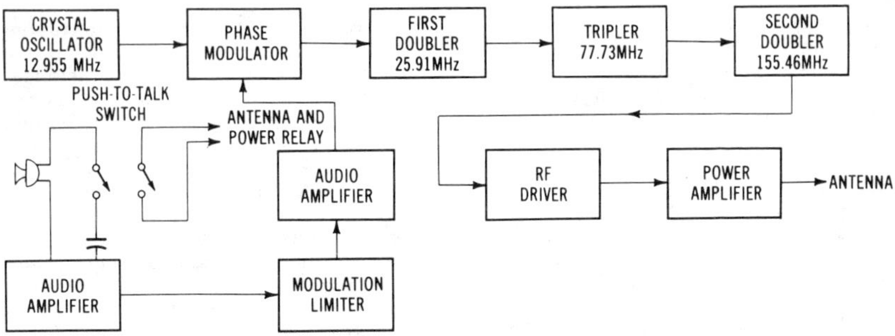

Fig. 15-10. Phase-modulated fm transmitter.

The crystal oscillator of the transmitter usually will not include an output tuning adjustment if it operates at the fundamental frequency. However, if the crystal also serves as a frequency doubler, an output resonant circuit will be present. The alignment usually begins by making certain that the proper crystal is plugged in and the operation is on the correct frequency. A very fine adjustment of the transmitter frequency can be made later. Connect a dummy load.

In the alignment procedure the ammeter is first inserted into the grid-current jack of the first doubler. If there are any peaking adjustments in the crystal oscillator or phase modulator, they are set for a maximum meter reading. The ammeter can now be moved to the jack associated with the grid circuit of the tripler. The first doubler resonant circuits are now adjusted for a maximum reading on the meter. In most designs the range of adjustment of the first-doubler tuning control is such that only one peak is obtained; this peak occurs at twice the frequency of the input. However, it is possible that the range of adjustment of a multiplier-tuning control could extend to the fundamental frequency or, perhaps, even to the third-harmonic frequency. An absorption wavemeter, held near the output of the doubler or the input to the tripler, will identify the proper harmonic. In this case the output of the doubler should be tuned to approximately 26 MHz.

The ammeter can now be moved to the grid-current jack at the input of the second doubler. The tripler resonant circuit is now peaked for a maximum meter reading. Again there is a possibility, in certain designs, that the tripler stage could be tuned to the second- or fourth-harmonic frequency of the input wave. The absorption wavemeter is again used for harmonic identification; the proper harmonic is approximately 78 MHz, not 52 MHz or 104 MHz.

The ammeter is now moved to the grid jack of the rf driver stage. The second doubler circuit can now be resonated for maximum grid-current reading. As before, there will probably be only one adjustment which will provide a maximum reading. However, it is again wise to check to make certain it is not a fundamental or third-harmonic output. In this case, the output frequency should be the transmit frequency of 155.46 MHz.

The ammeter is now transferred to the grid jack of the power-amplifier output stage. The driver resonant circuit and the input resonant circuit of the power amplifier are now peaked for a maximum grid-current reading. In many transmitters it is possible to adjust the power-output level of the rf driver. In this case, the rf drive level is adjusted until a certain grid-current reading is obtained, with the driver and the PA input resonant circuit tuned to resonance. Driver and power-amplifier circuits are tuned to the transmit frequency of 155.46 MHz.

(c) After the rf stage has been aligned, it is necessary to tune the final power amplifier. This involves resonating the plate tank circuit, as well as adjusting the load to transfer maximum power to the antenna system without exceeding the tube ratings. In most cases the supply voltage is known, or it can be measured. The tube manual can then be used to determine the proper cathode current (plate current plus screen-grid current for a multigrid-type tube) for optimum operation in the particular class of service and frequency of operation. Under most circumstances the appropriate operating conditions can be obtained by referring to the maintenance manual for the particular transmitter.

Many higher-powered rf amplifiers require that the initial alignment be done with a low supply voltage and a prescribed amount of loading (usually just enough light loading to prevent the development of excessively high rf voltages in the output tank circuit). The use of a lower than normal supply voltage minimizes the danger of having the tube draw excessive current during the initial tune-up procedure. Also, if the stage must be neutralized, this should be handled before the full supply voltage is applied. Many modern transmitters do not require neutralization.

As per (b), the input resonant circuit, if tunable, is peaked for a maximum grid-current reading. The magnitude of the rf excitation, if adjustable, is set until a particular value of grid-current occurs when the input resonant circuit is peaked. The ammeter can now be moved to the cathode-current jack of the power amplifier. The plate resonant circuit can now be adjusted to resonance by tuning for a maximum dip in the reading of the cathode-current meter. Maximum supply voltage can now be applied and the plate tank circuit retuned very finely for a maximum dip. The loading can now be increased gradually as the plate tank tuning is retouched slightly. Now there will be a higher cathode reading at the dip. Additional loading is now permissible, and the loading is increased until the rated cathode current is drawn when the plate tank circuit is set to resonance.

It is advisable to tune up the power amplifier working into a dummy antenna load to prevent radiating a signal and possible interference. The dummy load can well be an rf power-output meter that will also read the transmitter output as the adjustments are made. The dummy load, of course, should display the same resistance as the antenna system. Hence, when the antenna is connected to the output of the transmitter, only a minor readjustment of loading and plate circuit resonance is needed to establish optimum operating conditions. Check the transmit frequency, modulation and power input to ensure compliance with FCC regulations.

(d) The dc power input to the plate circuit is the product of the plate supply voltage and the plate current. Inasmuch as the cathode-current meter also indicates screen-grid and control-grid currents, the value for the plate current can be obtained by first subtracting the sum of the control-grid and screen-grid currents (obtained by measurement or approximated by reference to the tube manual) from the cathode-current meter reading.

In some cases *the stated dc power input* is based on the sum of screen grid and plate currents times the supply voltage.

(e) If the pa were self-oscillating, it would have an output even though the rf excitation were removed. The absorption wavemeter could be used to check for its presence. If the self-oscillation is on or near the frequency at which the amplifier is to operate, the stage would have to be neutralized to eliminate the problem.

If self-oscillation is occurring on a frequency far removed from the operating frequency, the trouble is parasitic self-oscillation. Parasitics are reduced using the procedure given in Questions 13-2-8 and 13-3-13 in Chapter 13. In most transmitters the parasitics have been eliminated in the design, and it is more likely that the oscillations result from improper neutralization of the stage.

(f) The reading would have to be established within 0.0003% (0.0005 – 0.0002). Thus the minimum and maximum readings that would be acceptable are

$$155.46 \text{ MHx} \pm (0.000003 \times 155.46)$$

or

$$155.45 \text{ MHz} \pm 466\text{Hz}$$

Keeping the frequency meter reading within these two extremes would ensure that the transmitter frequency would be within the required tolerance of 0.0005% using a heterodyne-frequency meter with an accuracy of 0.0002%.

(g) The test blocks (Fig. 15-11) show two typical arrangements for making the signal-frequency measurements. A heterodyne-frequency meter means that heterodyning facilities are available that present either an aural or visual display of a zero beat between two signals, one of known frequency and one of unknown frequency.

In arrangement A, the signal to be measured is picked up from the transmitter and applied to the input of the heterodyne-frequency meter. At the same time, the calibrated signal generator of the heterodyne-frequency meter can be switched to apply a calibrated

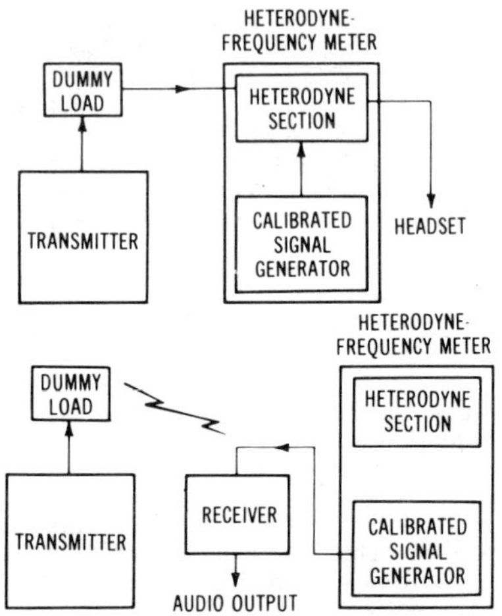

Fig. 15-11. Basic heterodyne-meter test arrangements.

comparison signal to the heterodyning portion of the meter. When the frequency meter is set to an appropriate fundamental or harmonic, there is a beat with the incoming signal, and an aural or visual display will show when the two are set to exactly the same frequency.

In our example, the fourth harmonic of the signal generator can be used in making the frequency check. To make the fourth harmonic available at the transmit frequency to be measured, it is necessary to set the signal generator on 38.865 MHz (155.4660/4).

The 20- to 40-MHz fundamental-frequency range is such that the fifth or sixth harmonic could have been used as well. However, it is customary to use the lowest order of harmonic in terms of output level, convenience, and less trouble with spurious beats.

A receiver can also be used for establishing a beat between the transmitter signal and a signal from the generator portion of the heterodyne-frequency meter. The output of the signal generator is applied to the input of the receiver. The same receiver can also pick up the transmit signal.

There should be proper isolation or separation between the transmitter and receiver so as not to overload the receiver. The receiver can be adjusted to tune in the transmitter signal. Then the heterodyne-frequency meter can be set to 38.865 MHz and a comparison made between the transmitter frequency and the fourth harmonic of the signal generator.

This latter arrangement is useful in checking the frequency of operation of mobile units from a base-station test location or, for that matter, to check the frequency of any incoming signals.

(h) A greater deviation is obtained by whistling rather than by speaking in a low voice into the microphone. In making oscilloscope observations, the use of a whistle permits a check to be made with a type of signal that approaches a sine-like waveform. Speech reproduces as a much more random and irregular waveform.

(i) An fm deviation meter can be used to make the measurement. If the transmitter is overmodulated when you are speaking into the microphone (deviation peaks in excess of the FCC deviation tolerance), there is a possibility of a fault in the speech amplifier-modulator section of the transmitter or, more than likely, the deviation control of the transmitter requires adjustment. This control sets the level of signal that is applied to the modulator as a function of the microphone-input signal level. It is adjusted so that the limiting will occur regardless of how high the input-signal level becomes, and under no circumstance should there be overmodulation. Limiting can be too severe. If it occurs regardless of how high the input, distortion results when you speak into the microphone at a normal voice level.

A combination of a modulation meter and an oscilloscope display are useful in making a modulation adjustment. The modulation meter is helpful in making certain that the maximum deviation is not exceeded, and the oscilloscope display shows the degree of limiting that the signal encounters. Usually the manufacturer of the transmitter suggests specific levels of audio-signal input, and an ac vtvm can be used in setting proper signal levels. Usually a 1000-Hz audio sine wave is used in making modulation adjustments.

15-2-2. What might be the effect on the transmitter frequency if a tripler stage in an otherwise perfectly aligned fm transmitter were slightly detuned?—A slightly detuned tripler stage might have some influence on the power level of the transmitted signal because of reduced drive to the following stages. It would have no effect on frequency, since frequency is determined by the basic carrier generator. However, carrier and side frequencies may be shifted in relative phase, producing distortion.

15-2-3. Under what usual condition of maintenance and/or repair should a transmitter be retuned?—A transmitter should be retuned when measurements indicate that it is not operating efficiently, or not in compliance with FCC technical requirements. Such adjustments can only be made by an FCC-licensed first- or second-class operator. Adjustments which influence frequency and modulation and, in some cases, power input or rf power output should be made only when the test instrumentation is of an accuracy that can measure well within established FCC tolerances. Retuning is often necessary when certain tubes or transistors or other components that are closely associated with resonant circuits and/or the modulation stages are replaced.

15-2-4. If an indirect fm transmitter without modulation is within carrier-frequency tolerance, but with modulation is out of tolerance, what are some of the possible causes?—Two causes of a shift in oscillator frequency are changes in the oscillator supply voltages with modulation and changes in oscillator loading with modulation.

Another possibility, if the oscillator is supplied from a regulated voltage source, is a fault with the regulator circuit, causing a supply voltage change under the changing conditions of modulation level. Of course, a similar change in supply voltage might result from a failure in the decoupling circuits along the supply line. With modulation, a fault in the phase modulator might reflect a change in loading to the oscillator. This might be the result of mistuning, too strong a modulating wave, or a bad modulator tube or semiconductor.

15-2-5. In an fm transmitter what would be the effect on antenna current if the grid bias on the current power amplifier were varied?—The rf power amplifier of an fm transmitter is operated class C. If the power-amplifier stage receives most or all of its class-C grid bias from an external power source, then a change in that grid bias (assuming a constant level of rf excitation) would influence the power output of the transmitter, and there would be a decrease in antenna current with an increase in the external grid bias.

If the power amplifier stage is signal-biased, some self-adjustment takes place because the bias is controlled by the level of the rf excitation to the stage. In general, as the rf excitation falls off, the grid bias declines, and initially there may be only a slight decline in output. However, when the loss of excitation is significant, the grid bias declines, and the stage begins to operate in a less efficient manner. Therefore, there is less power output and the antenna-current meter reading will decrease.

15-3. FM RECEIVERS

15-3-1. Draw a schematic diagram of each of the following stages of a superheterodyne fm receiver. Explain the principles of operation, and label the adjacent stages: (a) mixer with injected oscillator frequency, (b) if amplifier, (c) limiter, (d) discriminator, (e) differential-squelch circuit.—(a) A mixer-oscillator combination is shown in Fig. 15-12. Most fm communication receivers are double-conversion types, and they employ two mixer-oscillator combinations. In the first mixer, the rf signal is supplied to the grid circuit, and the output resonant circuit of the mixer is tuned to the high-if frequency. In the second mixer-oscillator combination, the input frequency to the mixer is the high-if frequency, while the output resonant circuit of the mixer is tuned to the low-if frequency. In the example shown in Fig. 15-12, it is assumed that the high-if frequency is being applied to the input of the second mixer. A triode mixer is used, and the output resonant transformer is tuned to the low-if frequency.

Inasmuch as most fm communications receivers are preset to one or more specific frequencies, the oscillators are crystal-controlled. For the operation of the first mixer, the crystal oscillator is often followed by a multiplier section to build up the frequency of the signal for proper mixing with the incoming rf signal. However, as shown in Fig. 15-8, for a second mixer-oscillator combination, the output of the triode crystal oscillator is supplied via a low-value capacitor (C1) to the control grid of the mixer. This process is referred to as *control-grid local-oscillator injection*. The mixer is biased on an essentially nonlinear portion of its characteristics by R2C2 so as to provide the most effective nonlinear conversion. The local oscillator and incoming signal beat together to develop four major signal components in the output (two fundamental frequencies, plus sum and difference frequencies). However, the

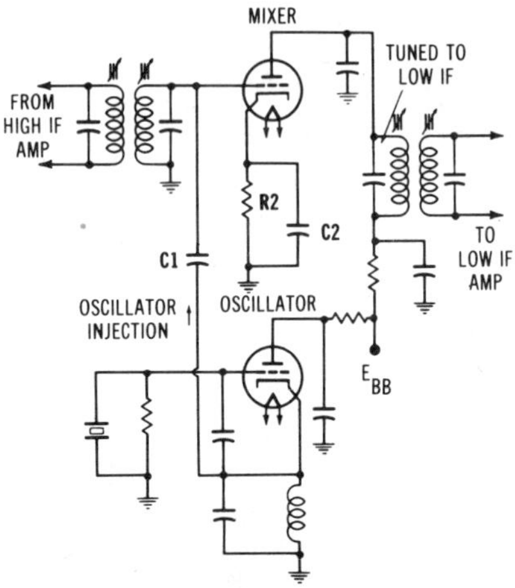

Fig. 15-12. A receiver second mixer/oscillator combination.

output resonant circuit is tuned only to the difference frequency, while the other frequency components are filtered out.

(b) A typical if amplifier, as found in fm communications receivers, is shown in Fig. 15-13. It is in the low if amplifier of the receiver that the selectivity of the receiver is established. Usually the input to the low-if amplifier is a sealed filter with an essentially flat response over the desired bandpass, and a very sharp drop-off (skirts) on the sides of the desired if bandpass. This use of a highly selective input filter with a good and constant selectivity characteristic minimizes off-channel signal interference, and makes the alignment of the receiver less critical and less subject to changes in tube, transistor, and other operating parameters. Selectivity is determined almost entirely by the sealed filter. This filter is usually a grouping of consecutive resonant circuits and appropriate M-derived sloping combinations.

A low-if amplifier, with a wider-band response and with less critical components and alignment requirements, follows the filter. The output if transformer is broadly tuned and not critical. The if amplifier is used for reasons of gain. The level of a very weak signal is built up to the required level for limiter operation (removal of amplitude variations) and proper drive of the fm detector.

(c) The function of the limiter is to remove any amplitude variations that appear on the incoming fm signal as a result of noise or interference. In fact, in most fm communications receivers, the gain and overall sensitivity of the receiver are such that limiting action occurs on even a very weak input signal and, on occasion, with the presence of noise in the absence of the signal.

A two-stage limiter is usually employed, as shown in Fig. 15-13B. The limiter stage is usually signal biased, operates at a relatively low plate voltage, and has low gain. Consequently, there is only a narrow-amplitude range between cutoff and grid limiting, and an incoming signal variation of very low amplitude will swing between cutoff and limiting or saturation. Furthermore, a second limiter stage follows the first and there is a further confining of the signal. Many limiters operate so tightly that regardless of the level of the incoming signal, weak or strong, they will deliver the same signal level at the output of the two-stage limiter. Thus, the follow-up fm detector can be designed for optimum operation on this level of applied signal.

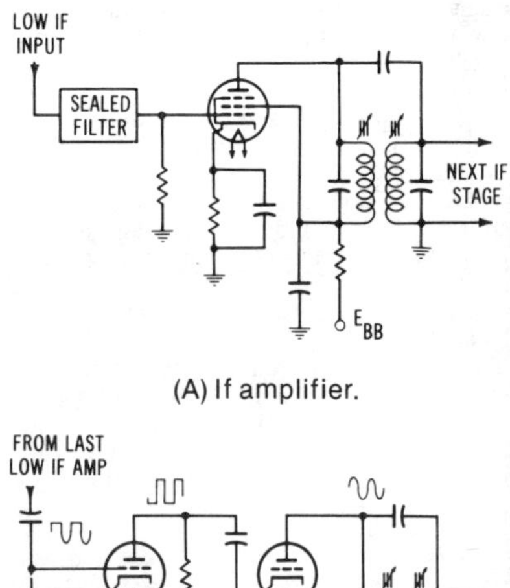

(A) If amplifier.

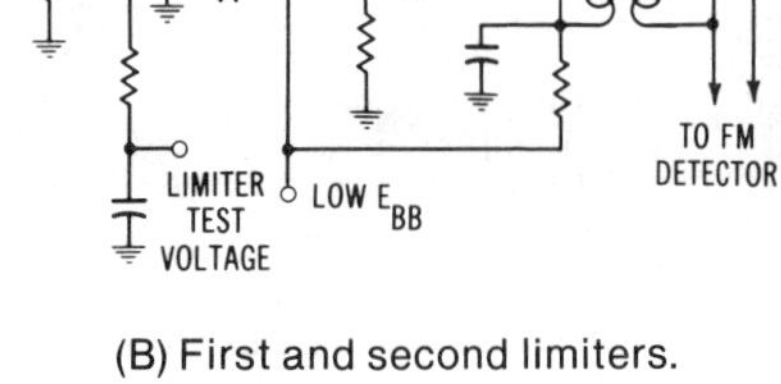

(B) First and second limiters.

Fig. 15-13. If amplifier and limiter circuits.

The waveforms show how the very limited-voltage range between cutoff and limiting causes clipping of the variations on an incoming signal. It might appear that the intelligibility of the modulation would be affected adversely. However, it must be recalled that the desired information is present in an fm signal in the form of a frequency change rather than an amplitude change. Despite the clipping of the waveform, the presence of the limiter resonant circuit restores the fm signal to sine waves of constant amplitude ahead of the discriminator.

(d) A basic discriminator circuit is shown in Fig. 15-14. The operation of a discriminator is based on the phase relationship between two fm rf components that appear at its input. One component in coupled directly by way of capacitor C1 and appears across discriminator inductor L1. Another rf component is coupled via the flux lines that link the primary and secondary resonant circuits. The secondary if voltage, when two coupled resonant circuits are employed, differs in phase 90° from the primary voltage, and therefore is also related by 90° to the if voltage that has been coupled directly across inductor L1. The 90° component is divided into two segments with reference to the secondary tap. One segment appears at the anode of diode D1 and the other appears at the anode of diode D2. In terms of the diode anodes, these two components are 180° apart.

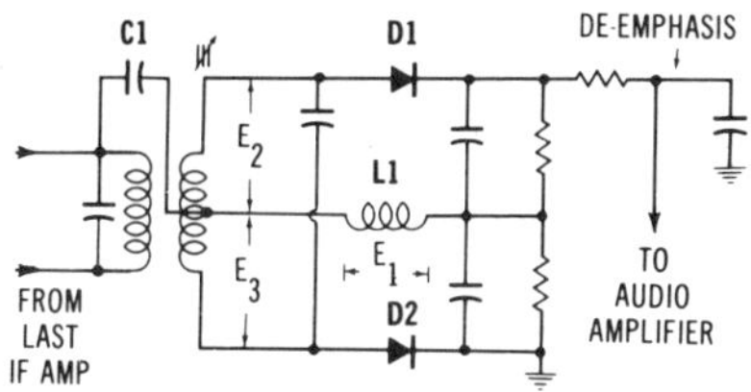

(A) Schematic.

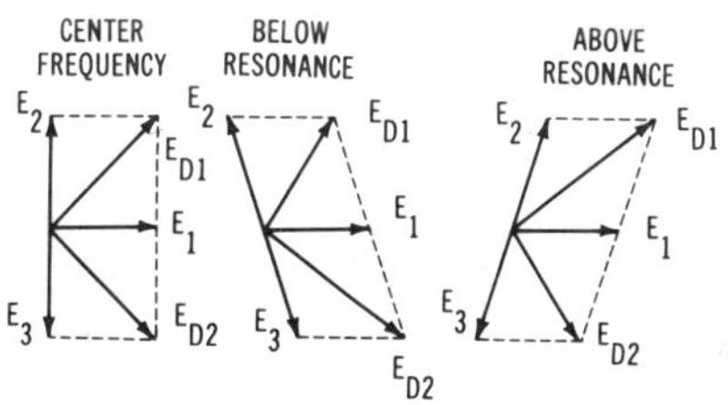

(B) Vectors.

Fig. 15-14. Basic discriminator circuit.

From the discussion, it is apparent that diode D1 is excited by two rf signals $(E_1 + E_2)$, while diode D2 is excited by a signal component composed of $(E_1 + E_3)$. At the frequency to which the discriminator transformer is tuned, these two signal components are of equal amplitude. Therefore, they draw the same peak diode current, and, inasmuch as the diode currents occur in opposite directions in the output circuit, there is no output from the discriminator.

When the if signal deviates, the signal applied to the discriminator transformer does not remain on one frequency (center frequency). Rather, it deviates with the modulation, swinging on each side of the center frequency. An off-resonance frequency applied to the discriminator transformer results in secondary components E_2 and E_3 which lead or lag the direct voltage (E_1) by more or less than 90°. Therefore, when the incoming signal deviates on one side of the center frequency, a higher resultant voltage is applied to the diode D1 than to diode D2. Conversely, when the signal swing is on the other side of the center frequency, diode D2 is driven by a higher net voltage than diode D1. Refer to the vectors in Fig. 15-14. When diode D1 draws the higher current, the net output of the discriminator is positive. Conversely, when the diode D2 draws the most current, the net output of the discriminator is negative.

How much positive or how much negative the output swings depends on just how far the incoming signal swings away from center frequency. Thus, as the incoming fm signal deviates in frequency, the output voltage of the discriminator follows the frequency change, and a conversion is made from frequency change to amplitude change. The greater the deviation of the incoming signal, the greater is the voltage range between maximum positive and maximum negative, as developed across the output.

The faster the rate of deviation of the incoming signal, the faster is the amplitude change at the output of the discriminator. In this manner, the modulating wave that is present as a frequency change of the incoming signal is converted back to the original amplitude variations of the modulating wave.

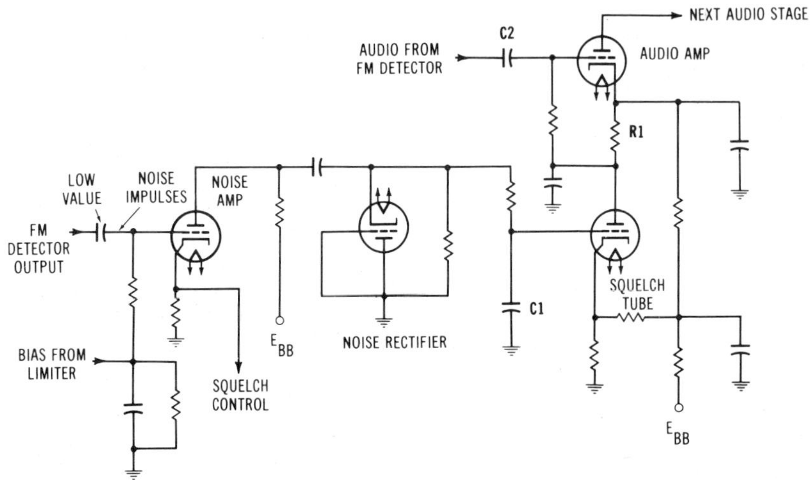

Fig. 5-15. Diagram, symbol, and truth table of the 7486 EXC/OR gate.

(e) A differential-squelch circuit is shown in Fig. 15-15. *In fm reception, an incoming carrier causes limiting action, and the amplitude variations and noise are quieted. However, when there is no incoming carrier, the background noise and interference are amplified and can be heard in the speaker.* This is an annoying condition because in the fm radiocommunication services, most receivers continuously monitor the transmit frequency. *The purpose of a squelch circuit is to quiet this background when there is no incoming signal.*

When there is no incoming signal, the noise output of the discriminator is high, and it is applied to a noise amplifier. The output of the noise amplifier is applied to a rectifier which develops a dc output voltage that appears on capacitor C1. This voltage is of a positive polarity and causes the squelch tube to conduct heavily. The current through resistor R1 in the cathode circuit of the first audio amplifier (positive at cathode side and negative at the grid side) biases off the first audio amplifier. Therefore the noise signal arriving at the input of the first audio amplifier from the discriminator through capacitor C2 is not amplified and passed on to the audio output stage and speaker. Thus the output is said to be "squelched."

When a signal is received, there is a limiting action, and a carrier is present at the discriminator. As a result, the noise output of the discriminator falls off drastically, and no noise component of significance is applied to the noise amplifier. Furthermore, a dc voltage resulting form limiter action biases the noise amplifier to cutoff or very low-gain operation. Therefore no noise component is rectified by the noise rectifier. Consequently, the positive charge on capacitor C1 disappears, and the squelch tube is cut off. This permits the first audio amplifier to operate in normal manner, and the demodulated signal from the discriminator, arriving at the grid of the first audio amplifier via capacitor C2, is amplified and passed on to the audio-output stage. Thus with the reception of a signal the receiver is said to be "unsquelched."

The squelch control must be set critically so that there will be no squelching action on an incoming weak signal. The squelch activity in the circuit of Fig. 15-11 is regulated by setting the bias applied to the cathode of the noise amplifier. Its level can be adjusted until the receiver just squelches in the absence of signal. An incoming weak signal then cuts back the noise-amplifier operation just enough to unsquelch the receiver.

15-3-2. Draw a diagram of a ratio detector, and explain its operation.—A ratio detector is shown in Fig. 15-16. Its operation is very similar to that of a discriminator. The schematic shows that,

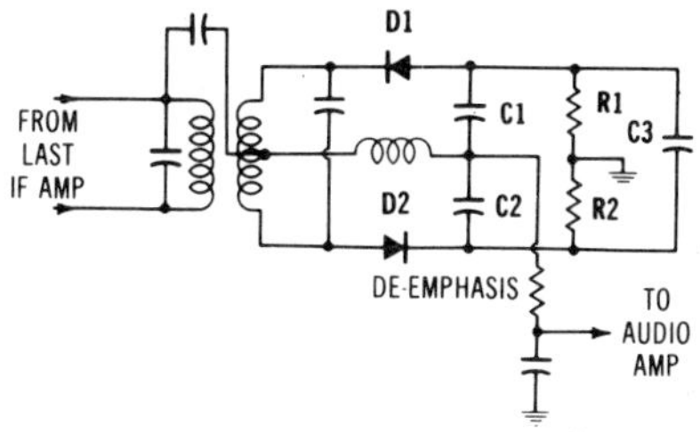

Fig. 15-16. Ratio detector.

except for the polarity of the two detector diodes and the output circuit, the two are quite similar. The ratio detector has an advantage in that it also provides some amplitude limiting.

The input circuit conditions and if voltages E_1, E_2, and E_3 have the same relations with respect to each other as in the conventional discriminator discussed in 15-3-1(d). Thus, the if input voltages influence the diode operation in a similar manner. However, the diode current is not the same. When the frequency (direction of deviation) is such that the D1 conducts more than D2, the charge placed on output capacitor C1 is more negative than that placed on capacitor C2. Oppositely, with a frequency swing to the other side of the center frequency, the negative charge on capacitor C2 is greater than the negative charge on capacitor C1. The output is removed at the junction of the two capacitors, and therefore the output voltage is the difference between the two charges. The output swings either negative or positive, depending on which capacitor has the higher negative charge. When capacitor C1 has the higher negative charge the output swings *positive*, while the output swings *negative* when capacitor C2 has the higher negative charge.

An unusual feature of the ratio detector is that although the output is the differential voltage of capacitors C1 and C2, the sum of the two charges is always a constant. In terms of dc current, there is a continuous motion of electrons down through resistors R1 and R2 connected in series. This current is a function of the strength of the incoming signal. Inasmuch as resistors R1 and R2 and capacitor C3 have a very long time constant, a dc charge which corresponds to the strength of the incoming signal develops on capacitor C3, and across C1 and C2.

This dc voltage is highly stable because of the long time constant. Therefore, if there are any amplitude variations of the average diode current, such as might occur with amplitude variations on the incoming signal they are filtered out and have no influence on the differential potential developed on capacitors C1 and C2. Hence, the differential voltage becomes a function of the frequency change only.

15-3-3. Explain how spurious signals can be received or created in a receiver. How can these be reduced in sets having sealed untunable filters?—In two-way radio-communications, the selectivity of the receiver is a very important factor. Channels are spaced very near to each other, and if the selectivity is poor, there can be interference from stations operating on adjacent channels. In a similar manner interference can be the result of signals generated within the receiver because of the two conversion processes and their associated oscillators and harmonic generators. There is also a multiplicity of mixing components of a spurious nature. These various signal components can combine with each other or with incoming off-frequency signals to produce interference components. Mistuning of sharply tuned if amplifiers can also result in instability and the generation of some undesired spurious signal components.

A strong local off-channel signal can swing one of the rf or if stages into nonlinear operation and lower the sensitivity (desensitize) of the receiver on the desired channel. Mixing components can also fall in the desired channel.

The above problems are greatly minimized with the use of stable and sealed untunable filters, which establish the basic selectivity of the receiver. As mentioned previously in Question 15-3-1, such filters are most often located ahead of the low if amplifier. These filters have a very sharp selectivity, which greatly emphasizes the narrow bandwidth of a desired incoming signal and has a maximum rejection to off-channel frequencies. Since these filters are sealed and isolated in a manner that prevents their operating characteristics from being affected by changes in the associated stages or changes in temperature, the selectivity becomes a stable and uniform response. The if amplifiers are now less critical of both alignment and circuit variables. They can also be of a broader bandwidth and made less likely to generate spurious frequency components.

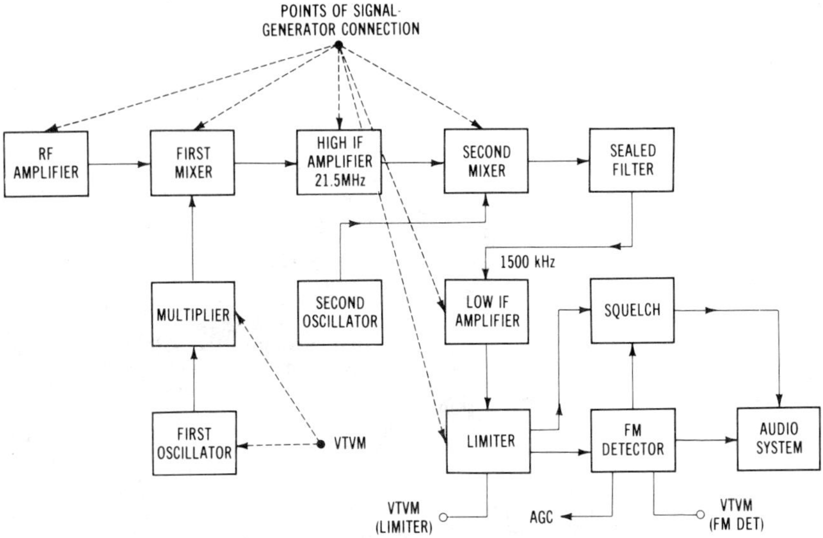

Fig. 15-17. Block presentation of a double-conversion receiver with alignment points indicated.

Of course, proper spacing of frequency relative to those of nearby transmitters can be made in a manner that minimizes adjacent channel operations and the formation of spurious heterodynes.

15-3-4. Describe, step by step, a proper procedure for aligning an fm double-conversion superheterodyne receiver.—Such a receiver is shown in block diagram form in Fig. 15-17. The recommended alignment procedure for most two-way radio fm receivers involves the use of a dc vtvm (or a suitable high-resistance vom or dvm), and an accurate signal generator that can make available the desired if frequencies and several rf spot frequencies.

Usually some limiter reference voltage is recommended, and in the alignment process a signal level should be used that maintains this reference signal level at the limiter. The voltmeter is attached to the limiter test point.

In a typical procedure, the signal generator, set to 1500 kHz, is connected to the input of the low if amplifier (usually at the output side of the untunable if filter which sets the selectivity of the receiver). In the alignment of the discriminator, the vtvm is connected across half of the discriminator output, and the primary is peaked for maximum output. Now the voltmeter is connected across the full output of the discriminator, and the secondary of the discriminator transformer is adjusted for a zero reading. True zero is indicated when there is an increase in output if the secondary is adjusted to either side of the correct zero setting (the swing will be negative on one side of zero and positive on the other).

The voltmeter is now connected to read the dc limiter voltage. As the alignment proceeds, the signal input from the signal generator must be held at a level that corresponds to the suggested limited voltage for alignment. The individual if transformers are now peaked for maximum output, proceeding from the discriminator toward the input of the low-if amplifier.

If the second mixer-oscillator converter unit contains adjustable resonant circuits, the signal generator (set to 1500 kHz) can be applied to the input of the mixer. The output-mixer resonant circuit and the second-oscillator resonant circuit can now be adjusted for maximum dc limiter voltage.

The signal generator can now be moved to the input of the high-if amplifier and set to the high-if frequency of 21.5 MHz (in the example of Fig. 15-17). Each of these included resonant circuits can now be peaked, beginning at the input of the second mixer and working back toward the output of the first mixer. Don't forget to maintain the proper test level of limiter voltage.

Some fm two-way radio receivers use double-tuned transformers with rather critical response characteristics having the required flat-topped desired bandpass and sharp drop-off skirts. Such a transformer must be aligned critically. Quite often the secondary is peaked with the primary loaded by a resistor, and then the primary is peaked with the secondary loaded by a resistor. *The precaution is sometimes necessary to avoid tuning the secondary to a different frequency than the primary and thus introducing a so-called split-tuned characteristic.*

In certain wideband applications split-tuning is an advantage. For example, an overcoupled wideband transformer has two resonant peaks on each side of the center frequency of its response. This is referred to as a split-tuned transformer. However, in many narrowband alignments, tuning of primary and secondary to two separate frequencies (split-tuning) is to be avoided. Too tight coupling can also cause split-tuning.

The signal generator can now be moved to the antenna or rf input of the receiver. Before applying the signal, it is first necessary to check the first local-oscillator operation. If vacuum tubes are used, oscillator operation can be ascertained by connecting the voltmeter to the oscillator grid, and on to the grids of the following multiplier stages. When an oscillator is in operation, it draws grid current, and therefore a negative dc voltage will be developed between grid and ground, or grid and cathode. Likewise, when the following multiplier stages are driven by a signal from the oscillator, there will also be grid-current readings. If the multiplier stages are tunable, they can be adjusted for a maximum reading on the meter when it is connected to the grid circuit of the following stage.

The voltmeter can now be reconnected to the limiter test point. This signal generator can be turned on, and an unmodulated rf signal of proper frequency applied to the receiver input. The level of applied signal should be set so that the reference voltage is read at the limiter circuit. The various rf controls can now be adjusted for a maximum reading. Always maintain the limiter reference voltage as the alignment proceeds. A local-oscillator injection control is often included; this should now be adjusted for a maximum limiter reading.

This concludes the overall alignment of the receiver. Usually the manufacturer will provide a SINAD or sensitivity factor. For example, it may state that with a 2-microvolt signal applied to the input of the receiver, a reading that corresponds to the limiter reference voltage will indicate proper alignment.

15-3-5. Discuss the cause and prevention of interference to radio receivers installed in motor vehicles.— Static and sparking noises are a major source of trouble in mobile installations. There is considerable sparking and interference associated with the spark plugs and ignition system of a vehicle. Generators, alternators, motors, and voltage regulators can present noise problems. Improper bonding, high-resistance rubbing grounds, and friction (rotating wheel noise) are also suspect. Spring-contact arrangements are available for reducing wheel static, and proper bonding and grounding of metal parts can reduce other scraping and motion static.

The principal noise reduction, however, involves the ignition and electrical system of the motor vehicle. The diagram of Fig. 15-18 shows how resistance and filter capacitors can do much in suppressing noise. Resistive spark plugs or resistive leads can be used to suppress spark interference. Filter capacitors can be used to reduce interference from the generator and voltage regulator.

15-3-6. What is receiver desensitization? How can receiver limiter current be used to determine if desensitization is occurring?—Desensitization is the reduction in the receiver sensitivity as the result of the reception of a very strong signal on an adjacent or nearby channel. This can occur at a repeater site where multiple transmitters and receivers are in operation. For example, a receiver is operating on a specific channel and is receiving a signal from one of its mobile transmitters. At the same time a transmitter at the site transmits on a channel other than the one to which the receiver is tuned. However, its signal is so overwhelming that it enters the receiver and operates the limiter and/or the agc system. In so doing the receiver sensitivity drops sharply (desensitization) and it is not able to receive the signal from its associated mobile transmitter. Inasmuch as the interfering transmit signal is so very strong, even though it is not on the

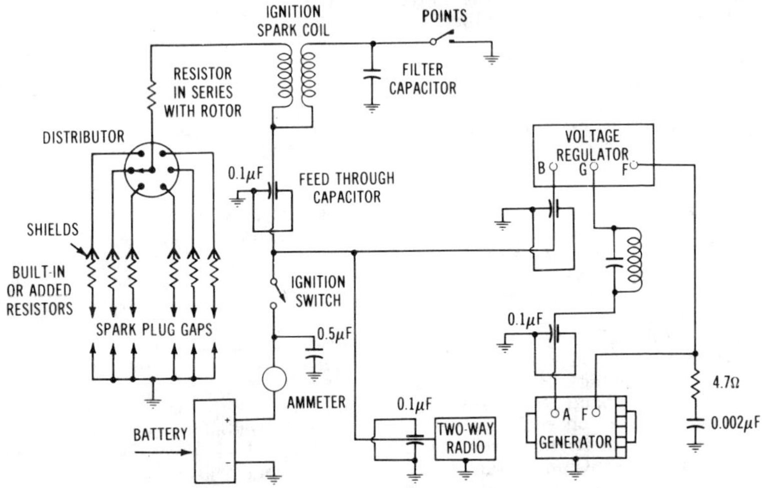

Fig. 15-18. Noise-reduction methods.

same channel, it is able to cause limiter current. Thus the presence of strong limiter current is a good indication that desensitization is occurring. One need only turn on the transmitter that is suspected of the desensitization and note if the limiter current is present when the transmitter is on.

15-3-7. Discuss the merits of using rf cavities as opposed to conventional LC filters for filtering between the transmitter and receiver of a repeater.—Radio-frequency cavities can be made very high Q and have a high order of stability. LC filters cannot be as sharply tuned and are subject to changes in characteristics as a result of changing operating conditions in the circuit and the presence of other components and stray fields in their vicinity. Radio-frequency cavities, because of their very construction, have a high order of shielding and can be designed to pass desired frequency components with maximum opposition to off-frequency components or can be designed to sharply reject a particular interfering frequency.

15-3-8. What is a continuous-tone-coded squelch system (CTCSS)? How is it used? How does CTCSS affect frequency deviation?—Study Section 7-10 in detail.

15-3-9. How is the Doppler effect used in police radar systems?—Doppler effect refers to the change in frequency of a wave when there is relative motion between the source of signal and the receiving point. If the distance between source and receiver decreases the received frequency becomes higher than the source frequency. If the distance is increasing, the received frequency becomes lower. In a Doppler radar system a pulse is sent out from a fixed position. When it strikes a moving target, such as a car, a reflected wave is sent back to the radar receiver. This returning radar component will change in frequency as a function of the velocity of the moving car. The radar receiver interprets this change in return frequency and displays it as a velocity of so many miles per hour.

15-3-10. Briefly explain the principles involved in frequency-shift keying (fsk). How is this signal detected?—*Frequency-shift keying permits the transmission of code and radioteletype signals that are less influenced by static and other noises.* In regular cw code transmission, the carrier is turned off during the spaces and on during the marks (dots and dashes). In fsk, by using one frequency for the spaces and a second frequency for the marks. Inasmuch as the mark and space frequencies can be positioned near each other, the receiver remains under control of a carrier whether marks or spaces are being transmitted.

As shown in Fig. 15-19, a frequency

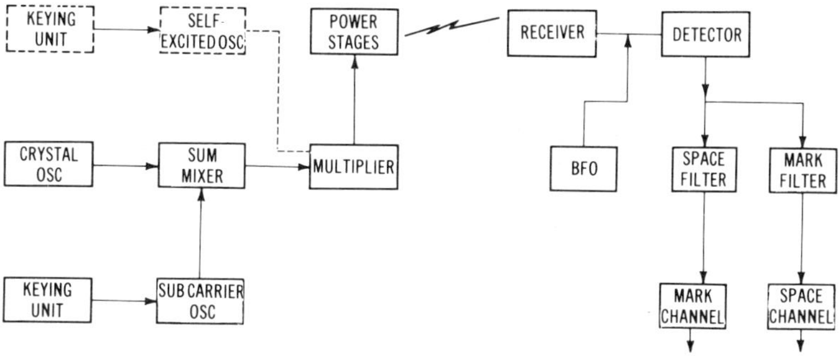

Fig. 15-19. Fsk transmission system.

shift can be accomplished by using the keying circuit to shift the frequency of a self-excited transmitter oscillator between two predetermined frequencies. Often the frequency shift is accomplished by a reactance modulator. The modulator is used to frequency shift a low-frequency subcarrier oscillator. The output of this oscillator is then mixed with a crystal-controlled component. A series of multipliers then builds up the frequency shift signal to the desired transmit frequency.

At the receiver, a bfo (beat-frequency oscillator) signal is beat with the incoming two-frequency signal. In so doing, two separate audio tones are developed at the detector output. The output signal is then passed to a discriminator-filter circuit which is able to segregate the two tones, forming a mark signal and a space signal for operation of radioteletype. For aural copying of a code signal, it is only necessary to use the mark-signal, it is only necessary to use the mark-signal component which carries the dots and dashes.

15-3-11. Compare fsk and afsk transmissions. What are typical baud rates and tone frequencies?—In frequency-shift keying (fsk) there is a continuous carrier. However, that carrier is shifted in frequency by a prescribed amount to delineate marks and spaces. Most often the frequency shift is 170 hertz with the regular carrier frequency representing the mark condition. A down shift of this frequency by 170 hertz represents the spaces of the fsk emissions. In compliance with FCC regulations fsk F1 emission is used below 50 MHz. Frequency shift maximum is 900 hertz.

At higher carrier frequencies audio frequency-shift keying (afsk) is more common. Again there is a continuous carrier with the modulation being a frequency-shifted tone. The popular frequency shift is again 170 hertz. The two tones are 2295 and 2125 hertz with the lower frequency representing marks and the higher representing spaces.

The two most common applications for fsk and afsk transmissions are associated with either the transmission of mechanical radioteletype data or print-out and display data for computer terminals. The 5-digit Baudot code is most often used for mechanical printing machines, while the ASCII 8-bit sequence is used in computer applications. Baudot data rates have specific values that extend between approximately 45 and 100 baud. These correspond to word-per-minute speeds that fall between 65 and more than 100. Baud rates for the ASCII code have specific values between 100 and more than 19,000 baud corresponding to words-per-minute speed that extends approximately between the very same limits.

15-3-12. To what service does DPLMRS refer? Briefly describe a mobile telephone system.—DPLMRS is the FCC designation for Domestic Public Land Mobile Radio Service. This service is concerned with the operation of telephones in automobiles and other vehicles. Presently there are a number of channels assigned to this service in the 150 MHz band. When such a telephone is installed in your car you can dial any telephone in your area without contacting a mobile service operator. Furthermore, when traveling into other cities you will be able to dial into their radio system and make calls to any telephone in the area. It is anticipated that the system will become

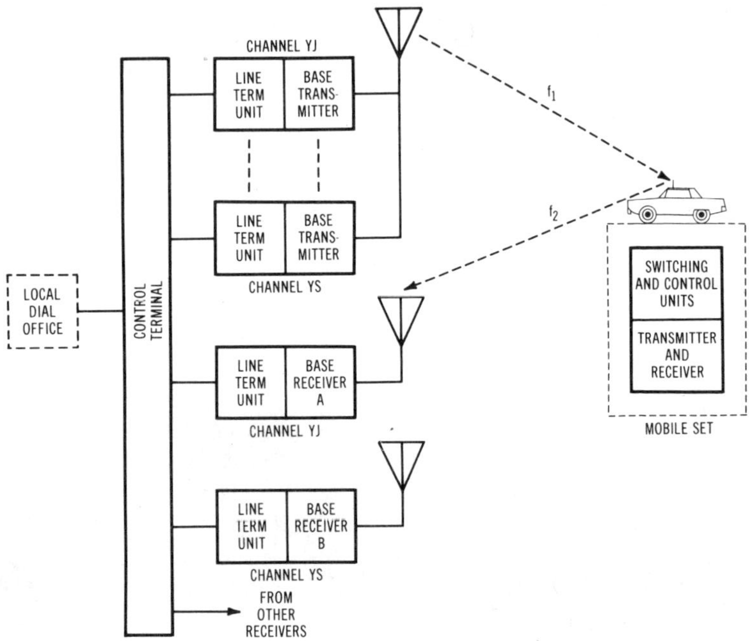

Fig. 15-20. Mobile telephone system.

nationwide and you will be able to dial any telephone in this country from any location.

A basic plan for such a system is shown in Fig. 15-20. For a large coverage area there is a centrally located group of high-powered transmitters, one for each channel, and in any given location up to a maximum of eight channels. Base receivers spotted conveniently around the area feed signals over wire lines to the central office. The most favorable mobile signal is switched into the telephone system for use. In small-coverage areas it is possible that a single low-powered base transmitter and single receiver at the same site are adequate.

Each mobile unit consists of the transmitter-receiver combination and control unit, plus the channel-switching facilities and signaling circuits. Each unit is equipped to operate on at least all of the channels active in a phone area. For subscribers who are on the road, the mobile set may be equipped with additional channels up to a specified maximum.

A special and very significant feature of the system is a feature referred to as multichannel access. A mobile unit automatically seeks an idle channel. Thus, when a mobile unit is activated, it will automatically seek an idle channel and establish a link from mobile to base station. The next call in either direction will then be completed over that channel. When this channel is occupied, all other mobiles in the area will automatically seek another idle base-station channel to establish still another connection. It is anticipated that each channel will be able to serve a maximum of 40 to 60 mobile units. The entire operation is automatic, and in most cases it is not necessary to call through a mobile service operator.

CHAPTER 16

Antennas and Lines

16-1. TRANSMISSION LINES AND WAVEGUIDES

16-1-1. What is meant by the characteristic (surge) impedance of a transmission line? To what physical quantities is it proportional?—A transmission line has a characteristic impedance which is determined by the unit-length inductance of its conductors and the capacitance between its conductors (as influenced also by the dielectric content of the material in the space between conductors). The value of this characteristic impedance is the impedance which must be connected to the output terminals of a section of line of any length to make the line appear to be infinitely long. An approximate expression for characteristic impedance is:

$$Z = \sqrt{\frac{L}{C}}$$

Thus it can be said that the characteristic impedance is proportional to the square root of the quotient of its inductance per unit length over its capacitance per unit length.

16-1-2. Why is the impedance of a transmission line an important factor with respect to matching "out of a transmitter" into an antenna?—When there is a match between the transmitter and the transmission line and between the transmission line and the antenna, there is a maximum transfer of power between the transmitter and the antenna. This efficient transfer of energy occurs when the transmitter output is matched to the *characteristic impedance* of the transmission line, and when the antenna system at the termination end of the line is matched to the characteristic impedance of the transmission line. When a transmission line is properly matched to an antenna system, there is a minimum radiation of radio-frequency energy from the line, eliminating a cause of energy loss or antenna-pattern distortion.

16-1-3. What is meant by standing waves, standing-wave ratio (swr), and characteristic impedance as referred to transmission lines? How can standing waves be minimized?—The definition for the characteristic impedance of a transmission line was given in Question 16-1-1. When a transmission line is terminated in its characteristic impedance, all of the energy conveyed along the line is delivered to the termination. If the line is not terminated in its characteristic impedance, only a part of the energy moving along the transmission line is delivered to the termination. The remainder of the electrical energy is reflected at the termination and moves back along the transmission line in the opposite direction. There is now an interaction between the direct rf energy and the reflected rf energy. In this case, the rf voltages and the rf current add algebraically to set up voltage and current standing waves on the transmission line. For example, at so-called voltage loops positioned along the line, the direct and reflected voltages are in phase and add to produce maxima. At other positions called voltage nodes, they are exactly out of phase and they subtract to produce minima (Fig. 16-1).

These voltages and currents can be measured with a suitable rf meter arrangement. The ratio between maximum and minimum voltage readings or between maximum and minimum current readings is called the standing-wave ratio (swr). Stated as an equation, this is

$$\text{swr} = \frac{E_{max}}{E_{min}} = \frac{I_{max}}{I_{min}}$$

The standing-wave ratio on the line is a measure of the degree of mismatch between the transmission line and the termination, and it can also be expressed as follows:

$$\text{swr} = \frac{Z_R}{Z_O} \text{ or } \frac{Z_O}{Z_R}$$

The larger quantity, Z_m or Z_o, is substituted in the numerator.

It should be noted that the ideal situation exists when the standing-wave ratio is unity. This indicates that the maximum and minimum readings are the same, and thus there is no standing wave on the line. In this case, there is no reflected energy, and the rf energy coming down the line has been delivered in its entirety to the terminating load. Thus, the standing wave on the transmission line can be minimized by supplying a properly matched termination to the receiving end.

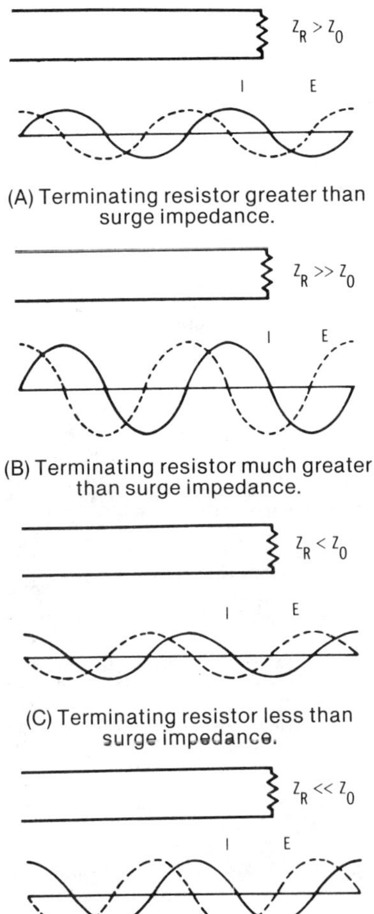

(A) Terminating resistor greater than surge impedance.

(B) Terminating resistor much greater than surge impedance.

(C) Terminating resistor less than surge impedance.

(D) Terminating resistor much less than surge impedance.

Fig. 16-1. Standing waves appearing on a transmission line.

16-1-4. If standing waves are desirable on a transmitting antenna, why are they undesirable on a transmission line?—The standing waves on an antenna are radiating rf energy. This is a preferred condition. However, a transmission line is a means of transferring rf power from a source of rf energy to an antenna. When standing waves appear on a transmission line, it is an indication that some radio-frequency energy is being radiated from the transmission line. This represents a loss of rf energy and means that some of the energy which should be transferred to the antenna is being lost in the line.

16-1-5. Explain the properties of a quarter-wavelength section of a radio-frequency transmission line. What is meant by stub tuning?—A radio wave in traveling a quarter-wavelength along a transmission line will encounter a 90-degree phase change. A full wavelength corresponds to 360 degrees. Inasmuch as the wave has travelled only one-quarter of this distance in a quarter-wavelength section of line this length of line is known as a 90-degree section. Because of the travel time the voltage output of the particular quarter-wavelength section of the radio-frequency transmission line will lag the voltage applied at the input of the same section by 90 degrees. A quarter-wavelength section of transmission line can be used as an impedance transformer. The transformation depends upon its characteristic or surge impedance. The basic equation is:

$$Z_O = \sqrt{Z_{in} Z_{out}}$$

When a quarter-wavelength section of transmission line is shorted, it displays a high impedance at its opposite end. In fact, this is a high resistive impedance and, in effect, a shorted quarter-wavelength section of transmission line acts as a parallel-resonant circuit. A simple equation can be used to calculate the physical length of a quarter-wavelength section of line. The equation is:

$$\text{Length in feet} = \frac{246 \times \text{velocity factor}}{\text{frequency in MHz}}$$

A stub tuner is a section of transmission line that is being used to match an rf load to a source of rf energy. In the case of a transmission line and antenna system, a stub-tuning arrangement can be used to match the characteristic impedance of a transmission line to the input impedance of an antenna. By so doing, there will be a maximum transfer of energy from line to antenna and a minimum standing-wave ratio on the transmission line. Typical stubs are shown in Fig. 16-2.

Such a stub can be used not only to function as an impedance transformer between the line and the antenna, but it is capable also of tuning out any reactance which is present in the input impedance of the antenna. A shorted quarter-wave stub has an impedance that rises from the point of the short to a maximum at the far end.

16-1-6. What would be the considerations in choosing a solid-dielectric cable over a hollow, pressurized cable for use as a transmission line?—The solid-dielectric line is more flexible and can be routed more conveniently. The hollow, pressurized cable (nitrogen or other gas) is a rigid type and involves higher cost with respect to mounting and pressurization. A hollow cable has very low loss and can be designed to handle extremely high power. It can be pressurized to keep out moisture and maintain a constant resistance. The solid-dielectric cable is adaptable to standard connectors and does not involve extensive plumbing as do the joints, bends, etc., of rigid cable.

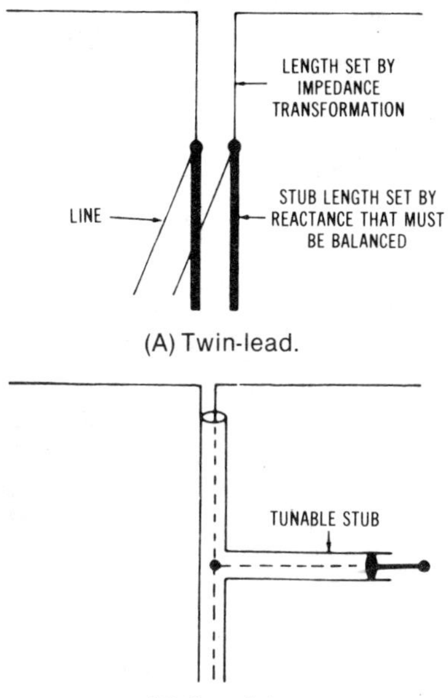

Fig. 16-2. Stub-tuning arrangements.

16-1-7. Describe briefly the construction and purpose of a *waveguide*. What precautions should be taken in the installation and maintenance of a waveguide to ensure proper operation?—Waveguides are constructed of durable weather-resistant material capable of reflecting microwave energy with a minimum of loss. The cross section of a waveguide can be rectangular or tubular (Fig. 16-3). There is no inner conductor, and the radio-frequency energy propagates along the waveguide enclosure much as a radio wave is propagated through space. However, the energy is confined within the guide walls. A good waveguide installation has very much less loss as a means of transferring microwave energy as compared to coaxial or parallel-conductor types of transmission line. Thus a waveguide functions as an efficient transmission line for conveying

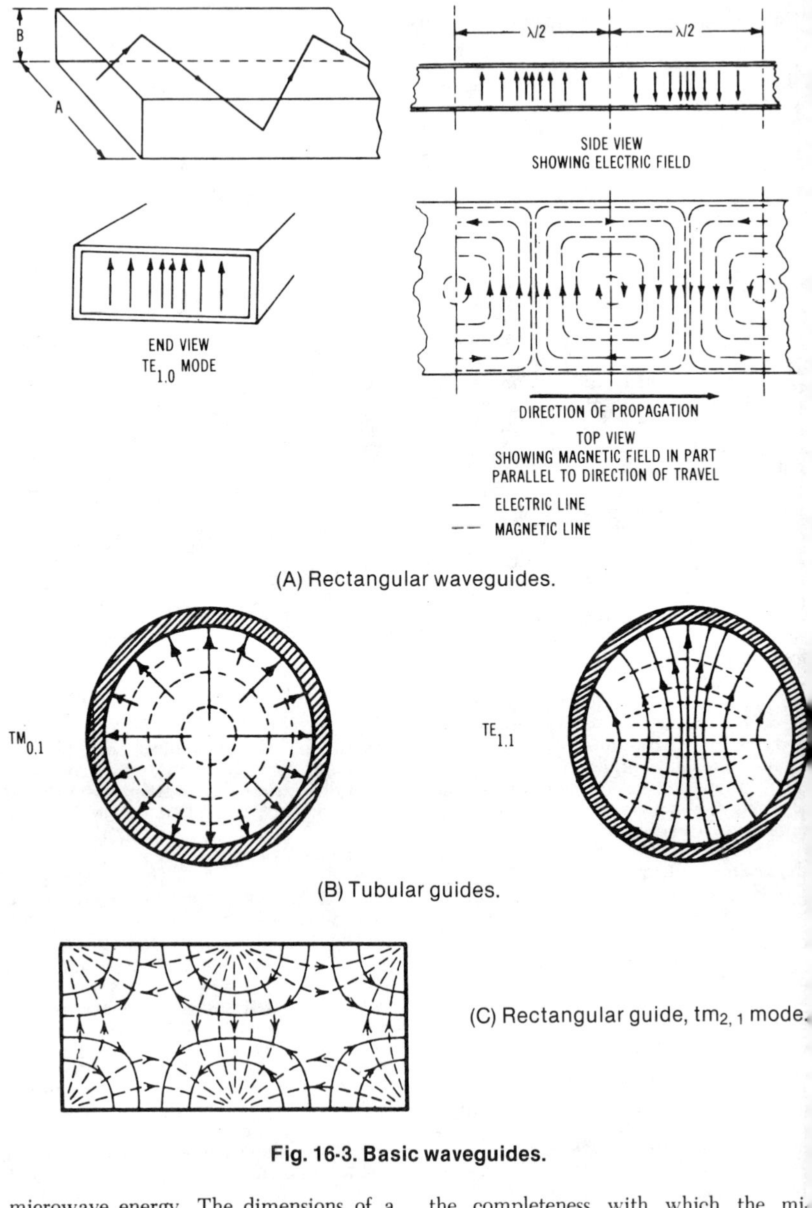

Fig. 16-3. Basic waveguides.

microwave energy. The dimensions of a waveguide are related to the wavelength of the signal to be conveyed; this makes the size of a waveguide impractical for use at lower frequencies.

The efficiency of a waveguide depends on the uniformity of the inner walls and the completeness with which the microwaves are reflected from the walls. Therefore the installation should be such that the interior is kept clean, smooth and dry. Water, dust, and corrosion will introduce discontinuities and loss of efficiency. To minimize loss, the installation should be planned for a short and direct routing of the waveguide. In the installation of a guide, special joints can be installed where connections or bends are to be made. These joints, called *choke joints*, prevent discontinuities from causing objectionable attenuation in the transfer of the microwave energy. Just like a conventional transmission line, a waveguide must be matched properly for most efficient operation. It must be correctly dimensioned relative to the microwave-frequency range that is to be conveyed.

16-1-8. Discuss the following with respect to waveguides: (a) relationship between frequency and size, (b) modes of operation, (c) coupling of energy into the waveguides, (d) general principles of operation.—(a) The higher the microwave frequency of operation is, the smaller are the required physical dimensions of a practical waveguide. As shown in Fig. 16-3A, a common dimension for a rectangular waveguide is an inside width (long wall) of approximately 0.8 wavelength (A), while the height (short wall) is 0.38 wavelength (B).

(b) The mode of operation of a waveguide has to do with the distribution of the electric and magnetic field patterns in a plane transverse to the direction of propagation of the microwave energy along the line. A *te mode* is one that has a transverse electric field, as shown in Fig. 16-3A. A *tm mode* is one with a transverse magnetic field, as in Fig. 16-3C. There are a variety of both te and tm modes according to the number of complete electric or magnetic fields in a half-period dimension of the microwave energy being conveyed.

(c) There are various ways of coupling energy into a waveguide. In some cases it is simply the correct positioning of an aperture which leads directly into the waveguide from a microwave resonant circuit such as a cavity, as shown in Fig. 16-4. It is also possible to insert a probe into the rectangular cavity, as shown in Fig. 16-4B. This would be a means of transferring energy from a short coaxial feed line into a waveguide.

The method of coupling depends on the propagation mode which is to be used in the waveguide. The arrangement shown in Fig. 16-4B is one used to establish the $te_{1,0}$ mode of Fig. 16-3A. It sets up a transverse electric field and a magnetic field which is oriented parallel to the direction of propagation along the line.

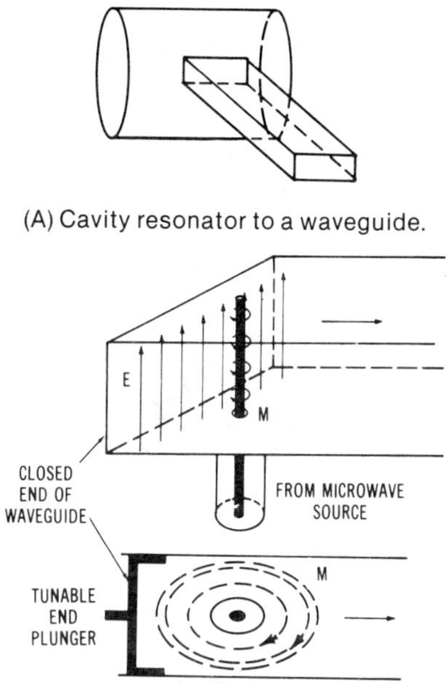

(A) Cavity resonator to a waveguide.

(B) Coaxial feed to a waveguide.

Fig. 16-4. Coupling microwave energy to a waveguide.

(d) The transmission of microwave energy along a waveguide is in the form of electromagnetic propagation, but the movement is confined within the guide. The propagation is by means of changing magnetic and electric lines of force, which have the same significance as the rf current and voltage distribution mentioned in the previous question on conventional two-conductor transmission lines. The microwave energy travels within the metal enclosure of a waveguide by bouncing off the inner walls. With proper design, there is no significant penetration of the walls, and the waves are reflected from the walls with little attenuation. The changing lines of force provide the continuous movement and completed energy loops needed to propagate within the guide. Either the electric- or magnetic-field lines must be oriented in part in the direction of propagation. These two fields are, of course, perpendicular to each other, one creating the other to propel the energy along the guide.

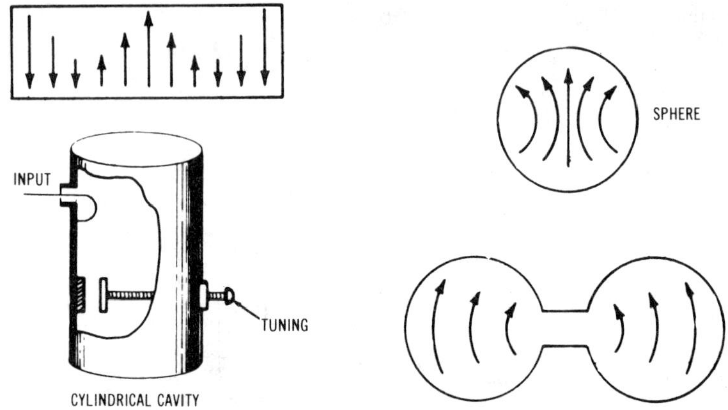

Fig. 16-5. Resonant cavity.

The dimensions of the waveguide are important in terms of the reflection of the energy from the walls. If dimension A, Fig. 16-3A, is half wavelength, the microwave energy will bounce back and forth side to side at its point of coupling, and it will not propagate along the line. To have propagation along the line, it is necessary to have the microwave energy hit the wall at an angle as shown in Fig. 16-3A. Since the angle of reflection is equal to the angle of incidence, the wave will then be bounced side to side down the line. This requires that dimension A be greater than a half wavelength or, stated another way, a half wavelength at the microwave signal frequency should be significantly shorter than dimension A. The value of 0.8 wavelength is typical.

16-1-9. Explain the principles of operation of a cavity resonator.—A properly dimensioned section of waveguide can be made to operate as a resonant circuit. Waveguide segments can be closed off in various shapes and forms, as shown in Fig. 16-5. These complete resonant enclosures are referred to as *cavities*. The simplest resonant cavity can be made by closing off both ends of a rectangular or circular waveguide. In so doing, there will be a circulating magnetic field similar to the circulating rf current in a conventional tuned circuit. Such a cavity can be made to have an extremely high Q at the microwave frequency for which it has been dimensioned.

As in the case of a conventional circuit, the energy is continuously exchanged between electric and magnetic fields (comparable to the electric field across the dielectric of the capacitor and the magnetic field surrounding the coil of a conventional LC resonant circuit). However, no closed electrical path is required, the energy being in the form of a confined electromagnetic wave.

16-1-10. How are cavities installed in vertical waveguides to prevent moisture from collecting? Why are long horizontal waveguides not desirable?—A good vertical mounting should be made so as to minimize the collection of moisture. Appropriate choke joints should be used, and mounting gaskets are available which block the intrusion of moisture. A choke joint ensures that a continuous electrical circuit is presented within the waveguide, although there is an actual physical break between the waveguide and any inserted device, such as a joint or a cavity.

Long, perfectly level horizontal runs of waveguide are to be avoided because they are especially prone to an accumulation of moisture which does not drain off readily and quickly.

16-1-11. How do frequency and line length affect transmission line attenuation?—Line attenuation increases with both frequency and line length.

16-1-12. Do transmission lines of the same characteristic impedance and the same length necessarily have the same attenuation? Explain.—They do not necessarily have the same attenuation.

Attenuation depends on conductor size and spacing as well as the dielectric material between the two conductors of the line. For example, RG-58/u has a greater attenuation per 100 feet of line than RG-8/u, even though both have the same characteristic impedance. A good quality line of the same type will often have less loss than one of poor quality.

16-1-13. In a 2-MHz marine installation, explain why the ground system of the boat affects the antenna impedance.—A poor ground system causes transmitter power to be wasted. The marine antenna is short in comparison to a quarter wavelength at transmit frequency. Therefore it has an extremely low resistance of 10 ohms or less. Thus everything should be done to keep the ground system resistance as low as possible. For a metal-hulled boat a large conductor should connect the ground of the transmitter to the hull. Large ground plates are often used for wooden-hulled boats to obtain an adequate ground for the 2-MHz marine transmitter. The marine antenna, because of its shortness in comparison to wavelength, has a very low impedance, the value of which is very much influenced by the ground system. In fact, 40% or more of the antenna impedance may be contributed by the ground system. This is a loss component.

16-1-14. Explain the use of a directional wattmeter in measuring vswr and transmitter power.—A directional wattmeter is inserted in the transmission line between the transmitter output and the antenna system. It is capable of measuring the power delivered forward as well as the reflected power in watts. Thus the efficiency of the transmitter match and the antenna system can be judged. Proper operating conditions are indicated when the reflected power measurement is very much lower than the direct power measurement. Some directional wattmeters also give an indication of the standing-wave ratio. The power delivered to the antenna can be calculated by subtracting the reflected power from the direct power. Refer to Sections 6-4 and 6-13.

16-1-15. Why is a microwave dish normally fed with waveguide instead of coaxial cable?—At microwave frequencies a waveguide has much less attenuation than a coaxial cable. When using a waveguide a simple arrangement, such as a horn, can be used to make a low-loss changeover between the waveguide output and the dish.

16-2. ANTENNAS

16-2-1. Explain the voltage and current relationships in a one-wavelength antenna, a half-wavelength (dipole) antenna, and a quarter-wavelength grounded antenna.—The current and voltage distributions for these fundamental antennas are shown in Fig. 16-6. When an antenna is excited with radio-frequency power, that energy moves along the antenna conductor(s) toward the free end(s). At the ends the energy is reflected just as it would be at the far open-end of a transmission line. Thus, a standing-wave pattern of voltage and current is set up on the antenna. In fact, many types of antennas become effective resonant circuits. As more and more rf energy is supplied to the antenna, oscillating rf waves break free and move out into space.

A voltage maximum occurs at the open-end points of reflection. Thus, in the case of the very common half-wavelength dipole (Fig. 16-6A), there are maximum voltage points at each antenna end. The half-wavelength antenna makes an efficient resonant circuit because it is an *electrical half wavelength* in dimension (its physical length is only slightly less than a half wavelength because of capacitive end effects). In a half-wavelength antenna, the voltage distribution goes from a maximum at one end through a minimum (90° away) to a second maximum of opposite polarity at the other end of the antenna. The minimum voltage point corresponds to a maximum current point, while current minima exist at the ends.

It can be anticipated from the discussed current and voltage relationship that the impedance will also vary along the length of the antenna. The impedance will be maximum where the voltage is maximum and the current is minimum, while the impedance will be minimum where the voltage is minimum and the current is maximum. Customarily, a half-wavelength antenna is a dipole type, meaning that it is divided into two quarter-wave sections and is fed at the center (Fig. 16-6A). At the center of the antenna we have noted that the impedance is minimum. Therefore, a suitable match can be arranged between the dipole antenna and the low impedance of a transmission line.

The voltage and current distribution is such that the impedance presented at the

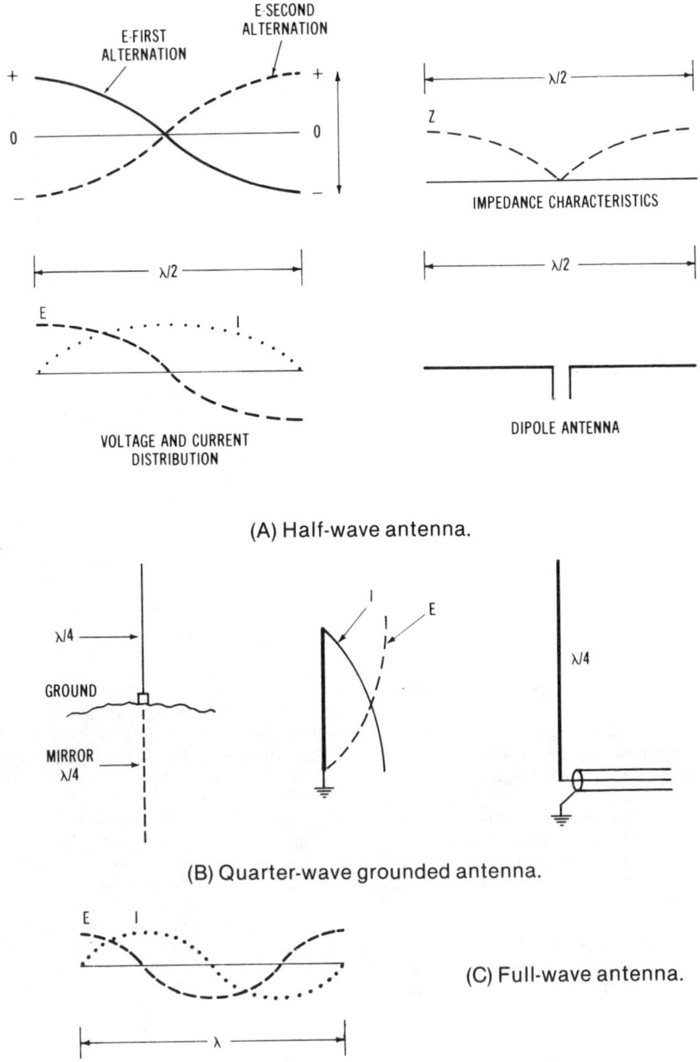

Fig. 16-6. Basic antennas and voltage-current distribution.

center feed point of the half-wavelength antenna is 72 ohms.

A quarter-wavelength antenna has the current and voltage distribution shown in Fig. 16-6B. The operational characteristics of a grounded quarter-wavelength antenna are similar to those of a half-wavelength antenna, with the exception that the ground acts as a mirror quarter-wavelength segment. Such a quarter-wavelength antenna is often fed near the ground point where it has a low impedance which can be matched conveniently to a transmission line. The impedance at the ground point is theoretically 36 ohms.

The voltage and current distribution for a one-wavelength antenna is shown in Fig. 16-6C. Again there are maximum voltage points at the open ends. This voltage maximum pattern is repeated along the length of the antenna. Therefore, a maximum exists at the center of the antenna as well as at the ends. Current maxima occur at the odd quarter-wavelength points along the antenna length.

From the preceding discussion it is apparent that current and voltage relations are significant in the matching and efficient transfer of energy between line and antenna. The physical length of the antenna is also a factor. If the antenna is to be an efficient rf radiator, its physical length must be such that electrical resonance can be established.

16-2-2. What effect does the magnitude of the voltage and current at a point on a half-wavelength antenna in free space (a dipole) have on the impedance at that point?—The impedance along a length of antenna is the quotient of the voltage and current at each particular point, i.e.,

$$Z = \frac{E}{I}$$

The impedance of a dipole antenna at its usual center feed point is 72-ohms resistive in free space. The actual impedance in a practical installation is also influenced by the presence of ground and nearby conducting surfaces. Dipoles can be shaped in ways other than thin, straight conductors, so as to change the center-point impedance.

If the dipole is fed at some point other than the center along its half wavelength, its impedance will be higher than at the center point. For example, if the antenna is fed at the end where the voltage is high and the current is minimum, the impedance is greater than 1000 ohms.

16-2-3. What kinds of fields emanate from a transmitting antenna, and what relationship do they have to each other?—The electromagnetic radiation from an antenna is in the form of changing electric and magnetic fields. As shown in Fig. 16-7, these fields are perpendicular to each other as the rf energy propagates through space.

There is also an induction field about an antenna — just as the one you would find about any type of coil. However, the induction field falls to an insignificant value a short distance from the antenna, as do the flux lines of a coil.

16-2-4. Can either of the two fields that emanate from an antenna produce an emf in a receiving antenna? If so, how?—The radio wave is propagated through space because of the interacting influence of its electric and magnetic components. Both are necessary in the activity that causes a current in the receiving antenna. The standard-length antenna has voltages (emf's) induced along its length as a result of the changing magnetic flux of the waves. The magnetic flux is, of course, a result of the changing electric and magnetic vectors of the electromagnetic wave. The voltages induced along the length of the antenna cause a current in the antenna and an associated external receiving system.

In the case of the usual straight receiving antenna, the electric vector has a significant bearing in terms of the current distribution. Conversely, for loop or frame antennas, which have dimensions that are very small in comparison to the wavelength of the incoming signal, the activity of the magnetic vector is of more significance in determining the current in the antenna.

16-2-5. Draw a sketch and discuss the horizontal- and vertical-radiation patterns of a quarter-wavelength vertical antenna.

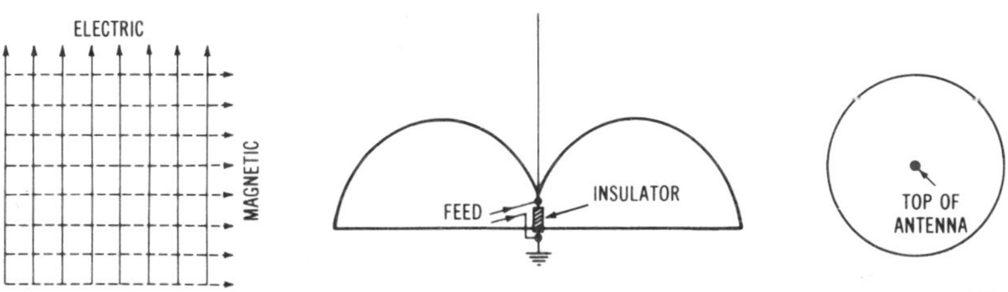

Fig. 16-7. Fields and patterns of a quarter-wave vertical antenna.

Would this also apply to a similar type of receiving antenna?—Refer to Fig. 16-7. The vertical quarter-wavelength antenna has an omnidirectional horiizontal-radiation pattern. This means that, under an ideal situation, signals of uniform strength are radiated from the antenna in all azimuthal directions. The quarter-wave vertical antenna above a good conductive ground has a low-angle vertical radiation pattern. This means that more energy is radiated directly over the surface of the earth toward the horizon, and much less energy is radiated skyward. This would also be representative of the sensitivity pattern of a similar quarter-wave receiving antenna. It would display the same sensitivity at all compass angles. It would display a maximum sensitivity toward signals skimming over the surface of the earth rather than to those signals which would be arriving at high angles (from the sky). This is referred to as the *law of reciprocity;* like antennas have similar radiation and sensitivity patterns.

16-2-6. Explain the following terms with respect to antennas (transmission or reception): (a) field strength, (b) power gain, (c) physical length, (d) electrical length, (e) polarization, (f) diversity reception, (g) corona discharge.—(a) Field strength has to do with the strength of an electromagnetic wave at a certain distance from a transmitting antenna. Customarily, electromagnetic radiation is expressed in volts, millivolts, or microvolts per meter. This represents the voltage which would be induced into an antenna by the rf wave for every meter its length.

(b) Power gain is a unit of comparison between antenna types in terms of their maximum radiated power. It is represented as a quotient as follows:

$$\text{Gain}_p = \frac{\text{Power Delivered by Antenna Under Test}}{\text{Power Delivered by Reference Antenna}}$$

There is no single standard reference antenna. Customarily, at high frequencies a half-wavelength dipole is often considered a test standard. In other cases the standard might be a quarter-wave or vertical or an isotropic (point-source) antenna.

Since the transmit and receive antennas are reciprocals, the power gain of the receiving antenna is a similar quotient. In this case, it is the power the antenna under test abstracts from the passing electromagnetic wave and delivers to the receiving load, as compared to the power which would be delivered when using the reference antenna.

(c) The physical length of an antenna is its linear length, so many inches or feet long. For a given-style antenna, say a half-wavelength antenna, the physical length of the antenna is usually somewhat shorter than the calculated "free-space" half-wavelength of the signal to be radiated or received.

(d) The electrical length represents the length of an antenna based on wavelength. A half-wavelength antenna has a length that displays an exact electrical half-wavelength resonance at the frequency to be radiated or received. As mentioned in (c), the actual physical length needed to obtain a given electrical length is usually slightly shorter than the calculated "free-space" half-wavelength of the signal to be received.

(e) The polarization of an electromagnetic wave refers to the direction of its electric vector. For example, a horizontally polarized wave is one that has a horizontal electric vector and a vertical magnetic vector. Conversely, a vertically polarized wave is one that has a vertical electric vector and a horizontal magnetic vector (Fig. 16-7).

A conventional dipole transmitting antenna must be mounted horizontally to radiate a horizontally polarized wave, and it must be mounted vertically to radiate a vertically polarized wave. Likewise, the receiving antenna must be mounted horizontally or vertically relative to an incoming horizontally or vertically polarized wave to extract the maximum signal.

(f) In a diversity-reception system there are two or more widely spaced antennas (spaced at least far enough to reduce mutual interaction) which pick up a particular signal and deliver outputs to a common receiving location. The diversity-receiving arrangement is such that the antenna delivering the strongest signal at the moment supplies a demodulated output for listening or display. With signal fading, the diversity-receiving system automatically selects the antenna which is receiving the maximum signal. In this way a diversity-reception system can minimize the influence of fading. The output signal-level change is therefore much less than that which would be present with a single receiving antenna.

(g) *Corona* is an electrical discharge

which occurs near the surface of a conductor when the voltage gradient exceeds a certain critical value. A fine point or conductor end would be most likely to produce a corona discharge. A mobile antenna in motion collects static charges that can build up and discharge in corona manner, producing receiver noise. This is most likely to occur at a sharp point such as the end of a whip antenna. For this reason, antenna ends are often capped or balled to hold down the voltage gradient.

16-2-7. Discuss series and shunt feeding of quarter-wave antennas, with respect to impedance matching.—Shunt- and series-fed quarter-wave vertical antennas and feed arrangements are shown in Fig. 16-8. In the shunt-fed arrangement, the

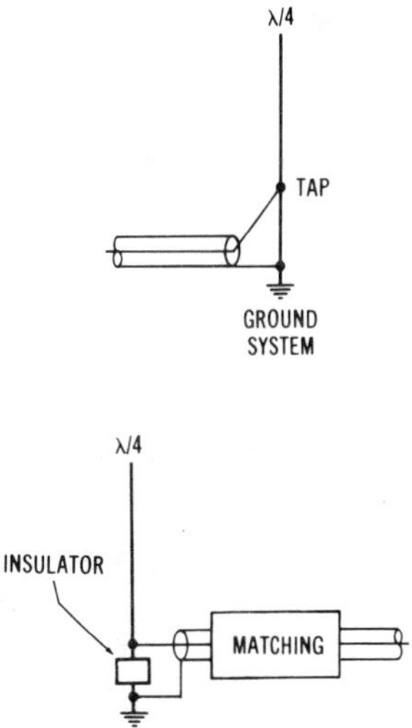

Fig. 16-8. Antenna feed systems for a quarter-wave antenna.

bottom of the quarter-wave antenna is grounded. An impedance match is made by proper positioning of the tap on the antenna. The higher the feed point is above ground, the higher is the impedance that the antenna presents to the transmission line or associated antenna-tuning unit.

In the series-fed arrangement, the bottom of the antenna is insulated from ground. The transmission line or the output of the tuning unit then connects between the bottom of the antenna and ground. The antenna resistance at this feed point is approximately 36 ohms for a quarter-wave antenna. The actual value will also be influenced to some extent by the ground conductivity. Thus, an impedance transformer or antenna tuner is often used to make an appropriate match from the transmission line to the antenna feed point.

16-2-8. Explain why a loading coil is sometimes associated with an antenna. Under this condition, would absence of the coil mean a capacitive antenna impedance?—The function of a loading coil is to increase the electrical length of the antenna when its physical length is too short to permit a resonant condition. Thus, as shown in Fig. 16-9, the presence of a loading coil permits the physical length of a quarter wavelength antenna to be significantly shorter than when used without a coil.

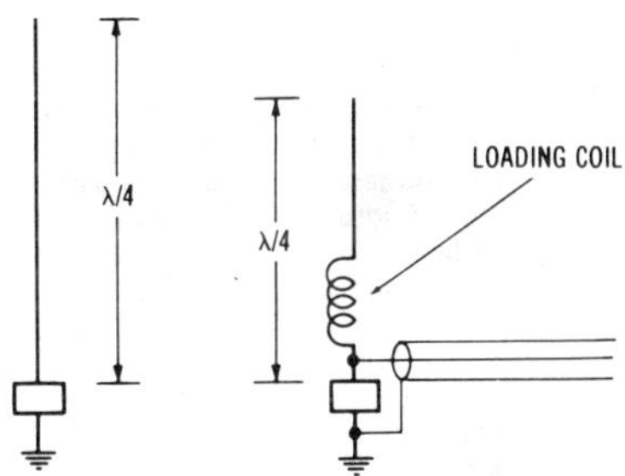

Fig. 16-9. Use of a loading coil to increase the electrical length of a short antenna.

Without the presence of the loading coil, the antenna length would therefore be shorter than the desired quarter-wavelength, and it would present a capacitive antenna impedance.

16-2-9. What constitutes the ground-plane when a quarter-wave grounded (whip) antenna one meter in length is mounted on the metal roof of an automobile? When it is mounted near the rear bumper of an automobile?—When a quarter-wave antenna that has a dimension of only one meter is used, the dimensions of the car roof are such that the roof functions as a good ground plane. In practice, a good ground-plane action is obtained if the roof dimensions are at least one-half wavelength (diameter of a circular ground plane or the minimum length of the sides of a squared ground plane). Such a ground plane provides a low angle of radiation, which is preferred for mobile two-way radio application.

A rear-bumper mount would display a characteristic less ideal than a more complete ground plane, and the horizontal-radiation pattern would not be as uniform as with the center of the roof mounting. The car bumper, and to some extent the car body, would provide some ground-plane radial characteristics. If the mount is made near the center of the bumper, the two bumper sides can act as approximately two quarter-wavelength radials.

16-2-10. Draw a sketch of a coaxial (whip) antenna. Identify the positions and discuss the purposes of the following components: (a) whip, (b) insulator, (c) skirt, (d) trap, (e) support mast, (f) coaxial line, (g) input connector.—Such an antenna construction is shown in Fig. 16-10. Fundamentally, the coaxial whip antenna is a half-wavelength type. The whip itself, which is a continuation of the inner conductor of the transmission line, is a quarter-wavelength long, and it functions as one side of the half-wavelength vertical dipole construction. The other quarter-wave of the basic dipole construction is in the form of the skirt. Signal is radiated by both the whip and the skirt.

(a) The whip is the top quarter-wave radiating section, and it is a continuation of the inner conductor of the coaxial line.

(b) The insulator permits center feed and keeps the whip and the inner conductor insulated from the skirt and the outer conductor of the coaxial line.

(c) The skirt forms the lower quarter-wave segment of the coaxial half-wave dipole. It is connected to the outer conductor of the coaxial line at the center feed point.

(d) The folded back quarter-wavelength of skirt acts as a quarter-wave shorted section of line, establishing a high impedance between the outer conductor of the coaxial line and the open end of the skirt. It functions as a parallel-resonant trap. It is this high impedance which prevents interaction between the dipole segment provided by the skirt and the coaxial line which feeds up through

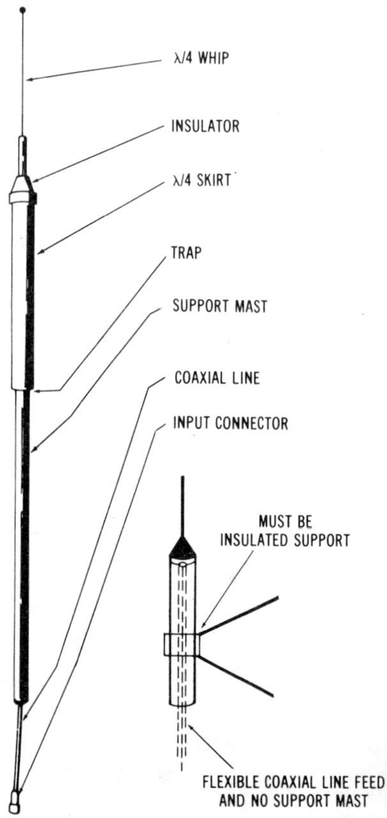

Fig. 16-10. Coaxial antenna.

its center, permitting the skirt to act as a radiator.

(e) If a rigid coaxial line is being used, the outer conductor can also serve as the supporting mast and can be fastened to a roof mount, wall mount, car body, etc. If a flexible coaxial line is being used, an insulated mounting arrangement is required, as shown in Fig. 16-10.

(f) The coaxial transmission line provides a means of transferring a signal from the transmitter to the coaxial antenna. The coaxial line is unbalanced, but the center-feed method using the skirt permits a proper impedance match and transformation to a balanced dipole radiator.

(g) In the rigid types of mount it is possible to install a coaxial receptacle at the base of the antenna. This receptacle can serve as the input connection. A

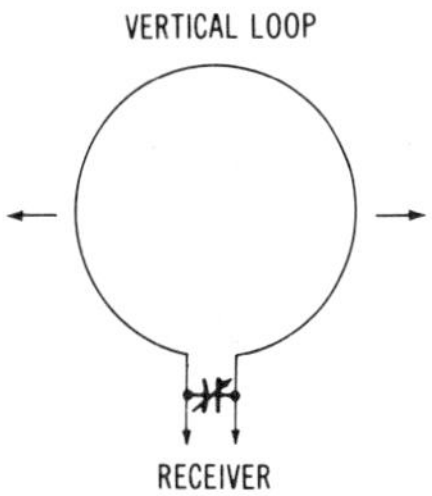

Fig. 16-11. The loop antenna.

coaxial transmission line of the flexible type and a suitable coaxial plug can then be used as a means of connecting the transmission line to the antenna.

16-2-11. Describe the directional characteristics of vertical-loop antennas.—The horizontal-radiation pattern of a vertical loop is a figure 8, with its two maxima in the direction of the plane of the loop (Fig. 16-11). The vertical-radiation pattern is circular.

16-2-12. Discuss the directivity and principal characteristics of the following types of antennas: (a) single-loop, (b) V-beam, (c) corner-reflector, (d) parasitic array, (e) stacked array.—(a) The single-loop antenna was discussed in the previous question. Most often it consists of several closely wound turns of wire formed into a rather large loop. This loop may be one to several feet in diameter, depending on frequency and application. Quite often it is the inductor of a resonant circuit, with an associated variable capacitor that can be used to tune the combination for resonant-circuit operation. The directional pattern is shown in Fig. 16-11.

(b) The V-beam antenna is a long, wire, directional antenna, as shown in Fig. 16-12A. The length of each leg is several wavelengths, with the angle between the two wires of the V depending on the leg wavelength and the desired characteristics. The V-beam antenna is unidirectional, with the maximum radiation and sensitivity in a direction that bisects the included angle between the two legs. It has a high gain and a reasonably sharp directional pattern which becomes sharper with an increase in the leg length.

(c) The corner reflector is used mainly as a high-gain unidirectional antenna in the vhf-uhf frequency spectrum. The element to which the transmission line is attached is a dipole or some form of modified dipole. This dipole is mounted at the bisector of two large area surfaces mounted at a corner angle between 45° and 90°. These conducting sheets reflect and confine the antenna radiation or sensitivity in the direction of the bisector of the corner (Fig. 16-12B). The conducting surfaces can be flat, solid conductors, wire mesh, or tabular elements.

(d) Parasitic antennas consist of a driven element (dipole or modified dipole) and one or more parasitic elements which have no direct electrical connection to the transmission line. Such parasitic elements are mutually coupled to the driven element, and as a function of their length and spacing they can be used to direct the radiated energy or to control the sensitivity pattern of the antenna.

A parasitic element which is *longer*

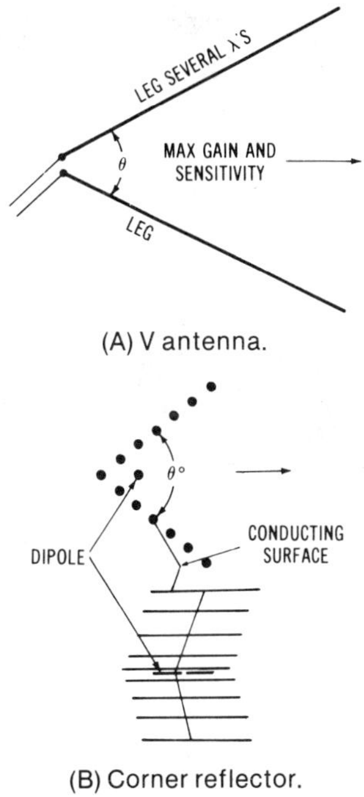

Fig. 16-12. V antenna and corner reflector antenna.

than the driven element acts as a *reflector* and directs any energy radiated from the dipole away from the reflector, as shown in Fig 16-13A. Parasitic elements which are cut *shorter* than the driven element also increase the radiation or improve the sensitivity of the antenna system. However, the radiation from such an antenna, as it comes off the driven element, is concentrated toward the *director* elements. The second antenna shown in Fig. 16-13A includes one reflector and three directors which are capable of concentrating the radiation from the antenna into a unidirectional, narrow-angle beam in the direction of the arrow. Only a small amount of this energy is radiated in the back direction, and for this reason the antenna is said to have a good front-to-back ratio.

(e) Additional antenna-system gain or sensitivity can be obtained by interconnecting driven elements in various groupings. One such combination is the broadside or stacked array, shown in Fig. 16-13B. In an arrangement for horizontal polarization, two or more half-wave elements can be stacked one above the other. Usually they are positioned with one-half wavelength separation and are fed in phase. To do this in the arrangement of Fig. 16-13B, it is necessary to center feed the transmission-line feed system between elements. In so doing, two half-wave elements are fed in phase, as shown by the polarity markings.

This type of antenna has the same horizontal figure-8 pattern as the single horizontal half-wave antenna. However, instead of a circular vertical-radiation pattern as with a single-element half-wave dipole, the vertical pattern also becomes a figure-8 pattern. The radiated energy of the antenna is stronger at low angles in a horizontal direction than in a skyward direction.

16-2-13. How is the operating power of an am transmitter determined, using values of antenna resistance and antenna current?—The rf power output of an am transmitter can be determined by using the power law.

$$P = I_A^2 R_A$$

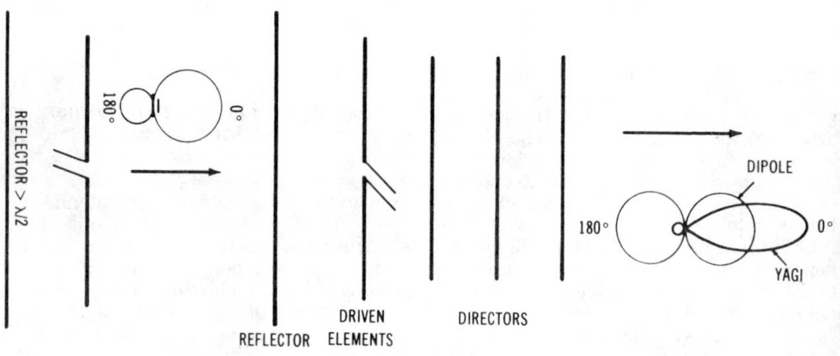

(A) Parasitic beam.

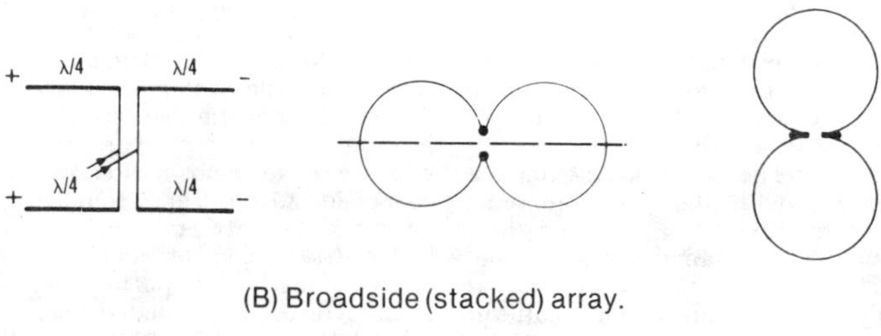

(B) Broadside (stacked) array.

Fig. 16-13. Two types of antennas.

If the antenna resistance is not known, it can be measured by an impedance bridge. The current can be determined with the insertion of an rf current meter.

The power output of an am transmitter is measured under various conditions. Sometimes the power output is to be known when the transmitter is unmodulated; this is referred to as the unmodulated carrier rf output. In other circumstances, according to the FCC recommended measurement procedure, the rf power output must be measured during maximum peaks of modulation. In other cases the measurement is made with standard 30% modulation of the carrier.

16-2-14. Why are insulators sometimes placed in antenna guy wires?—It could be possible that the guy wires themselves were of resonant length, and they would act as parasitic elements disturbing the normal pattern of the antenna system, and perhaps adversely affecting the antenna matching and efficiency. The insulators break up the guy wires into conductor segments that are very much smaller than the transmission wavelength.

16-2-15. Will the velocity of signal propagation differ in different materials? What effect, if any, would this have on wavelength or frequency?—The velocity of propagation does differ in different materials. This difference is expressed in a term called the *velocity factor*. This factor relates the velocity of radio-wave travel in a material in comparison to its travel in air. For example, if the velocity factor of a given type of transmission line is 0.9, it means the velocity of radio-wave travel along the line is 90% of the velocity of travel in free space. Assuming the same frequency, this means also that a physical wavelength segment of the transmission line would be 90% of the calculated free-space wavelength. Thus, in preparing transmission lines to be used as quarter-wave or half-wave stubs for matching sections, the velocity factor must be considered when dimensioning the cut. In the previous example a quarter-wave stub must then be 90% of the calculated free-space quarter-wavelength at the particular frequency to be handled by the stub arrangement.

The influence of velocity factor is one strictly of physical length as affected by velocity of wave travel; the frequency of the wave is unchanged.

16-2-16. Discuss the directivity and physical characteristics of a 5/8-wavelength vertical and collinear array of vertical dipoles.—These are two very common antennas used in the land-mobile radio services. Study the information in detail given in Sections 8-1 and 8-2. Both antennas have circular horizontal radiation patterns. They also have low-angle vertical patterns that spray the signal out over the area to be covered rather than at high angles. Consequently they have gain with the collinear array of vertical dipoles having a higher gain than a 5/8-wavelength vertical. Two vertical dipoles will have several dB gain above that of a 5/8-wavelength vertical. However, the collinear is a longer antenna. Its total length depends on the number of vertical dipoles that form the array.

16-2-17. What is effective radiated power? Given the transmitter power output, antenna transmission line loss, transmitter efficiency and antenna gain, show how erp is calculated.—The term effective radiated power means the product of the antenna power (transmitter output power less transmission line loss) times the antenna power gain. Where circular or elliptical polarization is employed, the term effective radiated power is applied separately to the horizontal and vertical components of radiation. For allocation purposes, the effective radiated power authorized is the horizontally polarized component of radiation only. By equation:

$$\text{erp} = (\text{transmitter power output} - \text{transmission-line power loss}) \times \text{antenna power gain.}$$

If the antenna rating is given in field gain the equation becomes:

$$\text{erp} = (\text{transmitter power output} - \text{transmission-line power loss}) \times (\text{antenna field gain})^2$$

16-2-18. If the power output of a station is quadrupled, what effect will this have on the field intensity at a given point?—*Field intensity* is a voltage measurement. It varies as the square root of power. Therefore if the broadcast station output power is quadrupled, the field intensity reading will double ($\sqrt{4}$).

16-2-19. What does the velocity factor of a transmission line mean and how might this be important?—The velocity factor relates the velocity of radio-wave travel in a transmission line as compared to its travel in free space. For example, if the velocity factor of a given type of transmission line is 0.66, it means the velocity of the radio-wave travel along the line is only 66% of the velocity of travel in free space. Assuming the same frequency, this means also that the physical length of a one wavelength segment of the transmission line would be 66% of the calculated free-space wavelength. This is important in preparing transmission lines to be used as quarter-wave or half-wave stubs or matching sections. For example, a quarter-wave stub must be 66 percent of the calculated free-space quarter-wavelength at the particular frequency to be handled by the stub arrangement. In directional antenna systems the electrical wavelength of the various sections of transmission line that feed the separate towers of a directional antenna system must be so many electrical degrees in length. The velocity factor of the transmission lines used must be considered in cutting the lines to a specified number of degrees of electrical length. An example will demonstrate the significance of velocity factor in determining the length of a transmission-line section.

What must be the *physical length* of a 200-degree length of line at 1 MHz (300 meters) if it has a velocity factor of 0.66? The equation is:

$$\text{Length} = \frac{200°}{360°}(300) \times 0.66$$
$$= 110 \text{ meters}$$

16-2-20. In what units is the field intensity of a station normally measured?—In millivolts per meter.

64. If the power output of a station has been increased so that the field intensity at a given point is doubled, what increase has taken place in antenna current?—Field intensity is a voltage measurement and will vary directly with the antenna current. Consequently if the field intensity is doubled it means that the antenna current has also been doubled.

16-2-21. If the antenna current of a station is 9.7 amperes for 5 kilowatts, what is the current necessary for a power of 1 kilowatt?—The resistance of the antenna system can be determined using the power law:

$$R = \frac{P}{I^2} = \frac{5000}{(9.7)^2} = 53.14 \text{ ohms}$$

The power law can be used again to determine the current needed for a power of only 1 kilowatt:

$$I = \sqrt{\frac{P}{R}} = \sqrt{\frac{1000}{53.14}} = 4.338 \text{ amperes}$$

16-2-22. The dc input power to the final amplifier stage is exactly 1500 volts and 700 milliamperes. The antenna resistance is 8.2 ohms and the antenna current is 9 amperes. What is the efficiency of the final amplifier?—The dc input power to the final stage is:

$$P = EI = 1500 \times 0.7 = 1050 \text{ watts}$$

The power to the antenna is:

$$P = I^2R = (9)^2 \times 8.2 = 664.2 \text{ watts}$$

The efficiency is:

$$\% \text{ efficiency} = \frac{P_o}{P_{in}} \times 100$$
$$= \frac{664.2}{1050} \times 1000 = 63.3\%$$

16-3. PROPAGATION AND INTERFERENCE

16-3-1. What is the relationship between the operating frequency and ground-wave coverage?—As the operating frequency increases, the distance between the radiating antenna and the point at which the ground wave drops to an insignificant level decreases. The ground-wave coverage per a given power output decreases with frequency.

16-3-2. In radio transmission, what bearing do the angle of radiation, density of the ionosphere, and frequency of emission have on the length of the skip zone?—The influence is shown in Fig. 16-14. As the angle of radiation relative to ground decreases, the length of the skip zone increases. In general, the higher the frequency, the greater is the penetration of the radio wave into the ionosphere, and the greater is the possible length of the skip zone. However, at too high a frequency, the penetration of the ionosphere is complete, and the radio wave does not return to the earth. Up to a point, the density of the ionosphere at a

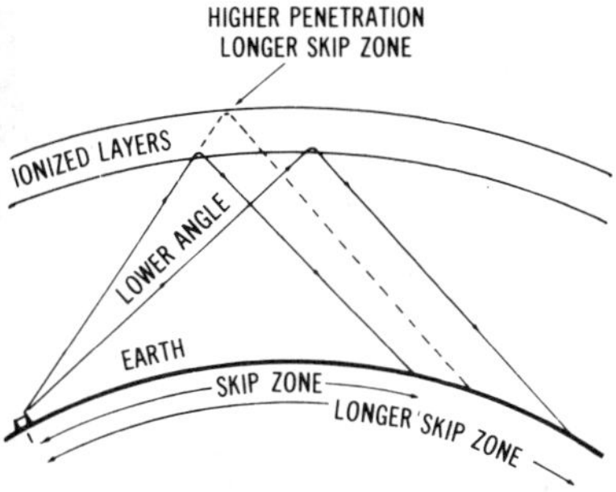

Fig. 16-14. Skip zone and radiation angle.

given time also controls the distance at which the reflected wave returns to the earth. The higher the density, the shorter is the skip zone. However, too dense an ionization will absorb all of the energy, and little energy will return to the earth.

In general, low angles of radiation are preferred for returning medium- and low-frequency signals to earth at preferred medium and long distances. Likewise, a low angle is preferred in the return of high-frequency radio waves in long-distance short-wave radiocommunications and broadcasting.

16-3-3. Why is it possible for a sky wave to meet a ground wave 180° out of phase?—It is this condition that causes fading. The out-of-phase condition is a result of the difference in path lengths traveled by a sky wave and a ground wave. A radio wave reverses in polarity for every half wavelength that it travels, thus it is very possible that the sky wave and ground wave will arrive out of phase at a particular point. More than likely, due to changing ionization the changing path lengths of the sky wave will cause addition and subtraction over a period of time, and the received radio signal will fade in and out as its net strength varies.

16-3-4. What radio frequencies are useful for long-distance communications that require continuous operation?—Low frequencies, 500 kHz and below, are preferred for continuous long-distance operation. To maintain such long-range communications, it is necessary that extremely large antennas and extremely high transmitter power be used.

Short-wave high-frequency channels are also kept open reasonably well using lower power and smaller (but more directional) antenna systems. However, frequency and antenna changes must be made to maintain communications, and even under these conditions the variables of ionosphere propagation can often make communications difficult or impossible. A reliable form of radio-relay communications has been developed incorporating the use of tropospheric scatter. Again such a system which operates in the vhf and higher-frequency ranges requires tremendous power and the use of complex intermediate-relay stations.

16-3-5. What type of modulation is usually contained in static and lightning radio waves?—Such noises are in the form of amplitude variations. Such interference begins at low frequencies and rises to a maximum in the medium frequency range, falling off in magnitude with a frequency increase into the high-frequency and vhf ranges.

16-3-6. What is knife-edge refraction and how can it be used?—Knife-edge refraction or diffraction represents the ability of a high-frequency wave to fill in behind a natural obstacle such as a mountain ridge, bridge, or tall structure. At the edge, the radio wave is refracted and broken up into rays that fan out from the knife-edge. Some of these fill in behind the obstacle and can provide radiocommunications to some degree behind the obstacle. Thus there is some coverage behind a mountain ridge and in metropolitan areas there is fill-in because of diffraction at the top of buildings and along the sides. The same applies to bridges and other high obstacles.

16-3-7. How is the operation of a two-way radio system affected by changes in antenna height? antenna gain? terrain? urban environment? frequency band?—Antenna height increases the range of operation of a two-way radio system and, in general, improves the reliability of communications within the service area. Antenna gain also improves range and reliability. In the two-way radio services an omnidirectional horizontal radiation pattern is most common because the mobiles are generally cruising on all sides of the base station. However, depending on the actual location of the base station antenna, it is sometimes helpful to have a directional horizontal pattern that provides a better reach into the region from which the bulk of the communications takes place, or perhaps into an area where communications are difficult. Of course antenna gain is also obtained by confining the vertical radiation of the antenna system to low angles. This is a way of obtaining better coverage at the same time that the horizontal pattern remains omnidirectional. Terrain has a definite influence on operating conditions of a two-way radio system. Usually the best performance is obtained when the mobile units are in the clear and on relatively high terrain. Reception in valleys and particularly behind ridges and mountains is sometimes difficult. Choice of the base-station height relative to the average terrain and preferred area of coverage is an important consideration. If the base-station antenna must be located at a relatively low site, antenna height is especially important. Also an efficient gain antenna is helpful. In an urban environment tall buildings, bridges, and other obstructions influence radio communications. Getting the base-station antenna high and clear and in a good position relative to the area to be covered are important factors. Dead spots and shadow areas occur among tall buildings and in the region of large metallic obstacles. The frequency band used too has an influence on operating conditions. The higher the frequency of operation is, the more beamlike is the propagated wave. Less energy is propagated beyond the horizon and the fill-in behind hills and buildings is not as extensive. In general, when the greatest possible range is to be obtained in suburban and rural areas, vhf propagation conditions are to be preferred over uhf propagation. Likewise, within the service areas, there are fewer dead spots and shadow areas. However, this does not rule out the use of uhf in these areas because in most situations the service areas to be covered are not great and uhf performance is fine. One might expect the same conditions to exist in crowded metropolitan areas of tall buildings and many obstructions. However, it has been found that uhf coverage in downtown areas is often significantly better than vhf because of the sharp reflections of uhf waves from buildings and other obstacles that send down strong signals to the street level where the mobiles cruise. Range is often not a consideration because the service area in miles is not particularly great.

16-3-8. Discuss the relationship between transmitter power and the field intensity of the reception point.—Field intensity or field strength refers to the magnitude of an electromagnetic field at a reception point. It is most often expressed in *microvolts per meter*. Field strength varies as the square root of the transmitted power. To double the field intensity it is necessary to quadruple the power.

16-3-9. How can heavy rain or snow disrupt a microwave link?—In the propagation of microwaves the water vapor content of the atmosphere absorbs the microwave energy. Thus there is greater attenuation in the path between transmitter and receiver. This added absorption, if severe, can reduce the microwave signal to the point that it will not be propagated over the entire link path.

16-3-10. How may transmitter and receiver antenna separation be used to reduce desensitization in a repeater system? Why is vertical separation better than an equal distance of horizontal separation in land-mobile operation?—The greater the separation is between the receiver antenna and the antenna of the transmitter that is causing desensitization, the weaker the interfering signal component becomes. The greater the isolation is between the two systems, the greater is the reduction in the desensitization. Signal exchange is reduced more from vertical separation of antennas than horizontal. Thus the reduction in desensitization can be made greater by a vertical separation of so many feet as compared to the same distance in horizontal separation. This is to be anticipated because of the characteristic low angle of vertical radiation inherent in the design of the gain antennas used at repeater sites. As you recall, the horizontal radiation patterns are largely omnidirectional. Refer to Sections 8-1, 8-2 and 8-6.

16-3-11. What are circulators? isolators? How are they used?—A circulator is a one-way signal device. Signal passes largely unattenuated in traveling one way through the circulator. A signal traveling in the opposite direction on the same, or a nearby frequency, will be attenuated. An isolator is a circulator that includes a dissipating resistor. Thus the energy from a signal coming in the attenuating direction can be dissipated by the terminating resistor. These devices can be used to reduce interference among the transmitters and receivers that employ a common site. Desensitization and intermodulation can be reduced by using such devices. Most often they are connected in the transmission line paths between the receivers and their antenna systems. Study Section 8-6.

16-3-12. Discuss the uses of directional antennas in eliminating land-mobile cochannel interference problems.—Cochannel interference refers to interference on the same channel. In the land-mobile radio services the use of directional mobile antennas is rarely feasible because the angular position of the mobile unit with respect to the base station is an unknown quantity. However, depending on the site of the base station, the use of a directional base-station antenna is sometimes possible. For example, if the base-station site is to the east of the primary coverage area of the system it is possible to use a directional antenna and point it west. This would send a stronger signal to all the mobile units in the area and make them less subject to cochannel interference from other stations. However, this type of installation is not feasible if the mobile units travel on all sides of the base station out to the limits of the service area to be covered. Sometimes base-station antennas are designed with a limited directivity toward a section of the service area in which there is the greatest cochannel interference problem. Thus some improvement can be made in the ability to reject cochannel signals at the same time service can be provided into other regions of the service area. However, antennas can also have a vertical directivity which refers to the vertical angle at which the antenna radiates most signal or is most sensitive to a received signal. Thus a base station antenna with low-angle vertical directivity helps to send as strong a signal as possible into the service area and again the receivers are made less sensitive to possible cochannel interference. Refer to Chapter 8 and especially Section 8-2.

16-3-13. What is intermodulation?—Intermodulation refers to the mixing of two or more radio signals to produce other frequencies that can cause interference among two-way radio systems. Study Section 8-6.

16-3-14. Discuss some methods for eliminating transmitter intermodulation. For eliminating receiver intermodulation.—Study Section 8-6.

16-3-15. What problems can electrical interference cause to CB stations and aeronautical ground stations?—CB stations and many aeronautical ground stations employ amplitude modulation. Electrical interference is predominantly amplitude modulation components that can cause interference in CB and aeronautical receivers. CB receivers in a mobile unit are subject to a variety of ignition and other amplitude-modulated noises in the vehicle. The same applies to aeronautical ground stations that are mobile. Electrical interference, if it is severe, can restrict the range of operations of a two-way radio system. Various types of noise-reduction systems are a part of modern mobile receivers and help considerably in reducing the effects of this type of interference. Refer to Question 15-3-5, Chapter 15.

16-3-16. Why is intermediate frequency pickup a problem when a 40-MHz transmitter is operated near a television receiver? What are some possible solutions to this problem?—The intermediate frequency of a typical television receiver is also in the 40-MHz frequency range. Consequently a 40-MHz transmitter signal picked up on the antenna and fed through to the if amplifier of the television receiver can cause interference. A number of things can be done to remedy the situation. One of these is the installation of a high-pass filter between the transmission line and the receiver antenna terminals that offers attenuation to the interfering signal at the same time it passes without attenuation the higher frequencies of the incoming television signals. Antenna system should be designed for maximum sensitivity on the television frequency band so as strong a television signal as possible is delivered to the receiver input. Receiver and high-pass filter should have a good ground to make the entire system more effective in accepting the desired signals and rejecting the undesired one. Resonant traps can be installed and precisely tuned to the undesired frequency to block it from the input of the receiver. Refer to Section 8-5.

16-3-17. Explain what occurs when the front end of a television receiver is overloaded by a nearby CB transmitter operating at 27 MHz. What are some possible symptoms which would be observed on the television?—Radio-frequency interference puts a close-spaced horizontal bar pattern on the television screen. This interference pattern can vary as the operator of the transmitter speaks into the microphone. The stronger the radio interference is, the more the desired picture is pushed into the background. With strong interference it can only be seen faintly in the background. If the overload is severe the picture may disappear completely because the tuner may be biased off by agc action.

16-3-18. On what television channels do the second and third harmonics of a 27-MHz Citizens Band transmitter fall?—Citizens Band transmissions usually affect channels 2, 5, and 6 most severely. The second and third harmonics of the Citizens Band frequency range fall on these channels. Refer to Section 8-5.

16-3-19. What is image reception? How would it cause a 120-MHz aircraft transmission to be received on an fm broadcast band receiver with a 10.7-MHz intermediate frequency?—Image reception is the pickup by the receiver of an image frequency. The image frequency differs from the desired frequency to be picked up by the receiver by twice the intermediate frequency. Receivers are, in general, designed to have a low image response, but if an incoming image frequency is strong enough it will penetrate the receiver and cause interference. For example, the local oscillator of an fm receiver tuned to 98.6 MHz in the fm band will have its local oscillator tuned to 109.3 MHz (98.6 + 10.7). If an incoming 120-MHz aircraft signal is strong enough it will mix with the 109.3-MHz local oscillator frequency to also produce a difference frequency of 10.7 MHz (120 − 109.3). Thus the if frequency from the desired and undesired signals is 10.7 MHz and interference results. Note that the difference between the interfering signal frequency and the desired signal frequency is twice the if frequency:

$$120 - (2 \times 10.7) = 98.6$$

16-3-20. Discuss the uses of high-pass and low-pass filters in elimination of 27-MHz Citizens Band interference to television reception.—The high-pass filter is installed at the receiver between the transmission line and the antenna terminal. The low-pass filter is installed at the transmitter between the transmission line and the CB radio. Study Section 8-5.

16-3-21. Explain how a 27-MHz Citizens Band transmission can be heard on an audio device through audio rectification. Discuss how this can be eliminated using bypassing and/or shielding.—Radio-frequency energy fed into an audio amplifier can encounter a component that will act as a detector such as a transistor, diode, integrated circuit, or electron tube. The rf signal once detected and demodulated will pass through the amplifier just as a normal audio signal. This audio component will be reproduced by the speaker. Usually the interference enters through one of the sensitive inputs of the audio system, particularly low-level and high-impedance inputs. If these inputs are suitably bypassed and a good grounding system is used for the audio components, the interference can be reduced. Make certain all the components are shielded and that each shield is connected to the ground system. Shielded leads should be used for the various inputs as well as the speakers. A power-line filter may help if the radio-frequency energy is entering in this manner. Refer to Section 8-5.

16-3-22. Discuss the uses of open and shorted stub traps to attenuate interfering

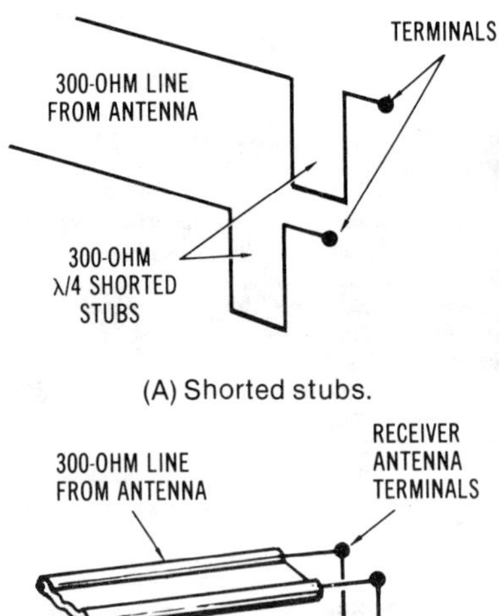

(A) Shorted stubs.

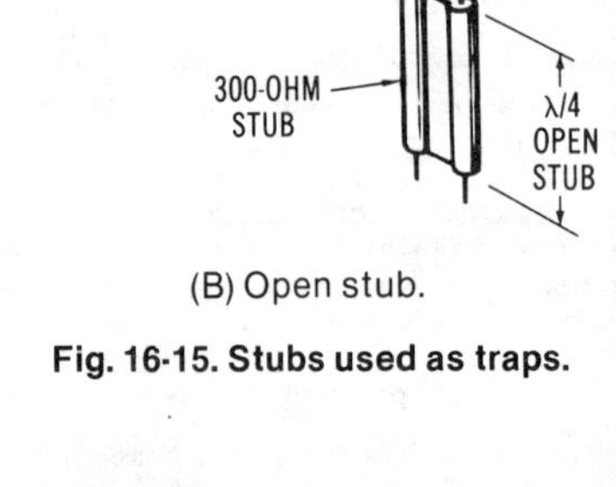

(B) Open stub.

Fig. 16-15. Stubs used as traps.

signals at a receiver.—A stub trap can be made of a section of transmission line. A quarter-wavelength section grounded acts as a parallel-resonant circuit; a quarter-wave section open, as a series-resonant circuit. Such traps can be installed in the transmission line path to the receiver input as shown in Fig. 16-15. A parallel-resonant trap is placed in series with the transmission line and offers maximum rejection of the frequency to which it is tuned. The length of the stub can be calculated using the following equation:

$$\text{Length in inches} = \frac{2952 \times V}{f}$$

where,

V equals velocity factor of line,

f equals frequency of interference in megahertz.

It might be necessary to adjust the length of the stub to obtain maximum rejection at the desired frequency. Perhaps a more common stub is the open-circuited arrangement of example B. This is connected right across the input terminals of the receiver. It acts as a short-circuit to the frequency at which it is tuned. Again some trimming may be necessary to obtain maximum rejection. The same equation is used to calculate the length of the stub. Coaxial stubs can be used too. Refer to Section 8-5.

16-3-23. How could improper grounding affect the harmonic radiation from a Citizens Band transmitter?—If the transmitter is not grounded properly there can be harmonic radiation through the case of the transmitter. Most Citizens Band transmitters include a harmonic rejection circuit. This circuit too is most effective when the transmitter has a proper ground.

CHAPTER 17

Test Equipment and Measurements

17-1. INDICATING INSTRUMENTS

17-1-1. Make a sketch showing the construction of the D'Arsonval-type meter, and label the various parts. Draw a circuit diagram of a vacuum-tube voltmeter and a wattmeter.—A sketch of a D'Arsonval meter is given in Fig. 17-1A. This basic meter movement has a permanent magnet and a moving coil. When the current to be measured flows through the coil of fine wire, the magnetic field set up around the coil reacts with the magnetic field of a permanent magnet. The reaction is such that the coil and its attached pointer rotate until a spiral spring exerts an equal but an opposite pull. The pointer moves across a calibrated current scale. The extent of the coil movement is a function of the magnitude of the introduced current.

A wattmeter diagram is shown in Fig. 17-1B. A wattmeter responds to both the voltage and the current of the power line into which it is connected. A basic vtvm circuit is shown in Fig. 17-2. A vtvm has a high input resistance. Because of its almost minute current demand, it places an unimportant load across any circuit to which it is connected.

17-1-2. Show by a diagram how a voltmeter and ammeter should be connected to measure power in a dc circuit.—Refer to Fig. 17-3. The power in the dc circuit will, of course, be the product of the voltage and current readings.

17-1-3. If a 0- to 1-dc milliammeter is to be converted into a voltmeter with a full-scale calibration of 100 volts, what value of resistance should be connected in series with a milliammeter?—A dc milliammeter can be made to measure voltage with the insertion of a proper current-limiting series resistance. The value of resistance needed can be determined by Ohm's law:

$$R_{series} = \frac{100}{0.001} = 100{,}000 \text{ ohms}$$

The value of the series resistance is the quotient of the maximum voltage to be measured divided by the maximum current that may be drawn by the meter movement and, if used, its meter shunt.

17-1-4. A 1-mA meter having a resistance of 25 ohms was used to measure an unknown current by shunting the meter with a 4-ohm resistor. The meter then indicated 0.4 milliampere. What was the value of the unknown current?—The answer is 2.9 milliamperes. The voltage drop across the meter is

$$E = I_M R = 0.0004 \times 25 = 0.01 \text{volt}$$

Therefore the current in the shunt becomes

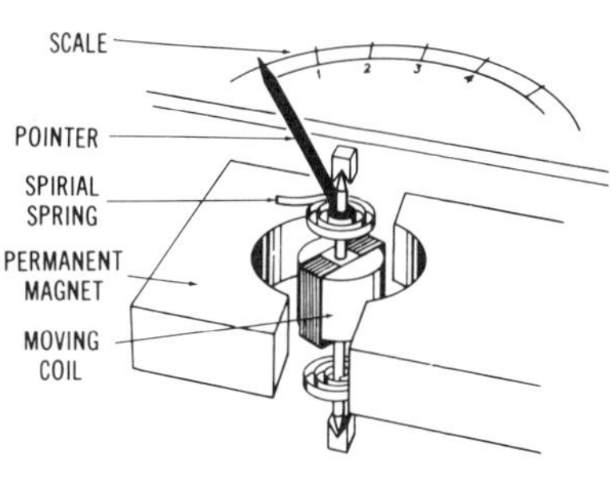

(A) D'Arsonval movement.

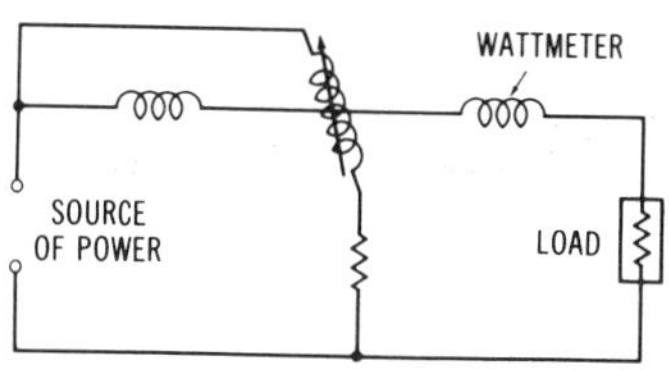

(B) Wattmeter circuit.

Fig. 17-1. Basic D'Arsonval meter movement and a wattmeter.

$$I_s = \frac{E}{R_s} = \frac{0.01}{4} = 2.5 \text{ mA}$$

The unknown current is the total current drawn through meter and shunt:

$$I_t = I_m + I_s = 0.4 + 2.5 = 2.9 \text{ mA}$$

17-1-5. An rf vtvm is available to locate resonance of a tunable primary tank circuit of an rf transformer. If the vtvm is measuring the voltage across the tuned secondary, how would resonance of the primary be indicated?—The meter would read a *maximum* at the resonant setting of the primary. When the primary is tuned to resonance, there is maximum circulating current, and maximum flux links the primary and secondary. Therefore, maximum voltage is induced across the secondary.

17-1-6. Define the following terms, and describe a practical situation in which they might be used: (a) rms voltage, (b) peak current, (c) average current, (d) power, (e) energy.—(a) Rms voltage is an abbreviation for *root-mean-square voltage*. It is the effective value of voltage for

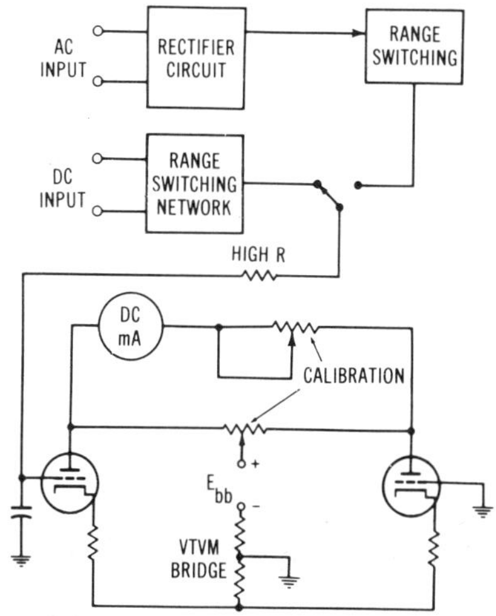

Fig. 17-2. Basic vtvm.

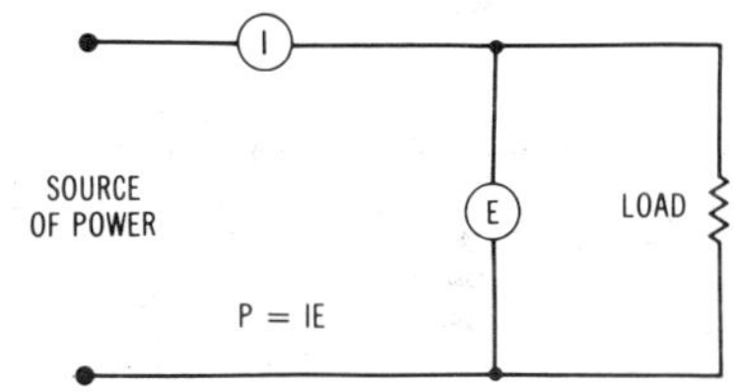

Fig. 17-3. Using ammeter and voltmeter for power measurement.

an ac sine wave, and it is 0.707 of the sine-wave peak voltage.

A 1-ampere peak ac sine wave produces the same heating effect and power dissipation as 0.707 ampere dc. Most ac voltmeters and ammeters are calibrated to read rms values; the rms values are important in calculating ac power.

(b) The *peak current* is the maximum amplitude of a sine wave (from base line to crest) or the peak value of current for a nonsinusoidal waveform. Peak current is significant in class-C amplifiers. Both the peak grid current and the peak plate current have practical importance because they influence the magnitude and the power of the rf energy developed in the associated plate tank circuit.

(c) An ac sine wave or a nonsinusoidal wave has peak- and average-current values. The *average current* is the net current when a repeating waveform is averaged over a period of time. The average value of a sine wave is 0.636 times the peak value. The average value of a nonsinusoidal wave cannot be defined so exactly because it is a function of shape, duration, and period. The dc grid- and plate-current meters of a class-C amplifier read the average grid and plate current, respectively. This average current is sometimes spoken of as the dc component of current which, in the case of a class-C amplifier, is important in terms of dc power input, rf drive, and modulation requirements.

(*d*) *Power* is the rate of electrical energy use; it can be the rate at which electrical energy is fed to a device or the rate at which electrical energy is taken from a source of power.

Using the previous class-C amplifier as an example, we find that there is a certain amount of *dc power input* in watts to the plate circuit of the amplifier. This is instrumental in making available a certain amount of *rf power output* in watts in the plate tank circuit.

(*e*) *Energy* is the capability of a device to supply electrical power. Electrical energy is measured in joules. Batteries, generator, and power supplies are examples of sources of electrical energy. We often speak of them as power sources. However, they are actually electrical-energy sources, and power is drawn only when they are put to work delivering so many joules per second or amperes to a load. A practical measure of consumed electrical energy is the kilowatt-hour (kWh).

17-1-7. Explain the operation of a resistance bridge. If the known resistances in such an instrument are 5 and 10 ohms, and if adjusting the third resistance to 50 ohms produces a perfect balance, what is the unknown resistance?—In a resistance bridge (Fig. 17-4) a source of voltage is connected across two opposite terminals of the bridge while a current meter is connected between the second pair of terminals. Such a bridge is balanced when the voltages at point A and point B are the same. With no difference of potential between points A and B there will be no current in the meter. To obtain the same voltage at points A and B it is necessary that the ratio of R_1/R_2 be the same as R_k/R_x. When so balanced, the voltages across resistor R_1 and R_k are the same, and so are the voltages across R_2 and R_x.

In the example of Fig. 17-4 the voltage across R_2 would be twice the voltage across R_1. Therefore, the voltage across

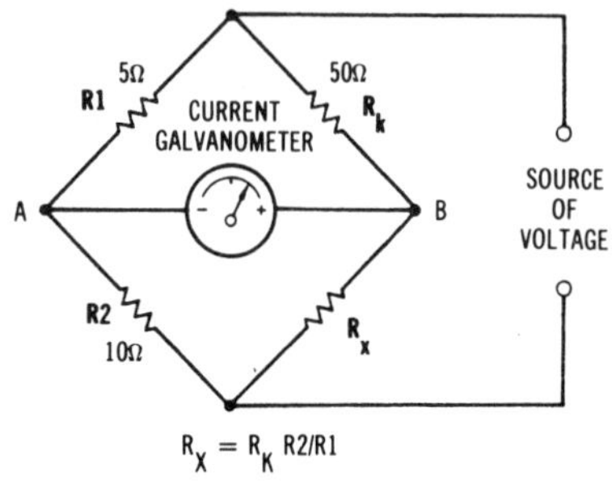

Fig. 17-4. A basic resistance bridge.

R_x must of necessity be two times greater than the voltage across R_k if the bridge is to be balanced. It follows then that resistance R_x is 100 ohms (50 × 2).

The same result can be obtained by rearranging the basic bridge equation as follows:

$$\frac{R_1}{R_2} = \frac{R_k}{R_x}$$

$$R_x = R_k \frac{R_2}{R_1}$$

$$= 50 \times \frac{2}{1} = 100 \text{ ohms}$$

17-1-8. Describe how horizontal and vertical deflection takes place in a cathode-ray oscilloscope. Include a discussion of the waveforms involved.—Fundamentally, an oscilloscope tube consists of a source of electrons which are directed and focused as a fine beam on a fluorescent screen. The screen glows at the point of impact with a brightness that depends on the density and the velocity of the beam of electrons. Refer to Fig. 17-5.

As the electrons proceed from the electron gun to the fluorescent screen, they pass between two pairs of deflection plates (horizontal and vertical). The electric stresses placed across the two pairs of plates are such that the beam can be deflected away from its dead-center position, striking the screen to the right, left, above, or below dead center.

For example, if the difference of potential between the two horizontal-deflection plates is zero and the difference of potential between the two vertical deflection plates is zero, the electron beam will strike the fluorescent screen dead center.

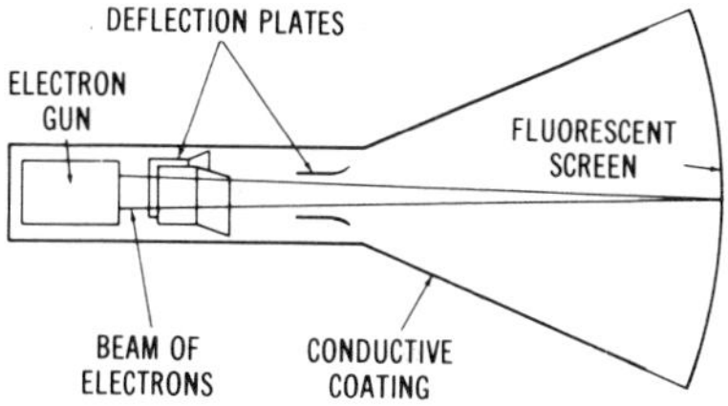

Fig. 17-5. Basic oscilloscope tube.

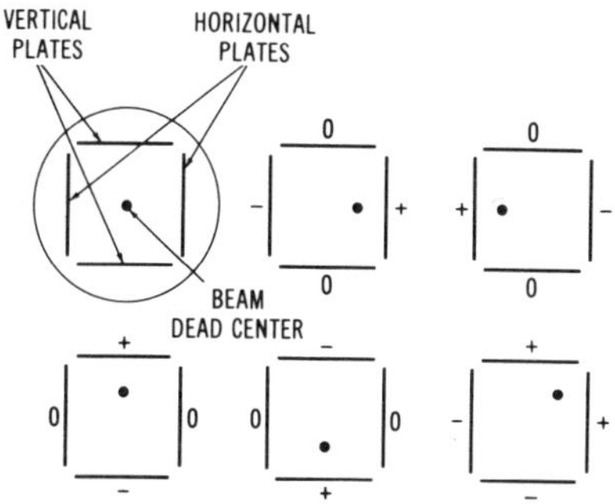

Fig. 17-6. Position of beam impact on fluorescent screen as a function of plate polarities.

However, looking into the face of the cathode-ray tube screen as shown in Fig. 17-6, let us now place a more positive voltage on the right deflection plate. In this case the beam, in passing between the plates, will be pulled toward the right plate and repelled away from the left plate. Therefore, it will strike the screen to the right of center. If the polarity on the two horizontal-deflection plates is reversed, the beam will strike to the left of center. How far the beam strikes to the left of center depends on how much more positive the left plate is than the right plate.

In a similar manner, a difference of potential applied to the vertical deflection plates (assuming that there is no difference of potential across the horizontal deflection plates) can position the beam directly above or directly below center.

If we now apply a positive voltage to the top plate and a similar positive voltage to the right plate, the beam will be deflected upward and to the right, appearing at a 45° angle relative to the horizontal and vertical axes. Actually, just where the beam strikes the screen in the right quadrant depends on the absolute and relative values of the voltages applied to the vertical- and horizontal-deflection plates. It is apparent, then, that with the use of proper dc voltage, the beam can be made to strike any given point on the fluorescent screen.

To go a step further, if instead of applying dc voltages to the deflection plates we apply ac voltages, the beam can be made to move or *scan* over the fluorescent screen. This is the basic principle of the test oscillograph as well as television, radar, and other electronic systems which employ cathode-ray tubes.

If we want to make the scanning beam move in a linear manner from the left side to the right side of the screen, we apply a sawtooth of voltage to the horizontal deflection plates, as shown in Fig. 17-7. With no voltage applied to the vertical deflection plates, the scanning beam will move across the screen as the sawtooth voltage rises linearly from a maximum negative through zero to a maximum positive. During the retrace portion of the sawtooth wave the scanning wave will snap back quickly from the right to the left because of the

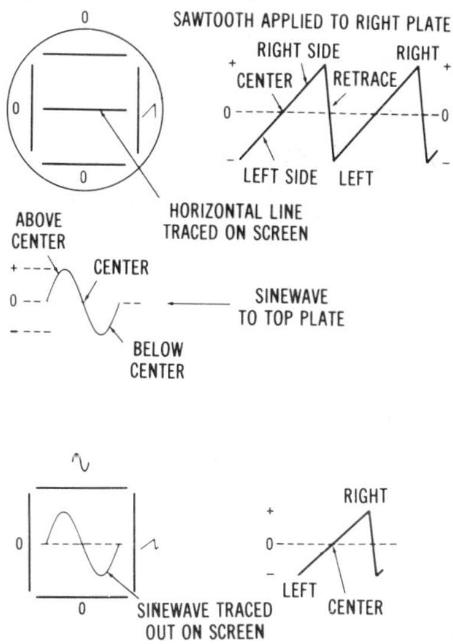

Fig. 17-7. Oscilloscope deflection waveforms and patterns.

very fast rate at which the voltage changes during this retrace time. In this manner, a horizontal "trace" line is formed on the screen as the beam scans back and forth.

Let us next consider the beam motion when, in addition to the sawtooth voltage applied to the horizontal deflection plates, we apply a sine wave of the same frequency to the vertical deflection plates. Under this condition, there will no longer be a straight horizontal line traced on the screen. As the horizontal sawtooth voltage increases linearly, it attempts to pull the beam in a straight line from left to right across the screen. However, at the same time, the sine wave applied to the vertical-deflection plate is causing the top plate to swing positive relative to the bottom plate. Thus, as the beam attempts to move from left to right across the screen, it is also pulled upward. Its upward deflection is a result of the positive alternation of the sine wave. When the sine wave is passing through zero the difference of potential between the vertical deflection plates will be zero, and at this very instant the beam will be crossing the horizontal axis at dead center. However, the sawtooth wave continues to pull the beam to the right, and at the same time the sine wave goes into its negative alternation, which makes the bottom plate positive relative to the top plate. As a result, the scanning beam is pulled below the horizontal axis, and the negative alternation of the sine wave is traced out on the screen. At the end of the cycle the beam is snapped back to the left very quickly by the retrace of the sawtooth, and the oscilloscope is ready to trace out the next sine-wave cycle. This cycle is traced out on top of the previous cycle, and the sine wave becomes visible. In a similar manner, pulses, sawtooth waves, and other nonsinusoidal waves can be traced out and made visible on the screen of an oscilloscope tube. Likewise, it is possible to display more than one cycle, and with appropriate circuits just a small portion of one cycle can be spread out for inspection on an oscilloscope.

17-1-9. Cathode-ray oscilloscopes are frequently used to register percentage modulation. Sketch the visual displays of (a) 0% modulation, (b) 50% modulation, (c) 100% modulation and (d) 120% modulation.—Waveform and trapezoidal patterns for the 0%, 50%, 100%, and 120% modulation rf carrier are shown in Fig. 17-8.

17-1-10. Explain what a voltmeter's sensitivity in ohms per volt means.—The ohms-per-volt sensitivity of a voltmeter provides an indication of the loading influence of the meter on a circuit being measured particularly when it is a high-resistance source. The loading influence or resistive component contributed by the meter can be calculated by multiplying the full-scale reading of the meter times the sensitivity of the voltmeter in ohms per volt. For example, if the full-scale meter reading is 100 and the sensitivity of a voltmeter is 20,000 ohms per volt, the total resistance of the meter will be 2 million ohms ($100 \times 20{,}000$). A total resistance of this value will have very little or no loading effect when measuring voltages across moderate- and low-resistance circuits. Particularly in high-resistance circuits, a high-resistance voltmeter permits more accurate measurements. Thus a voltmeter with a high sensitivity is preferred.

17-1-11. Draw a diagram of a simple ohmmeter and explain its operation.—A simple series ohmmeter is shown in Fig. 17-9. With the test probes shorted (representing zero resistance), the circuit current can be adjusted by varying potentiometer R1 (called the ohms-adjust control) to cause full scale deflection of the meter. In the example the meter is a 1-mA movement and the total resistance in the circuit (including the 100 ohms for the meter) is 3000 ohms. A 3-volt battery produces a 1-mA current in the circuit, thus causing full-scale deflection.

Since the full-scale deflection on the meter now represents zero resistance, that point is marked 0 ohms. If the short across the test probes is removed, the series circuit is open and the current in the meter drops to 0. This represents the case of an infinite resistance.

If intermediate values of resistance are connected across the probes, the total series resistance becomes greater than 3000 ohms and the meter reading will fall somewhere between 0 and infinity. For example, if a 3000-ohm resistance is connected across the probes the total series-circuit resistance becomes 6000 ohms and the meter current will be 0.5 mA, causing the meter pointer to deflect to a half-scale position. This position on the scale of the meter can then be calibrated as 3000 ohms.

Unknown resistance values above 3000 ohms will bring meter indications closer to the point marked infinity. Resistance values lower than 3000 ohms will give meter indications between mid-scale and full-scale deflection.

17-2. MEASUREMENTS

17-2-1. Draw a simplified circuit diagram of an absorption wavemeter (with galvanometer indicator). Explain its operation, and give some possible applications.—Such a wavemeter is shown in Fig. 17-10. It consists of a resonant tank circuit, a rectifier crystal and filter capacitor, and a dc meter. The dial of the variable capacitor is usually calibrated in frequency or in calibrations that can be referred to as associated frequency versus dial-setting chart.

The rf voltage across the resonant circuit is rectified by the crystal diode. Its pulsations are filtered out by the capaci-

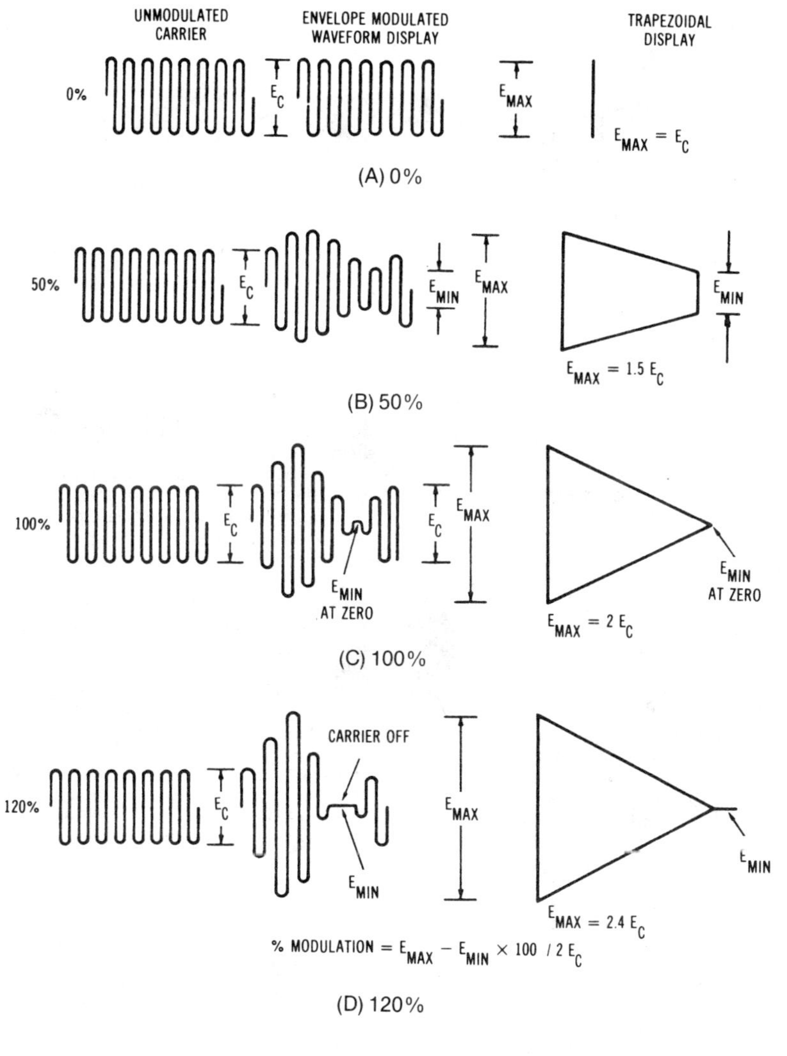

Fig. 17-8. Envelope and trapezoidal oscilloscope displays.

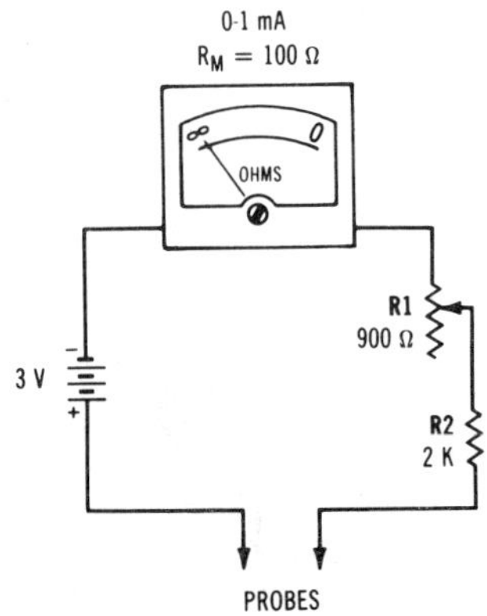

Fig. 17-9. Simple series ohmmeter circuit.

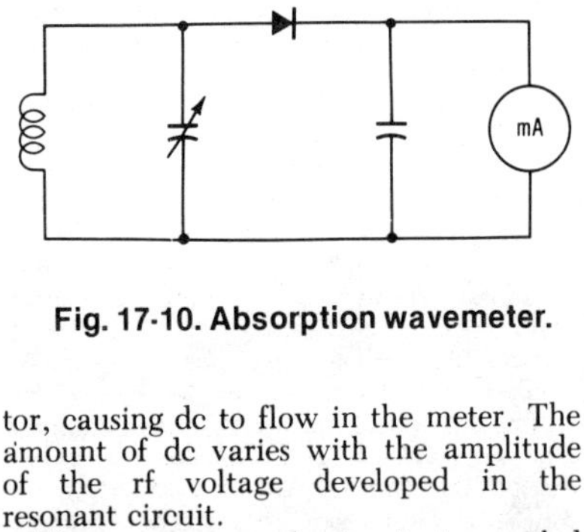

Fig. 17-10. Absorption wavemeter.

tor, causing dc to flow in the meter. The amount of dc varies with the amplitude of the rf voltage developed in the resonant circuit.

The resonant tank circuit is coupled just close enough to the source of rf energy to provide a meter reading. The capacitor is then tuned for maximum meter reading, indicating that the tank circuit has been tuned to resonance and maximum current is circulating within it. Because maximum current exists when the tank is tuned to resonant frequency of the rf energy source, the frequency calibration of the capacitor dial corresponds to the frequency of the rf energy source. If the wavemeter is carefully designed and calibrated, and if it is coupled just near enough to the rf source to provide a usable reading, a reasonably accurate frequency measurement can be obtained.

An absorption wavemeter can be used to determine the presence and frequency of a source of rf energy. It will indicate the frequency made available at the output of a multiplier stage, or the presence of spurious oscillations (self-oscillation or parasitics). It functions well as an rf indicator in making resonance, neutralization, and output adjustments.

17-2-2. Draw a simplified circuit diagram of a grid-dip meter. Explain its operation, and give some possible applications.—A grid dip meter is a frequency-calibrated oscillator (Fig. 17-11).

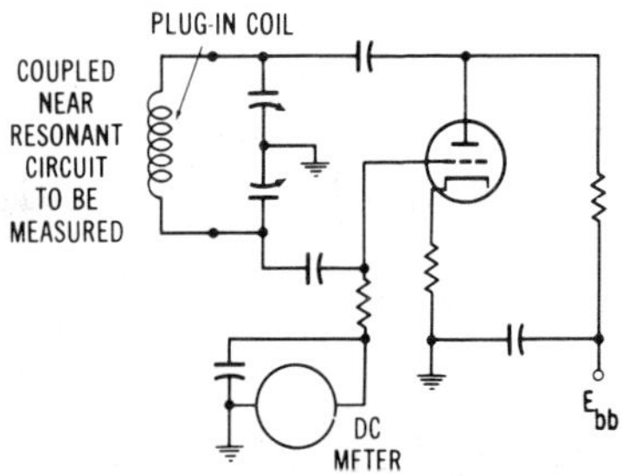

Fig. 17-11. Grid-dip meter.

Plug-in coils are usually employed to provide a wide frequency range. For measurement of grid current, a sensitive dc meter is located in the oscillator grid circuit. When the oscillator is in operation, there is a grid-current reading.

The grid-dip oscillator can be used to determine the resonant frequency of a tuned circuit in which there is no rf energy. Thus, one of the advantages of a grid-dip oscillator is that it can be used to measure resonant conditions in a transmitter or other electronic equipment that employs tuned circuits, without any power being applied to the unit.

When a resonant circuit is coupled to the grid-dip oscillating tank coil (or associated pickup coil), it absorbs some of the energy from the tank. Inasmuch as energy is removed from the oscillator, the oscillator feedback into the grid circuit decreases. As a result, the grid-current meter reading drops or dips. When the grid-dip oscillator is set on the exact frequency of the resonant circuit under measurement, the meter will dip to its minimum value. This setting indicates that the grid-dip oscillator is generating a signal that is the same frequency as the resonant frequency of the tuned circuit under check.

A grid-dip oscillator can be used to check resonant circuits throughout a two-way radio transmitter and receiver.

It can be used to search for the spurious resonant conditions which cause parasitic oscillations. Inasmuch as the grid-dip oscillator is itself an oscillator, it can be used as a signal generator for certain tests and measurements, and also for signal tracing.

Most grid-dip meters can also be used as heterodyne-frequency meters because they include a detector and a jack for the insertion of a headset. When the frequency of the grid-dip oscillator and the frequency of a source of signal are brought near to each other, a zero beat is heard in the headset, and the two frequencies can be matched. A grid-dip meter is also useful in the tuning and matching of antenna systems. Most grid-dip meters can also be operated as absorption wavemeters.

17-2-3. Draw a block diagram showing only those stages which would illustrate the principle of operation of a secondary frequency standard. Explain the functions of each stage.—A secondary frequency standard serves as a highly accurate frequency calibrator for signal generators and frequency meters. Often it comes as a built-in part of such instruments, while in other cases it serves as a laboratory standard for calibrating whatever signal generators and frequency meters are used. In the United States the primary-frequency standard is maintained by the National Bureau of Standards. Such primary signals are made available through the transmissions of station WWV. Thus, a secondary standard often includes a receiver for pickup of WWV, or facilities for using an external receiver that can be tuned to the WWV signal. A functional arrangement is shown in Fig. 17-12.

A crystal oscillator is usually employed as the calibrated base-frequency source of the secondary frequency standard. It

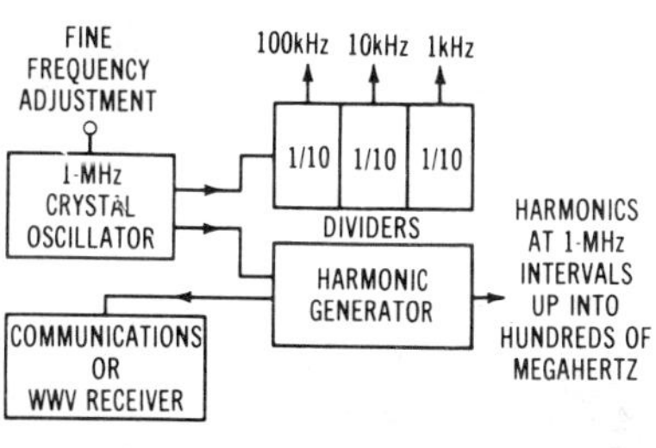

Fig. 17-12. Secondary frequency standard.

operates on one of the WWV frequencies or a subharmonic of one of the frequencies. In the example of Fig. 17-12, a 1-MHz crystal oscillator is being used as a secondary standard. It can be tuned to precisely 1 MHz by comparing its fifth harmonic to the 5-MHz WWV signal.

Additional accurate frequencies can be obtained by using harmonic generators and frequency dividers after the 1-MHz crystal oscillator. For example, a harmonic generator could supply harmonic checkpoints up into the hundreds of megahertz. A series of frequency dividers after the oscillator could provide accurate frequency components at 100 kHz, 10 kHz, and 1 kHz intervals.

17-2-4. Draw a block diagram of a heterodyne-frequency meter which would include the following stages: crystal oscillator, crystal oscillator harmonic amplifier, variable-frequency oscillator, mixer, detector and af amplifier, and af modulator. Show rf input, and rf, af, and calibration outputs. Assume a band-switching arrangement, and a dial having arbitrary units employing a vernier scale. (a) Describe the operation of the meter. (b) Describe, step by step, how the crystal should be checked against WWV, using a suitable receiver. (c) Under what conditions would the af modulator be used? (d) Describe, step by step, how the unknown frequency of a transmitter could be determined by use of headphones or by use of a suitable receiver. (e) What would be meant by calibrated check-points? When should they be used? (f) If in measuring a frequency the tuning dial should show an indication between two dial-frequency relationships in the calibration book, how could the frequency value be determined? (g) How could this meter be used as an rf generator? (h) Under what conditions would it be necessary to recalibrate the crystal oscillator?—(a) The basic principle of operation of the heterodyne-frequency meter involves the beating together of two radio-frequency signals, one whose frequency is known and the other whose frequency is to be determined. When two such frequencies are combined in a nonlinear device, such as a mixer, an audio tone is produced when the two frequencies are brought so near to each other that an audible difference frequency is produced. When the two frequencies are brought precisely together, the difference frequency will be zero (zero beat), and no output will be heard. Under this condition, the frequency of the signal to be measured matches the frequency calibration of the known-frequency oscillator. A functional plan of a heterodyne-frequency meter is shown in Fig. 17-13. It is the variable-frequency oscillator (including band-switching arrangement and vernier dial) which must be kept in accurate calibration. The crystal oscillator and its harmonic amplifier can serve as a secondary-frequency standard. The fifth harmonic of the 1-MHz crystal oscillator can be checked on the 5-MHz WWV signal. The 1-MHz and higher harmonics of the crystal section can then be used to check the variable-frequency dial calibrations over its operating range, up into the vhf and uhf range. There will be calibration points at 1-MHz intervals. Of course, it is possible to check the variable-frequency oscillator directly with WWV signals at those frequency outputs which correspond to the frequencies of WWV transmissions.

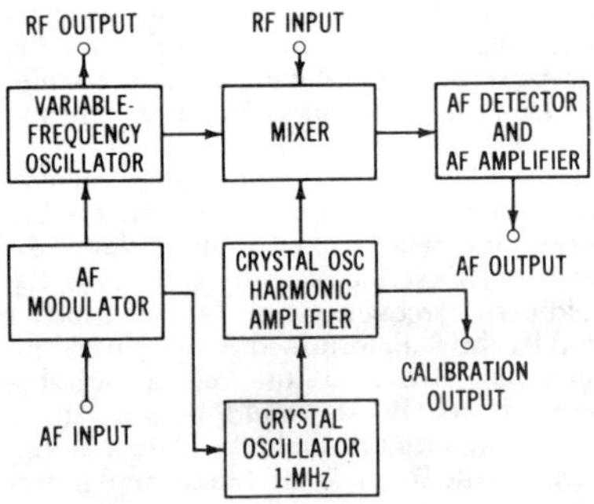

Fig. 17-13. Functional plan of a heterodyne-frequency meter.

The mixer is used to introduce the harmonic components of the crystal oscillator in such a manner that a beat note can be heard in the audio output of the heterodyne-frequency meter whenever the variable frequency oscillator is set to one of the crystal harmonics. The af amplifier is used to build up the level of the beat note or the level of any signal which has been demodulated by the af detector.

The heterodyne-frequency meter also includes an af modulator. Thus an external audio-tone signal or other modulating waves can be used to modulate the variable-frequency oscillator. The af detector permits the modulation to be removed from any incoming modulated signal.

The signal to be checked is also applied to the mixer. Here it combines with the variable-frequency oscillator signal or an appropriate harmonic to produce a zero beat which is amplified. The zero beat can be heard by inserting a headset into the af output jack. A zero beat indicates that the variable-frequency oscillator has been set to the frequency of the signal to be checked. The calibrated band switch and dial of the frequency meter can then be used to find the exact frequency of the incoming signal, using an associated calibration book.

(b) A 1-MHz crystal is being used in the heterodyne frequency meter of Fig. 17-13. A lead from the calibration output of the heterodyne-frequency meter can now be run to the input of a communications receiver that will be used to pick up the WWV signal. Tune the receiver to the 5-MHz WWV signal. Now turn on the 1-MHz crystal oscillator. A beat note will now be heard in the output of the WWV receiver. Adjust the calibration control of the 1-MHz crystal oscillator until there is an exact zero beat with the WWV signal.

(c) The af modulator can be used to apply modulation to the variable-frequency oscillator of the heterodyne-frequency meter. It can then be used as a signal generator supplying, when necessary, a modulated rf signal for test purposes.

(d) The heterodyne-frequency meter is turned on and permitted the prescribed warm-up period. An output taken from the transmitter is applied through a suitable attenuator-isolation network to the rf input of the heterodyne-frequency meter. The heterodyne-frequency meter is set to the proper frequency band. The calibration on this band can be checked using the appropriate crystal harmonics. The crystal can now be turned off and the variable-frequency oscillator varied in frequency until a zero beat is established with the unknown-frequency signal. This zero beat is heard by plugging the headphone into the af output jack. A true zero-beat point is indicated when the audio tone begins to rise in frequency when the variable-frequency oscillator is adjusted to either side of the zero beat.

A suitable receiver that is tunable to the transmitter-frequency range can also be used as a beat indicator. In this case, the transmitter signal is picked up on the receiver. Proper precautions must be taken so that the transmitter signal does not overload the receiver. A connection is made between the rf output of the heterodyne-frequency meter and the antenna input of the receiver. Loose coupling is usually preferred so as not to overload the receiver. The transmitter signal is tuned in on the receiver. The variable-frequency oscillator of the heterodyne-frequency meter is now set to the proper band, and its frequency is varied until a zero beat is heard in the output of the receiver. This indicates that the transmitter and the heterodyne-frequency meter are on the same frequency. The frequency of operation of the transmitter can now be read from the calibration dial of the heterodyne-frequency meter.

(e) When the crystal oscillator is turned on, zero-beats will be heard (headphones plugged into the af output jack) at regular points as the variable-frequency oscillator is tuned over its range. These are known as calibration checkpoints. A beat can be heard at every megahertz point in the example of Fig. 17-13. (It is possible with the addition of suitable circuits to have checkpoints every 100 kHz or even every 10 kHz.) Checkpoints are useful in checking the calibration of the variable-frequency oscillator on its various frequency bands. It is particularly helpful in checking two specific frequency points, one on each side of the variable-oscillator setting, after it has been made to match the frequency of the unknown-frequency signal. By knowing the dial readings for the two frequency checkpoints and also the dial reading that matches the unknown frequency, an interpolation procedure can be used to more exactly determine the unknown frequency.

(f) If the frequency to be measured falls between two dial-calibration points or between two checkpoints, an interpolation procedure can be followed. For example, if the dial reads 46 for 32 MHz and the dial reading for the unknown frequency is 39.2, the unknown frequency can be found. Assuming a straight line relation between the frequency and dial reading, the frequency calibration would be 125 kHz per division [1000 kHz ÷ (46 − 38)]. The unknown frequency to be measured can now be determined as follows:

1. Frequency is 1.2 divisions (39.2 − 38) above 31 MHz.
2. This corresponds to a frequency increase of 150 kHz (1.2 × 125).
3. Unknown frequency is 31.15 MHz (31 + 0.150).

(g) An rf output is made available by the variable-frequency oscillator. Consequently, the heterodyne-frequency meter can also be used as a calibrated signal source. It can be used for signal tracing or adjustment/alignment of a communications receiver. With a suitable audio oscillator, it can also be tone-modulated via the af modulator, if a modulated signal source is required to obtain a more accurate zero beat.

With the crystal oscillator turned on, a calibrating signal can also be made available for checking out the frequency calibration of a communications receiver. A modulated audible tone would be heard at each 1-MHz point over the frequency range of the communications receiver.

(h) The crystal-oscillator frequency should be checked on a scheduled basis and before use (after a suitable warm-up period) when very critical frequency measurements are to be made. Of course, the oscillator should be recalibrated whenever a crystal or an oscillator tube or transistor replacement is made, or when replacing any component of the oscillator-multiplier circuits that could influence the crystal-oscillator frequency. Inasmuch as crystals are temperature-sensitive, rather frequent WWV calibration may be necessary when the crystal is not a part of the thermostatically controlled oven.

17-2-5. Draw a block diagram of an fm deviation (modulation) meter which would include the following stages: mixer, if amplifier, limiter, discriminator, and peak-reading indicator. Explain the operation of this instrument, and draw a circuit diagram and explain how the discriminator would be sensitive to frequency changes rather than to amplitude changes.—A functional block diagram is shown in Fig. 17-14. The typical fm deviation (modulation) meter is

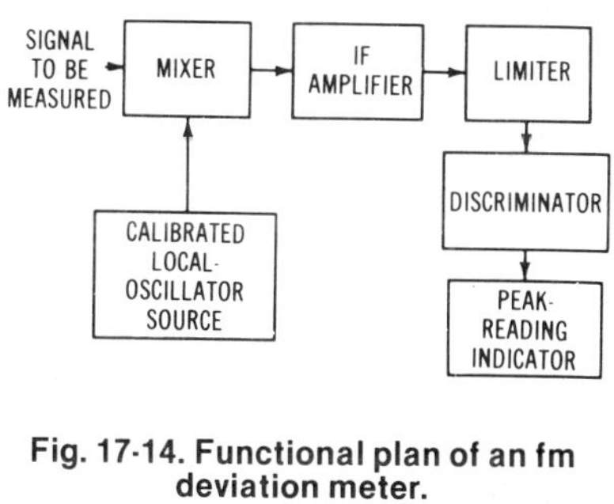

Fig. 17-14. Functional plan of an fm deviation meter.

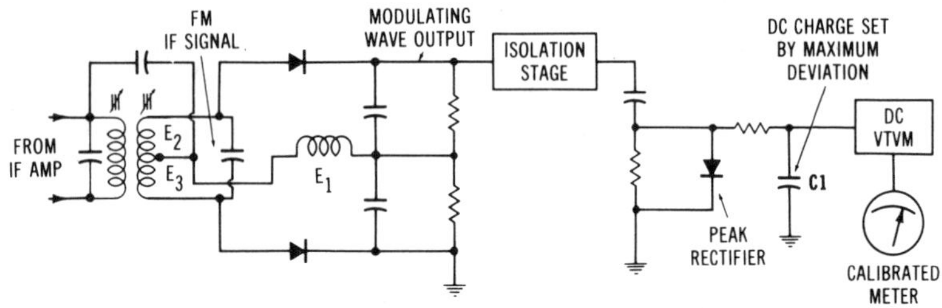

Fig. 17-15. Fm discriminator and peak voltmeter.

comparable to a high-quality fm receiver. The fm signal, modulated by a test signal, is applied to the mixer through a suitable isolation-attenuator arrangement. The modulation meter includes a local oscillator, and the mixer output is tuned to the difference if frequency. The if signal component is increased in amplitude and then applied to a limiter which removes any amplitude variations. The if output of the limiter is applied to the fm discriminator.

The audio test modulation of the fm signal is recovered at the output of the fm discriminator. To determine the peak deviation of the fm signal, it is necessary to develop a dc component of voltage which corresponds to the peak amplitude of the demodulated audio signal. This is accomplished by means of a peak rectifier. This dc component is then applied to a balanced vtvm stage and its associated sensitive current meter which is calibrated in terms of frequency deviation.

A simplified circuit diagram of a discriminator and peak voltmeter is given in Fig. 17-15. The output of the discriminator is a function of the applied frequency change. The limiter stages previous to the discriminator have removed any existing amplitude variations. Thus, amplitude changes which may have been present as a result of noise or interference are not able to contribute to any variations at the output of the discriminator. The discriminator responds to the phase relationship among the three voltages, E_1, E_2, and E_3. These phase changes are brought about by any frequency deviation of the incoming signal. The process of converting the phase changes to amplitude variations at the output of the discriminator was covered in detail in the discussion associated with Figs. 15-14 and 15-16.

As mentioned previously, the circuit must be made to respond to the peak amplitude of the demodulated audio variation. To do this, the output of the discriminator is fed through an isolation stage to a diode peak rectifier. The input time constant of the peak rectifier is such that when the rectifier conducts, the output capacitor charges to the peak amplitude of the demodulated wave. The long output time-constant keeps the dc charge steady between peaks, and a dc voltage is applied to the grid of the vtvm tube. As the peak amplitude of the modulating wave changes, so also does the peak charge placed on capacitor C1 change. It should be stressed that the output of the discriminator varies in amplitude in accordance with the fm deviation. The greater the deviation of the applied signal, the higher is the peak amplitude of the demodulated wave at the output of the discriminator.

17-2-6. Why is it important that a transmitter remain on frequency, and that the harmonics be attenuated?—It is important that transmitters remain on frequency because of the multiple use of the radio spectrum and, in the case of two-way radiocommunications, because of the use of crystal-controlled receivers which are preset to the transmit frequency. In the latter case, distortion and poor performance would result if the transmitter frequency would drift and be incorrect in relation to the crystal-controlled receiver frequency. Radio channels are allocated very near to each other, and a transmitter-frequency drift can cause interference with stations that are operating on adjacent channels.

Likewise, harmonic output is to be avoided because these harmonics may fall on the assigned frequencies of stations which operate on higher-frequency bands. Even very weak harmonic components can cause havoc in nearby re-

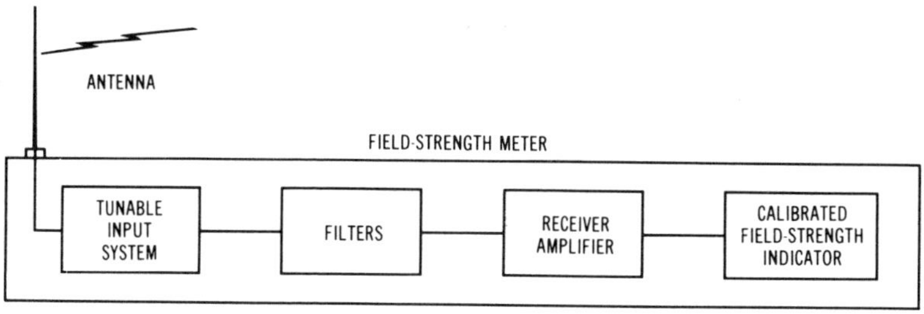

Fig. 17-16. Field-strength meter arrangement for measuring harmonics.

eivers which must pick up a signal from very distant station operating on the ame frequency.

.7-2-7. Describe a usual method (and quipment used) for measuring the harnonic attenuation of a transmitter.—A ield-strength meter, in the arrangement f Fig. 17-16, is best used to measure the evel of harmonic components. A fieldtrength meter is a combination receiver nd calibrated signal-strength meter. To neasure harmonics properly it is ecessary to use certain band-elimination nd bandpass filters. In a harmonic neasurement, the elimination filter is sed to tune out the carrier, preventing it rom overloading the field-strength neter. The bandpass filter is critically ıned to the frequency to be measured, nerefore it permits only the desired harnonic component to enter and produce a eld-strength reading.

In such an arrangement it is possible to easure exactly the level of undesired armonics. Furthermore these levels can e measured with relation to the strength f the fundamental carrier. Consequent-, it is possible to obtain a decibelmparison figure, making it appropriate r the manner in which the FCC ecifies the required harmonic tenuation.

7-2-8. What special problem is encounred when measuring the frequency of n fm carrier which contains a CTCSS ontinuous-tone-controlled squelch sysm) tone on a heterodyne frequency eter?—It will not be possible to obtain n exact zero beat because the fm carrier being deviated by the tone signal. For precise measurement the tone must be eactivated.

-2-9. Discuss the problems which ould be encountered in attempting to measure the frequency and deviation of a frequency-modulated signal with nonsymmetrical deviation.—In this situation the fm wave is not being deviated an equal amount on each side of the center frequency. In making such a measurement the maximum deviation is that figure obtained for the higher deviation peak (positive or negative). The difference between peak readings is known as the *transmitter deviation dissymmetry*. Again a precise measure of frequency is best obtained by deactivating the transmitter deviation.

17-2-10. What techniques can be used to measure the am modulation of a vhf aircraft transmitter?—The ideal method to measure am modulation is to use an oscilloscope if it has a frequency range that extends up to that of the vhf aircraft transmitter. An alternative is to use a vhf aircraft receiver to pick up the transmitted signal. The oscilloscope can then be connected into the if amplifier system of the receiver to obtain a modulation envelope display. The if frequency is much lower and most modern oscilloscopes can accommodate this frequency range. A spectrum monitor can also be used to check am modulation.

17-2-11. What is a nonradiating load (dummy antenna) for connection to the output of a transmitter?—A dummy load places a proper resistive load on the output of a transmitter that is to be serviced. Such a load is shielded and prevents signal radiation and interference to stations operating on the channel. The load is usually a noninductive resistor with a power rating capable of handling the power output level of the transmitter. Resistance of load matches the output impedance of the transmitter (usually 50 ohms).

CHAPTER 18

Sample Test

This is a sample test to act as a review and to give you some experience with the multiple-choice type of examination. Answers, some with explanation, are given in the next chapter.

Do each question without reference to the book material. Some questions are mathematical and will draw upon your knowledge of basic electricity and electronics. You may need some review of these basics. Most questions have to do with material covered in this handbook and will serve as a good review and will guide you in your further studies for the examination.

1. **The designation for the standard double-sideband amplitude modulation is**
   A. A1. C. A3.
   B. A2. D. A3J.

2. **The designation for the radiotelephone single-sideband and reduced carrier transmission is**
   A. A3. C. A3H.
   B. A3A. D. A3J.

3. **In a capacitive ac circuit**
   A. current lags voltage. C. voltage leads current.
   B. current leads voltage. D. there is no reactance.

4. **In an inductive ac circuit**
   A. there can be no resistance. C. voltage lags current.
   B. current leads voltage. D. there is reactance.

5. **Two series-connected .01-μF capacitors are charged to 100 volts, and then the source is removed. A third .01-μF capacitor is connected across the two. What is the voltage across the third capacitor?**
   A. 20. C. 100.
   B. 75. D. 33 1/3.

6. **A low-pass filter can be used to**
   A. attenuate lows.
   B. attenuate harmonics.
   C. amplify highs.
   D. pass a narrow frequency band.

7. **Two resistors of 1200 and 1800 ohms are connected in series across a 200-volt source. Determine the power in the circuit.**
   A. 33 1/3 W.
   B. 13 1/3 W.
   C. 66 2/3 W.
   D. 0.67 W.

8. **To keep power constant when you double the resistance in a circuit, the voltage must be**
   A. doubled.
   B. halved.
   C. squared.
   D. increased by 1.414.

9. **A base station transmitter draws 385 watts. It is on the air 120 hours. How much electrical energy is used?**
   A. 32 kWh.
   B. 46.2 kWh.
   C. 1200 kWh.
   D. 3900 joules.

10. **A sine wave of 10 volts rms is how much peak volts?**
   A. 10.
   B. 14.14.
   C. 7.07.
   D. 28.28.

11. **Find the impedance of a series circuit at 1000 Hz, with a resistance of 5000 ohms and an inductance of 1 henry.**
   A. 9850.
   B. 11,280.
   C. 985.
   D. 8027.

12. **A gain of 20 dB corresponds to an increase to**
   A. 1000 × watts.
   B. 10 × volts.
   C. 3.16 × watts.
   D. 31.6 × volts.

13. **A loss of 20 dB corresponds to a decline to**
   A. 10^{-2} × watts.
   B. 10^{-2} × volts.
   C. 10^{-3} × watts.
   D. 10^{-4} × volts.

14. **Find the power consumed in a 110-volt ac circuit of 0.35 power factor when the current demand is 7 amperes.**
   A. 770 W.
   B. 38.5 W.
   C. 269.5 W.
   D. 245 W.

15. **Ten cycles of a sine wave occur in 2 seconds. What is the frequency?**
   A. 20 Hz.
   B. 5 Hz.
   C. 0.5 Hz.
   D. 2 Hz.

16. **What is the wavelength of the sine wave in Question 15?**
   A. 60×10^6 meters.
   B. 15×16^6 meters.
   C. 15×10^3 meters.
   D. 60×10^3 meters.

17. **A wattmeter reads 400 watts. If the source voltage is 120 and the source current is 4 amperes, the power factor is**
   A. 0.48.
   B. 1.5.
   C. 1.6.
   D. 0.83.

18. **A common-base amplifier**
   A. has voltage gain.
   B. has a high input resistance.
   C. requires that input signal be applied to the collector.
   D. is similar to a vacuum-tube grounded-grid stage.

19. **To properly bias the collector junction of an npn transistor apply**
   A. plus to collector and minus to base.
   B. minus to collector and plus to emitter.
   C. plus to base and minus to collector.
   D. plus to emitter and minus to collector.

20. **A common-base transistor amplifier has**
A. a low input impedance.
B. no voltage gain.
C. out-of-phase input and output signals.
D. no current gain.

21. **A lead-acid storage battery is checked with a**
A. hydrometer.
B. voltmeter.
C. ammeter.
D. watt-hour meter.

22. **A bridge rectifier in a 60-Hz supply has a ripple frequency of**
A. 30.
B. 60.
C. 90.
D. 120.

23. **What transformer turns ratio is needed to match 2000 ohms to 200 ohms?**
A. 10.
B. 0.1.
C. 0.316.
D. 3.16.

24. **A 20-dB rise in a power gain of**
A. 10.
B. 20.
C. 100.
D. 1000.

25. **What happens when the rfc opens in a shunt-fed class-C amplifier?**
A. Plate current rises.
B. Supply voltage appears on resonant tank.
C. Neutralization is upset.
D. Plate voltage is removed.

26. **Electron transit time is taken advantage of in the design of the**
A. klystron.
B. lighthouse tube.
C. acorn.
D. all of the above.

27. **The most likely undesired frequency to appear at the output of a well-balanced 2.2-MHz push-pull class-C amplifier is**
A. 2.2 MHz.
B. 4.4 MHz.
C. 6.6 MHz.
D. 1.1 MHz.

28. **A pi-network provides**
A. harmonic suppression.
B. impedance matching.
C. efficient rf power transfer.
D. all of the above.

29. **The purpose of neutralization is to**
A. prevent radiation of spurious signals
B. prevent self-oscillation.
C. improve rf amplifier efficiency.
D. perform all of the above.

30. **If a plate-modualated class-C amplifier has an efficiency of 85% what is the dc input power if the rf carrier is 1000 watts?**
A. 850 W.
B. 1176 W.
C. 1150 W.
D. 1000 W.

31. **How much audio power is needed to 100% modulate the transmitter of Question 30?**
A. 850 W.
B. 500 W.
C. 588 W.
D. 425 W.

32. **What is the deviation ratio in the land-mobile services when maximum deviation is ± 5 kHz and highest audio frequency is 3000 Hz?**
A. 1.66.
B. 0.6.
C. 15.
D. 1.5.

33. **Modulation index is the ratio of**
A. carrier frequency to deviation.
B. deviation to modulating frequency.
C. modulating frequency to deviation.
D. deviation to number of side bands.

34. **An fm discriminator**
A. converts the if frequency change to audio frequency change.
B. cannot supply a dc output that responds to an incoming signal level.
C. converts frequency change to amplitude change.
D. is also a good limiter.

35. **Automobile ignition interference is reduced with**
A. resistance wire in series with plugs.
B. generator capacitor.
C. regulator capacitor.
D. all of the above.

36. **A transmitter operates on 25 MHz. If the frequency tolerance is 0.001%, how far may its carrier drift above and below 25 MHz?**
A. ± 2.5 kHz.
B. ± 25 kHz.
C. ± 400 Hz.
D. ± 250 kHz.

37. **Modulation index increases**
A. with deviation and modulating frequency.
B. with deviation and decreases with modulating frequency.
C. with amplitude of modulating wave and decreases with modulating frequency.
D. as in B and C above.

38. **A shorted quarter-wave section of transmission line**
A. has a high resistive impedance at the opposite end.
B. has an impedance equal to its surge impedance at the opposite end.
C. has a low resistive impedance at the opposite end.
D. displays a high reactance at the the opposite end.

39. **Velocity of wave propagation**
A. is the same at all frequencies.
B. rises with frequency.
C. is the product of wavelength and frequency.
D. is both A and C above.

40. **Atmospheric static decreases**
A. as frequency is increased above 10 MHz.
B. with the use of a long antenna.
C. when receiver noise is decreased.
D. as frequency is decreased below 30 MHz.

41. **At the top of a quarter-wave vertical there is a**
A. voltage maximum.
B. current maximum.
C. impedance maximum.
D. A and C above.

42. **A current of 0.5A is present in a 120-volt 60-Hz relay coil. If the dc resistance of the relay is 2 ohms, what is the relay impedance?**
A. 265.
B. 60.
C. 240.
D. 24.

43. **The base of a bipolar transistor is often compared loosely in function to what electrode of a vacuum tube?**
A. Plate.
B. Grid.
C. Cathode.
D. Filament.

44. **The emitter of a bipolar transistor is often compared loosely in function to what electrode of a vacuum tube?**
A. Plate.
B. Grid.
C. Cathode.
D. Filament.

45. **The collector of a bipolar transistor is often compared loosely in function to what electrode of a vacuum tube?**
A. Plate.
B. Grid.
C. Cathode
D. Filament.

46. **The source of a field-effect transistor is compared in function to what element of a bipolar transistor?**
A. Base.
B. Collector.
C. Emitter.
D. Collector junction.

47. **The gate of a field-effect transistor is compared in function to what element of a bipolar transistor?**
A. Base.
B. Collector.
C. Emitter.
D. Emitter junction.

48. **The drain of a field-effect transistor is compared in function to what element of a bipolar transistor?**
A. Base.
B. Collector.
C. Emitter.
D. Emitter junction.

49. **The pre-emphasis network of an fm system**
A. emphasizes highs.
B. emphasizes lows.
C. acts as a high-frequency notch filter.
D. maintains a level frequency response.

50. **A de-emphasis network is located at the**
A. transmitter.
B. receiver.
C. antenna input.
D. balanced modulator.

51. **Find resonant frequency of the circuit.**
A. 7.9 MHz.
B. 3.56 MHz.
C. 2.23 MHz.
D. 5 MHz.

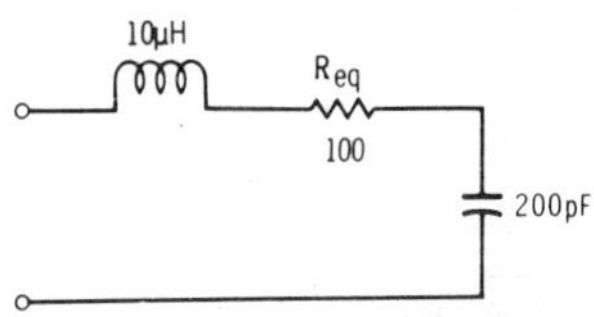

52. **For a series-resonant circuit at resonance.**
A. impedance is minimum.
B. current is maximum.
C. $Q = \frac{\omega L}{R}$.
D. all of the above apply.

53. **What is the impedance at resonance?**
A. More than 100 ohms.
B. Resistive.
C. 100 ohms.
D. B and C above.

54. **What is the impedance on the high-frequency side of resonance?**
A. Higher than 100 ohms.
B. Less than 100 ohms.
C. Inductive.
D. A and C above.

55. **Find resonant frequency of the circuit.**
A. 7.95 MHz.
B. 3.56 MHz.
C. 2.23 MHz.
D. 5 MHz.

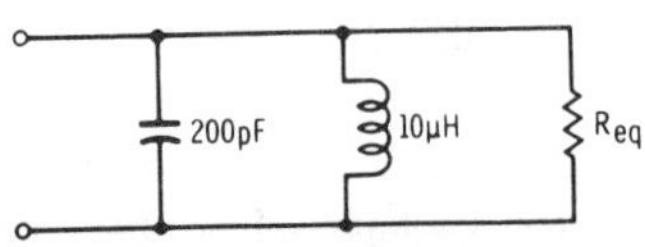

56. **For a parallel-resonant circuit at resonance**
A. impedance is maximum.
B. voltage is maximum.
C. current I is minimum.
D. all of the above pertain.

57. **What is the impedance at resonance?**
A. Minimum.
B. Resistive.
C. Less than R.
D. A and B above.

58. **What is the impedance on the high-frequency side of resonance?**
A. Lower than R.
B. Higher than R.
C. Capacitive.
D. A and C above.

59. **Purpose of the resonant circuits is**
A. to match antenna to receiver.
B. to prevent receiver overload.
C. to trap out interference.
D. for lightning protection.

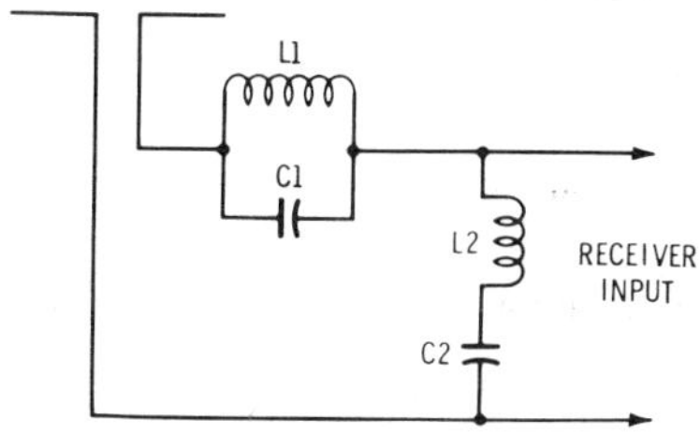

60. **In the previous circuit**
A. L1-C1 has maximum impedance at resonance.
B. L2-C2 has minimum impedance.
C. impedance is minimum for L1-C1 and maximum for L2-C2.
D. A and B above apply.

61. **Refer to the schematic of the NPN bipolar transistor**
A. Base-emitter junction is biased incorrectly.
B. Collector is biased incorrectly.
C. Schematic is correct.
D. A and B above.

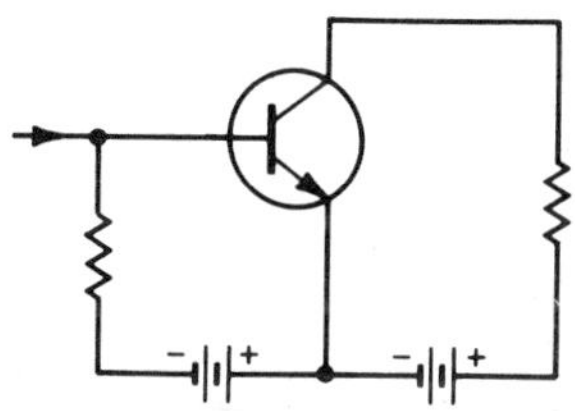

62. **Refer to the schematic of the n-channel junction FET**
A. Gate source junction is biased incorrectly.
B. Drain is biased incorrectly.
C. Schematic is correct.
D. A and B above.

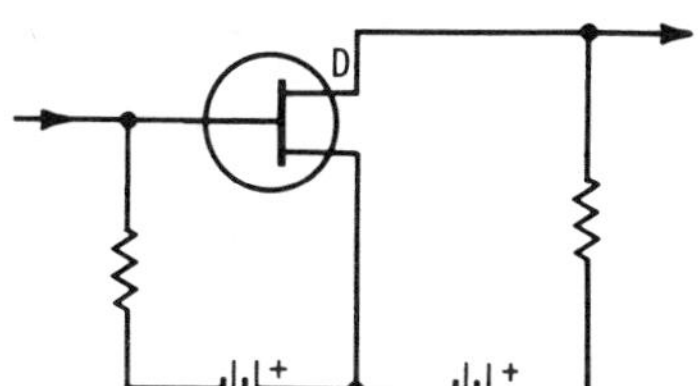

63. **What is the percentage power supply regulation if no load voltage is 24 volts and full load value is 20 volts?**
A. 24.
B. 20.
C. 1.7.
D. 2.4.

64. **If the resistive loading of a dc voltmeter on the 100-volt scale is 2 megohms, what is the meter sensitivity in ohms-per-volt?**
A. 2 Meg.
B. 1000.
C. 20,000.
D. 200.

65. **What is the impedance of a series RLC circuit if X_L equals 2000 ohms, X_C equals 1000 ohms and R equals 1000 ohms?**
A. 2000.
B. 1000.
C. 173.
D. 1414.

66. **An LED is to be operated from a 5-volt dc source. What value of series resistance is needed to limit current to approximately 15 mA?**
A. 220. C. 270.
B. 390. D. 330.

67. **Which circuit is a low-pass filter?**
A. C.
B. D.

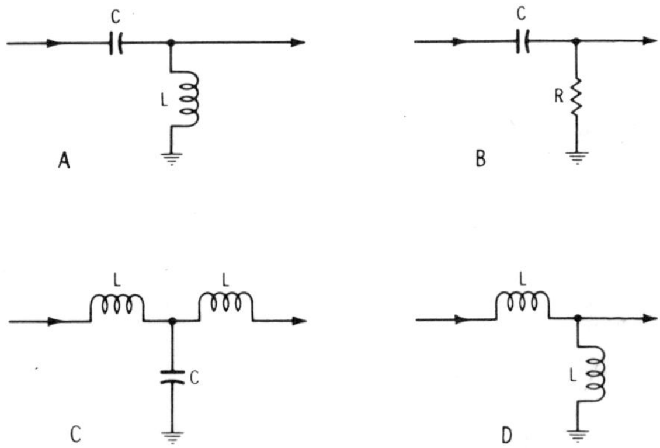

68. **Which is the NOR gate symbol?**
A. C.
B. D.

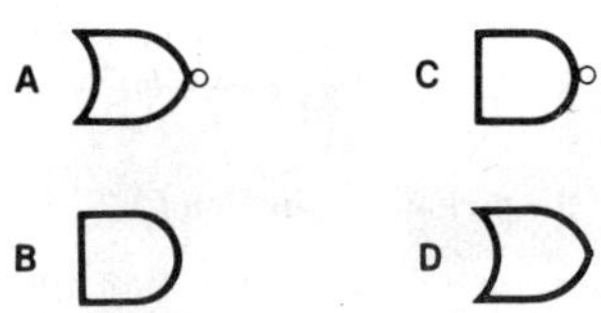

69. **What is the decade equivalent of binary 1101?**
A. 13. C. 3.
B. 14. D. 7.

70. **The operation of police radar depends upon the**
A. Miller effect. C. skin effect.
B. Doppler effect. D. corona effect.

71. **An ssb product detector**
A. requires two input signals. C. is a frequency multiplier.
B. is part of the second mixer. D. A and B above.

72. **Theoretical impedances of a dipole and a quarter-wave vertical are**
A. 36 and 50 ohms respectively. C. 72 and 36 ohms respectively.
B. the same. D. 52 and 36 ohms respectively.

73. **The frequency tolerance on a mobile band is 0.0005%. If frequency meter has tolerance of 0.00002%, what frequency range at 140 MHz can be certified as within th tolerance of 0.0005%?**
A. 139.8-140.2. C. 130.000328 140.000672.
B. 139.9872-140.0728. D. 139.928-140.072.

74. What does 2.8A3J emission mean?

A. Single-sideband modulation with carrier suppressed and 2.8-kHz bandwidth.
B. Single-sideband modulation with full carrier and 2.8-kHz bandwidth.
C. Amplitude modulation with 5.6-kHz bandwidth.
D. Amplitude modulation with 2.8-kHz bandwidth.

75. What is vswr when direct power as measured on a reflectometer is 100 watts and reflected power is 5 watts?

A. 0.905 to 1.
B. 1.105 to 1.
C. 1.05 to 1.
D. 2 to 1.

76. Station records must be retained by a licensee in the Land Mobile Radio Services for at least

A. 6 months.
B. 9 months.
C. 1 year.
D. 2 years.

77. Transmitters in the Land Mobile Radio Services must be measured to prove compliance with FCC technical standards

A. when the transmitter is installed initially.
B. when any change is made in the transmitter which may affect frequency, transmitter power input or modulation characteristics.
C. once a year.
D. A and B above.

The examination may contain block diagrams and circuit schematics that are in error or incomplete. In other questions there may be some related queries about diagrams or schematics. It is helpful if you can draw from memory a variety of circuits. Be able to describe their operation briefly. Some key basic circuits would be:

78. Be able to draw the various basic filter types (Figs. 10-3 and 10-4).
79. Basic bipolar common-emitter, common-base, and common-collector circuits (Sections 3-3 and 3-4).
80. Basic fet common-source, common-gate, and common-drain circuits (Section 3-12).
81. Vacuum-tube grounded-cathode, grounded-grid, and grounded-plate circuits (Figs. 11-10, 11-12 and 11-13).
82. MOSFET rf amplifier (Section 3-13).
83. Common-emitter rf amplifier (Fig. 11-32).
84. Common-gate rf amplifier (Fig. 11-31).
85. Common-emitter frequency doubler (Fig. 13-22).
86. Darlington amplifier (Fig. 11-33).
87. Full-wave rectifier supply (Fig. 12-3).
88. Bridge rectifier supply (Fig. 12-7).
89. Inverting operational amplifier (Section 3-16).
90. Noninverting operational amplifier (Section 3-16).
91. Zener diode voltage regulator supply (Fig. 12-11).
92. Varactor fm modulator (Fig. 15-8).
93. Fet rf amplifier (Fig. 11-30).
94. Transistorized voltage regulator using pass transistor and reference zener (Fig. 12-10).

95. Voltage regulated supply using a 3-terminal IC regulator (Fig. 12-9).
96. Series and shunt-fed plate arrangements (Fig. 11-18).
97. Push-pull audio amplifier (Figs. 13-7 and 13-8).
98. Fm discriminator (Fig. 15-10).
99. Triode crystal oscillator (Fig. 13-27).
100. Various bipolar rf oscillators and amplifiers. Refer to Chapters 3, 4, and 7. Study again the troubleshooting circuits at the end of Chapter 7.

CHAPTER 19

Answers to Sample Test

1. C.
2. B.
3. B.
4. D.
5. D. When capacitors are charged and the source of power is removed there are only so many coulombs of energy (CE) stored in the electrostatic fields. When other capacitors are added to the combination the energy is redistributed and the net charge voltage declines. In the example capacitance increased from 0.005 μF to 0.015 μF with the addition of the third capacitor. Therefore the net voltage across the combination and across this individual capacitor must drop from 100 volts to 33⅓ volts.

$$C_1E_1 = C_2E_2$$
$$\frac{C_1}{C_2} = \frac{E_2}{E_1}$$
$$\frac{0.005}{0.015} = \frac{E_2}{100}$$
$$E_2 = 33\tfrac{1}{3} \text{ volts}$$

6. B.
7. B. The power removed is a function of the total resistance and the applied voltage or:

$$P = \frac{E^2}{R} = \frac{200^2}{1200 + 1800} = 13\tfrac{1}{3} \text{ watt}$$

8. D.
9. B. Power = 385 × 120 = 46.2 kWh
10. B. Peak voltage equals $\sqrt{2} \times E_{rms}$ = 1.414 × 10 = 14.14 volts.
11. D. The standard impedance equation is used:

$$Z = \sqrt{R^2 + X^2}$$

The X_L depends upon frequency or:

$$X = 2\pi fL = 6.28 \times 1000 \times 1$$
$$= 6280 \text{ ohms}$$
$$Z = \sqrt{5000^2 + 6280^2} = 8027 \text{ ohms}$$

12. B. A 20-dB gain corresponds to a tenfold increase in voltage.
13. A. The loss of 20 dB corresponds to a 1/100 decline in power. This is written as $10^{-2} \times$ watts.
14. C. In an ac circuit the power consumed is a function of the power factor. The true power becomes:

$$P = EI \times \text{Power Factor} = 110 \times 7 \times 0.35 = 269.5 \text{ watts}$$

15. B. The period of the wave is 0.2 second (2/10). Frequency is

$$\frac{1}{\text{Period}}$$

or:

$$F = \frac{1}{T} = \frac{1}{0.2} = \text{five hertz}$$

16. A. Wavelength is the quotient of the velocity over frequency or:

$$\lambda = \frac{300 \times 10^6}{5} = 60 \times 10^6 \text{ meters}$$

17. D. The true power as read by the wattmeter is 400 watts. However, the product of the voltage and current is 480 watts. This is the apparent power. The power factor is their quotient.

$$\text{Power Factor} = \frac{\text{True Power}}{\text{Apparent Power}} = \frac{400}{4 \times 120} = 0.83$$

18. A. A common-base amplifier does have voltage gain. However, it has a very low input resistance. In operation, it is similar to a vacuum-tube grounded-grid circuit.
19. A.
20. A.
21. A.
22. D.
23. D. Keep in mind that the turns ratio equals the square root of the impedance ratio. Conversely, the impedance ratio equals the square of the turns ratio as shown:

$$\text{Turns Ratio} = N_P/N_S = \sqrt{Z_p/Z_S}$$

$$\text{Turns Ratio} = \sqrt{\frac{2000}{200}} = 3.16$$

$$\text{Impedance Ratio} = Z_p/Z_S = (N_P/N_S)^2$$

24. C.
25. D.
26. A.
27. C. 6.6 MHz is an odd harmonic (3rd)
28. D.
29. D.
30. B. The equation is:

$$\text{Dc Input Power} = \frac{\text{Rf Power Output}}{\text{Efficiency}} = \frac{1000}{0.85} = 1176 \text{ watts}$$

31. C. 100% modulation requires that the audio power be one-half of the dc power input or 588 watts (1176/2).
32. A. The deviation ratio is determined by the maximum permissible deviation and the highest modulating frequency or:

$$\text{Deviation Ratio} = \frac{\text{Permissible Deviation}}{\text{Highest Audio Frequency}} = \frac{5000}{3000} = 1.66$$

33. B.
34. C.
35. D.
36. D. If the frequency tolerance is 0.001% then the maximum frequency drift is:
Permissible Frequency Shift = $0.00001 \times 25{,}000{,}000 = \pm 250$ Hz
37. D.
38. A.
39. D.
40. A.
41. D.
42. C. It is the impedance that determines the current level. Therefore:

$$Z = \frac{E}{I} = \frac{120}{0.5} = 240 \text{ ohms}$$

43. B.
44. C.
45. A.
46. C.
47. A.
48. B.
49. A.
50. B.
51. B. The resonance equation is:

$$f_o = \frac{1}{2\pi\sqrt{LC}} = \frac{1}{6.28\sqrt{10 \times 10^{-6} \times 200 \times 10^{-12}}} = 3.56 \text{ MHz}$$

52. D. A series resonant circuit has minimum impedance at resonance and draws maximum current from the source. The Q is the ratio of reactance to the equivalent series resistance.
53. D. The reactive components cancel and the only impedance is that contributed by the resistance itself. Therefore the series resonant circuit itself is resistive.
54. D. On the high-frequency side of resonance the inductive reactance is greater than the capacitive reactance and, therefore, it has a limiting influence on the current. The circuit is said to be inductive on the high frequency side of resonance.
55. B. The same equation is used to calculate resonance for both series- and parallel-resonant circuits.

56. D. Impedance of a parallel-resonant circuit is maximum at the resonant frequency. Therefore current demand is minimum but the voltage drop across the resonant circuit is maximum.
57. B. The impedance is resistive and the actual impedance is equal to the effective parallel resistance. The Q of a parallel-resonant circuit in terms

of the effective parallel resistance is $\frac{\text{Resistance}}{\text{Reactance}}$ (just the opposite of the series-resonant case).

58. D. On the high-frequency side of resonance the capacitive reactance is lower than the inductive reactance and a higher current results because of the drop in impedance. A parallel-resonant circuit on the high-frequency side of resonance is capacitive. Remember that the impedance of a series-resonant circuit off resonance increases, while the impedance of a parallel-resonant circuit off resonance decreases.
59. C. This is a method of blocking interfering signals from the input of a receiver.
60. D. The parallel combination of L1-C1 has a high impedance at the resonant frequency. If it is tuned to the interfering frequency signal it will block the transfer of that signal to the receiver input. The impedance of a series-resonant circuit is minimum at resonance. Therefore if L2-C2 is tuned to the interfering signal frequency there is additional rejection of the undesired signal because L2-C2 places a low impedance shunt across the receiver input at the undesired frequency.
61. A.
62. C.
63. B.

$$\%\ \text{Reg.} = \frac{E_{\text{No Load}} - E_{\text{Full}}}{E_{\text{Full}}} \times 100$$
$$= \frac{24 - 20}{20} = 20\%$$

64. C.

2,000,000 divided by 100 equals 20,000 ohms-per-volt

65. D.

$$Z = \sqrt{R^2 + (X_L - X_C)^2}$$
$$Z = \sqrt{1000^2 + (2000 - 1000)^2}$$
$$= 1414 \text{ ohms}$$

66. D.

$$R = \frac{E}{I} = \frac{5}{.015} = 333 \text{ ohms}$$

67. C.
68. A.
69. A.

1-eight
1-four
1-one
———
thirteen

70. B.
71. A.
72. C.
73. C. Tolerance of meter requires that the actual overall tolerance must be scaled down to determine the range of readings that would be acceptable for certification. In the example this value becomes 0.00048% (0.0005%-0.00002%).
Permissible deviation from correct frequency is 140 times 0.0000048 equals 672 Hz. This is added to and subtracted from 140 MHz to set range.

140.000000	140.000000
− 672	+ 672
139.999328	140.000672

74. A.
75. B.

$$\text{VSWR} = \frac{1 + K}{1 - K}$$

$$\text{VSWR} = \frac{1 + \frac{5}{100}}{1 - \frac{5}{100}} = 1.105$$

76. C.
77. D.

Index